Trends in Mathematics is a series devoted to the publication of volumes arising from conferences and lecture series focusing on a particular topic from any area of mathematics. Its aim is to make current developments available to the community as rapidly as possible without compromise to quality and to archive these for reference.

Proposals for volumes can be sent to the Mathematics Editor at either

Birkhäuser Verlag
P.O. Box 133
CH-4010 Basel
Switzerland

or

Birkhäuser Boston Inc.
675 Massachusetts Avenue
Cambridge, MA 02139
USA

Material submitted for publication must be screened and prepared as follows:

All contributions should undergo a reviewing process similar to that carried out by journals and be checked for correct use of language which, as a rule, is English. Articles without proofs, or which do not contain any significantly new results, should be rejected. High quality survey papers, however, are welcome.

We expect the organizers to deliver manuscripts in a form that is essentially ready for direct reproduction. Any version of $\TeX$ is acceptable, but the entire collection of files must be in one particular dialect of $\TeX$ and unified according to simple instructions available from Birkhäuser.

Furthermore, in order to guarantee the timely appearance of the proceedings it is essential that the final version of the entire material be submitted no later than one year after the conference. The total number of pages should not exceed 350. The first-mentioned author of each article will receive 25 free offprints. To the participants of the congress the book will be offered at a special rate.

A Tribute to C.S. Seshadri

A Collection of Articles on Geometry and Representation Theory

V. Lakshmibai
V. Balaji
V.B. Mehta
K.R. Nagarajan
K. Paranjape
P. Sankaran
R. Sridharan
Editors

Birkhäuser Verlag
Basel · Boston · Berlin

Organizing editor:

V. Lakshmibai
Department of Mathematics
Northeastern University
Boston, MA 02115
USA
lakshmibai@neu.edu

Co-editors:

V. Balaji (Chennai Mathematical Institute, Chennai, India)
V.B. Mehta (Tata Institute of Fundamental Research, Bombay, India)
K.R. Nagarajan (Chennai Mathematical Institute, Chennai, India)
K. Paranjape (Institute of Mathematical Sciences, Chennai, India)
P. Sankaran (Institute of Mathematical Sciences, Chennai, India)
R. Sridharan (Chennai Mathematical Institute, Chennai, India)

2000 Mathematical Subject Classification 00B30, 14-XX, 53-XX

A CIP catalogue record for this book is available from the
Library of Congress, Washington D.C., USA

Bibliographic information published by Die Deutsche Bibliothek
Die Deutsche Bibliothek lists this publication in the Deutsche Nationalbibliografie; detailed bibliographic data is available in the Internet at <http://dnb.ddb.de>.

ISBN 3-7643-0444-2 Birkhäuser Verlag, Basel – Boston – Berlin

Authorized edition by Birkhäuser Verlag, P.O. Box 133, CH-4010 Basel, Switzerland
Member of the BertelsmannSpringer Publishing Group
for exclusive distribution worldwide except India
Printed on acid-free paper produced from chlorine-free pulp. TCF ∞
Printed in Germany
ISBN 3-7643-0444-8

9 8 7 6 5 4 3 2 1 www.birkhauser.ch

Contents

C.S. Seshadri

Preface

Seshadri turned seventy on March 1st, 2002. Some of his colleagues, friends and students in India and abroad decided to celebrate this occasion by holding a symposium to highlight Seshadri's contributions to mathematics and to felicitate him. A one-day symposium was organised by The Chennai Mathematical Institute and the Institute of Mathematical Sciences, Chennai on March 1st, 2002, and held at The Hotel Residency.

Invitations were sent to a large number of friends and admirers of Seshadri. Many of them attended the function and those who could not, sent in their greetings.

The proceedings began with an inauguration by Prof. R. Balasubramaniam, the director of the Institute of Mathematical Sciences, Chennai.

The academic session started with some of Seshadri's colleagues and students giving talks on various aspects of his mathematical contributions spread over nearly five decades. The speakers were: S. Ramanan, V. Balaji, V. Mehta, N. Nitsure, D.S. Nagaraj, C. Musili and V. Lakshmibai. This was followed by talks by K.R. Nagarajan and M. Mukund on the educational programme at CMI, initiated by Seshadri. The session concluded with a talk by M.S. Raghunathan on the National Board for Higher Mathematics and the role of CMI in mathematics education.

Late in the afternoon, the invitees gathered to felicitate Seshadri, his mother Chudamani and his wife Sundari. Messages of greetings from many of his friends were read out. Several distinguished guests also spoke bringing out the multi-faceted personality of Seshadri.

Seshadri, in his reply, recalled among other things the genesis of CMI and the key role played by the late Dr. S. Parthasarathy and Shri A.C. Muthiah in its development. The day, which was marked by warmth and goodwill, ended with a banquet in honour of Seshadri.

The help and cooperation extended by several individuals made the function a memorable one. Sripathy and his colleagues at the offfice of the CMI deserve special mention for their unstinted and enthusiastic support.

About this volume:

We thought that it would be appropriate to publish a volume in honour of Seshadri, containing the texts of the talks at the symposium as well as invited research and expository articles from some of his colleagues. The response to our invitation was spontaneous, and I want to thank all the authors who have contributed to this volume.

The articles published here were refereed, and I am indebted to the referees for their gracious cooperation. I thank the Hindustan Book Agency, New Delhi for their immense help in bringing out this volume. Finally, I am much obliged to my colleagues in the editorial board for their efforts which made this volume possible.

V. Lakshmibai
Chief Editor

C.S. Seshadri
A Biographical Sketch

Conjeevaram Srirangachari SESHADRI was born on February 29th, 1932, in Kanchipuram (the "Golden City" turned into "Conjeevaram" during the British Raj).

He was the eldest among eleven children of his parents, Sri C. Srirangachari (a well-known advocate in Chengleput) and Srimati Chudamani.

Seshadri's early schooling was at the St. Joseph's Elementary School and at Sri Ramakrishna School in Chengleput. His final years of schooling were at the St. Joseph's High School at Chengleput.

Seshadri recalls with feeling the lasting influence his uncle T.A. Rangaswamy had in his formative years in school, especially in kindling his interest in mathematics.

He joined the Loyola College, Chennai in 1948, and he graduated from there in 1953 with a BA (Hons) degree in mathematics.

During his early years at college, he was encouraged by his teacher Prof S. Narayanan. Later, Fr. C. Racine played a decisive role in Seshadri's taking up mathematics as a profession.

Seshadri's research career began in 1953 when he joined the Tata Institute of Fundamental Research, Mumbai, as a student.

He received his Ph.D. degree in 1958 from the Bombay University for his thesis entitled "Generalised multiplicative meromorphic functions on a complex manifold". His thesis adviser was Prof K. Chandrasekharan who shaped the mathematical career of Seshadri as he did for many others.

Seshadri spent the years 1957–60 in Paris, where he came under the influence of many luminaries of the French school, like Chevalley, Cartan, Schwartz, Grothendieck and Serre.

He returned to the TIFR in 1960 and was a member of the faculty of the School of Mathematics until 1984, where he was responsible for establishing an active school of algebraic geometry. He moved to the Institute of Mathematical Sciences, Chennai in 1984.

In 1989, Seshadri became the director of the Chennai Mathematical Institute, then called the SPIC Mathematical Institute, founded by A.C. Muthiah, and in the formation of which the late S. Parthasarathy played a key role. This institute is now well known not only as a research institution of repute but also as a pioneering centre for undergraduate and graduate education in mathematics and computer science.

Seshadri is a recipient of numerous distinctions. He received the Bhatnagar Prize in 1972 and was elected a fellow of the Royal Society, London in 1988. He has held distinguished positions in various centres of mathematics, all over the world.

Seshadri is also well known as an accomplished exponent of Carnatic Music.

Seshadri married Sundari in 1962, and they have two sons, Narasimhan and Giridhar.

List of Publications of C.S.Seshadri

[1] Generalised Multiplicative Meromorphic Functions on a Complex Analytic Manifold, J.I.M.S., **21** (1957), 149–175.

[2] Triviality of Vector Bundles over the Affine Space K, Proc. Mat. Acad. Sci., U.S.A. **44** (1958), 456–458.

[3] Exposés in the Chevalley Seminar – Seminaire Chevalley, 3e année, 1958–59.

[4] Algebraic Vector Bundles over the Product of an Affine Curve and the affine line, Proc. A.M.S. **10** (1959), 670–673.

[5] Varieté de Picard d'une Varieté complete, Annali di Mat. Italy IV, Vol. LVII (1962), 117–142.

[6] On a theorem of Weitzenböck in invariant theory, Memoirs of the College of Science, Kyoto Univ. **1**, No. 3 (1962).

[7] Some results on the quotient space by an algebraic group of automorphism, Math. Ann., **149** (1963), 286–301.

[8] Quotient space by an Abelian variety, Math. Ann., **152** (1963), 185–194.

[9] ......(with M.S. Narasimhan) Holomorphic vector bundles on a compact Riemann surface, Math. Ann., **155** (1964), 69–80.

[10] ......(with M.S. Narasimhan) Stable bundles and unitary bundles on a compact Riemann surface, Proc. Nat. Acad. Soc., **52** (1964), 207–211.

[11] ......(with M.S. Narasimhan) Stable bundles and unitary bundles on a compact Riemann surface, Annals of Math., **82** (1965), 540–567.

[12] Universal Property of the Picard variety of a complete variety, Math. Ann., **156** (1965), 293–296.

[13] Space of unitary vector bundles on a compact Riemann surface, Annals of Math., **85** (1967), 303–335.

[14] Mumford's conjecture for $GL(2)$ and applications, Proc. Inter. Colloquium on Algebraic Geometry, Bombay (1968), 347–371.

[15] Moduli of π-vector bundles over an algebraic curve, Proc. C.I.M.E. Session, Varenna (1969).

[16] Quotient spaces modulo reductive algebraic groups and applications to moduli of vector bundles on algebraic curves, Proc. Inter. Congress of Mathematicians, Nice (1970).

[17] Quotient spaces modulo reductive algebraic groups, Annals of Math., **95**, No.3 (1972), 511–556.

[18] Errata to 'Quotient spaces module reductive algebraic groups', Annals of Math. **96** (1972), 599.

[19] ......(with V. Lakshmibai and C. Musili) Cohomology of line bundles on G/B, Annales Scientifiques ENS, 4 Series, 4.7, fasc. 1, (1974), 89–138.

[20] Correction to 'Cohomology of line bundles on G/B', Annales Scientifiques ENS, 4 Series, t.7 (1974).

[21] Theory of moduli, Proc. Symposia in Pure Mathematics, Algebraic Geometry, Arcata, 1974, A.M.S. (1975), 265–304.

[22] Cohomology of line bundles on G/B - unpublished, talk given at the Institute for Advanced Study (1976).

[23] Geometric reductivity over arbitrary base, Advances in Maths., **26** (1977), 225–274.

[24] Moduli of vector bundles on curves with parabolic structures, Bull. Amer. Math. Soc., **83** (1977), 124–126.

[25] Desingularisation of the moduli varieties of vector bundles on curves, Proc. Tokyo Symposium on Algebraic Geometry (1977).

[26] Geometry of G/P-I (Theory of standard monomials for miniscule representation), C.P. Ramanujam - A Tribute, TIFR Publication, Springer Verlag (1978), 207–239.

[27] ......(with V. Lakshmibai) - Geometry of G/P-II (The work of De Concini and Procesi and the basic conjectures), Proc. Indian Acad. Sci., **87 A**, No.2 (1978), 1–54.

[28] ......(with T. Oda) - Compactifications of the generalised Jacobian variety, Trans. Amer. Math. Soc. **253** (1979), 190.

[29] ......(with V. Lakshmibai and C. Musili) Geometry of G/P-III (Standard Monomial Theory for a quasi-miniscule P), Proc. Indian Acad. Sci. **87 A**, (1979), 93–177.

[30] ......(with V. Lakshmibai and C. Musili) Geometry of G/P-IV (Standard Monomial Theory for classical types), Proc. Indian Acad. Sci. **88 A**, Part III No.4 (1979), 279–362.

[31] ......(with V. Lakshmibai and C. Musili) - Geometry of G/P, Bull. Amer. Math. Soc. (1979).

[32] ......(with V.B. Mehta) - Moduli of vector bundles on curves with parabolic structures, Math. Ann. **228** (1980), 205–239.

[33] Vector bundles on curves - Lectures at the Ecole Normale Superieure, 1980 - Asterisque, 96.

[34] ......(with C. Musili) Standard Monomial Theory, Seminaire de'Algebre Paul Dubreil et Marie-Poule Malliavin, Lecture Notes in Mathematics, Springer Verlag No.867, 441–476.

[35] Standard Monomial Theory and the work of Demazure, Proceedings of a Symposium on Algebraic and Analytic Varieties, Tokyo (1981).

[36] ......(with C. Musili) - Schubert varieties and the variety of complexes - Volume dedicated to Prof. Shaferevich on his 60th birthday, Birkhauser, 329–359.

[37] ......(with V. Lakshmibai) - Singular Locus of a Schubert Variety, Bulletin (New series) of the Amer. Math. Soc. Vol. II (1984), 363–366.

[38] Line bundles on Schubert varieties - International Colloquium on 'Vector bundles on algebraic varieties', TIFR, Bombay (1984).

[39] Introduction to the theory of Standard Monomials - Brandeis Lecture Notes 4 (1985).

[40] ......(with V. Lakshmibai) - Geometry of G/P-V, Journal of Algebra, **100** (1986).

[41] Standard Monomial Theory and the Geometry of Schubert Varieties - Srinivasa Ramanujan Medal Lecture, 1985, Proc. Indian Nat. Sci. Acad. **52**, A, No.2 (1986).

[42] ......(with V. Lakshmibai) Theorie normale standard pour SL - C.R. Acad. Sci. Paris, t.305, Serie I (1987), 183–185.

[43] ......(with V. Balaji) Cohomology of a moduli space of vector bundles - The Grothendieck Festschrift Volume 1, Birkhauser (1990), 87–120.

[44] ......(with V. Balaji) Poincare polynomials of some Moduli Varieties - Algebraic Geometry and Analytic Geometry, ICM-90 Satellite Conference Proceedings, Springer Verlag (1991), 1–25.

[45] Vector bundles on curves - dedicated to Prof. Steinberg, Contemporary Mathematics, Volume **153** (1993), 163–200.

[46] The work of P. Littelmann and Standard Monomial Theory - Recent Trends in Mathematics and Physics: A Tribute to Harish Chandra, Narosa (1995), 178–197.

[47] ......(with D.S. Nagaraj) Degenerations of the moduli spaces of vector bundles on curves I, Proc. Indian Acad. Sci. (Math. Sci.) **107** (1997), 101–137.

[48] Methods of Construction of the Picard Variety, International Conference on Teaching and Research in Mathematics (A Tribute to Fr. C. Racine, S.J.), Racine Research Centre, Department of Mathematics, Loyola College, Chennai-34, India, 53–72.

[49] ......(with D.S. Nagaraj) Degenerations of the moduli spaces of vector bundles on curves II (Generalized Gieseker moduli spaces), Proc. Indian Acad. Sci. (Math. Sci.) **109** (1999), 165–201.

[50] "Chevalley: Some Reminiscences", Transformation Groups, **4** (1999), 119–125.

[51] Degenerations of the moduli spaces of vector bundles on curves, ICTP Lecture Notes 1 (2000), 205–265.

[52] ...... (with V. Balaji) Semistable principal bundles - 1 (in characteristic zero), Journal of Algebra, **258** (2002), 321–347.

List of Participants at the Symposium and the Felicitation function

- M.S. Ananth (IITM, Chennai);
- V.K. Balachandran (formerly of the Ramanujan Institute, Univrsity of Madras);
- Indranil Biswas (TIFR, Mumbai);
- V. Chandrasekhar (Chennai);
- E.R. Desikan (Chennai);
- P.P. Divakaran (CMI, Chennai);
- Sudhir Ghorpade (IITB, Mumbai);
- N.S. Gopalakrishnan (formerly of the University of Pune);
- N. Shanthi (Chennai);
- Ilangovan (TIFR, Mumbai);
- S.K. Joshi (Hon. Scientist Emeritus CSIR);
- P. Jothilingam (University of Pondicherry);
- Kalyan Sinha (ISI, Delhi);
- Dr. Kamala (a well wisher and a generous donor for CMI);
- Krishna Kumar (University of Calicut);
- V. Lakshmibai (Northeastern University, Boston);
- Vikram Mehta (TIFR, Mumbai);
- Gadhadhar Misra (ISI, Bangalore);
- C. Musili (University of Hyderabad);
- A. Muthukumar (Trustee Secretary, CMI);
- Nitin Nitsure (TIFR, Mumbai);
- K.S. Padmanabhan (formerly of the Ramanujan Institute, Univ. of Madras);
- R. Parimala (TIFR, Mumbai);
- S. Parvathi (Ramanujan Institute, University of Madras);
- S. Ponnusamy (IIT, Chennai);
- A.N. Radhakrishnan (Adviser, Shriram Group of Companies and CMI);
- N. Raghavendra (TCS, Hyderabad);
- M.S. Raghunathan (TIFR, Mumbai);
- C.S. Rajan (TIFR, Mumbai);
- S. Ramanan (TIFR, Mumbai);
- M.R. Rangarajan (formerly of the Loyola College, Chennai);
- K.V. Ramanathan (formerly of the TIFR, Mumbai);
- S. Ramaswamy (University of Pondicherry);
- P.S. Rema (formerly at the Ramanujan Institute, Univ. of Madras),
- Alladi Sitaram (ISI, Bangalore);
- Milind Sohoni (IIT, Mumbai);

- R. Sribala (Ramanujan Institute, University of Madras);
- Suchitra Srinivas, Industrial Economist;
- R. Thiagarajan (Chairman, Shriram Group of Companies);
- Usha Rubugunday (a well wisher and a generous donor for CMI);
- T.C. Vasudevan (Vivekananda College, Chennai);
- S. Venkitaramanan (Vice Chairman, Governing Council, CMI);
- S. Viswanathan, Editor, Industrial Economist.

Members and staff of CMI and IMSc. along with their families. Seshadri's family.

Excerpts from Greetings and Messages to Seshadri on his 70th birthday

Ram and Yvonne Abhyankar, U.S.A
Dear Seshadri:

It is indeed a great pleasure for us to wish you a very very happy 70th birthday. It has been a wonderful experience having known you ever since we first met for lunch in the Cite University in Paris in June 1958. Also there are the many happy memories of our very numerous get–togethers in TIFR and in other places in the world. We do look forward to seeing you and Sundari soon in Chennai or some other place.

Arun Ram, U.S.A.
Dear Sridharan,

Perhaps one of Professor Seshadri's greatest contributions is the non-quantifiable effect from the interest that he has shown in the work of young people throughout his career. Many of us are deeply indebted to him. He gave us a reason to believe in our own work: "if Prof. Seshadri is interested then perhaps it has some worth after all".

Michael Atiyah, U.K.
Although I cannot be with you to celebrate the 70th birthday of my old friend Prof Seshadri I would like to send my greetings for this happy occasion. He and I have had for many years a common interest in the study of vector bundles on algebraic curves, and I was much influenced by his early papers on the subject, written in collaboration with Narasimhan. The development of algebraic geometry and the related parts of theoretical physics have shown how important this work was. It pioneered the whole area.

Bhanu, U.S.A.
I have known Seshadri as a friend. What impressed me most is his humility, sense of humor and eagerness to learn. I have had the privilege of taping his carnatic music rendering. That brought out a couple of more endearing qualities of his – spontaneity and enjoying what he does. Behind every man is a woman – I can vouch for that in his case. Sundari is great – friendly, thoughtful, caring, and spirited. All the best to the couple from a friend of theirs from a long distance and seeking their blessings for her family.

Best wishes,

Michel Brion, France
Dear Lakshmi,

If it is not too late, please give my best wishes to Seshadri for his 70th birthday. And please tell him how much I admire his mathematical achievements and how much I enjoyed the opportunities we had to meet. His enthusiasm and kindness are deeply appreciated by all of us.

K. Chandrasekharan, Switzerland
My dear Sridharan,

Seshadri's career witnesses the leap of India in the world of mathematics from the era of Srinivasa Ramanujan into the post Ramanujan era of multivalent structures that inter-penetrate and make of mathematics a many-splendoured thing.

The lamp that Ramanujan lit burns ever so brightly, but a hundred other lights now dot the landscape and bring beauty and richness and colour to the scene.

It is a joy to notice that Sehadri's originality is not just confined to his own research work with its inspiring influence. His accomplishment of the integration of undergraduate education with graduate courses and postgraduate research makes him, and the institute he heads, exceptional in the realm of science. It opens up, especially in India, a myriad possibilities for the gifted young.

One cannot but applaud the founders of the institute, and its benefactors, and wish Seshadri and all his colleagues new conquests in the glorious struggle that is scientific research. Very sincerely yours, KC
"In the prison of his days
Teach the free man how to praise. "– W.H. Auden, Feb. 1939.

Dale Cutkosky, U.S.A.
Dear Sridharan,

Seventieth birthday wish to Professor Seshadri,

One of the most influential mathematicians of our time. I wish him many happy returns, and many more wonderful theorems.

Corrado De Concini, Italy
Dear Seshadri,

.... I heard that you are going to be 17 and a half very soon. This is not possible!!. In any case all my best wishes and thank you for all you gave me both mathematically and as a human being. I am sorry I could not come but I had engaged myself for a long time to go to Berkeley. In any case all my best wishes and a special hello to Sundari.

David Eisenbud, U.S.A.
Dear Seshadri,

I have enjoyed our friendship for a long time! Among the works of yours that have influenced me directly let me mention first your work on compactified jacobians and second your work on standard monomials, both essential contributions.

You may be more surprised to hear that your enthusiastic and searching glance often comes to my mind when I am looking for inspiration in a problem!

I also remember your beautiful singing with great pleasure, and a conversation we once had about it stands out in my mind: you said that as you walk down the street you were very frequently turning over not only mathematical problems, but also problems of how to make the most interesting musical phrases in a particular context; I remember thinking: "That's how music should be studied!"

With warm regards on your seventieth birthday!

Gopalakrishnan, Pune, India

Dear Sridharan,

I am happy to note that Prof. Seshadri will be completing 70 years and I feel happy to send my heartfelt greetings on this occasion. Prof. Seshadri's mathematical career spread over 5 decades started with his brilliant solution of an outstanding conjecture viz Serre's conjecture for 2 variables. This he did while at TIFR which may be called the gateway of modern mathematics, India. His further research solving some deep problems in geometry, and producing some brilliant students culminated in the award of FRS to him. In a quiet and unassuming manner he interacted with the mathematical community in India with an abiding concern for the development of mathematics in India. May God shower on him health and happiness and many more achievements in the years to come.

Phillip. A. Griffiths, U.S.A.

Dear Prof. Sridharan,

Professor C.S. Seshadri has made absolutely fundamental contributions to algebraic geometry. His originality and insight have given us precious new understanding of this classical, beautiful and central area of mathematics. Speaking for all my colleagues at the Institute for Advanced Study, we send you appreciation and best wishes on this joyous occasion.

P. Jothilingam, Pondicherry, India

Dear Professor Sridharan,

I remember with deep gratitude the unstinted support Professor C.S. Seshadri extended to all his younger colleagues at the T.I.F.R. ..... I pray that God almighty bestow good health and longevity on Prof. C.S. Seshadri.

Ravi Kannan, U.S.A.

It has been a great pleasure to know Prof. Seshadri for over two decades. Besides being such an outstanding mathematician, ... Prof. Seshadri has been an inspiration to many of us in very different areas and of course is a very friendly, approachable and kind person....

Peter Littelmann, Germany

Dear Seshadri,

It is a great pleasure for me to send you my congratulations and all my best wishes to your 70th birthday.

I am sorry not to be present personally, but while the ceremony is taking place I'll sit in a train to Paris, where I am going to give some lectures to students. These lectures will for sure also include an introduction into the theory of standard monomials, so, you see, since you are partially responsible for the content of the lectures, after all, my not being present is hence at least partially also your fault.

I will never forget the September 1983, when I arrived in the US as an exchange student. The Brandeis Math Department was organizing a "beginning of the semester" party, and while I was trying to figure out who is who, David Eisenbud came by and explained to me the tradition at Brandeis of having a Lecture Notes series based on the courses given by distinguished visiting professors. Slowly I began to understand that he was actually looking for a victim to take notes of your course on the "*Geometry of G/P and standard monomial theory*". I don't really know why he was asking me, maybe because a former undergradute student of Hanspeter Kraft was supposed to know at least what G/P means.

Anyway, Pradeep Shukla and I accepted the job, and I still remember this time with a tremendous joy. In retrospect, one reason for this is of course that you introduced me during that year into an exciting field in mathematics which turned out to have very interesting and surprising connections with various other fields in mathematics. I am sure that in the speeches on your research contributions somebody will mark the beginning of standard monomial theory and its implications.

But the most important reason for the joy was, and still is, the way you introduced me into the subject. I remember during the writing up of the Brandeis Lecture Notes your nearly infinite patience with me and your careful answers to all the questions I had. In some sense, this has not changed during the years. Always when we meet, which, I hope, will be also very often during the next 70 years, the talks with you are always an enormous source of inspiration.

Dear Seshadri, again all my best wishes for the future and at least as much good and exciting mathematics as we have had by you during the last 70 years.

M.G.K. Menon, New Delhi, India

Dear Prof. Sridharan,

It would have given me great pleasure to be present for the occasion, since I regard Prof. Seshadri very highly. We were colleagues at TIFR from 1955 when I joined the Institute. I always knew of him as an outstanding mathematician who was an important influence in the development of the School of Mathematics at the Institute, the finest school of this country in this field. Apart from that, he is a great human being gentle, courteous, honest, sincere and frank. His wife, Sundari and he are a highly cultured couple, particularly with their understanding of Carnatic Music. It has always been a privilege for me to have known him as a friend. ..... My very best wishes have always been with him.

..... Please convey to Prof. Seshadri my very best wishes for a wonderful seventieth birthday celebration and many many more happy and productive years, which he well deserves; and affectionate regards to him and to Sundari.

David Mumford, U.S.A.

A Toast to C.S.Seshadri

Seshadri, my old friend, congratulations on your 70th birthday (although that is a lie: we know it is only your $17\frac{1}{2}$th birthday). I am not far behind – maybe I'll catch up. In the US, the tradition at these occasions is to be quite informal, so I hope this is admissible in India too. Our lives have been so long intertwined, both in our mathematics and with our families. The math started it: it must have been around 1960 that I first heard of your ingenious work on the triviality of vector bundles on the plane. But I guess what thrilled both of us – it certainly thrilled me – was when our work on vector bundles on curves arrived at the same idea from two such different directions. What a strange thing it is that three people (you, me and M.S.), on *opposite* sides of the world (which, by the way, seemed a lot bigger in those days) using *totally* different techniques should construct the *same* compact moduli space!

Then we met, first in Cambridge and afterwards when my family and I came to TIFR, in 1967–68. ..... Our mathematics flourished. I found an old letter from me to you from about 1970 which says:

"Thanks for your letter which is beautiful and very encouraging. I was personally getting quite discouraged about extending my result when your letter arrived. The blowing up trick is marvelous ···"

You kept solving, one after the other, all these questions our ideas had led to! It was a heady period and we continued to alternate between Cambridge and that unreal mathematical pleasure dome on the Arabian Sea, the TIFR. With all our colleagues, these were the two poles of algebraic geometry at that time.

..... Good luck with your work, with research and with the Chennai Mathematical Institute. You are an extraordinarily brave man to embark on such a significant venture at this stage in your career. Good luck too to those who share your life, Sundari, Narasi and Giddhu and their growing families, and to your music, your other love. My only regret right now is not to be able to raise a glass of duty-free scotch in person and toast you and your future.

Pavaman Murthy, U.S.A.

Dear Sridharan,

...... I am truly fortunate to have been associated with Seshadri as his first student. I am sure that many colleagues and students of his will speak about his impressive achievements in algebraic geometry and his contributions to mathematics education in India. On this happy occasion, I would like to record my deep gratitude to Seshadri for the guidance, encouragement and valuable advice he has given me in my formative years at the T.I.F.R.

My earlier work on projective modules grew entirely out of understanding and using ideas from his work on the Serre problem in two variables. Later, our families became good friends. Prema and I very much cherish this friendship and convey our love to Seshadri, Sundari and family.

M.S. Narasimhan, Italy

Dear Sridharan,

I am very happy to learn that a symposium will be held in Chennai on 1 March, 2002, to celebrate Seshadri's seventieth birthday. I am sure that there will be others who will talk about his profound contributions to mathematics, his crucial role in the development of mathematical research in India and his recent efforts to promote mathematical education. So I would like to confine my remarks to my personal association with him extending over more than 50 years.

Although we were students together in Loyola college in the Intermediate and B.A. Honours classes, I got to know him well when we joined the Tata Institute in 1953. In those heady days as research students at the Institute we interacted very closely, reading and learning together an enormous amount of mathematics, perhaps in an amateurish but enthusiastic fashion. As I look back, this has played an important role in our future mathematical development. I still treasure the copy of his first pulished paper which he gave me with an inscription recalling the "many things learnt in company". Later, in the 60's, we collaborated in a work with which, I am sure, both of us are pleased and proud of. While we did not formally collaborate afterwards, I have profited by his scholarship in algebraic geometry and he has generously shared his insights with me. In fact, the task of my initiation to and assimilation of a large part of the work of Grothendieck was facilitated by conversations with him. I have been fortunate to have been associated with him over this long period. His friendship and thoughtful advice have enriched my life and I am grateful to him for this. I would like to convey to him my best wishes for many returns of leap year birthdays.

With best regards,

Raghavan Narasimhan, U.S.A.

Dear Sridharan,

It gives me great pleasure to pay homage to C.S. Seshadri on the occasion of the 70th year following his birth; as you all know, he has made it hard to speak more conventionally about his birthday except once in four years.

After K. Chandrasekharan, Seshadri and M.S. Narasimhan were the people most responsible for the remarkable atmosphere at the Tata Institute. Narasimhan and Seshadri were very influential in creating the feeling of openness and easy accessibility which was of such importance to young people coming to TIFR from all over the country. They exerted this influence even when they were away from Bombay in far away Paris. First, the outstanding mathematical research they were doing set an example; and then they communicated to people in Bombay the excitement of current research all over the world. Their influence after they returned to Bombay was enormous, and, I believe, continues to be felt even today.

Seshadri has taken this attitude with him to Madras, and has tried to create a similar group in a place where the ground had not already been broken. No one can fail to admire the success he has achieved in this very difficult attempt.

May I offer to Seshadri and his wife Sundari my very best wishes for many years of personal (and, in his case, mathematical) happiness?

Ann Newstead, U.K.

Dear Seshadri,

It is now 37 years since I first met you when you and Narasimhan visited the department here for two months. Although I have seen Narasimhan many times since, I never met you again until a conference in ICTP 35 years later! I was therefore delighted to recognise you across the Adriatico terrace when Peter and I arrived and we have happy recollections now of that month including especially happy memories of the evening when you joined us at our wedding anniversary dinner along with several other old friends. I hope very much that you are enjoying the celebrations which are being made in your honour and I look forward to another meeting in the not too distant future.

Peter Newstead, U.K.

Dear Seshadri.

It is a pleasure to write to you on the occasion of your 70th birthday. Your papers with Narasimhan were among the first research papers I ever read and most of my mathematical work derives from them. I have enjoyed all our (far too few) subsequent meetings. I hope we shall meet again before too long and I wish you all the best for the future.

Madhav Nori, U.S.A.

Dear Seshadri,

Poetry not being my forte, my sentiments on this happy occasion are best expressed by greetings telegram number sixteen "Many happy returns of the occasion". To add but two lines: right now, I'm lecturing on Quot schemes, GIT, the moduli of bundles on curves, exactly what I learnt from you many years ago right after stepping into the Tata Institute. Wishing you and your family and your extended family that includes the multitude of algebraic geometers educated by you, great joy on this occasion and hereafter, your old student, Madhav.

K.R. Parthasarathy, New Delhi, India

Dear Sridharan,

Best wishes for a useful symposium which could inspire the young colleagues of Seshadri to stay in India and bring up a new generation of mathematicians.

R. Parthasarathy, Mumbai, India

..... Seshadri, I bow to you. You have single-handedly created a volume of work for generations of mathematicians to lean on.

Gopal Prasad, U.S.A.

Dear Professor Sridharan,

I am very happy to learn that Professor Seshadri's seventieth birthday will be celebrated on March 1. .....I would like to wish Professor Seshadri a long and fruitful life in the service of mathematics. ....

Professor Seshadri was a role-model for my generation in the School of Mathematics at TIFR. His varied interests in and outside of mathematics are awe–inspiring. His gentle and friendly disposition and affectionate ways left a very nice impression on us. My wife and I were especially lucky to have also known Sundari. On this very happy occasion, both of us would like to congratulate her, Narasi and Giridhar, whom we fondly remember.

As a senior and very influential member of the Faculty of School of Mathematics at TIFR, Professor Seshadri was responsible for shaping the policies of the School for many years and was instrumental in its spectacular growth...........

Vladimir Popov, Russia

Dear Professor Seshadri,

On the occasion of your 70th birthday, let me express my admiration and highest appreciation of your remarkable contribution to mathematics. Your classical results on Picard varieties, moduli of vector bundles on algebraic curves, Serre conjecture, geometric invariant theory and Mumford conjecture belong to the golden fund of algebraic geometry and algebraic group theory. You are the leader of a scientific school that recently created modern standard monomial theory, the beautiful breakthrough linking representation theory, algebraic geometry and quantum groups.

I wish you very good health and new remarkable scientific achievements, and I wish further prosperity to your scientific school.

Claudio Procesi, Italy

Dear Sridharan,

I met Seshadri for the first time in Arcata in 1974 and we had immediately some mathematical exchanges on the Mumford conjecture, later he visited Roma with his family and I visited TIFR. We became friends and the collaboration has never stopped.

I followed in particular the developments of the program of Seshadri on standard monomials, in which his intuition and technical ability has been the driving force and which has finally led to a complete and beautiful theory.

I also appreciate very much his enthusiasm in supporting young researchers and creating suitable environment for research to grow. Finally I learned a bit from him to appreciate Indian classical music. So my best greetings for the anniversary and hoping to meet you soon.

B.P. Purnaprajna and Sunita, U.S.A.

Dear Sridharan,

.........Even though there will be many experts who will be talking about his math contributions I cannot resist saying that the things (in my area of interest) that he did 30 years ago are now being pursued with great interest, one of which is this very clever idea to measure the ampleness of a linear series (the now famous Seshadri constant)............

V.S. Ramamurthy, Secretary, DST, New Delhi, India

Dear Professor Sridharan,

Kindly convey my hearty congratulations to Professor Seshadri on his turning the age of seventy on March 1, 2002. It is befitting to felicitate him on this occasion. Professor Seshadri's contribution to mathematics is well-known. He has rendered yeoman service in the field of mathematics education.

I wish the symposium and the felicitation function all success.

Raja Ramanna, Bangalore, India

Dear Prof. Seshadri,

Please accept my good wishes on completing 70 years. Your contribution to mathematics has brought great honours to India and I am sure you will continue to do this for many years to come.

With best wishes,

Ravi Rao, Mumbai, India

Dear Prof. Sridharan,

I have certainly been quite influenced by Professor Seshadri's simple idea to show freeness of projective modules over a polynomial ring in two variables over a field by viewing it as a problem of completing a unimodular polynomial vector. Please convey my appreciation of all that Professor Seshadri has done in and for mathematics. I wish him a happy 70th birthday.

Jayant Shah, U.S.A.

Dear Seshadri,

Emma and I send you our best wishes on your 70th birthday. Your dedication to mathematics and music sets the highest standard for the next generation to follow.

T.A. Springer, Netherlands

Dear Professor Seshadri,

It is a pleasure to send you my congratulations on your seventieth birthday. I hope that it will be a memorable day for you.

Although our mathematical interests are not exactly the same, they are not too far apart either. Hence, over the years, I have come into contact with much of your admirable work, for example the work on standard monomial theory (with your collaborators) and with its many ramifications. I have very good memories of our contacts, over a period of more than the thirty years. The last one was last June, in Rome.

My wife joins me in the congratulations. We wish you many good and fruitful future years.

B.V. Sreekantan, New Delhi, India

Dear Sridharan,

I am very happy to know that a symposium and a felicitation function is being organized on March 1st on the occasion of the 70th birthday of Prof. C.S. Seshadri. It is most appropriate. ..... Please convey on behalf of Ratna and myself our best wishes to Seshadri and Sundari on this joyous occasion and our prayers for their continued well-being in the decades to come...

Bhama Srinivasan, U.S.A.

I am very happy to hear about the symposium to celebrate Prof. Seshadri's 70th birthday. As we all know Prof. Seshadri has made important contributions to algebraic geometry. However, he is also to be specially commended for his efforts to build a school of mathematics in Chennai, and for the encouragement he has given to many young people in this endeavor. While many of us have chosen to practice our profession abroad, he has been an inspiration to many mathematicians in India. The flourishing Chennai Mathematics Institute is the result of his vision of a great future for mathematics in India.........

P.S. Thiagarajan, Singapore

Dear Seshadri,

Warm felicitations! I wish you the very best in the years to come.

I hope you remember (fat chance) that we celebrated in style your 60th birthday as well. Looking back, you must be proud of the progress that the Institute has made in the last 10 years; we have so many more problems now!

But seriously, progress indeed has been made and it is mainly due to your sense of values, quiet determination and the enormous store of good will you have accumulated over the years by being what you are.

I am proud of my association with you and in the years to come, I greatly look forward to further assisting you to realize your vision of CMI.

Again, on this happy occasion, Vishu, Usha and I wish you and your family the very best.

Warm regards,

K. Varadarajan, Canada

Dear colleagues,

..... Many of us have had a long association with him and benefited immensely from his deep and profound knowledge of mathematics. He is responsible for building a very strong school of Algebraic Geometry in India.

After establishing a strong school in Algebraic Geometry at T.I.F.R., he felt that similar centres should be built elsewhere in India. He moved to Madras (presently Chennai) solely for the purpose of building such a centre in Madras. We can now definitely say that his efforts have borne fruit and we have a group of young people dedicated to many important branches of modern mathematics and producing work of the highest quality. All of you are aware that for his significant achievements in mathematics Seshadri was elected a Fellow of the Royal Society of London (U.K.).

It was nice of him to have involved me indirectly in the process of building a good school in Chennai. I will always remain indebted to him for giving me an opportunity to get involved in the growth of mathematics in India.

More recently, he felt the need to train talented undergraduates in mathematics by providing them with facilities to learn mathematics "the right way" through interaction with top quality mathematicians. I am sure, that his efforts in this direction will have beneficial effects in the near future.

Finally let me conclude by wishing him a long and healthy life so that all of us, as well as future generations can benefit from the wealth of knowledge he possesses.

V.S. Varadarajan, U.S.A.

Dear Sridharan,

It is my great privilege to have known Seshadri for many years. I have always had great admiration for his profound contributions to mathematics, but even more for his efforts to lift mathematical research in India to high levels. In this matter he is a leader, leading by personal example and sacrifice. Such persons are rare in any generation and India is very fortunate that he decided to share his great gifts with the young people of his country. I wish him continued health and strength so that he can continue his mission in life according to his lights. My wife Veda joins me in wishing Seshadri and Sundari many more years of happiness.

R. Vijayaraghavan, Mumbai, India

Dear Prof Seshadri,

..... You deserve every honour that the mathematical community can confer. You are also an eminent musician and scholar of good repute. Such functions give joy and happiness to your friends and colleagues.

I wish the function all success. All your good friends and colleagues will echo my wish with a vaishnavite saying "Innum oru nootrandu irum".

My greetings to Sundari and family.

Michel Waldschmidt, France

Dear Seshadri,

It seems it will be more convenient that I address you in English in spite of the fact that your French is excellent compared to my broken English.

I understand that you are turning seventy on March 1, 2002 and that your colleagues at Chennai Mathematical Institute and The Institute of Mathematical Sciences, Chennai, are organizing a one day symposium on March 1, 2002. I would have liked very much to be with you at this occasion; unfortunately it will not be possible for me to come to Chennai before September 2002. But I am pleased to send you this personal message, even if it does not replace my physical presence.

A few days ago I attended a meeting of the Committee for Developing Countries (COPED) in the French Academy of Sciences, devoted to mathematics. I took this opportunity to quote the agreement between Ecole Normale Superieure ENS Ulm and your undergraduate program as a successful example of cooperation between our two countries.

I wish to congratulate you not only for your mathematical achievements, but also for your founding of the SPIC Institute (now Chennai Mathematical Institute) first, and then your undergraduate program. This action of yours will have an everlasting influence on the development of mathematics in India.

I wish also to convey the best wishes to you from my daughter, who appreciated very much to meet you last summer in Chennai. Finally together with my wife we address our cordial greetings to you and all your family.

Felicitation messages were also received from:

- T.S. Bhanu Murthy, Chennai;
- R. Bhatia and family, ISI, Delhi;
- Vyjayanthi Chari, UCR;
- R.C. Cowsik, University of Bombay;
- Jayanta Ghosh, ISI, Kolkata;
- R.V. Gurjar, Mumbai;
- P.K. Iyengar, Mumbai;
- Anil Kakodkar, Chairman, Atomic Energy;
- Ravi Kulkarni and colleagues from HRI, Allahabad;
- S. Kumaresan, University of Bombay;
- B.V. Limaye, Nirmala Limaye, Mumbai;
- V.C. Nanda, University of Punjab;
- Narsi Seshadri, Zurich;
- A.J. Parameswaran, TIFR;
- I.B.S. Passi, HRI, Allahabad;
- S. Raghavan, Chennai;
- N. Ramabhadran, Chennai;
- K. Ramachandra, TIFR;
- C.N.R. Rao, Jawaharlal Nehru Centre for Advanced Scientific Research, Bangalore;
- Sawant and colleagues at the Office of the School of Mathematics, TIFR;
- Balwant Singh, IIT, Mumbai;
- Sujatha, TIFR;
- K.K. Thilakan, Mumbai;
- P.A. Vishwanath, Mysore.

Seshadri and Sundari, 1966

Seshadri and family with M.S. Narasimhan, Erika Mumford and children, 1967

Seshadri, his mother Chudamani and his wife Sundari being felicitated

At the symposium

Part I

Mathematical Contributions of C.S. Seshadri

Talks given at the Symposium

The early work of Seshadri

S. RAMANAN

It is with immense pleasure that I associate myself with the celebration of Seshadri's birthday. Many of our foibles and shortcomings go back to our childhood days. I think that Seshadri's habit of leaving sentences incomplete, can be traced back to his birthday, which is February 29th, and often the month fails to complete itself!

I have been asked to talk about his early work. I confine myself therefore to the first ten years of his research career. Apparently, when Seshadri was a graduate student at the Tata Institute, Prof K. G. Ramanathan suggested to him a problem connected with the functional equation of the Zeta function, and Seshadri solved the problem. Some more work on those lines was supposed to follow, but Seshadri's interests changed in the meanwhile and the paper never got published.

His first published paper [1], which constitued his thesis as well, was on 'Generalised multiplicative meromorphic functions on a complex manifold'. In 1954, Laurent Schwartz (known for his theory of distributions) visited the Tata Institute, and both Seshadri and M.S. Narasimhan were influenced considerably by him. There is some evidence of it in that paper, since there are some analytical arguments of the kind typical of Schwartz's approach to differentiable manifolds.

In any case, the paper addressed some complex analytical problems related to Riemann surfaces. Two important problems dealt with in that paper are the following.

The first is the fact that all vector bundles on the Riemann sphere are direct sums of line bundles. Seshadri essentially gives a proof of this, unaware that Grothendieck had stated and proved it. In fact, it is nowadays believed to have been known in some form to many mathematicians before.

The second relates to the Hodge decomposition. The cohomology of a compact Kähler manifold admits a decomposition (although these days we emphasize the filtration which gives rise to this direct sum decomposition). Indeed, de Rham's theorem can be stated for a complex manifold M as follows. Consider the complex of sheaves

$$0 \to \mathbb{C} \to \mathcal{O} \to \Omega^1 \to \Omega^2 \to \cdots \to \Omega^n \to 0$$

where Ω^i denotes the sheaf of holomorphic forms of degree i, and n is the dimension of the complex manifold. Then the cohomology spaces of M with coefficients in $\mathbb{C}$ are canonically isomorphic to the hypercohomology of the above complex. The significance of this formulation is that the boundary homomorphisms are holomorphic operators, and so in a certain sense, the topology of M has been described, purely by complex analytic data. The key point now is that Hodge decomposition is equivalent to saying that if we replace in the above complex, all the operators by 0, we still get the same cohomology. From this point of view, the filtration becomes more natural than the decomposition.

It is trivial to see that de Rham's theorem can be generalised when the coefficients are changed from $\mathbb{C}$ to a local system $\mathcal{L}$. However Hodge decomposition is not valid for an arbitrary local system. Indeed, take a Riemann surface M of genus at least 2, and a nontrivial extension of the type

$$0 \to \xi \to E \to \xi^{-1} \to 0$$

where ξ is a line bundle with ξ^2 isomorphic to the canonical bundle. (Such a line bundle is called a *theta characteristic*). It is easy to see (and can for example be deduced from a theorem of Andr/'e Weil which asserts that indecomposable bundles of degree 0 come from local systems) that E is associated to a local system. Also if ξ has nonzero sections, E also does, but the local system does not have any section. This shows that a Hodge type decomposition is not true for this local system. Seshadri proves that if the local system is *unitary*, then the Hodge theory goes through. There has been a resurgence in recent years, in understanding what happens in the non-unitary case, thanks to the work of Hitchin, Donaldson, Uhlenbeck, Yau, Corlette and Simpson.

In the 50's, the foundations of Modern Algebraic Geometry were being laid and there was much interplay between algebraic and analytic geometry. One of the theorems that Grauert proved in this era was that the topological classification and analytic classification of vector bundles on a Stein manifold were equivalent. Since $\mathbb{C}^n$ is contractible, this implies that all holomorphic bundles on the complex manifold $\mathbb{C}^n$ are trivial. A natural question arose whether there are similar theorems for affine algebraic varities. at least over $\mathbb{C}$. In particular, one may ask if all algebraic vector bundles on the affine space $\mathbb{C}^n$ are algebraically trivial.

The algebraic translation of this question is whether projective modules over polynomial rings are free. This goes by the name 'Serre's conjecture'. Seshadri visited Paris for a few years, and came to know of this problem. Since submodules of free modules are free over $\mathbb{C}[x]$, this assertion is clearly true for one variable. Thus the first nontrivial case is for two variables, and Seshadri settled it in the affirmative. Serre knew that projective modules M over polynomial rings A were stably free (i.e. the direct sum of M with a free module is free) and so it is enough to show that if M is a projective A-module such that $M \oplus A$ is free, then M is itself free. This leads to the following easily posed question: If $(a_1, \cdots, a_k)$ is a primitive vector, then does it occur as the 1st row of a unimodular (k, k) matrix? Apparently, Seshadri proved this for $A = \mathbb{C}[X, Y]$ and Serre told him that this finishes the problem for $\mathbb{C}[X, Y]$. The final paper [2] deals with the case of modules over $A[X]$ where A is a principal ideal domain. Later [3] Seshadri also handled the case of $A[X]$ where A is the ring of functions over an affine curve. I had already joined the Tata Institute as a graduate student by this time and vividly remember the excitement this achievement caused there, and even more importantly, the confidence that it gave us all that work of international quality could be done by Indian mathematicians. As is well known, this spawned a lot of work in India and abroad, the culmination being the Suslin-Quillen theorem which settled the question affirmatively in any number of variables.

The classical Abel-Jacobi problem concerns itself with the understanding of divisor classes on a compact Riemann surface. With the laying of the foundation of the new algebraic geometry, it began to be posed as an existence problem for the Picard variety for any algebraic variety M. Chevalley and Matsusaka had independently constructed it and had proved the essential properties when M has only normal singularities. Seshadri observed that the construction in the case of arbitrary varieties was relatively easy, using purely formal machinery, but to show that all algebraically trivial Cartier divisors are covered by this construction, was quite difficult. Rosenlicht had constructed such a Jacobian for singular curves, and Seshadri, in a real tour de force [4], showed that the general case can be deduced from this. It is true to say that Serre, H. Cartan and Chevalley realised at this point that they had on their hands a mathematician of exceptional ability. In fact, Cartan suggested that Seshadri be asked to give a talk at the Chevalley seminar, and I believe that Seshadri was genuinely proud of this. He checked the universal property in a subsequent paper [8]. This paper starts with the following remark! The proof of the universal property of the Picard variety G of a complete variety U, is not complete.

The next paper I want to touch upon concerns itself with the quotient space problem [5]. Many existence problems, including that of the Picard variety, can be reduced to the existence of the quotient of an algebraic variety under the action of an algebraic group. The quotient comes with a natural sheaf of rings and the problem is to investigate if it is actually a variety. Seshadri considered this question and this paper contains the following theorem. If G acts on a normal variety M over $\mathbb{C}$ such that the graph of the action is closed and the isotropy subgroups at all points have the same dimension, then every point has an open neighbourhood U and an action by a finite group F on U such that the quotient of M by G is locally like the quotient of U by F.

Seshadri used this result subsequently in [6] where he considered free actions by Abelian varieties. He showed that in this case the quotient is always a variety. In the course of the proof he proved that a Weil divisor in $V \times T$, with V nonsingular and T normal, is locally principal, if does not contain any set of the form $V \times \{t\}$. He gave a global proof of this local result and C.P, Ramanujam, in an appendix, gave it a purely local treatment. The latter is now known in the literature as Samuel-Ramanujam theorem.

Finally, I should mention the elegant proof [7] that Seshadri gave of the theorem of Weizenbock. In the Theory of Invariants, one is, in the first place, interested in the following question. Given a representation of an algebraic group G on a vector space V. there is also a natural action of G on the algebra of polynomial functions on V, namely the symmetric algebra $S(V^*)$ over the dual. Let I be the set of polynomials fixed by all elements of G. It is obvious that I is a subalgebra of $S(V^*)$. One would like to know first (in the spirit of Hilbert) if it is finitely generated as an algebra, and then to write down in specific cases, a set of generators, and relations, etc. This was fourteenth in the list of Hilbert's problems. This had been shown to be true for finite groups by E. Noether, for reductive groups by Weyl and for rational representations of the additive group of the field by

Weizenbock. It was therefore widely believed that it is true for all algebraic groups. However Nagata showed by a counter-example finite generation does not hold in general. Seshadri gave a simple proof of the theorem of Weitzenböck. There was subsequent work due to G. Hochschild and G.D. Mostow ("Unipotent groups in invariant theory", Proc. Nat. Acad. Sci. U.S.A. 70, 646-648) generalizing his idea to the unipotent radical of parabolic subgroups of reductive groups.

As I have said, Seshadri's early work had the themes 'Vector bundles, Invariant Theory, Algebraic groups, Quotient spaces'. He devoted much of his later years to these areas and I am sure that others will have more to say on his later work.

References

[1] Generalised multiplicative meromorphic functions on a complex analytic manifold, Jour. Ind. Math. Soc. (**21**) 1957, 149-178.

[2] Triviality of vector bundles over the affine space K^2, Proc. Nat. Acad. Sci. U.S.A. (**44**) 1958, 456-458.

[3] Algebraic vector bundles over the product of an affine curve and the affine line, Proc. Amer. Math. Soc. (**10**) 1959, 670-673.

[4] La variété de Picard d'une variété complète, Annali di matematica pura ed applicata (**55**) 1962, 117-142.

[5] On the quotient space by an algebraic group of automorphisms, Math. Annalen (**149**) 1963, 296-301.

[6] Quotient space by an Abelian variety, Math. Annalen (**152**) 1963, 185-194.

[7] On a theorem of Weitzenböck in invariant theory, Mem of the Coll. of Sc., Kyoto University (**1**) 1962, 403-409.

[8] Universal property of the Picard variety of a complete variety, Math. Annalen (**158**) 1965, 293-296.

Chennai Mathematical Institute
92 G.N. Chetty Road, Chennai 600 017,
INDIA.
E-mail address: sramanan@cmi.ac.in

Geometry of Moduli Spaces (The work of C.S.Seshadri)

V.Balaji

The aim of this brief article is to survey three fundamental papers of C.S.Seshadri on the moduli space of semi-stable bundles and their desingularisation.

The Narasimhan-Seshadri Theorem

Let X be a Riemann surface of genus $g \geq 2$ or equivalently a smooth projective curve of genus $g \geq 2$. Let $U(n)$ be the group of $n \times n$-unitary matrices. It is fairly classical that the cohomology group $H^1(X, U(1))$ gets the structure of a torus, identified with the quotient $H^1(X, \mathbf{R})/H^1(X, \mathbf{Z})$. The torus is given a complex structure by fixing one on the topological surface X and making the identification:

$$H^1(X, \mathbf{R}) \simeq H^1(X, \mathcal{O}) \simeq H^{0,1}(X)$$

the dual of the space of holomorphic differentials on X. This may be termed a version of the *abelian class field theory.* One can identify the group cohomology $H^1(X, U(1))$ with the homomorphisms

$$\pi_1(X) \longrightarrow U(1).$$

The theme in the seminal paper of Narasimhan-Seshadri ([**NS2**]) is the generalisation to the non-abelian case namely the study of the space $H^1(X, U(n))$, which parametrises conjugacy classes of representations:

$$\pi_1(X) \longrightarrow U(n).$$

It is well known that the fundamental group $\pi_1(X)$ of X has generators $\{a_1, a_2, ..., a_g, b_1, b_2, ...b_g\}$ with a single relation given by:

$$\Pi[a_i, b_i] = 1.$$

Consider the map $U(n)^g \times U(n)^g \longrightarrow U(n)$ given by $(a_i, b_i) \longrightarrow \Pi[a_i, b_i]$. Let $\mathcal{U}$, (a subset of $U(n)^{2g}$), be the inverse image of 1. One sees that the space $H^1(X, U(n))$ is the quotient of $\mathcal{U}$ by $U(n)$, making the space $H^1(X, U(n))$ compact Hausdorff. Further, its subset of equivalence classes of all *irreducible* representations gets a structure of a real analytic manifold of dimension $2\{n^2(g-1)+1\}$ whose tangent space at a point $p \in H^1(X, U(n))$ can be identified with $H^1(\pi(X), ad(\rho))$, where ρ is an irreducible representation in the class corresponding to the point p. In fact, what is proven is that the space of equivalence classes of *irreducible* unitary representations of $\pi_1(X)$ of rank n has a natural structure of a complex manifold of the dimension $\{n^2(g-1)+1\}$ which was the expected dimension of the moduli space of vector bundles of rank n and degree d. This is primarily the theme in the first paper ([**NS1**]) of Narasimhan-Seshadri.

In the classical abelian case, the Abel-Jacobi map identifies the space of representations $H^1(X, U(1))$ with the Jacobian of the curve X. The Jacobian can be seen geometrically as the moduli space of holomorphic line bundles on X of degree 0. The corresponding non-abelian analogy is the theme in the second paper ([**NS2**]) of Narasimhan-Seshadri.

The basic object in the non-abelian theory is what is known as a *stable bundle*, obtained in an altogether different context by D.Mumford ([**GIT**]). A holomorphic vector bundle E over X is said to be *stable* if, for all holomorphic sub-bundles F, we have the inequality:

$$degree(F)/rank(F) < degree(E)/rank(E),$$

where, by degree(E) we mean the first Chern class $c_1(E) \in H^2(X, \mathbf{Z}) \simeq \mathbf{Z}$.

The second theorem of Narasimhan-Seshadri studies the geometry of the space of isomorphism classes *stable holomorphic bundles of rank n and degree d* which we denote by $M(n,d)^s$. They prove that $M(n,d)^s$ has a natural structure of a connected complex manifold and its underlying manifold structure can be identified with that of the manifold of *irreducible unitary representations* of $\pi_1(X)$ of *type* τ (a technical notion needed to take care of bundles of non-zero degree). They also show that the space $M(n,d)^s$ is *compact*, if n and d are coprime. Further, the topology on the space of stable bundles is independent of the complex structure on the curve X.

To describe this further, recall that the curve X can be realised as the base space of a principal bundle $Z \longrightarrow X$ with structure group $\pi_1(X)$, where Z is the universal cover of X. Let $\rho : \pi_1(X) \longrightarrow U(n)$ be a representation. Then, via ρ, we can get the *associated vector bundle* to the principal bundle $Z \longrightarrow X$ given by, $E_\rho \simeq (Z \times^\rho \mathbf{C}^n)$, and one sees that E_ρ gets a natural holomorphic structure. It is then proved that a vector bundle E on X of degree 0 is *stable* if and only if it is isomorphic to a bundle E_ρ associated to an *irreducible* unitary representation ρ of $\pi_1(X)$. The main theme of this paper ([**NS2**]) can be described as establishing a functorial correspondence $\rho \longrightarrow E_\rho$ between the categories of $\{$*irreducible unitary representations*$\}$ and $\{$*stable holomorphic bundles of degree* 0$\}$. Moreover, this functorial correspondence is realised geometrically by the identification of the moduli space of stable bundles $M(n,d)^s$ with the manifold of irreducible representations mentioned above.

The links between holomorphic bundles and representations of the fundamental group was apparent in the earlier work of A.Weil, namely, that if V is an indecomposable bundle on X, then V is isomorphic to a bundle arising from a linear representation ρ of the fundamental group $\pi_1(X)$ if and only if $deg(V) = 0$. This is equivalent to the condition that V admits a flat holomorphic connection.

From a different standpoint, the Narasimhan-Seshadri theorem illustrates in a natural manner the philosophy underlying Grothendieck's Tannakian formalism wherein "fiber functors" play the role of points. Let V be a vector bundle of degree 0 on X. Then the basic theorem in ([**NS2**] in fact proves that V is isomorphic to a unitary bundle V_ρ if and only if, V is isomorphic to a direct sum of stable bundles of degree 0 i.e a *polystable* bundle of degree 0. By this identification, it follows that the *tensor product* of two polystable bundles of degree 0 is again polystable. Let V be polystable of degree 0. Let $\mathcal{C}(V)$ be the full sub-category consisting of subquotients of tensor powers of V and V^*. If we fix a base point $x \in X$, then this category is naturally equipped with a *fibre functor* ω_x, namely, associate to each bundle E its fibre E_x at x.

Then it can be shown that $(\mathcal{C}(V), \omega_x)$, gives an example of a *Tannakian category*. The Grothendieck-Tannaka theorem realises this category as a category of representations of a pro-algebraic group scheme $M(V)$. The image of $M(V)$, i.e $Im(M(V)) \in GL(V_x)$ is the monodromy group of the representation ρ i.e the Zariski closure of $\rho(\pi(X,x))$ in $GL(n)$. (cf [**BBN**], [**No**] for some applications of this outlook)

The Narasimhan-Seshadri theorem has had a profound impact on the subject. It has been developed and generalised on many fronts. Just to name a few, starting with the seminal paper of Atiyah and Bott, unexpected links with mathematical physics were perceived, coming from the so-called Yang-Mills equations, where they prove that the irreducible unitary representations realises the minimum, in the Morse theoretic sense, of the Yang-Mills functional. This led to a radically different differential geometric approach to these problems and vast ramifications in mathematical physics, starting with the work of S.Donaldson, who gave a purely differential geometric proof of the Narasimhan-Seshadri theorem, and in the subsequent work of Kobayashi-Uhlenbeck-Yau. In the Hodge theoretic direction the non-abelian theory was developed in the fundamental work of N.Hitchin and C.Simpson on Higgs bundles; earlier, working over bundles with more general structure groups A.Ramanathan had generalised the Narasimhan-Seshadri theorem for stable principal bundles on curves.(cf [**Sim 0**], [**R2**])

Compactification of the moduli spaces

A holomorphic bundle E is termed *semi-stable* if the inequality given for stable bundles admits equality as well. In [**Ses1**] Seshadri then proceeds to compactify the moduli spaces $M(n,d)^s$ by extending Mumford's Geometric Invariant Theory([**GIT**]). Let $M(n,d)$, the moduli space of semi-stable bundles on X of rank n and degree d under the S-equivalence i.e $V \sim V'$ if and only if the associated graded objects $gr(V) \simeq gr(V')$. Seshadri first proves that the category of semi-stable bundles of fixed *slope* μ ($\mu = \frac{degree}{rank}$) is an abelian category which is artinian and noetherian and the stable bundles are the *simple objects* in this category; the S-equivalence corresponds to the usual Jordan-Holder equivalence. In particular, if the degree of the underlying bundles (and hence slope) is zero , each semi-stable bundle on X is S-equivalent to a unitary bundle on X. In the categorical realisation these objects correspond to the *semi-simple objects* and the functorial correspondence is also extended to include all unitary representations.

It is of interest to relate this to the Tannakian philosophy mentioned above. One first proves that given a *semi-stable* bundle, not necessarily polystable, the category $(\mathcal{C}(V), \omega_x)$ defined above is a Tannakian category as well. The key ingredient is the *tensor product theorem*, namely, that the tensor product of two semi-stable vector bundles of degree zero is again semi-stable. This can be proved using the Narasimhan-Seshadri theorem and the fact that semi-stable bundles have a Jordan-Holder filtration in terms of stable bundles. The property that a semi-stable bundle V of degree zero is polystable (i.e "semi-simple") is then equivalent to the *reductivity* of the monodromy group of V.

The moduli space is shown to be *topologically compact* and the projectivity deduced as a consequence. In a subsequent paper ([**Ses1**]) Seshadri proves this by a direct method of embedding the Hilbert scheme in a suitable product of Grassmannians which gives the projectivity of the moduli spaces over fields of arbitrary characteristic as well (see the talk of V.Mehta in this collection). Philosophically, in a certain sense, the *projectivity* of the moduli space of semi-stable bundles, *thought of as a GIT quotient*, is equivalent, as a model case, to the Mumford conjecture on the equivalence of *geometric reductivity* and *reductivity* for $GL(n)$.

I remember, during Seshadri's 60th birthday celebrations, David Mumford pointing out that this construction of a compactification was a perfect representative example and a forerunner of all subsequent GIT constructions of compactification in various moduli problems. For a detailed exposition of these ideas and subsequent generalisations (cf. [**V**]). For further development (cf. [**R1**], [**BS**], [**BP**], [**Nit**], [**Sim 1**], ...).

Desingularisation of moduli spaces

As we observed in the last section, the moduli spaces of stable bundles namely $M(n,d)^s$ was compactified by using GIT and the semi-stable bundles under suitable equivalence relation were used to compactify the moduli space of stable bundles. In the case when the rank and degree of the bundles namely n, d were co-prime, it is easy to see that the notions of semi-stability and stability coincide. Further, it is shown in ([**NS2**]) that the moduli space $M(n,d)^s$ is a smooth variety for all (n,d). Combining this with our earlier remark, it follows that when $gcd(n,d) = 1$, $M(n,d)^s$ is smooth projective variety.

Let X be a smooth projective curve of genus $g \geq 2$ defined over an algebraically closed field k. Let $M(n,0)$, be as above, the moduli space of semi-stable bundles on X of rank n and degree 0 under the equivalence $V \sim V'$ if and only if the associated graded objects $gr(V) \simeq gr(V')$. Recall that semi-stable objects are the *semi-simple* objects in our category of vector bundles and the stable bundles the simple objects; this equivalence corresponds to the usual Jordan-Holder equivalence. One may also recall that the moduli space $M(n,0)^s$ of stable bundles is an *open smooth subvariety* of $M(n,0)$ and it was shown by Narasimhan and Ramanan that

$M(n,0)^s$ is precisely the smooth locus of $M(n,0)$ except in the exceptional case when $g = 2$ and $r = 2$ in which case it was shown that $M(2,0) \simeq \mathbf{P}^3$.(cf [**NR1**]).

The general problem of desingularisation of $M(n,d)$ can be described as seeking a construction of a birational model for $M(n,d)$ as a smooth compactification of $M(n,d)^s$ in *a moduli theoretic fashion*, for example, as representing a suitable moduli functor. The existence of a desingularisation over fields of char 0 follows by the general theorem of Hironaka, but an explicit realisation (moduli-theoretic) of such a model for the purposes of application was still not done. This problem was solved for the case $n = 2, d = 0$, first by Narasimhan and Ramanan and almost contemporaneously by Seshadri by a completely different approach.(cf [**Ses2**], [**NR2**]).

We now describe the model constructed by Seshadri. The length of the present section is somewhat longer than the earlier ones, primarily because we felt that this work of Seshadri is somewhat less known. Moreover, since the question of desingularisation is not completely answered, this offers some very challenging problems.

Notation: For the rest of the talk we shall denote by M the variety $M(2,0)$ and M^s the subvariety $M(2,0)^s$.

Let PV_4 denote the category of parabolic semi-stable bundles (V,Δ), where V is of rank 4 and degree 0, and Δ is a parabolic structure at a marked point $x_0 \in X$ with small weights $(\alpha_1, alpha_2)$. (for a definition of these concepts see the next article by N.Nitsure)

The candidate for the smooth compactification of M^s is the variety N which parametrises isomorphism classes of vector bundles $(V,\Delta) \in PV_4$ such that the algebra of endomorphisms, $End(V)$, is a specialisation of the (2×2)-matrix algebra $\mathcal{M}_2$.

One defines the variety of specialisations of $(n \times n)$-matrix algebra $\mathcal{M}_n$ as follows: Let W be n^2-dimensional vector space. A structure of a k-algebra on W is just giving a multiplication, namely a k-linear map

$$A : W \times W \longrightarrow W$$

or equivalently an element of the vector space $Hom(W \otimes W, W)$. Fix an element $e_0 \in W$. Consider the closed subset $Z \subset Hom(W \otimes W, W)$, which consists of algebra structures, with e_0 being the identity element.

Denote by $\mathcal{A}'_n$ the subspace of $Hom(W \otimes W, W)$ which give algebra structures A on W such that $(W, A) \simeq \mathcal{M}_n$. One has a natural left-action of the group $GL(W)$ on $Hom(W \otimes W, W)$, namely

$$g.A_1 = A_2 \; iff \; (W, A_1) \simeq (W, A_2)$$

Note that $\mathcal{A}'_n$ is an orbit under $GL(W)$. Let $H = Stab(e_0)$ be the isotropy subgroup of $GL(W)$at e_0. Then the variety Z is an orbit under H. Let $\mathcal{A}_n$ be the Zariski closure of Z in $Hom(W \otimes W, W)$. Then $\mathcal{A}_n$ gets a natural structure of a closed subvariety of the affine space $Hom(W \otimes W, W)$ and for every $A \in \mathcal{A}_n$ we call the algebra (W, A) a *specialisation* of $\mathcal{M}_n$.

We will briefly collect a few facts about the variety N:

- The variety N *represents a natural moduli functor* and hence there exists a universal family of vector bundles of rank 4 equipped with parabolic structures on the product $X \times N$. We observe that this in contrast to the fact that there is no family of semistable vector bundles on $X \times M$. We further remark that, this makes the variety N similar to the moduli space of vector bundles of coprime rank and degree which supports universal families.
- The *specialisation* property of the endomorphism algebra of bundles in N show that there exists a family of quadratic forms on a fixed three dimensional vector space, $\{Q_t\}_{t \in N}$, parametrised by N. This stratifies N into subvarieties $\{N_i\}$ $i = 1, 2, 3$ defined as follows:

$$N_i = \{t \in N | rankQ_t \leq 3 - i\}$$

  Thus,

$$N_3 \subset N_2 \subset N_1 \subset N.$$

  We observe that $N - N_1 \simeq M^s$ and the other strata can be completely described in terms of the singular locus of M below.

The proof of the smoothness of N is quite involved and is related functorially to the question of smoothness of the variety of specialisations $\mathcal{A}_2$. The *projectivity* of N is very ingeniously handled by realising N as a closed subvariety of a suitable moduli space of parabolic bundles which one already knows is projective. The appearance of parabolic bundles here is almost magical! Most of the construction and properties generalise for all ranks n except the key property of smoothness. In fact, it can be shown that this variety is not smooth for $n \geq 3$ (cf [**BR**]).

One may add a word here regarding the general philosophy which Seshadri had on desingularising the moduli spaces of vector bundles.

The idea is to embed diagonally $GL(2)$ as a subgroup of $GL(4)$ and realise the $GL(2)$-bundles as obtained from reduction of structure group of the corresponding $GL(4)$-bundles. Then one *compactifies* the homogeneous space $GL(4)/GL(2)$ in a an equivariant fashion. The isotropies of the orbit closures gives one the so-called specialisations of $GL(2)$ and also the candidates for structure groups of the *limiting bundles*. The problem then would be to isolate certain components of the variety of specialisations wherin one would wish the limiting bundles to have their structure group. This general method would lead to some compactification, but the difficulty in general would be to prove the smoothness of the variety of specialisations as well as the smoothness of the morphism which associates the bundle to its structure group. The general problem would then be the following: let G a reductive algebraic group. Classify the $\mathcal{H}$-torsors, where $\mathcal{H}$ is a smooth subgroup scheme of G coming from the specialisations of H.

The desingularisation has been studied in depth by Seshadri and his students in a series of papers which give results parallelling the ones known for the usual moduli of vector bundles of coprime rank and degree. (cf [**BS1**] [**BS2**] [**B1**] [**BV**]). The construction also carries through over an arbitrary ring as is shown in a recent work of another student of Seshadri's (cf [**T**]).

One of the open problems in the subject is to give such moduli theoretic construction of desingularisation of moduli spaces of bundles of higher rank. One is led to suspect that this construction really works because of some coincidences in terms of *isomorphisms between algebraic groups of low rank* and that this construction may actually give desingularisations of other moduli spaces of principal bundles with structure groups like the Spin groups. This is yet to be understood and answered.

References

[AB] M.Atiyah and R.Bott : Yang-Mills Equations over Riemann Surfaces, *Proc. Trans.R.Soc.Lond.*, **A 308**, (1982), 523-615.

[B1] V.Balaji: Intermediate Jacobian of some moduli spaces of vector bundles on curves, *American Jour Math*, **112**, (1990), 611-630.

[BS1] V.Balaji and C.S.Seshadri: Cohomology of a moduli space of vector bundles, *The Grothendieck Festchrift Volume 1, Birkhauser*, (1990) pp. 87-120.

[BS2] V.Balaji and C.S.Seshadri: Cohomology of some moduli spaces of vector bundles (to appear) in the *Proceedings of the Indo-French Conference in 'Geometry'*, held in T.I.F.R., Bombay, 1989)

[BV] V.Balaji and P.A.Vishwanath: Deformations of certain moduli spaces of vector bundles, *Amer. Jour. Math.*, Vol. 115, No. 2 (1993) 279-303.

[BBN] V.Balaji, I.Biswas and D.S.Nagaraj: Principal bundles over projective manifolds with parabolic structure over a divisor. *Tôhoku Math. Jour.* **53** (2001), 337–368.

[BS] V.Balaji and C.S.Seshadri: Semistable Principal Bundles-I (in characteristic zero), *Journal of Algebra.*, **258**, 321-347.

[BP] V.Balaji and A.J.Parameswaran : Semistable principal bundles-II (in positive characteristics). (to appear in *Transformation Groups*)

[F] G.Faltings: Stable G-bundles and projective connections, *J. Alg. Geom.* **2** (1993), 507-568.

[GIT] D.Mumford, J.Fogarty and F.Kirwan: Geometric Invariant theory, Ergebnisse 34, Springer 3rd Edition.

[NS1] M.S. Narasimhan and C.S.Seshadri: Holomorphic vector bundles on a compact Riemann surface, *Math.Annalen*, **155**,(1964), 69-80.

[NS2] M.S. Narasimhan and C.S.Seshadri: Stable and unitary vector bundles on a compact Riemann surface, *Annals of Mathematics* , **82**, (1965) pp 540-567.

[NR1] M.S.Narasimhan and S.Ramanan: Moduli of vector bundles on a compact Riemann surface, *Annals of Mathematics*, **89**, (1969), 14-51.

[NR2] M.S.Narasimhan and S.Ramanan: Geometry of Hecke Cycles-I, *C.P.Ramanujam - A Tribute*, (Bombay TIFR), (1978).

[Nit] N.Nitsure, Moduli spaces of semistable pairs on a curve, *Proc.London Math Soc*, **62**, (1991), 275-300.

[No] M. V. Nori: The fundamental group scheme, *Proc.Ind.Acad.Sci (Math.Sci)* **91** (1982), 73–122.

[R1] A. Ramanathan: Stable principal bundles on a compact Riemann surface - Construction of moduli space (*Thesis*, Bombay University 1976).

[R2] A. Ramanathan: Stable principal bundles on a compact Riemann surface, *Math. Ann.* **213** (1975), 129-152.

[Ses1] C.S.Seshadri : Spce of unitary bundles on a compact Riemann surface, *Annals of Mathematics*, **85**, (1967), 303-336.

[Ses1] C.S.Seshadri : Mumford's Conjecture for $GL(2)$ and applications, *Algebraic Geometry, Bombay Colloquium* (1968), 347-371.

[Ses3] C.S. Seshadri: Fibrés vectoriels sur un courbes algébriques, *Asterisque* **96** (1982).

[Ses2] C.S.Seshadri: Desingularisation of moduli varieties of vector bundles on curves, *International Symp. on Algebraic geometry*, (ed.) M.Nagata (Kyoto), (1977), 155-184.

[Sim 0] C. Simpson: Higgs bundles and Local systems, *Pub. I.H.E.S.* **75** (1992), 5-95.

[Sim 1] C. Simpson: Moduli of representations of the fundamental group of a smooth projective variety-I, *Pub. I.H.E.S.* **79** (1994) pp 47-129.

[Sim 2] C. Simpson: Moduli of representations of the fundamental group of a smooth projective variety-II, *Pub. I.H.E.S.* **80** (1995), 5-79.

[T] T.E.Venkata Balaji, Limits of rank 4 Azumaya algebras and applications to desingularisation *Proc Ind.Acad.Sci* **112** (2002), 485-537.

[V] E.Vieweg, Quasi-Projective Moduli of Polarised Manifolds, Ergebnisse Vol 30, Springer Verlag, (1995)

[BR] Lieven Le Bruyn and Z.Reichstein, Smoothness in algebraic geography. *Proc. London Math. Soc.* (3) **79** (1999), no. 1, 158–190.

Chennai Mathematical Institute
92 G.N. Chetty Road, Chennai 600 017,
INDIA.
E-mail address: balaji@cmi.ac.in

Seshadri's Contributions to Moduli and Geometric Invariant Theory

Vikram B.Mehta

Dedicated to C.S.Seshadri on his 70th birthday

In this very brief article, I shall attempt to highlight some of Seshadri's work on the topics mentioned in the title. I am going to concentrate on two of his papers, namely "Mumford's Conjecture for $GL(2)$ and Applications" and "Quotient Spaces Modulo Reductive Algebraic Groups". This is not to suggest that his other papers on the subject, such as "Quotient Spaces by an Abelian Variety" and "Quotient Space by an Algebraic Group of Automorphisms" are not important. On the contrary, this only means that those papers are outside the present author's sphere of competence. Further, what follows is an extremely informal, an almost leisurely stroll through the two papers.

Let me begin by outlining the first of the above mentioned papers. Recall that at that time, the conjecture of Mumford (reductive groups are geometrically reductive) had not yet been proved in general. In order to construct the moduli spaces of rank two vector bundles on a smooth curve, one has to construct the good quotient of N copies of $H^{t,2}$ by a large linear group $GL(t)$ for $t >> 0$. Here $H^{t,2}$ denotes the Grassmannian of 2 dimensional quotients of a t–dimensional vector space. In characteristic 0, Mumford could achieve this, as $GL(n)$ is geometrically reductive for any n. Seshadri had already proved that $GL(2)$ is geometrically reductive in characteristic p. However, one has to divide by $GL(t)$ for $t >> 0$ as mentioned earlier. Before I outline Seshadri's ingenious method , let us introduce some notation. Let W denote $M_{t,2}$, the space of all $t \times 2$ matrices, and let W^N be the space of all N tuples from W. $SL(2)$ acts on the right and $GL(t)$ on the left. Let H be the subgroup of $GL(2)^N$ defined by elements

$(g_1, g_2 ...... g_N)$ such that $\Pi det g_i = 1$. Then $H/SL(2)^N$ is a torus group and hence H is geometrically reductive.

Now consider the following diagram :

$$\begin{array}{ccc} W^N & \longrightarrow & \widehat{X} \\ \downarrow & & \downarrow \\ (*) & \longrightarrow & \widehat{Y} \end{array}$$

Here $\hat{X}$ is the cone over N copies of $H^{t,2}$; and we have to take the good quotient of $\hat{X}$ by $GL(t)$. But it was known classically that one can take the good quotient of W^N by $SL(t)$ (the left vertical arrow), and the bottom horizontal arrow is a good quotient because the group is H, which is geometrically reductive. Hence the right vertical arrow is a good quotient by $SL(t)$, which is what we wanted. Here the main point is that an abstract problem of taking good quotients by $SL(t)$ has been reduced to a known quotient by the same group, and then by further dividing out by $GL(2)$.

The second main point in the same paper is the following: in order to construct the space of all semistable bundles on a curve, one has to show that the "covariant" from $R^{s,s}$ to a product of Grassmannians is an imbedding. The injectivity and differential injectivity are easily proved. What is missing is the properness. Here Seshadri introduces an amazing device namely, a multiple-valued mapping. Using this and a great deal of ingenuity, Seshadri proves that the covariant is proper. He goes to the heart of the problem as he has realized at once that this is the main problem of constructing semistable moduli. This has been borne out by the later works of Gieseker, Maruyama, Simpson on higher dimensional varieties, as well as for principal bundles on curves by Ramanathan over the complex numbers, and by other people in characteristic p. This was also the perhaps the first time that a moduli problem had been compactified purely algebraically. For rank n stable bundles over curves over the complex numbers, this had already been achieved by Seshadri, but purely analytically, using the Narasimhan-Seshadri theorem. Compactifying gives a much better insight into the moduli problem, as had been already observed by Seshadri.

Let me turn to the second paper of Seshadri mentioned in the introduction. Here he addresses the problem of constructing the space of stable bundles of any rank in characteristic p. We briefly mention the statement and the applications of the main theorems. The first one says that $X^{s,s}$ mod G is complete, where X is a projective variety with a linear action of G. This is proved by showing that any G invariant section of L over $X^{s,s}$ extends to the whole of X. So if the quotient did exist, it would be complete as a variety. The extension property has been very useful in the works of Narasimhan-Ramadas and Mehta-Ramadas on rank two bundles. The notion of X mod G being complete was also used by Gieseker while constructing $\overline{M}_g$, the space of stable curves.

He now proceeds by making a reduction to the stable case. That is, if G acts linearly on X, then he considers $X \times G/B = Z$ say. By choosing the polarisation carefully, he proves that on Z, stable = semistable and the canonical rational map from Z/G to X/G is a morphism. Ramanathan's work on the moduli of principal bundles makes very heavy use of this result. But even stable = semistable is not enough. Seshadri goes on to "eliminate finite isotropies". In other words, if G acts on X with finite isotropies, then one can construct a variety Z with a G action and a map Z to X which is finite and surjective; and he proves that Z/G exists. So if we put H = Aut (Z/X), then H would act on Z/G. Hence if Z/G is quasiprojective, then X/G would exist. This is a technically difficult theorem, which has been used extensively by Kollar and Viehweg (among others) in the construction of moduli of higher dimensional varieties. These two results, namely reducing to the case semistable = stable and removing finite isotropies, are really the heart of the paper.

Now Seshadri proves the following theorem : let G act linearly on X, with stable = semistable and G acts freely on X^s. Assume X^s/G exists. Further assume that the linearizing line bundle L descends to a line bundle M on $Y = X^s/G$. Then M is pseudo-ample on Y. This has to use the methods of etale cohomology, which was being developed at that time. This result has the remarkable by-product of providing another criterion of ampleness of line bundles on varieties, and the phrase the "Seshadri constant" has become a common one.

The last theorem in this paper proves that for any linear action of G on X, the quotient of X^s mod G exists and is a quasi projective variety. One can now construct the moduli space of stable bundles in characteristic p by combining all the above results and those in the first paper.

Let me add that Oda had already proved the conjecture for $GL(2)$ in characteristic 2. Shortly after the second paper of Seshadri mentioned above, Formanek and Procesi proved the conjecture for $GL(n)$ for any n, and Haboush proved it in general. But these two papers of Seshadri went a long way in convincing people that the conjecture must be true.

I hope I have done some justice to the two papers of Seshadri. I have left out many theorems for reasons of space. Their power and beauty become apparent not only in solving a difficult problem, but also in the amazing way in which the methods and ideas have proved to be of vital importance in the work of other people. This is the true hallmark of a great paper. Let me conclude by wishing Seshadri many, many more years of active mathematical research.

References

[M] D. Mumford, Geometric invariant theory, Springer-Verlag, 1965.

[S-1] C.S. Seshadri, *Space of unitary vector bundles on a compact Riemann surface*, Ann. of Math. 82 (1965), 303-336.

[S-2] C.S. Seshadri, *Mumford's conjecture for $GL(2)$ and applications*, Algebraic geometry (Papers presented at the Bombay Colloquium, 1968), Oxford University Press.

[S-3] C.S. Seshadri, *Quotient spaces modulo reductive algebraic groups*, Ann. of Math. 95 (1972), 511-556.

Seshadri's work on moduli spaces - The case of singular curves

D.S. NAGARAJ

1. Introduction

The following is an expanded version of my talk at the symposium to celebrate C.S.Seshadri's seventieth birthday. The aim of the talk was to summarise the work of C.S. Seshadri on the Moduli spaces of vector bundles on singular curves. In this note I will try to explain the motivation for the studying such moduli spaces and what is contribution of C.S. Seshadri. At the end of the article I mention a few open problems.

2. Preliminaries

The standard reference for most of the notations used in this article is [H]. A quasi projective scheme X over an algebraically closed field k is said to be *equidimensional* if all the irreducible components of X have the same dimension. For a point p of X the stalk of the structure sheaf $\mathcal{O}_X$ is a local ring and is denoted by $\mathcal{O}_{X,p}$ and the maximal ideal is denoted by $m_{X,p}$. For any closed subscheme Y of X the ideal sheaf of Y is denoted by $\mathcal{I}_Y$. On an integral scheme X, a coherent sheaf $\mathcal{F}$ of $\mathcal{O}_X$ -modules is said to be *torsion free* if $\mathcal{F}$ does not contain any sub $\mathcal{O}_X$ -modules which is supported on a proper closed subset of X. If η is the generic point of the integral scheme X then the dimension of the vector space $\mathcal{F} \otimes k(\eta)$ is called the *rank* of the coherent sheaf $\mathcal{F}$, where $k(\eta) = \mathcal{O}_{X,\eta}$. A coherent sheaf $\mathcal{F}$ on an equidimensional reduced projective scheme X is said to be *torsion free* if $\mathcal{F} \otimes \mathcal{O}_{X_i}$ is of same rank for all irreducible component X_i of X and $\mathcal{F}$ has no sub $\mathcal{O}_X$ -module whose support is contained in a smaller dimensional subscheme. A torsion free sheaf $\mathcal{F}$ on an equidimensional reduced projective scheme X is said to be *locally free* (or a *vector bundle*) if for every point p of X the stalk $\mathcal{F}_p$ is a free $\mathcal{O}_{X,p}$ -module. A locally free sheaf of rank one on X is called a *invertible sheaf* (or a *line bundle*) on X.

EXAMPLES Let X be a integral scheme and Y be a closed proper subscheme. Assume dimension of X is bigger than zero.

1) The sheaves $\mathcal{O}_X^n$, $\mathcal{I}_Y$ are examples of torsion free sheaves.

2) Let p be a closed point of X and n be a positive integer. The coherent sheaves $\mathcal{O}_X/\mathcal{I}_p^n$ and $\mathcal{O}_X \oplus \mathcal{O}_X/\mathcal{I}_p^n$ are examples of sheaves which are *not* torsion free

In what follows, an equidimensional projective scheme of dimension one over an algebraically closed field k is called a *curve.* A closed point q of a curve X is said to be *non-singular* or *smooth* if the maximal ideal m_q of the local ring $\mathcal{O}_{X,q}$ is generated by a single element. A point p of a curve X which is not smooth is called a *singular* point of X. If the curve X has at least one singular point then we say that X is *singular curve*; otherwise X is said to be *non-singular curve.*

If X is a non-singular curve, then any torsion free sheaf is a locally free sheaf and hence corresponds to a vector bundle. On the other hand, if X is a singular curve and p is a singular point on X then the ideal sheaf $\mathcal{I}_p$ is an example of a torsion free sheaf which is not locally free.

Let X be a singular curve and q be a singular point of X. If the completion of the local ring $\mathcal{O}_{X.q}$ with respect to the maximal ideal is isomorphic to the quotient ring $k[[U,V]]/(UV)$ of the power series ring in two variables then the point q is said to be a *nodal point* (or an *ordinary double point*) of X. If a curve X has only nodal singularities then X is said to be a *nodal curve.*

REMARK Let X be a irreducible nodal curve, p be a node of X and $\mathcal{F}$ be a torsion free sheaf on X. Then the stalk of $\mathcal{F}$ at p is isomorphic $\mathcal{O}_{X,p}^a \oplus m_p^b$ for some non-negative integer a and b (See. Proposition 2, Huitième Partie [Se1]).

EXAMPLES: 3) Any connected curve with more than one irreducible component is an example of a singular curve.

4) Let $n \geq 1$ be an integer. In the projective plane over k the curve defined by the homogeneous polynomial $X^n + Y^n + Z^n$ is an example of a non-singular curve (if n and $char(k)$ are coprime).

5) In the projective plane over k the curve defined by the equation $Y^2Z + Y^3 + X^2Z$ in the projective plane over k is an example of an irreducible singular curve. (Here we assume $char(k) \neq 2, 3$).

6) In the projective plane over k the curve defined by the equation $Y^2Z + Y^3$ is also an example of an irreducible singular curve. (Here again we assume $char(k) \neq 2, 3$).

7) In the projective plane over k the curve defined by the equation XY is an example of a singular curve which not irreducible.

In the above examples 5),6) and 7) the point corresponding (0,0,1) is a singular point of the curve. In all these three examples the curve has no other singularities. The singular point is a node in the case of 5) and 7). Hence the curves in 5) and 7) are examples of nodal curves. The curve in 6) is an example of a singular curve which is not a nodal curve.

Let X be a connected reduced curve. The set $Pic(X)$ of isomorphism classes of line bundles X is a group under the operation of tensor product (in fact it is a *group scheme*). If X is non-singular $Pic(X)$ is called the *Picard group* (*Picard group scheme*) of X and connected component $Pic^0(X)$ containing $\mathcal{O}_X$ is in fact a complete variety and is known as the *Jacobian* of X. When X is singular the connected component $Pic^0(X)$ containing $\mathcal{O}_X$ is a variety but need not be complete, in that case it is called the generalized Jacobian.

EXAMPLE 8) Let X be an integral nodal curve with exactly one node, say p. Let $\pi : \tilde{X} \to X$ be the normalization of X. Then there is an exact sequence of coherent sheaves on X

$$0 \to \mathcal{O}_X \to \pi_*(\mathcal{O}_{\tilde{X}}) \to T \to 0$$

with T a torsion sheaf supported at p and $T_p \simeq k$. This gives rise to an exact sequence (of group schemes)

$$1 \to k^* \to Pic(X) \to Pic(\tilde{X}) \to 1.$$

Thus we see that in this case $Pic^0(X)$ is a k^* bundle over $Pic(\tilde{X})$ and hence is not a complete variety.

3. Motivation

There are two motivations for the study of moduli spaces of vector bundles on a singular curve and compactifications of these moduli spaces.

If X is a singular curve of arithmetic genus g, then its normalization $\tilde{X}$ is a nonsingular curve of genus less than g and a vector bundle on X is related to a vector bundle on $\tilde{X}$ with some extra structure. Thus the moduli space of vector bundles X and its compactifications are related to the moduli space of vector bundles $\tilde{X}$. Assume that there is an algebraic family of smooth curves of genus g specializing to X and also assume that one is able to construct a corresponding family of moduli spaces of bundles specializing to a compactification of the moduli space of vector bundles on X. Then to study the geometry of the moduli space of vector bundles on a smooth curve of genus g, one can set up an inductive procedure. This method is successfully used to compute the Chern classes of the tangent bundle of the moduli space of vector bundles on a smooth curve (See. [G], [Se2]) and to compute the dimension of the vector space of sections of a line bundle on the moduli space of a smooth curve (See. [NR], [R], [su1], [su2]).

The other motivation is that in a way the study of the moduli space of bundles on singular curves is similar to the study of the moduli space of bundles on higher dimensional nonsingular varieties. More precisely, in the case of singular curves as well as higher dimensional non singular varieties, the moduli spaces of vector bundles are in general not proper and hence in order to get natural proper varieties one is forced to look at moduli spaces of more general objects namely torsion free sheaves. One hopes that the understanding of the moduli space of torsion free sheaves on singular curves may lead to a better understanding of the moduli spaces of torsion free sheaves on higher dimensional non-singular varieties. This aspect is yet to get enough attention from the people working in this area.

4. Seshadri's work

Seshadri's interest in this subject can be traced back to early 1970's through the work of his students Cyril D'souza and C.J. Rego on the problem of compactifying the generalized Jacobian of an irreducible singular curve. C.J. Rego also worked on the problem of compactifying the moduli space of vector bundles of higher rank on a singular curve.

The first paper of Seshadri in this area is the joint work with T.Oda [OS]. They constructed finitely many compactifications of the generalized Jacobian of a curve which is not necessarily irreducible but with only ordinary double points as singularities. These finitely many compactifications are given in terms of the topology of the graph associated with such a curve. The structure of these comapctifications is described in terms of torus embeddings. Note that the generalized Jacobian of a curve of the above type is a torus bundle over the Jacobian of the normalization. These compactifications are *good* in the following sense: 1) these are proper varieties, 2) points of these compactifications corresponds to torsion free sheaves of rank one, 3) generalized Jacobian acts on them, and 4) if the nodal curve is a specialization of non-singular curve then these compactifications (union of finitely many copies) are specializations of the Jacobian of the non-singular curve.

The first account of Seshadri's work on the problem of compactification of the moduli space of vector bundles of rank greater than one on a singular curve appeared in the Astérisque volume [Se1]. As it is mentioned in the introduction of this volume, the things which figure there were motivated by Gieseker's lectures at TIFR and an attempt to understand the whole aspect of Gieseker's work. In this work, among other things, he discusses the local properties of the moduli space of semi-stable sheaves on a reduced curve with arbitrary singularities. By specializing to the case of a nodal curve, it is shown that the moduli space of semi stable torsion free sheaves of rank two on such a curve is reduced.

Inspired by the work of Usha Bhosle [B1] on the Generalized Parabolic Bundles(GPB), Seshadri comes back to the problem of studying the moduli space of torsion free sheaves on singular curves in the mid 1990's.

On X, a smooth curve a *Generalized Parabolic Bundle*(*GPB*) on X with support $\{p_1, p_2\} \subset X$ is a pair (E, Q), where E is a vector bundle of rank r and $Q \subset E_{p_1} \oplus E_{p_2}$ is a subspace of dimension r.

In a joint work with Nagaraj [NS1] he takes up the case of the moduli space of rank two torsion free sheaves on a connected reducible curve $X = X_1 \cup X_2$ with one node p and no other singularities. The main results of this work are:

The study of torsion free sheaves on X is equivalent the study of triples (V_1, V_2, ϕ) consisting of vector bundle V_i on X_i, $i = 1, 2$ and a linear map $\phi : V_{1,p} \to V_{2,p}$. For a suitable choice of polarization on X, the rank two sheaves of the type $m_p \oplus m_p$ are not semi-stable, where m_p is the maximal ideal of X at the node p. Thus the moduli space of rank two torsion free sheaves of odd degree is the union of two irreducible smooth components which intersect transversally along a smooth subvariety Y. The variety Y is isomorphic to the product $Y_1 \times Y_2$, where Y_i is a moduli space of rank two parabolic vector bundles on X_i, $i = 1, 2$.

The above moduli space has good specialization properties. Further, a "Fixed determinant moduli space" exists and has good specialization properties.

Attempts to generalize the above work to the higher rank case has had several inherent difficulties: namely the number of irreducible components increases with the rank and the singularities of the moduli space will not be (analytic) normal crossings (See. [Su2]). On the other hand, D.Gieseker had constructed a good compactification of the moduli space of rank two vector bundles on a nodal curve. His construction involved only vector bundles on the nodal curve and its semistable equivalents. Using his compactification, Gieseker proved a conjecture of Newstead and Ramanan on the vanishing of certain Chern classes of the moduli space of vector bundles of rank two and degree one. Generalizing Gieseker's work to higher rank seems complicated. With Nagaraj, Seshadri (See. [NS2]) found a way to overcome the difficulties and constructed a compactification of the moduli space of vector bundles on a nodal curve X, using only vector bundles on X and its semi stable equivalents. These are called the Generalized Gieseker moduli spaces. The singularities of the Generalized Gieseker moduli space are (analytic) normal crossings and it has good specialization properties. There is a morphism from the Generalized Gieseker moduli space to the moduli space of torsion free sheaves on X. This morphism also specializes well.

Seshadri in his lectures at ICTP [Se2] described the concrete realization of the Gieseker moduli space in rank two and degree one and showed that some of the arguments of Gieseker can be simplified. In these lectures he also indicated a proof of the following: any fiber of the morphism from Gieseker moduli space (of rank two) to the moduli space of torsion free sheaves is the wonderful compactification of the projective linear group in the sense of De Concini and Procesi.

5. Open problems

There are several unsolved problems in this area, and I believe Seshadri has some idea how to go about to solve them.

One such problem is to find good compactification of the moduli space of vector bundles on a higher dimensional non singular variety X, using vector bundles on X and vector bundles on flat degenerations (which may not be irreducible) of X. This problem is mentioned in the introduction of [NS2]. This, in a way would be doing some justice to the second motivation for studying the compactification moduli space of vector bundle on a singular curve which I mentioned earlier. More specifically, is it possible to construct a compactificaton of moduli space of rank two vector bundles (with fixed Chern classes) on the projective plane using vector bundles on the projective plane and vector bundles on $X \bigcup \cup_{i=1}^{n} X_i$, where X is a blow up of $\mathbb{P}^2$ along n smooth points p_i $(1 \leq i \leq n)$ and $X_i \simeq \mathbb{P}^2$ $(1 \leq i \leq n)$ with $X_i \cup X_j = \emptyset$ if $i \neq j$ and for $1 \leq i \leq n$ the intersection $X \cap X_i$ is the exceptional curve over p_i? If this is the case then is it true that n is bounded in terms of Chern classes?

Another problem is to find a good compactification of the moduli space of G bundles on a singular curve, where G is a reductive group. Let me quote from [NS2]: "Our set seems also to give a tool for a systematic investigation of compactifications of moduli spaces of principal bundles with reductive structure groups, on curves with singularities (ordinary double points)". For the group of $G = SL(2)$ on a connected complete curve with two irreducible components and exactly one node, the problem has been solved in [NS1]. For the group $G = SL(n)$ on a connected complete curve with two irreducible components and exactly one node, this problem has been recently solved by X. Sun [SU2]. The case of $G = Sp(n)$ and $O(n)$ has been treated by U.Bhosle [B2] and G.Faltings [F2].

Acknowledgement: I thank V. Balaji, C.S. Seshadri and R. Sridharan for their valuable suggestions. I also thank Suresh Nayak.

References

[B1] Bhosle, Usha N. *Generalised parabolic bundles and applications to torsion free sheaves on nodal curves.* Ark. Mat. 30 (1992) no.2. 187-215.

[B2] Bhosle, Usha N. *Orthogonal and symplectic bundles on curves and representations.* J. Ramanujan Math. Soc. 13 (1998) no.2. 195-215.

[C] Cyril D'souza. *Compactifications of generalised Jacobians.* Proc. Indian. Acad. Sci. 88 A, Part III (1979) no.5. 419-457.

[F1] G. Faltings. *A proof for the Verlinde formula.* J. Alg. Geom. 3 (1994) 347-374.

[F2] G. Faltings. *Moduli-stacks for bundles on semistable curves.* Math. Ann. 304 (1996) 489-515.

[G] D. Gieseker. *A degeneration of the moduli space of stable bundles.* J. Diff. Geom. 19 (1984) 173-206.

[H] R. Hartshorne. *Algebraic geometry.* Springer GTM. 52 (1977)

[K] Ivan Kausz. *A Gieseker type degeneration of moduli stack of vector bundles on curves* math.AG/0201197 (2002).

[NR] M.S. Narasimhan and T.R. Ramadas. *Factorisation of generalised theta functions I.* Invent. Math. 114 (1993) 565-623.

[NS1] D.S. Nagaraj and C.S. Seshadri. *Degenerations of the moduli spaces of vector bundles on curves I.* Proc, Indian. Acad. Sci.(Math.Sci.) 107 (1997) 101-137.

[NS2] D.S. Nagaraj and C.S. Seshadri. *Degenerations of the moduli spaces of vector bundles on curves II (Generalized Gieseker moduli spaces).* Proc, Indian. Acad. Sci. (Math.Sci.) 109 (1999) 165-201.

[OS] Tadao Oda and C.S. Seshadri. *Compactifications of the generalized Jacobian variety* Tran. AMS. 253 (1979) 1-89.

[R] T.R. Ramadas. *Factorisation of generalised theta functions. II. The Verlinde formula.* Topology 35 (1996), no. 3, 641–654.

[Re1] C.J. Rego. *The compactified Jacobian.* Ann. Sci. École Norm. Sup. (4) 13 (1980), no. 2, 211–223.

[Re2] C.J. Rego. *Compactification of the space of vector bundles on a singular curve.* Comment. Math. Helv. 57 (1982), no. 2, 226–236.

[Se1] C.S. Seshadri. *Fibrés vectoriels sur les courbes algébriques.* Astérisque. 96 (1982).

[Se2] C.S. Seshadri. *Degenerations of the moduli spaces of vector bundles on curves.* School on Algebraic Geometry. ICTP. Lect. Notes.1. Abdus Salam Int. Cent. Theoret. Phys., Trieste, (2000) 205-265.

[Su1] Xiaotao Sun. *Degeneration of Moduli spaces and generalized theta functions.* J. Alg. Geom. 9 (2000) 459-527.

[Su2] Xiaotao Sun. *Factorization of generalized theta functions at reducible case.* math.AG/0004111 (2000).

[Su3] Xiaotao Sun. *Degeneration of SL(n)-Bundles on a reducible curve.* math.AG/0112072 (2001).

Institute of Mathematical Sciences
C I T Campus, Taramani Chennai-600 113.
E-mail address: dsn@imsc.res.in

Parabolic Bundles

NITIN NITSURE

1. A Brief Parabolic History

The concept of parabolic bundles was invented by Seshadri in the late 1960's. His idea was to investigate vector bundles arising from arbitrary unitary representations of fundamental groups of punctured Riemann surfaces, so as to generalize the Narasimhan-Seshadri theorem.

The initial discovery that Seshadri made was that such representations naturally give rise to vector bundles with *some additional structures*. These additional structures were named by Seshadri as parabolic structures. The additional structures, as we will describe fully later, involve flags in some of the fibers E_s of the bundle E, and such flags naturally correspond to parabolic subgroups in $GL(E_s)$, from which the name 'parabolic' comes. Also, it is no accident that parabolic structures also have something to do with parabolic cusps for the action of discrete groups on the upper half plane.

The ideas of André Weil from his 1938 paper 'Généralisation des fonctions abéliennes' played an important role in inspiring parabolic bundles, as Seshadri has emphasized.

Throughout a period of 10 years from his invention of parabolic bundles, Seshadri was completely occupied with his research programmes on geometric invariant theory, and the geometry of G/B. So apart from a lecture course [S1] in Varenna in 1969, in which he treated a special case which he called as 'π-bundles', and a brief mention in his address to the International Congress of Mathematicians in Nice in 1970 [S2], the world did not get to hear anything on the subject, with one important exception: Seshadri used the (yet to be constructed) coarse moduli scheme of parabolic bundles to provide an explicit desingularization of the moduli of rank 2 semi-stable vector bundles with trivial determinant on any compact riemann surface of genus $g \geq 2$ (see [S3]). So even before the basic definitions and results on parabolic bundles were published, they already had an interesting application!

At last, in 1977, came Seshadri's research announcement [S4] in the Bulletin of American Mathematical Society. It contains the basic definitions, the statement of the existence of a coarse moduli scheme for semi-stable parabolic bundles, and the statement of a theorem which identifies stable parabolic bundles with irreducible unitary representations. Three years after this announcement, the existence of a coarse moduli scheme for semi-stable parabolic bundles was established in Seshadri's joint paper [M-S] with Vikram Mehta in 1980.

With this paper of Mehta and Seshadri, published more than a decade after the birth of parabolic bundles, a new stage began in the career of parabolic bundles – the subject was now thrown open to the mathematical public. Meanwhile its creator, Seshadri himself, withdrew from the scene to pursue his other mathematical interests.

Today, about 35 years after Seshadri's original discovery of the notion of parabolic bundles, the subject has grown enormously, and has become important in many, very unexpected, ways. It has found its way from Algebraic Geometry into Differential Equations, and Physics. An ever expanding chain of ideas has a crucial parabolic link somewhere. From a quiet beginning, this brain-child of Seshadri has now become an indispensable part of mathematics!

This article, based on a talk on the occasion of Seshadri's 70 th birthday, is arranged as follows. We first give the definition of parabolic bundles, then state the basic theorems on connection with unitary representations and existence of moduli. After that, we give a telegraphic list of some of the interesting developments in the subject. This list, far from comprehensive, is just meant as a glimpse of the wide variety of interesting mathematics coming out of parabolic bundles.

2. Definition of Parabolic Bundles

Let X be a compact Riemann surface (or more generally, an algebraic curve over a field k in any characteristic, which we assume for simplicity to be algebraically closed), let $S \subset X$ be finite subset, and E a holomorphic vector bundle on X.

A **quasi-parabolic structure** on E over S by definition consists of a flag $E_s = F_1 \supset F_2 \supset \ldots \supset F_r \supset 0$ in the fiber over each $s \in S$. The number r as well as the dimensions of the F_i depend on s.

Parabolic structure : Moreover, for each $s \in S$ we are given **weights** $0 \leq \alpha_1 < \ldots < \alpha_r < 1$.

Parabolic bundle : This is a vector bundle together with a parabolic structure on it.

Parabolic degree of a parabolic bundle is defined by

$$pardeg(E) = deg(E) + \sum_{s \in S} \sum_{i} \alpha_i \, dim(F_i/F_{i+1}).$$

Parabolic slope : This is the ratio

$$par\mu(E) = pardeg(E)/rank(E).$$

Given any subbundle $E' \subset E$ of a parabolic bundle, we get an induced parabolic structure on E'. The flag in E'_s is obtained by intersection with the given flag in E_s. The weights are taken to be the 'largest possible' allowed by the intersections.

Parabolic semi-stability : Every nonzero subbundle must satisfy the inequality $par\mu(E') \leq par\mu(E)$.

Parabolic stability : $par\mu(E') < par\mu(E)$ for all nonzero proper subbundles.

A **poly-stable** parabolic bundle is a semi-stable parabolic bundle which can be written as a direct sum of stable parabolic bundles.

A **parabolic Harder-Narasimhan flag** exists in any parabolic bundle, with usual properties.

An alternative (equivalent) definition of parabolic bundles is possible in terms of $E|_{X-S}$ and a **periodic filtration** of $j_*(E|_{X-S})$ at each $s \in S$ indexed by reals, where $j : X - S \hookrightarrow X$ is the inclusion.

The **dual** of parabolic bundle and **tensor product** of two parabolic bundles makes sense as a parabolic bundle in a natural way.

A **parabolic homomorphism** $f : E \to F$ between two parabolic bundles is a homomorphism of vector bundles which satisfies the following at each $s \in S$: $f_s(F_i) \subset F'_{i+1}$ whenever $\alpha_i > \alpha'_{i+1}$.

It can be shown in characteristic zero that poly-stable parabolic bundles on (X, S) with parabolic degree zero form a Tannakian category in the sense of Grothendieck, with fiber functor given by taking the usual fiber at a chosen point of $X - S$.

3. How unitary representations give parabolic bundles

Let X be a compact Riemann surface, and let $\pi_1(X)$ be its fundamental group. By the Narasimhan-Seshadri theorem, every stable vector bundle of rank n and degree 0 arises from an irreducible unitary representation $\pi_1(X) \to U(n)$, and this gives a bijective correspondence between the set of all isomorphism classes of rank n stable vector bundles of degree zero and the set of all conjugacy classes of irreducible unitary representation $\pi_1(X) \to U(n)$.

The Narasimhan-Seshadri theorem also says that more generally, stable bundles of rank n degree d arise from irreducible unitary representations $\pi_1(X - P) \to U(n)$ (where $X - P$ is X with one point P removed) for which the local monodromy (image of a small loop around the puncture at P) is the scalar matrix $\exp(2\pi i d/n)I$.

Seshadri's construction of parabolic bundles can be described as follows. Consider any unitary representation $\rho : \pi_1(X - S) \to U(n)$ where S is an arbitrary finite set, and there is no restriction on the local monodromies. If $j : X - S \hookrightarrow X$ is the inclusion, then the representation $\rho : \pi_1(X - S) \to U(n)$ defines a holomorphic vector bundle F on $X - S$, which comes equipped with a flat connection and a natural hermitian metric. Seshadri defines a subsheaf $E \subset j_*(F)$ which is a vector bundle which prolongs F to all of X. The sheaf E is defined in terms of growth rates of norms of sections with respect to the hermitian metric. The connection on F prolongs to a logarithmic connection on E (with – at worst – logarithmic singularities along S) whose monodromy representation is ρ, and whose residual eigenvalues at each $s \in S$ are in the interval $[0, 1) \subset \mathbb{R}$. (Following a more general later construction of Deligne, one now says that E is given by the 'Deligne construction'.)

The fiber of E at $s \in S$ comes with a natural filtration $E_s = F_1 \supset F_2 \supset \ldots \supset F_r \supset 0$ where the number r is the number of distinct eigenvalues $\exp(2\pi i \alpha_i)$ of the local monodromy around s. (Here, the α_i are the residual eigenvalues at s of the logarithmic connection on E.) If $0 \leq \alpha_1 < \ldots < \alpha_r < 1$ are the distinct residual eigenvalues, then $dim(F_i/F_{i+1})$ equals the multiplicity of α_i. The filtration F_i gives a quasi-parabolic structure on E, with weights α_i. The filtration is again defined in terms of growth rates of norms of sections with respect to the hermitian metric on $E|_{X-S}$.

Thus, we get a parabolic bundle E for any representation $\rho : \pi_1(X - S) \to U(n)$. Such an E necessarily has parabolic degree 0, in view of the general relation between ordinary degree and trace of residue for a bundle with a logarithmic connection. For any morphism (intertwining operator) between two unitary of representations, one gets a morphism of parabolic bundles, which makes the construction functorial.

Using the hermitian metric on $E|_{X-S}$ and maximum modulus principle for plurisubharmonic functions, it is shown that the parabolic bundles associated to an irreducible unitary representation is parabolic stable. The converse is also true, but it needs the existence of a moduli space which is complete.

4. Moduli of parabolic bundles

Let X be a smooth projective curve over a field k, and let $S \subset X$ be a finite set of k-rational points. Consider parabolic bundles with a fixed rank n and ordinary degree d, and fixed parabolic data (namely, weights with multiplicities at each $s \in S$).

THEOREM (Mehta-Seshadri [M-S], 1980) *There exists a coarse moduli space for semi-stable parabolic bundles with the above data. It is a normal projective variety over k, whose points are isomorphism classes of poly-stable parabolic bundles. Stable parabolic bundles form a smooth open subvariety of the moduli.*

With the existence of moduli, a proof based on the 'continuity method' deployed earlier by Narasimhan-Seshadri now also works for parabolic bundles, showing that every stable parabolic bundle of parabolic degree 0 on (X, S) comes from an irreducible unitary representation of $\pi_1(X - S)$, giving a bijective correspondence between the set of isomorphism classes of stable parabolic bundles and the set of conjugacy classes of irreducible unitary representations.

5. Some highlights, in telegraphic form

(Precise references can be quickly obtained by the interested reader from internet sources such as MathSciNet, arXiv:math, etc.)

- π-bundles (Seshadri Varenna lectures, 1968-69).
- Definition of parabolic bundles (Seshadri 1977).
- Moduli construction (Mehta-Seshadri 1980).
- Application to desingularization of moduli (Seshadri 1977).
- Application to desingularization of moduli in all characteristics (T.E.V. Balaji 2001).
- Generalization to higher dimension, moduli construction. (Maruyama, Yokogawa 1992).
- Generalized parabolic bundles (Bhosle 1992).
- Application to factorization of generalized theta functions (Narasimhan-Ramadas 1993).
- Parabolic Higgs bundles, moduli (Yokogawa 1993).

• Harmonic metrics and parabolic stability (Simpson 1990, Biquard 1997, Li-Narasimhan 1999).

• Determination of cohomology of moduli of parabolic bundles by gauge theory (Nitsure 1984).

• Quasi-Parabolic Siegel formula and explicit calculation of the Betti numbers of the moduli of parabolic bundles using Weil conjectures. (Nitsure 1996, Holla 1999).

• Application of cohomology of moduli to 3-manifolds and knot theory (Boden 1993).

• Application of parabolic bundles to quantum cohomology, orbifold cohomology (Teleman-Woodward, Fantechi-Göttsche 2001, etc).

• Application to classical matrix problems (Biswas, Belkale 1999).

• Definition of parabolic principal bundles (due to Narasimhan, in Balaji-Biswas-Nagaraj 2001).

• And a wide variety of on-going stuff: One can just try the search 'parabolic bundles' on any internet search engine and see the profusion of entries!

For an exposition of the basic theory of vector bundles on curves, including parabolic bundles, one can see Seshadri's lecture notes [S5].

References

[S1] C.S. Seshadri : Moduli of π-vector bundles over an algebraic curve. In *Questions on Algebraic Varieties*, C.I.M.E. III Ciclo Varenna 1969, 139-260 Edizioni Cremonese, Rome (1970).

[S2] C.S. Seshadri : Quotient spaces, modulo reductive algebraic groups and applications to moduli of vector bundles on algebraic curves. Actes du Congrès International des Mathématiciens (Nice, 1970), Tome 1, 479-482. Gauthier-Villars, Paris (1971).

[S3] C.S. Seshadri : Desingularization of the moduli varieties of vector bundles on curves. Proceedings of the International Symposium on Algebraic Geometry (Kyoto Univ., Kyoto, 1977), 155-184, Kinokuniya Book Store, Tokyo (1978).

[S4] C.S. Seshadri : Moduli of vector bundles on curves with parabolic structures. Bull. Amer. Math. Soc. **83** 124-126 (1977).

[S5] C.S. Seshadri : Fibrés vectoriels sur les courbes algébriques. (Notes by J.-M. Drezet) Astérisque **96** (1982).

[M-S] V.B. Mehta and C.S. Seshadri : Moduli of vector bundles on curves with parabolic structures. Math. Annln. **248**, 205-239 (1980).

School of Mathematics
Tata Institute of Fundamental Research
Mumbai 400 005

Seshadri's Work on Flag & Schubert Varieties

V. Lakshmibai and C. Musili

While Hodge is the founder of *classical Standard Monomial Theory*, Seshadri is the founder of *modern Standard Monomial Theory*. In [**7, 8**], Hodge constructed a nice basis for the coordinate ring of a Grassmann variety (as well as a Schubert subvariety) for its canonical Plücker imbedding in a projective space. The basis is indexed by "standard Young tableaux". The modern Standard Monomial Theory was initiated by Seshadri in the early 1970's. The aim of Standard Monomial Theory (SMT) is the construction of nice bases of the spaces of sections of line bundles on the generalized flag variety as well as its Schubert varieties, in particular bases for finite dimensional representations of a semisimple algebraic group. SMT has led to very many developments in the areas of Combinatorics, Algebraic Geometry, Representation Theory, Quantum Groups, etc. We shall now give a brief account of the development of this theory; this could be broken into the following phases.

Modern Interpretation of Hodge's work

Let $G_{d,n}$ denote the Grassmanian variety of d-dimensional subspaces of k^n, k being an algebraically closed field. For the Plücker embedding of $G_{d,n}$, let R denote its homogeneous co-ordinate ring. Hodge constructed a k-basis for R in terms of what he called *standard monomials* in the Plücker co-ordinates, characteristic of k being 0. Following Seshadri's suggestion, Musili (cf. [**35**]) gave a modern interpretation of Hodge's work, extending Hodge's result to all characteristics. As a by-product, Musili obtained such geometric results as vanishing of higher cohomology groups, arithmetic normality, arithmetic Cohen-Macaulayness of Schubert varieties in $G_{d,n}$ etc. (see [**36**]).

A detailed account of this short write-up on Standard Monomial Theory appears in [**36, 14**] in this volume.

Minuscule case – Beginning of a general SMT

After the above work of Musili, Seshadri could see that the Grassmannian is a special case of a more general class of G/P's, namely the minuscule G/P's, i.e., maximal parabolic subgroups whose associated fundamental weights ω satisfy $(\omega, \beta^\vee) \leq 1$, for all positive roots β, and that all the results of Musili (cf. [**35**]) could be formulated for this more general class. Thus Seshadri extended the results of Musili in a characteristic-free way to the case of a minuscule G/P (see [**38, 36**]). While Musili's work (cf. [**35**]) uses the well-known quadratic Plücker relations for proving the generation by standard monomials, Seshadri's work uses geometric techniques.

SMT for classical type

The SMT for the minuscule case, though a good beginning for a general SMT, was not strong enough, since, for a general semi-simple group there are only a few minuscule fundamental representations. Even for a classical group of type other than A_n, this is the case. Besides there may be no minuscule fundamental representation at all (e.g. E_8). Around this time, De Concini and Procesi (cf.[**4**]) constructed in a characteristic-free way, bases for rings of invariants appearing in Classical Invariant Theory. Seshadri observed that while the (categorical) quotients i.e., Spec R^G for the actions of the linear groups (resp. symplectic groups) are "open parts" of suitable Schubert varieties in the Grassmannian (resp. orthogonal Grassmannian) both of which cases are minuscule, the categorical quotients for the actions of the orthogonal groups are "open parts" of suitable Schubert varieties in the symplectic Grassmannian (and hence non-minuscule!). This gave the clue for an SMT far more general than the minuscule case. While in the minuscule case, the basis is indexed by Weyl group elements, inspired by De Concini-Procesi's work Seshadri could see that the indexing for the nonminuscule case could be given by certain pairs of Weyl group elements. Together with Lakshmibai, Seshadri formulated conjectures (cf. [**21**]) for *classical type* G/P's, i.e., P is a maximal parabolic subgroup such that the associated fundamental weight ω satisfies $(\omega, \alpha^\vee)(= 2\frac{(\omega,\alpha)}{(\alpha,\alpha)}) \leq 2$, for all positive roots α; and more generally conjectures were formulated in [**21**] for G/Q's where Q is an intersection of maximal parabolic subroups of classical type. The formulation of these conjectures could be considered as the real beginning of a general SMT. Part of these conjectures was proved in [**19**] by Seshadri in collaboration with Lakshmibai and Musili; the conjectures were proved in full generality in [**22**] by Seshadri in collaboration with Lakshmibai.

Main conjectures leading to the completion of SMT

Once SMT was established for classical type (i.e., for G/Q such that every maximal parabolic subgroup P containing Q is of classical type) (cf. [**22**]), it became clear that there should be a general SMT valid for all semi-simple groups i.e., on G/Q, where Q is any parabolic subgroup. Following the method pursued so far, one had first to consider the case of G/P, where P is a maximal parabolic subgroup not of classical type. Then G is necessarily an exceptional group. Since for a fundamental weight ω, one can have $(\omega, \alpha^\vee) \geq 3$ (G being exceptional), the indexing set of a basis of $H^0(G/P, L_\omega)$ would be *certain* n-tuples $(\tau_1, \ldots, \tau_n)$, $\tau_i \in W/W_P$ (for the case of ω being of classical type, the indexing set consists of certain pairs - called "admissible pairs" - of Weyl group elements (see [**22**], and also [**19**]), and the problem was at first to write a precise recipe for these "*admissible n-tuples*" as was done for the case of admissible pairs. The simplest case is G_2/P, where P is associated to the right end root. This case was worked out in [**11**], and later for E_6 in [**20**]. Working out SMT for the other exceptional groups in the same manner became quite complicated and the attention turned to formulating good conjectures for a general SMT. Guided by the partial results for exceptional groups as well as some extensive analysis, computations with exceptional groups and several crucial discussions with Seshadri, Lakshmibai formulated the conjectures for a general SMT (cf. [**24**]). There are two parts in these conjectures - part one gives an indexing set for the standard monomial basis thus giving rise to a conjectural (positive-termed) character formula, and part two gives a conjectural basis for Weyl (and Demazure) modules.

Besides the already established SMT, another evidence for the conjectures in [**24**], was SMT for the simplest Kac-Moody case, namely $SL(2, \infty)$ (cf. [**25**]). A priori the conjectures in [**24**] could be formulated for Schubert varieties in the Kac-Moody set up. The importance of the case $SL(2, \infty)$ is that the multiplicities $(\omega, \alpha^\vee)$ even with respect to a fundamental weight ω tend to ∞. One could work out SMT for this case since the geometry of the Schubert varieties for this case is not hard. It supported the conjectures in [**24**], besides it showed that SMT would also hold in the Kac-Moody set up.

The work of Littelmann - Completion of SMT

Solving the conjectures of [**24**] required new ideas. This is what Littelmann has done in his work (cf. [**26, 27**]). Littelmann's idea is to view the indexing set for the standard monomial basis as certain paths (which he calls *LS-paths - Lakshmibai-Seshadri paths*) and work with these paths. He defines the *root operators* e_α, f_α (α simple) on the set of all paths, and proves the indexing part of the conjectures in [**24**] using these operators.

The construction of a nice basis for $H^0(G/B, L_\lambda)$ turned out to be more difficult. Littelmann came out with a beautiful new idea to construct the basis elements, using Lusztig's Frobenius map for quantum groups at a root of unity (cf. [**31**]). See [**27, 14**] for details.

K-theory and SMT

While "the generation by standard monomials" is proved in [**35, 38, 18, 19**] by using certain quadratic relations, the strategy of proof in G/P-V (cf. [**22**]) as well as [**28**] involves a certain filtration for the ideal sheaf of the boundary of a given Schubert variety. Such a filtration is closely linked to the Grothendieck ring of G/B (see [**30, 15**]). The proof of SMT in G/P-V could be termed K-theoretic. In fact in [**30**], it is shown that in the case of a classical G/Q (i.e., Q is a maximal parabolic subgroup of classical type), SMT on G/Q is equivalent to giving an explicit filtration for $\mathcal{O}_X \otimes L$ (here X is a Schubert variety in G/Q and L is a line bundle on G/Q), which in turn determines the structure constants (in the multiplicative structure) in the product $[\mathcal{O}_X] \cdot [L]$, in $K_T(G/Q)$, the Grothendieck ring of T-equivariant coherent sheaves on G/Q. For a general G/Q, SMT appears to be stronger than the K-theory of G/Q (see [**15, 30**] for details).

A geometric approach to Standard monomial theory:

In the construction of the standard monomial basis in [**27**], the theory of Quantum groups (at a root of unity) is used in a crucial way. One may ask "Is it possible to develop a Standard Monomial Theory purely geometrically (without using any Quantum group theory)"? In [**2**], Brion and Lakshmibai answer this question in the affirmative. Nevertheless, it is not known if the basis of [**2**] is the same as Littelmann's standard monomial basis. For details, see [**2**].

Conclusion:

SMT has led to many interesting developments in Combinatorics, Algebraic Geometry and Representation Theory and Quantum Groups.

SMT provides a systematic tool for the study of the geometry of Schubert varieties, since it leads to a concrete determination of the ideal (or ideal sheaf) of a Schubert variety in its imbedding in G/Q. In particular, SMT can be used to prove the vanishing of higher cohomology of line bundles on Schubert varieties which come from line bundles on the ambient flag variety. In the initial stages of development of SMT this was one of its primary motivations. Later the technique of Frobenius splitting due to Mehta and Ramanathan achieved the proof of these vanishing theorems in a remarkably efficient manner (cf.[**34**]). An important application of SMT concerns the determination of the singular locus of a Schubert variety. The first breakthrough in this classical problem is in the work of Seshadri with Lakshmibai (cf. [**23**]). Lakshmibai worked extensively on the singular locus problem and obtained further results. (cf. [**12, 13**]). Much progress has been made in this area (of singularities of Schubert varieties) over the past two decades, thanks to the efforts of several mathematicians working on the combinatorics and geometry of Schubert varieties. For a detailed account of these results, see [**1**]. Interesting classes of varieties like the classical determinantal varieties, ladder determinantal varieties, varieties of complexes, varieties associated to certain quivers and varieties defined by the rings of invariants in Classical Invariant Theory, can be considered as certain open subsets of Schubert varieties and hence SMT can be used to study these varieties (cf. [**5, 6, 16, 21**]). SMT has close connections with representation theory; SMT yields a (positive-termed) character formula for finite dimensional representations of a semisimple algebraic group (more generally for the Demazure modules under a symmetrizable Kac-Moody set up) which is, in spirit, quite different from the Weyl character formula. SMT has also led to a generalisation of Littlewood-Richardson rule (cf. [**26**]), good filtrations (cf. [**28**]) etc.

It should be remarked that Littelmann's contribution to SMT has really given a quantum leap to SMT leading to important Combinatorial, Geometric and Representation-theoretic results. Littemann has not only solved the conjectures of SMT in full generality, but he has also brought in new ideas leading to connections of SMT with quantum group theory. His introduction of "path operators" relates the combinatorics of the graph of L-S paths and Kashiwara's crystal graph - in fact the two graphs are isomorphic (cf. [**9**]). Further, his standard monomial basis relates SMT and the representation-theory of quantum groups (cf. [**27**]). In addition, his basis has also close connections with the canonical basis of Kashiwara as well as Lusztig (for quantum groups) (cf. [**3, 32, 33**]). Also, his standard monomial basis has led to important connections between SMT and K-theory (cf. [**15, 30, 37**]). Interesting combinatorial results have been obtained (cf. [**39, 40**]) using Littelman's standard monomial basis.

References

[1] S. Billey and V. Lakshmibai, On the singular locus of a Schubert variety, *Progress in Mathematics* 182, Birkhaüser, 2001.

[2] M. Brion and V. Lakshmibai: A geometric approach to Standard Monomial Theory, preprint math.AG/0111054 (2001) (to appear in *Representation Theory* (electronic journal)).

[3] P. Caldero and P. Littelmann, Adapted algebras and standard monomials, math. RT/0201026.

[4] C. DeConcini and C. Procesi, A characteristic-free approach to invariant theory, *Advances in Math.*, **21** (1976), 330-354.

[5] N. Gonciulea: Singular loci of varieties of complexes, *J. Alg.*, **232** (2001).

[6] N. Gonciulea and V. Lakshmibai: Singular loci of Schubert varieties and Ladder determinantal varieties, *J. Alg.*, **232** (2000), 360-395.

[7] W.V.D.Hodge: Some enumerative results in the theory of forms, *Proc. Camb. Phil. Soc.* **39** (1943), 22-30.

[8] W.V.D.Hodge and D.Pedoe: Methods of Algebraic Geometry, **Vol.II**, *Cambridge University Press* (1952).

[9] A. Joseph, Quantum groups and their primitive ideals, *Springer Verlag*, New York (1995).

[10] S. Kumar and P. Littelmann: Frobenius splitting in characteristic zero and the quantum Frobenius map, *J. Pure and Applied Alg.*, **152** (2000), 201-216.

[11] V. Lakshmibai: Standard monomial theory for G_2, *J. Alg.*, **98** (1986), 281-318.

[12] V. Lakshmibai: Singular loci of Schubert varieties, *Bull. A.M.S.*, Vol. **16** (1987), 83-90.

[13] V. Lakshmibai: Tangent spaces to Schubert Varieties, *Math. Research Letters*, **2** (1995),473-477.

[14] V. Lakshmibai: The development of SMT II (*appears in this volume*).

[15] V. Lakshmibai and P. Littelmann: Ricchardson varieties and equivariant K-theory, math.AG/0201075 (to appear in the volume dedicated to Steinberg on his 80-th birthday).

[16] V. Lakshmibai and P. Magyar Degeneracy schemes, Quiver schemes, and Schubert varieties, *IRNM*, no:**12** (1998), 627-640.

[17] V. Lakshmibai, C. Musili, and C. S. Seshadri: Cohomology of line bundles on G/B, *Ann. Sci. E.N.S.*, **t.7**, (1974), 89-138.

[18] V. Lakshmibai, C. Musili, and C. S. Seshadri: Geometry of $G/P - III$, *Proc. Ind. Acad. Sci.*, **87A** (1978), 93-177.

[19] V. Lakshmibai, C. Musili, and C. S. Seshadri: Geometry of $G/P - IV$, *Proc. Ind. Acad. Sci.*, **88A** Geometry of $G/P - IV$ (1979) 279-362.

[20] V. Lakshmibai and K.N. Rajeswari: Towards a standard monomial theory for exceptional groups, *Contemporary Math., A.M.S.*, **88** (1989),

[21] V. Lakshmibai and C.S. Seshadri: Geometry of G/P - II, *Proc. Ind. Acad. Sci.*, **87A** (1978), 1-54.

[22] V. Lakshmibai and C.S. Seshadri: Geometry of G/P-V, *J. Alg*, **100, no:2** (1986), 462–557.

[23] V. Lakshmibai and C.S. Seshadri: Singular locus of a Schubert variety, *Bull. A.M.S.*, Vol. **11** (1984), 363-366.

[24] V. Lakshmibai and C.S. Seshadri: Standard monomial theory, *Proc. Hyderabad Conference on Algebraic Groups*,(S. Ramanan et al.,eds.), Manoj Prakashan, Madras (1991) 279-323.

[25] V. Lakshmibai and C.S. Seshadri: Standard monomial theory for $\widehat{SL}_2$, *Advanced Studies in Mathematical Physics*, Vol. **7** (1989), p.178-234, "Infinite Dimensional Lie Algebras and Groups", Luminy, Proceedings of the conference held at CIRM, Luminy, Marseille, July 4-8, 1988.

[26] P. Littelmann: A Littlewood-Richardson formula for symmetrizable Kac-Moody algebras. *Invent. Math.* **116**, 329-346 (1994).

[27] P. Littelmann: Contracting modules and standard monomial theory. *JAMS* **11**, 551–567 (1998).

[28] P. Littelmann: The path model, the quantum Frobenius map and Standard Monomial Theory, Algebraic groups and their representations, eds., R. Carter and J. Saxl, *Kluwer Academic Publishers* (1998).

[29] P. Littelmann: Bases for representations, L-S paths and Verma-Flags (appears in this volume).

[30] P. Littelmann and C.S. Seshadri: A Pieri-Chevalley formula for K(G/B) and Standard Monomial Theory, *The proceedings of the SCHUR MEMORIAM WORKSHOP in Rehovot, 2000* (to appear), to be published by Birkhäuser.

[31] G. Lusztig: Introduction to quantum groups. *Progress in Mathematics* **110**, Birkhäuser (1993).

[32] R.J. Marsh: Algorithms to obtain the canonical basis in some fundamental modules of quantum groups, *Journal of Algebra*, 186(3) (1996), 831-860.

[33] R.J. Marsh: Canonical bases for fundamental modules of quantized enveloping algebras of type A, *Communications in Algebra*, 25(5) (1997), 1365-1386.

[34] V. Mehta and A. Ramanathan: Frobenius splitting and cohomology vanishing for Schubert varieties , *Ann. Math.* **122** (1985), 27–40.

[35] C. Musili: Postulation formula for Schubert varieties , *J. Ind. Math. Soc.*, **36** (1972) 143-171.

[36] C. Musili: The development of Standard Monomial Theory-I, *this volume*, pp 385– 420 .

[37] H. Pittie and A. Ram, A Pieri-Chevalley formula for $K(G/B)$, *Electron. Res. Announc. Amer. Math. Soc.*, Vol. **5** (1999), 102-107.

[38] C.S. Seshadri: Geometry of G/P - I, *C.P. Ramanujam: A Tribute*, **207** (Springer-Verlag), published by Tata Institute (1978), 207–239.

[39] J. Stembridge: A Local Characterization of Simply-Laced Crystals, *Trans. Amer. Math. Soc.*, to appear.

[40] J. Stembridge: Combinatorial Models for Weyl Characters, *Advances in Math.*, 168 (2002), 96-131.

Northeastern University, Mathematics Department
Boston, MA-02115, U.S.A.
E-mail address: lakshmibai@neu.edu

Department of Math. and Stat., University of Hyderabad
P.O. Central University, Hyderabad 500 046, INDIA.
E-mail address: cmsm@uohyd.ernet.in

Seshadri and the Chennai Mathematical Institute

K.R.Nagarajan

In 1958, Irving Kaplansky was giving a course in Commutative Algebra at the University of Chicago. During the lectures he referred to the now famous result of Seshadri on projective modules over polynomial rings in two variables over a field. I was then a student attending the course, and the glow of happiness and pride I felt at that time is still fresh in my memory. This was my first introduction to Seshadri.

In talking about Seshadri's contribution to mathematics education, it would be perhaps easier for me to start at the "end", that is, describe his achievements at CMI during the last decade, and go backwards in time. However, keeping to the maxim that the easiest is not always the best, I shall start by sketching the developments in the early years, when Seshadri was involved in building up a school of Algebraic Geometry at the TIFR. I am conscious of the fact that there are several distinguished colleagues of Seshadri here who are much better fitted for this task.

Seshadri joined the TIFR in 1953, and very soon after that he and M.S.Narasimhan were looked upon as role models for the developing group of young mathematicians. Looking at his present style of interactions with the students and aspiring younger colleagues, one can easily visualize how it would have been in those early days, very much more energetic, intense and fierce.

Seshadri returned to TIFR from France in the early sixties, imbibing the French taste in mathematics, philosophy, and other things. He still "exposes" mathematics and occasionally talks about integrity domains. My first meeting with Seshadri was at the TIFR; that was soon after he got married. I remember with happiness that it was he who advised me to read Serre's Faiseaux Algebraic Coherents;

listened to both my music and my mathematics with a total concentration and encouragement. This continues to this day.

For Seshadri, teaching and research form a synthetic whole. His teaching should be understood in a very broad sense, not just as instruction by formal lectures. He can listen to a question and quickly position the problem in a very broad canvas, illuminate it, and lead the learner to a wider understanding of it. There have been many great mathematicians and many great teachers, in the past. Among them a few have achieved a synthesis of the two roles, and Seshadri is one of them.

The coming of Grothendieck's school of Algebraic Geometry to India should be attributed primarily to the efforts of Seshadri. His long courses on Schemes and Algebraic Geometry at the TIFR inspired many people, and after a few years, TIFR came to be recognized as one of the best centers in the world for learning and research in Algebraic Geometry. As an index of Seshadri's inspiration to the "younger" generation, we can list some of his students - M.P. Murthy, C. Musili, V. Lakshmibai, Rego, Cyril D'Souza, Krishnaprasad, M.V. Nori, N. Sundaram, V. Balaji, S. Kannan.

Seshadri's availability to the mathematicians and aspiring students was not confined to the TIFR alone. For instance, starting from 1967 he had visited Madurai (where I was working) several times. I still have the notes of his lectures – and it was from one of these lectures that I learnt that the correct and geometric way of describing a curve is by drawing a square (with fibers)!!

After his visits to Brandeis and UCLA, Seshadri came to the Institute of Mathematical Sciences (IMSc.) in 1984. He came to Chennai essentially supported by his self assurance – like a sage of old starting a new Ashram. Students and other mathematicians joined him soon. P.S. Thiagarajan, who later joined Seshadri at CMI, R. Balasubramian and several other mathematicians for short periods, formed a very active group at the IMSc. In a little more than five years at IMSc., four Ph.D's were produced and Seshadri continued with his instruction by courses on basic Geometry, Curves and Jacobians. Attracted by the ambiance created by Seshadri, there were then several distinguished visitors, D.Mumford, M.Artin, T.Oda, J.L.Verdier, R.Hotta, J.Thomson, M.V.Nori, V.S.Varadarajan, S.R.S.Varadhan, who stayed on for short stretches of time.

It was perhaps during these years at IMSc. that Seshadri crystallized his ideas about the need to reach out into undergraduate education in Mathematics. I remember a conversation we had in 1989 about this; and one has to remember that that was much before any actual move was made. I learnt later that efforts towards setting up a center for mathematics education at the undergraduate level, were in the air much earlier, initiated by mathematicians at the TIFR and NBHM. Perhaps the time was not ripe.

Seshadri's dream of starting an instructional programme beginning with undergraduates, grew out of the philosophy that great learning can be had only from great masters. This meant that one should first have a group of mathematicians who are scholars in the true sense, learning and creating new ideas in mathematics continuously, and then this group should disseminate its knowledge and inspire young minds. Also, Seshadri has the firm conviction, that research and learning is not complete without the activity of continuous instruction. We are reminded of the injunctions of the Taittiriya Upanishad - Every one of the aspects of human excellence is associated with, and gains fulfillment by "Swadhyaya Pravachana"–self learning and instruction.

The Chennai Mathematical Institute is a somewhat unique institution in India. Originally it was part of the SPIC Science Foundation realized by the vision of Mr. A.C. Muthiah, Chairman of SPIC group. The inclusion of mathematics as one of the divisions of the foundation is an example of a right thing happening at the right time. The late Dr S. Parthasarathy played a crucial role in the bringing together of, a group of outstanding mathematicians under the leadership of Seshadri, and an industrialist with vision. The mathematics division was called the SPIC Mathematical Institute and later changed its name to Chennai Mathematical Institute. Seshadri and P.S. Thiagarajan with their younger colleagues undertook the task of building the CMI first into an active and first rate centre for research in Mathematics and Computer Science. Then in 1998 the undergraduate educational programme was launched. Anyone who has worked in higher educational institutions would appreciate the difficulties in planning the curriculum, the details of admission of students, evaluation, and the logistics of actual instruction. And here Seshadri had the unstinted cooperation of his colleagues at CMI and in a generous measure from the IMSc. In fact even now a substantial part of the instruction is being handled by the mathematicians and computer scientists of the IMSc. I would like to offer the proposition that such enthusiastic involvement in the teaching programme is due to the universal perception shared by all scholars that teaching young minds is an integral part of the scholarship. Seshadri must have been very sure of this, to have embarked on this programme which needs so many different inputs for its success.

The three year programme of BSc (Hons) in mathematics consists of six semesters with a total of 26 units (courses). Out of these, about half are in mathematics, a third in computer science, two courses in humanities/languages and the rest are optional, available to the advanced undergraduates of the 5th and 6th semesters. According to Seshadri, the ideal programme ought to offer and make available to the young minds all that is best in sciences and humanities, create an atmosphere of open enquiry and learning, and at the same time guide and encourage the students to concentrate and achieve depth of learning in one or two areas. The younger colleagues at CMI and from IMSc. proposed a programme which fully reflected Seshadri's ideas. In order to put this into action there were several constraints. First and foremost at the basic level there had to be affiliation to a university and it was here CMI has been most fortunate. Prof. H.P. Dikshit who was then the V.C of Madhya Pradesh Bhoj (Open) University not only approved the programme but also encouraged it by his enthusiastic support. At the second level, Seshadri had the whole hearted cooperation from a large group of mathematicians and computer scientists from the IMSc., TIFR, and ISI. The courses of instruction were by experts and scholars and were innovative. Thirdly, it was decided that students would be selected on the basis of an entrance examination held at various centres in India, and this has proved to be a wise decision.

This process that was set in motion in 1999 has now succeeded remarkably. Seshadri says that in the early days of the programme he and his colleagues felt that the success of the programme would, in the Indian context, be measured by the number of graduates who would get admitted in well known foreign universities. For the record let me mention that there are now three of them abroad, two in the U.S. , and one in Germany, working for the PhD.s in mathematics/computer science!! We must remember that only one batch has graduated so far.

The French associations of Seshadri has continued into the teaching programme also. CMI has an arrangement with ENS Paris for mutual exchange of visits by students and scholars. Each year three or four students of the passing out batch of the CMI would visit ENS for about six weeks, and reciprocally, three or four Ph.D. students of ENS would visit CMI and deliver short courses of lectures. Seshadri often talks about his philosophy of education and his dream of creating an ideal university. He usually sums up by saying that it should be somewhat like the Ecole Normale. What has been happening at the CMI should not be seen as an accident, though Seshadri often says that things must be simply initiated and allowed to evolve. I know and many of my friends will agree, that this evolution has already taken place in Seshadri's mind.

After the introduction of the undergraduate programme, Seshadri has taken the natural step of initiating graduate programmes in mathematics and computer science. We are now only in the first year of the programme. Further, in keeping with the spirit of his philosophy, Seshadri is also planning to introduce a BSc. programme in Physics; with the long term goal of a single undergraduate programme, where courses in all disciplines will be available to the students, and they can major in any one of them.

In one sense Seshadri's model for a centre for mathematics education could have been realized only at a place like CMI, because of the large amount of freedom available to innovate. At the same time there has always been a substantial constraint for expansion due to the lack of sufficient monetary support. As mentioned earlier, the enlightened support given by A.C. Muthiah for nearly a decade was crucial and thereafter the NBHM under the leadership of M.S. Raghunathan, and the ISRO have supported the activities substantially. Recently the Shriram Group of Companies, Chennai, headed by R. Thyagarajan have taken great interest and have in-effect continued the private funding. Seshadri has said on many occasions that a substantial private presence in the basic support structure is a major factor in the success of CMI. Support for research and education in the pure sciences especially mathematics is inadequate in our country. The leaders of finance and industry by and large are yet to fully realize the importance of basic research in pure sciences for the future growth of the country.

Most of us mathematicians remember the singular dedication of our teachers and their remarkable achievements in isolation, and remember their yearning for an enlightened environment. It has taken many decades for the mind-set to change and we at the CMI are always hopeful that our dreams will be realized. Seshadri is here to lead us.

Chennai Mathematical Institute
92 G.N. Chetty Road, Chennai 600 017, INDIA.
E-mail address: nagaraj@cmi.ac.in

C.S. Seshadri's impact on education: A Computer Scientist's Perspective

MADHAVAN MUKUND

On the occasion of Prof. Seshadri's 70th birthday, it is my privilege to be able to talk about his influence on teaching and research in Computer Science in India. I have been in close contact with Prof. C.S. Seshadri for about fifteen years now: five years as a Ph.D. student at the Institute of Mathematical Sciences (IMSc) and Chennai Mathematical Institute (CMI) and over ten years as a member of the Computer Science faculty at CMI. I have thus interacted with him both as a student and as a colleague.

Computer Science has the dubious distinction of being one of the most misunderstood academic disciplines. There is tremendous confusion about the scope of the term Computer Science, not only among laymen, but also among fellow scientists—a term that I use in the broad sense to include mathematicians, physical scientists and even engineers.

Computer Science has become synonymous with the design and manufacture of computer systems. Thus, it is taken for granted that the work of every Computer Scientist is directly linked to the production of some tangible hardware or software component. A natural consequence of this attitude is that Computer Science is seen as a subject to be taught in engineering colleges.

Most people are surprised when they are told that Computer Science is, at its heart, a mathematical activity that is largely independent of the use of computers! The subject originated in the early part of the twentieth century through the work of Alonzo Church, Alan Turing and others who sought to formalize the notion of "effective computability" in response to questions raised by, among others, David Hilbert. These theoretical studies led to the model of the stored memory computer, commonly known as the von Neumann model, which was then realized in electronics. Even as advances in electronics have led to faster, more powerful and more compact physical realizations of computers, the body of work initiated by Church and Turing has continued in parallel, conceptually well ahead of the state of the art of physical computing devices.

Given the general prejudice that Computer Science is a subject that belongs in engineering colleges, it was tremendously courageous of Prof. Seshadri to broaden the scope of the Mathematics group at IMSc in the mid-1980s to include Computer Science. His vision has led to the establishment of one of the most active research groups in Computer Science in the country (now divided between IMSc and CMI). Several distinguished Computer Scientists across the country have been a part of this group at various points in their academic careers. The most prominent example at the time of writing is Manindra Agrawal, now very much in the news for his outstanding result that primality can be checked in polynomial time.

The start of the teaching programme at CMI revealed another important facet of Prof. Seshadri's thinking. While the research activity at CMI and IMSc is primarily theoretical, and thus more "obviously" linked to mathematics, a teaching programme must necessarily expose students to more practical computing skills. Prof. Seshadri was very quick to recognize that for undergraduate students, this rather mundane side of Computer Science was as valuable as the esoteric theoretical side that he was by now used to. He has allowed the Computer Science group complete freedom to design a curriculum that balances the mathematical theory of computing with more practical issues that arise in the design of computer software and hardware.

To summarize, Prof. Seshadri has been extremely supportive of Computer Science at CMI. His vision and encouragement have led to the growth of an active research group and a vibrant teaching programme. The research group in Computer Science at IMSc and CMI is one of the best in India and has an established international reputation. In the years to come, I am confident that the teaching programme will also come to be recognized in the same light.

Chennai Mathematical Institute
92 G.N. Chetty Road, Chennai 600017, India
E-mail address: madhavan@cmi.ac.in
URL: http://www.cmi.ac.in/~madhavan

CMI's Undergraduate Programme and NBHM

M.S. RAGHUNATHAN

Over the last decade and a half scientists have voiced their concern over the decline in interest in basic sciences among young people in many forums. The sorry state of science education, especially in the undergraduate years has been identified as one of the principal causes for this decline. In the West, undergraduates are taught largely by active research scientists and they are able to convey the excitement of science effectively. This rarely happens in our country where undergraduate education is handled mostly by affiliated colleges.

It is to the great credit of Professor C.S. Seshadri, that he initiated the B.Sc(Hons) programme at CMI, a programme in which active research mathematicians interact with undergraduates intensely. NBHM has welcomed this initiative enthusiastically recognising it as the first attempt at a practical initiative to do something about a problem which was much talked about. It is gratifying to see that already in the first few years, the programme has proved itself successful. We are particularly happy that NBHM's Olympiad activity has provided a good number of talented students for the CMI's programme.

As a body charged with the promotion of Mathematics, NBHM has been supporting CMI with financial grants almost from its inception. We were happy to enhance this support substantially when CMI embarked on the undergraduate programme. We are happy to note that the programme is living up to Professor Seshadri's aims of providing undergraduate education in Mathematics comparable to the best available anywhere in the world. We wish the programme continued success and offer our congratulations to Professor Seshadri on its performance hitherto. NBHM is also very happy to note that The Institute of Mathematical Sciences is participating vigorously in the programme and is in fact providing very substantial inputs. Their collaboration has been crucial for the success of the programme. We offer our congratulations also to the Faculty of that institute as well as of those of CMI for the splendid job they have done.

CHAIRMAN, NATIONAL BOARD FOR HIGHER MATHEMATICS

Part II

Invited Articles

Translates of Polynomials

Shreeram S. Abhyankar, William J. Heinzer
and Avinash Sathaye

Abstract. We undertook this study of affine pencils especially to celebrate the 70th birthday of our good friend C. S. Seshadri. The first named author met Seshadri in Paris in 1958 and had the pleasure of seeing him frequently ever since. We are very happy to say to him: JEEVEMA SHARADAH SHATAM.

1. Introduction

Let $f = 0$ be a hypersurface in the n-dimensional affine space over a field k with $n > 1$, i.e., $f \in R \setminus k$ where R is the polynomial ring $k[X_1, \ldots, X_n]$. We want to consider the pencil of hypersurfaces $f - c = 0$ with c varying over k, and wish to consider the sets $\mathrm{singset}(f) = \{c \in k : f - c \text{ is singular}\}$ and $\mathrm{redset}(f) = \{c \in k : f - c \text{ is reducible}\}$. In the first case $f - c$ is singular means the local ring $R_P/((f-c)R_P)$ is nonregular for some $P \in \mathrm{spec}(R)$ with $f - c \in P$, and in the second case $f - c$ is reducible means $f - c = gh$ with g, h in $R \setminus k$.

As consequences of the two famous theorems of Bertini, which may be called Bertini I or Singular Bertini, and Bertini II or Reducible Bertini, it can be shown that, under suitable conditions, singset(f) and redset(f) are finite. One of our aims is to give short direct proofs of these consequences. We shall do this in Sections 3 and 4. In Section 2 we shall recall the Theorems of Bertini and also the Theorem of Lüroth. Before outlining the contents of the rest of the paper, let us fix some notation.

2000 Mathematical Subject Classification: 12F10, 14H30, 20D06, 20E22. Abhyankar's work was partly supported by NSF Grant DMS 99-88166 and NSA grant MDA 904-99-1-0019.

By $|S|$ we denote the cardinality of a set S. By $U(S)$ we denote the multiplicative group of all units in a ring S, and by $S^\times$ we denote the set of nonzero elements in it. By $\mathrm{QF}(S)$ we denote the quotient field of a domain S. By $A = R/(fR)$ we denote the affine coordinate ring of $f = 0$, and we identify k with its image under the residue class epimorphism $\phi : R \to A$; if f is irreducible in R then by $L = \mathrm{QF}(A)$ we denote the function field of $f = 0$. By an affine domain over a field k' we mean an overdomain of k' which is a finitely generated ring extension of k'. By a DVR we mean a real discrete valuation ring; if the said ring has quotient field L' then we call it a DVR of L'; if it also contains a subfield k' then we call it a DVR of L'/k'. Note that a finitely generated free abelian group is isomorphic to $\mathbb{Z}^r$ for a unique nonnegative integer r which is called its rank; we shall apply this to the multiplicative group $U(A')/U(k')$ for an overdomain A' of a field k'.

Given polynomials $g_1, \ldots, g_m$ in one or more variables with coefficients in some field, we write $\gcd(g_1, \ldots, g_m) = 1$ or $\neq 1$ to mean that they do not or do have a nonconstant common factor. In particular we shall apply this to the partial derivatives $f_{X_1}, \ldots, f_{X_n}$ of f.

Let $R^* = k^*[X_1, \ldots, X_n]$ where k^* is an algebraic closure of k. Let $\mathrm{singset}(f)^* = \{c \in k^* : R^*_P/((f-c)R^*_P)$ is nonregular for some $P \in \mathrm{spec}(R^*)$ with $f - c \in P\}$, and $\mathrm{redset}(f)^* = \{c \in k^* : f - c = gh$ for some g, h in $R^* \setminus k^*\}$. We shall also consider the multiple set and the primary set of f defined by putting $\mathrm{multset}(f)^* = \{c \in k^* : f - c = gh^2$ for some $g \in R^* \setminus \{0\}$ and $h \in R^* \setminus k^*\}$, and $\mathrm{primset}(f) = \{c \in k : f - c = gh^\mu$ for some $g \in k^\times$ and $h \in R \setminus k$ and integer $\mu > 1\}$.

In Section 5 we shall discuss the notion of composite pencils, and we shall find some bounds for $|\mathrm{redset}(f)|$. A series of Examples, to be outlined in Remark 5, will illustrate the spread of the various values which $|\mathrm{redset}(f)|$ can take.

In Section 6 we shall extend our study to more general pencils $f - cw$ where $w \in R^\times$ with $\gcd(f, w) = 1$. In Remark 8 we shall link-up the redset of a general pencil to Klein's parametrization of a special rational surface, and in Question 4 we shall pose a related problem.

In Section 7 we shall employ a refined version of $\mathrm{redset}(f)$ to give necessary conditions for the ring $R[1/f]$ to be isomorphic to the ring $R[1/f']$ where $f' = 0$ is another hypersurface. This is when f and f' are irreducible. In the reducible case we shall give necessary conditions in terms of the quotient group $U(R[1/f])/U(k)$ and the function fields of the irreducible components of $f = 0$. Now an isomorphism of the rings $R[1/f]$ and $R[1/f']$ can be geometrically paraphrased as biregular equivalence of the complements of the hypersurfaces $f = 0$ and $f' = 0$ in affine n-space; moreover, an automorphism of R sending f to f' induces such an isomorphism. In Questions 5 to 8 of Section 7, these facts provide a link-up of the results of that section to: Abhyankar's theorem on exceptional nonruled varieties, the birational invariance of the arithmetic genus of a nonsingular projective variety via the domination part of Abhyankar's desingularization theory, the epimorphism theorems and problems discussed by Abhyankar in his Kyoto Notes, and the work of Zariski, Fan, Teicher, and others on the topology of complements. We were originally motivated to consider the said isomorphisms by a question which Roger Wiegand asked us in 1988.

In Section 8 we shall deduce the finiteness of the redset of a hypersurface from that of a plane curve by the intervention of Zariski's famous Lemma 5. It is by means of this Lemma that Abhyankar proved the Galois case of the Jacobian Problem. More precisely he deduced the Galois case from the birational case which was itself proved by using Zariski's Main Theorem.

In Section 9, we shall find a bound for the singset of plane curve f in terms of its deficiency set defset(f) which is the set of all constants c for which the algebraic rank $\rho_a(f - c)$ is different from the pencil-rank $\rho_\pi(f)$. In the complex case, $\rho_a(f)$ of an irreducible f coincides with its first homology rank. In the general case, $\rho_a(f)$ is defined in terms of the genera and numbers of branches of the various irreducible components of f, and the pencil-rank $\rho_\pi(f)$ is the general value of $\rho_a(f - c)$ taken over all constants c. We shall show how the pencil rank is related to the Zeuthen-Segre invariant. We shall also discuss Jung's formula which relates the rank to the Zeuthen-Segre invariant.

In Section 10 we shall extend our study of the deficiency set to that of a general pencil.

To put things in proper perspective, in Section 11 we shall briefly talk about linear systems and pencils on normal varieties, and say a few words about the Zeuthen-Segre invariant of a nonsingular projective algebraic surface.

2. Theorems of Bertini and Lüroth

Considering a linear systems of codimension one subvarieties of an algebraic variety, and calling it irreducible if its generic member is irreducible, in Mantra form, Bertini's Theorems may be stated thus:

BERTINI I OR SINGULAR BERTINI. Outside the singularities of the variety and outside its base points, members of an irreducible linear system do not have variable singularities.

BERTINI II OR REDUCIBLE BERTINI. If a linear system, without fixed components, is not composite with a pencil, then it is irreducible.

These were obtained by Bertini in his 1882 paper [**Ber**]. They were revisited by Zariski in [**Za1**] and [**Za2**].

We shall also use the equally hoary:

THEOREM OF LÜROTH. If a curve has a rational parametrization then it has a faithful rational parametrization.

This is in his 1875 paper [**Lur**]. We need the refined version given by Abhyankar-Eakin-Heinzer in [**AEH**]. Also see Igusa [**Igu**] and Nagata [**Na2**].

3. Singset

Let us now prove our:

SINGSET THEOREM. If k is of characteristic zero then singset(f) is finite.

Proof. Let I be the ideal in R^* generated by $f_{X_1}, \ldots, f_{X_n}$. For any $P \in \operatorname{spec}(R^*)$ with $I \subset P$, consider the residue class map $\Phi_P : R^* \to R^*/P$. Since all the partials of f belong to P, it follows that $D(\Phi_P(f)) = 0$ for every $\Phi_P(k^*)$-derivation of $\mathrm{QF}(R^*/P)$. Therefore, since k^* is of characteristic zero, we have $\Phi_P(f) = \Phi_P(\kappa(P))$ for a unique $\kappa(P) \in k^*$. Clearly $P + (f - c)R^* = R^*$ whenever $\kappa(P) \neq c \in k^*$. Let $P_1, \ldots, P_s$ be the minimal primes of I in R^*, where $s = 0 \Leftrightarrow I = R^*$. Then for all $c \in k^* \setminus \{\kappa(P_1), \ldots, \kappa(P_s)\}$ we have

$I + (f - c)R^* = R^*$. Since k is of characteristic zero, it follows that $\mathrm{singset}(f) \subset \{\kappa(P_1), \ldots, \kappa(P_s)\}$, and hence $\mathrm{singset}(f)$ is finite.

4. Redset

Next we prove our:

Redset Theorem. If f is irreducible in R and k is relatively algebraically closed in L, then $\mathrm{redset}(f)$ is finite.

Proof. By the following Lemma we can find a finite number of DVRs $V_1, \ldots, V_t$ of L/k such that $A \cap V_1 \cap \cdots \cap V_t = k$. For every $z \in L^\times$ let $W_i(z) = \mathrm{ord}_{V_i}(z)$, and let $W : L^\times \to \mathbb{Z}^t$ be the map given by putting $W(z) = (W_1(z), \ldots, W_t(z))$. Let G be the set of all $g \in R \setminus k$ such that $gh = f - c$ for some $h \in R \setminus k$ and $c \in k^\times$. Since the degree of g is clearly smaller than the degree of f, the set G is contained in a finite dimensional k-vector-subspace of R. Therefore for every $i \in \{1, \ldots, t\}$, the set $W_i(\phi(g))_{g \in G}$ is bounded from below. Since h also belongs to G and clearly $W_i(\phi(g)) = -W_i(\phi(h))$, it follows that the set $W_i(\phi(g))_{g \in G}$ is also bounded from above. Since, for $1 \le i \le t$, the set $W_i(\phi(g))_{g \in G}$ is bounded from both sides, it follows that $W(\phi(G))$ is a finite set. Also clearly $\phi(G) \subset U(A)$. Let $g_1 h_1 = f - c_1$ and $g_2 h_2 = f - c_2$ with g_1, h_1, g_2, h_2 in $R \setminus k$ and c_1, c_2 in $k^\times$ be such that $W(\phi(g_1)) = W(\phi(g_2))$. Then $\phi(g_1)/\phi(g_2) \in A \cap V_1 \cap \cdots \cap V_t = k$ and hence $\phi(g_2) = c\phi(g_1)$ for some $c \in k^\times$. Consequently $g_2 - cg_1$ is divisible by f in R and hence, because $\deg(g_2 - cg_1) < \deg(f)$, we must have $g_2 = cg_1$. Therefore, by subtracting the equation $g_2 h_2 = f - c_2$ from the equation $g_1 h_1 = f - c_1$ we get $c_2 - c_1 = g_1 h_1 - g_2 h_2 = g_1(h_1 - ch_2)$ which implies that $c_2 - c_1 \in k$ is divisible in R by the positive degree polynomial g_1. Consequently we must have $c_2 = c_1$. Therefore, because the set $W(\phi(G))$ is finite, we conclude that $\mathrm{redset}(f)$ is finite.

Lemma. Given any affine domain A' over a field k', let k'' be the algebraic closure of k' in $L' = \mathrm{QF}(A')$, and let A'' be the integral closure of A' in L'. Then there exists a finite number of DVRs $V_1, \ldots, V_t$ of L'/k' such that $A'' \cap V_1 \cap \cdots \cap V_t = k''$. Moreover, if $k'' = k'$ then $A' \cap V_1 \cap \cdots \cap V_t = k'$ and $U(A')/U(k')$ is a finitely generated free abelian group of rank r with $r \le \max(0, t-1)$.

Proof. Let m be the transcendence degree of L' over k'. If $m = 0$ then we can take $t = 0$, and if also $k' = k''$ then clearly $U(A')/U(k') = 1$ and hence $r = 0$. So assume $m > 0$. Then by Noether Normalization A' is integral over a polynomial ring $\widehat{A} = k'[Y_1, \ldots, Y_m] \subset L'$. Let $\widehat{L} = k'(Y_1, \ldots, Y_m)$ and $\widehat{V} = \{g/h : g, h \in \widehat{A}$ with $h \neq 0$ and $\deg(g) \leq \deg(h)\}$. Then $\widehat{V}$ is a DVR of $\widehat{L}/k'$ with $\widehat{A} \cap \widehat{V} = k'$. Let V'' be the integral closure of $\widehat{V}$ in L'. Then $V'' = V_1 \cap \cdots \cap V_t$ where $V_1, \ldots, V_t$ are DVRs of L'/k' with $t > 0$, and by the following Sublemma we have $A'' \cap V'' = k''$. It follows that $A'' \cap V_1 \cap \cdots \cap V_t = k''$, and if $k'' = k'$ then $A' \cap V_1 \cap \cdots \cap V_t = k'$. Now assuming $k'' = k'$, let $W : U(L') \to \mathbb{Z}^t$ be the homomorphism (from a multiplicative group to an additive group) given by $W(z) = (W_1(z), \ldots, W_t(z))$ with $W_i(z) = \mathrm{ord}_{V_i}(z)$. Then $U(A') \cap \ker(W) = U(k')$ and hence we get a monomorphism $\overline{W} : U(A')/U(k') \to \mathbb{Z}^t$. Therefore $U(A')/U(k')$ is a finitely generated free abelian group of rank r with $r \leq t$. Suppose if possible that $r = t$. Then we can find $z_1, \ldots, z_t$ in $U(A')$ such that the $t \times t$ matrix $W_i(z_j)$ has a nonzero determinant. Now the column vectors of this matrix are $\mathbb{Q}$-linearly independent vectors in $\mathbb{Q}^t$ and hence the column vector $(1, 0, \ldots, 0)$ can be expressed as a $\mathbb{Q}$-linear combination of them, i.e., we can find rational numbers $a_1, \ldots, a_t$ such that $\sum_{1 \leq j \leq t} W_i(z_j)a_j = 1$ or 0 according as $i = 1$ or $2 \leq i \leq t$. Next we can find integers $a, b_1, \ldots, b_t$ with $a > 0$ such that $aa_j = b_j$ for $1 \leq j \leq t$. Let $z = \prod_{1 \leq j \leq t} z_j^{b_j}$. Then $z \in U(A')$ with $W_1(z) = a > 0 = W_i(z)$ for $2 \leq i \leq t$. Consequently $z \in A' \cap V_1 \cdots \cap V_t$ but $z \notin k'$ which is a contradiction. Therefore we must have $r \leq t - 1$.

Sublemma. Let $\widehat{A}$ and $\widehat{V}$ be normal domains with a common quotient field $\widehat{L}$. Let A'' and V'' be the respective integral closures of $\widehat{A}$ and $\widehat{V}$ in an algebraic field extension L' of $\widehat{L}$. Then $A'' \cap V''$ is the integral closure of $\widehat{A} \cap \widehat{V}$ in L'.

Proof. Clearly the integral closure of $\widehat{A} \cap \widehat{V}$ in L' is contained in $A'' \cap V''$. Conversely, given any $z \in L'$ let $H(Z)$ be the minimal monic polynomial of z over $\widehat{L}$. By Kronecker's Theorem (or obviously), if $z \in A''$ then $H(Z) \in \widehat{A}[Z]$, and if $z \in V''$ then $H(Z) \in \widehat{V}[Z]$. Consequently, if $z \in A'' \cap V''$ then $H(Z) \in (\widehat{A} \cap \widehat{V})[Z]$. Thus $A'' \cap V''$ is integral over $\widehat{A} \cap \widehat{V}$. Therefore $A'' \cap V''$ is the integral closure of $\widehat{A} \cap \widehat{V}$ in L.

REMARK 1. With notation as in the proof of the Redset Theorem, in the $n = 2$ case, the Lemma also follows by taking $V_1, \ldots, V_t$ to be the valuation rings of the places at infinity of the plane curve $f = 0$. Moreover, in this case, in the proof of the Redset Theorem, the finiteness of $W(\phi(G))$ can be deduced from the boundedness from below of the sets $W_1(\phi(G)), \ldots, W_t(\phi(G))$ by invoking the fact that the number of zeros of a rational function on the curve $f = 0$ equals the number of its poles.

REMARK 2. Assuming k to be of characteristic zero, the $n > 2$ case of the Redset Theorem can be deduced from the $n = 2$ case by invoking the famous Lemma 5 of Zariski's paper [**Za1**] thus. By applying a linear k-automorphism to R and multiplying f by an element in $k^\times$, we can arrange f to be a monic polynomial of positive degree in X_1 with coefficients in $S = k[X_2, \ldots, X_n]$. Now $\phi(X_2), \ldots, \phi(X_n)$ is a transcendence basis of L/k and hence, invoking Zariski's Lemma 5 and applying a k-linear automorphism to S, we can arrange the field $k(\phi(X_3), \ldots, \phi(X_n))$ to be relatively algebraically closed in L. Let $\widetilde{R} = \widetilde{k}[X_3, \ldots, X_n]$ with $\widetilde{k} = k(X_3, \ldots, X_n)$, and let $\widetilde{L} = \mathrm{QF}(\widetilde{A})$ with $\widetilde{A} = \widetilde{R}/(f\widetilde{R})$. By Gauss Lemma, f is irreducible in $\widetilde{R}$ and hence, by the $n = 2$ case of the Redset Theorem, the set $\{c \in \widetilde{k} : f - c \text{ is reducible in } \widetilde{R}\}$ is finite. Therefore, again by Gauss Lemma, the set $\{c \in k : f - c \text{ is reducible in } R\}$ is finite.

REMARK 3. Assume f is irreducible in R. Now the Redset Theorem says that if the ground field k is relatively algebraically closed in the function field L then redset(f), i.e., the set of all constants c for which the polynomial $f - c$ factors, is finite. We want to observe that this set, although determined by f, is not determined by the affine coordinate ring A. Indeed if redset(f) is nonempty, say it contains a constant c (which is necessarily nonzero because f is irreducible), then every constant γ in A factors. Namely, since c is in redset(f), we can write $f - c = gh$ with g, h in $R \setminus k$, and multiplying both sides by $-\gamma/c$ we get $(-\gamma/c)f + \gamma = g'h'$ with $g' = g$ and $h' = (-\gamma/c)h$. Thus we have factored γ not only in A, i.e., modulo the ideal fR, but also "modulo" the constant multiples kf of f. Note that if $\gamma \neq 0$ then both g' and h' belong to $R \setminus k$, but if $\gamma = 0$ then h' does not. To take care of this extra "desire" and to make sure that the "multiplier $e = -(\gamma/c)$" of f is also nonzero, by taking $e = ug - (\gamma/c)$ with any $u \in R^\times$ we get $ef + \gamma = g'h'$

with $g' = g \in R \setminus k$ and $h' = uf - (\gamma/c)h \in R \setminus k$. As example, $f = X_1X_2 + 1$ is irreducible with $\text{redset}(f) = \{1\}$, and for any $\gamma \in k$ we have $(X_1X_2 - \gamma)f + \gamma = X_1[X_1X_2^2 + (1-\gamma)X_2]$.

5. Generic Members and Composite Pencils

Let $R^\sharp = k(Z)[X_1, \ldots, X_n]$ where Z is an indeterminate over R. By the generic member of the pencil $(f-c)_{c\in k}$ we mean the hypersurface $f^\sharp = 0$ with $f^\sharp = f - Z \in R^\sharp$. Let $\text{singset}(f^\sharp)$ be the set of all $c \in k(Z)$ such that $R^\sharp_P/((f^\sharp - c)R^\sharp_P)$ is nonregular for some $P \in \text{spec}(R^\sharp)$ with $f^\sharp - c \in P$, and let $\text{redset}(f^\sharp)$ be the set of all $c \in k(Z)$ such that $f^\sharp - c$ is reducible in $R^\sharp$.

By Gauss Lemma $f^\sharp$ is irreducible in $R^\sharp$, i.e., $0 \notin \text{redset}(f^\sharp)$. Let $\phi^\sharp : R^\sharp \to R^\sharp/(f^\sharp R^\sharp)$ be the residue class epimorphism. Clearly $R \cap \ker(\phi^\sharp) = \{0\}$ and $\phi^\sharp(f) = \phi^\sharp(Z)$, and hence there exits a unique isomorphism $\psi^\sharp : k(X_1, \ldots, X_n) \to \text{QF}(R^\sharp/(f^\sharp R^\sharp))$ such that for all $r \in R$ we have $\psi^\sharp(r) = \phi^\sharp(r)$. Thus the triple $\phi^\sharp(k(Z)) \subset \phi^\sharp(R^\sharp) \subset \text{QF}(\phi^\sharp(R^\sharp))$ is isomorphic to the triple

$$k^\sharp = k(f) \subset A^\sharp = k(f)[X_1, \ldots, X_n] \subset L^\sharp = k(X_1, \ldots, X_n)$$

and hence we may regard the above three displayed sets as the ground field, the affine coordinate ring, and the function field of $f^\sharp = 0$.

The ring $A^\sharp$ is regular because it is the localization of the regular ring R at the multiplicative set $k[f]^\times$. Thus we have the:

Generic Singset Theorem. $0 \notin \text{singset}(f^\sharp)$.

In view of the above isomorphism of triples, by the Redset Theorem we get the:

Generic Redset Theorem. If $k^\sharp$ is relatively algebraically closed in $L^\sharp$ then $\text{redset}(f^\sharp)$ is finite.

In [**AEH**] we proved the following:

Refined Lüroth Theorem. Assume that $k^\sharp$ is not relatively algebraically closed in $L^\sharp$. Let $\widehat{B}^\sharp$ be the integral closure of $k[f]$ in $L^\sharp$, let $\widehat{k}^\sharp$ be the algebraic closure of $k^\sharp$ in $L^\sharp$, and let $\nu = [\widehat{k}^\sharp : k^\sharp]$. Then ν is an integer with $\nu > 1$, and there exist $\widehat{f} \in R \setminus k$ and $\Lambda \in k[Z] \setminus k$ with $\widehat{B}^\sharp = k[F]$ and $\widehat{k}^\sharp = k(F)$ such that $f = \Lambda(\widehat{f})$ and $\deg_Z \Lambda = \nu$.

REMARK 4. In the above situation, every member of the pencil $(f-c)_{c\in k^*}$ consists of ν members (counted properly) of the pencil $(\widehat{f}-\widehat{c})_{\widehat{c}\in k^*}$, and so we say that **the pencil $(f-c)_{c\in k^*}$ is composite with the pencil $(\widehat{f}-\widehat{c})_{\widehat{c}\in k^*}$**. In greater detail, for any $c \in k^*$ we have $\Lambda(Z) - c = \widehat{c}_0 \prod_{1\le i\le\nu}(Z-\widehat{c}_i)$ where $\widehat{c}_0, \widehat{c}_1, \ldots, \widehat{c}_\nu$ in k^* with $\widehat{c}_0 \neq 0$, and by substituting $\widehat{f}$ for Z we get $f - c = \widehat{c}_0 \prod_{1\le i\le\nu}(\widehat{f}-\widehat{c}_i)$ and so the hypersurface $f = c$ is the union of the hypersurfaces $\widehat{f} = \widehat{c}_1, \ldots, \widehat{f} = \widehat{c}_\nu$. Since $\deg_Z\Lambda = \nu > 1$, if k is of characteristic 0 then by taking a root ζ of the Z-derivative of $\Lambda(Z)$ in k^* we see that $\Lambda(Z) - \Lambda(\zeta)$ has a multiple root in k^* and hence $f - \Lambda(\zeta)$ has a nonconstant multiple factor in R^*, and therefore $\mathrm{multset}(f)^* \neq \emptyset$. Note that if k is of characteristic $p > 0$ then this does not work as can be seen by taking $\Lambda(Z) = Z^p + Z$. Without assuming any condition on the pair $(k^\sharp, L^\sharp)$ we see that, for any $c \in k^*$, every multiple factor of $f - c$ in R^* divides f_{X_j} for $1 \le j \le n$, and hence: $\mathrm{multset}(f)^* \neq \emptyset \Rightarrow \gcd(f_{X_1}, \ldots, f_{X_n}) \neq 1$. Again without assuming any condition on the pair $(k^\sharp, L^\sharp)$ we see that $c \mapsto c - Z$ gives an injection of $\mathrm{redset}(f)$ into $\mathrm{redset}(f^\sharp)$, and hence $|\mathrm{redset}(f)| \le |\mathrm{redset}(f^\sharp)|$. Thus, in view of the above two Theorems, we get the:

COMPOSITE PENCIL THEOREM. We have the following:

(I) $\gcd(f_{X_1}, \ldots, f_{X_n}) = 1 \Rightarrow \mathrm{multset}(f)^* = \emptyset$.

(II) characteristic of k is 0 and $\mathrm{multset}(f)^* = \emptyset \Rightarrow k^\sharp$ is relatively algebraically closed in $L^\sharp$.

(III) $k^\sharp$ is relatively algebraically closed in $L^\sharp \Rightarrow \mathrm{redset}(f)$ and $\mathrm{redset}(f^\sharp)$ are finite with $|\mathrm{redset}(f)| \le |\mathrm{redset}(f^\sharp)|$.

(IV) $k^\sharp$ is not relatively algebraically closed in $L^\sharp \Rightarrow \mathrm{redset}(f)^* = k^*$.

Next we prove the:

MIXED PRIMSET THEOREM. If $k^\sharp$ is relatively algebraically closed in $L^\sharp$ then $\mathrm{primset}(f) = \emptyset$.

Proof. If $\mathrm{primset}(f) \neq \emptyset$ then for some $c \in k$ and integer $\mu > 1$ we have $f - c = gh^\mu$ with $g \in k^\times$ and $h \in R \setminus k$, and this implies $[k(h) : k(f)] = \mu$ and hence $k^\sharp$ is not relatively algebraically closed in $L^\sharp$.

Let us now prove the following theorem which is some kind of a mixture of the Redset Theorem and the Generic Redset Theorem.

MIXED REDSET THEOREM. If $k^\sharp$ is relatively algebraically closed in $L^\sharp$ then $U(A^\sharp)/U(k^\sharp)$ is a finitely generated free abelian group of rank r with $|\mathrm{redset}(f)| \leq r$.

Proof. Assuming that $k^\sharp$ is relatively algebraically closed in $L^\sharp$, by the Lemma we see that $U(A^\sharp)/U(k^\sharp)$ is a finitely generated free abelian group of some rank r. Suppose if possible that $|\mathrm{redset}(f)| > r$ and take distinct elements $c_1, \ldots, c_s$ in $\mathrm{redset}(f)$ with integer $s > r$. By the Mixed Primset Theorem, for $1 \leq i \leq s$ we have

$$f - c_i = g_i h_i \quad \text{where} \quad g_i, h_i \text{ in } R \setminus k \quad \text{with} \quad \gcd(g_i, h_i) = 1.$$

Since $s > r$, there exist integers $a_1, \ldots, a_s$ at least one of which is nonzero, say $a_\iota \neq 0$, such that

$$\prod_{1 \leq i \leq s} g_i^{a_i} \in U(k(f)). \tag{2}$$

For the rest of the proof we shall give two alternative arguments.

FIRST ARGUMENT. Clearly the images of g_i and h_i in $U(A^\sharp)/U(k^\sharp)$ are inverses of each other and hence replacing g_i by h_i for those i for which $a_i < 0$, we can arrange matters so that $a_i \geq 0$ for all i, and hence in particular $a_\iota > 0$. Again since the images of g_i and h_i in $U(A^\sharp)/U(k^\sharp)$ are inverses of each other, by (2) we get

$$\prod_{1 \leq i \leq s} h_i^{a_i} \in U(k(f)). \tag{3'}$$

Any element in $k(f)^\times$ can be written as $u(f)/v(f)$ where $u(Z), v(Z)$ in $k[Z]^\times$ with $u(Z)\widehat{u}(Z) + v(Z)\widehat{v}(Z) = 1$ for some $\widehat{u}(Z), \widehat{v}(Z)$ in $k[Z]$, and substituting $Z = f$ we get $u(f)\widehat{u}(f) + v(f)\widehat{v}(f) = 1$; it follows that if $u(f)/v(f) \in R$ then $v(f) \in k^\times$ and hence $u(f)/v(f) \in k[f]$. Thus $R \cap k(f) = k[f]$ and therefore by (2) and (3′) we get

$$\prod_{1 \leq i \leq s} g_i^{a_i} = g(f) \quad \text{and} \quad \prod_{1 \leq i \leq s} h_i^{a_i} = h(f) \tag{4'}$$

with $g(Z), h(Z)$ in $k[Z]^\times$. Multiplying the two equations in (4′) we obtain

$$\prod_{1 \leq i \leq s} (f - c_i)^{a_i} = g(f)h(f)$$

and hence by the k-isomorphism $k[Z] \to k[f]$ with $Z \mapsto f$ we get

$$\prod_{1\le i\le s} (Z-c_i)^{a_i} = g(Z)h(Z)$$

and therefore we have

$$(5') \quad g(Z) = \overline{g} \prod_{1\le i\le s} (Z-c_i)^{\alpha_i} \quad \text{and} \quad h(Z) = \overline{h} \prod_{1\le i\le s} (Z-c_i)^{\beta_i}$$

with $\overline{g}, \overline{h}$ in $k^\times$ and nonnegative integers α_i, β_i such that $\alpha_i + \beta_i = a_i$. Substituting $Z = f$ and $f - c_i = g_i h_i$ in (5′) and then comparing it with (4′) we get

$$(6') \quad \prod_{1\le i\le s} g_i^{a_i} = \overline{g} \prod_{1\le i\le s} (g_i h_i)^{\alpha_i} \quad \text{and} \quad \prod_{1\le i\le s} h_i^{a_i} = \overline{h} \prod_{1\le i\le s} (g_i h_i)^{\beta_i}.$$

Clearly $\gcd(f - c_i, f - c_j) = 1$ for all $i \neq j$, and hence by (1) and (6′) we get

$$(7') \qquad a_i = \alpha_i \text{ and } a_i = \beta_i \text{ for } 1 \le i \le s.$$

Since $\alpha_\iota + \beta_\iota = a_\iota > 0$ with $\alpha_\iota \ge 0$ and $\beta_\iota \ge 0$, we must have either $\alpha_\iota < a_\iota$ or $\beta_\iota < a_\iota$ which contradicts (7′). Therefore $|\mathrm{redset}(f)| \le r$.

Second Argument. We shall make this more condensed. By (2) we get

$$(3'') \qquad \prod_{1\le i\le s} g_i^{a_i} = \overline{u} \prod_{1\le i\le \sigma} u_i(f)^{\alpha_i} \quad \text{with} \quad \overline{u} \in k^\times$$

where $\alpha_1, \dots, \alpha_\sigma$ are nonzero integers and $u_1(Z), \dots, u_\sigma(Z)$ are pairwise coprime monic members of $k[Z] \setminus k$. As in the First Argument, for any u, v in $k[Z] \setminus k$ with $\gcd(u, v) = 1$ we have $u(f), v(f)$ in $R \setminus k$ with $\gcd(u(f), v(f)) = 1$. Therefore, by looking at divisibility by a nonconstant irreducible factor of g_i in R, in view of (1) and (3″), we see that for each $i \in \{1, \dots, s\}$ with $a_i \neq 0$ there is some $\theta(i) \in \{1, \dots, \sigma\}$ with $u_{\theta(i)}(Z) = Z - c_i$ and $\alpha_{\theta(i)} = a_i$. In particular we get $u_{\theta(\iota)}(Z) = Z - c_\iota$ and $\alpha_{\theta(\iota)} = a_\iota$. Again in view of (1) and (3″), by looking at divisibility by a nonconstant irreducible factor of h_ι in R, we get a contradiction. Therefore $|\mathrm{redset}(f)| \le r$.

Remark 5. Assume $n = 2$. Let t be the number of places at infinity of the irreducible plane curve $f^\sharp = 0$ and let $V_1, \dots, V_t$ be their valuation rings. Then t is a positive integer and by the Lemma and the Mixed Redset Theorem we see that if $k^\sharp$ is relatively algebraically closed in $L^\sharp$, then $R^\sharp \cap V_1 \cap \cdots \cap V_t = k^\sharp$ and

$|\text{redset}(f)| \le t-1$. We shall show that this bound is the best possible. In fact, in the following Examples 1 to 5, assuming k to contain sufficiently many elements, we shall show that, for any $t > 0$, $|\text{redset}(f)|$ can be equal to any integer μ with $0 \le \mu \le t-1$; Example 1 works for $0 \le \mu = t-1$ and $t > 0$; Example 2 works for $1 \le \mu < t-1$ and $t > 2$; Example 3 works for $\mu = 0$ and $t > 3$; Example 4 works for $\mu = 0$ and $t > 0$; Example 5 works for $\mu = 1$ and $t > 1$. As common notation for these examples, let $n = 2$, let $m \ge 0$ be an integer, and let $a(X_1) = (X_1 - a_1)\dots(X_1 - a_m)$ where $a_1, \dots, a_m$ are pairwise distinct elements in k. Moreover, for the irreducible $f \in R \setminus k$ to be constructed, let $\tau(f)$ denote the number of places at infinity of the plane curve $f = 0$, and let $\tau(f^\sharp)$ denote the number of places at infinity of the plane curve $f^\sharp = 0$.

EXAMPLE 1. Let $f = a(X_1)X_2 + X_1 - z$ with $z \in k \setminus \{a_1, \dots, a_m\}$. Then f is irreducible in R and
$\text{redset}(f) = \text{redset}(f)^* = \{a_1 - z, \dots, a_m - z\}$.
Moreover, $k^\sharp$ is relatively algebraically closed in $L^\sharp$ and we have $|\text{redset}(f)| = |\text{redset}(f)^*| = m$ with $\tau(f) = \tau(f^\sharp) = m + 1$.

Proof. Clearly $(X_1 - a_i)$ divides $f - (a_i - z)$ in R for $1 \le i \le m$, and hence $\{a_1 - z, \dots, a_m - z\} \subset \text{redset}(f)$. By Gauss Lemma we also see that $\text{redset}(f)^* \subset \{a_1 - z, \dots, a_m - z\}$. Therefore f is irreducible in R, and we have $\text{redset}(f) = \text{redset}(f)^* = \{a_1 - z, \dots, a_m - z\}$. In particular $|\text{redset}(f)| = |\text{redset}(f)^*| = m < \infty$, and hence by the Composite Pencil Theorem we see that $k^\sharp$ is relatively algebraically closed in $L^\sharp$. For the degree form (which consists of the highest degree terms) of f we have $\text{defo}(f) = X_1^m X_2$, and hence the points at infinity of the curve $f = 0$ are $(X_0, X_1, X_2) = (0, 1, 0)$ and $(X_0, X_1, X_2) = (0, 0, 1)$. Homogenizing f we obtain

$$(X_1 - a_1X_0)\dots(X_1 - a_mX_0)X_2 + X_0^m X_1 - zX_0^{m+1}$$

Putting $X_1 = 1$ we get

$$f_1 = (1 - a_1X_0)\dots(1 - a_mX_0)X_2 + X_0^m - zX_0^{m+1}$$

and hence $(0, 1, 0)$ is a simple point. Putting $X_2 = 1$ we get

$$f_2 = (X_1 - a_1X_0)\dots(X_1 - a_mX_0) + X_0^m X_1 - zX_0^{m+1}$$

and hence $(0, 0, 1)$ is an m-fold point with m distinct tangents. Therefore $\tau(f) = m + 1$, and hence by taking $(z + Z, k^\sharp)$ for (z, k) we get $\tau(f^\sharp) = m + 1$.

EXAMPLE 2. Assume $m \geq 1$ and let μ be any integer with $1 \leq \mu \leq m$. Let $b(X_1) = (X_1 - b_1) \dots (X_1 - b_m)$ where $b_1, \dots, b_m$ be pairwise distinct elements in k such that $b_i = a_i$ for $1 \leq i \leq \mu$, and $b_i \notin \{a_1, \dots, a_m\}$ for $\mu < i \leq m$. If $m \neq 1$ then let γ be any nonzero element of k, and if $m = 1$ then let γ be any nonzero element of k such that $Z^2 + \gamma Z + 1 = (Z - \gamma_1)(Z - \gamma_2)$ with $\gamma_1 \neq \gamma_2$ in k. Let $f = a(X_1)X_2^2 + \gamma b(X_1)X_2 + X_1 - z$ with $z \in k \setminus \{a_1, \dots, a_\mu\}$. Then f is irreducible in R and $\text{redset}(f) = \text{redset}(f)^* = \{a_1 - z, \dots, a_\mu - z\}$. Moreover, $k^\sharp$ is relatively algebraically closed in $L^\sharp$ and we have $|\text{redset}(f)| = |\text{redset}(f)^*| = \mu$ with $\tau(f) = \tau(f^\sharp) = m + 2$.

Proof. Clearly $(X_1 - a_i)$ divides $f - (a_i - z)$ in R for $1 \leq i \leq \mu$, and hence $\{a_1 - z, \dots, a_\mu - z\} \subset \text{redset}(f)$. Again by Gauss Lemma we see that if for some $c \in k^* \setminus \{a_1 - z, \dots, a_\mu - z\}$ we have $c \in \text{redset}(f)$ then there exists $\eta \in (k^*)^\times$ and a disjoint partition $\{1, \dots, m\} = U \coprod V$ such that upon letting

$$u(X_1) = \prod_{i \in U} (X_1 - a_i) \quad \text{and} \quad v(X_1) = \prod_{i \in V} (X_1 - a_i)$$

we have

$$f - z - c = [u(X_1)X_2 + \eta] \, [v(X_1)X_2 + (1/\eta)(X_1 - z - c] \, .$$

Equating the coefficients of X_2 we get

$$(*) \qquad \gamma b(X_1) = (1/\eta)(X_1 - z - c)u(X_1) + \eta v(X_1).$$

This gives a contradiction because the LHS is divisible by $(X_1 - a_1)$, but exactly one term in the RHS is so divisible. Therefore $\text{redset}(f)^* \subset \{a_1 - z, \dots, a_m - z\}$. Consequently f is irreducible in R, and we have $\text{redset}(f) = \text{redset}(f)^* = \{a_1 - z, \dots, a_m - z\}$. In particular $|\text{redset}(f)| = |\text{redset}(f)^*| = m < \infty$, and hence by the Composite Pencil Theorem we see that $k^\sharp$ is relatively algebraically closed in $L^\sharp$. Now $\text{defo}(f) = X_1^m X_2^2$, and hence the points at infinity of the curve $f = 0$ are $(X_0, X_1, X_2) = (0, 1, 0)$ and $(X_0, X_1, X_2) = (0, 0, 1)$. Homogenizing f we obtain

$$\begin{aligned}&(X_1 - a_1X_0) \dots (X_1 - a_mX_0)X_2^2 \\ &+ \gamma(X_1 - b_1X_0) \dots (X_1 - b_mX_0)X_2X_0 + X_1X_0^{m+1} - zX_0^{m+2}.\end{aligned}$$

Putting $X_1 = 1$ we get

$$\begin{aligned} f_1 =&(1-a_1X_0)\dots(1-a_mX_0)X_2^2 \\ &+\gamma(1-b_1X_0)\dots(1-b_mX_0)X_2X_0 + X_0^{m+1} - zX_0^{m+2} \end{aligned}$$

and hence $(0,1,0)$ is a double point with two distinct tangents. Putting $X_2 = 1$ we get

$$\begin{aligned} f_2 =&(X_1-a_1X_0)\dots(X_1-a_mX_0) \\ &+\gamma(X_1-b_1X_0)\dots(X_1-b_mX_0)X_0 + X_1X_0^{m+1} - zX_0^{m+2} \end{aligned}$$

and hence $(0,0,1)$ is an m-fold point with m distinct tangents. Therefore $\tau(f) = m+2$, and hence by taking $(z+Z, k^\sharp)$ for (z,k) we get $\tau(f^\sharp) = m+2$.

EXAMPLE 3. Assume $m \geq 2$. Let α_i and α_{ij} be the elements in k such that

$$a(X_1) = X_1^m + \sum_{1\leq j\leq m} \alpha_j X_1^{m-j}$$

and

$$a(X_1)/(X_1 - a_i) = X_1^{m-1} + \sum_{1\leq j\leq m-1} \alpha_{ij} X_1^{m-1-j} \quad \text{for} \quad 1 \leq i \leq m.$$

Let $b(X_1) = X_1^m + \sum_{i\in\{1,2,m\}} \beta_i X_1^{m-i}$ where $(\beta_i)_{i\in\{1,2,m\}}$ are elements k such that:

(i) $b(a_i) \neq 0$ for $1 \leq i \leq m$;

(ii) if $m > 3$ then $\beta_1 \neq \alpha_1$, and $\beta_2 \neq \beta_1\alpha_{i1} - \alpha_{i1}^2 + \alpha_{i2}$ for $1 \leq i \leq m$;

(iii) if $m = 3$ then $\beta_1 \neq \alpha_1$, and $\beta_2 \neq \beta_1\alpha_{i1} - \alpha_{i1}^2 + \alpha_{i2} + 1$ for $1 \leq i \leq m$;

(iv) if $m = 2$ then $\beta_1 \neq 1 + \alpha_1$, and $\beta_2 \neq \beta_1\alpha_{i1} - \alpha_{i1}^2 - \alpha_{i1} - a_i$.

Let $f = a(X_1)X_2^2 + b(X_1)X_2 + X_1 - z$ with $z \in k$. Then f is irreducible in R and $\text{redset}(f) = \text{redset}(f)^* = \emptyset$. Moreover, $k^\sharp$ is relatively algebraically closed in $L^\sharp$ and we have $|\text{redset}(f)| = |\text{redset}(f)^*| = 0$ with $\tau(f) = \tau(f^\sharp) = m+2$.

Proof. Let $\mu = 0$. By (i) we have $\gcd(a(X_1), b(X_1)) = 1$ and hence the proof is identical with the proof of Example 1 except we have to get a contradiction to (*) where $\gamma = 1$ and c is any element of k^*. So assume (*) and let $y = z + c$. Then comparing degrees and coefficients of X_1^m in the LHS and RHS of (*) we see that $\eta = 1$ and either $(u(X_1), v(X_1)) = (1, a(X_1))$ or

$(u(X_1), v(X_1)) = (a(X_1)/a_i(X_1), a_i(X_1))$ for some $i \in \{1, \ldots, m\}$. Let us expand the said LHS and RHS and compare the coefficients of X_1^{m-1} and X_1^{m-2} in them. Then in case of $m > 3$ we have

$$\begin{aligned}
&(u(X_1), v(X_1)) = (1, a(X_1)) \\
&\Rightarrow X_1^m + \sum_{j \in \{1,2,m\}} \beta_j X_1^{m-j} = (X_1 - y) + \left[X_1^m + \sum_{1 \le j \le m} \alpha_j X_1^{m-j} \right] \\
&\Rightarrow \beta_1 = \alpha_1
\end{aligned}$$

and

$$\begin{aligned}
&(u(X_1), v(X_1)) = (a(X_1)/a_i(X_1), a_i(X_1)) \\
&\Rightarrow X_1^m + \sum_{j \in \{1,2,m\}} \beta_j X_1^{m-j} = (X_1 - y) \left[X_1^{m-1} + \sum_{1 \le j \le m} \alpha_{ij} X_1^{m-1-j} \right] \\
&\qquad\qquad + (X_1 - a_i) \\
&\Rightarrow \beta_1 = \alpha_{i1} - y \text{ and } \beta_2 = -y\alpha_{i1} + \alpha_{i2} \\
&\Rightarrow \beta_2 = \beta_1 \alpha_{i1} - \alpha_{i1}^2 + \alpha_{i2}
\end{aligned}$$

and hence we get a contradiction by (ii). Likewise in case of $m = 3$ we have

$$\begin{aligned}
&(u(X_1), v(X_1)) = (1, a(X_1)) \\
&\Rightarrow X_1^3 + \beta_1 X_1^2 + \beta_2 X_1 + \beta_3 = (X_1 - y) + (X_1^3 + \alpha_1 X_1^2 + \alpha_2 X_1 + \alpha_3) \\
&\Rightarrow \beta_1 = \alpha_1
\end{aligned}$$

and

$$\begin{aligned}
&(u(X_1), v(X_1)) = (a(X_1)/a_i(X_1), a_i(X_1)) \\
&\Rightarrow X_1^3 + \beta_1 X_1^2 + \beta_2 X_1 + \beta_3 = (X_1 - y)(X_1^2 + \alpha_{i1} X_1 + \alpha_{12}) \\
&\qquad\qquad + (X_1 - a_i) \\
&\Rightarrow \beta_1 = \alpha_{i1} - y \text{ and } \beta_2 = -y\alpha_{i1} + \alpha_{i2} + 1 \\
&\Rightarrow \beta_2 = \beta_1 \alpha_{i1} - \alpha_{i1}^2 + \alpha_{i2} + 1
\end{aligned}$$

and hence we get a contradiction by (iii). Finally in case of $m = 2$ we have

$$\begin{aligned}&(u(X_1), v(X_1)) = (1, a(X_1))\\ &\Rightarrow X_1^2 + \beta_1 X_1 + \beta_2 = (X_1 - y) + (X_1^2 + \alpha_1 X_1 + \alpha_2)\\ &\Rightarrow \beta_1 = 1 + \alpha_1\end{aligned}$$

and

$$\begin{aligned}&(u(X_1), v(X_1)) = (a(X_1)/a_i(X_1), a_i(X_1))\\ &\Rightarrow X_1^2 + \beta_1 X_1 + \beta_2 = (X_1 - y)(X_1 + \alpha_{i1}) + (X_1 - a_i)\\ &\Rightarrow \beta_1 = \alpha_{i1} - y + 1 \text{ and } \beta_2 = -y\alpha_{i1} - a_i\\ &\Rightarrow \beta_2 = \beta_1\alpha_{i1} - \alpha_{i1}^2 - \alpha_{i1} - a_i\end{aligned}$$

and hence we get a contradiction by (iv).

EXAMPLE 4. Let $f = a(X_1)X_2^2 + X_2 + X_1^{m+1} - z$ with $z \in k$. Then f is irreducible in R and $\operatorname{redset}(f) = \operatorname{redset}(f)^* = \emptyset$. Moreover, $k^\sharp$ is relatively algebraically closed in $L^\sharp$ and we have $|\operatorname{redset}(f)| = |\operatorname{redset}(f)^*| = 0$ with $\tau(f) = \tau(f^\sharp) = m + 1$.

Proof. By Gauss Lemma we see that if $c \in \operatorname{redset}(f)^*$ then for some $u(X_1)$, $v(X_1)$, $p(X_1)$, $q(X_1)$ in $R^\times$ we have

$$f - c = [u(X_1)X_2 + p(X_1)][v(X_1)X_2 + q(X_1)]$$

and equating coefficients of powers of X_2 we get

$$(4_1) \qquad u(X_1)v(X_1) = a(X_1)$$

and

$$(4_2) \qquad p(X_1)q(X_1) = r(X_1) \quad \text{with} \quad r(X_1) = X_1^{m+1} - z - c$$

and

$$(4_3) \qquad u(X_1)q(X_1) + v(X_1)p(X_1) = 1.$$

For a while suppress the variable X_1. Note that then $\deg(a) = m$ and $\deg(r) = m + 1$. First suppose m is even. Then $m + 1$ is odd. Multiplying (4_3) by u and using (4_1) we get

$$(4_4') \qquad u^2q + ap = u.$$

By (4_2) we know that $\deg(pq)$ is odd and hence $(\deg(q), \deg(p)) =$ (even, odd) or (odd, even). Therefore $(\deg(u^2q), \deg(ap)) =$ (even, odd) or (odd, even), and hence $\deg(u^2q+ap) = \max(\deg(u^2q), \deg(ap)) > \deg(u)$ which contradicts $(4_4')$. Next suppose m is odd. Then $m+1$ is even. Multiplying (4_3) by p and using (4_2) we get

$$(4_4'') \qquad ur + vp^2 = p.$$

By (4_1) we know that $\deg(uv)$ is odd and hence $(\deg(u), \deg(v)) =$ (even, odd) or (odd, even).

Therefore $(\deg(ur), \deg(vp^2)) =$ (even, odd) or (odd, even), and hence $\deg(ur + vp^2) = \max(\deg(ur),\deg(vp^2)) > \deg(p)$ which contradicts $(4_4'')$. Thus we have shown that $\text{redset}(f)^* = \emptyset$. Consequently f is irreducible in R and $\text{redset}(f) = \emptyset$. In particular $|\text{redset}(f)| = |\text{redset}(f)^*| = 0 < \infty$, and hence by the Composite Pencil Theorem we see that $k^\sharp$ is relatively algebraically closed in $L^\sharp$. Now $\text{defo}(f) = X_1^m X_2^2$, and hence the points at infinity of the curve $f = 0$ are $(X_0, X_1, X_2) = (0,1,0)$ and $(X_0, X_1, X_2) = (0,0,1)$. Homogenizing f we obtain

$$(X_1 - a_1X_0)\dots(X_1 - a_mX_0)X_2^2 + X_2X_0^{m+1} + X_1^{m+1}X_0 - zX_0^{m+2}.$$

Putting $X_1 = 1$ we get

$$f_1 = (1 - a_1X_0)\dots(1 - a_mX_0)X_2^2 + X_2X_0^{m+1} + X_1^{m+1}X_0 - zX_0^{m+2}$$

and hence $(0,1,0)$ is a simple point. Putting $X_2 = 1$ we get

$$f_2 = (X_1 - a_1X_0)\dots(X_1 - a_mX_0) + X_0 + X_1^{m+1}X_0 - zX_0^{m+2}$$

and hence $(0,0,1)$ is an m-fold point with m distinct tangents. Therefore $\tau(f) = m+1$, and hence by taking $(z+Z, k^\sharp)$ for (z,k) we get $\tau(f^\sharp) = m+1$.

EXAMPLE 5. Assume $m \geq 2$. Let $f = X_2^m a(X_1/X_2) + z$ with $z \in k^\times$. Then f is irreducible in R and $\text{redset}(f) = \text{redsset}(f)^* = \{z\}$. Moreover, $k^\sharp$ is relatively algebraically closed in $L^\sharp$ and we have $|\text{redset}(f)| = |\text{redset}(f)^*| = 1$ with $\tau(f) = \tau(f^\sharp) = m$.

Proof. Obviously $z \in \text{redset}(f)$. Moreover, for any $c \neq z$, homogenizing $f - c$ and then putting $X_2 = 1$ we get the polynomial $f_2 = a(X_1) + (z-c)X_0^m$ which is clearly irreducible and hence so in $f - c$. Also obviously $f - c = 0$ has m places at infinity.

6. More General Pencils

Let us now study more general pencils of hypersurfaces $f-cw=0$ with c varying over k, where $w \in R^\times$ is such that $\gcd(f,w)=1$. Let $R^\flat = R^\sharp = k(Z)[X_1,\dots,X_m]$.

The assumption $\gcd(f,w)=1$ says that the pencil $(f-cw)_{c\in k}$ is without fixed components. However, it may have base loci of dimension $< n-1$, i.e., there may be primes of height > 1 in R which contain f and w both, and nothing can be said about the singularities of $f-cw=0$ at a base point. To indicate these primes, for any $g_1,\dots,g_m$ in R we let $\mathcal{V}(g_1,\dots,g_m)$ denote the variety defined by $g_1=\dots=g_m=0$, i.e. we put

$\mathcal{V}(g_1,\dots,g_m)=\{P\in \operatorname{spec}(R) \text{ with } (g_1,\dots,g_m)R\subset P\}$;

now $\mathcal{V}(f,w)$ is the set of all base loci of our pencil; for $n=2$ the set $\mathcal{V}(f,w)$ is finite and its members are the base points of the pencil. Likewise for any $g_1,\dots,g_m$ in R^* we put

$\mathcal{V}(g_1,\dots,g_m)^*=\{P\in \operatorname{spec}(R^*) \text{ with } (g_1,\dots,g_m)R^*\subset P\}$,

and for any $g_1,\dots,g_m$ in $R^\flat$ we put

$\mathcal{V}(g_1,\dots,g_m)^\flat=\{P\in \operatorname{spec}(R^\flat) \text{ with } (g_1,\dots,g_m)R^\flat\subset P\}$.

Let $\operatorname{singset}(f,w)=\{c\in k : R_P/((f-cw)R_P)$ is nonregular for some $P\in\operatorname{spec}(R)\setminus\mathcal{V}(f,w)$ with $f-cw\in P\}$, and let $\operatorname{singset}(f,w)^*=\{c\in k^* : R^*_P/((f-cw)R^*_P)$ is nonregular for some $P\in\operatorname{spec}(R^*)\setminus\mathcal{V}(f,w)^*$ with $f-cw\in P\}$. Also let $\operatorname{redset}(f,w)=\{c\in k :$ $f-cw=gh$ for some g,h in $R\setminus k\}$, and let $\operatorname{redset}(f,w)^*=\{c\in k^* :$ $f-cw=gh$ for some g,h in $R^*\setminus k^*\}$. Finally let $\operatorname{multset}(f,w)^*=$ $\{c\in k^* : f-cw=gh^2$ for some $g\in(R^*)^\times$ and $h\in R^*\setminus k^*\}$, and let $\operatorname{primset}(f,w)=\{c\in k : f-cw=gh^\mu$ for some $g\in k^\times$ and $h\in R\setminus k$ with integer $\mu>1\}$.

Sometimes we need to projectivise the pencil $f-cw=0$ by allowing c to vary over $k\cup\{\infty\}$ and declaring that $f-\infty w$ means w. To take care of this we put $\operatorname{redset}(f,w)_+=\operatorname{redset}(f,w)$ or $\operatorname{redset}(f,w)\cup\{\infty\}$ according as we cannot or can write $w=gh$ with g,h in $R\setminus k$. Likewise we put $\operatorname{redset}(f,w)^*_+=\operatorname{redset}(f,w)^*$ or $\operatorname{redset}(f,w)^*\cup\{\infty\}$ according as we cannot or can write $w=gh$ with g,h in $R^*\setminus k^*$. Similarly we put $\operatorname{multset}(f,w)^*_+=\operatorname{multset}(f,w)^*$ or $\operatorname{multset}(f,w)^*\cup\{\infty\}$ according as we cannot or can write $w=gh^2$ for some $g\in(R^*)^\times$ and $h\in R^*\setminus k$. Finally we put $\operatorname{primset}(f,w)_+=\operatorname{primset}(f,w)$ or $\operatorname{primset}(f,w)\cup\{\infty\}$ according as we cannot or can write $w=gh^\mu$ with $g\in k^\times$ and $h\in R\setminus k$ with integer $\mu>1$. Let

$$d=\max(\deg(f),\deg(w))$$

and observe that, without assuming $\deg(f) > 0$ but assuming $d > 0$, the condition $\gcd(f, w) = 1$ implies that

$$(') \qquad \deg(f - cw) < d \text{ for at most one } c \in k^* \cup \{\infty\}$$

and

$$('') \qquad \gcd(f - c_1 w, f - c_2 w) = 1 \text{ for all } c_1 \neq c_2 \text{ in } k^* \cup \{\infty\}.$$

By the generic member of the pencil $(f - cw)_{c \in k}$ we mean the hypersurface $(f, w)^\flat = 0$ with $(f, w)^\flat = f - Zw \in R^\flat$. Let $\text{singset}(f, w)^\flat = \{c \in k(Z) : R^\flat_P/(f - (Z + c)w)R^\flat_P) \text{ is nonregular for some } P \in \text{spec}(R^\flat) \setminus \mathcal{V}(f, w)^\flat \text{ with } f - (Z + c)w \in P\}$. Let $\text{redset}(f, w)^\flat = \{c \in k(Z) \text{ such that } f - (Z + c)w = gh \text{ for some in } g, h \text{ in } R^\flat \setminus k(Z)\}$.

By Gauss Lemma $(f, w)^\flat$ is irreducible in $R^\flat$, i.e., $0 \notin \text{redset}(f, w)^\flat$. Let $\phi^\flat : R^\flat \to R^\flat/((f, w)^\flat R^\flat)$ be the residue class epimorphism. Clearly $R \cap \ker(\phi^\flat) = \{0\}$ and $\phi^\flat(f) = \phi^\flat(Z)\phi^\flat(w)$, and hence there exists a unique isomorphism $\psi^\flat : k(X_1, \ldots, X_n) \to \text{QF}(R^\flat/((f, w)^\flat R^\flat))$ such that for all $r \in R$ we have $\psi^\flat(r) = \phi^\flat(r)$. Thus the triple $\phi^\flat(k(Z)) \subset \phi^\flat(R^\flat) \subset \text{QF}(\phi^\flat(R^\flat))$ is isomorphic to the triple

$$k^\flat = k(f/w) \subset A^\flat = k(f/w)[X_1, \ldots, X_n] \subset L^\flat = k(X_1, \ldots, X_n)$$

and hence we may regard the above three displayed sets as the ground field, the affine coordinate ring, and the function field of $(f, w)^\flat = 0$.

Let us reiterate that $\text{singset}(f, w)$, $\text{singset}(f, w)^*$, and $\text{singset}(f, w)^\flat$ denote singularities outside the base points. For any $P \in \text{spec}(A^\flat)$ with $f - (Z + c)w \in P$ and $(f, w)A^\flat \not\subset P$, upon letting $Q = R \cap P$, we see that the local ring $A^\flat_P$ is regular because it equals the localization of the regular local ring R_Q at its multiplicative subset $k[f/w]^\times$ or $k[w/f]^\times$ depending on whether $w \notin P$ or $f \notin P$. Thus we have the:

General Generic Singset Theorem. $0 \notin \text{singset}(f, w)^\flat$.

In view of the above isomorphism of triples, by the General Redset Theorem which we shall state and prove in a moment, we get the:

General Generic Redset Theorem. If $k^\flat$ is relatively algebraically closed in $L^\flat$, then $\text{redset}(f, w)^\flat$ is finite.

General Redset Theorem. If f is irreducible in R with $\deg(f) \geq \deg(w)$ and k is relatively algebraically closed in L, then $\operatorname{redset}(f, w)$ is finite.

Proof. By the Lemma we can find a finite number of DVRs $V_1, \ldots, V_t$ of L/k such that $A[1/\phi(w)] \cap V_1 \cap \cdots \cap V_t = k$. For every $z \in L^\times$ let $W_i(z) = \operatorname{ord}_{V_i}(z)$, and let $W : L^\times \to \mathbb{Z}^t$ be the map given by putting $W(z) = (W_1(z), \ldots, W_t(z))$. Let G be the set of all $g \in R \setminus k$ such that $gh = f - cw$ for some $h \in R \setminus k$ and $c \in k^\times$. Since the degree of g is clearly smaller than the degree of f, the set G is contained in a finite dimensional k-vector-subspace of R. Therefore for every $i \in \{1, \ldots, t\}$, the set $W_i(\phi(g))_{g \in G}$ is bounded from below. Since h also belongs to G and clearly $W_i(\phi(g)) = W_i(\phi(w)) - W_i(\phi(h))$ with integer $W_i(\phi(w))$ which depends only on i and w, it follows that the set $W_i(\phi(g))_{g \in G}$ is also bounded from above. Since, for $1 \leq i \leq t$, the set $W_i(\phi(g))_{g \in G}$ is bounded from both sides, it follows that $W(\phi(G))$ is a finite set. Also clearly $\phi(G) \subset U(A[1/\phi(w)])$. Let $g_1 h_1 = f - c_1 w$ and $g_2 h_2 = f - c_2 w$ with g_1, h_1, g_2, h_2 in $R \setminus k$ and c_1, c_2 in $k^\times$ be such that $W(\phi(g_1)) = W(\phi(g_2))$. Then $\phi(g_1)/\phi(g_2) \in A[1/\phi(w)] \cap V_1 \cap \cdots \cap V_t = k$ and hence $\phi(g_2) = c\phi(g_1)$ for some $c \in k^\times$. Consequently $g_2 - cg_1$ is divisible by f in R and hence, because $\deg(g_2 - cg_1) < \deg(f)$, we must have $g_2 = cg_1$. Therefore, by subtracting the equation $g_2 h_2 = f - c_2 w$ from the equation $g_1 h_1 = f - c_1 w$ we get $(c_2 - c_1)w = g_1 h_1 - g_2 h_2 = g_1(h_1 - ch_2)$ which implies that $(c_2 - c_1)w$ is divisible in R by the positive degree polynomial g_1 with $\gcd(w, g_1) = 1$. Consequently we must have $c_2 = c_1$. Therefore, because the set $W(\phi(G))$ is finite, we conclude that $\operatorname{redset}(f, w)$ is finite.

The following version of Lüroth's Theorem was first proved by Igusa [**Igu**], and then it was deduced by Nagata [**Na2**] as a consequence of Abhyankar's paper [**A01**].

General Refined Lüroth Theorem. Assume that $k^\flat$ is not relatively algebraically closed in $L^\flat$. Let $\widehat{k^\flat}$ be the algebraic closure of $k^\flat$ in $L^\flat$, and let $\nu = [\widehat{k^\flat} : k^\flat]$. Then ν is an integer with $\nu > 1$, and there exist $\Gamma \in k[Z]^\times$, $\Omega \in k[Z]^\times$, $\widehat{f} \in R \setminus k$, and $\widehat{w} \in R \setminus k$, with $\gcd(\Gamma, \Omega) = 1$, $\max(\deg(\Gamma), \deg(\Omega)) = \nu$, $\gcd(\widehat{f}, \widehat{w}) = 1$, $\max(\deg(\widehat{f}), \deg(\widehat{w})) > 0$, and $\deg(\widehat{w}) = \min\{\deg(\widehat{f} - \widehat{c}\widehat{w}) : \widehat{c} \in k^* \cup \infty\}$, such that $\widehat{k^\flat} = k(\widehat{f}/\widehat{w})$ and $f/w = \Gamma(\widehat{f}/\widehat{w})/\Omega(\widehat{f}/\widehat{w})$.

REMARK 6. In the above situation, every member of the pencil $(f - cw)_{c\in k^*}$ consists of ν members (counted properly) of the pencil $(\widehat{f} - \widehat{c}\widehat{w})_{\widehat{c}\in k^*}$, and so we say that **the pencil** $(f - cw)_{c\in k^*}$ **is composite with the pencil** $(\widehat{f} - \widehat{c}\widehat{w})_{\widehat{c}\in k^*}$. In greater detail, upon letting $\Lambda(Y,Z) = \Gamma(Z) - Y\Omega(Z)$ we get irreducible $\Lambda(Y,Z)$ in $k[Y,Z]$ of Y-degree 1 and Z-degree ν such that $\Lambda(f/w, \widehat{f}/\widehat{w}) = 0$. For most $c \in k^*$ we have $\Lambda(c,Z) = \widehat{c}_0 \prod_{1\le i\le\nu}(Z - \widehat{c}_i)$ where $\widehat{c}_0, \widehat{c}_1, \ldots, \widehat{c}_\nu$ in k^* with $\widehat{c}_0 \neq 0$ and so the "object" $f/w = c$ is the union of the "objects" $\widehat{f}/\widehat{w} = \widehat{c}_1, \ldots, \widehat{f}/\widehat{w} = \widehat{c}_\nu$. Alternatively, by letting $\gamma(Z,T) = T^\nu\Gamma(Z/T)$, $\omega(Z,T) = T^\nu\Omega(Z/T)$, and $\lambda(Y,Z,T) = T^\nu\Lambda(Y,Z/T)$, we get polynomials which are homogeneous of degree ν in (Z,T), and for which we have $f/w = \gamma(\widehat{f},\widehat{w})/\omega(\widehat{f},\widehat{w})$ and $\lambda(Y,Z,T)/\omega(Z,T) = (\gamma(Z,T)/\omega(Z,T)) - Y$. Now $\lambda(c,Z,T) = \widehat{c}_0\prod_{1\le i\le\nu}(Z - \widehat{c}_iT)$, and so the hypersurface $f - cw = 0$ is the union of the hypersurfaces $\widehat{f} - \widehat{c}_1\widehat{w} = 0, \ldots, \widehat{f} - \widehat{c}_\nu\widehat{w} = 0$. In the above phrase "most $c \in k^*$" we were referring to the tacit assumption that $\lambda(c,Z,T)$ is not divisible by T; in the contrary case, if T^μ is the highest power of T which divides $\lambda(c,Z,T)$ then μ of the roots, say $c_1, \ldots, c_\mu$, have "gone to infinity" and the hypersurface $f - cw = 0$ is composed of the hypersurfaces $\widehat{w} = 0, \ldots, \widehat{w} = 0, \widehat{f} - \widehat{c}_{\mu+1}\widehat{w} = 0, \ldots, \widehat{f} - \widehat{c}_\nu\widehat{w} = 0$ with $\widehat{w} = 0$ occurring μ times. By factoring $\omega(Z,T)$ we get $\omega(Z,T) = \widetilde{c}_0T^\epsilon\prod_{\epsilon+1\le i\le\nu}(Z - \widetilde{c}_iT)$ where $\widetilde{c}_0, \widetilde{c}_{\epsilon+1}\ldots, \widetilde{c}_\nu$ are elements in k^* with $\widetilde{c}_0 \neq 0$. The assumption $\gcd(\Gamma,\Lambda) = 1$ implies that $\widetilde{c}_i \neq \widehat{c}_j$ for $\epsilon + 1 \le i \neq \nu$ and $\mu + 1 \le j \le \nu$, and if $\epsilon > 0$ then the phrase "most $c \in k^*$" can be changed to the phrase "all $c \in k^*$." In other words, for two different members of the "old pencil" $(f - cw)_{c\in k^*\cup\{\infty\}}$ the corresponding ν-tuples in the "new pencil" $(\widehat{f} - \widehat{c}\widehat{w})_{\widehat{c}\in k^*\cup\{\infty\}}$ are disjoint. Upon multiplying the polynomials Γ and Ω by suitable constants it can be arranged that $f = \gamma(\widehat{f},\widehat{w})$ and $w = \omega(\widehat{f},\widehat{w})$; now for all $c \in k^*$ we have $f - cw = \lambda(c,\widehat{f},\widehat{w})$. The picture is completed by noting that every member of the projectivised pencil $(f - cw)_{c\in k^*\cup\{\infty\}}$ consists of ν members of the projectivised pencil $(\widehat{f} - \widehat{c}\widehat{w})_{\widehat{c}\in k^*\cup\{\infty\}}$, counted with multiplicities.

The above correspondence $(\widehat{c}_1, \ldots, \widehat{c}_\nu) \mapsto c$ can be elucidated by noting that the $(Z - \widehat{c}_1)$-adic, $\ldots$, $(Z - \widehat{c}_\nu)$-adic valuations of $k^*(Z)$ are the extensions of the $(Y - c)$-adic valuation of $k^*(Y)$ with $Y = \Gamma(Z)/\Omega(Z)$. Here the $(Z - \infty)$-adic and the $(Y - \infty)$-adic valuations are the negative degree functions, and the multiplicity of the root $\widehat{c}_i$

is the ramification exponent of the $(Z - \widehat{c}_i)$-adic valuation. As said above, since $\gcd(\Gamma, \Omega) = 1$, for any two different c's in $k^* \cup \{\infty\}$ the two corresponding sets $(\widehat{c}_1, \ldots, \widehat{c}_\nu)$ are disjoint. By $(')$ we see that $\deg(\widehat{f} - \widehat{c}\widehat{w}) < \widehat{d}$ for at most one $\widehat{c} \in k^* \cup \{\infty\}$ and this can equal a $\widehat{c}_i$ for at most one $c \in k^* \cup \{\infty\}$.

Since $[k^*(Z) : k^*(Y)] = \nu > 1$, if the characteristic of k is 0 then, for some $c \in k^*$, there exits $\widehat{c} \in k^*$ such that $(Z - \widehat{c})$-adic valuation lies above the $(Y - c)$-adic valuation and has ramification exponent greater than one, and hence $f - cw$ has a nonconstant multiple factor in R, and therefore $\text{multset}(f, w)^* \neq \emptyset$. To find c explicitly, upon letting $\Theta(Y)$ be the Z-resultant of $\Lambda(Y, Z)$ and its Z-derivative $\lambda_Z(Y, Z)$ we can show that $\Theta(Y)$ has a root in k^* which is not a root of the coefficient of Z^ν in $\Lambda(Y, Z)$; now for c we can take such a root.

Without assuming any condition on the pair $(k^\flat, L^\flat)$, if $c \in k^*$ is such that $f - cw = gh^2$ with $g \in R^* \setminus \{0\}$ and $h \in R^* \setminus k^*$ then $(f/w) - c = (g/w)h^2$ and hence taking the partials of both sides relative to X_i and multiplying by w^2 we get $fw_{X_i} - wf_{X_i} = w^2[(g/w)_{X_i}h^2 + 2hh_{X_i}(g/w)] = h[(gw_{X_i} - wg_{X_i})h + 2gwh_{X_i}]$; therefore $\text{multset}(f, w)^* \neq \emptyset \Rightarrow \gcd(fw_{X_1} - wf_{X_1}, \ldots, fw_{X_n} - wf_{X_n}) \neq 1$.

Again without assuming any condition on the pair $(k^\flat, L^\flat)$ we see that $c \mapsto c - Z$ gives an injection of $\text{redset}(f, w)$ into $\text{redset}(f, w)^\flat$, and hence $|\text{redset}(f, w)| \leq |\text{redset}(f, w)^\flat|$.

Thus, in view of the General Generic Redset Theorem, the General Refined Lüroth Theorem, and the above two observations $(')$ and $('')$, we get the:

General Composite Pencil Theorem. We have the following:

(I) $\gcd(fw_{X_1} - wf_{X_1}, \ldots, fw_{X_n} - wf_{X_n}) = 1 \Rightarrow \text{multset}(f, w)^* = \emptyset$.

(II) characteristic of k is 0 and $\text{multset}(f, w)^* = \emptyset \Rightarrow k^\flat$ is relatively algebraically closed in $L^\flat$.

(III) $k^\flat$ is relatively algebraically closed in $L^\flat \Rightarrow \text{redset}(f, w)$ and $\text{redset}(f, w)^\flat$ are finite with $|\text{redset}(f, w)| \leq |\text{redset}(f, w)^\flat|$.

(IV) $k^\flat$ is not relatively algebraically closed in $L^\flat \Rightarrow$
$|(k^* \cup \{\infty\}) \setminus \text{redset}(f, w)^*_+| \leq 1$ and $\deg(f - cw) < d$ for any
$c \in (k^* \cup \{\infty\}) \setminus \text{redset}(f, w)^*_+$.

By slightly changing the proof of the Singset Theorem we shall now prove the:

GENERAL SINGSET THEOREM. If k is of characteristic zero then $\mathrm{singset}(f, w)$ is finite.

Proof. Let I be the ideal in R^* generated by the n elements $fw_{X_i} - wf_{X_i}$ with $1 \leq i \leq n$, and note that these elements when divided by w^2 give us the partials $(f/w)_{X_i}$. For any $P \in \mathrm{spec}(R^*)$ with $I \subset P$ and $w \notin P$, consider the residue class map $\Phi_P : R^* \to R^*/P$. Since all the partials of f/w when multiplied by w^2 belong to P, it follows that $D(\Phi_P(f)/\Phi_P(w)) = 0$ for every $\Phi_P(k^*)$-derivation of $\mathrm{QF}(R^*/P)$. Therefore, since k^* is of characteristic zero, we have $\Phi_P(f)/\Phi_P(w) = \Phi_P(\kappa(P))$ for a unique $\kappa(P) \in k^*$. For any $c \in k^*$ we clearly have: $f - cw \in P \Leftrightarrow c = \kappa(P)$. Let $P_1, \ldots, P_s$ be the minimal primes of I in R^* which do not contain w. Then for any $c \in k^* \setminus \{\kappa(P_1), \ldots, \kappa(P_s)\}$ and $Q \in \mathrm{spec}(R^*)$ with $f - cw \in Q$ and $(f, w)R^* \not\subset Q$, we must have $I \not\subset Q$. Since k is of characteristic zero, it follows that $\mathrm{singset}(f, w) \subset \{\kappa(P_1), \ldots, \kappa(P_s)\}$, and hence $\mathrm{singset}(f, w)$ is finite.

REMARK 7. To explain the ideas behind the proofs of the Singset and General Singset Theorems, suppose k to be of characteristic zero. If the equations $f_{X_1} = \cdots = f_{X_n} = 0$ have only a finite number of common solutions $(a_{i1}, \ldots, a_{in})_{1 \leq i \leq s}$ in $(k^*)^n$ then upon letting $c_i = f(a_{i1}, \ldots, a_{in})$ we clearly get $\mathrm{singset}(f) = \{c_1, \ldots, c_s\}$ provided $k = k^*$ and hence $\mathrm{singset}(f) \subset \{c_1, \ldots, c_s\}$ without that proviso. Without assuming the common solutions to be finite, this is generalized by letting $P_1, \ldots, P_s$ to be the minimal primes of $(f_{X_1}, \ldots, f_{X_n})R^*$, i.e., upon letting $\mathcal{V}(P_1)^*, \ldots, \mathcal{V}(P_s)^*$ to be the irreducible components of the affine variety $f_{X_1} = \cdots = f_{X_n} = 0$, then showing that f is constant on $\mathcal{V}(P_i)^*$ and letting c_i be that constant value. To apply this to get the singset of the more general pencil $f - cw$ outside its base points we take only those irreducible components of the variety $fw_{X_1} - wf_{X_1} = \cdots = fw_{X_n} - wf_{X_n} = 0$ which are not contained in the hypersurface $w = 0$. Note that, since in these two theorems the characteristic is assumed to be zero, the ideal I is a nonzero ideal. Observe that here the characteristic assumption is essential as is shown by taking $f = X_1^p + \cdots + X_n^p$ with characteristic $p > 0$ and $k = k^*$ and noting that then $\mathrm{singset}(f) = k$. However, no assumption on the characteristic is required in the Generic Singset Theorem and the General Generic Singset Theorem.

In the Mixed Primset Theorem we showed that if the special pencil $f-c$ is noncomposite, i.e., if $k^\sharp$ is relatively algebraically closed in $L^\sharp$, then primset(f) is empty. However, if the general pencil $f-cw$ is noncomposite, i.e., if $k^\flat$ is relatively algebraically closed in $L^\flat$, then primset(f,w) may be nonempty. More precisely, we prove the:

GENERAL MIXED PRIMSET THEOREM Assume that k is perfect and $k^\flat$ is relatively algebraically closed in $L^\flat$. For any $c \in k \cup \{\infty\}$ let $d(c) = \deg(f-cw)$. For any $c \in \text{primset}(f,w)_+$ by definition we have $f - cw = g(c)h(c)^{\mu(c)}$ with $g(c) \in k^\times$, $h(c) \in R \setminus k$, and integer $\mu(c) > 1$; we shall assume that the triple $(g(c), h(c), \mu(c))$ is so chosen that $\mu(c)$ is maximal. Call $c' \in k \cup \{\infty\}$ inseparable if $(f-c'w)_{X_i} = 0$ for $1 \le i \le n$; otherwise call c' separable. Then we have the following:

(I) primset$(f,w)_+$ has at most one inseparable member.

(II) If primset$(f,w)_+$ has an inseparable c' then it has at most one $c \neq c'$ with $d(c) = d$. If primset$(f,w)_+$ has an inseparable member then $|\text{primset}(f,w)_+| \le 3$.

(III) primset$(f,w)_+$ has at most three members c with $d(c) = d$.

(IV) If primset$(f,w)_+$ has three distinct members c_i with $d(c_i) = d$ for $1 \le i \le 3$ then $\{\mu(c_1), \mu(c_2), \mu(c_3)\} = \{2,3,5\}$.

(V) $|\text{primset}(f,w)| \le |\text{primset}(f,w)_+| \le 4$.

Proof. To prove (I) and (II), assuming primset$(f,w)_+$ to have an inseparable member, and replacing f and w by suitable k-linear combinations of them, we may suppose ∞ to be that inseparable member. Note that now k must be of characteristic $p > 0$ and $w = (w')^p$ with $w' \in R \setminus k$. If primset$(f,w)_+$ has another inseparable member then replacing it by a suitable k-linear combination of it and w we may suppose the other inseparable member to be 0; now $f = (f')^p$ with $f' \in R \setminus k$ and $k(f/w) \subset k(f'/w')$ with $[k(f'/w') : k(f/w)] = p$ contradicting the noncompositness of our pencil. This proves (I). In particular, we must have $f_{X_j} \neq 0$ for some $j \in \{1, \dots, n\}$. If primset$(f,w)_+$ has two separable members $c_1 \neq c_2$ with $d(c_1) = d = d(c_2)$ then for $1 \le i \le 2$ we get $f_{X_j} = (f - c_i w)_{X_j} = g(c_i)h(c_i)_{X_j} h(c_i)^{\mu(c_i)-1}$ and hence, because clearly $\gcd(h(c_1)^{\mu(c_1)-1}, h(c_2)^{\mu(c_2)-1}) = 1$, we see that

$h(c_1)^{\mu(c_1)-1}h(c_2)^{\mu(c_2)-1}$ divides f_{X_j} in R, and therefore

$$\begin{aligned} d-1 &\geq \deg(f_{X_j}) \\ &\geq \deg(h(c_1)^{\mu(c_1)-1}) + \deg(h(c_2)^{\mu(c_2)-1}) \\ &= d - (d/\mu(c_1)) + d - (d/\mu(c_2)) \end{aligned}$$

and dividing the first and the last expressions by d and rearranging terms suitably we obtain

$$(1/\mu(c_1)) + (1/\mu(c_2)) \geq 1 + (1/d).$$

Since our pencil is noncomposite, we must have $\gcd(\mu(c_1), \mu(c_2)) = 1$, and hence the left hand side is at most $5/6$ which is less than the right hand side, giving a contradiction. This proves (II).

To prove (III) to (V), let if possible c_i in $\text{primset}(f, w)_+$ with $d(c_i) = d$ for $1 \leq i \leq u$ be distinct members where $u = 3$ or 4. If $(f/w)_{X_j} = 0$ for $1 \leq j \leq n$ then k must be of characteristic $p > 0$ and we must have $f/w = (f'/w')^p$ for some f', w' in $R^\times$ contradicting the noncompositness of our pencil. Therefore $(f/w)_{X_j} \neq 0$ for some $j \in \{1, \ldots, n\}$. By the equation $f - c_i w = g(c_i)h(c_i)^{\mu(c_i)}$ we see that $f - c_i w$ and $(f - c_i w)_{X_j}$ are both divisible by $h(c_i)^{\mu(c_i)-1}$ in R. Clearly $w^2(f/w)_{X_j} = f w_{X_j} - w f_{X_j}$ and hence if $c_i = \infty$ then $w^2(f/w)_{X_j}$ is divisible by $h(c_i)^{\mu(c_i)-1}$ in R. If $c_i \in k$ then $w^2(f/w)_{X_j} =$ $w^2((f - c_i w)/w)_{X_j} = (f - c_i w)w_{X_j} - w(f - c_i w)_{X_j}$ and hence again $w^2(f/w)_{X_j}$ is divisible by $h(c_i)^{\mu(c_i)-1}$ in R. Clearly $\gcd(h(c_i), h(c_{i'}) =$ 1 for all $i \neq i'$, and hence $w^2(f/w)_{X_j}$ is divisible by $\prod_{1 \leq i \leq u} h(c_i)^{\mu(c_i)-1}$ in R. Therefore

$$\begin{aligned} 2d - 1 &\geq \deg(w^2(f/w)_{X_j}) \\ &\geq \sum_{1 \leq u} \deg(h(c_i)^{\mu(c_i)-1}) \\ &= \sum_{1 \leq i \leq u} [d - (d/\mu(c_i))] \end{aligned}$$

and dividing the first and the last expressions by d and rearranging terms suitably we obtain

$$\sum_{1 \leq i \leq u} (1/\mu(c_1)) \geq (u-2) + (1/d).$$

Again as above, two different μ_i must have gcd 1, for otherwise, the system is easily seen to be composite. In case of $u = 4$, the LHS of the above inequality has each term at most $1/2$ and at most one term equal to $1/2$, leading to a maximum value less than $2 = (u-2)$, which is a contradiction in view of the RHS. This proves (III) and hence also (V). In case of $u = 3$, we see that $2, 3, 5$ are the only choices for mutually coprime $\mu(c_i)$ giving a sum bigger than $1 = (u-2)$ for the LHS, which proves (IV).

REMARK 8. Referring to (IV) above, for $n = 2$ and k any field of characteristic zero, there is indeed an example of a pencil with three such members in the primset. Namely we use the fact that the surface $X^2 + Y^3 + Z^5 = 0$ is rational. The specific parametrization is in Klein's Lectures on the Icosahedron [**Kle**]. To quote it explicitly, from Article 13 of Chapter I of [**Kle**], let

$$f = H_1^2 \quad \text{and} \quad w = H_2^3 \quad \text{and} \quad v = 1728H_3^5$$

where

$$\begin{aligned} H_1 = X_1^{30} + X_2^{30} - 10005X_1^{10}X_2^{10}(X_1^{10} + X_2^{10}) \\ +522X_1^5X_2^5(X_1^{20} - X_2^{20}) \end{aligned}$$

and

$$H_2 = -[X_1^{20} + X_2^{20} + 494X_1^{10}X_2^{10}] + 228X_1^5X_2^5(X_1^{10} - X_2^{10})$$

and

$$H_3 = X_1X_2[X_1^{10} - X_2^{10} + 11X_1^5X_2^5].$$

Let us put

$$P = X_1^{10} - X_2^{10} \quad \text{and} \quad Q = X_1^5X_2^5.$$

Then

$$\begin{aligned} H_1 &= (X_1^{10} + X_2^{10})^3 - 10008X_1^{10}X_2^{10}(X_1^{10} + X_2^{10}) \\ &\qquad + 522X_1^5X_2^5(X_1^{20} - X_2^{20}) \\ &= (X_1^{10} + X_2^{10})(P^2 - 10004Q^2 + 522PQ) \end{aligned}$$

and hence

$$\begin{aligned} f &= (P^2 + 4Q^2)(P^2 - 10004Q^2 + 522PQ)^2 \\ &= (P^2 + 4Q^2)[P^4 + 1044P^3Q + ((522)^2 - 20008)P^2Q^2 \\ &\qquad - (1044)(10004)PQ^3 + (10004)^2Q^4] \\ &= P^6 + 1044P^5Q + [(522)^2 - 20004]P^4Q^2 - (1044)(10000)P^3Q^3 \\ &\quad + [(10004)^2 + (1044)^2 - 4(20008)]P^2Q^4 \\ &\quad - 4(1044)(10004)PQ^5 + (20008)^2Q^6. \end{aligned}$$

Also

$$H_2 = -P^2 - 496Q^2 + 228PQ$$

and hence

$$\begin{aligned} w = &-[P^6 + 3(496)P^4Q^2 + 3(496)^2P^2Q^4 + (496)^3Q^6] \\ &+ 3(228)PQ[P^4 + 2(496)P^2Q^2 + (496)^2Q^4] \\ &- 3(228)^2P^2Q^2[P^2 + 496Q^2] + (228)^3P^3Q^3 \\ = &-P^6 + 684P^5Q - 3[496 + (228)^2]P^4Q^2 \\ &+ [6(228)(496) + (228)^3]P^3Q^3 - 3[(496)^2 + (228)^2(496)]P^2Q^4 \\ &+ 3(228)(496)^2PQ^5 - (496)^3Q^6. \end{aligned}$$

Moreover

$$H_3 = X_1X_2(P + 11Q)$$

and hence

$$\begin{aligned} H_3^5 &= Q(P + 11Q)^5 \\ &= P^5Q + 5(11)P^4Q^2 + 10(11)^2P^3Q^3 + 10(11)^3P^2Q^4 \\ &\qquad + 5(11)^4PQ^5 + (11)^5Q^6. \end{aligned}$$

Adding the coefficients of $P^6, P^5Q, \ldots, Q^6$ in f and w, and comparing the sums to the corresponding coefficients in H_3^5, we get $f + w = 1728H_3^5 = v$.

Now we shall prove the:

General Mixed Redset Theorem. Assume that k is perfect and $k^\flat$ is relatively algebraically closed in $L^\flat$. Note that now by the Lemma there exists a finite number of DVRs $V_1, \ldots, V_t$ of $L^\flat/k^\flat$ with $A^\flat[1/w] \cap V_1 \cap \cdots \cap V_t = k^\flat$ and, for any such t, we have that t is a positive integer and $U(A^\flat[1/w])/U(k^\flat)$ is a finitely generated free abelian group of rank $r \leq t - 1$. By the General Mixed Primset Theorem, upon letting $\rho = |\text{primset}(f, w)|$, we get an integer ρ with $0 \leq \rho \leq 4$. We claim that $|\text{redset}(f, w)| \leq r + \rho$.

Proof. As in the proof of the Mixed Redset Theorem, suppose if possible that we have distinct elements $c_1, \ldots, c_s$ in k with integer $s > r + \rho$ such that

$$(1^*) \qquad f - c_i w = g_i h_i \quad \text{where} \quad g_i, h_i \text{ in } R \setminus k.$$

Let $\tau = s - \rho$. Then τ is an integer with $\tau > r$, and upon relabelling $c_1, \ldots, c_s$ we can arrange matters so that

$$(2^*) \qquad \gcd(g_i, h_i) = 1 \quad \text{for} \quad 1 \le i \le \tau.$$

Since $\tau > r$, there exist integers $a_1, \ldots, a_\tau$ at least one of which is nonzero, say $a_\iota \neq 0$, such that

$$(3^*) \qquad \prod_{1 \le i \le \tau} g_i^{a_i} \in U(k(f/w)).$$

As in the Second Argument of the proof of the Mixed Redset Theorem, by (3*) we get

$$(4^*) \qquad \prod_{1 \le i \le \tau} g_i^{a_i} = \overline{u} \prod_{1 \le i \le \sigma} u_i(f/w)^{\alpha_i} \qquad \text{with} \qquad \overline{u} \in k^\times$$

where $\alpha_1, \ldots, \alpha_\sigma$ are nonzero integers and $u_1(Z), \ldots, u_\sigma(Z)$ are pairwise coprime monic members of $k[Z] \setminus k$. For any u, v in $k[Z] \setminus k$ with $\gcd(u, v) = 1$ we have $w^{\deg(u)} u(f/w), w^{\deg(v)} v(f/w)$ in $R \setminus k$ with $\gcd(w^{\deg(u)} u(f/w),\ w^{\deg(v)} v(f/w)) = 1$. Therefore, by looking at divisibility by a nonconstant irreducible factor of g_i in R, in view of (1*), (2*), and (4*), we see that for each $i \in \{1, \ldots, \tau\}$ with $a_i \neq 0$ there is some $\theta(i) \in \{1, \ldots, \sigma\}$ with $u_{\theta(i)}(Z) = Z - c_i$ and $\alpha_{\theta(i)} = a_i$. In particular we get $u_{\theta(\iota)}(Z) = Z - c_\iota$ and $\alpha_{\theta(\iota)} = a_\iota$. Again in view of (1*), (2*), and (4*), by looking at divisibility by a nonconstant irreducible factor of h_ι in R, we get a contradiction.

Remark 9. In the $n = 2$ case of the above theorem we can take $V_1, \ldots, V_t$ to be valuation rings of the places at infinity of the curve $f^\flat = 0$ together with the valuation rings of its places at finite distance centered at points where it meets the curve $w = 0$.

Remark 10. To draw a deduction chart for the various incarnations of the Redset Theorem, as already observed, we have:

Lemma $\Rightarrow$ Redset Theorem $\Rightarrow$ Generic Redset Theorem
$\Rightarrow$ Composite Pencil Theorem

and

Lemma $\Rightarrow$ General Redset Theorem
$\Rightarrow$ General Generic Redset Theorem
$\Rightarrow$ General Composite Pencil Theorem

and we have:

$$\text{Lemma} \Rightarrow \text{Mixed Redset Theorem}$$

and

$$\text{Lemma} \Rightarrow \text{General Mixed Redset Theorem}$$

where A $\Rightarrow$ B means B can be deduced from A. Also clearly:

$$\text{Composite Pencil Theorem} \Rightarrow \text{Redset Theorem when } k = k^*$$

and

$$\text{General Composite Pencil Theorem} \Rightarrow \text{General Redset Theorem when } k = k^*.$$

REMARK 11. The condition $\deg(f) \geq \deg(w)$ in the General Redset Theorem is necessary as well as reasonable. It is reasonable because if $\deg(f) < \deg(w)$ and we pass to the projective n-space by homogenizing f and w, then f becomes reducible by acquiring the hyperplane at infinity counted $\deg(w) - \deg(f)$ times, and we have not yet obtained any control over the rest of the pencil. To see that it is necessary, take w to be a polynomial in f of degree $\nu > 1$ and note that then, for every $c \in k^*$, $f - cw$ is clearly a product of ν members of $R^* \setminus k^*$.

It is time to whet the appetite of the reader by raising a few questions.

QUESTION 1. As the bounds found in the Mixed Redset Theorem were sharpened in Remark 5 and Examples 1 to 5, can you do similar things to sharpen the bounds found in the General Mixed Redset Theorem?

QUESTION 2. In the Mixed Redset Theorem it was shown that $k^\sharp$ relatively algebraically closed in $L^\sharp \Rightarrow |\text{redset}(f)| \leq \text{rank}(U(A^\sharp)/U(k^\sharp))$. Can you similarly obtain a good bound in the context of the Redset Theorem by showing that, assuming f to be irreducible in R, k relatively algebraically closed in $L \Rightarrow |\text{redset}(f)| \leq \text{rank}(U(A)/U(k))$? Can you relate the conditions "k relatively algebraically closed in L" and "$k^\sharp$ relatively algebraically closed in $L^\sharp$"? In case of $n = 2$, how far can you relate the numbers of places at infinity of the curves $f = 0$ and $f^\sharp = 0$?

QUESTION 3. Can you, in a manner similar to Question 2, relate the hypotheses and conclusions of the General Redset Theorem and the General Mixed Redset Theorem?

QUESTION 4. Let the unique component set of (f, w) be defined by putting $\text{uniset}(f, w) = \{c \in k : f - cw = gh^{\mu}$ for some $g \in k^{\times}$ and irreducible $h \in R \setminus k$ with integer $\mu > 1\}$, and let us put $\text{uniset}(f, w)_+ = \text{uniset}(f, w)$ or $\text{uniset}(f, w) \cup \{\infty\}$ according as we cannot or can write $w = gh^{\mu}$ with $g \in k^{\times}$ and irreducible $h \in R \setminus k$ with integer $\mu > 1$. Note that clearly $\text{uniset}(f, w) \subset \text{primset}(f, w)$ and $\text{uniset}(f, w)_+ \subset \text{primset}(f, w)_+$. Also note that the Generic Mixed Redset Theorem and its proof remain valid if we let $\rho = |\text{uniset}(f, w)|$. In this manner we get a possibly stronger from of the Generic Mixed Redset Theorem. Can you show that this is indeed a stronger from? In other words, can you show that $|\text{uniset}(f, w)_+| \leq 3$ or $|\text{uniset}(f, w)_+| \leq 2$? Hint: vis-a-vis Remark 8, study all parametrizations of the Klein surface. You may also redo Question 3 by putting in the above stronger form of the Generic Mixed Redset Theorem.

7. Complements of Hypersurfaces

We shall now give several criteria, i.e., necessary conditions, for the ring $R[1/f]$ to be isomorphic to the ring $R[1/f']$ where f' is another member of $R \setminus k$. In geometric terms, this amounts to giving criteria for the complements of the hypersurfaces $f = 0$ and $f' = 0$ in the affine n-space to be biregularly equivalent to each other. Note that $R[1/f]$ may be viewed as the affine coordinate ring of the hypersurface $f(X_1, \ldots, X_n)X_{n+1} - 1 = 0$ in the affine $(n+1)$-space. Also note that if the hypersurfaces $f = 0$ and $f' = 0$ are automorphic, i.e., if there exists an automorphism of R which sends f to f', then the rings $R[1/f]$ and $R[1/f']$ must be isomorphic, and hence our criteria also provide necessary conditions for the hypersurfaces $f = 0$ and $f' = 0$ to be automorphic.

For any $c \in k$ let us write $f - c = f_{c,0} f_{c,1}^{e(c,1)} \ldots f_{c,q(c)}^{e(c,q(c))}$ with $f_{c,0} \in k^{\times}$, integers $1 \leq e(c, 1) \leq \cdots \leq e(c, q(c))$, and pairwise coprime irreducible elements $f_{c,1}, \ldots, f_{c,q(c)}$ in $R \setminus k$. Note that then $c \in \text{redset}(f) \Leftrightarrow e(c, 1) + \cdots + e(c, q(c)) > 1$. Let $e(c)$ denote the sequence $(e(c, 1), \ldots, e(c, q(c)))$. Let $\text{refset}(f) =$ the refined redset of f be defined to be the family of sequences $e(c)_{c \in \text{redset}(f)}$.

Let $f'-c = f'_{c,0}f'^{e'(c,1)}_{c,1} \dots f'^{e'(c,q'(c))}_{c,q'(c)}$ and $e'(c) = (e'(c,1),\dots,e'(c,q'(c)))$ be the corresponding factorization and the corresponding sequence for $f'-c$. Let us write $\mathrm{refset}(f) = \mathrm{refset}(f')$ to mean that there exists a bijection $\theta : \mathrm{redset}(f) \to \mathrm{redset}(f')$ such that for all $c \in \mathrm{redset}(f)$ we have $e(c) = e'(\theta(c))$, i.e., $q(c) = q'(c)$ and $e(c,i) = e'(\theta(c),i)$ for $1 \le i \le q(c)$. Thus, geometrically speaking, $\mathrm{refset}(f) = \mathrm{refset}(f')$ means there is a multiplicities preserving bijection between the reducible members of the pencils $f-c=0$ and $f'-c=0$.

Let us write $R[1/f] \approx R[1/f']$ to mean that there exists a ring isomorphism of $R[1/f]$ onto $R[1/f']$, and let us write $R[1/f] \approx_k R[1/f']$ to mean that there exists a ring k-isomorphism of $R[1/f]$ onto $R[1/f']$. As our first criterion we want to show that (i) if $R[1/f] \approx_k R[1/f']$ then $q(0) = q'(0)$, and (ii) if f and f' are irreducible in R and $R[1/f] \approx_k R[1/f']$ then $\mathrm{refset}(f) = \mathrm{refset}(f')$. But since it takes only a little more effort, we might as well prove the somewhat stronger:

Refset Criterion. (i) If $R[1/f] \approx R[1/f']$ then $q(0) = q'(0)$, and (ii) if f and f' are irreducible in R and $R[1/f] \approx R[1/f']$ then $\mathrm{refset}(f) = \mathrm{refset}(f')$.

Since R is a UFD and clearly $k = \{0\} \cup U(R)$ is relatively algebraically closed in the quotient field $k(X_1,\dots,X_n)$ of R, by applying the following Corollary of Lemma with $(k',A') = (k, R[1/f])$, this follows by taking $S = R$ and changing capital letters to lower case letter1/s in the even stronger:

UFD Criterion. Let S be a UFD, and let F, F' be in $S \setminus U(S)$. For any $C \in \{0\} \cup U(S)$ let us write $F-C = F_{C,0}F^{E(C,1)}_{C,1} \dots F^{E(C,Q(C))}_{C,Q(C)}$ with $F_{C,0} \in U(S)$, integers $1 \le E(C,1) \le \dots \le E(C,Q(C))$, and pairwise coprime irreducible elements $F_{C,1},\dots,F_{C,Q(C)}$ in $S \setminus U(S)$. Let $F'-C = F'_{C,0}F'^{E'(C,1)}_{C,1} \dots F'^{E'(C,Q'(C))}_{C,Q'(C)}$ be the corresponding factorization of $F'-C$. Assume that there exists a ring isomorphism $\Delta : S[1/F] \to S[1/F']$ such that $\Delta(U(S)) = U(S)$. Then (i) $Q(0) = Q'(0)$. Moreover (ii) if F and F' are irreducible in S then there exists a bijection $\Theta : U(S) \to U(S)$ such that for all $C \in U(S)$ we have $Q(C) = Q'(\Theta(C))$ and $E(C,i) = E'(\Theta(C),i)$ for $1 \le i \le Q(C)$.

Proof. Clearly $S[1/F]$ is UFD in which the irreducible nonunits are associates of the irreducible nonunits of S except $F_{0,1},\dots,F_{0,Q(0)}$

which have become units. Also clearly $U(S[1/F])/U(S)$ is a free abelian group of rank $Q(0)$ generated by $F_{0,1},\ldots,F_{0,Q(0)}$. Similarly $U(S[1/F'])/U(S)$ is a free abelian group of rank $Q'(0)$ generated by $F'_{0,1},\ldots,F'_{0,Q(0)}$. For any ring isomorphism $\Delta : S[1/F] \to S[1/F']$ we obviously have $\Delta(U(S[1/F]) = U(S[1/F'])$. Therefore, since we are assuming $\Delta(U(S)) = U(S)$, we get an induced isomorphism $U(S[1/F])/U(S) \to U(S[1/F'])/U(S)$ and hence $Q(0) = Q'(0)$. This proves (i). Now assume that F and F' are irreducible in S. Then because of the said induced isomorphism we must have either $\Delta(F) = \alpha F'$ with $\alpha \in U(S)$ or $\Delta(F) = \alpha/F'$ with $\alpha \in U(S)$. If $\Delta(F) = \alpha F'$ then, letting $\Theta : U(S) \to U(S)$ be the bijection given by $C \mapsto \Delta(C)/\alpha$, we have

$$\Delta(F_{C,0}) \prod_{1\le i\le Q(C)} \Delta(F_{C,i}^{E(C,i)}) = \Delta(F - C) = \alpha(F' - \Theta(C))$$

$$= \alpha F'_{\Theta(C),0} \prod_{1\le i\le Q'(\Theta(C))} {F'}_{\Theta(C),i}^{E'(\Theta(C),i)}$$

and hence we get $Q(C) = Q'(\Theta(C))$ and $E(C,i) = E'(\Theta(C),i)$ for $1 \le i \le Q(C)$. If $\Delta(F) = \alpha/F'$ then, letting $\Theta : U(S) \to U(S)$ be the bijection given by $C \mapsto \alpha/\Delta(C)$, we have

$$\begin{aligned}
&\Delta(F_{C,0}) \prod_{1\le i\le Q(C)} \Delta(F_{C,i}^{E(C,i)}) \\
&= \Delta(F - C) = (-\Delta(C)/F')(F' - \Theta(C)) \\
&= (-\Delta(C)/F')F'_{\Theta(C),0} \prod_{1\le i\le Q'(\Theta(C))} {F'}_{\Theta(C),i}^{E'(\Theta(C),i)}
\end{aligned}$$

and hence again we get $Q(C) = Q'(\Theta(C))$ and $E(C,i) = E'(\Theta(C),i)$ for $1 \le i \le Q(C)$. In the last two sentences we have used the obvious facts that the elements $\Delta(F_{C,1}),\ldots,\Delta(F_{C,Q(C)})$ are pairwise coprime irreducible nonunits in $S[1/F']$ and so are the elements $F'_{\Theta(C),1},\ldots,F'_{\Theta(C),Q'(C)}$; moreover, the elements $\Delta(F_{C,0})$, $\alpha F'_{\Theta(C),0}$, and $(-\Delta(C)/F')F'_{\Theta(C),0}$ are units in $S[1/F']$.

Corollary of Lemma. Let A' be an affine domain over a field k', let k'' be the algebraic closure of k' in $L' = \mathrm{QF}(A')$, and let $\overline{k} = k'' \cap A'$. Then $\overline{k}$ can be located only using the ring structure of A' by noting that it is only subfield of A' which equals the intersection of A' with a finite number of DVRs of L'.

Proof. Clearly $\overline{k}$ is a subfield of A' and by the Lemma there exists a finite number of DVRs $V_1, \ldots, V_t$ of L'/k' with $A' \cap V_1 \cap \cdots \cap V_t = \overline{k}$. Let $V_1', \ldots, V_{t'}'$ be any finite number of DVRs of L' such that $A' \cap V_1' \cap \cdots \cap V_{t'}'$ is a subfield $\widetilde{k}$ of A'. Let $\widehat{k} = \overline{k} \cap \widetilde{k}$. We want to show that then $\widehat{k} = \overline{k}$. Clearly $\widehat{k}$ is a subfield of $\overline{k}$ and we have $\widehat{k} = \overline{k} \cap V_1' \cap \cdots \cap V_{t'}'$. For any i the intersection $\widehat{k} \cap V_i'$ is either a DVR of $\overline{k}$ or equals $\overline{k}$. Let $i_1 < \cdots < i_s$ be those values of i for which the said intersection is a DVR, and let $W_j = \overline{k} \cap V_{i_j}'$. Now $W_1, \ldots, W_s$ are a finite number of DVRs of the field $\overline{k}$ and their intersection is the field $\widehat{k}$. Now it is a well-known fact that if $W_1, \ldots, W_s$ are any finite number of valuation rings with a common quotient field $\overline{k}$ then $\overline{k}$ is the quotient field of their intersection, and hence in our situation we must have $\widehat{k} = \overline{k}$. A proof of the said fact can be found in Theorem (11,11) on page 38 of [**Na1**]. Since our valuations are real, we can deduce the fact from the Approximation Theorem (see Theorem 18 on page 45 of volume II of [**ZaS**]) thus. Since the statement is obvious when $s = 0$, suppose $s > 0$. Let $M(W_i)$ denote the maximal ideal of W_i. By the said theorem we can find $u \in W_1 \setminus M(W_1)$ such that $u \in M(W_i)$ for all $i > 1$. Now $u \in W_1 \cap \cdots \cap W_s$ and, since the valuations are real, for any $v \in W_1$ we can find an integer $m > 0$ such that $vu^m \in W_1 \cap \cdots \cap W_s$ and hence $v \in \mathrm{QF}(W_1 \cap \cdots \cap W_s)$. Therefore $\mathrm{QF}(W_1 \cap \cdots \cap W_s) = \mathrm{QF}(W_1) = \overline{k}$.

EXAMPLE 6. Take $f = X_1^{a_1} \ldots X_n^{a_n} - 1$ and $f' = X_1^{a_1'} \ldots X_n^{a_n'} - 1$ where $1 \leq a_1 \leq \cdots \leq a_n$ and $1 \leq a_1' \leq \cdots \leq a_n'$ are integers such that $\gcd(a_1, a_j) = 1$ and $\gcd(a_1', a_{j'}') = 1$ for some $j \in \{2, \ldots, n\}$ and $j' \in \{2, \ldots, n\}$, and $a_i \neq a_{i'}$ for some $i \in \{1, \ldots, n\}$ and $i' \in \{1, \ldots, n\}$. Then clearly $\mathrm{redset}(f) = \{-1\}$ and $\mathrm{redset}(f') = \{-1\}$ with $\mathrm{refset}(f) = \{(a_1, \ldots, a_n)\}$ and $\mathrm{refset}(f') = \{(a_1', \ldots, a_n')\}$. Consequently $\mathrm{refset}(f) \neq \mathrm{refset}(f')$. It follows that for all b, b' in $k^\times$ we have $\mathrm{refset}(bf) \neq \mathrm{refset}(b'f')$ and therefore by the Refset Criterion we get $R[1/(bf)] \not\approx R[1/(b'f')]$. Hence in particular no automorphism of R can send the ideal fR to the ideal $f'R$.

Now suppose $n = 2$. Let $\epsilon : R \to k[T, T^{-1}]$ be the k-homomorphism given by $(X_1, X_2) \mapsto (T^{-a_2}, T^{a_1})$. Then clearly $\ker(\epsilon) = fR$. Since $\gcd(a_1, a_2) = 1$, for some integers b_1, b_2 we have $b_1a_1 - b_2a_2 = 1$; by adding a positive multiple of a_2 to b_1 and adding the same multiple of a_1 to b_2 we can arrange b_1 to be positive and then automatically b_2 will also become positive; clearly $\epsilon(X_1^{b_2}X_2^{b_1}) = T$. Also we have $\beta_1 a_1 - \beta_2 a_2 = -1$ where $\beta_1 = -b_1$ and $\beta_2 = -b_2$; by adding a positive multiple of a_2 to β_1 and adding the same multiple of a_1 to β_2 we can arrange β_1 to be positive and then automatically β_2 will also become positive; clearly $\epsilon(X_1^{\beta_2} X_2^{\beta_1}) = T^{-1}$. Thus ϵ is surjective. Similarly we find a surjective k-homomorphism $\epsilon' : R \to k[T, T^{-1}]$ with $\ker(\epsilon') = f'R$. However, as shown above, ϵ and ϵ' do not differ from each other by an automorphism of R. Thus $f = 0$ and $f' = 0$ are "hyperbolas" which are not automorphic to each other.

Next we come to the:

NONRULED CRITERION. Recall that an irreducible $\overline{f} \in R \setminus k$ is said to be ruled if, after identifying k with a subfield of $R/(\overline{f}R)$, there exits a subfield $\widetilde{L}$ of $\overline{L} = \mathrm{QF}(R/(\overline{f}))$ such that $k \subset \widetilde{L} \subset \widetilde{L}(t) = \overline{L}$ where t is transcendental over $\widetilde{L}$. Let us relabel $f_{0,1}, \ldots, f_{0,q(0)}$ so that $f_{0,i}$ is nonruled or ruled according as $1 \le i \le r$ or $r+1 \le i \le q(0)$, and let us relabel $f'_{0,1}, \ldots, f_{0,q'(0)}$ so that $f'_{0,i}$ is nonruled or ruled according as $1 \le i \le r'$ or $r' + 1 \le i \le q'(0)$; note that now the exponent sequences $e(0,1), \ldots, e(0, q(0))$ and $e'(0,1), \ldots, e'(0, q'(0))$ need not be nondecreasing. Assume that $R[1/f] \approx_k R[1/f']$. Then $r = r'$ and, after relabelling $f_{0,1}, \ldots, f_{0,r}$ suitably, we have $\mathrm{QF}(R/(f_{0,i}R)) \approx_k \mathrm{QF}(R/(f'_{0,i}R))$ for $1 \le i \le r$.

Proof. Recall that $\mathfrak{V}(R) = \{R_P : P \in \mathrm{spec}(R)\}$. For $0 \le i \le n$ let $R_i = k[X_0/X_i, \ldots, X_n/X_i]$ where $X_0 = 1$, and let the projective n-space $\mathfrak{P}$ over k be defined to be the nonsingular projective model $\cup_{0 \le i \le n} \mathfrak{V}(R_i)$ of $K = k(X_1, \ldots, X_n)$ over k. For the language of models see Abhyankar's books [**A02, A06, A09**]. In particular note that for any valuation ring V of K/k, i.e., valuation ring with quotient field K and having k as a subfield, the center of V on $\mathfrak{P}$ is the unique member of $\mathfrak{P}$ dominated by V; we identify k with a subfield of $V/M(V)$ where $M(V)$ is the maximal ideal of V, and we let $\mathrm{restrdeg}_k(V)$ denote the transcendence degree of $V/M(V)$ over k. By a prime divisor of K/k we mean a DVR V of K/k such that $\mathrm{restrdeg}_k(V) = n - 1$; we call the prime divisor ruled if there exits a subfield $\widetilde{L}$ of $V/M(V)$ such that $k \subset \widetilde{L} \subset \widetilde{L}(t) = V/M(V)$ where t is transcendental over $\widetilde{L}$. Upon letting $V_i = R_{f_{0,i}R}$ and $V'_i = R_{f'_{0,i}R}$ we get nonruled prime divisors $V_1, \ldots, V_r$ and $V'_1, \ldots, V'_{r'}$ of K/k. Note that $\mathfrak{V}(R[1/f]) \subset \mathfrak{P}$ and $\mathfrak{V}(R[1/f']) \subset \mathfrak{P}$. By Proposition 3 on page 336 of [**A01**] we see that $V_1, \ldots, V_r$ are exactly all the nonruled prime divisors of K/k whose center on $\mathfrak{P}$ is not in $\mathfrak{V}(R[1/f])$, and

$V'_1, \dots, V'_{r'}$ are exactly all the nonruled prime divisors of K/k whose center on $\mathfrak{P}$ is not in $\mathfrak{V}(R[1/f'])$. Since $R[1/f] \approx_k R[1/f']$, it follows that $r = r'$ and, after relabelling $f_{0,1}, \dots, f_{0,r}$ suitably, we have $\mathrm{QF}(R/(f_{0,i}R)) \approx_k \mathrm{QF}(R/(f'_{0,i}R))$ for $1 \le i \le r$.

EXAMPLE 7. Take $f = X_1^a + \cdots + X_n^a - 1$ and $f' = X_1^{a'} + \cdots + X_n^{a'} - 1$ where a and a' are positive integers with $a > a'$ and $a > n > 1$; assume that a and a' are nondivisible by the characteristic of k in case the latter is nonzero. The polynomials f and f' are clearly irreducible in R. It is expected that:

(7*) f is nonruled and $\mathrm{QF}(R/(fR)) \not\approx_k \mathrm{QF}(R/(f'R))$.

Assuming (7*), by the Nonruled Criterion we would get $R[1/f] \not\approx_k R[1/f']$.

QUESTION 5. Can you prove the above statement (7*)? Hint for $n = 2$: show that the genus of the nonsingular plane curve $f = 0$ is $(a-1)(a-2)/2$ and that of $f' = 0$ is $(a'-1)(a'-2)/2$; see [**A06**]. Hint for $n > 2$: show that the arithmetic genus of the nonsingular hypersurface $f = 0$ is $(a-1)\dots(a-n)/n!$ and that of $f' = 0$ is $(a'-1)\dots(a'-n)/n!$; see [**A06**]. Further hint for $n > 2$: by using the domination part of the desingularization theory, show that the arithmetic genus of a nonsingular projective model is a birational invariant; see [**A09**].

To avoid the problem of showing that f is nonruled in the above Question 5, let us establish the:

GENERIC CRITERION. Assume that f and f' are irreducible in R. Also assume that $R[1/f] \approx_k R[1/f']$. Then the generic members of the pencils $R^\sharp((f-Z)R^\sharp) \approx_{k(Z)} R^\sharp((f'-Z)R^\sharp)$ where we recall that Z is an indeterminate over R and $R^\sharp = k(Z)[X_1, \dots, X_n]$. Hence in particular the said generic members are birationally equivalent to each other, i.e., $\mathrm{QF}(R^\sharp((f-Z)R^\sharp)) \approx_{k(Z)} \mathrm{QF}(R^\sharp((f'-Z)R^\sharp))$.

Proof. Given a k-isomorphism $\delta : R[1/f] \to R[1/f']$, as in the proof of the UFD Criterion, for some $\alpha \in k^\times$ we have either $\delta(f) = \alpha f'$ or $\delta(f) = \alpha/f'$. Consequently we have either $\delta(k[f]^\times) = \delta(k[f']^\times)$ or $\delta(k[f]^\times) = \delta(k[1/f']^\times)$ respectively. Recall that $k^\sharp = k(f)$ and $A^\sharp = k(f)[X_1, \dots, X_n]$. Let $k'^\sharp = k(f')$ and $A'^\sharp = k(f')[X_1, \dots, X_n]$.

Clearly $A^\sharp$ is the localization of $R[1/f]$ at $k[f]^\sharp$, and $A'^\sharp$ is the localization of $R[1/f']$ at $k[f']^\sharp$ as well as at $k[1/f']^\times$. Therefore δ has a unique extension to an isomorphism $\delta^\sharp : A^\sharp \to A'^\sharp$. As noted before, the pair $(k(Z), R^\sharp((f-Z)R^\sharp))$ is isomorphic to the pair $(k^\sharp, A^\sharp)$, and similarly the pair $(k(Z), R^\sharp((f'-Z)R^\sharp))$ is isomorphic to the pair $(k'^\sharp, A'^\sharp)$. It follows that $R^\sharp((f-Z)R^\sharp) \approx_{k(Z)} R^\sharp((f'-Z)R^\sharp)$ and hence $\mathrm{QF}(R^\sharp((f-Z)R^\sharp)) \approx_{k(Z)} \mathrm{QF}(R^\sharp((f'-Z)R^\sharp))$.

QUESTION 6. In connection with Example 7, can you show that $\mathrm{QF}(R^\sharp((f-Z)R^\sharp)) \not\approx_{k(Z)} \mathrm{QF}(R^\sharp((f'-Z)R^\sharp))$? Assuming this, by the Generic Criterion, we would get $R[1/f] \not\approx_k R[1/f']$.

REMARK 12. In geometric terms, considering the birational equivalence of the complements of two hypersurfaces $f=0$ and $f'=0$ in the affine n-space, and relating it to their refined redsets as well as to the birational equivalence of their irreducible components and the biregular equivalence of the generic members of their associated pencils $f-c=0$ and $f'-c=0$, we have the following:

(I) The first part of the Redset Criterion says that if the affine complements of two hypersurfaces are biregularly equivalent then they have the same number of irreducible components.

(II) The second part of the Redset Theorem says that if the affine complements of two irreducible hypersurfaces are biregularly equivalent then they have the same refined redsets.

(III) the Nonruled Criterion says that if the affine complements of two hypersurfaces are biregularly equivalent then their nonruled irreducible components are birationally equivalent.

(IV) the Generic Criterion says that if the affine complements of two irreducible hypersurfaces are biregularly equivalent then the generic members of their associated pencils are biregularly equivalent and hence birationally equivalent.

QUESTION 7. In the complex case, can you topologize the conclusions of Remark 12 by replacing biregular equivalence by some kind of topological type? Can you somehow relate this to the topology of complements as exemplified by the work of Zariski, Fan, Teicher, and others as discussed in [**GaT**]? Can you also tie it to Abhyankar's algebraization of the tame fundamental groups of complements as described in [**A03**]?

QUESTION 8. We have already noted the fact that the hypersurfaces $f = 0$ and $f' = 0$ being automorphic implies the biregular equivalence of their complements. Can you exploit this fact to link-up the results of Remark 12 with the epimorphism theorems and problems discussed by Abhyankar in his Kyoto Notes [**A04**]

8. Redset of a Plane Curve and Zariski's Lemma

We shall now show how, in case of characteristic zero, the finiteness of the redset of a hypersurface can be deduced from that of a plane curve via the famous Lemma 5 of Zariski's Bertini II paper [**Za1**]. It may be noted that, Abhyankar [**A05, A06, A08**] reduced the Galois case of the Jacobian Problem to the birational case by means of Zariski's Lemma and then settled the birational case by using Zariski's Main Theorem for which reference may be made to [**A09**].

REMARK 13. As pointed out in Remark 1, for $n = 2$, the Lemma next to the Redset Theorem follows by taking $V_1, \ldots, V_t$ to be the valuation rings of the places at infinity of the irreducible plane curve $f = 0$, and then the argument in the proof of the Redset Theorem shows that if k is relatively algebraically closed in L then $\text{redset}(f)$ is finite. To see how, for characteristic zero, the $n = 2$ case of the Redset Theorem implies the $n > 2$ case we can proceed thus. Let ι be the smallest positive integer $\leq n$ such that $f \in k[X_1, \ldots, X_\iota]$. If $\iota = 1$ then by doing nothing, and if $\iota > 1$ then by applying a $k-$automorphism to R of the form $X_j \mapsto X_j$ for all $j \in \{1, \ldots, n\} \setminus \{\iota\}$ and $X_\iota \mapsto X_\iota + X_1^l$ with $l >$ twice the total degree of f, we can arrange matters so that f is essentially monic in X_1, i.e., for some $e' \in k^\times$ and integer $e > 0$ we have $f = e'X_1^e +$ terms of X_1-degree $< e$. For any positive integer $i \leq n$ let $R_i = k_i[X_1, \ldots, X_i]$ with $k_i = k(X_{i+1}, \ldots, X_n)$. Now clearly $f \in R_\iota \setminus k_\iota$ and by Gauss Lemma, f is irreducible in R_ι. Let us identify k_ι with a subfield of $L_\iota = \text{QF}(R_\iota/(fR_\iota))$. If $\iota = 2$ and k_ι were relatively algebraically closed in L_ι then by the $n = 2$ case of the Redset Theorem we would see that $\{c \in \widetilde{k} : f - c = gh$ with g, h in $R_\iota \setminus k_\iota\}$ is finite and hence so is $\text{redset}(f)$. In the next Example 8 we shall show that f irreducible in R_ι does not imply k_ι relatively algebraically closed in L_ι. However, in case of characteristic zero, by taking $(L', Z_1, \ldots, Z_s) = (L, \phi(X_2), \ldots, \phi(X_n))$ in the Second Version of Zariski's Lemma given below, we can arrange k_2 to be relatively algebraically closed in L_2.

EXAMPLE 8. The geometric significance of the condition that the ground field k of the irreducible polynomial f be relatively algebraically closed in its function field L is due to the well-known fact that, in case of characteristic zero, it is equivalent to assuming f to be absolutely irreducible, i.e., irreducible in R^*. To illustrate this take $n = 2$ and $f = X_1^2 + X_2^u$ with integer $u > 0$. Assume that -1 is not a square in k; for instance k could be the field of real numbers. Let $R_1 = k_1[X_1]$ and $R_1^* = k_1^*[X_1]$ where k_1^* is an algebraic closure of $k_1 = k(X_2)$. Now clearly f is always irreducible in R_1, but it is irreducible in $R_1^* \Leftrightarrow u$ is odd. Let $\phi_1 : R_1 \to R_1/(fR_1)$ be the canonical epimorphism, and let us identify k_1 with a subfield of $L_1 = \mathrm{QF}(R_1/(fR_1))$. Note that now $L_1 = k(\overline{X}_1, X_2)$ with $\overline{X}_1 = \phi_1(X_1)$. If $u = 2v$ is even then $(\overline{X}_1/X_2^v)^2 = -1$ and hence k_1 is not relatively algebraically closed in L_1. If $u = 2v + 1$ is odd then $(\overline{X}_1/X_2^v)^2 = -X_2$ and hence $L_1 = k(\overline{X}_1/X_2^v)$ and therefore k_1 is relatively algebraically closed in L_1.

REMARK 14. Before coming to Zariski's Lemma, let us note that he calls maximally algebraic (m.a. for short) what we have called relatively algebraically closed., i.e., a field is m.a. in an overfield if every element of the overfield which is algebraic over the field belongs to the field; likewise he calls a field quasi-maximally algebraic (q.m.a. for short) in an overfield if every element of the overfield which is separable algebraic over the field belongs to the field; this is sometimes called relatively separably closed. Recall that an overfield is said to be regular over (or a regular extension of) a field if the overfield is a finitely generated extension of the field such that the field is m.a. in the overfield and the overfield is separably generated over the field.

The well-known fact about an irreducible polynomial mentioned in Example 8, also applies to any prime ideal P in R, after identifying k with a subfield of the function field $\mathrm{QF}(R/P)$ of the variety $\mathcal{V}(P)$. Namely, P is absolutely prime $\Leftrightarrow$ $\mathrm{QF}(R/P)$ is regular over k. Recall that P is absolutely prime means PR^* is prime, and note that then: P is absolutely prime $\Leftrightarrow$ $P\overline{k}[X_1, \ldots, X_n]$ is prime for every field extension $\overline{k}$ of k $\Leftrightarrow$ $P\overline{k}[X_1, \ldots, X_n]$ is prime for every algebraic field extension $\overline{k}$ of k $\Leftrightarrow$ $P\overline{k}[X_1, \ldots, X_n]$ is prime for every finite algebraic field extension $\overline{k}$ of K. Let us call an ideal Q in R quasiprime if it is primary; note that this implies $Q \neq R$; also note fR is quasiprime $\Leftrightarrow f = gh^\mu$ for some $g \in k^\times$ and irreducible $h \in R \setminus k$ with integer $\mu > 0$. Let us call P absolutely quasiprime to mean that PR^* is quasiprime, and note that then: P is absolutely quasiprime $\Leftrightarrow P\overline{k}[X_1, \dots, X_n]$ is quasiprime for every field extension $\overline{k}$ of $k \Leftrightarrow P\overline{k}[X_1, \dots, X_n]$ is quasiprime for every algebraic field extension $\overline{k}$ of $k \Leftrightarrow P\overline{k}[X_1, \dots, X_n]$ is quasiprime for every finite algebraic field extension $\overline{k}$ of k. As a well-known variation of the above well-known fact we have that P is absolutely quasiprime $\Leftrightarrow k$ is q.m.a. in QF(R/P). Proofs of all these assertions can be found in [**ZaS**].

Out of the following four versions of Zariski's Lemma, the first two constitute Lemma 5 of [**Za1**], the third is Proposition I.6.1 of [**Za3**], and the fourth is Theorem 2.4 of [**Mat**] or Proposition 9.31 of [**FrJ**].

Zariski's Lemma. For any finitely generated field extension L' of k we have the following.

First Version. Assume that k is of characteristic zero and m.a. in L'. Let elements Z_1, Z_2 in L' be algebraically independent over k. Then for all except a finite number of c in k we have that $k(Z_1 + cZ_2)$ is m.a. in L'.

Second Version. Assume that k is of characteristic zero and m.a. in L'. Let elements $Z_1, \dots, Z_s$ in L', with $s > 1$, be algebraically independent over k. Then by applying a k-linear automorphism to $k[Z_1, \dots, Z_s]$ it can be arranged that $k(Z_1, \dots, Z_{s-1})$ is m.a. in L'. In other words, there exists a nonsingular $n \times n$ matrix $C = (C_{ij})$ over k such that upon letting $Z_i^C = \sum_{1 \le j \le s} C_{ij} Z_j$ we have that $k(Z_1^C, \dots, Z_{s-1}^C)$ is m.a. in L'. Moreover, the constants C_{ij} are nonspecial in the sense that for every $\Lambda \in k$ there is $H_i(\Lambda) \subset k$ with $|k \setminus H_i(\Lambda)| < \infty$ for $1 \le i < s$ such that: if $C_{ij} = 1$ for $1 \le i \le s$, $C_{ij} = 0$ for $1 < i + 1 < j \le s$, $C_{12} \in H_1(1)$, and $C_{i.i+1} \in H_i(C_{i-1,i})$ for $1 < i < s$. then $k(Z_1^C, \dots, Z_{s-1}^C)$ is m.a. in L'.

Third Version. Assume that k is q.m.a. in L'. Let $z_1, \dots, z_r$ be a finite number of elements in L' such that $\text{trdeg}_k k(z_1, \dots, z_r) > 1$. Let $u_1, \dots, u_r$ be indeterminates over L'.
Then $k(u_1, \dots, u_r, u_1 z_1 + \cdots + u_r z_r)$ is q.m.a. in $L'(u_1, \dots, u_r)$.

Fourth Version. Assume that L' is regular over k. Let elements Z_1, Z_2 in L' be algebraically independent over k, and assume that $D(Z_2) \neq 0$ for some derivation D of L'/k. Then for all except a finite number of c in k we have that L' is regular over $k(Z_1 + cZ_2)$.

9. Singset of a Plane Curve and the Zeuthen-Segre Invariant

We have found bounds on redset and primset for the special pencil $(f-c)_{c\in k}$ but in case of the singset we have only stated that it is finite under appropriate conditions. Actually, from the proof of the Singset Theorem it does follow that, in case of characteristic zero, $|\text{singset}(f)| \le$ the number of irreducible components of the variety of partials $\mathcal{V}(f_{X_1},\dots,f_{X_n})^*$.

Assuming $n=2$ with $(X_1,X_2)=(X,Y)$, and k is of characteristic zero with $k=$ its algebraic closure k^*, it is possible to give a more quantitative estimate of the singset using the rank $\rho(f)$ of f as introduced in Section 11 of [**AbA**]. As we shall see, this rank $\rho(f)$ is related to the Zeuthen-Segre invariant. For the convenience of the reader let us review the definition of $\rho(f)$.

Let g,h in R. For $Q=(u,v)$ in the affine plane $\mathcal{A}=k^2$ we define the intersection multiplicity $I(g,h;Q)$ to be the k-vector-space dimension of $S/(g,h)S$ where S is the localization of R at the maximal ideal $(X-u,Y-v)R$. Note that: if $(g,h)R\not\subset(X-u,Y-v)R$ then $I(g,h;Q)=0$; if $(g,h)R\subset qR$ for some $q\in R\setminus k$ with $q(u,v)=0$ then $I(g,h;Q)=\infty$; otherwise $I(g,h;Q)$ is a positive integer. We define further intersection multiplicities by putting

$$I(g,h;\mathcal{A})=\sum_{Q\in\mathcal{A}} I(g,h;Q) \text{ and } I(g,h;f)=\sum_{\{Q\in\mathcal{A}:f(u,v)=0\}} I(g,h;Q)$$

and

$$I(g,h;\mathcal{A}\setminus f)=\sum_{\{Q\in\mathcal{A}:f(u,v)\neq 0\}} I(g,h;Q)$$

with the usual conventions about infinity, and we note that these are nonnegative integers or ∞. We also put

$$\widehat{I}(g,h;\mathcal{A})=\max_{\mu\in k} I(g,h-\mu;\mathcal{A})$$

and we note that $\widehat{I}(g,h;\mathcal{A})$ is a nonnegative integer or ∞, and: $\widehat{I}(g,h;\mathcal{A})=\infty \Leftrightarrow \gcd(g,h-c)\neq 1$ for some $c\in k$. Next we put

$$\alpha(g,h;\mathcal{A})=\{\lambda\in k : I(g,h-\lambda;\mathcal{A})<\widehat{I}(g,h;\mathcal{A})\}$$

and

$$\beta(g,h;\mathcal{A})=\sum_{0\neq\lambda\in\alpha(g,h;\mathcal{A})}[\widehat{I}(g,h;\mathcal{A})-I(g,h-\lambda;\mathcal{A})]$$

and we note that then $\beta(g,h;\mathcal{A})$ is a nonnegative integer or ∞. Finally we define

$$\rho(f) = I(f_X, f_Y; \mathcal{A} \setminus f) + \beta(f_Y, f; \mathcal{A})$$

and we note that this is a nonnegative integer or ∞.

By augmenting $\mathcal{A}$ by points on the line at infinity we get the projective plane $\mathcal{P}$ over k. For any $Q \in \mathcal{P} \setminus \mathcal{A}$ we define $I(g,h;Q)$ in an obvious manner and we put

$$I(g,h;\mathcal{P}) = \sum_{Q \in \mathcal{P}} I(g,h;Q).$$

For any $Q = (u,v) \in \mathcal{A}$ we put

$$\overline{\chi}(f;Q) = (\text{number of branches of } f \text{ at } Q) - 1$$

and we note that if $f(u,v) = 0$ then $f(X+u, Y+v)$ is a product of $\overline{\chi}(f;Q)+1$ irreducible nonconstant power series in $k[[X,Y]]$ in case $f(u,v) = 0$, and if $f(u,v) \neq 0$ then $\overline{\chi}(f;Q) = -1$. We let

$$\overline{\chi}(f;\mathcal{A}) = \sum_{Q=(u,v) \in \mathcal{A} \text{ with } f(u,v)=0} \overline{\chi}(f;Q)$$

and we note that this is a nonnegative integer. We define the integer $\overline{\chi}(f;\mathcal{P}) \geq \overline{\chi}(f;\mathcal{A})$ in an analogous manner and we put

$$\overline{\chi}(f;\infty) = \overline{\chi}(f;\mathcal{P}) - \overline{\chi}(f;\mathcal{A}).$$

If f is irreducible then by $\gamma(f)$ we denote its genus, and we note that by the genus formula for f, given any $g \in R$ with $g - c \notin fR$ for all $c \in k$, we have

$$2\gamma(f) - 2 = \deg(d\phi(g)) = \sum_{V \in \mathfrak{R}(f,\mathcal{P})} \operatorname{ord}_V(d(\phi(g))$$

where $\deg(\phi(g))$ is the degree of the divisor of the differential of $\phi(g)$ in the function field L/k of f, with canonical epimorphism $\phi : R \to A = R/(fR)$ and $L = \mathrm{QF}(A)$, and where $\mathfrak{R}(f,\mathcal{P})$ is the set of all DVRs of L/k; we also put $\mathfrak{R}(f,\mathcal{A}) = \{V \in \mathfrak{R}(f,\mathcal{P}) : A \subset V\}$ and $\mathfrak{R}(f,\infty) = \mathfrak{R}(f,\mathcal{P}) \setminus \mathfrak{R}(f,\mathcal{A})$; moreover, for any $Q \in \mathcal{P}$ we let $\mathfrak{R}(f,Q)$ denote the set of $V \in \mathfrak{R}(f,\mathcal{P})$ having center Q on f, and we note that then $\mathfrak{R}(f,Q)$ is a finite set which is empty if and only if "Q does not lie on f."

In the general case, by writing $f = f_1 \dots f_s$ with irreducible $f_1, \dots, f_s$, we generalize the definition of $\gamma(f)$ by putting

$$\gamma(f) = 1 + \sum_{1 \leq i \leq s} (\gamma(f_i) - 1).$$

Relabelling $f_1, \ldots, f_s$ suitably we can arrange that the first r of them are pairwise nonassociates, and every f_i is an associate of f_j for some $j \leq r$; now we put $\text{rad}(f) = f_0 f_1 \ldots f_r$ where $0 \neq f_0 \in k$ is chosen so that the coefficients of the highest lexicographic terms of f and $\text{rad}(f)$ are equal; the lexicographic order is such that $\deg(X^a Y^b) \geq \deg(X^{a'} Y^{b'}) \Leftrightarrow$ either $b = b'$ and $a \geq a'$ or $b > b'$. Now we define the algebraic rank of f by putting

$$\rho_a(f) = 2\gamma(\text{rad}(f)) + \overline{\chi}(\text{rad}(f); \mathcal{P}).$$

If $k = \mathbb{C}$ then by $\rho_t(f)$ we denote the rank of the first homology group of f, i.e., of the point-set $\{(u, v) \in \mathbb{C}^2 : f(u, v) = 0\}$.

Consider the condition:

$$(*) \qquad f \text{ is } Y\text{-monic of } Y\text{-degree } N > 0,$$

i.e., $\deg_{X,Y}(f - f_0 Y^N) < N$ with $f_0 \in k^\times$ and integer $N > 0$. Also consider the conditions:

$$(1^*) \qquad \gcd(f_Y, f - c) = 1 \text{ for all } c \in k,$$

and

$$(2^*) \qquad \gcd(f_Y, f - c) = 1 \text{ for all } c \in k, \text{ and } f \text{ is irreducible},$$

and

$$(3^*) \quad \gcd(f_Y, f - c) = 1 \text{ for all } c \in k,\ f \text{ is irreducible, and } k = \mathbb{C},$$

and

$$(4^*) \qquad \text{rad}(f) = f.$$

In (11.2) of [**AbA**] it is shown that:

$$(9.1) \quad (*) + (1^*) \Rightarrow \rho(f) = (1 - N) + I(f, f_Y; \mathcal{A}) - I(f_X, f_Y; f)$$

where all the terms are integers. In (11.5) of [**AbA**] it is shown that:

$$(9.2) \qquad (*) + (2^*) \Rightarrow \rho(f) = \rho_a(f)$$

and just after (11.5) it is asserted that:

$$(9.3) \qquad (*) + (3^*) \Rightarrow \rho(f) = \rho_t(f).$$

In a moment we shall generalize (9.1) by showing that:

$$(9.4) \quad (*) + (4^*) \Rightarrow \rho_a(f) = (1 - N) + I(f, f_Y; \mathcal{A}) - I(f_X, f_Y; f)$$

where obviously all the terms are integers. From this we shall deduce that:

(9.5)

$$(*) \Rightarrow \rho_a(f) = (1 - N) + \deg_Y[f] + I(f, f_Y/\widehat{[f]}; \mathcal{A}) - I(f_X, f_Y/\widehat{[f]}; f)$$

with all terms integers, where

$$[f] = \gcd(f, f_X, f_Y) \quad \text{and} \quad \widehat{[f]} = \gcd(f_X, f_Y)$$

with the gcds made unique by requiring them to be Y-monic. While proving (9.5) we shall also show that:

$$(9.6) \qquad (*) \Rightarrow \begin{cases} f = [f]\mathrm{rad}(f) \\ \text{and } \widehat{[f]} = [f]\widetilde{f} \text{ where } \widetilde{f} \in R \text{ with } \mathcal{V}(f, \widetilde{f}) = \emptyset \\ \text{and } \widehat{[f]} = \prod_{c \in \mathrm{multset}(f)^*} [f - c] \\ \text{and hence } |\mathrm{multset}(f)^*| \leq \deg_Y \widehat{[f]}. \end{cases}$$

As a consequence of (9.5) we shall show that:

$$(9.7) \qquad \begin{cases} \text{there exists a unique integer } \rho_\pi(f) \text{ together with} \\ \text{a unique finite subset defset}(f) \text{ of } k \text{ such that} \\ \rho_a(f - c) = \rho_\pi(f) \text{ for all } c \in k \setminus \mathrm{defset}(f) \text{ and} \\ \rho_a(f - c) \neq \rho_\pi(f) \text{ for all } c \in \mathrm{defset}(f). \end{cases}$$

In reference to (9.7) we put

$$\rho_\pi(f) = \text{the pencil-rank of the pencil } (f - c)_{c \in k}$$

and

$$\mathrm{defset}(f) = \text{the deficiency set of } f.$$

From (9.7) we shall deduce that:

$$(9.8) \qquad (^*) \Rightarrow \rho_\pi(f) = (1 - N) + \widehat{I}(f, f_Y/\widehat{[f]}; \mathcal{A}).$$

From (9.8) we shall deduce that:

$$(9.9) \qquad (^*) \Rightarrow \rho_\pi(f) \geq \rho_a(f - c) \text{ for all } c \in k \setminus \mathrm{multset}(f)^*.$$

From (9.8) we shall deduce the jungian formula for the pencil-rank saying that:

$$(9.10) \qquad \rho_\pi(f) = 1 - |\mathcal{V}_\infty(f)| + \sum_{c \in \mathrm{defset}(f)} [\rho_\pi(f) - \rho_a(f - c)]$$

with the points at infinity of f defined by

$$\mathcal{V}_\infty(f) = \text{set of height-one members of } \mathcal{V}(f^+)$$

where f^+ is the degree form of f consisting of its highest degree terms. We are using the adjective jungian in view of the fundamental contribution of Jung [**Jun**] to the theory of algebraic rank. The Zeuthen-Segre invariant of f is the integer $\zeta(f)$ defined by putting

$$\zeta(f) = -|\mathcal{V}_\infty(f)| - \rho_\pi(f) + \sum_{c \in \text{defset}(f)} [\rho_\pi(f) - \rho_a(f - c)]$$

and from (9.10) it immediately follows that:

(9.11) $$\zeta(f) = -1.$$

Finally from (9.11) we shall deduce the:

(9.12) DEFSET THEOREM. If (*) then $\text{singset}(f) \setminus \text{multset}(f)^* \subset \text{defset}(f)$, and $|\text{defset}(f)| \leq 1 + \rho_a(f) + \deg_Y \widehat{[f]}$ with $|\text{singset}(f)| \leq 1 + \rho_a(f) + 2\deg_Y \widehat{[f]}$.

Before turning to the proof of items (9.4) to (9.12), let us establish some common

NOTATION AND CALCULATION. Given any $f' \in R^\times$, write $f' = \theta f'_1 \dots f'_{s'}$ where $f'_1, \dots, f'_{s'}$ are irreducible members of $R \setminus k$ and $\theta \in k^\times$, and let $\phi'_i : R \to A'_i = R/(f'_i R)$ be the canonical epimorphism and identify k with a subfield of $L'_i = \text{QF}(A'_i)$. Let $Z_i = \{V \in \mathfrak{R}(f'_i, \infty) : \text{ord}_V \phi'_i(f) \geq 0\}$ and $P_i = \{V \in \mathfrak{R}(f'_i, \infty) : \text{ord}_V \phi'_i(f) < 0\}$, where the letters Z and P are meant to suggest zeros and poles, and note that these are obviously finite sets. Let

$$Z(f, f') = \sum_{1 \leq i \leq s'} \sum_{V \in Z_i} \text{ord}_V \phi'_i(f) \text{ and } P(f, f') = \sum_{1 \leq i \leq s'} \sum_{V \in P_i} \text{ord}_V \phi'_i(f)$$

and note that $Z(f, f')$ is a nonnegative integer or ∞ according as $\gcd(f, f') = 1$ or $\gcd(f, f') \neq 1$, and $P(f, f')$ is always a nonpositive integer. Clearly for each $V \in Z_i$ there is a unique $c_i(V) \in k$ such that $\text{ord}_V \phi'_i(f - c_i(V)) > 0$. Let $D(f, f') = \cup_{1 \leq i \leq t} \{c_i(V) : V \in S_i\}$ and Let $E(f, f') = D(f, f') \cup \text{singset}(f)$. Note that clearly $D(f, f')$ is a finite subset of k and hence by the Singset Theorem so is $E(f, f')$. Assuming (*), write $f = f_1 \dots f_s$ where $f_1, \dots, f_s$ are irreducible Y-monic members of $R \setminus k$, and let $\phi_i : R \to A_i = R/(f_i R)$ be the canonical epimorphism and identify k with a subfield of $L_i = \text{QF}(A_i)$.

Without assuming (*), consider the conditions:
(1′)

$$\begin{cases} \text{for a given } Q=(u,v)\in\mathcal{A} \text{ with } f(u,v)=0, \text{ and} \\ \text{for every irreducible factor } g \text{ of } f' \text{ in } R\setminus k \text{ with } g(u,v)=0 \\ \text{we have } f_Y\in gR \text{ with } f\notin gR \text{ and } f_X\notin gR, \end{cases}$$

(2′) $$f'=f_Y \text{ and } Q=(u,v)\in\mathcal{A} \text{ with } f(u,v)=0$$

and

(3′) $$\gcd(f,f')=1.$$

Then we have (I) to (III) stated below.

(I) If (*)+(1′) then

$$I(f,f';Q)-I(f_X,f';Q)=I(X-u,f';Q)$$

where all the terms are integers.

(II) If (*)+(2′) then

$$I(X-u,f';Q)=\overline{\chi}(f;Q)+\sum_{1\le i\le s}\sum_{V\in\mathfrak{R}(f_i,Q)}\operatorname{ord}_V(d\phi_i(X))$$

where all the terms are integers.

(III) If (*)+(3′) then for all $\lambda\in k$ we have

$$I(f,f';\mathcal{A})=-Z(f,f')-P(f-\lambda,f')$$

where all the terms are integers.

Proof of (I). If (*)+(1′) then we have, with all terms integers,

$$\begin{aligned}
\text{LHS of (I)} &= \sum_{1\le i\le s'} [I(f,f_i';Q) - I(f_X,f_i';Q)] \\
&= \sum_{1\le i\le s'} \sum_{V\in\mathfrak{R}(f_i',Q)} [\mathrm{ord}_V(\phi_i'(f)) - \mathrm{ord}_V(\phi_i'(f_X))] \\
&= \sum_{1\le i\le s'} \sum_{V\in\mathfrak{R}(f_i',Q)} \mathrm{ord}_V(\phi_i'(X-u)) \\
&= \sum_{1\le i\le s'} I(X-u,f_i';Q) \\
&= \text{RHS of (I)}.
\end{aligned}$$

Proof of (II). If (*)+(2′) then we have, with all terms integers,

$$\begin{aligned}
\text{LHS of (II)} &= -1 + I(X-u,f;Q) \\
&= -1 + \sum_{1\le i\le s} \sum_{V\in\mathfrak{R}(f_i,Q)} \mathrm{ord}_V(\phi_i(X-u)) \\
&= -1 + \sum_{1\le i\le s} \sum_{V\in\mathfrak{R}(f_i,Q)} [\mathrm{ord}_V(d\phi_i(X)) + 1] \\
&= \text{RHS of (II)}.
\end{aligned}$$

Proof of (III). If (*)+(3′) then we have, with all terms integers,

$$\begin{aligned}
\text{LHS of (III)} &= -Z(f,f') - P(f,f') \\
&\quad \text{(since number of zeros of a function equals number of its poles)} \\
&= \text{RHS of (III)} \\
&\quad (\text{since } \mathrm{ord}_V\phi_i'(f) = \mathrm{ord}_V\phi_i'(f-\lambda) \text{ for all } V\in P_i \text{ and } \lambda\in k.)
\end{aligned}$$

Proof of (9.4). Assume (*)+(4*). Write $f = f_1 \dots f_r$ with pairwise distinct irreducible $f_1, \dots, f_r$ in $R \setminus k$. Let $\phi_i : R \to A_i = R/(f_i)R$ be the canonical epimorphism and identify k with a subfield of

$L_i = \mathrm{QF}(A_i)$. Letting $\overline{\sum}$ stand for summation over $\{Q = (u, v) \in \mathcal{A} : f(u, v) = 0\}$ we get, with all terms integers,

$$
\begin{aligned}
&\text{RHS of (9.4)}\\
&= 1 - N + \overline{\sum}[I(f, f_Y; Q) - I(f_X, f_Y; Q)]\\
&= 1 - N + \overline{\sum}\left[\overline{\chi}(f; Q) + \sum_{1\le i\le s}\ \sum_{V\in\mathfrak{R}(f_i,Q)} \mathrm{ord}_V(d\phi_i(X))\right]\\
&\qquad\qquad\qquad\text{(by taking } f' = f_Y \text{ in (I) and (II))}\\
&= 1 - N + \overline{\chi}(f; \mathcal{A}) + \sum_{1\le i\le s}\ \sum_{V\in\mathfrak{R}(f_i,\mathcal{A})} \mathrm{ord}_V(d\phi_i(X))
\end{aligned}
$$

$$
\begin{aligned}
&= 1 - N + \overline{\chi}(f; \mathcal{A}) + (2\gamma(f) - 2)\\
&\qquad - \sum_{1\le i\le s}\ \sum_{V\in\mathfrak{R}(f_i,\infty)} \mathrm{ord}_V(d\phi_i(X))\\
&\qquad\text{(by genus formula for } f_i \text{ and definition of } \gamma(f))\\
&= 2\gamma(f) + \overline{\chi}(f; \mathcal{A}) - N - 1\\
&\qquad - \sum_{1\le i\le s}\ \sum_{V\in\mathfrak{R}(f_i,\infty)} [\mathrm{ord}_V(\phi_i(X)) - 1]\\
&\qquad\qquad\text{(because } \mathrm{ord}_V\phi_i(X) \neq 0)\\
&= 2\gamma(f) + \overline{\chi}(f; \mathcal{A}) - N - 1 - [-N - (\overline{\chi}(f; \infty) + 1)]\\
&= 2\gamma(f) + \overline{\chi}(f; \mathcal{P})\\
&= \text{LHS of (9.4).}
\end{aligned}
$$

Proof of (9.5) and (9.6). Assume (*). Let $\overline{f} = \mathrm{rad}(f)$. Then by the argument in the proof of the Singset Theorem we see that

(i) $[f] = f/\overline{f}$ and $\widehat{[f]} = [f]h$ where $h \in R$ with $\mathcal{V}(f, h) = \emptyset$.

and

$$\widehat{[f]} = \prod_{c\in\mathrm{multset}(f)^*} [f - c] \quad \text{and hence} \quad |\mathrm{multset}(f)^*| \le \deg_Y \widehat{[f]}.$$

which proves (9.6). Applying (9.4) to $\overline{f}$ we get

$$\rho_a(f) = 1 - \deg_Y \overline{f} + I(\overline{f}, \overline{f}_Y; \mathcal{A}) - I(\overline{f}_X, \overline{f}_Y; \overline{f}).$$

The first two terms of the above RHS combine to give $1-N+\deg_Y[f]$ which are the first three terms of the RHS of (9.5). It remains to compare the intersection multiplicity terms, i.e., the proof of (9.5) will be completed by showing that

(i*) $I(\overline{f},\overline{f}_Y;\mathcal{A}) - I(\overline{f}_X,\overline{f}_Y;\overline{f}) = I(f,f_Y/\widehat{[f]};\mathcal{A}) - I(f_X,f_Y/\widehat{[f]};f)$.

For any $Q=(u,v)\in\mathcal{A}$ with $f(u,v)=0$, by taking $(\overline{f},\overline{f}_Y)$ for (f,f') in (I) we get

(ii) $$I(\overline{f},\overline{f}_Y;Q) - I(\overline{f}_X,\overline{f}_Y;Q) = I(X-u,\overline{f}_Y;Q),$$

and, in view of (i), by taking $(f,f_Y/\widehat{[f]})$ for (f,f') in (I) we get

(iii) $$I(f,f_Y/\widehat{[f]};Q) - I(f_X,f_Y/\widehat{[f]};Q) = I(X-u,f_Y/\widehat{[f]};Q).$$

By summing over $Q=(u,v)\in\mathcal{A}$ with $f(u,v)=0$, the LHS of (ii) gives the LHS of (i*), and the RHS of (iii) gives the RHS of (i*). Therefore the proof of (9.5) will be complete by proving that, for any $Q=(u,v)\in\mathcal{A}$ with $f(u,v)=0$, the RHS of (ii) equals the RHS of (iii). Let $\psi: R\to R/(X-u)R$ be the canonical epimorphism, and let W be the localization of $\psi(R)$ at $\psi((Y-v)R)$. Now

$$\begin{aligned}
\text{RHS of (ii)} &= \operatorname{ord}_W\psi(\overline{f}_Y)\\
&= -1+\operatorname{ord}_W\psi(\overline{f}) \quad (\text{because } \overline{f}(u,v)=0)\\
&= -1+\operatorname{ord}_W\psi(f/\widehat{[f]}) \quad (\text{by (i)})\\
&= [-1+\operatorname{ord}_W\psi(f)] - \operatorname{ord}_W\psi(\widehat{[f]})\\
&= \operatorname{ord}_W\psi(f_Y) - \operatorname{ord}_W\psi(\widehat{[f]}) \quad (\text{because } f(u,v)=0)\\
&= \operatorname{ord}_W\psi(f_Y/\widehat{[f]})\\
&= \text{RHS of (iii)}
\end{aligned}$$

and this completes the proof.

Proof of (9.7) to (9.12). For any k-automorphism σ of R we clearly have $\rho_a(f-c)-\rho_a(\sigma(f)-c) = |\mathcal{V}_\infty(\sigma(f)-c)| - |\mathcal{V}_\infty(f-c)|$ for all $c\in k$, and hence: (9.7) is true for $f \Leftrightarrow$ it is true for $\sigma(f)$. After having defined $\rho_\pi(f)$ and defset(f), for any k-automorphism σ of R we clearly have $\jmath(\sigma(f))=\jmath(f)$, and hence: (9.9) is true for $f\Leftrightarrow$ it is true for $\sigma(f)$. Also it is well-known that $\sigma(f)$ satisfies (*) for some k-automorphism σ of R. Therefore in proving (9.7) to (9.12), without loss of generality we may and **we shall assume that f satisfies (*).** Let $f'=f_Y/\widehat{[f]}$. Clearly f_X and f_Y are unchanged if we replace

f by $f-c$ with $c \in k$, and hence so are $\widehat{[f]}$ and f'. Given any $c \in k$, by taking $f-c$ for f in (9.5) we get

(1) $\rho_a(f-c) = 1-N+\deg_Y[f-c]+I(f-c,f';\mathcal{A})-I(f_X,f';f-c)$

with all terms integers. By (9.6) we have $\gcd(f-c,f')=1$ and hence by taking $(0,f-c)$ for (λ,f) in (III) we get

$$(2) \qquad I(f-c,f';\mathcal{A}) = -Z(f-c,f') - P(f,f')$$

where all the terms are integers. In view of (9.6), by the definitions of the sets $D(f,f')$, $\mathrm{multset}(f)^*$, $\mathrm{singset}(f)$ and $E(f,f')$ we see that

$$(3) \qquad \begin{cases} Z(f-c,f') > 0 \text{ or } Z(f-c,f') = 0 \\ \text{according as } c \in D(f,f') \text{ or } c \notin D(f,f'), \end{cases}$$

and

$$(4) \qquad \deg_Y[f-c] > 0 \text{ or } = 0 \text{ according as } c \in \mathrm{multset}(f)^* \text{ or } c \notin \mathrm{multset}(f)^*,$$

and
(5)
$I(f_X,f';f-c) \geq 0$ and if $c \notin \mathrm{singset}(f)$ then $I(f_X,f';f-c)=0$,

and

(6) $E(f,f') = D(f,f') \cup \mathrm{singset}(f)$ with $\mathrm{multset}(f)^* \subset \mathrm{singset}(f)$.

By (1) to (6) we get

$$(7) \qquad c \notin E(f,f') \Rightarrow \rho_a(f-c) = 1-N-P(f,f')$$

with all terms integers. Since $E(f,f')$ is a finite set and the above RHS is independent of c, this proves (9.7) and establishes the existence of $\rho_\pi(f)$ and $\mathrm{defset}(f)$. Now by (1) to (7) we see that

$$(8) \qquad \rho_\pi(f) = 1-N-P(f,f')$$

and

$$(9) \qquad \mathrm{defset}(f) \subset E(f,f')$$

and

$$(10) \qquad \rho_\pi(f) = (1-N) + \widehat{I}(f,f';\mathcal{A})$$

with all terms integers, which proves (9.8). By (9.5) and (9.8) we get

$$(11) \qquad \begin{cases} \rho_\pi(f) - \rho_a(f-c) = & -\deg_Y[f-c] + I(f_X,f';f-c) \\ & + \left(\widehat{I}(f,f';\mathcal{A}) - I(f-c,f';\mathcal{A})\right). \end{cases}$$

Since the second parenthesis term and the third big parenthesis term in the above RHS are obviously nonnegative, in view of (4) we see that

$$\rho_\pi(f) - \rho_a(f-c) \geq -\deg_Y[f-c] \tag{12}$$

and

$$\rho_\pi(f) \geq \rho_a(f-c) \text{ for all } c \in k \setminus \text{multset}(f)^* \tag{13}$$

which proves (9.9). By (1), (2) and (8) we get

$$\rho_\pi(f) - \rho_a(f-c) = -\deg_Y[f-c] + I(f_X, f'; f-c) + Z(f-c, f') \tag{14}$$

with all terms integers. Now, with all terms integers, we have

$$\begin{aligned}
&\text{RHS of (9.10)}\\
&= \sum_{c\in E(f,f')} [\rho_\pi(f) - \rho_a(f-c)]\\
&\qquad\qquad \text{(by (9))}\\
&= -\deg_Y[\widehat{f}] + \sum_{c\in E(f,f')} [I(f_X, f'; f-c) + Z(f-c, f')]\\
&\qquad\qquad \text{(by (9.6), (4), (6) and (14))}\\
&= -\deg_Y[\widehat{f}] + I(f_X, f'; \mathcal{A}) + \sum_{c\in E(f,f')} Z(f-c, f')\\
&\qquad\qquad \text{(by (5) and (6))}\\
&= -\deg_Y[\widehat{f}] + I(f_X, f'; \mathcal{A}) + \sum_{1\leq i\leq s'} \sum_{V\in Z_i} \text{ord}_V \phi_i'(f - c_i(V))\\
&\qquad\qquad \text{(by (7) and (9))}\\
&= -\deg_Y[\widehat{f}] + I(f_X, f'; \mathcal{A}) + \sum_{1\leq i\leq s'} \sum_{V\in Z_i} [\text{ord}_V \phi_i'(f_X) + \text{ord}_V \phi_i'(X)]\\
&\qquad\qquad \text{(because } f_Y \in f_i'R \text{ and } \text{ord}_V \phi_i'(f - c_i(V)) \neq 0 \neq \text{ord}_V \phi_i'(X))\\
&= -\deg_Y[\widehat{f}] + I(f_X, f'; \mathcal{A}) + Z(f_X, f') + Z(X, f')\\
&= -\deg_Y[\widehat{f}] + [-Z(f_X, f') - P(f_X, f')] + Z(f_X, f') + Z(X, f')\\
&\qquad\qquad \text{(by taking } (0, f_X) \text{ for } (\lambda, f) \text{ in (III))}\\
&= -\deg_Y[\widehat{f}] + Z(X, f') - P(f_X, f'))
\end{aligned}$$

and we have

$$
\begin{aligned}
&-\deg_Y \widehat{[f]} + Z(X, f') - P(f_X, f')) \\
&= -\deg_Y \widehat{[f]} + Z(X, f') - \sum_{1 \le i \le s'} \sum_{V \in P_i} \operatorname{ord}_V \phi'_i(f_X) \\
&= -\deg_Y \widehat{[f]} + Z(X, f') - \sum_{1 \le i \le s'} \sum_{V \in P_i} [-\operatorname{ord}_V \phi'_i(X) + \operatorname{ord}_V \phi'_i(f)] \\
&\qquad \text{(because } f_Y \in f'_i R \text{ and } \operatorname{ord}_V \phi'_i(f) \neq 0 \neq \operatorname{ord}_V \phi'_i(X))
\end{aligned}
$$

$$
\begin{aligned}
&= -\deg_Y \widehat{[f]} + Z(X, f') + P(X, f') - P(f, f') \\
&= -\deg_Y \widehat{[f]} + Z(X, f') + P(X, f') - P(f, f') \\
&= -\deg_Y \widehat{[f]} - \deg_Y f' - P(f, f') \\
&= -\deg_Y f_Y - P(f, f') \\
&= 1 - N - P(f, f') \\
&= \rho_\pi(f) \\
&\quad \text{(by (8))}
\end{aligned}
$$

and so we get

$$\rho_\pi(f) = 1 - |\mathcal{V}_\infty(f)| + \sum_{c \in \operatorname{defset}(f)} [\rho_\pi(f) - \rho_a(f - c)] \tag{15}$$

and hence

$$\zeta(f) = -1 \tag{16}$$

which proves (9.10) and (9.11). By (9.6) we see that

$$c \notin \operatorname{multset}(f)^* \Rightarrow I(f_X, f'; f - c) = I(f_X, f_Y; f - c)$$

and hence

$$c \in \operatorname{singset}(f) \setminus \operatorname{multset}(f)^* \Rightarrow I(f_X, f'; f - c) > 0$$

and therefore, because the third big parenthesis term in the RHS of (11) is nonnegative, in view of (4) and (11) we conclude that

$$\operatorname{singset}(f) \setminus \operatorname{multset}(f)^* \subset \operatorname{defset}(f). \tag{17}$$

Rewriting the jungian excess formula (9.9) we get

$$\rho_\pi(f) = \sum_{c \in \operatorname{defset}(f)} [\rho_\pi(f) - \rho_a(f - c)] \tag{18}$$

and adding $\rho_a(f) - \rho_\pi(f)$ to both sides we obtain

$$\rho_a(f) = \sum_{c\in\operatorname{defset}(f)\setminus\{0\}} [\rho_\pi(f) - \rho_a(f-c)]. \tag{19}$$

By (4), (12) and (19) we get

$$\begin{aligned}\rho_a(f) + &\sum_{0\neq c\in\operatorname{multset}(f)^*} \deg_Y[f-c] \\ &\geq \sum_{0\neq c\in\operatorname{defset}(f)\setminus\operatorname{multset}(f)^*} [\rho_\pi(f) - \rho_a(f-c)]\end{aligned}$$

and hence by (9.6) we have

$$\rho_a(f) + \deg_Y\widehat{[f]} \geq \sum_{0\neq c\in\operatorname{defset}(f)\setminus\operatorname{multset}(f)^*} [\rho_\pi(f) - \rho_a(f-c)]$$

and therefore, since every term in the above RHS is nonnegative, we obtain

$$|\operatorname{defset}(f)\setminus\{0\}| \leq \rho_a(f) + \deg_Y\widehat{[f]}$$

and hence

$$|\operatorname{defset}(f)| \leq 1 + \rho_a(f) + \deg_Y\widehat{[f]} \tag{20}$$

By (17) we have

$$|\operatorname{singset}(f)| \leq |\operatorname{defset}(f)| + |\operatorname{multset}(f)^*|$$

and hence, in view of (9.6), by (20) we get

$$|\operatorname{singset}(f)| \leq 1 + \rho_a(f) + 2\deg_Y\widehat{[f]}. \tag{21}$$

The proof of (9.12) is completed by (17), (20) and (21).

QUESTION 9. Is it always true that $\operatorname{redset}(f) \subset \operatorname{defset}(f)$?

QUESTION 10. More explicitly, is it possible to find a nonsingular reducible $f \in R\setminus k$ satisfying (*) such that $[f] = 1$ and $\rho_a(f) = \rho_\pi(f)$. If such an f is found then $0 \in \operatorname{redset}(f) \setminus \operatorname{defset}(f)$ and we have a negative answer to Question 9. In this connection it would be worthwhile to study the famous examples of Artal-Bartolo, Cassou-Nogues, and Luengo-Velasco [**ACL**] in which they construct curves whose redsets and singsets are empty and whose ranks are arbitrarily high.

QUESTION 11. In connection with the Defset Theorem (9.12), consider the Example given by the pencil $(h(Y) - c)_{c\in k}$ where

$f = h(Y) \in k[Y]$ is monic of degree $N > 1$; using (9.5) we can easily show that: (i) $\rho_\pi(f) = 1-N$, (ii) $\rho_a(f-c) = 1-N+\deg_Y[f-c]$ for all $c \in k$, and (iii) $\rho_\pi(f) = \sum_{c\in k}[\rho_\pi(f)-\rho_a(f-c)]$; thus $\rho_a(f-c) \geq \rho_\pi(f)$ for all $c \in k$, and $\rho_a(f-c) > \rho_\pi(f)$ for precisely those c for which $c = h(c')$ with $h_Y(c') = 0$. It is easy to make similar examples for pencils obtained by replacing Y by any nonconstant polynomial in R. However we ask: is it true that if the pencil $(f-c)_{c\in k}$ is noncomposite, i.e., if $k(f)$ is m.a. in $k(X,Y)$, then $\rho_\pi(f) \geq \rho_a(f-c)$ for all $c \in k$? If this has a positive answer then, for noncomposite pencils, the term $2\deg_Y\widehat{[f]}$ may be dropped from the estimate given in (9.12).

QUESTION 12. Can rank and defset be generalized to $n > 2$?

MORAL. Thus it is all a matter of measuring change in a quantity relative to the corresponding change in another quantity on which the previous quantity depends. This after all is the Newton-Leibnitz idea of derivative. By avoiding limits, which are algebraically awkward to take, it also gives rise to the twisted derivative introduced in [**A07**] and exploited in numerous succeeding papers summarized in [**A10**] for calculating Galois groups and hence fundamental groups. The same principle of measuring change gives rise to the invariants $I, \beta, \gamma, \rho, \rho_a, \rho_\pi, \jmath, \zeta$. Amongst these, β measures the change in I, the quantities $\beta, \gamma, \rho, \rho_a, \rho_\pi$ are different incarnations of the same underlying reality, and the almost identical quantities $\jmath$ and ζ measure the change in ρ. Or, as may be easier to remember, the genus γ measures the change in the intersection multiplicity I, while the Zeuthen-Segre invariant ζ measures the change in the genus γ.

10. Defset of a General Pencil

In this section we continue to assume $n = 2$ with $(X_1, X_2) = (X, Y)$, and k is of characteristic zero with $k =$ its algebraic closure k^*. We shall now extend our study of rank and defset to a general pencil $(f - cw)_{c\in k\cup\{\infty\}}$ where $w \in R^\times = R \setminus \{0\}$ with $\gcd(f, w) = 1$ and $d = \max(\deg(f), \deg(w))$. Recall that by convention $f - \infty w = w$.

Let $\mathcal{L}_\infty = \mathcal{P} \setminus \mathcal{A} = \{(\infty, v) : v \in k \cup \{\infty\}\}$ and call this the line at infinity. In homogeneous coordinates (X, Y, Z) we think of $\mathcal{L}_\infty$ as given by $Z = 0$, and also as the (Y/Z)-axis. As abbreviations, for the two special points of this line, i.e., for the two infinite points of the fundamental triangle, we put $(\infty) = (\infty, 0)$ and $((\infty)) = (\infty, \infty)$ and note that in homogeneous coordinates they are $(1, 0, 0)$ and $(0, 1, 0)$.

Let $R_0 = k[[X,Y]]$ and note that $k(X,Y) \cap R_0$ is the localization of R at the maximal ideal $(X,Y)R$ and R_0 is its completion. Let $e \geq 0$ be an integer. Define the e-th homogenization of any $g \in R$ by putting $g_{[e]} = g_{[e]}(X,Y,Z) = Z^{e+\deg(g)} g(X/Z, Y/Z)$ in case $g \neq 0$, and $g_{[e]} = g_{[e]}(X,Y,Z) = 0$ in case $g = 0$. For any $Q = (u,v) \in \mathcal{P}$, define the k-monomorphism $T^{[Q,e]} : R \to R_0$ with $g \mapsto g^{[Q,e]} = g^{[Q,e]}(X,Y)$ by putting

$$(10.1) \qquad g^{[Q,e]}(X,Y) = \begin{cases} g(X+u, Y+v) & \text{if } Q \in \mathcal{A} \\ g_{[e]}(1, Y+v, X) & \text{if } ((\infty)) \neq Q \in \mathcal{L}_\infty \\ g_{[e]}(X,1,Y) & \text{if } Q = ((\infty)) \end{cases}$$

and call this the Taylor map at $[Q,e]$.

Maclaurin and Taylor were two disciples of Newton of calculus-fame, the former expanding things around the origin and the latter around other points! Newton having accomplished many other things, the adjective "of calculus-fame" is clearly directed more towards the disciples!!.

For any $Q \in \mathcal{P}$ let $T^{[Q]} : R \to R_0$ be given by $g \mapsto g^{[Q]} = g^{[Q,0]}$. Note that if $g^{[Q]}(0,0) \neq 0$ then $\overline{\chi}(g;Q) = -1$, and if $g^{[Q]}(0,0) = 0 \neq g$ then $g^{[Q]}$ is product of $1 + \overline{\chi}(g;Q)$ irreducibles in R_0. Also note that for any g, h in R we have $I(g,h;Q) = [R_0/(g^{[Q]}, h^{[Q]})R_0 : k]$. For any $g, h, \overline{g}, \overline{h}$ in R we put

$$(10.2) \qquad I(g,h;\overline{g} \setminus \overline{h}) = \sum_{\{Q \in \mathcal{A} : \overline{g}^{[Q]}(0,0) = 0 \neq \overline{h}^{[Q]}(0,0)\}} I(g,h;Q).$$

Let

$$t = |\mathcal{V}(f,w)|$$

and let $Q_1 = (u_1, v_1), \ldots, Q_t = (u_t, v_t)$ in $\mathcal{A}$ be the t points of $\mathcal{V}(f,w)$ called the finite base points of the pencil. For any $g, h, \overline{g}$ in R we put

$$(10.3) \qquad I(g,h;\overline{g} \setminus \mathcal{V}(f,w)) = \sum_{\{Q \in \mathcal{A} \setminus \{Q_1, \ldots, Q_t\} : \overline{g}^{[Q]}(0,0) = 0\}} I(g,h;Q).$$

To avoid repeating the considerations of the previous section and for simplicity of calculation, most of the time we shall suppose that:

(r*) the elements $f, w, 1$ are linearly independent over k,

i.e., by replacing f, w by suitable k-linear combinations of them, our pencil cannot be converted into the special pencil considered in the previous section.

From the previous section recall condition

$$(*) \qquad f \text{ is } Y\text{-monic of } Y\text{-degree } N > 0$$

and consider condition

$$(**) \qquad (*) \text{ and } w \text{ is } Y\text{-monic of } Y\text{-degree } 0 < M < N.$$

We claim that:
(10.4)

$$(\mathrm{r}^*) \Rightarrow \begin{cases} \text{there exists a } k\text{-automorphism } \sigma \text{ of } R \\ \text{together with } c_i^* \neq c_2^* \text{ in } k \cup \{\infty\} \text{ such that} \\ \text{if we let the pair } (\sigma(f - c_1^* w), \sigma(f - c_2^* w)) \text{ be called } (f, w) \\ \text{then condition } (**) \text{ is satisfied.} \end{cases}$$

Namely, by a well-known argument there exists a k-automorphism σ_1 of R such that $\sigma_1(f), \sigma_1(w)$ are both Y-monic. If their degrees are distinct, then we already have (**). If not, then we can find $c_1^*, c_2^* \in k$ such that $\sigma_1(f - c_1^* w)$ has bigger degree than $\sigma_1(f - c_2^* w)$, which must be positive because of (r*). By applying a suitable k-automorphism σ_2 of R to these two polynomials, we can arrange that condition (**) is satisfied. Now take σ to be the composition of σ_2 with σ_1.

Analogous to (9.5) we shall prove that:

$$(10.5) \qquad (**) \Rightarrow \begin{cases} \text{for any } c \in k \text{ we have:} \\ \rho_a(f - cw) \\ = (1 - N) \\ \quad + \deg_Y [f - cw] \\ \quad + I\left(f - cw, \frac{f_Y w - f w_Y}{\widehat{[f,w]}}; (f - cw) \setminus w\right) \\ \quad - I\left(f_X w - f w_X, \frac{f_Y w - f w_Y}{\widehat{[f,w]}}; (f - cw) \setminus w\right) \\ \quad + \sum_{1 \le i \le t} \left(I(X - u_i, \frac{f - cw}{[f - cw]}; Q_i) - 1\right) \end{cases}$$

with all terms integers, where

$$\widehat{[f, w]} = \gcd(f_X w - f w_X, f_Y w - f w_Y)$$

with, as before, the gcd made unique by requiring it to be Y-monic. While proving (10.5) we shall show that:
(10.6)

$$(**) \Rightarrow \text{for any } c \in k \cup \{\infty\} \text{ we have } [f - cw] = \gcd(f - cw, \widehat{[f, w]}).$$

As a consequence of (10.5) we shall show that:
(10.7)

$$(\text{r}^*) \Rightarrow \begin{cases} \text{there exists a unique integer } \rho_\pi(f,w) \text{ together with} \\ \text{a unique finite subset defset}(f,w) \text{ of } k \cup \{\infty\} \text{ such that} \\ \rho_a(f-cw) = \rho_\pi(f,w) \text{ for all } c \in (k\cup\{\infty\}) \setminus \text{defset}(f,w) \\ \text{and } \rho_a(f-cw) \neq \rho_\pi(f,w) \text{ for all } c \in \text{defset}(f,w). \end{cases}$$

In reference to (10.7), we put

$$\rho_\pi(f,w) = \text{the pencil-rank of the pencil } (f-cw)_{c\in k\cup\{\infty\}}$$

and

$$\text{defset}(f,w) = \text{the deficiency set of } (f,w).$$

From (10.7) we shall deduce that:
(10.8)

$$(**) \Rightarrow \begin{cases} \rho_\pi(f,w) \\ = (1-N) \\ \quad +\max_{c\in k} I\left(f-cw, \frac{f_Y w - f w_Y}{\widehat{[f,w]}}; (f-cw)\setminus w\right) \\ \quad + \sum_{1\le i\le t} \min_{c\in k\setminus \text{multset}(f,w)^*} \left(I(X-u_i, \frac{f-cw}{[f-cw]}; Q_i) - 1\right) \end{cases}$$

with all terms integers. From (10.8) we shall deduce that:

$$(10.9) \quad (**) \Rightarrow \begin{cases} \rho_\pi(f,w) \ge \rho_a(f-cw) \\ \text{for all } c \in k \setminus (\text{multset}(f,w)^* \cup \text{conset}(f,w)^*) \end{cases}$$

where, without assuming $k = k^*$, the contact set of (f,w) is defined by putting

$$\text{conset}(f,w)^* = \cup_{Q=(u,v)\in\mathcal{V}(f,w)^*} \text{conset}(f,w;Q)^*$$

with the set $\text{conset}(f,w;Q)^*$ of size at most one defined by

$$\begin{cases} \text{conset}(f,w;Q)^* \\ = \{c \in k^* : I(X-u, f-cw; Q) \\ \qquad > I(X-u, f-c'w; Q) \text{ for some } c' \in k^* \setminus \text{multset}(f,w)^*\}. \end{cases}$$

From (10.8) we shall deduce that:
(10.10)

$$(**) \Rightarrow \begin{cases} \rho_a(f) + \rho_a(w) \\ = 1 - t + \sum_{c\in\text{defset}(f,w)\setminus\{0,\infty\}} [\rho_\pi(f,w) - \rho_a(f-cw)]. \end{cases}$$

Clearly (10.10) is equivalent to saying that:

$$(10.11) \qquad (**) \Rightarrow \zeta(f,w) = -1$$

where

$$\zeta(f,w) = -t - 2\rho_\pi(f,w) + \sum_{c\in \text{defset}(f,w)} [\rho_\pi(f,w) - \rho_a(f-cw)].$$

Finally, from (10.11) we shall deduce the:

(10.12) GENERAL DEFSET THEOREM. If (**) then we have:

(i) $(a(f,w)\setminus b(f,w)) \subset \text{defset}(f,w)$ where $a(f,w) = k \cap \text{singset}(f,w)$ and $b(f,w) = \text{multset}(f,w)^* \cup \text{conset}(f,w)^*$.

(ii) $|\text{defset}(f,w)| \le \rho_a(f) + \rho_a(w) + \deg_Y \widehat{[f,w]} + 2t + 1$.

(iii) $|\text{singset}(f,w)| \le \rho_a(f) + \rho_a(w) + 2\deg_Y \widehat{[f,w]} + 3t + 2$.

Before turning to the proof of items (10.5) to (10.12), let us establish some common

NOTATION AND CALCULATION. Given any $f' \in R^\times$, write $f' = \theta f'_1 \dots f'_{s'}$ where $f'_1,\dots,f'_{s'}$ are irreducible members of $R\setminus k$ and $\theta \in k^\times$, and let $\phi'_i : R \to A'_i = R/(f'_iR)$ be the canonical epimorphism and identify k with a subfield of $L'_i = \text{QF}(A'_i)$. Let $Z_i = \{V \in \mathfrak{R}(f'_i,\infty) : \text{ord}_V\phi'_i(f) \ge \text{ord}_V\phi'_i(w)\}$ and $P_i = \{V \in \mathfrak{R}(f'_i,\infty) : \text{ord}_V\phi'_i(f) < \text{ord}_V\phi'_i(w)\}$, where the letters Z and P are meant to suggest zeros and poles, and note that these are obviously finite sets.

Consider the conditions:

$$\text{(0')} \qquad \gcd(w,f') = 1,$$

(1')

$$\begin{cases} \text{for a given } Q=(u,v)\in\mathcal{A} \text{ with } f(u,v)=0, \text{ and} \\ \text{for every irreducible factor } g \text{ of } f' \text{ in } R\setminus k \text{ with } g(u,v)=0 \\ \ne w(u,v) \text{ we have } w^2(f/w)_Y \in gR \text{ with } f\notin gR \text{ and} \\ w^2(f/w)_X \notin gR, \end{cases}$$

(2') $f' = w^2(f/w)_Y$ and $Q=(u,v)\in\mathcal{A}$ with $f(u,v) = 0 \ne w(u,v)$

and

$$\text{(3')} \qquad \gcd(f,f') = 1.$$

Assuming (0'): Let

$$Z((f,w),f') = \sum_{1\le i\le s'}\sum_{V\in Z_i} [\text{ord}_V\phi'_i(f) - \text{ord}_V\phi'_i(w)]$$

and

$$P((f,w),f') = \sum_{1\le i\le s'} \sum_{V\in P_i} [\mathrm{ord}_V \phi_i'(f) - \mathrm{ord}_V \phi_i'(w)]$$

and note that $Z((f,w),f')$ is a nonnegative integer or ∞ according as $\gcd(f,f')=1$ or $\gcd(f,f')\neq 1$, and $P((f,w),f')$ is always a nonpositive integer. Clearly for each $V\in Z_i$ there is a unique $c_i(V)\in k$ such that $\mathrm{ord}_V\phi_i'(f-c_i(V)w) > \mathrm{ord}_V\phi_i'(w)$. Let $D((f,w),f') = \cup_{1\le i\le t}\{c_i(V) : V\in Z_i\}$ and let $E((f,w),f') = D((f,w),f')\cup \mathrm{singset}((f,w))$. Note that clearly $D((f,w),f')$ is a finite subset of k and hence by the General Singset Theorem so is $E((f,w),f')$.

Without assuming $(0')$ but assuming (**): Write $f=f_1\dots f_s$ where $f_1,\dots,f_s$ are irreducible Y-monic members of $R\setminus k$, and let $\phi_i : R\to A_i = R/(f_iR)$ be the canonical epimorphism and identify k with a subfield of $L_i = \mathrm{QF}(A_i)$.

With these conditions in mind, we have (I) to (IV) stated below.

(I) If (**)+$(1')$ then

$$I(f,f';Q) - I(w^2(f/w)_X, f';Q) = I(X-u,f';Q)$$

where all the terms are integers.

(II) If (**)+$(2')$ then

$$I(X-u,f';Q) = \overline{\chi}(f;Q) + \sum_{1\le i\le s}\sum_{V\in\mathfrak{R}(f_i,Q)} \mathrm{ord}_V d\phi_i(X)$$

where all the terms are integers.

(III) If (**)+$(0')+(3')$ then for all $\lambda\in k$ we have

$$I(f,f';\mathcal{A}) - I(w,f';\mathcal{A}) = -Z((f,w),f') - P((f-\lambda w,w),f')$$

where all the terms are integers.

(IV) If (*) then

$$1-\deg_Y f = 2 + \overline{\chi}(f;\infty) + \sum_{1\le i\le s}\sum_{V\in\mathfrak{R}(f_i,\infty)} \mathrm{ord}_V d\phi_i(X)$$

with all terms integers.

Proof of (I). If (**)+(1′) then we have, with all terms integers,

$$\begin{aligned}\text{LHS of (I)} &= \sum_{1\le i\le s'} \left[(I(f,f'_i;Q) - I(w^2(f/w)_X, f'_i;Q)\right] \\ &= \sum_{1\le i\le s'} \sum_{V\in\mathfrak{R}(f'_i,Q)} \left[\operatorname{ord}_V \phi'_i(f) - \operatorname{ord}_V \phi'_i(w^2(f/w)_X)\right] \\ &= \sum_{1\le i\le s'} \sum_{V\in\mathfrak{R}(f'_i,Q)} \operatorname{ord}_V \phi'_i(X-u) \\ &= \sum_{1\le i\le s'} I(X-u, f'_i;Q) \\ &= \text{RHS of (I)}.\end{aligned}$$

Proof of (II). If (**)+(2′) then we have, with all terms integers,

$$\begin{aligned}\text{LHS of (II)} &= -1 + I(X-u,f;Q) \\ &= -1 + \sum_{1\le i\le s} \sum_{V\in\mathfrak{R}(f_i,Q)} \operatorname{ord}_V \phi_i(X-u) \\ &= -1 + \sum_{1\le i\le s} \sum_{V\in\mathfrak{R}(f_i,Q)} [\operatorname{ord}_V d\phi_i(X) + 1] \\ &= \text{RHS of (II)}.\end{aligned}$$

Proof of (III). If (**)+(0′) + (3′) then we have, with all terms integers,

$$\begin{aligned}&\text{LHS of (III)} \\ &= -Z(f,f') - P(f,f') + Z(w,f') + P(w,f') \\ &\quad \text{(since number of zeros of a function equals number of its poles)} \\ &= \text{RHS of (III)} \\ &\quad (\text{since } \operatorname{ord}_V \phi'_i(f) = \operatorname{ord}_V \phi'_i(f-\lambda w) \text{ for all } V \in P_i \text{ and } \lambda \in k.)\end{aligned}$$

Proof of (IV). If (*) then we have, with all terms integers,

$$\begin{aligned}\text{RHS of (IV)} &= 2+\overline{\chi}(f;\infty)+\sum_{1\le i\le s}\sum_{V\in\mathfrak{R}(f_i,\infty)}(\mathrm{ord}_V\phi_i(X)-1)\\ &\qquad(\text{because } \mathrm{ord}_V\phi_i(X)<0)\\ &= 2+\overline{\chi}(f;\infty)-\deg_Y f-\sum_{1\le i\le s}|\mathfrak{R}(f_i,\infty)|\\ &= 1-\deg_Y f\\ &= \text{LHS of (IV)}.\end{aligned}$$

Proof of (10.5) and (10.6). Assume (**). Then by the argument in the proof of the Singset Theorem we see that

(i)

for any $c\in k\cup\{\infty\}$ we have $f-cw=[f-cw]\overline{f}$ where $\overline{f}=\mathrm{rad}(f-cw)$

and

$$\text{for any } c\in k\cup\{\infty\} \text{ we have } [f-cw]=\gcd(f-cw,\widehat{[f,w]})$$

which proves (10.6). Take $f'=\frac{f_Yw-fw_Y}{\widehat{[f,w]}}$. For any $c\in k$ and $Q=(u,v)\in\mathcal{A}$ with $f(u,v)-cw(u,v)=0\neq w(u,v)$, by using (I) we get

(ii) $\quad I(f-cw,f';Q)-I(f_Xw-fw_X,f';Q)=I(X-u,f';Q)$

with all terms integers. Let $\overline{f}=\mathrm{rad}(f-cw)$. Let $\psi:R\to R/(X-u)R$ be the canonical epimorphism, and let W be the localization of $\psi(R)$ at $\psi((Y-v)R)$. Then we have, with all terms integers,

$$\begin{aligned}&I(X-u,w^2(\overline{f}/w)_Y;Q)\\ &\quad= \mathrm{ord}_W\psi(w)+\mathrm{ord}_W\psi(\overline{f})-1\\ &\quad= \mathrm{ord}_W\psi(\overline{f})-1\\ &\qquad\quad(\text{since } w(u,v)\neq 0)\\ &\quad= \mathrm{ord}_W\psi(f-cw)-\mathrm{ord}_W\psi([f-cw])-1\\ &\quad= \mathrm{ord}_W\psi(w^2((f-cw)/w)_Y)-\mathrm{ord}_W\psi([f-cw])\\ &\quad= \mathrm{ord}_W\psi(w^2((f-cw)/w)_Y)-\mathrm{ord}_W\psi(\widehat{[f-w]})\\ &\qquad\quad(\text{since } \mathcal{V}(f-cw)\cap\mathcal{V}(\widehat{[f,w]}/[f-cw])=\emptyset)\\ &\quad= \mathrm{ord}_W\psi(f')\\ &\quad= I(X-u,f';Q)\end{aligned}$$

and hence, with all terms integers, we have

(iii) $$I(X-u, w^2(\overline{f}/w)_Y; Q) = I(X-u, f'; Q).$$

Now, upon letting $\widetilde{\sum}$ and $\widehat{\sum}$ stand for summations over $\{Q=(u,v)\in \mathcal{A} : f(u,v)-cw(u,v)=0\neq w(u,v)\}$ and $\{Q=(u,v)\in\mathcal{A} : f(u,v)-cw(u,v)=0\}$ respectively, we have, with all terms integers,

$$\begin{aligned}
&\text{RHS of (10.5)}\\
&= 1-N+\deg_Y[f-cw]+\sum_{1\le i\le t}\left(I(X-u_i,\overline{f};Q_i)-1\right)\\
&\qquad +\widetilde{\sum} I(X-u,w^2(\overline{f}/w)_Y;Q)\\
&\qquad\qquad \text{(by (i), (ii) and (iii))}\\
&= 1-\deg_Y\overline{f}+\widehat{\sum}\left[\overline{\chi}(\overline{f};Q)+\sum_{1\le i\le\overline{s}}\ \sum_{V\in\mathfrak{R}(\overline{f}_i,Q)}\operatorname{ord}_V d\overline{\phi}_i(X)\right]\\
&\qquad \text{(by taking } \overline{f} \text{ for } f \text{ in (II) and writing } \overline{f}=\overline{f}_1\dots\overline{f}_{\overline{s}} \text{ with}\\
&\qquad\quad \text{irreducible } Y\text{-monic members } \overline{f}_1,\dots,\overline{f}_{\overline{s}} \text{ of } R\setminus k \text{ and}\\
&\qquad\quad \text{letting } \overline{\phi}_i : R\to R/(\overline{f}_iR) \text{ be the canonical epimorphism)}\\
&= \rho_a(\overline{f})\\
&\quad \text{(by (IV))}\\
&= \text{LHS of (10.5)}.
\end{aligned}$$

Proof of (10.7) to (10.12). In view of (10.4), as in the proof of (9.7) to (9.12), while proving (10.7) to (10.12) we may and **we shall assume that** (f,w) **satisfies (**).** Take

$$f'=\frac{f_Yw-fw_Y}{\widehat{[f,w]}}\qquad\text{and}\qquad f''=f_Xw-fw_X.$$

Let

$$a(f,w)=k\cap\operatorname{singset}(f,w)\text{ and } b(f,w)=\operatorname{multset}(f,w)^*\cup\operatorname{conset}(f,w)^*.$$

Now in the RHS of (10.5), for all $c\in k\setminus\operatorname{multset}(f,w)^*$ the second line is zero, for $c\in k\setminus D((f,w),f')$ the third line attains a maximum, for $c\in\operatorname{singset}(f,w)$ the fourth line is zero, and for $c\in k\setminus b(f,w)$ the fifth line attains a minimum. This proves that

(1) $$\operatorname{defset}(f,w)\subset\left(D((f,w),f')\cup\operatorname{singset}(f,w)\cup\operatorname{conset}(f,w)^*\right)$$

and establishes (10.7) and (10.8). Clearly (10.9) is evident from (10.8). Also note that if $c \in \text{singset}(f,w)\setminus b(f,w)$ then the above consideration of the RHS lines of (10.5) establishes that $c \in \text{defset}(f,w)$, and hence

$$(2) \qquad (a(f,w)\setminus b(f,w)) \subset \text{defset}(f,w)$$

which proves part (i) of (10.12). Clearly $|\text{conset}(f,w)^*| \le t$ and hence by (10.9) and assuming (10.10) we get

(3)

$$|\text{defset}(f,w)| \le \rho_a(f)+\rho_a(w)+t-1+|\text{multset}(f,w)^*|+|\text{conset}(f.w)^*|+2$$

where the last number 2 is added for possible $0,\infty$ in $\text{defset}(f,w)$ not accounted by (10.10). This gives

$$(4) \qquad |\text{defset}(f,w)| \le \rho_a(f)+\rho_a(w)+\deg_Y \widehat{[f,w]}+2t+1$$

showing that (10.10) $\Rightarrow$ part (ii) of (10.12). By (2) we see that we get

$$|\text{singset}(f,w)| \le |\text{defset}(f,w)| + \deg_Y \widehat{[f,w]} + t + 1$$

and hence by (4) we get

$$(5) \qquad |\text{singset}(f,w)| \le \rho_a(f)+\rho_a(w)+2\deg_Y \widehat{[f,w]}+3t+2$$

showing that (10.10) $\Rightarrow$ part (iii) of (10.12). Thus, it only remains to prove (10.10). We shall do this in STEPS (6) to (12).

STEP (6). Using the proof of (10.5) we shall now show that:

$$(10.5^*) \quad (***) \Rightarrow \begin{cases} \rho_a(w) & = (1-M)+\deg_Y[w] \\ & \quad +I\,(w,f';w\setminus f) - I\,(f'',f';w\setminus f) \\ & \quad +\sum_{1\le i\le t}\left(I\left(X-u_i,\frac{w}{[w]};Q_i\right)-1\right) \end{cases}$$

where condition

$$(***) \qquad f \text{ and } w \text{ are } Y\text{-monic of positive } Y\text{-degree } N \neq M.$$

is obviously weaker than condition (**). So let $\overline{w} = \text{rad}(w)$. Then by the argument in the proof of the Singset Theorem we see that: $[w] = \gcd(w, \widehat{[f,w]})$ and $w = [w]\overline{w}$.

The idea of the proof is to redo the calculations in (10.5) reversing the role of f,w and for this purpose, let us note that under our current notation, we have:

$$\text{(i)} \qquad w^2(f/w)_Y = -f^2(w/f)_Y \text{ and } w^2(f/w)_X = -f^2(w/f)_X$$

Note that our arguments in (I) thru (IV) remain valid under exchange of f,w, if we change the degree condition (**) by the weaker condition (***). Indeed, the degrees, N, M never enter the calculations in (I) to (IV).

For $Q = (u,v) \in \mathcal{A}$ with $w(u,v) = 0 \neq f(u,v)$, by using calculations of (I) we get

$$\text{(ii)} \qquad I(w, f'; Q) - I(f'', f'; Q) = I(X - u, f'; Q)$$

with all terms integers. Let $\psi : R \to R/(X-u)R$ be the canonical epimorphism, and let W be the localization of $\psi(R)$ at $\psi((Y-v)R)$. Then we have, in view of (i) and with all terms integers,

$$\begin{aligned}
I(X-u, f^2(\overline{w}/f)_Y; Q) &= \operatorname{ord}_W \psi(f) + \operatorname{ord}_W \psi(\overline{w}) - 1 \\
&= \operatorname{ord}_W \psi(\overline{w}) - 1 \\
&\qquad (\text{since } f(u,v) \neq 0) \\
&= \operatorname{ord}_W \psi(w) - \operatorname{ord}_W \psi([w]) - 1 \\
&= \operatorname{ord}_W \psi(f^2(w/f)_Y) - \operatorname{ord}_W \psi([w]) \\
&= \operatorname{ord}_W \psi(f^2(w/f)_Y) - \operatorname{ord}_W \psi(\widehat{[f,w]}) \\
&\qquad (\text{since } \mathcal{V}(w) \cap \mathcal{V}(\widehat{[f,w]}/[w]) = \emptyset \\
&= \operatorname{ord}_W \psi(f') \\
&= I(X-u, f'; Q)
\end{aligned}$$

and hence, with all terms integers, we have

$$\text{(iii)} \qquad I(X-u, f^2(\overline{w}/f)_Y; Q) = I(X-u, f'; Q).$$

Now, upon letting $\widetilde{\sum}$ and $\widehat{\sum}$ stand for summations over

$$\{Q = (u,v) \in \mathcal{A} : w(u,v) = 0 \neq f(u,v)\} \quad \text{and}$$
$$\{Q = (u,v) \in \mathcal{A} : w(u,v) = 0\}$$

respectively, we have, with all terms integers,

RHS of (10.5*)

$$= 1 - M + \deg_Y[w] + \sum_{1\le i\le t} (I(X - u_i, \overline{w}; Q_i) - 1)$$

$$+ \widetilde{\sum} I(X - u, f^2(\overline{w}/f)_Y; Q)$$

(by (ii) and (iii))

$$= 1 - \deg_Y \overline{w} + \widehat{\sum} \left[\overline{\chi}(\overline{w}; Q) + \sum_{1\le i\le \overline{s}} \sum_{V\in\mathfrak{R}(\overline{w}_i,Q)} \mathrm{ord}_V d\overline{\phi}_i(X) \right]$$

(by taking $\overline{w}$ for w in (II) and writing $\overline{w} = \overline{w}_1 \dots \overline{w}_{\overline{s}}$ with irreducible Y-monic members $\overline{w}_1, \dots, \overline{w}_{\overline{s}}$ of $R \setminus k$ and letting $\overline{\phi}_i : R \to R/(\overline{w}_i R)$ be the canonical epimorphism)

$$= \rho_a(\overline{w})$$

(by (IV))

$= \mathrm{RHS}$ of (10.5)*.

Step (7). Combining (10.5*) with (10.5) we see that for any $c \in k$ we have:

$$\rho_a(f - cw) + \rho_a(w) + t - 1 = \sum_{1\le j\le 4} F_j(c) \tag{7*}$$

where

$$F_1(c) = (1 - N - M) + \deg_Y[f - cw] + \deg_Y[w]$$

and

$$F_2(c) = I\left((f - cw)w, f'; (f - cw)w\right) - I\left(f'', f'; (f - cw)w\right)$$

and

$$F_3(c) = \sum_{1\le l\le t} (I(f'', f'; Q_l) - I((f - cw)w, f'; Q_l))$$

and

$$F_4(c) = \sum_{1\le l\le t} \left(I\left(X - u_l, \frac{(f - cw)w}{[f - cw][w]}; Q_l\right) - 1\right).$$

Fix some $c_\pi \in k$ such that $\rho_a(f - c_\pi w) = \rho_\pi(f, w)$ and such that c_π gives the various extremal values as described in (10.8). Explicitly, we assume that c_π is chosen so that it satisfies the following additional conditions (i) to (iv).

(i) For each $V \in \mathfrak{R}(f'_i, \infty)$, with $1 \le i \le s'$, we have $\text{ord}_V(\phi'_i(f - c_\pi w)) = \min(\text{ord}_V(\phi'_i(f), \text{ord}_V(\phi'_i(w)).$

(ii) For each $V \in \mathfrak{R}(f'_i, Q_l)$, with $1 \le i \le s'$ and $1 \le l \le t$, we have $\text{ord}_V(\phi'_i(f - c_\pi w) = \min(\text{ord}_V(\phi'_i(f), \text{ord}_V(\phi'_i(w)).$

(iii) For $1 \le l \le t$ we have $I(X - u_i, f - c_\pi w; Q_i) = \min(I(X - u_i, f; Q_i), I(X - u_i, w; Q_i)).$

(iv) $c_\pi \notin \text{defset}(f, w)$ and hence in particular $I(f'', f'; \ (f - c_\pi w) \setminus w) = 0$ by the General Singset Theorem.

For various numerical functions $F(c)$ to be considered, with c varying in k, let $H(F(c))$ denote the variation $\sum_{c \in D}(F(c_\pi) - F(c))$ where D is a finite subset of k which is defined below and which is a large enough "defset" to be applicable to all the relevant F's.

(i*) For each $V \in \mathfrak{R}(f'_i, \infty)$, with $1 \le i \le s'$, we define $c_V \in k$ thus. In case $\text{ord}_V \phi'_i(f) < \text{ord}_V \phi'_i(w)$, we take $c_V = c_\pi$. In case $\text{ord}_V \phi'_i(f) \ge \text{ord}_V \phi'_i(w)$, we take c_V to be the unique element of k such that $\text{ord}_V \phi'_i(f - c_V w) > \text{ord}_V \phi'_i(f - c_\pi w)$. Let D_{i} be the set of all these elements c_V.

(ii*) For each $V \in \mathfrak{R}(f'_i, Q_l)$, with $1 \le i \le s'$ and $1 \le l \le t$, we define $c_V \in k$ thus. In case $\text{ord}_V \phi'_i(f) < \text{ord}_V \phi'_i(w)$, we take $c_V = c_\pi$. In case $\text{ord}_V \phi'_i(f) \ge \text{ord}_V \phi'_i(w)$, we take c_V to be the unique element of k such that $\text{ord}_V \phi'_i(f - c_V w) > \text{ord}_V \phi'_i(f - c_\pi w)$. Let D_{ii} be the set of all these elements c_V.

(iii*) For $1 \le l \le t$ we define $c_l \in k$ thus.
In case $I(X - u_l, f - c_\pi w; Q_l) = \max\{I(X - u_l, f - cw; Q_l) : c \in k\}$, we take $c_l = c_\pi$.
In case $I(X - u_l, f - c_\pi w; Q_l) < \max\{I(X - u_l, f - cw; Q_l) : c \in k\}$, we take c_l to be the unique element of k such that $I(X - u_l, f - c_l w; Q_l) = \max\{I(X - u_l, f - cw; Q_l) : c \in k\}$. Let $D_{\text{iii}} = \{c_1, \ldots, c_t\}$.

(iv*) Let D_{iv} be the union of $\text{defset}(f, w) \cap k$ and $\text{singset}(f, w)$.

Let $D = D_{\text{i}} \cup D_{\text{ii}} \cup D_{\text{iii}} \cup D_{\text{iv}}$.

Now, in view of (7*), equation (10.10) is equivalent to the equation

$$\rho_a(f - c_\pi w) + \rho_a(w) + t - 1 = \sum_{1 \le j \le 4} H(F_j(c))$$

and hence to the equation

$$\text{(10.10*)} \qquad \sum_{1 \le j \le 4} H(F_j(c)) = \sum_{1 \le j \le 4} F_j(c_\pi).$$

In words, (10.10) says that the function* $\sum_{1 \le j \le 4} F_j(c)$, *or equivalently the function* $\rho_a(f - cw) + \rho_a(w) + t - 1$, *replicates itself, i.e.,* *it has a constant value at most points and that value equals the total variation of the function.*

STEP (8). In view of (9.6) and (10.6) we see that

$$(8^*) \qquad H(F_1(c))) = -\deg_Y \widehat{[f,w]} + \deg_Y [w].$$

STEP (9). For each $V \in \mathfrak{R}(f_i', \infty)$, with $1 \le i \le s'$, we clearly have

$$(9a) \qquad -\mathrm{ord}_V \phi_i'(f'') = \mathrm{ord}_V \phi_i'(X) - \mathrm{ord}_V \phi_i'((f - c_V w)w).$$

Now

$$\begin{cases} \mathrm{ord}_V \phi_i'(f) < \mathrm{ord}_V \phi_i'(w) \\ \Rightarrow \mathrm{ord}_V \phi_i'((f - cw)w) = \mathrm{ord}_V \phi_i'(fw) \text{ for all } c \in k \end{cases}$$

and thus in this case, using (9a), we get

$$\begin{aligned} &- H\left(\mathrm{ord}_V \phi_i'((f - cw)w)\right) - \mathrm{ord}_V \phi_i'(f'') \\ &= 0 - \mathrm{ord}_V \phi_i'(f'') \\ &= \mathrm{ord}_V \phi_i'(X) - \mathrm{ord}_V \phi_i'((f - c_\pi w)w). \end{aligned}$$

Likewise

$$\begin{cases} \mathrm{ord}_V \phi_i'(f) \ge \mathrm{ord}_V \phi_i'(w) \\ \Rightarrow \mathrm{ord}_V \phi_i'((f - cw)w) \\ = \mathrm{ord}_V \phi_i'(fw) \text{ except for exactly one } c = c_V \in k \end{cases}$$

and thus in this case, again using (9a), we get

$$\begin{aligned} &- H\left(\mathrm{ord}_V \phi_i'((f - cw)w)\right) - \mathrm{ord}_V \phi_i'(f'') \\ &= -\mathrm{ord}_V (\phi_i'((f - c_\pi w)w)) + \mathrm{ord}_V (\phi_i'((f - c_V w)w)) - \mathrm{ord}_V \phi_i'(f'') \\ &= \ \mathrm{ord}_V \phi_i'(X) - \mathrm{ord}_V \phi_i'((f - c_\pi w)w). \end{aligned}$$

Consequently we always have

$$(9b) \qquad \begin{cases} -H\left(\mathrm{ord}_V \phi_i'((f - cw)w)\right) - \mathrm{ord}_V \phi_i'(f'') \\ = \mathrm{ord}_V \phi_i'(X) - \mathrm{ord}_V \phi_i'((f - c_\pi w)w). \end{cases}$$

Clearly

$$(9c) \qquad \begin{aligned} H(F_2(c)) = H(I\left((f - cw)w, f'; (f - cw)w\right)) \\ -H(I\left(f'', f'; (f - cw)w\right)). \end{aligned}$$

Since $I\left(f'', f'; (f - c_\pi w)w \setminus w\right) = 0$ by our choice of c_π, we also have

$$(9d) \qquad I\left(f'', f'; (f - c_\pi w)w\right) = I\left(f'', f'; w\right).$$

Now

$$\begin{aligned}
&H(I(f'',f';(f-cw)w))\\
&=\sum_{c\in D}(I(f'',f';(f-c_\pi w)w)-I(f'',f';(f-cw)w))\\
&=-\sum_{c\in D}I(f'',f';(f-cw)w\setminus w)\quad\text{by (9d)}\\
&=-I((f'',f';\mathcal{A}\setminus w))\\
&=I(f'',f';w)-I(f'',f';\mathcal{A})\\
&=I(f'',f';w)+\sum_{1\le i\le s'}\sum_{V\in\mathfrak{R}(f_i',\infty)}\operatorname{ord}_V\phi_i'(f'')
\end{aligned}$$

and

$$\begin{aligned}
&H(I\left((f-cw)w,f';(f-cw)w\right))\\
&=-\sum_{1\le i\le s'}\sum_{V\in\mathfrak{R}(f_i',\infty)}H(\operatorname{ord}_V\phi_i'((f-cw)w)\\
&=\sum_{1\le i\le s'}\sum_{V\in\mathfrak{R}(f_i',\infty)}[\operatorname{ord}_V\phi_i'(f'')+\operatorname{ord}_V\phi_i'(X)-\operatorname{ord}_V\phi_i'((f-c_\pi w)w)]\\
&\qquad\text{by (9b)}
\end{aligned}$$

and hence by (9c) we get

$$\begin{aligned}
H(F_2(c))=&-\deg_Y(f')-I(f'',f';w)\\
&-\sum_{1\le i\le s'}\sum_{V\in\mathfrak{R}(f_i',\infty)}\operatorname{ord}_V\phi_i'((f-c_\pi w)w).
\end{aligned}$$

Clearly have $\deg_Y(f')=\deg_Y(f_Y w-fw_Y)-\deg_Y(\widehat{[f,w]})$ and, since $N\neq M$, we also have

$$\deg_Y(f')=N+M-1-\deg_Y(\widehat{[f,w]}).$$

Combining the above two displayed equations we conclude that

$$\begin{aligned}
H(F_2(c))=&1-N-M+\deg_Y\widehat{[f,w]}-I(f'',f';w)\\
&-\sum_{1\le i\le s'}\sum_{V\in\mathfrak{R}(f_i',\infty)}\operatorname{ord}_V\phi_i'((f-c_\pi w)w)
\end{aligned}$$

where the last line is clearly equal to $I((f-c_\pi w)w, f', \mathcal{A})$ and hence by using (8*) we get

(9e)

$$\begin{cases} H(F_1(c)) + H(F_2(c)) \\ = 1 - N - M + \deg_Y[w] - I(f'', f'; w) + I((f-c_\pi w)w, f'; \mathcal{A}). \end{cases}$$

Clearly $\deg_Y[f-c_\pi w] = 0$ and hence by (9d) and (9e) we conclude that

$$H(F_1(c)) + H(F_2(c)) = F_1(c_\pi) + F_2(c_\pi). \tag{9*}$$

STEP (10). Upon letting

$$F_{3,l}(c) = I(f'', f'; Q_i) - I((f-cw)w, f'; Q_l)$$

we get

$$\begin{aligned} &H(F_{3,l}(c)) \\ &= -H(I((f-cw)w, f'; Q_l)) \\ &= -\sum_{1\le i\le s'} \sum_{V\in\mathfrak{R}(f_i', Q_l)} H(\operatorname{ord}_V \phi_i'((f-cw)w) \\ &= \sum_{1\le i\le s'} \sum_{V\in\mathfrak{R}(f_i', Q_l)} \big[\operatorname{ord}_V \phi_i'(X-u_l) - \operatorname{ord}_V \phi_i'((f-c_\pi w)w)) \\ &\qquad\qquad + \operatorname{ord}_V \phi_i'(f'')\big] \qquad \text{as in (9a)} \\ &= I(X-u_l, f'; Q_l) - I((f-c_\pi w)w, f'; Q_l) + I(f'', f'; Q_l) \end{aligned}$$

and hence

$$\begin{cases} F_3(c) = \sum_{1\le l\le t} F_{3,l}(c) \text{ with} \\ H(F_{3,l}(c)) = F_{3,l}(c_\pi) + I(X-u_l, f'; Q_l). \end{cases} \tag{10*}$$

STEP (11). Upon letting

$$F_{4,l}(c) = \left(I\left(X-u_l, \frac{(f-cw)w}{[f-cw][w]}; Q_l\right) - 1\right)$$

we clearly have

$$\begin{cases} F_{4,l}(c) = & I(X-u_l, f-cw; Q_l) + I(X-u_l, w; Q_l) - 1 \\ & -I(X-u_l, [f-cw]; Q_l) - I(X-u_l, [w]; Q_l). \end{cases} \tag{11a}$$

Let $\mu_l = I(X - u_l, f - c_\pi w; Q_l)$ and $\theta_l = \max\{I(X - u_l, f - cw; Q_l) : c \in k\}$ and $\nu_l = I(X - u_l, w; Q_l)$. Then

$$\begin{aligned}
&H(F_{4,l}(c))\\
&= H(I(X - u_l, f - cw; Q_l)) - H(I(X - u_l, [f - cw]; Q_l))\\
&= \mu_l - \theta_l - H(I(X - u_l, [f - cw]; Q_l))\\
&= \mu_l - \theta_l - I(X - u_l, [w]; Q_l) + I(X - u_l, \widehat{[f, w]}; Q_l)\\
&= \mu_l - \theta_l - I(X - u_l, [w]; Q_l) + [\nu_l - 1 + \theta_l - I(X - u_l, f'; Q_l)]\\
&= I(X - u_l, (f - c_\pi w)w; Q_l) - 1 - I(X - u_l, f'; Q_l)\\
&\qquad\qquad\qquad\qquad - I(X - u_l, [w]; Q_l)
\end{aligned}$$

and hence, because of (11a) and the obvious fact that $I(X - u_l, [f - c_\pi w]; Q_l) = 0$, we get

$$\text{(11b)} \qquad \begin{cases} F_4(c) = \sum_{1 \le l \le t} F_{4,l}(c) \text{ with} \\ H(F_{4,l}(c)) = F_{4,l}(c_\pi) - I(X - u_l, f'; Q_l). \end{cases}$$

By (10*) and (11b) we see that

$$\text{(11*)} \qquad H(F_3(c)) + H(F_4(c)) = F_3(c_\pi) + F_4(c_\pi).$$

STEP (12). By (9*) and (11*) we get (10.10*).

Conclusion. Genus plus excess branch number is the rank of a curve. The total variation of the rank as a curve moves thru a pencil is independent of the pencil. A suitable modification of this variation is the Zeuthen-Segre invariant of the surface. For the plane it equals minus one. The defset of a polynomial is the set of translation constants which produce nongeneral rank. The defset gives a bound on the singset, i.e., the set of translation constants which produce singular curves or more generally singular hypersurfaces. The redset of a polynomial is the set of translation constants which produce reducible hypersurfaces. Bounds for the redset are found in terms of the group of units of the affine coordinate ring.

11. Linear Systems and Pencils on Normal Varieties

To say a word about the Zeuthen-Segre invariant of a surface, let us very briefly talk about linear systems and pencils on normal varieties.

So assume $k = k^*$, let E be an irreducible n-dimensional normal algebraic variety over k, for $0 \le i \le n$ let E_i be the set of all irreducible i-dimensional subvarieties of E, for any $C \in \cup_{0 \le i \le n} E_i$ let

$k(C)$ be the local ring of C on E, let $\widehat{E}_i = \{k(C) : C \in E_{n-i}\}$, and finally let $\widehat{E} = \cup_{0\le i\le n}\widehat{E}_i$. Then, in the language of models (see [**A09**]), $\widehat{E}_i$ is the set of all i-dimensional members of the n-dimensional normal model $\widehat{E}$ of $k(E)/k$.

Recall that a premodel $\widehat{E}$ of a finitely generated field extension K/k is a collection of local domains whose quotient field is K and which have k as a subring; $\widehat{E}$ is irredundant (resp: complete) means any valuation ring of K/k dominates at most (resp: at least) one of its members; $\widehat{E}$ is a model if it is an irredundant premodel which can be expressed as a finite union $\widehat{E} = \cup_{0\le j\le m}\mathfrak{V}(B_j)$ where $B_j = k[x_{j0}, \dots, x_{jm}]$ = an affine domain over k and where $\mathfrak{V}(B_j)$ = the set of all localizations $(B_j)_P$ with P varying over spec(B_j). The normality assumption says that there is an injection $R \to B_j$ such that B_j is the integral closure of the image in a finite algebraic field extension of the quotient field of the image, or equivalently that every member of $\widehat{E}$ is normal. The normality assumption implies that $\widehat{E}_1$ is a set of DVRs of $k(E)/k = K/k$. Recall that E or $\widehat{E}$ is nonsingular means every member of $\widehat{E}$ is a regular local ring, and so nonsingular $\Rightarrow$ normal. If m can be taken to be 0 then we call $\widehat{E}$ (resp: E) to be an affine model (resp: affine variety). If we can find nonzero elements $z_0, \dots, z_m$ in $k(E)$ such that $x_{ji} = z_i/z_j$ for all i, j in $\{0, \dots, m\}$ then we call $\widehat{E}$ (resp: E) to be a projective model (resp: projective variety).

Now any $C \in E_{n-i}$ can be recovered from $k(C) \in \widehat{E}_i$ by observing that (closed) points P in C are characterized by saying that $k(P)$ are those members of $\widehat{E}_n$ for which $k(C)$ belongs to $\mathfrak{V}(k(P))$. Thus we may dispense with the geometric beginning of commencing with an algebraic variety, and start (and end) with a model. This economy of thought is the beauty of the language of models.

Let $\mathcal{D}$ be the group of all divisors on E, i.e., the set of all maps $E_{n-1} \to \mathbb{Z}$ with finite support. Let $\mathcal{D}_+$ be the set of all effective divisors, i.e., nonzero divisors D with $D(E_{n-1}) \subset \mathbb{N}$ = the set of all nonnegative integers. The degree deg(D) of any divisor D is defined to be $\sum D(C)$ taken over all C in E_{n-1}. The divisor (z) of any $z \in k(S)^\times$ is defined by the equation $(z)(C) = \mathrm{ord}_{k(C)}z$. For any k-vector subspace H of $k(E)$, by $\mathbb{P}(H)$ we denote the associated projective space, i.e., the set of all 1-dimensional subspaces of H; for any nonzero z, z' in any $y \in \mathbb{P}(H)$ we clearly have $(z) = (z')$ and this common divisor is denoted by (y). By a linear system on E we mean a subset

$\mathcal{C}$ of $\mathcal{D}_+$ for which there exists a finite dimensional k-vector-subspace H of $k(E)$ together with $D' \in \mathcal{D}$ such that $y \mapsto (y) + D'$ gives a bijection $\mathbb{P}(H) \to \mathcal{C}$, and for which there does not exist $D_0 \in \mathcal{D}_+$ such that for all $D \in \mathcal{C}$ we have $D \geq D_0$; the second proviso means that we assume our linear systems to be devoid of fixed components. It is easily seen that the dimension of H depends only on $\mathcal{C}$, and we call this dimension decreased by one to be the dimension of $\mathcal{C}$. If the dimension of $\mathcal{C}$ is one then we call $\mathcal{C}$ a pencil.

Now assume that k is of characteristic zero with $n = 2$, and E is an irreducible nonsingular projective algebraic surface. For any $C \in E_1$ let $\gamma(C)$ be its genus, and let us generalize this to any $D \in \mathcal{D}_+$ by putting

$$\gamma(D) = 1 + \sum_{C \in E_1} D(C)(\gamma(C) - 1).$$

For any $Q \in E_0$ the completion $\widetilde{Q}$ of $k(Q)$ is clearly isomorphic to $k[[X, Y]]$. For any $D \in \mathcal{D}_+$ we let $\chi(D; Q)$ denote the number of branches of D at Q, i.e., upon letting M stand for maximal ideal,

$$\left(\prod\nolimits_{\{C \in E_1 : D(C) > 0 \text{ and } k(Q) \subset k(C)\}} [k(Q) \cap M(k(C))]^{D(C)}\right) \widetilde{Q} = U_0 U_1 \dots U_{\chi(D;Q)} \widetilde{Q}$$

where U_0 is a unit in $\widetilde{Q}$, and $U_1, \dots, U_{\chi(D;Q)}$ are irreducible nonunits in $\widetilde{Q}^\times$. Let

$$\overline{\chi}(D; Q) = \chi(D; Q) - 1$$

and put

$$\overline{\chi}(D; E) = \sum_{\{Q \in E_0 : \chi(D;Q) > 1\}} \overline{\chi}(D : Q).$$

Let $\mathrm{rad}(D) \in \mathcal{D}_+$ be defined by

$$(\mathrm{rad}(D))(C) = \begin{cases} 1 & \text{if } D(C) > 0 \\ 0 & \text{if } D(C) = 0. \end{cases}$$

Also put

$$\rho_a(D) = 2\gamma(\mathrm{rad}(D)) + \overline{\chi}(\mathrm{rad}(D); E).$$

Finally define the base point set of a pencil $\mathcal{C}$ on E by putting

$$B(\mathcal{C}) = \cap_{D \in \mathcal{C}} S_0(D)$$

where the curve-support and point-support of D are given by

$$S_1(D) = \{C \in E_1 : D(C) \neq 0\} \text{ and}$$
$$S_0(D) = \{Q \in E_0 : k(Q) \subset \cup_{C \in S_1(D)} k(C)\}.$$

If there exists a finite subset defset($\mathcal{C}$) and an integer $\rho_\pi(\mathcal{C})$ such that for any $D \in \mathcal{C}$ we have: $\rho_\pi(\mathcal{C}) = \rho_a(D) \Leftrightarrow D \in \mathcal{C}\setminus\text{defset}(\mathcal{C})$, then these two objects are clearly unique and we call them the **deficiency set** and the **pencil-rank** of the pencil $\mathcal{C}$ on E. When this is so,

we define the Zeuthen-Segre invariant of $\mathcal{C}$ to be the integer $\zeta(\mathcal{C})$ given by

$$\zeta(\mathcal{C}) = -|B(\mathcal{C})| - 2\rho_\pi(\mathcal{C}) + \sum_{D\in\text{defset}(\mathcal{C})} [\rho_\pi(\mathcal{C}) - \rho_a(D)].$$

EPILOGUE. Assume it has been shown that defset($\mathcal{C}$) and $\rho_\pi(\mathcal{C})$ exist for every pencil on E, and $\zeta(\mathcal{C})$ depends only on E and not on $\mathcal{C}$. Let $\zeta(E)$ be the Zeuthen-Segre invariant of the surface E, i.e., the common value of $\zeta(\mathcal{C})$ for all pencils $\mathcal{C}$ on E. Taking any two distinct members F and G of $\mathcal{C}$, and adding $\rho_a(F) + \rho_a(G) - \zeta(E)$ to both sides of the above equation we get the jungian formula

$$\rho_a(F) + \rho_a(G) = -\zeta(E) - |B(\mathcal{C})| + \sum_{D\in\text{defset}(\mathcal{C})\setminus\{F,G\}} [\rho_\pi(\mathcal{C}) - \rho_a(D)].$$

Thinking of F and G as "curves on the surface" E, their "common points" or "set-theoretic intersection" is given by $I^*(F,G) = S_0(F) \cap S_0(G)$, and clearly we have $B(\mathcal{C}) = I^*(F,G)$. Moreover, F and G have no common component, i.e., there is no $C \in E_1$ with $F(C) \neq 0 \neq G(C)$. Also $\mathcal{C}$ is "generated" by F and G; so every member of $\mathcal{C}$ can symbolically be written as $F - cG$ with $c \in k \cup \{\infty\}$ with $F - \infty G = G$. In the situation of the previous section, this symbolism becomes more real by taking E to be the projective plane $\mathcal{P}$ over k, and taking $F = Z^d f(X/Z, Y/Z)$ and $G = Z^d w(X/Z, Y/Z)$. As a thought for the future, going back to the section on More General Pencils, ρ_π of the pencil could be defined as ρ_a of the generic member $(f,w)^\flat$ with affine coordinate ring $k(f/w)[X,Y]$ over ground field $k(f/w)$. Similar trick could be played when E is any nonsingular surface.

References

[A01] S. S. Abhyankar, *On the valuations centered in a local domain,* American Journal of Mathematics, vol. 78 (1956), pages 321-348.

[A02] S. S. Abhyankar, *Ramification Theoretic Methods in Algebraic Geometry,* Princeton University Press (1959).

[A03] S. S. Abhyankar, *Tame coverings and fundamental groups of algebraic varieties, Parts I to VI,* American Journal of Mathematics, vol. 81 (1959), pages 46-94, vol. 82 (1960), pages 120-178, vol. 82 (1960), pages 179-190, vol. 82 (1960), pages 341-362,, vol. 82 (1960), pages 365-373, vol. 82 (1960), pages 374-388.

[A04] S. S. Abhyankar, *On the semigroup of a meromorphic curve, Part I,* Proceedings of the International Symposium on Algebraic Geometry, Kyoto (1977), pages 240-414.

[A05] S. S. Abhyankar, *Lectures on Expansion Techniques in Algebraic Geometry,* Tata Institute of Fundamental Research, Bombay (1977).

[A06] S. S. Abhyankar, *Algebraic Geometry for Scientists and Engineers,* American Mathematical Society (1990).

[A07] S. S. Abhyankar, *Galois theory on the line in nonzero characteristic,* Bulletin of the American Mathematical Society, vol. 27 (1992), pages 68-133.

[A08] S. S. Abhyankar, *Some remarks on the jacobian question,* Proceedings of the Indian Academy of Sciences, vol. 104 (1994), pages 515-542.

[A09] S. S. Abhyankar, *Resolution of Singularities of Embedded Algebraic Surfaces,* Springer-Verlag (1998).

[A10] S. S. Abhyankar, *Resolution of singularities and modular Galois theory,* Bulletin of the American Mathematical Society, vol. (2001), pages 131-169.

[AbA] S. S. Abhyankar and A. Assi, *Jacobian pairs,* Contemporary Mathematics, To Appear.

[AEH] S. S. Abhyankar, P. Eakin, and W. J. Heinzer, *On the uniqueness of the coefficient ring in a polynomial ring,* Journal of Algebra, vol. 23 (1972), pages 310-342.

[ACL] E. Artal-Bartolo, P. Cassou-Nogues, and I. Luengo-Velasco, On polynomials whose fibres are irreducible with no critical points, Mathematische Annalen, vol. 299 (1994), pages 477-490.

[Ber] E. Bertini. *Sui sistemi lineari,* Rend. Ist. Lomb. II.s, vol. 15 (1882).

[FrJ] M. D. Fried and M. Jarden, *Field Arithmetic,* Springer-Velag (1986).

[GaT] D. Garber and M. Teicher, *The fundamental group's structure of the complement of some configurations of real line arrangements,* Complex Analysis and Algebraic Geometry, Edited by T. Peternell and F. O. Schreyer, Walter de Gruyter, Berlin (2000), pages 173-223.

[Igu] J. Igusa. *On a theorem of Lüroth,* Mem. Coll. Sci., Univ. Kyoto, Ser. A, vol. 26 (1951), pages 251-253.

[Jun] H. W. E. Jung, *Algebraische Flächen,* Hannover (1925).

[Kle] F. Klein. *Lectures on the Icosahedron and the Solution of Equations of the Fifth Degree,* Dover Reprint (1956).

[Lur] A. Lüroth, *Beweis eines Satzes über rationale Curven,* Math. Annalen, vol. 9 (1875), page 135.

[Mat] T. Matsusaka, *The theorem of Bertini on linear systems in modular fields,* Memoirs of the College of Science, University of Kyoto, vol. 26 (1950), pages 51-62.

[Na1] M. Nagata. *Local Rings,* Interscience (1962).

[Na2] M. Nagata. *A theorem on valuation rings and its applications,* Nagoya Math. Journal, vol. 29 (1967), pages 85-91.

[Za1] O. Zariski, *Pencils on an algebraic variety and a new proof of a theorem of Bertini,* Transactions of the American Mathematical Society, vol. 50 (1941), pages 48-70.

[Za2] O. Zariski, *The theorem of Bertini on the variable singular points of a linear system of varieties,* Transactions of the American Mathematical Society, vol. 56 (1944), pages 130-140.

[Za3] O. Zariski, *Introduction to the Problem of Minimal Models in the Theory of Algebaic Surfaces,* Publication 4 of the Mathematical Society of Japan (1958).

[ZaS] O. Zariski and P. Samuel, *Commutative Algebra,* Vols I and II, Van Nostrand (1960).

Mathematics Department, Purdue University
West Lafayette, IN 47907, U.S.A.
E-mail address: ram@cs.purdue.edu

Mathematics Department, Purdue University
West Lafayette, IN 47907, U.S.A.
E-mail address: heinzer@math.purdue.edu

Mathematics Department, University of Kentucky
Lexington, KY 40506, U.S.A.
E-mail address: sohum@ms.uky.edu

Orbits of Certain Endomorphisms of Nilmanifolds and Hausdorff Dimension

C.S.Aravinda and P.Sankaran

Dedicated to Professor C.S.Seshadri on the occasion of his seventieth birthday

Abstract. Let R be an element of $GL(n, \mathbb{R})$ having integer entries and let $\rho : \mathbb{T}^n \longrightarrow \mathbb{T}^n$ denote the induced map on the torus $\mathbb{T}^n = \mathbb{R}^n/\mathbb{Z}^n$. It is well known that ρ is ergodic with respect to the Haar measure on $\mathbb{T}^n$ if and only if none of the eigenvalues of R is a root of unity. Dani has shown that there exists a subset S of $\mathbb{T}^n$ such that for any $x \in S$ and any semisimple surjective endomorphism ρ of $\mathbb{T}^n$ such that the corresponding linear endomorphism has no eigenvalue on the unit circle, the closure of the orbit $\{\rho^k(x) | k \geq 0\}$ contains no periodic points and that the set S is 'large' in the sense that for any nonempty open set U of $\mathbb{T}^n$ the set $U \cap S$ has Hausdorff dimension n. In this paper, we shall prove an analogous result for certain endomorphisms of nilmanifolds and infranil manifolds.

1. Introduction

Let N be a simply connected nilpotent Lie group. Assume that N has a discrete subgroup Λ so that $M = N/\Lambda$ is compact. (cf. Theorem 2.12 and Remark 2.14, [**8**].) Then $\Lambda \cap N^1$, where $N^1 = [N, N]$, is a uniform lattice in N^1. Let $\overline{\mathbb{T}} = \overline{N}/\overline{\Lambda} \cong N/(N^1 \cdot \Lambda)$ where $\overline{N} = N/N^1$ is the abelianization of N and $\overline{\Lambda}$ is the image of Λ under the natural projection $N \longrightarrow \overline{N}$. One has a smooth bundle projection $p : M \longrightarrow \overline{\mathbb{T}}$ with fibre $N^1/\Lambda \cap N^1$ which is again a nilmanifold. We put the Riemannian metric on M induced by a right invariant metric on N.

Any automorphism R of N such that $R(\Lambda) \subset \Lambda$ induces an automorphism $\overline{R}$ of $\overline{N}$ such that $\overline{R}(\overline{\Lambda}) \subset \overline{\Lambda}$. R and $\overline{R}$ induce self-maps

$\rho : M \longrightarrow M$ and $\overline{\rho} : \overline{\mathbb{T}} \longrightarrow \overline{\mathbb{T}}$. The map $\rho : M \longrightarrow M$ is a bundle map covering the map $\overline{\rho} : \overline{\mathbb{T}} \longrightarrow \overline{\mathbb{T}}$. We call ρ an endomorphism of M. A theorem of W.Parry [**7**] says that ρ is ergodic if and only if $\overline{\rho}$ is, that is, if and only if none of the eigenvalues of $\overline{R} : \overline{N} \longrightarrow \overline{N}$ is a root of unity.

DEFINITION 1.. Let $\mathcal{C}$ be a collection of self-maps of a topological space X. We say that a nonempty subset $S \subset X$ is *exceptional relative to* $\mathcal{C}$ if the following condition holds: for any $f \in \mathcal{C}$ and any $x \in S$, the closure of the f-orbit $\{f^k(x) | k \geq 0\}$ does not contain any f-periodic point.

When a class $\mathcal{C}$ of self maps of a smooth Riemannian manifold M contains an ergodic map, one expects any exceptional set to be 'small'. Indeed such a set is necessarily of measure zero and is of first category, i.e, its complement contains a residual set. However, relative to the class of all semisimple surjective endomorphisms of the standard n-torus $\mathbb{T}^n$ such that the corresponding linear endomorphisms of $\mathbb{R}^n$ have no eigenvalue on the unit circle, Dani has proved the existence of an exceptional set $S \subset \mathbb{T}^n$ which is 'large' in the sense that for any nonempty open subset U of $\mathbb{T}^n$, the set $S \cap U$ has Hausdorff dimension n. In fact, Dani shows that S is an α-winning set of the Schmidt game (cf. §2).

The purpose of this note is to extend Dani's result to a certain class of endomorphisms of nilmanifolds and of infranil manifolds. Examples of nilmanifolds are U/Γ where U is the group of unipotent upper triangular n-by-n matrices over $\mathbb{R}$ (resp. $\mathbb{C}$) and Γ is the subgroup of those having entries in $\mathbb{Z}$ (resp. in $\mathbb{Z}[i]$). Nilmanifolds occur naturally in Riemannian geometry. For example, they arise in the study of the geometry near a cusp in a finite volume (non-compact) Riemannian manifold whose sectional curvatures are bounded above by a negative constant. See [**5**] for a detailed study of certain nilmanifolds arising this way.

Let $\mathcal{C}_{\mathbb{T}}$ denote the collection of all surjective endomorphisms of an n-torus $\mathbb{T}$ such that the corresponding linear automorphisms of $\mathbb{R}^n$ are semisimple and have no eigenvalue on the unit circle. We prove

THEOREM 2.. Let $M = N/\Lambda$ be an n-dimensional nilmanifold. Let $\mathcal{C}$ be the set of all surjective endomorphisms ρ of M such that $\overline{\rho} \in \mathcal{C}_{\overline{\mathbb{T}}}$. Then there exists a positive number $\mu \leq 1/2$ and an exceptional set $S \subset M$ relative to $\mathcal{C}$ which is α-winning for any $\alpha \in (0, \mu]$. In particular, $S \cap U$ has Hausdorff dimension n for any nonempty open set $U \subset M$.

Let N be a simply connected nilpotent Lie group. Let $Aff(N)$ denote the group of all affine transformations of N which acts on the right of N, that is, $Aff(N)$ is the group generated by the right translations by elements of N and the automorphisms of N. Let Γ be a subgroup of $Aff(N)$ such that action of Γ on N is free and properly discontinuous with the quotient N/Γ being compact. The quotient $\widehat{M} := N/\Gamma$ is called an infranil manifold. It is known that the subgroup $\Lambda \subset \Gamma$ consisting of all right translations of N is a finite index normal subgroup. We shall identify an element of Λ with the image of the identity element $e \in N$ under it. Then it is a result of Auslander [**2**] that Λ is a uniform lattice in N.

Denote by M the nilmanifold N/Λ. One has a finite normal covering $\pi : M \longrightarrow \widehat{M}$ with deck transformation group $G := \Gamma/\Lambda$. Starting with the metric on M induced by the right invariant metric on N, the averaging process leads to a G-invariant metric on M. We put this G-invariant metric on M and the induced metric on $\widehat{M}$ so that π is a local isometry.

Let $R : N \longrightarrow N$ be an automorphism which preserves the Γ action, that is, for any $x \in N$ and $\gamma \in \Gamma$, there exists a $\gamma' \in \Gamma$ such that $R(x\gamma) = R(x)\gamma'$. Then $R(\Lambda) \subset \Lambda$. R induces a self-map $\hat{\rho} : \widehat{M} \longrightarrow \widehat{M}$ which we call an endomorphism of $\widehat{M}$. Note that $\hat{\rho}$ is covered by the endomorphism ρ of M defined by R. We are now ready state

THEOREM 3.. Let $\widehat{\mathcal{C}}$ be the collection of all endomorphisms $\hat{\rho}$ of $\widehat{M}$ induced by automorphisms $R : N \longrightarrow N$ such that the corresponding toral endomorphism $\overline{\rho} : \overline{\mathbb{T}} \longrightarrow \overline{\mathbb{T}}$ is in $\mathcal{C}_{\overline{\mathbb{T}}}$. Then there exist a positive number $\nu \leq 1/2$ and an exceptional set $\widehat{S}$ relative to $\widehat{\mathcal{C}}$ which is α-winning for any $\alpha \in (0, \nu]$. In particular, for any nonempty open set $U \subset \widehat{M}$, the set $\widehat{S} \cap U$ has Hausdorff dimension n.

Remark: Theorem 3 includes an important class of manifolds, namely, compact connected flat Riemannian manifolds; these are the infranil manifolds when the simply connected nilpotent Lie group N is abelian, i.e., $N = \mathbb{R}^n$. Then the compact flat manifolds are precisely those finitely covered by an n-torus $\mathbb{T}^n$.

Our proofs use Dani's result and some basic observations about the nature of pre-image (resp. image) of α-winning sets under a smooth bundle projection (resp. a covering projection).

Acknowledgements: We thank Professor S.G.Dani for his valuable comments on an earlier version of this paper. The authors gratefully acknowledge financial support from Department of Science & Technology, Government of India.

2. The Schmidt game

In this section we give a brief description of the Schmidt game [**10**] and recall here for the benefit of the reader some basic facts about winning sets.

Let X be a complete metric space. Let $0 < \alpha, \beta < 1$ and let S be a subset of X. The Schmidt game on X is played by two players $\mathcal{A}$ and $\mathcal{B}$ as follows: $\mathcal{B}$ first chooses any closed ball B_0 in X of radius r where $r \in \mathbb{R}$ is positive. Then $\mathcal{A}$ chooses a closed ball $A_1 \subset B_0$ of radius αr. Then $\mathcal{B}$ chooses a closed ball $B_1 \subset A_1$ of radius $r_1 = \beta\alpha r$. Inductively, after $\mathcal{B}$ has chosen closed a ball B_k of radius $r_k = (\alpha\beta)^k r$, $k \geq 1$, $\mathcal{A}$ chooses a closed ball $A_{k+1} \subset B_k$ of radius αr_k and $\mathcal{B}$ chooses a closed ball B_{k+1} of radius $r_{k+1} = \beta\alpha r_k$. Since $\lim_{k\to\infty} r_k = 0$, and since X is a complete metric space, $\cap_{k\geq 1} A_k$ is a singleton set $\{x\}$. We say that S is an (α, β)-winning set (for $\mathcal{A}$) if for any choices of $\mathcal{B}$, $\mathcal{A}$ can always make her choices so that $x \in S$. We say that S is α-winning if it is (α, β)-winning for any $\beta \in (0, 1)$. Of course, the whole space X is an α-winning set for any $\alpha \in (0, 1)$. Schmidt has shown that if $\beta \leq 2-\alpha^{-1}$, then the only (α, β)-winning set is X itself. However, if $\beta > 2-\alpha^{-1}$, then there can be proper subsets of X which are (α, β)-winning. Schmidt introduced these notions in his study of badly approximable numbers and related Diophantine problems. Schmidt games have proved to be very useful in showing that certain sets which naturally arise in dynamical systems and which are small in the conventional sense are indeed large in the sense of Hausdorff dimension.

We state without proofs some basic properties of α-winning sets:

Proposition 4.. (i) Suppose S, S' are α-winning sets in X, X' then $S \times S' \subset X \times X'$ is also an α-winning set.
(ii) Let X_i, $1 \leq i \leq k$, be closed subspaces of a compact metric space X whose interiors cover X. If $S_i \subset X_i$ is α_i-winning in X_i then $S = \bigcup_{1\leq i\leq k} S_i$ is α winning where $\alpha = \min\{\alpha_i \mid 1 \leq i \leq k\}$.
(iii) (Prop. 3.3[**1**], cf. §11,[**10**]) Any α-winning set S in a complete Riemannian manifold M is large, i.e., for any nonempty open set $U \subset M$ the Hausdorff dimension of $S \cap U$ is n where $n = \dim M$.
(iv) ([**4**], cf. Th. 1, [**10**]) Let $f : X \longrightarrow Y$ be a homeomorphism

between complete metric spaces and λ be a bi-Lipschitz constant for f, that is, $\lambda \geq 1$ is a real number such that for any $x, y \in X$, one has $\lambda^{-1} d_X(x,y) \leq d_Y(f(x), f(y)) \leq \lambda d_X(x,y)$. If $S \subset X$ is (α, β)-winning set in X with $\beta\lambda^2 < 1$, then $f(S)$ is an $(\alpha/\lambda^2, \beta\lambda^2)$-winning set in Y. In particular, if S is α-winning, then $f(S)$ is α/λ^2-winning. □

REMARK 5.. If X, Y are smooth compact Riemannian manifolds and $f : X \longrightarrow Y$ is a diffeomorphism, then a bi-Lipschitz constant for f always exists (see Ch. 5 of [**4**], Prop. 3.3 of [**1**]). Thus, in view of Prop. 4(iv), the 'largeness' of a subset of a compact Riemannian manifold does not depend on the specific choice of the metric.

REMARK 6.. Let Λ be any lattice in the Euclidean space $\mathbb{R}^n$ and let $\mathbb{T} = \mathbb{R}^n/\Lambda$. Consider the collection $\mathcal{C}_{\mathbb{T}}$. Then there exists a subset $S \subset \mathbb{T}$ which is exceptional relative to $\mathcal{C}_{\mathbb{T}}$ such that S is α-winning for any $\alpha \in (0, 1/2]$. In particular, $S \cap U$ has Hausdorff dimension n for any nonempty open set $U \subset \mathbb{T}$. To see this, note that if $\Lambda = \mathbb{Z}^n \subset \mathbb{R}^n$, then this is just a restatement of Dani's result [**3**]. In the general case, we observe that Dani's proof still goes through when $\mathbb{Z}^n$ is replaced by an arbitrary uniform lattice Λ. We can also see this easily using Dani's result for the torus $\mathbb{R}^n/\mathbb{Z}^n$ and Remark 5.

We conclude this section with the following

LEMMA 2.1.. *Let $p : M \longrightarrow M'$ be a smooth fibre bundle where M and M' are compact Riemannian manifolds. Assume that $S' \subset M'$ is α-winning for any $\alpha \in (0, \nu]$ for some $\nu \in (0, 1/2]$. Then there exists a $\mu > 0$ such that $S := p^{-1}(S')$ is α-winning for any $\alpha \in (0, \mu]$.*

Proof: Let F be the fibre of the bundle, endowed with a Riemannian metric. Let $V \subset M'$ be any trivializing open set which is diffeomorphic to the unit ball B in $\mathbb{R}^n$. Let K be the image in V of the closed disk of radius 1/2 under the diffeomorphism $B \cong V$. Let $h_V : p^{-1}(V) \longrightarrow V \times F$ be a diffeomorphism which is a trivialization. Then $S'_K := S' \cap K$ is an α- winning set in K for any $\alpha \in (0, \nu]$ and so, by Prop. 4(i), $S'_K \times F$ is an α-winning set in $K \times F$ for any $\alpha \in (0, \nu]$. Now, it follows from Prop. 4(iv) that $p^{-1}(S'_K) = h_V^{-1}(S'_K \times F)$ is α-winning in $p^{-1}(K)$ for any $\alpha \in (0, \nu/\lambda^2]$ where λ is a bi-Lipschitz constant for the diffeomorphism h_V. One can cover M' by the interiors of (closed) disk neighbourhoods $K_i, 1 \leq i \leq k$, where $K_i \subset V_i \subset M'$, the V_i being trivializing open sets for the bundle $p : M \longrightarrow M'$. Choose bi-Lipschitz constants λ_i for trivializations $p^{-1}(V_i) \longrightarrow V_i \times F$. Then $p^{-1}(S'_{K_i})$ is α-winning in $p^{-1}(K_i)$ for any $\alpha \in (0, \nu/\lambda_i^2]$ and for each

i such that $1 \le i \le k$. Hence, by Prop. 4(ii), it follows that $S = \bigcup_{1\le i\le k} p^{-1}(S'_{K_i})$ is α-winning for any $\alpha \in (0, \mu]$ where μ is the smallest of the numbers ν/λ_i^2, $1 \le i \le k$. □

3. Proofs of Main Theorems

Let $f : X \longrightarrow X$ be a self-map and let $x \in X$. Denote by $C_f(x)$ the closure of the set $\{f^k(x) \mid k \ge 0\}$.

LEMMA 3.1.. *Let $p : X \longrightarrow X'$ be any continuous map. Let $\mathcal{C}$ and $\mathcal{C}'$ be any collections of self-maps of X and X' respectively such that given $f \in \mathcal{C}$, there exists an $f' \in \mathcal{C}'$ such that $p \circ f = f' \circ p$. Suppose that $S' \subset X'$ is exceptional relative to $\mathcal{C}'$, then $S := p^{-1}(S')$ is exceptional relative to $\mathcal{C}$.*

Proof: Let $x \in S$, $f \in \mathcal{C}$. Suppose that $y \in C_f(x)$ is f-periodic, say, $f^k(y) = y$. Let $x' = p(x) \in S'$ and let $y' = p(y)$. Choose $f' \in \mathcal{C}'$ such that $p \circ f = f' \circ p$. Then, $p(C_f(x)) \subset C_{f'}(x')$. In particular $y' \in C_{f'}(x')$. Note that $f'^k(y') = f'^k \circ p(y) = p \circ f^k(y) = p(y) = y'$ which contradicts our hypothesis that S' is exceptional. This shows that S is exceptional relative to $\mathcal{C}$. □

LEMMA 3.2.. *Let $p : X \longrightarrow \widehat{X}$ be a finite covering projection where X, $\widehat{X}$ are compact metric spaces. Let $\mathcal{C}$ and $\widehat{\mathcal{C}}$ be any collections of self-maps of X and $\widehat{X}$ respectively such that for any $\hat{f} \in \widehat{\mathcal{C}}$, there exists an $f \in \mathcal{C}$ such that $p \circ f = \hat{f} \circ p$. Let S be an exceptional set relative to $\mathcal{C}$. Then $\widehat{S} := p(S) \subset \widehat{X}$ is an exceptional set relative to $\widehat{\mathcal{C}}$.*

Proof: Let $\hat{x} \in \widehat{S}$, $\hat{f} \in \widehat{\mathcal{C}}$ and let $\hat{y} \in C_{\hat{f}}(\hat{x})$. We shall assume that $\hat{y}$ is an $\hat{f}$-periodic point and arrive at a contradiction.

Since $\hat{y} \in C_{\hat{f}}(\hat{x})$, there exists a sequence of positive integers n_k such that $\hat{f}^{n_k}(\hat{x})$ converges to $\hat{y}$. Choose $x \in S$, $f \in \mathcal{C}$ such that $p(x) = \hat{x}$ and $p \circ f = \hat{f} \circ p$. Consider the set $\{f^{n_k}(x) | k \ge 1\} \subset C_f(x)$. Because $x \in S$, this set is infinite. Since X is compact, it has a subsequence $f^{n_{k_i}}(x)$ converging to an element $y \in X$ and $y \in C_f(x)$ since $C_f(x)$ is closed. The element y must be in the fibre over $\hat{y} \in \widehat{X}$ by continuity of p. Since $\hat{y}$ is $\hat{f}$-periodic, and since f covers $\hat{f}$, it follows that y must be f-periodic. That is, $y \in C_f(x)$ is an f-periodic point which is a contradiction since $x \in S$, $f \in \mathcal{C}$ and S is an exceptional set relative to $\mathcal{C}$. □

We are now ready to prove the main theorems stated in the introduction. We keep the notations of §1.

Proof of Theorem 2: Consider the fibration $p : M \longrightarrow M'$ where M' is the torus $\overline{N}/\overline{\Lambda}$ and let $\mathcal{C}' = \mathcal{C}_{\overline{\mathbb{T}}}$. By Remark 6, there exists an exceptional set $S' \subset M'$ relative to $\mathcal{C}'$ which is α-winning for any $\alpha \in (0, 1/2]$. Note that for the collection $\mathcal{C}$ of the theorem, the hypotheses of Lemma 3.1 are satisfied. It follows that $S = p^{-1}(S')$ is exceptional. By Lemma 2.1, we see that S is α-winning for any $\alpha \in (0, \mu]$ for some $\mu \in (0, 1/2]$. □

Proof of Theorem 3: Consider the finite covering projection $\pi : M \longrightarrow \widehat{M}$. Recall that the deck transformation group $G = \Gamma/\Lambda$ acts on M by isometries.

The hypotheses of Lemma 3.2 holds when we take $\mathcal{C}$ to be as in Theorem 2, that is, the set of all surjective endomorphisms ρ of M such that $\overline{\rho} \in \mathcal{C}_{\overline{\mathbb{T}}}$.

By Theorem 2 there exists a positive number $\mu \leq 1/2$ and a set S which is exceptional relative to $\mathcal{C}$ such that S is α-winning for any $\alpha \in (0, \mu]$ when M is given the right invariant metric. By Prop. 4 (iv) applied to the identity map of M, the same statement holds if we replace μ by a suitably smaller positive number ν when we put the G-invariant metric on M. By Lemma 3.2, $\widehat{S} := \pi(S)$ is an exceptional set relative to $\widehat{\mathcal{C}}$.

Since G acts on M by isometries, the covering projection π is a local isometry. Choose a finite collection $\{K\}$ of normal (closed) disk neighbourhoods whose interiors cover M and such that $\pi|K : K \longrightarrow \pi(K)$ is an isometry for each K. Then, $S \cap K$ is α-winning in K and $\pi(S \cap K)$ is α-winning for each $\alpha \in (0, \nu]$ by Theorem 1, [**10**]. It follows from Prop. 4 (ii) that $\widehat{S} = \pi(S)$ is an α-winning set for any $\alpha \in (0, \nu]$. This completes the proof. □

Concluding Remarks: (*i*) It would be interesting to know whether there are large exceptional sets for the class consisting of an arbitrary expanding endomorphism (cf. [**9**]) of a nilmanifold or an infranil manifold.

(*ii*) Kleinbock [**6**] has obtained a generalization of Dani's theorem showing existence of large sets in a very general setting. However, Kleinbock's result still does not subsume our results because the exceptional set in his set up is not known to be an α-winning set. This is a stronger conclusion in our case which enables us to deduce the maximality of Hausdorff dimension for exceptional sets of a *family* of endomorphisms and not just for exceptional sets of a single endomorphism.

References

[1] C.S.Aravinda, Bounded geodesics and Hausdorff dimension, Math. Proc. Camb. Phil. Soc.,**116**,(1994), 505-511.
[2] L.Auslander, Bieberbach's theorems on space groups and discrete uniform subgroups of Lie groups, Ann. Math. **71**, (1960), 579-590.
[3] S.G.Dani, On orbits of endomorphisms of tori and the Schmidt game, Ergod. Th. & Dynam. Sys. **8**,(1988), 523-529.
[4] S.G.Dani, On badly approximable numbers, Schmidt games, and bounded orbits of flows, in *Number theory and Dynamical Systems, York-1987*, Lond. Math. Soc. Lect. Notes Sr., **134**(1989), 69-86.
[5] P. Eberlein, Geometry of 2-step nilpotent groups with a left invariant metric, Ann. Sci. Ecole Nor. Sup.,**27**(1994), 611-660.
[6] D.Y.Kleinbock, Nondense orbits of flows on homogeneous spaces, Ergod. Th. & Dynam. Sys., **18**,(1998), 373-396.
[7] W.Parry, Ergodic properties of affine transformations and flows on nilmanifolds, Amer. Jour. Math, **91**, (1969), 757-771.
[8] M.S.Raghunathan, *Discrete subgroups of Lie Groups,* Ergeben. Math. Grenzgeb. **68**, Springer-Verlag, New York, 1972.
[9] M.Shub, Expanding maps, Proc. Symp. Pure Math., XVI, (1970), 273-277.
[10] W.M.Schmidt, On badly approximable numbers and certain games, Trans. Amer. Math. Soc., **123**, (1966), 178-199.

Chennai Mathematical Institute
92, G.N.Chetty Road, Chennai 600 017, INDIA
E-mail address: aravinda@cmi.ac.in

Institute of Mathematical Sciences
CIT Campus, Chennai 600 113, INDIA
E-mail address: sankaran@imsc.res.in

A Family of Quantum Stabilizer Codes Based on the Weyl Commutation Relations over a Finite Field

V. ARVIND AND K.R. PARTHASARATHY

Dedicated to C.S. Seshadri on his 70th birthday

Abstract. Using the Weyl commutation relations over a finite field $\mathbb{F}_q$ we introduce a family of error-correcting quantum stabilizer codes based on a class of symmetric matrices over $\mathbb{F}_q$ satisfying certain natural conditions. When $q = 2$ the existence of a rich class of such symmetric matrices is demonstrated by a simple probabilistic argument depending on the Chernoff bound for i.i.d. symmetric Bernoulli trials. If, in addition, these symmetric matrices are assumed to be circulant it is possible to obtain concrete examples by a computer program. The quantum codes thus obtained admit elegant encoding circuits.

1. Introduction

Let A be a finite abelian group with operation denoted by $+$ and identity 0. We identify A with the alphabet of symbols transmitted on a classical communication channel. Consider the n-fold cartesian product A^n of copies of A. Elements of A^n are called *words* of length n. A commonly used group is $\{0, 1\}$ with addition modulo 2. Let $\hat{A}$ denote the character group of A, the multiplicative group of all homomorphisms from A into the multiplicative group of complex numbers of modulus unity. For $\mathbf{a} = (a_1, a_2, \ldots, a_n)^T \in A^n$ we define its *weight* $w(\mathbf{a})$ to be $\#\{i \mid a_i \neq 0\}$. We say that a subgroup $\mathcal{C}_n$ of A^n is a *t-error correcting group code* if for every non-zero element $\mathbf{x} = (x_1, x_2, \ldots, x_n)^T$ in $\mathcal{C}_n$, $w(\mathbf{x}) \geq 2t + 1$. In other words, if messages transmitted through a noisy channel are encoded into words from $\mathcal{C}_n$ and during transmission of a word errors at the output occur in at most t positions, then the message can be decoded without any error. There is a vast literature on the construction of t-error correcting group codes and the reader may find an introduction to this subject and pointers to literature in [**9, 8**].

A broad class of quantum error correcting codes known as stabilizer codes was introduced by Gottesman [**5**] and Calderbank et al [**3**] (also see [**4, 14, 13**]). Nonstabilizer codes have received some attention recently [**14, 16, 1**] but, to date, most of the work in the area of quantum codes deals with stabilizer codes, primarily due to the influential paper of Calderbank et al. [**3**]. A theory of quantum stabilizer codes based on additive classical codes over $\mathbb{F}_4$ is developed in [**3**]. A large part of the Calderbank et al. paper deals with the explicit construction of a wide variety of quantum stabilizer codes with different parameters, all arising from classical additive codes over $\mathbb{F}_4$.

In the present paper, our main aim is to give a new description of the theory of error-correcting quantum stabilizer codes. First we introduce some definitions. We choose and fix an N-dimensional complex Hilbert space $\mathcal{H}$ and consider the unit vectors of $\mathcal{H}$ as pure states of a finite level quantum system. If A is a finite abelian group with N elements and $\{e_x \mid x \in A\}$ is an orthonormal basis of $\mathcal{H}$ indexed by elements of A we express it in the Dirac notation as $|x\rangle = e_x$. If $\mathbf{x} = (x_1, x_2, \ldots, x_n)^T \in A^n$ is a word of length n, we write

$$|\mathbf{x}\rangle = |x_1 x_2 \ldots x_n\rangle = e_{x_1} \otimes e_{x_2} \otimes \ldots \otimes e_{x_n}$$

where the right-hand side is a product vector in the n-fold tensor product $\mathcal{H}^{\otimes^n}$ of n copies of $\mathcal{H}$. Thus, with the chosen orthonormal basis, every word $\mathbf{x}$ in A^n is translated into a basis state $|\mathbf{x}\rangle$ of $\mathcal{H}^{\otimes^n}$.

A *quantum code* is a subspace $\mathcal{C}_n$ in $\mathcal{H}^{\otimes^n}$. Note that a pure state in $\mathcal{H}^{\otimes^n}$ described by a unit vector $|\psi\rangle$ in $\mathcal{H}^{\otimes^n}$ has density matrix $|\psi\rangle\langle\psi|$. A *density matrix* ρ in $\mathcal{H}^{\otimes^n}$ is a non-negative operator of unit trace. In quantum probability, a projection operator E in $\mathcal{H}^{\otimes^n}$ is interpreted as an event concerning the quantum system and a density matrix ρ as a state of the quantum system. The probability of the event E in the state ρ is given by $\mathrm{Tr}\rho E$. Messages to be transmitted through a quantum channel are encoded into pure states in $\mathcal{H}^{\otimes^n}$. When a pure state $|\psi\rangle$, or equivalently, a density matrix $|\psi\rangle\langle\psi|$ is transmitted the channel output is hypothesized to be a state of the form

$$\rho = \sum_i L_i |\psi\rangle\langle\psi| L_i^\dagger \tag{1.1}$$

where the operators $\{L_i\}$ belong to a linear subspace $\mathcal{A}$ of the algebra of all operators on $\mathcal{H}^{\otimes^n}$. As in classical coding theory, the theory of quantum error correcting codes does not depend on the nature of the quantum channel but only on the subspace $\mathcal{A}$ of operators from which the errors $\{L_i\}$ arise. Thus, we may assume, in general, that the operators $\{L_i\}$ depend on $|\psi\rangle$, but in order to ensure that the output ρ is a density matrix it is assumed that $\langle\psi|\sum_i L_i^\dagger L_i|\psi\rangle = 1$. By the spectral theorem ρ can be expressed as

$$\rho = \sum_j p_j |\psi_j\rangle\langle\psi_j|$$

where $|\psi_j\rangle$ is an orthonormal set in $\mathcal{H}^{\otimes^n}$ and $\{p_j\}$ is a probability distribution with $p_j > 0$ for each j. In other words, the output state ρ is not necessarily pure even though the input state is pure. The operators L_i are called *error operators* and the linear space $\mathcal{A}$ from which they come is called the *error space.*

Suppose there is a finite family $\{M_j\}$ of operators in $\mathcal{H}^{\otimes^n}$ satisfying the condition $\sum_j M_j^\dagger M_j = I$ and for any output state ρ with $|\psi\rangle$ in the code $\mathcal{C}_n$,

$$\sum_j M_j \rho M_j^\dagger = \sum_{i,j} M_j L_i |\psi\rangle\langle\psi| L_i^\dagger M_j \dagger = |\psi\rangle\langle\psi|.$$

Then we say that the quantum code $\mathcal{C}_n$ together with the family $\{M_j\}$ of 'decoding operators' corrects any error induced by $\{L_i\}$ from $\mathcal{A}$. In this context we have the following fundamental theorem of Knill and Laflamme [**6**] which gives necessary and sufficient conditions for the existence of such a family of decoding operators.

THEOREM 1.2. [**6**] *Let $\mathcal{A}$ be a family of operators in $\mathcal{H}^{\otimes^n}$ and let $\mathcal{C}_n \subset \mathcal{H}^{\otimes^n}$ be a quantum code with an orthonormal basis $|\psi_1\rangle, |\psi_2\rangle, \dots,$ $|\psi_d\rangle$. Then there exists a finite family $\{M_j\}$ of operators in $\mathcal{H}^{\otimes^n}$ satisfying the conditions:*

(i) $\sum_j M_j^\dagger M_j = I$; *and*
(ii)

$$\sum_j M_j L|\psi\rangle\langle\psi|L^\dagger M_j^\dagger = \langle\psi|L^\dagger L|\psi\rangle|\psi\rangle\langle\psi| \ \forall\ |\psi\rangle \in \mathcal{C}_n, L \in \mathcal{A}$$

if and only if the following condition holds:
$\langle\psi_p|L_1^\dagger L_2|\psi_q\rangle = \delta_{p,q} c(L_1, L_2)$ *for all* $L_1, L_2 \in \mathcal{A}$, $1 \le p, q \le d$, *where* $c(L_1, L_2)$ *is a scalar independent of* p *and* q *and* $\delta_{p,q}$ *is 1 if* $p = q$ *and 0 otherwise.*

REMARK 1.3. The proof of the above theorem is constructive and therefore yields the decoding operators in terms of $\mathcal{A}$ and the basis $|\psi_1\rangle, \ldots, |\psi_d\rangle$ of $\mathcal{C}_n$. In this case we say that $\mathcal{C}_n$ is an $\mathcal{A}$-error correcting quantum code.

Now we specialize the choice of $\mathcal{A}$. Consider all unitary operators in $\mathcal{H}^{\otimes^n}$ of the form $U = U_1 \otimes U_2 \otimes \ldots \otimes U_n$ where each U_i is a unitary operator on $\mathcal{H}$ and all but t of the U_i's are equal to I. Such a U when operating on $|\psi\rangle = |\psi_1\rangle \otimes \ldots \otimes |\psi_n\rangle \in \mathcal{H}^{\otimes^n}$ produces $U|\psi\rangle$ which is an n-fold tensor product that differs from $|\psi\rangle$ in at most t places. Denote by $\mathcal{A}_t$ the linear span of all such unitary operators U. A quantum code $\mathcal{C}_n$ is called a *t-error correcting quantum code* if $\mathcal{C}_n$ is an $\mathcal{A}_t$-correcting quantum code.

2. Quantum codes and subgroups of the error group

Let $(A, +)$ be a finite abelian group with N elements and identity denoted by 0. Denote by $\hat{A}$ the character group of A and $\mathcal{H}$ the N-dimensional Hilbert space $L^2(A)$ of all complex-valued functions on A, spanned by $\{|x\rangle\}_{x \in A}$ (where the vector $|x\rangle$ denotes the indicator function 1_x of the singleton $\{x\}$). Define the unitary operators U_a and V_χ on $\mathcal{H}$ for every $a \in A$ and $\chi \in \hat{A}$ by

$$U_a|x\rangle = |x + a\rangle, \qquad V_\chi|x\rangle = \chi(x)|x\rangle$$

where $x \in A$. Then

$$\chi(a)U_a V_\chi = V_\chi U_a \ \forall\ a \in A, \chi \in \hat{A}.$$

These are the Weyl commutation relations between the unitary operators representing A by translations and $\hat{A}$ by multiplications. The family of operators $\{U_a V_\chi \mid a \in A \chi \in \hat{A}\}$ is irreducible.

If $\mathbf{a} \in A^n$ then any element $\boldsymbol{\chi} \in \hat{A}^n$ can be identified with an element of $\hat{A}^n$ so that

$$\boldsymbol{\chi}(\mathbf{a}) = \prod_{i=1}^{n} \chi_i(a_i) \ \chi_i \in \hat{A}, a_i \in A$$

where $\boldsymbol{\chi} = (\chi_1, \ldots, \chi_n)$ and $\mathbf{a} = (a_1, \ldots, a_n)$. Put $U_{\mathbf{a}} = U_{a_1} \otimes \ldots \otimes U_{a_n}$ and $V_{\boldsymbol{\chi}} = V_{\chi_1} \otimes \ldots \otimes V_{\chi_n}$. Then $\{U_{\mathbf{a}} V_{\boldsymbol{\chi}} \mid \mathbf{a} \in A^n, \boldsymbol{\chi} \in \hat{A}^n\}$ is again an irreducible family of unitary operators satisfying the Weyl commutation relations

$$\boldsymbol{\chi}(\mathbf{a}) U_{\mathbf{a}} V_{\boldsymbol{\chi}} = V_{\boldsymbol{\chi}} U_{\mathbf{a}} \ \forall\ \mathbf{a} \in A^n, \boldsymbol{\chi} \in \hat{A}^n.$$

In the Hilbert space of all linear operators on $\mathcal{H}^{\otimes^n}$ equipped with the scalar product $\langle X \mid Y \rangle = \mathrm{Tr} X^\dagger Y$ the set $\{N^{-n/2} U_{\mathbf{a}} V_{\boldsymbol{\chi}} \mid \mathbf{a} \in A^n, \boldsymbol{\chi} \in \hat{A}^n\}$ is an orthonormal basis. The weight $wt(\mathbf{a}, \boldsymbol{\chi})$ of a pair $(\mathbf{a}, \boldsymbol{\chi}) \in A^n \times \hat{A}^n$ is defined to be $\#\{i \mid 1 \leq i \leq n, (a_i, \chi_i) \neq (0,1)\}$, where $\mathbf{a} = (a_1, a_2, \ldots, a_n)$ and $\boldsymbol{\chi} = (\chi_1, \ldots, \chi_n)$. The irreducibility of $\{U_{\mathbf{a}} V_{\boldsymbol{\chi}} \mid \mathbf{a} \in A^n, \boldsymbol{\chi} \in \hat{A}^n\}$ implies that $\{U_{\mathbf{a}} V_{\boldsymbol{\chi}} \mid \mathbf{a} \in A^n, \boldsymbol{\chi} \in \hat{A}^n, wt(\mathbf{a}, \boldsymbol{\chi}) \leq t\}$ spans $\mathcal{A}_t$. The Knill-Laflamme theorem for $\mathcal{A}_t$-correcting quantum codes assumes the following form which can be readily derived from Theorem 1.2.

THEOREM 2.1. *$\mathcal{C}_n \subset L^2(A)^{\otimes^n}$ is a t-error correcting quantum code if and only if $\mathcal{C}_n$ has an orthonormal basis $|\psi_1\rangle, |\psi_2\rangle, \ldots, |\psi_d\rangle$ satisfying the following conditions:*
For every $(\mathbf{a}, \boldsymbol{\chi}) \in A^n \times \hat{A}^n$ such that $wt(\mathbf{a}, \boldsymbol{\chi}) \leq 2t$

(i) *$\langle \psi_i | U_{\mathbf{a}} V_{\boldsymbol{\chi}} | \psi_j \rangle = 0$ if $i \neq j$, and*
(ii) *$\langle \psi_i | U_{\mathbf{a}} V_{\boldsymbol{\chi}} | \psi_i \rangle$ is a scalar independent of $|\psi_i\rangle$ for $i = 1, 2, \ldots, d$.*

Let l be the least positive integer such that $la = 0$ for all $a \in A$ and let $\omega = e^{\frac{2\pi i}{l}}$. We define the *error group* as the following finite group of unitary operators in $L^2(A)^{\otimes^n}$.

$$\mathcal{E} = \{\omega^i U_{\mathbf{a}} V_{\boldsymbol{\chi}} \mid 0 \leq i \leq l-1, \mathbf{a} \in A^n, \boldsymbol{\chi} \in \hat{A}^n\}.$$

The group $\mathcal{E}$ has a natural action on the Hilbert space $L^2(A)^{\otimes^n}$ defined by:

$$U_{\mathbf{a}} |\mathbf{x}\rangle = |\mathbf{x} + \mathbf{a}\rangle, \qquad V_{\boldsymbol{\chi}} |\mathbf{x}\rangle = \chi(\mathbf{x}) |\mathbf{x}\rangle.$$

Subspaces of $L^2(A)^{\otimes^n}$ that are point-wise fixed by some subgroup of the error group $\mathcal{E}$ are called *stabilizer codes.*

Let $\mathcal{S}$ be a subgroup of $\mathcal{E}$. Denote by $\mathcal{C}(\mathcal{S})$ the subspace of $L^2(A)^{\otimes^n}$ that is point-wise stabilized by $\mathcal{S}$. More precisely,

$$\mathcal{C}(\mathcal{S}) = \{|\psi\rangle \in L^2(A)^{\otimes^n} \mid U|\psi\rangle = |\psi\rangle \ \forall\, U \in \mathcal{S}\}.$$

LEMMA 2.2. *$\mathcal{C}(\mathcal{S}) \neq 0$ if and only if $\mathcal{S}$ is an abelian subgroup of $\mathcal{E}$ such that $\omega^i I \notin \mathcal{S}$ for $i \neq 0$. Furthermore, when $\mathcal{C}(\mathcal{S}) \neq 0$ the dimension of $\mathcal{C}(\mathcal{S})$ is $\#A^n / \#\mathcal{S}$.*

Proof. Suppose $\omega^i I \in \mathcal{S}$ for some $i \neq 0$. For any $|\psi\rangle \in \mathcal{C}(\mathcal{S})$ we have $\omega^i I |\psi\rangle = |\psi\rangle$, which implies $|\psi\rangle = 0$. Hence $\mathcal{C}(\mathcal{S}) = 0$.

It follows from the Weyl commutation relations that two elements $\omega^i U_{\mathbf{a}} V_{\boldsymbol{\alpha}}$ and $\omega^j U_{\mathbf{b}} V_{\boldsymbol{\beta}}$ in $\mathcal{S}$ commute if and only if $\boldsymbol{\alpha}(\mathbf{b}) = \boldsymbol{\beta}(\mathbf{a})$. Now, let $|\psi\rangle \in \mathcal{C}(\mathcal{S})$. We have

$$|\psi\rangle = \omega^i U_{\mathbf{a}} V_{\boldsymbol{\alpha}} \omega^j U_{\mathbf{b}} V_{\boldsymbol{\beta}} |\psi\rangle = \omega^j U_{\mathbf{b}} V_{\boldsymbol{\beta}} \omega^i U_{\mathbf{a}} V_{\boldsymbol{\alpha}} |\psi\rangle.$$

Applying the commutation relations we can see that the above equation holds for a $|\psi\rangle \neq 0$ if and only if $\boldsymbol{\alpha}(\mathbf{b}) = \boldsymbol{\beta}(\mathbf{a})$. Thus, $\mathcal{C}(\mathcal{S}) \neq 0$ if and only if $\mathcal{S}$ is abelian and $\omega^i I \notin \mathcal{S}$ for $i \neq 0$.

Now, let $\mathcal{S}$ be an abelian subgroup of $\mathcal{E}$ such that $\omega^i I \notin \mathcal{S}$ for $i \neq 0$. Define the projection operator

$$P = \frac{1}{\#\mathcal{S}} \sum_{U \in \mathcal{S}} U.$$

Since $\mathrm{Tr} U_{\mathbf{a}} V_{\boldsymbol{\beta}} = 0$ unless $(\mathbf{a}, \boldsymbol{\beta}) = (0, 1)$ it follows that $\mathrm{Tr}(P) = \#A^n/\#\mathcal{S}$. It is easy to see that P is the projection onto $\mathcal{C}(\mathcal{S})$. Thus, the dimension of $\mathcal{C}(\mathcal{S})$ is $\mathrm{Tr}(P) = \#A^n/\#\mathcal{S}$. This completes the proof. ∎

Next, we state Theorem 2.1 in a form that will give the criteria for constructing t-error correcting quantum stabilizer codes. Let $Z(\mathcal{S})$ denote the centralizer of $\mathcal{S}$ in $\mathcal{E}$, i.e.,

$$Z(\mathcal{S}) = \{U \in \mathcal{E} \mid UU' = U'U \ \forall \ U' \in \mathcal{S}\}.$$

THEOREM 2.3. *Let $\mathcal{S}$ be an abelian subgroup of the error group $\mathcal{E}$ such that $\omega^i I$ is not in $\mathcal{S}$ for $i \neq 0$. Then $\mathcal{C}(\mathcal{S})$ is a t-error correcting quantum code if $wt(\mathbf{a}, \boldsymbol{\alpha}) > 2t$ for each $\omega^i U_{\mathbf{a}} V_{\boldsymbol{\alpha}} \in Z(\mathcal{S}) \setminus \mathcal{S}$.*

Proof. Suppose $wt(\mathbf{a}, \boldsymbol{\alpha}) > 2t$ for each $\omega^i U_{\mathbf{a}} V_{\boldsymbol{\alpha}} \in Z(\mathcal{S}) \setminus \mathcal{S}$. Now, by the previous lemma $\mathcal{C}(\mathcal{S})$ is a subspace of $L^2(A)^{\otimes^n}$ of dimension $\#A^n/\#\mathcal{S} = d$. Let $|\psi_1\rangle, \ldots, |\psi_d\rangle$ be an orthonormal basis of $\mathcal{C}(\mathcal{S})$. Consider a $(\mathbf{a}, \boldsymbol{\chi}) \in A^n \times \hat{A}^n$ with the property that $wt(\mathbf{a}, \boldsymbol{\chi}) \leq 2t$. We check the Knill-Laflamme conditions (Theorem 2.1). There are two cases:

(a) If $\omega^i U_{\mathbf{a}} V_{\boldsymbol{\chi}} \in \mathcal{S}$ for some $i \geq 0$ then

$$\langle \psi_j | \omega^i U_{\mathbf{a}} V_{\boldsymbol{\chi}} | \psi_k \rangle = \langle \psi_j \mid \psi_k \rangle = \delta_{jk}, \ 1 \leq j, k \leq d.$$

Thus, $\langle \psi_j | U_{\mathbf{a}} V_{\boldsymbol{\chi}} | \psi_k \rangle = \omega^{-i} \delta_{jk} 1 \leq j, k \leq d$, where δ_{jk} is the Kronecker delta function.

(b) If $\omega^i U_{\mathbf{a}} V_{\boldsymbol{\chi}} \notin \mathcal{S}$ for each $i \geq 0$, then since $wt(\mathbf{a}, \boldsymbol{\chi}) \leq 2t$, $\omega^i U_{\mathbf{a}} V_{\boldsymbol{\chi}} \notin Z(\mathcal{S})$ for each $i \geq 0$ by the assumption. Let $|\psi\rangle \in \mathcal{C}(\mathcal{S})$ and $\omega^r U_{\mathbf{b}} V_{\boldsymbol{\beta}}$ be some element of $\mathcal{S}$. Then we can write $\langle \psi | U_{\mathbf{a}} V_{\boldsymbol{\chi}} | \psi \rangle$ as $\langle \omega^r U_{\mathbf{b}} V_{\boldsymbol{\beta}} \psi | U_{\mathbf{a}} V_{\boldsymbol{\chi}} | \omega^r U_{\mathbf{b}} V_{\boldsymbol{\beta}} \psi \rangle$, which can be simplified to get the following

$$\langle \psi | U_{\mathbf{a}} V_{\boldsymbol{\chi}} | \psi \rangle = \overline{\boldsymbol{\beta}(\mathbf{a})} \boldsymbol{\chi}(\mathbf{b}) \langle \psi | U_{\mathbf{a}} V_{\boldsymbol{\chi}} | \psi \rangle. \tag{2.4}$$

Since $\omega^i U_{\mathbf{a}} V_{\boldsymbol{\chi}} \notin Z(\mathcal{S})$ for each $i \geq 0$, for some $\omega^r U_{\mathbf{b}} V_{\boldsymbol{\beta}} \in \mathcal{S}$ we must have $\boldsymbol{\beta}(\mathbf{a}) \neq \boldsymbol{\chi}(\mathbf{b})$. This choice of $\omega^r U_{\mathbf{b}} V_{\boldsymbol{\beta}} \in \mathcal{S}$ yields $\langle\psi|U_{\mathbf{a}} V_{\boldsymbol{\chi}}|\psi\rangle = 0$.

■

At this point it is useful to introduce a standard notation using which it is convenient to describe quantum stabilizer codes. Let $\mathcal{S}$ be an abelian subgroup of $\mathcal{E}$ with centralizer $Z(\mathcal{S})$. The *minimum distance* $d(\mathcal{S})$ is defined to be the minimum of

$$\{wt(\mathbf{a}, \boldsymbol{\alpha}) \mid \omega^i U_{\mathbf{a}} V_{\boldsymbol{\alpha}} \in Z(\mathcal{S}) \setminus \mathcal{S}\}.$$

When A is the additive abelian group of the finite field $\mathbb{F}_q$ we define an $[[n, k, d]]_q$ quantum stabilizer code to be a q^k-dimensional subspace $\mathcal{C}(\mathcal{S})$ of $L^2(\mathbb{F}_q)^{\otimes^n}$, where $\mathcal{S}$ is an abelian subgroup of $\mathcal{E}$ with $d(\mathcal{S}) \geq d$ and cardinality q^{n-k}.

By Theorem 2.3 it follows that an $[[n, k, d]]_q$ quantum stabilizer code is a $\lfloor (d-1)/2 \rfloor$-error correcting quantum code.

REMARK 2.5. Let $\mathcal{S}$ be an abelian subgroup of $\mathcal{E}$ such that $\omega^i I \notin \mathcal{S}$ for every $i \neq 0$. This is equivalent to demanding that $\mathcal{S}$ is an abelian subgroup of $\mathcal{E}$ such that for any $\mathbf{a} \in A^n$ and $\boldsymbol{\chi} \in \hat{A}^n$ the operator $\omega^i U_{\mathbf{a}} V_{\boldsymbol{\chi}}$ can be in $\mathcal{S}$ for at most one $i : 0 \leq i \leq l-1$. Thus $\mathcal{S}$ has the form

$$\mathcal{S} = \{p(\mathbf{a}, \boldsymbol{\chi}) U_{\mathbf{a}} V_{\boldsymbol{\chi}} \mid (\mathbf{a}, \boldsymbol{\chi}) \in S\}$$

where $S \subset A^n \times \hat{A}^n$ is a subgroup satisfying $\boldsymbol{\chi}(\mathbf{a}') = \boldsymbol{\chi}'(\mathbf{a})$ for any $(\mathbf{a}, \boldsymbol{\chi}), (\mathbf{a}', \boldsymbol{\chi}') \in S$ and p is a function on S with values in $\{\omega^i \mid 0 \leq i \leq l-1\}$.

3. Quantum stabilizer codes in the finite field setting

In order to construct stabilizer quantum codes, we need to study abelian subgroups $\mathcal{S}$ of $\mathcal{E}$ such that elements in $Z(\mathcal{S}) \setminus \mathcal{S}$ have large weight. We choose A to be a finite field $\mathbb{F}_q$, $q = p^r$ for some prime p. In particular, the Hilbert space in which we seek stabilizer codes is $L^2(\mathbb{F}_q)^{\otimes^n}$. Since $\mathbb{F}_q$ is an abelian group under its addition operation with each nonzero element of order p, it follows that every nontrivial character of $\mathbb{F}_q$ is of order p. Choose a nontrivial character $\tilde{\omega} \in \hat{\mathbb{F}}_q$. Then every other character $\omega' \in \hat{\mathbb{F}}_q$ is of the form ω_a where $\omega_a(x) = \tilde{\omega}(ax)$ for all $x \in \mathbb{F}_q$. To see this, it suffices to notice that $\omega_a, a \in \mathbb{F}_q$ are all distinct and thus exhaust $\hat{\mathbb{F}}_q$. Likewise, every character in $\hat{\mathbb{F}}_q^{\,n}$

is of the form $\omega_{\mathbf{a}}$ where $\omega_{\mathbf{a}}(\mathbf{x}) = \tilde{\omega}(\mathbf{a}\cdot\mathbf{x})$ for all $\mathbf{x} \in \mathbb{F}_q^n$, where $\mathbf{a}\cdot\mathbf{x}$ is the inner product $\sum_i a_i x_i$, for $\mathbf{a} = (a_1, \ldots, a_n)^T$ and $\mathbf{x} = (x_1, \ldots, x_n)^T$.

If we identify $\hat{\mathbb{F}_q}^n$ with $\mathbb{F}_q^n$, we can index the elements of the error group $\mathcal{E}$ as $\omega^i U_{\mathbf{a}} V_{\mathbf{b}}$, $0 \le i \le p-1$, and $\mathbf{a}, \mathbf{b} \in \mathbb{F}_q^n$, where $V_{\mathbf{b}}$ now stands for the operator $V_{\boldsymbol{\chi}}$ with $\boldsymbol{\chi} = \omega_{\mathbf{b}}$. Thus, $\mathcal{E}$ is rewritten as

$$\mathcal{E} = \{\omega^i U_{\mathbf{a}} V_{\mathbf{b}} \mid 0 \le i \le p-1, \mathbf{a}, \mathbf{b} \in \mathbb{F}_q^n\}.$$

Notice that $\mathcal{E}$ is a finite group of cardinality pq^{2n}. The Weyl commutation relations take the following form

$$\tilde{\omega}(\mathbf{b}\cdot\mathbf{a}) U_{\mathbf{a}} V_{\mathbf{b}} = V_{\mathbf{b}} U_{\mathbf{a}} \ \forall\ \mathbf{a}, \mathbf{b} \in \mathbb{F}_q^n.$$

If $\mathcal{S}$ is a subgroup of $\mathcal{E}$ it is readily seen that $\mathcal{S}$ is abelian if and only if for any two elements $\omega^i U_{\mathbf{a}} V_{\mathbf{b}}$ and $\omega^j U_{\mathbf{c}} V_{\mathbf{d}}$ in $\mathcal{S}$ we have $\tilde{\omega}(\mathbf{a}\cdot\mathbf{d} - \mathbf{b}\cdot\mathbf{c}) = 1$. Indeed, for q prime this condition is equivalent to $\mathbf{a}\cdot\mathbf{d} = \mathbf{b}\cdot\mathbf{c}$ for all $\omega^i U_{\mathbf{a}} V_{\mathbf{b}}$ and $\omega^j U_{\mathbf{c}} V_{\mathbf{d}}$ in $\mathcal{S}$.

For $(\mathbf{a}, \mathbf{b}) \in \mathbb{F}_q^n \times \mathbb{F}_q^n$, define $wt(\mathbf{a}, \mathbf{b}) = \#\{i \mid (a_i, b_i) \neq (0,0)\}$, where $\mathbf{a} = (a_1, \ldots, a_n)$ and $\mathbf{b} = (b_1, \ldots, b_n)$. Let $S \subset \mathbb{F}_q^n \times \mathbb{F}_q^n$ be a subgroup for which $\mathbf{a}\cdot\mathbf{d} = \mathbf{b}\cdot\mathbf{c}$ for all $(\mathbf{a}, \mathbf{b}), (\mathbf{c}, \mathbf{d}) \in S$. Define

$$S^{\perp_s} = \{(\mathbf{a}, \mathbf{b}) \in \mathbb{F}_q^n \times \mathbb{F}_q^n \mid \mathbf{a}\cdot\mathbf{d} - \mathbf{b}\cdot\mathbf{c} = 0 \text{ for all } (\mathbf{c}, \mathbf{d}) \in S\}.$$

Lemma 2.2 and the Knill-Laflamme conditions can be restated as follows.

LEMMA 3.1. *Let* $S \subset \mathbb{F}_q^n \times \mathbb{F}_q^n$ *be a subgroup for which* $\mathbf{a}\cdot\mathbf{d} = \mathbf{b}\cdot\mathbf{c}$ *for all* $(\mathbf{a}, \mathbf{b}), (\mathbf{c}, \mathbf{d}) \in S$. *Suppose* $\tilde{p} : S \to \{\omega^i \mid 0 \le i \le p-1\}$ *is a function such that* $\mathcal{S} = \{\tilde{p}(\mathbf{a}, \mathbf{b}) U_{\mathbf{a}} V_{\mathbf{b}} \mid (\mathbf{a}, \mathbf{b}) \in S\}$ *is an abelian subgroup of* $\mathcal{E}$. *Then* $\mathcal{C}(\mathcal{S}) \subset L^2(\mathbb{F}_q)^{\otimes^n}$ *is a quantum stabilizer code of dimension* $q^n/\#S$. *Furthermore, if* $wt(\mathbf{a}, \mathbf{b}) > 2t$ *for all nonzero elements* $(\mathbf{a}, \mathbf{b}) \in S^{\perp_s} \setminus S$ *then* $\mathcal{C}(\mathcal{S})$ *is a* t*-error correcting quantum stabilizer code.*

Thus the problem is to find subgroups S of $\mathbb{F}_q^n \times \mathbb{F}_q^n$ such that $\mathbf{a}\cdot\mathbf{d} = \mathbf{b}\cdot\mathbf{c}$ for all $(\mathbf{a}, \mathbf{b}), (\mathbf{c}, \mathbf{d}) \in S$ and $wt(\mathbf{a}, \mathbf{b})$ is large whenever $(\mathbf{a}, \mathbf{b}) \in S^{\perp_s} \setminus S$. The other problem is to ensure that we can build an abelian subgroup $\mathcal{S}$ of $\mathcal{E}$ by picking a suitable $\tilde{p} : S \to \{\omega^i \mid 0 \le i \le p-1\}$ such that $\mathcal{S} = \{\tilde{p}(\mathbf{a}, \mathbf{b}) U_{\mathbf{a}} V_{\mathbf{b}} \mid (\mathbf{a}, \mathbf{b}) \in S\}$. To this end, we formulate an approach.

Let $\mathcal{V}$ be an m-dimensional vector space over $\mathbb{F}_q$ for a positive integer m. Let $L : \mathcal{V} \to \mathbb{F}_q^n$ and $M : \mathcal{V} \to \mathbb{F}_q^n$ be two linear transformations. Thus, L and M can be written as $n \times m$ matrices over $\mathbb{F}_q$. We restrict attention to abelian subgroups of $\mathcal{E}$ that are of the form

$$\{\tilde{p}(\mathbf{v}) U_{L\mathbf{v}} V_{M\mathbf{v}} \mid \mathbf{v} \in \mathcal{V}\}.$$

The condition

$$\mathbf{v}_2^T L^T M \mathbf{v}_1 = \mathbf{v}_1^T L^T M \mathbf{v}_2 \ \forall \ \mathbf{v}_1, \mathbf{v}_2 \in \mathcal{V} \tag{3.2}$$

ensures that the above subgroup is abelian. Furthermore,

$$\frac{\tilde{p}(\mathbf{v}_1 + \mathbf{v}_2)}{\tilde{p}(\mathbf{v}_1)\tilde{p}(\mathbf{v}_2)} = \tilde{\omega}(\mathbf{v}_2^T L^T M \mathbf{v}_1) = \tilde{\omega}(\mathbf{v}_1^T L^T M \mathbf{v}_2) \ \forall \ \mathbf{v}_1, \mathbf{v}_2 \in \mathcal{V} \tag{3.3}$$

ensures that $\{\tilde{p}(\mathbf{v}) U_{L\mathbf{v}} V_{M\mathbf{v}} \mid \mathbf{v} \in \mathcal{V}\}$ is closed under multiplication.

Equation (3.2) will hold if we choose L and M such that $M^T L = L^T M$ (i.e. L and M are such that $M^T L$ is symmetric).

Writing $\tilde{p}(\mathbf{v}) = \tilde{\omega}(\tilde{q}(\mathbf{v}))$, for some function $\tilde{q} : \mathcal{V} \to \mathbb{F}_q$, Equation (3.3) assumes the form

$$\tilde{q}(\mathbf{v}_1 + \mathbf{v}_2) - \tilde{q}(\mathbf{v}_1) - \tilde{q}(\mathbf{v}_2) = \mathbf{v}_2^T L^T M \mathbf{v}_1 = \mathbf{v}_1^T L^T M \mathbf{v}_2 \ \forall \ \mathbf{v}_1, \mathbf{v}_2 \in \mathcal{V}.$$

For $p \neq 2$ we can choose $\tilde{q}$ to be the quadratic form $\frac{1}{2}\mathbf{v}^T M^T L \mathbf{v}$. For $p = 2$ the problem of recovering a suitable quadratic form as a solution to the above equation is more difficult.

For the purpose of this article, we look for special solutions: we demand that $L^T M$ be expressible as $D + D^T$ for some matrix D over $\mathbb{F}_q$, which implies that $L^T M$ is a symmetric matrix with diagonal entries as scalar multiples of 2. For example, we can choose D to be an upper triangular matrix. Then $q(\mathbf{v}) = \mathbf{v}^T D \mathbf{v}$ is a solution to Equation (3.3). We summarize this below.

LEMMA 3.4. *Let $\mathcal{V}$ be a finite dimensional vector space, and $L : \mathcal{V} \to \mathbb{F}_q^n$ and $M : \mathcal{V} \to \mathbb{F}_q^n$ be two linear transformations such that $M^T L$ is symmetric and of the form $D + D^T$ for a linear map $D : \mathcal{V} \to \mathbb{F}_q^n$. Then*

$$\mathcal{S} = \{\tilde{\omega}(\mathbf{v}^T D \mathbf{v}) U_{L\mathbf{v}} V_{M\mathbf{v}} \mid \mathbf{v} \in \mathcal{V}\}$$

is an abelian subgroup of the error group $\mathcal{E}$ on $L^2(\mathbb{F}_q)^{\otimes^n}$.

An element $\omega^i U_{\mathbf{x}} V_{\mathbf{y}}$ of $\mathcal{E}$ is in $Z(\mathcal{S})$ if and only if $\mathbf{v}^T M^T \mathbf{x} = \mathbf{v}^T L^T \mathbf{y}$ for all $\mathbf{v} \in \mathcal{V}$. Equivalently, $\omega^i U_{\mathbf{x}} V_{\mathbf{y}} \in Z(\mathcal{S})$ if and only if $M^T \mathbf{x} = L^T \mathbf{y}$.

From the Knill-Laflamme conditions as stated in Theorem 2.3, $\mathcal{C}(\mathcal{S})$ is a t-error correcting quantum code with $\mathcal{S}$ defined as above if for any $(\mathbf{x}, \mathbf{y}) \in \mathbb{F}_q^n \times \mathbb{F}_q^n$, the condition $M^T \mathbf{x} = L^T \mathbf{y}$ implies that either $\mathbf{x} = L\mathbf{v}$ and $\mathbf{y} = M\mathbf{v}$ for some $\mathbf{v} \in \mathcal{V}$ or $wt(\mathbf{x}, \mathbf{y}) > 2t$.

If q is a prime different from 2 there is a partial converse to Lemma 3.4: Suppose $\mathcal{C}(\mathcal{S})$ is some stabilizer code in $L^2(\mathbb{F}_q)^{\otimes n}$ where $\mathcal{S} = \{\tilde{p}(\mathbf{a},\mathbf{b})U_{\mathbf{a}}V_{\mathbf{b}} \mid (\mathbf{a},\mathbf{b}) \in S\}$ for some additive subgroup S such that $\mathbf{a}\cdot\mathbf{d} = \mathbf{b}\cdot\mathbf{c}$ for all $(\mathbf{a},\mathbf{b}),(\mathbf{c},\mathbf{d}) \in S$. Let $\#S = q^r$ and $(\mathbf{a}_1\mathbf{b}_1),(\mathbf{a}_2,\mathbf{b}_2)\ldots,(\mathbf{a}_r\mathbf{b}_r)$ be an independent generating set for S. Then $S_1 = \{\mathbf{a} \in \mathbb{F}_q^n \mid \exists \mathbf{b} \in \mathbb{F}_q^n : (\mathbf{a},\mathbf{b}) \in S\}$ and $S_2 = \{\mathbf{b} \in \mathbb{F}_q^n \mid \exists \mathbf{a} \in \mathbb{F}_q^n : (\mathbf{a},\mathbf{b}) \in S\}$ are linear subspaces of $\mathbb{F}_q^n$. Let $\mathbf{e}_1,\mathbf{e}_2,\ldots,\mathbf{e}_r$ be the standard basis for $\mathbb{F}_q^r$. Define $L : \mathbb{F}_q^r \to \mathbb{F}_q^n$ and $M : \mathbb{F}_q^r \to \mathbb{F}_q^n$ by letting $L\mathbf{e}_i = \mathbf{a}_i$ and $M\mathbf{e}_i = \mathbf{b}_i$ for $i = 1,2,\ldots,r$. Since , $\mathbf{a}\cdot\mathbf{d} = \mathbf{b}\cdot\mathbf{c}$ for all $(\mathbf{a},\mathbf{b}),(\mathbf{c},\mathbf{d}) \in S$, it follows that L^TM is symmetric. Suppose $\mathbb{F}_q$ is of characteristic $p \neq 2$. For $(\mathbf{a},\mathbf{b}) \in S$, let $\mathbf{v} \in \mathbb{F}_q^r$ be such that $L\mathbf{v} = \mathbf{a}$ and $M\mathbf{v} = \mathbf{b}$ and define $p'(\mathbf{v}) = \frac{1}{2}\mathbf{v}^TL^TM\mathbf{v}$. It is easy to check that $p(\mathbf{a},\mathbf{b}) = p'(\mathbf{v}) + \mathbf{c}\cdot\mathbf{v}$, for some $\mathbf{c} \in \mathbb{F}_q^r$. More precisely, we have the following proposition.

PROPOSITION 3.5. *Suppose q is a prime different from 2 and $\mathcal{C}(\mathcal{S})$ is some stabilizer code in $L^2(\mathbb{F}_q)^{\otimes n}$ of dimension q^{n-r}. Then there are linear transformations $L : \mathbb{F}_q^r \to \mathbb{F}_q^n$ and $M : \mathbb{F}_q^r \to \mathbb{F}_q^n$ such that L^TM is symmetric and there is a $\mathbf{c} \in \mathbb{F}_q^n$ such that $\mathcal{S} = \{\tilde{\omega}(\frac{1}{2}\mathbf{v}^TL^TM\mathbf{v} + \mathbf{c}\cdot\mathbf{v})U_{L\mathbf{v}}V_{M\mathbf{v}} \mid \mathbf{v} \in \mathbb{F}_q^r\}$.*

We can derive the following proposition from Lemma 3.4.

PROPOSITION 3.6. *Let $L : \mathbb{F}_q^{n-1} \to \mathbb{F}_q^n$ be an injective linear map with range $C = \{\mathbf{a} \in \mathbb{F}_q^n \mid \sum_{i=1}^n a_i = 0\}$ and $M = M'L$ for some symmetric linear map $M' : \mathbb{F}_q^n \to \mathbb{F}_q^n$ of the form $M' = D + D^T$. Then*

(i) *$\mathcal{S} = \{\tilde{\omega}(\mathbf{a}^TD\mathbf{a})U_{\mathbf{a}}V_{M'\mathbf{a}} \mid \mathbf{a} \in C\}$ is an abelian subgroup of $\mathcal{E}$.*
(ii) *$\mathcal{C}(\mathcal{S})$ is t-error correcting if for any $(\mathbf{x},\mathbf{y}) \in \mathbb{F}_q^n \times \mathbb{F}_q^n$, the condition $\mathbf{y} - M'\mathbf{x} \in C^\perp$ implies that either $\mathbf{x} \in C$ and $\mathbf{y} = M'\mathbf{x}$ or $wt(\mathbf{x},\mathbf{y}) > 2t$.*

Let $\mathcal{S} = \{\tilde{\omega}(\mathbf{a}^TD\mathbf{a})U_{\mathbf{a}}V_{L\mathbf{a}} \mid \mathbf{a} \in C\}$, where $L = D + D^T$, L and D are $n \times n$ matrices over $\mathbb{F}_q$, and C is a subspace of $\mathbb{F}_q^n$. As already observed $\mathcal{S}$ is an abelian subgroup of $\mathcal{E}$. Our next goal is to give an orthonormal basis for $\mathcal{C}(\mathcal{S})$. Notice that $\mathcal{C}(\mathcal{S})$ is a $q^n/\#C$-dimensional subspace of $L^2(\mathbb{F}_q)^{\otimes n}$. Since C is an additive subgroup of $\mathbb{F}_q^n$, it suggests that an orthonormal basis for $\mathcal{C}(\mathcal{S})$ can be indexed by the cosets of C in $\mathbb{F}_q^n$. It suffices to describe unit vectors $|\psi_{C+\mathbf{x}}\rangle \in L^2(\mathbb{F}_q)^{\otimes n}$ that have disjoint support in $\mathbb{F}_q^n$, and show that each $|\psi_{C+\mathbf{x}}\rangle$ is fixed by $\mathcal{S}$, where $\mathbf{x}$ runs over a set of distinct coset representatives of C in $\mathbb{F}_q^n$. Define

$$(3.7) \qquad |\psi_{C+\mathbf{x}}\rangle = \frac{1}{\sqrt{\#C}} \sum_{\mathbf{a}\in C} \tilde{\omega}(\mathbf{a}^T D\mathbf{a})\tilde{\omega}(\mathbf{a}^T L\mathbf{x})|\mathbf{a}+\mathbf{x}\rangle$$

for each coset $C + \mathbf{x}$ as $\mathbf{x}$ runs over a set of distinct coset representatives of C in $\mathbb{F}_q^n$. The vectors $|\psi_{C+\mathbf{x}}\rangle$ have unit norm, and as they have mutually disjoint supports, they form an orthonormal set of $q^n/\#C$ vectors in $L^2(\mathbb{F}_q)^{\otimes^n}$. It can be easily verified that $\mathcal{S}$ fixes each $|\psi_{C+\mathbf{x}}\rangle$. We summarize our observations below.

PROPOSITION 3.8. *Let* $\mathcal{S} = \{\tilde{\omega}(\mathbf{a}^T D\mathbf{a})U_{\mathbf{a}}V_{L\mathbf{a}} \mid \mathbf{a} \in C\}$, *where* $L = D + D^T$, *L and D are $n \times n$ matrices over $\mathbb{F}_q$, and C is a subspace of $\mathbb{F}_q^n$. Then the collection of vectors $\{|\psi_{C+\mathbf{x}}\rangle\}$ defined as*

$$|\psi_{C+\mathbf{x}}\rangle = \frac{1}{\sqrt{\#C}} \sum_{\mathbf{a}\in C} \tilde{\omega}(\mathbf{a}^T D\mathbf{a})\tilde{\omega}(\mathbf{a}^T L\mathbf{x})|\mathbf{a}+\mathbf{x}\rangle$$

for each coset $C + \mathbf{x}$ as $\mathbf{x}$ runs over a set of distinct coset representatives of C in $\mathbb{F}_q^n$, is an orthonormal basis for $\mathcal{C}(\mathcal{S})$. In particular, $\dim \mathcal{C}(\mathcal{S}) = q^{n-\dim C}$.

REMARK 3.9. $\mathcal{C}(\mathcal{S})$ is an $[[n,k,d]]_q$ quantum code if it has dimension q^k and $d(\mathcal{S}) \geq d$. In line with classical coding theory we can define the *rate* of an $[[n,k,d]]_q$ quantum code as k/n and *relative distance* as d/n. It is clearly desirable to design quantum codes with large rates and relative distance. An $[[n,k,d]]_q$ quantum code $\mathcal{C}(\mathcal{S})$ is a *pure* code if the corresponding centralizer subgroup $Z(\mathcal{S})$ has the property that $wt(\mathbf{a},\mathbf{b}) \geq d$ for each $\omega^i U_{\mathbf{a}} V_{\mathbf{b}} \in Z(\mathcal{S})$, $(\mathbf{a},\mathbf{b}) \neq (0,0)$. (Notice that this is a stronger property than guaranteed by Theorem 2.3). Given a pure quantum stabilizer code, the following simple method can be used for deriving new quantum codes.

Suppose we have an $[[n,k,d]]_q$ pure quantum code, with a small k and large d. From such a code we can construct an $[[n-1,k+1,d-1]]_q$ quantum code that is again pure, by the technique of puncturing S to yield an additive subgroup S' of $\mathbb{F}_q^{n-1} \times \mathbb{F}_q^{n-1}$, of size still q^{n-k} and distance at least $d(\mathcal{S}) - 1$. The idea of punctured classical codes (see McWilliams and Sloane [**9**]) can be adapted to punctured pure quantum stabilizer codes following [**3**] where it is shown for $q = 2$. A repeated application of puncturing will give $[[n-k',k+k',d-k']]_q$ codes for different choices of k'.

4. A class of stabilizer codes

First choose and fix the following subspace C of $\mathbb{F}_q^n$:

$$C = \{(a_1, \ldots, a_n)^T \in \mathbb{F}_q^n \mid \sum_i a_i = 0\}.$$

The subspace C is invariant under the cyclic shift permutation $\sigma : i \mapsto (i+1) mod\ n$. Thus, $C^\perp = \{(a, \ldots, a) \in \mathbb{F}_q^n \mid a \in \mathbb{F}_q\}$ is also invariant under σ. An $n \times n$ matrix L over $\mathbb{F}_q$ is said to be *circulant* if for $i = 2, \ldots, n$, the ith row of L is obtained by applying σ^{i-1} to the first row.

Let $\mathcal{S} = \{\tilde{\omega}(\mathbf{a}^T D\mathbf{a})U_\mathbf{a} V_{L\mathbf{a}} \mid \mathbf{a} \in C\}$, where $L = D + D^T$ is an $n \times n$ matrix over $\mathbb{F}_q$ and C is as chosen above. We further specialize our construction by choosing L to be an $n \times n$ symmetric circulant matrix with entries from $\{0, 1\}$ and with all diagonal entries 0. Let $\mathbf{e}_1 = (1, 0, \ldots, 0) \in \mathbb{F}_q^n$. For such an L observe that $\mathbf{a}^T L\mathbf{e}_1 = \mathbf{a}^T D^T \mathbf{e}_1$ for $\mathbf{a} \in C$. Then the orthonormal basis $\{|\psi_{C+\mathbf{x}}\rangle\}$ for $\mathcal{C}(\mathcal{S})$ (as described in Equation (3.7)) can be written in the following form:

(4.1)
$$|\psi_{C+c\mathbf{e}_1}\rangle = \frac{1}{\sqrt{\#C}} \sum_{\mathbf{a} \in C} \tilde{\omega}((\mathbf{a}^T + c\mathbf{e}_1^T)D(\mathbf{a} + c\mathbf{e}_1))|\mathbf{a} + c\mathbf{e}_1\rangle, \ \mathbf{c} \in \mathbb{F}_q.$$

In particular, for $q = 2$ the above stabilizer code has a neat encoding circuit that we describe in Figure 8 in the appendix.

As an example of stabilizer codes given by Equation (4.1), we now describe a $[[5, 1, 3]]_q$ quantum code for every finite field $\mathbb{F}_q$. In particular, for $q = 2$, the $[[5, 1, 3]]_2$ code is the Laflamme code which was originally obtained by a computer search [**7**]. Let L_5 be the following symmetric circulant matrix in $\mathbb{F}_q^{5\times 5}$.

$$\begin{pmatrix} 0 & 0 & 1 & 1 & 0 \\ 0 & 0 & 0 & 1 & 1 \\ 1 & 0 & 0 & 0 & 1 \\ 1 & 1 & 0 & 0 & 0 \\ 0 & 1 & 1 & 0 & 0 \end{pmatrix}.$$

It can be checked that $S = \{(\mathbf{a}, L_5\mathbf{a}) \mid \mathbf{a} \in C\}$ is an additive subgroup of $\mathbb{F}_q^5 \times \mathbb{F}_q^5$ such that $d(\mathcal{S}) \geq 3$. Thus, by Theorem 2.3, $\mathcal{C}(\mathcal{S})$ is a $[[5, 1, 3]]_q$ quantum code for every finite field $\mathbb{F}_q$. The encoding circuit for the $[[5, 1, 3]]_2$ can be obtained easily from the general encoding circuit already described for codes given by Equation (4.1).

For a vector $\mathbf{c} \in \mathbb{F}_2^n$, let $\sigma\mathbf{c} \in \mathbb{F}_2^n$ denote the vector obtained by a cyclic shift of $\mathbf{c}$. An $n \times n$ circulant matrix with first column $\mathbf{c} \in \mathbb{F}_2^n$ can be conveniently written as

$$\begin{pmatrix} \mathbf{c} & \sigma\mathbf{c} & \ldots & \sigma^{n-1}\mathbf{c} \end{pmatrix}$$

We give two more examples of quantum codes defined using circulant matrices.

First, there is a $[[13, 1, 5]]_2$ quantum code defined by a 13×13 circulant matrix L_{13} over $\mathbb{F}_2$, whose first column is

$$\mathbf{c} = (0, 0, 1, 1, 0, 0, 0, 0, 0, 0, 1, 1, 0)^T.$$

It can be checked (with the help of a computer program) that $S = \{(\mathbf{a}, L_{13}\mathbf{a}) \mid \mathbf{a} \in C\}$ is an additive subgroup of $\mathbb{F}_2^{13} \times \mathbb{F}_q^{13}$ such that $d(\mathcal{S}) \geq 5$. Thus, by Theorem 2.3, $\mathcal{C}(\mathcal{S})$ is a pure $[[13, 1, 5]]_2$ quantum code.

Similarly, there is a $[[21, 1, 7]]_2$ quantum code defined by a 21×21 circulant matrix L_{21} over $\mathbb{F}_2$, whose first column is

$$\mathbf{c} = (0, 1, 1, 0, 1, 1, 1, 0, 0, 0, 0, 0, 0, 0, 0, 1, 1, 1, 0, 1, 1)^T$$

It can be checked using a computer program that $S = \{(\mathbf{a}, L_{21}\mathbf{a}) \mid \mathbf{a} \in C\}$ is an additive subgroup of $\mathbb{F}_2^{21} \times \mathbb{F}_q^{21}$ such that $d(\mathcal{S}) \geq 7$. Thus, by Theorem 2.3, $\mathcal{C}(\mathcal{S})$ is a pure $[[21, 1, 7]]_2$ quantum code.

If $k = 1$, it is interesting to note that for $n = 5, 13$ and 21, the best achievable minimum distance [**3**] is $d = 3, 5$, and 7 respectively.

5. Existence of good stabilizer codes

Using a probabilistic argument we show that there is a number $\alpha > 0$ and a natural number n_α such that for each $n > n_\alpha$ there exists a $[[n, 1, \lfloor \alpha n \rfloor]]_2$ pure quantum stabilizer code. Now, as observed in Remark 3.9, given β such that $0 < \beta < \alpha$, by the method of punctured codes we can obtain a family of $[[\lfloor(1-\beta)n\rfloor, \lfloor \beta n \rfloor, \lfloor(\alpha-\beta)n\rfloor]]_2$ quantum codes for all $n > n_\alpha$. These are good quantum codes with constant rate $\beta/(1-\beta)$ and constant relative distance $(\alpha-\beta)/(1-\beta)$.

We first recall a particular form of the Chernoff bounds for bounding the probability that a random variable deviates far from its expectation.

THEOREM 5.1. [**10**, Theorem 4.2, page 70] *Let $X_1, X_2, \ldots, X_n$ be independent Bernoulli random variables such that for each i,* $\Pr[X_i = 1] = p$ *and* $\Pr[X_i = 0] = 1 - p$, *for $0 < p < 1$. Let $X = \sum_i X_i$ and* *let μ denote the expectation* $\mathbf{E}[X]$. *Then for* $0 < \delta < 1$

$$\Pr[X < (1-\delta)\mu] < e^{-\mu\delta^2/2}.$$

Our existence proof for stabilizer codes will be guided by Lemma 3.4.

As before, we first choose and fix the following subspace C of $\mathbb{F}_q^n$:

$$C = \{(a_1, \ldots, a_n) \in \mathbb{F}_q^n \mid \sum_i a_i = 0\}.$$

DEFINITION 5.2. *An $n \times n$ matrix R over $\mathbb{F}_2$ is said to be α-*good *if the following conditions are true.*

(i) *The sum of every $\lfloor \alpha n \rfloor$ columns of R has weight at least αn.*
(ii) *The sum of every $\lfloor \alpha n \rfloor$ rows of R has weight at least αn.*
(iii) *The sum of every $\lfloor \alpha n \rfloor$ columns of R has weight at most $(1-\alpha)n$.*
(iv) *The sum of every $\lfloor \alpha n \rfloor$ rows of R has weight at most $(1-\alpha)n$.*

As in classical coding theory [**9**], given a vector $\mathbf{a} = (a_1, a_2, \ldots, a_n)^T \in \mathbb{F}_q^n$ we denote $\#\{i \mid a_i \neq 0\}$ by $w(\mathbf{a})$. The next proposition describes a way of constructing stabilizer codes from good matrices.

THEOREM 5.3. *For $0 < \alpha < 1$, suppose R is an $n \times n$ α-good matrix over $\mathbb{F}_2$. Let L be the following $2n \times 2n$ symmetric matrix over $\mathbb{F}_2$:*

$$\begin{pmatrix} 0 & R \\ R^T & 0 \end{pmatrix}$$

If we write $L = D + D^T$, where D is the upper triangular matrix with zeros on the principal diagonal, and define the abelian subgroup $\mathcal{S}$ of $\mathcal{E}$ as $\mathcal{S} = \{\tilde{\omega}(\mathbf{a}^T D\mathbf{a}) U_{\mathbf{a}} V_{L\mathbf{a}} \mid \mathbf{a} \in C\}$, then $\mathcal{C}(\mathcal{S})$ is a $[[2n, 1, \lfloor \alpha n \rfloor]]_2$ pure stabilizer code.

Proof. From Lemma 3.4, we know that $\mathcal{C}(\mathcal{S})$ is a $[[2n, 1, \lfloor \alpha n \rfloor]]_2$ stabilizer code if for any $(\mathbf{x}, \mathbf{y}) \in \mathbb{F}_2^{2n} \times \mathbb{F}_2^{2n}$, the condition $\mathbf{y} - L\mathbf{x} \in C^\perp$ implies that either $\mathbf{x} \in C$ and $\mathbf{y} = L\mathbf{x}$ or $wt(\mathbf{x}, \mathbf{y}) > \alpha n$. It is easy to check that the assumptions about R in Definition 5.2, in fact, guarantees a stronger property: for any nonzero vector $\mathbf{x} \in \mathbb{F}_2^n$ such that $w(\mathbf{x}) \leq \alpha n$, the assumptions (i) and (iii) imply that $\alpha n \leq w(R\mathbf{x}) \leq (1-\alpha)n$. Similarly, assumptions (ii) and (iv) imply that $\alpha n \leq w(R^T\mathbf{x}) \leq (1-\alpha)n$. Putting these together, it follows that $\alpha n \leq w(L\mathbf{x}) \leq (1-\alpha)n$ if $w(\mathbf{x}) \leq \alpha n$ for nonzero $\mathbf{x} \in \mathbb{F}_2^{2n}$.

Since $C^\perp = \{(1,1,\ldots,1)^T,(0,0,\ldots,0)^T\}$, we can see that the above observation implies that $wt(\mathbf{x},\mathbf{y}) > \alpha n$ if $(0,0) \neq (\mathbf{x},\mathbf{y}) \in \mathbb{F}_2^{2n} \times \mathbb{F}_2^{2n}$, and $\mathbf{y} - L\mathbf{x} \in C^\perp$. It follows that $\mathcal{C}(\mathcal{S})$ is a $[[2n,1,\lfloor \alpha n \rfloor]]_2$ pure stabilizer code. This completes the proof. ∎

We now show the existence of $n \times n$ matrices over $\mathbb{F}_2$ that fulfill the conditions of Theorem 5.3.

LEMMA 5.4. *Let R_{ij}, $1 \le i,j \le n$ be independent identically distributed random variables taking values in $\{0,1\}$ such that* $\Pr[R_{ij} = 1] = 1/2$, $1 \le i,j \le n$. *Let R be the uniformly distributed $n \times n$ random matrix over $\mathbb{F}_2$ whose ij^{th} entry is the random variable R_{ij}. There exist constants $\alpha > 0$ and $n_\alpha > 0$ such that*

$$\Pr[R \textit{ is } \alpha\textit{-good }] > 0.$$

Proof. Let BAD denote the event that R is not α-good. Let $\mathbf{r}_1,\mathbf{r}_2,\ldots,\mathbf{r}_n$ be the rows of R and $\mathbf{c}_1,\mathbf{c}_2,\ldots,\mathbf{c}_n$ be the columns of R. For any subset $S \subseteq \{1,2,\ldots,n\}$ with $1 \le \#S \le \alpha n$, we define E_S, D_S, A_S, B_S as the events $w(\bigoplus_{i\in S}\mathbf{r}_i) < \alpha n$, $w(\bigoplus_{i\in S}\mathbf{r}_i) > (1-\alpha)n$, $w(\bigoplus_{i\in S}\mathbf{c}_i) < \alpha n$, $w(\bigoplus_{i\in S}\mathbf{c}_i) > (1-\alpha)n$ respectively. Then BAD can be written as follows

$$\text{BAD} = \bigcup_{S\subset[n],1\le\#S\le\alpha n} A_S \cup B_S \cup D_S \cup E_S. \tag{5.5}$$

We analyze A_S for a fixed S. Let $\bigoplus_{i\in S}\mathbf{c}_i = (x_1,x_2,\ldots,x_n)^T$. Since R_{ij} $1 \le i,j \le n$ are all independent random variables taking values in $\mathbb{F}_2$, $x_1,x_2,\ldots,x_n$ are n independent uniformly distributed random variables taking values in $\mathbb{F}_2$. We will use Chernoff bounds as given in Theorem 5.1 to analyze the random variable $\#\{i \mid x_i = 1, 1 \le i \le n\}$. Let $X = \sum_{i=1}^n x_i$. Then $\mathbf{E}[X] = n/2$. Applying Theorem 5.1 we get

$$\Pr[A_S] = \Pr[X < \alpha n] \le e^{-\frac{n(1-2\alpha)^2}{4}}.$$

Notice that under $\mathbb{F}_2$ addition $1+x_1, 1+x_2, \ldots, 1+x_n$ are also n independent uniformly distributed random variables taking values in $\mathbb{F}_2$. Thus, by Chernoff bounds we again obtain $\Pr[B_S] \le e^{-\frac{n(1-2\alpha)^2}{4}}$. Likewise, $\Pr[E_S]$ and $\Pr[D_S]$ are also bounded above by $e^{-\frac{n(1-2\alpha)^2}{4}}$.

Putting these together with the definition of BAD in Equation (5.5) we get

$$\Pr[\text{BAD}] \le 4e^{-\frac{n(1-2\alpha)^2}{4}} \cdot \sum_{i=1}^{\lfloor \alpha n \rfloor} \binom{n}{i} \le 4e^{-\frac{n(1-2\alpha)^2}{4}} 2^{nH(\alpha)}$$

where $H(\alpha) = -\alpha(\log \alpha) - (1-\alpha)\log(1-\alpha)$. To ensure that $\Pr[\text{BAD}] < 1$, it suffices to pick $\alpha < 1/4$ such that $H(\alpha) < (\log e)3/8 - 2/n$, which can be done by choosing n larger than some constant n_α and $\alpha > 0$ sufficiently small. ■

From Theorem 5.3, Lemma 5.4, and Remark 3.9 we can immediately deduce the following.

COROLLARY 5.6. *There are constants $\alpha > 0$ and $n_\alpha > 0$ such that for each $n > n_\alpha$ there is a $[[n, 1, \lfloor \alpha n \rfloor]]_2$ pure quantum stabilizer code. Furthermore, for any β such that $0 < \beta < \alpha$, and $n > n_\alpha$ there is a $[[\lfloor (1-\beta)n \rfloor, \lfloor \beta n \rfloor, \lfloor (\alpha - \beta)n \rfloor]]_2$ pure quantum stabilizer code.*

REMARK 5.7. The above existence argument can be easily extended to stabilizer codes over any finite field. More precisely, for $\mathbb{F}_q$ there are constants $\alpha > 0$ and $n_\alpha > 0$ such that for each $n > n_\alpha$ there is a $[[n, 1, \lfloor \alpha n \rfloor]]_q$ pure quantum stabilizer code. Also, given a β such that $0 < \beta < \alpha$, and $n > n_\alpha$ there is a $[[\lfloor (1-\beta)n \rfloor, \lfloor \beta n \rfloor, \lfloor (\alpha - \beta)n \rfloor]]_q$ pure quantum code.

Acknowledgment We thank Piyush P Kurur for helping with the computer search for codes given in Section 4. We are indebted to the referee for his meticulous refereeing and useful comments.

References

[1] V. Arvind, Piyush P Kurur, and K.R. Parthasarathy, *Nonstabilizer quantum codes from abelian subgroups of the error group,* quant-ph/0210097.

[2] Alexei Ashikhmin, Emanuel Knill, *Nonbinary quantum stabilizer codes,* IEEE Transactions on Information Theory, vol. 47, No. 7, 3065-3072, 2001.

[3] A. R. Calderbank, Eric M. Rains, Peter W. Shor and N. J. A. Sloane, *Quantum error correction via codes over GF(4),* IEEE Trans. Inform. Theory, Vol. 44 (1998), pp. 1369-1387.

[4] D. G. Cory, W. Mass, M. Price, E. Knill, R. Laflamme, W. H. Zurek, T. F. Havel, S. S. Somaroo, *Experimental quantum error correction,* Phys.Rev.Lett. 81 (1998) 2152-2155.

[5] D. Gottesman, *A class of quantum error-correcting codes saturating the quantum Hamming bound,* Phys. Rev. A 54, 1862-1868 (1996). Also quant-ph/9604038.

[6] Emanuel Knill, Raymond Laflamme, *A theory of quantum error-correcting codes*, Phys.Rev.Lett. 84 (2000) 2525-2528.

[7] Raymond Laflamme, Cesar Miquel, Juan Pablo Paz, Wojciech Hubert Zurek, *Perfect quantum error correction code,* Physical Review Letters, 98, 77, 1996.

[8] J. H. van Lint, Introduction to Coding Theory, *Springer Verlag,* 3rd edition, 1998.

[9] F. J. Macwilliams and N. J. A. Sloane, Theory of Error-Correcting Codes, *North Holland,* Amsterdam, 1978.

[10] R. Motwani and P. Raghavan, Randomized Algorithms, *Cambridge University Press,* 1995.

[11] Michael A. Nielsen and Isaac L. Chuang, Quantum Computation and quantum Information, *Cambridge,* 1999.

[12] K. R. Parthasarathy, *Lectures on Quantum Computation and Quantum Error-Correcting Codes*, Institute of Mathematical Sciences, Chennai.

[13] E. M. Rains, *Nonbinary quantum codes,* IEEE Transactions on Information Theory, vol. 45, no. 6, pp. 1827-1832, 1999.

[14] E. M. Rains, R. H. Hardin, P. W. Shor, and N. J. A. Sloane, *A nonadditive quantum code,* Phys.Rev.Lett. 79 (1997) 953-954.

[15] Vwani P. Roychowdhury, Farrokh Vatan, *On the existance of nonadditive quantum codes,* QCQC (1998) 325-336.

[16] Vwani P. Roychowdhury, Farrokh Vatan, *On the Structure of Additive Quantum Codes and the Existence of Nonadditive Codes,* quant-ph/9710031.

Institute of Mathematical Sciences
C.I.T Campus, Chennai 600113, India.

E-mail address: {arvind,krp}@imsc.res.in

Appendix A. The encoding circuit

We describe the encoding circuit for the stabilizer code with orthonormal basis defined by Equation 4.1 in the figures below. The elements that we use to build our encoding circuit are the Hadamard gate, the C-NOT gate and the Z gate. We first describe these gates. Then we give the four main components B, $H^{\otimes^n}$, D, and E and finally, the complete encoding circuit in Figure 8.

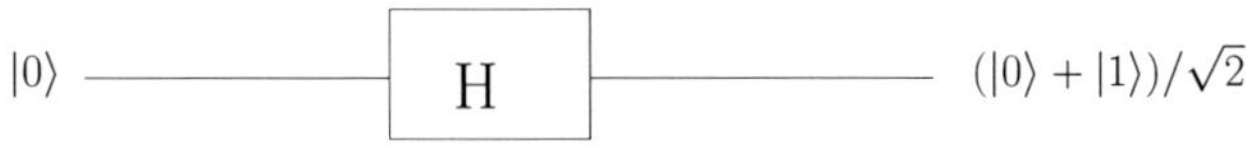

FIGURE 1. *The Hadamard gate.*

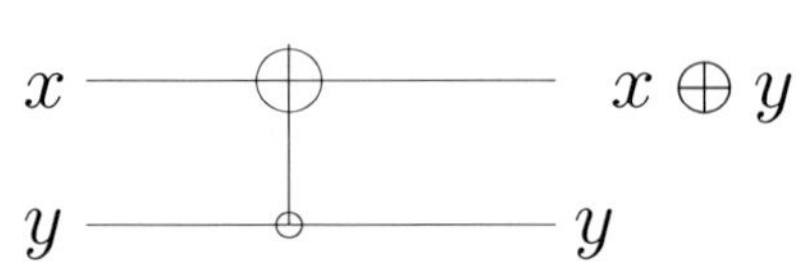

FIGURE 2. *The controlled NOT gate.*

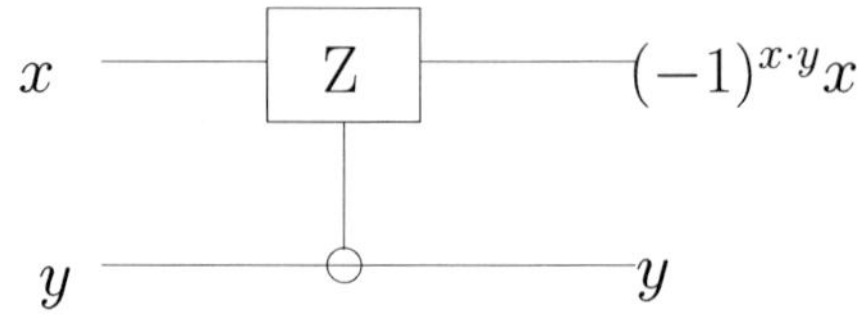

FIGURE 3. *The controlled Z gate.*

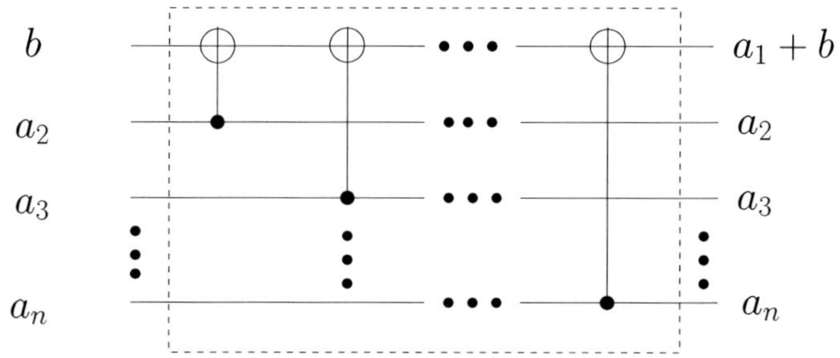

FIGURE 4. *B: computing* $\mathbf{a} + b\mathbf{e}_1$.

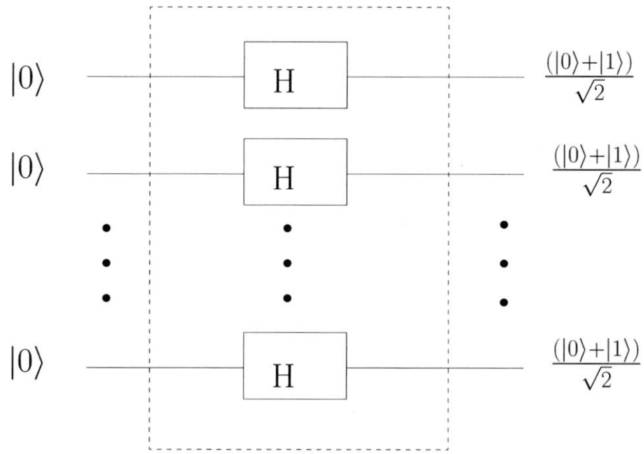

FIGURE 5. $H^{\otimes^{n-1}}$.

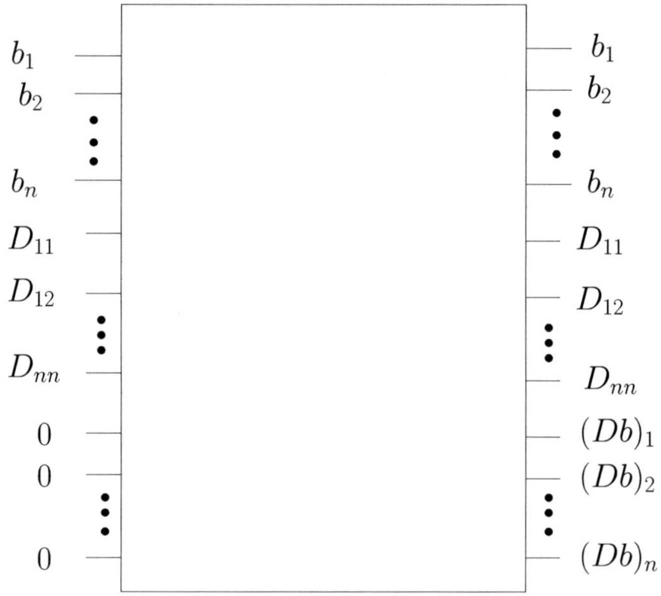

FIGURE 6. *The linear map D.*

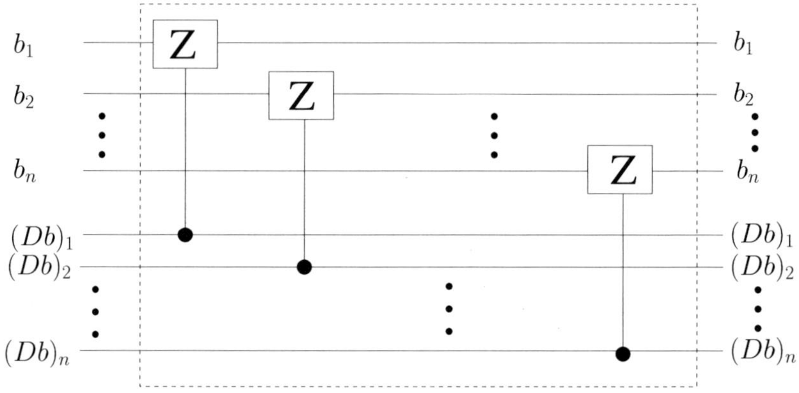

FIGURE 7. *Component E.*

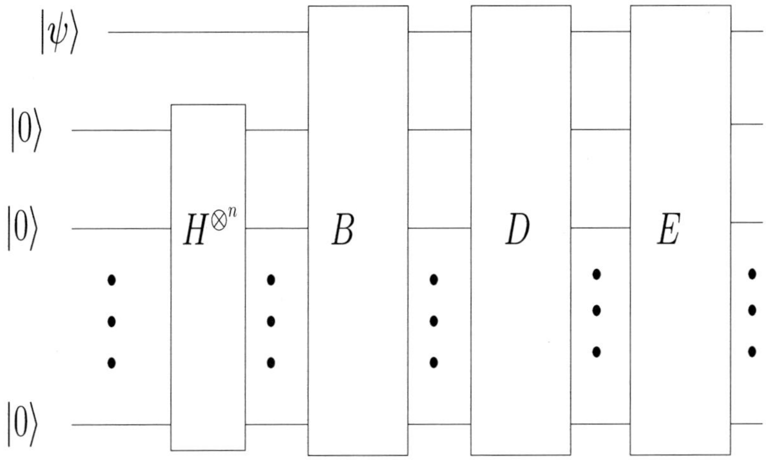

FIGURE 8. *The encoding circuit.*

Principal bundle, parabolic bundle, and holomorphic connection

INDRANIL BISWAS

Dedicated to C.S. Seshadri on his 70th birthday

Abstract. Let E_G be a principal G–bundle over a rationally connected variety, where G is a complex algebraic group. Then any holomorphic connection on E_G is flat. We describe a necessary and sufficient condition for a parabolic vector bundle over a Riemann surface to admit a logarithmic connection compatible with the parabolic structure.

1. Introduction

The notion of parabolic vector bundles was introduced by Seshadri in [**Se2**]. Parabolic vector bundles have turned out to be useful in various situations.

Let X be a compact connected Riemann surface and $S \subset X$ a finite subset. Let E be a holomorphic vector bundle over X. A parabolic structure on E over S is a decreasing filtration $\{F_i^s\}$ of the fiber E_s over each $s \in S$ together with an increasing sequence of numbers $0 \leq \lambda_i^s < 1$ corresponding to the subspaces occurring in the filtration. The parabolic degree is defined using the degree of V together with the parabolic weights λ_i^s multiplied by the dimension of the corresponding graded pieces.

A theorem due to A. Weil says that a holomorphic vector bundle V over X admits a holomorphic connection if and only if each indecomposable component of V is of degree zero [**We**], [**At2**].

To extend this theorem to the context of parabolic vector bundles, we need to define holomorphic connection for a parabolic vector bundle. The notion of a logarithmic connection introduced in [**De**] is well suited for this purpose. So a holomorphic connection on the parabolic vector defined above would be a logarithmic connection D on the underlying vector bundle E which is singular over S, and furthermore, the residue of D at each $s \in S$ is required to have some compatibility properties with respect to the filtration of the fiber E_s (see Section 7 for the precise definition).

In Theorem 7.1, extending the above criterion for connection to the context of parabolic bundles we prove:

A parabolic vector bundle admits a holomorphic connection if and only if every direct summand of it is of parabolic degree zero.

The proof of Theorem 7.1 is carried out using a bijective correspondence between parabolic bundles and π–bundles. We first prove a corresponding criterion for the existence of an equivariant holomorphic connection on a π–bundle (see Theorem 7.4). Using this result, the bijective correspondence between parabolic bundles and π–bundles yields the above criterion for the existence of a holomorphic connection on a parabolic vector bundle.

We recall that π–bundles were introduced in [**Go**]. In [**Se1**], the moduli space of semistable π–bundles was constructed as a normal projective variety, and a bijective correspondence between polystable π–bundles and π–unitary representations was established.

In [**AB**], the above criterion of [**We**] for the existence of a holomorphic connection was extended to principal G–bundles, where G is a complex reductive algebraic group. It would be interesting to be able to actually extend this criterion to principal P–bundles, where P is a parabolic subgroup of G.

In [**KMM1**], the notion of a rationally connected variety was introduced. Roughly, a rationally connected variety contains plenty of rational curves (see Section 3 for the precise definition).

Let G be a complex algebraic group. Let M be a smooth projective rationally connected variety.

In Theorem 3.1 we prove the following:

Any holomorphic connection D on a principal G–bundle over M is flat, that is, the curvature of D vanishes.

Acknowledgements

The author is very grateful to Christophe Mourougane for very useful discussions regarding Theorem 3.1. The author thanks firstly the organizers of the symposium in honor of Seshadri for the invitation and, secondly, the Tata Institute of Fundamental Research for a travel support that enabled him to attend the symposium.

2. Principal bundle and connection

Let M be a connected smooth projective variety over $\mathbb{C}$. Let G be an algebraic group over $\mathbb{C}$.

A *Principal G–bundle* over M is a smooth complex variety E equipped with an action of G on the right and a surjective morphism $p : E \longrightarrow M$ satisfying the following conditions

(1) the map p is smooth;
(2) the map p is a morphism of G–spaces, with the action of G on M being the trivial one;
(3) the map $E \times G \longrightarrow E \times_M E$ defined by $(z\,,g) \longmapsto (z\,,zg)$ is an isomorphism.

Note that we do *not* assume E to be locally trivial in Zariski topology.

Let Z be a variety equipped with a left action of G. For a principal G–bundle E over M, consider the "twisted" diagonal action of G on $E \times Z$. The "twisted" diagonal action of any $g \in G$ sends $(y, z) \in E \times Z$ to $(yg, g^{-1}z)$. The quotient space $(E \times Z)/G$ is a fiber bundle over M with Z as the typical fiber. This fiber bundle, which will be denoted by $E(Z)$, is known as the *associated fiber bundle* — associated to E for the G–variety Z.

Let $\mathfrak{g}$ denote the Lie algebra of G. The group G acts on $\mathfrak{g}$ by the adjoint representation. For a principal G–bundle E, the associated vector bundle $E(\mathfrak{g})$ is called the adjoint vector bundle, and it is denoted by $\mathrm{ad}(E)$.

If $\rho : G \longrightarrow H$ is a homomorphism of algebraic groups, then ρ induces a left action of G on H. For a principal G–bundle E, the associated fiber bundle $E(H)$ is a principal H–bundle. This construction is known as the *extension of structure group*. Note that there is a morphism $E \longrightarrow E(H)$ that sends any $z \in E$ to $(z\,, e) \in (E \times H)/G$, where $e \in H$ is the identity element.

Let H_1 be a closed subgroup of G. For a principal G–bundle E, consider the quotient space $E/H_1 = E(G/H_1)$. The projection of E to E/H_1 will be denoted by q. A section

$$\sigma : M \longrightarrow E(G/H_1)$$

of the fiber bundle has the property that the inverse image $q^{-1}(\sigma(M))$ is a principal H_1 bundle over M. In other words, $q^{-1}(\sigma(M))$ is closed under the action of H_1 and the restriction of p to $q^{-1}(\sigma(M))$ makes it into a principal H_1 bundle. Here p, as before, denotes the projection of E to M. Conversely, if $E_{H_1} \subset E$ is a subvariety which is closed under the action of H_1 (for the action of G on E), and H_1 acts transitively on the fibers of the projection of E_{H_1} to M, then E_{H_1} is a principal H_1 bundle, and furthermore, sending any $x \in M$ to the H_1 orbit $E_{H_1} \cap p^{-1}(x)$ in the G space $p^{-1}(x)$ we get a section of the fiber bundle $E(G/H_1)$ over M. This construction of a principal H_1 bundle as a subvariety of E is known as a *reduction of structure group* of E.

Now we will define a holomorphic connection on a principal bundle.

Let E be a principal G–bundle over M. For any analytic open subset $U \subset M$, consider the inverse image $p^{-1}(U)$, which is a complex manifold equipped with an action of G. Let $\mathcal{A}(U)$ denote the space of holomorphic vector fields on $p^{-1}(U)$ that are left invariant by the action of G on $p^{-1}(U)$. Note that $\mathcal{A}(U)$ is closed under multiplication by holomorphic functions pulled back from U. Since the action of G is transitive on the fibers of p, the sheaf on M that associates to any U the vector space $\mathcal{A}(U)$ is a coherent analytic sheaf. Let $\mathrm{At}(E)$ denote the corresponding holomorphic, hence algebraic, vector bundle over M for the coherent sheaf.

The vector bundle $\mathrm{At}(E)$ defined above is known as the *Atiyah bundle* for E [**At2**].

The sheaf of sections of $\mathrm{At}(E)$ has the structure of a Lie algebra. Since $\mathcal{A}(U)$ is closed under the Lie bracket operation, there is an induced Lie algebra structure on the sheaf of section of $\mathrm{At}(E)$.

Let $\mathcal{A}_{\mathrm{rel}}(U) \subset \mathcal{A}(U)$ be the subspace defined by the vertical (relative) tangent bundle, that is, $\mathcal{A}_{\mathrm{rel}}(U)$ is in the kernel of the differential dp of the projection p. Using the fact that the holomorphic tangent bundle TG over G is identified with the trivial vector bundle with fiber $\mathfrak{g}$ by right translations of G, the vector space $\mathcal{A}_{\mathrm{rel}}(U)$ is identified with the space of all holomorphic sections of the adjoint bundle $\mathrm{ad}(E)$ over U.

Consequently, we have an exact sequence of vector bundles

$$0 \longrightarrow \mathrm{ad}(E) \stackrel{\iota}{\longrightarrow} \mathrm{At}(E) \stackrel{\sigma}{\longrightarrow} TM \longrightarrow 0 \tag{2.1}$$

where σ is defined using dp [**At2**]. This sequence is known as the *Atiyah exact sequence.*

A *holomorphic connection* on E is a splitting of the Atiyah exact sequence, that is, a homomorphism

$$D : TM \longrightarrow \mathrm{At}(E)$$

such that $\sigma \circ D$ is the identity automorphism of TM [**At2**]. For another such splitting D' of (2.1), the difference $D - D'$ is a homomorphism from TM to $\mathrm{ad}(E)$. Conversely, for any holomorphic section s of $\Omega^1_M \bigotimes \mathrm{ad}(E)$, clearly $D + s$ is a splitting of (2.1) if D is so. In other words, the space of all holomorphic connections on E is an affine space for the vector space $H^0(M, \Omega^1_M \bigotimes \mathrm{ad}(E))$.

Recall that the sheaf of sections of $\mathrm{At}(E)$ is equipped with a natural Lie algebra structure. The image of $\mathrm{ad}(E)$ by the inclusion map ι in (2.1) is a subalgebra. This algebra structure on $\mathrm{ad}(E)$ coincides with the fiberwise Lie algebra structure induced by $\mathfrak{g}$ on the associated bundle $\mathrm{ad}(E)$. Since the adjoint action of G on $\mathfrak{g}$ is Lie algebra structure preserving, the fibers of $\mathrm{ad}(E)$ are Lie algebras. The sequence (2.1) is in fact an exact sequence of Lie algebra sheaves, with the Lie algebra structure on the sheaf of sections of TM being the given by Lie bracket.

The image $D(TM)$ of a splitting D of (2.1) need not be a subalgebra of $\mathrm{At}(E)$. Since σ in (2.1) is Lie algebra structure preserving,

$$[D(v)\,, D(w)] \in \mathrm{ad}(E)_x$$

where $v, w \in T_xM$ and $x \in M$. (Actually we should take the commutator of two local sections passing through $D(v)$ and $D(w)$, but the evaluation at x of such a commutator does not depend on the choice of the sections.) In other words, the obstruction for D being Lie algebra structure preserver is a section

$$K(D) \in H^0(M, \Omega^2_M \otimes \mathrm{ad}(E))\,. \tag{2.2}$$

The $\mathrm{ad}(E)$ valued holomorphic two–form $K(D)$ is called the *curvature* of M. If $K(D) = 0$, then D is called a *flat connection.*

Note that if M is a Riemann surface, then any holomorphic connection is automatically flat, as $\Omega^2_M = 0$.

Let $\rho : G \longrightarrow H$ be a homomorphism of algebraic groups and E a principal G–bundle over M. Consider the map

$$\hat{\rho} : E \longrightarrow E(H)$$

defined earlier that sends any z to $(z\,, e)$. If ν is a G invariant vector field on $p^{-1}(U) \subset E$, where p is the projection of E to M, then $d\hat{\rho}(\nu)$ is a H invariant vector field on $q^{-1}(U) \subset E(H)$, where q is the projection of the principal H–bundle $E(H)$ to M, and $d\hat{\rho}$ is the differential of the above map $\hat{\rho}$. Therefore, we have a homomorphism

$$\tilde{\rho} : \mathrm{At}(E) \longrightarrow \mathrm{At}(E(H))$$

of Atiyah bundles. If $D : TM \longrightarrow \mathrm{At}(E)$ is a holomorphic connection on E, then $\tilde{\rho} \circ D$ is a holomorphic connection on the principal H–bundle $E(H)$.

Giving a holomorphic connection D on a G–bundle E is equivalent to giving a holomorphic $\mathfrak{g}$–valued one–form

$$\omega \in H^0(E, \Omega^1_E \otimes \mathfrak{g}) \tag{2.3}$$

satisfying the following two conditions:

(1) ω is G–equivariant for the adjoint action of G on $\mathfrak{g}$ and the obvious action on E;
(2) the restriction of the form ω to any fiber of the projection p to M coincides with the Maurer–Cartan form.

Given such a form ω, the kernel of ω defines a homomorphism from TM to $\mathrm{At}(E)$ splitting the Atiyah exact sequence. Conversely, given a splitting of the Atiyah exact sequence, there is a unique one–form ω satisfying the above two conditions whose kernel is the image of TM for the splitting homomorphism.

3. Connection on a rationally connected variety

If E is a principal G bundle over M admitting a holomorphic connection, then all the rational characteristic classes of E vanish [**At2**]. It is a well known question whether E admits a flat holomorphic connection given that it admits a holomorphic connection. The above result of [**At2**] shows that there are no topological obstructions for a G–bundle to admit a flat holomorphic connection assuming that it admits a holomorphic connection.

Let the projective manifold M be rationally connected. We recall that a projective manifold X is called *rationally connected* if *any one* of the following five *equivalent conditions* holds (see [**KMM1**, p. 433, Theorem 2.1] and [**KMM1**, p. 434, Definition–Remark 2.2]):

(1) given finitely many arbitrary points $x_1, \cdots, x_m \in X$ there is an irreducible rational curve on X which contains $x_1, \cdots, x_m$;
(2) given two arbitrary points $x_1, x_2 \in X$ there is an irreducible rational curve on X which contains x_1, x_2;
(3) for a sufficiently general point $(x_1, x_2) \in X \times X$ there is an irreducible rational curve on X which contains x_1, x_2;

(4) given two arbitrary points $x_1, x_2 \in X$ there is a connected curve on X containing x_1, x_2 which is a union of rational curves;
(5) there exists a morphism

$$f : \mathbb{CP}^1 \longrightarrow X$$

with f^*TX ample.

It is known that if X is a Fano manifold, that is, the top exterior product $\bigwedge^{\text{top}} TM$ is an ample line bundle, then X is rationally connected [**KMM2**, p. 766, Theorem 0.1].

Let E be a principal G–bundle over M. The following theorem says that any holomorphic connection on E is automatically flat. A similar theorem was proved in [**BM**] under the stronger assumption that $G = \mathrm{GL}(n, \mathbb{C})$ and M is a Fano manifold whose Picard number, that is, the rank of the Néron–Severi group $H^{1,1}(M) \cap H^2(X, \mathbb{Z})$, is one (see [**BM**, p. 129, Theorem 2.2]).

THEOREM 3.1. *Let D be a holomorphic connection on a principal G–bundle E over a rationally connected complex projective manifold M. Then the curvature $K(D)$, defined in* (2.2), *vanishes identically.*

Proof. Since M is rationally connected, given a general point $x \in M$, there is a morphism

$$f : \mathbb{CP}^1 \longrightarrow M$$

such that $f(z_0) = x$ and f^*TM is ample, where z_0 is a fixed base point of $\mathbb{CP}^1$. This follows from Corollary 1.3 (p. 430) and Claim (2.3.1) (p. 434) of [**KMM1**]; the above assertion is in Proof (2.6) (p. 436) of [**KMM1**].

The holomorphic connection D on E induces a holomorphic connection f^*D on the G–bundle f^*E over $\mathbb{CP}^1$. So we have a holomorphic connection on the adjoint vector bundle $\mathrm{ad}(f^*E)$.

We noted earlier that a holomorphic connection on a Riemann surface is flat. Since $\mathbb{CP}^1$ is simply connected, a vector bundle over $\mathbb{CP}^1$ with a flat connection must be trivial. In particular, $\mathrm{ad}(f^*E)$ is a trivial vector bundle over $\mathbb{CP}^1$.

Now, $\bigwedge^2 f^*TM$ is ample since f^*TM is ample. Since $\bigwedge^2 f^*\Omega^1_M$ is the dual of an ample bundle and $\mathrm{ad}(f^*E)$ is trivial, we have

$$H^0(\mathbb{CP}^1, (\bigwedge^2 f^*\Omega^1_M) \otimes \mathrm{ad}(f^*E)) = 0 .$$

Note that $f^*K(D)$ is a section of $f^*(\Omega^2_M \bigotimes \mathrm{ad}(E)) = (\bigwedge^2 f^*\Omega^1_M) \bigotimes \mathrm{ad}(f^*E)$. Consequently, the curvature $K(D)$ vanishes at $y \in M$ if $y \in f(\mathbb{CP}^1)$.

Therefore, $K(D) = 0$ as it vanishes on a general point of X. This completes the proof of the theorem.

In the next section we will consider connections on a Riemann surface.

4. Connection on a Riemann surface

Let X be a connected smooth projective curve over $\mathbb{C}$, or equivalently, a compact connected Riemann surface. Let E be a holomorphic vector bundle over X. The vector bundle E is called *decomposable* if there are holomorphic subbundles W_1 and W_2 of E of positive rank such that $E \cong W_1 \bigoplus W_2$; and E is called *indecomposable* if it is not decomposable.

Every vector bundle can be expressed as a direct sum of indecomposable vector bundle, and furthermore, the decomposition is unique up to a permutation of the direct summands [**At1**].

A *holomorphic connection* on E is a first order holomorphic differential operator

$$D \in H^0(X, \mathrm{Diff}^1_X(E, \Omega^1_X \otimes E))$$

whose symbol is the identity automorphism of E. It is easy to see that this condition on the symbol is equivalent to the condition that D satisfies the Leibniz identity which says that

$$D(fs) = fD(s) + \partial(f) \otimes s$$

where f is a locally defined holomorphic function on X and s is a locally defined holomorphic section of E. A holomorphic connection on E is same as a holomorphic connection on the corresponding principal $\mathrm{GL}(n, \mathbb{C})$–bundle, where $n = \mathrm{rank}(E)$.

To explain this, let $E_{\mathrm{GL}(n)}$ be a principal $\mathrm{GL}(n, \mathbb{C})$ bundle over X, and let V be the rank n vector bundle associated to $E_{\mathrm{GL}(n)}$ by the standard representation of $\mathrm{GL}(n, \mathbb{C})$. The Atiyah bundle $\mathrm{At}(E_{\mathrm{GL}(n)})$ defined in Section 2 is identified with a certain subbundle $\mathrm{At}(V)$ of the vector bundle $\mathrm{Diff}^1_X(V, V)$ of differential operators. The subbundle $\mathrm{At}(V)$ in question is the inverse image of $TX \bigotimes \mathrm{Id}_V$ of the symbol homomorphism

$$\mathrm{Diff}^1_X(V, V) \longrightarrow TX \otimes \mathrm{End}(V).$$

In this identification, the homomorphism σ in (2.1) is the symbol map. Finally $\mathrm{At}(V)\bigotimes\Omega^1_X$ is identified with the subbundle of $\mathrm{Diff}^1_X(V\,,\Omega^1_X\bigotimes V)$ defined by the inverse image of the line subbundle of $\mathrm{End}(V)$ generated by Id_V for the symbol homomorphism

$$\mathrm{Diff}^1_X(V\,,\Omega^1_X\otimes V)\ \longrightarrow\ TX\otimes\Omega^1_X\otimes\mathrm{End}(V)\ \cong\ \mathrm{End}(V)$$

(see [**At2**] for the details).

Let D be a holomorphic connection on E. So we have degree$(E) = 0$. Assume that $E\cong W_1\bigoplus W_2$, where W_1 and W_2 are holomorphic subbundles of E. Let q denote the projection of E to W_1 defined using the decomposition. The composition

$$W_1\hookrightarrow E\stackrel{D}{\longrightarrow}\Omega^1_X\otimes E\stackrel{\mathrm{Id}\otimes q}{\longrightarrow}\Omega^1_X\otimes W_1$$

defines a holomorphic connection on W_1. Consequently, degree$(W_1) = 0$.

Therefore, if a holomorphic vector bundle E over X admits a holomorphic connection, then every indecomposable component of E is of degree zero. The converse is also true. In other word, a holomorphic vector bundle E over X admits a holomorphic connection if and only if every indecomposable component of E is of degree zero [**At2**], [**We**].

Let G be a connected reductive algebraic group over $\mathbb{C}$. We will describe a condition on G–bundles that ensures the existence of a holomorphic connection.

A parabolic subgroup of G is a closed algebraic subgroup $P\subset G$ such that G/P is complete. Therefore, G itself will be considered as a parabolic subgroup of G. Let $R_u(P)$ denote the unipotent radical of a parabolic subgroup P. The Levi factor $P/R_u(P)$ will be denoted by $L(P)$.

It is possible to find a subgroup of P that maps isomorphically onto $L(P)$. To explain this, fix a maximal torus T of G. Let $L'\subset P$ be the largest T invariant reductive subgroup. Then the restriction of the quotient map $P\longrightarrow L(P)$ to L' is an isomorphism of $L(P)$ with L'.

Henceforth, we will consider the Levi factor $L(P)$ as a subgroup of P by fixing a maximal torus T of G. We have $P = R_u(P)L(P)$.

Let

$$L_0(P)\ :=\ \frac{L(P)}{[L(P)\,,L(P)]}$$

be the quotient by the commutator. So, $L_0(P)$ is the maximal abelian quotient of $L(P)$. Note that $L_0(P)$ is isomorphic to a product of copies of $\mathbb{C}^*$. Since there are no nontrivial (multiplicative) characters on $[L(P), L(P)]$, the characters of $L(P)$ are in bijective correspondence with the characters of $L_0(P)$. Consequently, for a $L(P)$–bundle $E_{L(P)}$ on X, the corresponding $L_0(P)$–bundle $E_{L(P)}(L_0(P))$, obtained by extension of structure group, is topologically trivial if and only if for every character

$$\chi : L(P) \longrightarrow \mathbb{C}^*$$

degree$(E_{L(P)}(\mathbb{C})) = 0$, where $E_{L(P)}(\mathbb{C})$ is the line bundle over X associated to $E_{L(P)}$ for the action of P on $\mathbb{C}$ that factors through χ.

Let E be a principal G–bundle over X. Let

$$E_{L(P)} \hookrightarrow E$$

be a reduction of structure group of E to the Levi factor $L(P)$ of a parabolic subgroup P. Take a character

$$\chi : L(P) \longrightarrow \mathbb{C}^*$$

of $L(P)$. Let

$$\xi := \frac{E_{L(P)} \times \mathbb{C}}{P} = E_{L(P)}(\mathbb{C})$$

be the holomorphic line bundle over X, where the action of P on $\mathbb{C}$ is χ composed with the multiplication action of $\mathbb{C}^*$ on $\mathbb{C}$.

LEMMA 4.1. *If the G–bundle E admits a holomorphic connection then* degree$(\xi) = 0$.

Proof. Let $\mathfrak{l}(P)$ denote the Lie algebra of $L(P)$. Consider the two $L(P)$ modules, namely $\mathfrak{l}(P)$ and $\mathfrak{g}$, equipped with the adjoint action of $L(P)$. Since $L(P)$ is reductive, the inclusion map

$$\iota : \mathfrak{l}(P) \hookrightarrow \mathfrak{g}$$

of $L(P)$–modules admits a splitting. Take a splitting

$$\phi : \mathfrak{g} \longrightarrow \mathfrak{l}(P)\,. \tag{4.1}$$

So, ϕ is $L(P)$ equivariant and $\phi \circ \iota = \mathrm{Id}_{\mathfrak{l}(P)}$.

Let τ be the inclusion map of $E_{L(P)}$ in E. Let

$$\omega \in H^0(E, \Omega^1_E \otimes \mathfrak{g})$$

be a connection form, as in (2.3), on E. Then the form

$$\phi \circ \tau^* \omega \in H^0(E_{L(P)}, \Omega^1_{L(P)} \otimes \mathfrak{l}(P))$$

where ϕ as in (4.1), defines a holomorphic connection on the $L(P)$–bundle $E_{L(P)}$. Indeed, since ϕ is $L(P)$ equivariant, the form $\phi \circ \iota^* \omega$

evidently satisfies both the conditions needed to define a holomorphic connection.

The connection $\omega' := \phi \circ \tau^*\omega$ on $E_{L(P)}$ induces a holomorphic connection on the principal $\mathbb{C}^*$–bundle $E_{L(P)}(\mathbb{C}^*)$ obtained by extending the structure group using the homomorphism χ. Since $E_{L(P)}(\mathbb{C}^*)$ admits a holomorphic connection and ξ is the line bundle associated to $E_{L(P)}(\mathbb{C}^*)$ by the standard action, we conclude that ξ admits a holomorphic connection. Consequently, degree$(\xi) = 0$. This completes the proof of the lemma.

It was proved in [**AB**] that the converse of Lemma 4.1 is also valid. In other words, the following theorem is valid ([**AB**, p. 342, Theorem 4.1]).

THEOREM 4.2. *Let G be a connected reductive algebraic group over $\mathbb{C}$ and E a principal G–bundle over a compact connected Riemann surface X. The principal bundle E admits a holomorphic connection if and only if for every reduction*

$$E_{L(P)} \subset E\,,$$

where $L(P)$ is the Levi factor of some parabolic subgroup P, and for every character χ of $L(P)$, the degree of the associated line bundle $(E_{L(P)} \times \mathbb{C})/L(P)$ is zero.

If we set $G = \mathrm{GL}(n, \mathbb{C})$, then the criterion in Theorem 4.2 coincides with the earlier stated criterion, that is, the rank n vector bundle over X associated to E by the standard representation admits a holomorphic connection if and only if all the indecomposable components are of degree zero.

Let P be a parabolic subgroup of the reductive group G. We can ask the following question: what is a necessary and condition for the existence of a holomorphic connection on a given principal P–bundle over X?

Let E_P be a principal P–bundle over X. Let $E_P(L(P))$ be the principal $L(P)$–bundle obtained by extending the structure group using the projection of P to $P/R_u(P) = L(P)$. Now, a holomorphic connection on E_P induces a holomorphic connection on $E_P(L(P))$. Note that since $L(P)$ is reductive, we can use the criterion in Theorem 4.2 to decide if $E_P(L(P))$ admits a holomorphic connection.

QUESTION 4.3. *Assume that $E_P(L(P))$ admits a holomorphic connection. Does this imply that the P–bundle E_P admits a holomorphic connection?*

The question posed above has an affirmative answer if $G = \mathrm{GL}(n,\mathbb{C})$ and the principal $L(P)$–bundle $E_P(L(P))$ is polystable. Polystability is defined in the next section.

5. Semistable principal bundle

Let M be a connected complex projective manifold of dimension d. Fix an ample line bundle $\mathcal{O}_M(1)$ on M. The degree of any torsionfree coherent sheaf W on M will be defined as

$$\mathrm{degree}(W) := \int_M c_1(W)(c_1(\mathcal{O}_M(1)))^{d-1} \in \mathbb{Z}\,.$$

A torsionfree coherent sheaf W over M is called *semistable* if for any nonzero subsheaf $W' \subset W$ the inequality

$$\frac{\mathrm{degree}(W')}{\mathrm{rank}(W')} \leq \frac{\mathrm{degree}(W)}{\mathrm{rank}(W)}$$

is valid. If the strict inequality "<" is valid for subsheaves with the extra condition that W' is a proper subsheaf with W/W' torsionfree, then W is called *stable* (see [**Ko**]).

The quotient degree(W)/rank(W) is called *slope* of W and is usually denoted by $\mu(W)$. The sheaf W is called *polystable* if it is a direct sum of stable sheaves of same slope. In particular, any polystable sheaf is semistable.

A. Ramanathan extended the notion of (semi)stability to principal bundles [**Ra1**], [**Ra2**], [**Ra3**]. We will briefly recall the definition.

Let G be a connected complex reductive algebraic group. The center of G will be denoted by $Z(G)$.

Take a principal G–bundle E over X. Let

$$E_P \subset E|_U$$

be a reduction of structure group of E to a proper parabolic subgroup $P \subset G$ over a nonempty Zariski open subset $U \subset M$ with the codimension of the complement $M \setminus U$ being at least two. Note that since codimension of $M\setminus U$ in M is at least two, given a coherent sheaf F on U, the direct image $\iota_* F$ is coherent, where ι is the inclusion map of U in M. The degree of F is defined to be the degree of $\iota_* F$. Let χ be a character of P dominant with respect to a Borel subgroup contained in P. So, $E_P(\mathbb{C})$ is a line bundle over U, where P acts on $\mathbb{C}$ through χ.

The G–bundle E is called *semistable* (respectively, *stable*) if in *every* such situation describe above, the degree of the line bundle $E_P(\mathbb{C})$ is nonpositive (respectively, strictly negative).

The G–bundle E is called *polystable* if there is a Levi factor $L(P) \subset P \subset G$ of a parabolic subgroup P and a reduction, over M, of structure group $E_{L(P)}$ of E to the subgroup $L(P)$ such that

(1) the $L(P)$–bundle $E_{L(P)}$ is stable;
(2) for any character $\chi : L(P) \longrightarrow \mathbb{C}^*$ with the property that the restriction of χ to the center $Z(G)$ is trivial, the degree of the associated line bundle $E_{L(P)}(\mathbb{C})$ is zero.

A $\mathrm{GL}(n, \mathbb{C})$–bundle E_{GL} is stable or semistable or polystable if and only if the vector bundle of rank n associated to E_{GL} by the standard representation of $\mathrm{GL}(n, \mathbb{C})$ is stable or semistable or polystable respectively.

A G–bundle E is semistable if and only if its adjoint vector bundle $\mathrm{ad}(E)$ is semistable [**RR**, p. 285, Theorem 3.18], [**ABi**, p. 214, Proposition 2.10]. Similarly, E is polystable if and only if $\mathrm{ad}(E)$ is polystable [**RR**, p. 285, Theorem 3.18], [**ABi**, p. 212, Theorem 2.6].

We will now recall the main result of [**MR2**].

Assume the dimension of M to be at least two. Fix an integer $t \in [1, d-1]$, where $d = \dim M$, and

$$(\alpha_1, \cdots, \alpha_t) \in \mathbb{Z}^t$$

with $\alpha_i \geq 2$ for all $i \in [1, t]$. For any integer $m \geq 1$, by a complete intersection of type (m) we will mean a complete intersection of divisors

$$D_1 \cap D_2 \cap \cdots \cap D_t$$

with D_i in the complete linear system of $\mathcal{O}_M(\alpha_i^m) = \mathcal{O}_M(1)^{\otimes \alpha_i^m}$. Recall the $\mathcal{O}_M(1)$ is the fixed ample line bundle over M.

Note that for m sufficiently large, the general complete intersection of type (m) is a smooth irreducible subvariety of M. The restriction of $\mathcal{O}_M(1)$ to any subvariety of M remains ample. This restriction will be used in defining the degree of a coherent sheaf on the subvariety.

Now we are in a position to state the main result of [**MR2**].

THEOREM 5.1 (MR2, p. 167, Theorem 4.3). *A holomorphic vector bundle V over M is stable if and only if there is an integer m_0 such that for any $m \geq m_0$, the restriction of E to the general complete intersection of type (m) is a stable vector bundle.*

This can be extended to principal bundles as shown in the following theorem.

THEOREM 5.2. *A principal G–bundle E over M is stable if and only if there is an integer m_0 such that for any $m \geq m_0$, the restriction of E to the general complete intersection of type (m) is stable.*

Proof. Let H be an algebraic group with $\mathfrak{h}$ its Lie algebra and $\mathfrak{h}_0$ the center of $\mathfrak{h}$. If F_H is a principal H–bundle over M, then $\mathfrak{h}_0$ is naturally contained in $H^0(M, \mathrm{ad}(F_H))$, as H acts trivially on $\mathfrak{h}_0$ by the adjoint action on $\mathfrak{g}$.

Let $\mathfrak{z}$ denote the center of the Lie algebra $\mathfrak{g}$ of G. Let F be a polystable G–bundle over M. We will show that F is stable if and only if

$$H^0(M, \mathrm{ad}(F)) = \mathfrak{z} \tag{5.1}$$

If F is stable then $H^0(M, \mathrm{ad}(F)) = \mathfrak{z}$ [**Ra1**, p. 136, Proposition 3.2]. For the converse direction, recall from the definition of polystability that F admits a reduction of structure group

$$F_{L(P)} \subset F$$

to the Levi factor $L(P)$ of some parabolic subgroup P. If F is *not* stable, then P is a proper parabolic subgroup of G. In that case, the connected component of the center of $L(P)$ containing the identity element is strictly larger than the connected component of the center of G containing the identity element. Since

$$\dim H^0(M, \mathrm{ad}(F_{L(P)})) \geq \dim Z_{L(P)}$$

where $Z_{L(P)}$ is the connected component of the center of $L(P)$ containing the identity element, we conclude that if (5.1) is valid, then F must be stable.

We recall the Enriques–Severi–Zariski lemma (see [**MR1**, p. 218, Proposition 3.2], [**Ha**, p. 244, Corollary 7.8]). Given any holomorphic vector bundle V over M, there is an integer n_0 such that for all $n \geq n_0$, the restriction map

$$H^0(M, V) \longrightarrow H^0(X, V|_X)$$

is surjecitve for the general complete intersection $X \subset M$ of type (n).

Since a polystable vector bundle is a direct sum of stable vector bundles, the above quoted Theorem 5.1 remains valid if stability is replaced by polystability.

Let E be a stable G–bundle over M. So the adjoint bundle $\mathrm{ad}(E)$ is polystable [**RR**, p. 285, Theorem 3.18], [**ABi**, p. 212, Theorem 2.6]. Therefore, there is an integer k_0 such that for any $m \geq k_0$, the restriction of $\mathrm{ad}(E)$ to the general complete intersection of type (m) is polystable Theorem 5.1. This, in turn, implies that the restriction of E to such a complete intersection is polystable [**RR**, p. 285, Theorem 3.18], [**ABi**, p. 212, Theorem 2.6].

Now applying the Enriques–Severi–Zariski lemma to the vector bundle $\mathrm{ad}(E)$ and using the earlier observation that a polystable G–bundle is stable if (5.1) is valid for it we conclude that there is an integer m_0 such that for any $m \geq m_0$, the restriction of E to the general complete intersection of type (m) is stable.

The converse is easy. If there is an integer m_0 such that for any $m \geq m_0$, the restriction of E to the general complete intersection of type (m) is stable, then the restriction of $\mathrm{ad}(E)$ to any such complete intersection is polystable. This implies that the vector bundle $\mathrm{ad}(E)$ is polystable [**MR2**, p. 167, Theorem 4.3].

Now, if (5.1) is *not* valid for E, that is, $\mathrm{ad}(E)$ has more sections, then (5.1) must fail for the general complete intersection of type (m) for m sufficiently large. But this contradicts the assumption that the restriction of E to the general complete intersection of type (m) is stable for m sufficiently large. This completes the proof of the theorem.

In the next section we will consider vector bundles with parabolic structure.

6. Parabolic bundles on a Riemann surface

Let X be a compact connected Riemann surface, and $S \subset X$ be a finite subset.

A *parabolic structure* over S on a holomorphic vector bundle E over X consists of the following data (see [**Se2**], [**MS**]):

(1) a strictly decreasing filtration of subspaces

$$E_s = F_1^s \supset F_2^s \supset \cdots \supset F_{l_s}^s \supset F_{l_s+1}^s = 0$$

for each $s \in S$ which is known as the *quasi–parabolic filtration*;

(2) a sequence of rational numbers

$$0 \leq \lambda_1^s < \lambda_2^s < \cdots < \lambda_{l_s}^s < 1\,,$$

where λ_i^s corresponds to the subspace F_i^s.

A *parabolic vector bundle* is a holomorphic vector bundle over X equipped with a parabolic structure. The divisor S over which the parabolic structure is defined is called the *parabolic divisor*.

If we denote by E_* the above defined parabolic vector bundle, then the *parabolic degree* of E_* is defined to be

$$\text{par-deg}(E_*) := \text{degree}(E) + \sum_{s\in S} \sum_{i=1}^{l_s} \lambda_s^i \dim(F_i^s / F_{i+1}^s) .$$

So, in general, the parabolic degree is a rational number not necessarily an integer. The quotient

$$\text{par-}\mu(E_*) := \frac{\text{par-deg}(E_*)}{\text{rank}(E)} \tag{6.1}$$

is called the *parabolic slope* of E_*.

Given a parabolic vector bundle E_* as above, any subbundle F of E has an induced parabolic structure. Indeed the quasiparabolic filtration of the fiber F_s, $s \in S$, is the intersection of F_s with the quasiparabolic filtration on E_s.

A parabolic vector bundle E_* is called *semistable* if for all subbundles F of E of positive rank the inequality

$$\text{par-}\mu(E_*) \geq \text{par-}\mu(F_*)$$

is valid, where F_* is the induced parabolic structure on F and par-μ is defined in (6.1). If the strict inequality "$>$" is valid for all proper subbundles F of E, then E_* is called *stable*. The parabolic vector bundle E_* is called *polystable* if it is a direct sum of stable parabolic bundles of same parabolic slope, the parabolic structure on the direct sum being the induced one — induced by the direct summands. We will describe the direct sum operation in detail later in this section.

In [**MY**], M. Maruyama and K. Yokogawa define a parabolic vector bundle in a slightly different way. But their definition is equivalent to the above definition of [**MS**]. We will recall the definition of [**MY**].

In [**MY**], a quasiparabolic structure on V is a filtration

$$V = F_1(V) \supset F_2(V) \supset \cdots \supset F_l(V) \supset F_{l+1}(V) = V \otimes \mathcal{O}_X(-S) \tag{6.2}$$

of V by coherent subsheaves. Note that the length of this filtration is not bounded by the rank of V. In [**MY**], a parabolic structure is a quasiparabolic structure as above together with a system of parabolic weights

$$0 \leq \lambda_1 < \lambda_2 < \cdots < \lambda_l < 1 \tag{6.3}$$

where the rational number λ_i corresponds to the subsheaf $F_i(V)$. We will show how a parabolic vector bundle defined earlier (in the sense of [**MS**]) gives a parabolic vector bundle in the sense of [**MY**].

Take the parabolic vector bundle E_* defined earlier (as in [**MS**]). Let

$$\Lambda := \bigcup_{s\in S}\bigcup_{1\leq i\leq l_s} \lambda^s_i \subset \mathbb{Q}$$

be the union. For any $\lambda \in \Lambda$ and $s \in S$, set $s(\lambda) \in [1, l_s]$ such that the two conditions

(1) $\lambda^s_{s(\lambda)} \geq \lambda$, and
(2) $\lambda^s_{s(\lambda)-1} < \lambda$ if $s(\lambda) \geq 2$

are satisfied. Note that the above two condition uniquely fixes $s(\lambda)$. Let $F^s(\lambda) \subset E$ be the subsheaf defined by the exact sequence

$$0 \longrightarrow F^s(\lambda) \longrightarrow E \longrightarrow E_s/F^s_{s(\lambda)} \longrightarrow 0$$

where $F^s_{s(\lambda)}$ is in the quasiparabolic filtration of the fiber E_s. Let

$$F(\lambda) := \bigcap_{s\in S} F^i(\lambda) \subset E$$

be the subsheaf defined by the intersection.

The parabolic vector bundle in the sense of [**MY**] associated to the parabolic vector bundle E_* (in the sense of [**MS**]) has parabolic weights Λ, and for any weight $\lambda \in \Lambda$, the subsheaf corresponding to it is $F(\lambda)$.

It is easy to see that the parabolic vector bundle in the sense of [**MY**] constructed this way determines the parabolic vector bundle (in the sense of [**MS**]) we started with. It other words, the parabolic vector bundle we started with can be reconstructed back from it. Conversely, any parabolic vector bundle in the sense of [**MY**] corresponds to some parabolic vector bundle in the sense of [**MS**].

Given a parabolic vector bundle, in [**MY**] M. Maruyama and K. Yokogawa associates to it a filtration of subsheaves parametrized by $\mathbb{R}$. This filtration, which we will recall below, is very useful.

Take any $t \in \mathbb{R}$. Let $[t] \in \mathbb{Z}$ be such that $0 \leq t - [t] < 1$. If the fractional part $t - [t]$ is bigger than every number in the set of parabolic weights Λ, then define $\alpha_t := 1$. Otherwise, let $\alpha_t \in \Lambda$ be the smallest number (in Λ) with $t - [t] \leq \alpha_t$. Define

$$E_t := F(\alpha_t) \otimes \mathcal{O}_X(-[t]S)$$

if $\alpha_t \in \Lambda$, and define $E_t := E \bigotimes \mathcal{O}_X(-([t]+1)S)$ if $\alpha_t = 1$. Note that in terms of the quasiparabolic filtration as in (6.2), and with parabolic weights as in (6.3), we have

$$E_t = F_i(E) \otimes \mathcal{O}_X(-[t]S)$$

where $\lambda_{i-1} < t \leq \lambda_i$ with $\lambda_0 = \lambda_l - 1$ and $\lambda_{l+1} = 1$.

From the construction of the filtration $\{E_t\}_{t\in\mathbb{R}}$ it follows immediately that

(1) the filtration is decreasing as t increases;
(2) it is left continuous, which means that for each $t \in \mathbb{R}$, there is $\epsilon(t) > 0$ such that $E_t = E_{t-\epsilon}$ for all $\epsilon \in (0, \epsilon(t))$;
(3) $E_{t+1} = E_t \bigotimes \mathcal{O}_X(-S)$ for all $t \in \mathbb{R}$;
(4) given any finite interval $[a, b] \subset \mathbb{R}$, the set

$$\{t \in [a, b] \mid E_{t_0} \neq E_{t_0+\delta} \text{ for all } \delta > 0\}$$

is finite;
(5) the filtration has a jump at t_0, that is, $E_{t_0} \neq E_{t_0+\delta}$ for all $\delta > 0$, if and only if $t_0 - [t_0]$ is a parabolic weight.

For any $t \in \mathbb{R}$, by E_{t+} we will denote the limit of $E_{t+\epsilon}$ as $\epsilon > 0$ converges to 0. Note that for each $t \in \mathbb{R}$, there is $\epsilon(t) > 0$ such that $E_{t+} = E_{t+\epsilon}$ for all $\epsilon \in (0, \epsilon(t))$.

The parabolic structure on E can easily be recovered from the filtration $\{E_t\}_{t\in\mathbb{R}}$. Note that $E = E_0$ and a number $0 \leq \lambda < 1$ is a parabolic weight if and only if $E_\lambda \neq E_{\lambda+}$. If λ is a parabolic weight, then the corresponding term in the quasiparabolic filtration is recovered using the quotient $E_\lambda/E_{\lambda+}$. More precisely, $E_\lambda/E_{\lambda+}$ coincides with the graded piece of the quasiparabolic filtration.

Let $\tau : X\backslash S \longrightarrow X$ be the inclusion map. Consider the quasicoherent sheaf $\tau_*\tau^*E$ on X given by the direct image of the restriction of E to $X \setminus S$. Note that $\tau_*\tau^*E$ is *not* coherent if the set S is nonempty. Note that each E_t, $t \in \mathbb{R}$, is naturally contained in $\tau_*\tau^*E$. Furthermore, $\tau_*\tau^*E$ is generated by the collection of subsheaves E_t, $t \in \mathbb{R}$.

Now we are in a position to define the direct sum, dual and tensor product operations on parabolic vector bundles.

Given two parabolic vector bundle E_* and V_*, with S as the common parabolic divisor, and E and V respectively the underlying vector bundles, consider

$$W := \tau_*\tau^*E \oplus \tau_*\tau^*V = \tau_*\tau^*(E \oplus V)$$

The *direct sum* $E_* \bigoplus V_*$ is defined to be the parabolic vector bundle that corresponds to the filtration $\{W_t\}_{t\in\mathbb{R}}$ in W defined as

$$W_t := E_t \oplus V_t$$

where $\{E_t\}_{t\in\mathbb{R}}$ and $\{V_t\}_{t\in\mathbb{R}}$ are the filtrations for E_* and V_* respectively. Note that the underlying vector bundle for the parabolic vector bundle $E_* \bigoplus V_*$ coincides with $E \bigoplus V$.

Now define

$$U := (\tau_*\tau^*E) \otimes_{\mathcal{O}_X} (\tau_*\tau^*V) = \tau_*\tau^*(E \otimes V).$$

So each $E_t \bigotimes V_{t'}$ is a subsheaf of U. For any $t \in \mathbb{R}$, let U_t denote the subsheaf of U generated by all the subsheaves $E_{t_1} \bigotimes V_{t_2}$ with $t_1 + t_2 \geq t$. It is easy to see that each U_t is a coherent subsheaf of U. In fact, U_t is a subsheaf of $E_{t-1} \bigotimes V_{-1}$. This follows immediately from the property that $E_{t+1} = E_t \bigotimes \mathcal{O}_X(-S)$.

The parabolic vector bundle defined by the filtration $\{U_t\}_{t\in\mathbb{R}}$ of U is the *parabolic tensor product* $E_* \bigotimes V_*$.

Note that the underlying vector bundle for the parabolic vector bundle $E_* \bigotimes V_*$ *need not* coincide with $E \bigotimes V$. In fact, if E_* has a parabolic weight α and V_* has a parabolic weight β such that $\alpha + \beta \geq 1$, then the underlying vector bundle for the parabolic vector bundle $E_* \bigotimes V_*$ *does not* coincide with $E \bigotimes V$. In that case $E \bigotimes V$ is a proper subsheaf of the underlying vector bundle.

For any $t \in \mathbb{R}$, set

$$\mathcal{E}_t := (E_{-t-1+})^* \subset \mathcal{E}' := \tau_*\tau^*E^*$$

(the right limit $E_{t'+}$ was defined earlier). The parabolic vector bundle corresponding to the filtration $\{\mathcal{E}_t\}_{t\in\mathbb{R}}$ of $\mathcal{E}'$ is defined to be *parabolic dual* of E_*. This parabolic dual will be denoted by E^*_*.

If E_* has at least one nonzero parabolic weight, then the underlying vector bundle for the parabolic vector bundle E^*_* *does not* coincide with E^*. The underlying vector bundle for E^*_* is a subsheaf of E^*.

For two parabolic vector bundles E_* and V_* the *parabolic homomorphism* bundle from E_* to V_* is the parabolic vector bundle

$$\mathrm{Hom}_P(E_*, V_*) := E^*_* \otimes V_*.$$

See [**Bi3**] for details of the above constructions.

The parabolic tensor product is self–dual, that is, $(E^*_*)^* = E_*$. The tensor product is associative, that is,

$$E_* \otimes (V_* \otimes W_*) = (E_* \otimes V_*) \otimes W_*$$

where E_*, V_* and W_* are any parabolic vector bundles. Furthermore, the tensor product is distributive, that is,

$$E_* \otimes (V_* \oplus W_*) = (E_* \otimes V_*) \oplus (E_* \otimes W_*).$$

In the next section we will consider holomorphic connections on a parabolic vector bundle.

7. Connection on a parabolic vector bundle

We will now define a holomorphic connection in the context of parabolic bundles.

A *logarithmic connection* on a vector bundle V over X with singularity over S is a first order differential operator

$$D : V \longrightarrow \Omega^1_X \otimes \mathcal{O}_X(S) \otimes V$$

satisfying the Leibniz identity [**De**]. We recall that the Leibniz identity says

$$D(f\beta) = fD(\beta) + \partial(f) \otimes \beta$$

where f is a locally defined holomorphic function on X and β is a locally defined holomorphic section of V.

The Poincaré adjunction formula says that the fiber of the line bundle $\mathcal{O}_X(S)$ over any $s \in S$ is identified with the tangent space T_sX at s. In other words, the fiber $(\Omega^1_X \bigotimes \mathcal{O}_X(S))_s$ is identified with $\mathbb{C}$. Given a logarithmic connection D, consider the composition

$$V \xrightarrow{D} \Omega^1_X \otimes \mathcal{O}_X(S) \otimes V \longrightarrow (\Omega^1_X \otimes \mathcal{O}_X(S) \otimes V)_s = V_s .$$

From the Leibniz identity it follows immediately that the composition is $\mathcal{O}_X$–linear. Consequently, it defines an endomorphism of the fiber V_s.

The above constructed endomorphism is called the *residue* of D at s [**De**], and it is denoted by $\mathrm{Res}(D\,, s)$.

Let E_* be a parabolic structure on E as described in Section 6. A *holomorphic connection* on E_* is a logarithmic connection D on E, singular over S, satisfying the following conditions:

(1) the residue $\mathrm{Res}(D\,, s)$ is semisimple;
(2) for any $s \in S$, the residue $\mathrm{Res}(D\,, s)$ preserves the quasi-parabolic filtration of E_s;
(3) for each $s \in S$ and $i \in [1\,, l_s]$, the action of $\mathrm{Res}(D\,, s)$ on F^s_i/F^s_{i+1} is multiplication by the corresponding parabolic weight λ^s_i. (Note that since $\mathrm{Res}(D\,, s)$ preserves the quasi-parabolic filtration, it acts on each quotient F^s_i/F^s_{i+1}.)

We will call a parabolic bundle E'_* to be a *direct summand* of E_* if there is another parabolic bundle E^1_* such that E_* is isomorphic to the parabolic vector bundle $E'_* \bigoplus E^1_*$ (the direct sum operation was defined in the previous section). So, in particular E is isomorphic to

$E' \bigoplus E^1$, where E' and E^1 are the underlying vector bundles for E'_* and E^1_* respectively.

Note that if E^2 and E^3 are two holomorphic subbundles of E with $E = E^2 \bigoplus E^3$, then it is not necessary that $E_* = E^2_* \bigoplus E^3_*$, where E^2_* and E^3_* have the induced parabolic structures from E_*.

The following theorem is the parabolic analog of the criterion of [**We**], [**At2**] (described Section 4) for the existence of a holomorphic connection on a vector bundle.

THEOREM 7.1. *A parabolic vector bundle E_* admits a holomorphic connection if and only if every direct summand of E_* is of parabolic degree zero.*

Proof. The proof involves an essential use of π–bundles introduced by C. S. Seshadri (see [**Se1**]). We will recall the definition of [**Se1**]. Nowadays, it is customary to call π–bundles as Γ–linearized bundles, where Γ is the Galois group in the set up. We will use the current terminology.

Let Y be a compact connected Riemann surface. Fix a finite subgroup $\Gamma \subset \mathrm{Aut}(Y)$ of the automorphism group Y.

A Γ*–linearized vector bundle* W over Y is a holomorphic vector bundle equipped with an action Γ compatible with the obvious action of Γ on Y. In other words, Γ acts on the total space of the vector bundle W and for any $g \in \Gamma$, the action of g on W is a vector bundle isomorphism of W with $(g^{-1})^*W$.

Given two Γ–linearized vector bundles, say W_1 and W_2, a Γ*–homomorphism* from W_1 to W_2 is a $\mathcal{O}_Y$–linear homomorphism $h : W_1 \longrightarrow W_2$ that commutes with the actions of Γ, that is, $h \circ g = g \circ h$ for all $g \in \Gamma$.

A Γ*–holomorphic connection* on W is a holomorphic connection that is preserved by the action of Γ. In other words, for a Γ–holomorphic connection D on W and for any $g \in \Gamma$, the isomorphism of W with $(g^{-1})^*W$, defined by the action of g, takes the connection D to the connection $(g^{-1})^*D$.

We will prove the following proposition.

PROPOSITION 7.2. *If the vector bundle W admits a holomorphic connection, then it admits a Γ–holomorphic connection.*

Proof. The space of all holomorphic connections on W is a convex space. Indeed, as we noted in Section 2, the space of all holomorphic connections on W is an affine space for the vector space

$H^0(Y, \Omega^1_Y \bigotimes \mathrm{End}(W))$. An affine space has an obvious convex structure.

Given a holomorphic connection D on W, consider the average

$$D' := \frac{1}{\#\Gamma} \sum_{g \in \Gamma} g^* D$$

where $\#\Gamma$ denotes the order of the group Γ and the average is taken using the convex space structure of the space of all holomorphic connections on W.

The holomorphic connection D' is evidently a Γ–holomorphic connection. This completes the proof of the proposition.

A Γ–linearized vector bundle W will be called *decomposable* if there are two Γ–linearized vector bundles W_1 and W_2 of positive ranks such that W and $W_1 \bigoplus W_2$ are isomorphic as Γ–linearized vector bundles. We will call W to be *indecomposable* if it is not decomposable.

The following lemma is proved in [**Bi3**].

LEMMA 7.3. *If W is an indecomposable Γ–linearized vector bundle of degree zero, then W admits a Γ–holomorphic connection.*

A Γ–linearized vector bundle W_1 is called a *direct summand* of the Γ–linearized vector bundle W if there is a Γ–linearized vector bundle W_2 such that W and $W_1 \bigoplus W_2$ are isomorphic as Γ–linearized vector bundles.

In Section 4 we saw that a holomorphic connection on W induces a holomorphic connection on each direct summand of it. If W_1 and W_2 both admit holomorphic connections, then obviously $W_1 \bigoplus W_2$ also admits a holomorphic connection.

Consequently, the following theorem follows from Lemma 7.3 and Proposition 7.2.

THEOREM 7.4. *A Γ–linearized vector bundle W admits a Γ–holomorphic connection if and only if every direct summand of it is of degree zero.*

Continuing with the proof of Theorem 7.1, given a parabolic vector bundle E_* over X, in [**Bi1**] a (ramified) Galois covering

$$p : Y \longrightarrow X$$

is constructed. Let Γ denote the Galois group for p. From E_*, a Γ–linearized vector bundle W on Y is constructed. See [**Bi1**, Section 3] for the details.

Let W denote the Γ–linearized bundle over Y (for the automorphism group Γ) constructed in [**Bi1**, Section 3] from E_*. Now [**Bi1**, (3.12)] says that

$$\text{par-deg}(E_*) = \frac{\text{degree}(W)}{\#\Gamma}\,. \tag{7.1}$$

Also, there is a one–to–one correspondence between subbundles of E and Γ invariant subbundles of W [**Bi1**, p. 318].

Assume that every direct summand of E_* is of parabolic degree zero. Since subbundles of E are in one-to-one correspondence with the Γ invariant subbundles of W, using (7.1) it follows that every direct summand of the Γ–linearized vector bundle W is of degree zero. Therefore, we conclude from Theorem 7.4 that W admits a Γ–holomorphic connection.

Take a Γ–holomorphic connection D on W. Fix a point $y \in p^{-1}(X \setminus S)$. Let

$$\rho\,:\,\pi_1(Y,y) \longrightarrow \mathrm{GL}(n,\mathbb{C})$$

be the monodromy representation of the flat connection $\nabla := D + \overline{\partial}_W$, where $n = \text{rank}(W)$ and $\overline{\partial}_W$ is the Dolbeault operator defining the holomorphic structure of W. Since ∇ is left invariant by the action of Γ on W, the representation ρ clearly descends to a representation of $\pi_1(X \setminus S, p(y))$. This gives a connection on the restriction of E to $X \setminus S$. It can also be directly seen that the condition that ∇ is Γ invariant ensures that it descends to a flat connection on E over $X \setminus S$.

This connection on E over $X \setminus S$ extends to a connection on E_* [**Bi2**, Lemma 4.11]. Indeed, Lemma 4.11 of [**Bi2**] says that Γ invariant forms on Y descend as logarithmic forms on X. From this it follows immediately that the above holomorphic connection on E over $X \setminus S$ gives a holomorphic connection on E_*.

Conversely, if E_* has a holomorphic connection, then W has a Γ–holomorphic connection. Indeed, the pull back of a holomorphic connection on E_* is a holomorphic connection on the restriction of W to $Y \backslash p^{-1}(S)$ that is left invariant by the action of Γ on $W|_{Y\backslash p^{-1}(S)}$. It is easy to see that this connection extends to W over Y. We remarked earlier that direct summands of E_* are in one-to-one correspondence with direct summands of W. Therefore, using (7.1) it follows that if E_* admits a holomorphic connection, then any direct summand of E_* is of parabolic degree zero. This completes the proof of Theorem 7.1.

A polystable parabolic vector bundle of parabolic degree zero clearly has the property that any direct summand of it is of parabolic degree zero. Such a parabolic bundle admits a holomorphic connection which is unitary [**MS**, Theorem 4.1].

In [**BBN1**] the parabolic analog for principal bundles were introduced (also [**BBN2**] and [**BBN3**]). The definition of a parabolic G–bundle is modeled on the original definition of Nori of a principal G–bundle as a functor from the category of left G–modules to vector bundles satisfying certain conditions (see [**No1**], [**No2**]). If the tensor product, dual and homomorphisms of vector bundles are replaced by the corresponding construction in the category of parabolic vector bundles (see Section 6 for the definitions) in Nori's definition of a principal bundles, the definition of a parabolic G–bundle is obtained. Therefore, [**No1**] and [**No2**] provided the inspiration as well as the set–up for the definition of parabolic G–bundles. In any case, this would be clear to any reader of [**BBN1**].

One might ask the questions addressed here in the context of parabolic principal bundles. We hope to return to this in future.

References

[At1] M. F. Atiyah: On the Krull–Schmidt theorem with application to sheaves. *Bull. Soc. Math. Fr.* **84** (1956), 307–317.

[At2] M. F. Atiyah: Complex analytic connections in fibre bundles. *Trans. Amer. Math. Soc.* **85** (1957), 181–207.

[AB] H. Azad and I. Biswas: On holomorphic principal bundles over a compact Riemann surface admitting a flat connection. *Math. Ann.* **322** (2002), 333–346.

[ABi] B. Anchouche and I. Biswas: Einstein–Hermitian connections on polystable principal bundles over a compact Kähler manifold. *Amer. Jour. Math.* **123** (2001), 207–228.

[BBN1] V. Balaji, I. Biswas and D. S. Nagaraj: Principal bundles over projective manifolds with parabolic structure over a divisor. *Tôhoku Math. Jour.* **53** (2001), 337–367.

[BBN2] V. Balaji, I. Biswas and D. S. Nagaraj: Principal bundles with parabolic structures. *Elect. Res. Announc. Amer. Math. Soc.* **7** (2001), 37–44.

[BBN3] V. Balaji, I. Biswas and D. S. Nagaraj: Ramified G-bundles as parabolic bundles. Preprint (2001).

[Bi1] I. Biswas: Parabolic bundles as orbifold bundles. *Duke Math. Jour.* **88** (1997), 305–325.

[Bi2] I. Biswas: Parabolic ample bundles. *Math. Ann.* **307** (1997), 511–529.

[Bi3] I. Biswas: On the principal bundles with parabolic structure. *Jour. Math. Kyoto Univ.* (to appear).

[BM] I. Biswas and C. Mourougane: Holomorphic connection on a Fano manifold with Picard number one. *Jour. Ramanujan Math. Soc.* **14** (1999), 125–130.

[De] P. Deligne: *Equations Différentielles à Points Singuliers Réguliers.* Lecture Notes in Math. No.: 163, Springer-Verlag, Berlin, 1970.

[Go] A. Grothendieck: Sur la mémoire de Weil "Généralisation des fonctions abéliennes". *Seminaire Bourbaki,* expose 141, 1955–1956.

[Ha] R. Hartshorne: *Algebraic Geometry.* Graduate Texts in Mathematics, No. 52, Springer–Verlag, New-York, Heidelberg, Berlin, 1977.

[Ko] S. Kobayashi: *Differential Geometry of Complex Vector Bundles.* Publications of the Math. Society of Japan 15. Iwanami Shoten Publishers and Princeton University Press, 1987.

[KMM1] J. Kollár, Y. Miyaoka and S. Mori: Rationally connected varieties. *Jour. Alg. Geom.* **1** (1992), 429–448.

[KMM2] J. Kollár, Y. Miyaoka and S. Mori: Rational connectedness and boundedness of Fano manifolds. *Jour. Diff. Geom.* **36** 765–779.

[MY] M. Maruyama and K. Yokogawa: Moduli of parabolic stable sheaves. *Math. Ann.* **293** (1992), 77–99.

[MR1] V. B. Mehta and A. Ramanathan: Semistable sheaves on projective varieties and their restriction to curves. *Math. Ann.* **258** (1982), 213–224.

[MR2] V. B. Mehta and A. Ramanathan: Restriction of stable sheaves and representations of the fundamental group. *Invent. Math.* **77** (1984), 163–172.

[MS] V. B. Mehta and C. S. Seshadri: Moduli of vector bundles on curves with parabolic structures. *Math. Ann.* **248** (1980), 205–239.

[No1] M. V. Nori: On the representations of the fundamental group. *Compositio Math.* **33** (1976), 29–41.

[No2] M. V. Nori: The fundamental group scheme. *Proc. Ind. Acad. Sci. (Math. Sci.) 91* (1982), 73–122.

[RR] S. Ramanan and A. Ramanathan: Some remarks on the instability flag. *Tôhoku Math. Jour.* **36** (1984), 269–291.

[Ra1] A. Ramanathan: Stable principal bundles on a compact Riemann surface. *Math. Ann.* **213** (1975), 129–152.

[Ra2] A. Ramanathan: Moduli for principal bundles over algebraic curves: I. *Proc. Ind. Acad. Sci. (Math. Sci.)* **106** (1996), 301–328.

[Ra3] A. Ramanathan: Moduli for principal bundles over algebraic curves: II. *Proc. Ind. Acad. Sci. (Math. Sci.)* **106** (1996), 421–449.

[Se1] C. S. Seshadri: Moduli of π–vector bundles over an algebraic curve. *Questions on Algebraic Varieties*, (C.I.M.E., III Ciclo, Varenna, 1969) pp. 139–260

[Se2] C. S. Seshadri: Moduli of vector bundles on curves with parabolic structures. *Bull. Amer. Math. Soc.* **83** (1977), 124–126.

[We] A. Weil: Généralisation des fonctions abéliennes. *Jour. Math. Pures Appl.* **17** (1938), 47–87.

School of Mathematics, Tata Institute of Fundamental Research
Homi Bhabha Road, Mumbai 400005, India
E-mail address: indranil@math.tifr.res.in

The cone of effective one–cycles of certain G–varieties

Michel Brion

To C. S. Seshadri for his 70th birthday

Abstract. Let X be a normal projective variety admitting an action of a semisimple group with a unique closed orbit. We construct finitely many rational curves in X, all having a common point, such that every effective one–cycle on X is rationally equivalent to a unique linear combination of these curves with non–negative rational coefficients. When X is nonsingular, these curves are projective lines, and they generate the integral Chow group of one–cycles.

Introduction

One associates to any complete algebraic variety X the group $N_1(X)$ of one–cycles on X modulo numerical equivalence; this is a free abelian group of finite rank [**6**, Example 19.1.4]. In the corresponding real vector space $N_1(X)_{\mathbb{R}}$, the convex cone generated by the classes of (closed irreducible) curves in X is the *cone of effective one–cycles*, denoted by $NE(X)$.

Dually, one has the group $N^1(X)$ of Cartier divisors on X modulo numerical equivalence, endowed with a non–degenerate pairing $N^1(X) \times N_1(X) \to \mathbb{Z}$ via intersection numbers of Cartier divisors with curves. Again, $N^1(X)$ is a free abelian group of finite rank; in the corresponding real vector space $N^1(X)_{\mathbb{R}}$, the dual cone of $NE(X)$ is the *cone of numerically effective divisors* (called *nef* for brevity). The interior of the nef cone is the ample cone, by Seshadri's criterion [**7**, Theorem I.7.1].

The cone $NE(X)$ encodes much information about morphisms with source X. Specifically, any surjective separable morphism $f : X \to Y$ with connected fibers, where Y is a normal projective variety, is uniquely determined by the subcone $NE(f) \subseteq NE(X)$ generated by the classes of curves lying in fibers of f. Moreover, $NE(f)$ is a face of the convex cone $NE(X)$.

However, the structure of $NE(X)$ may be quite complicated: this cone may not be closed, its closure may not be polyhedral, and certain faces may not arise from morphisms with source X, due to the existence of nef divisors having no globally generated positive multiple. Only the part of $NE(X)$ where the canonical divisor is negative is well–understood, if X is a nonsingular projective variety in characteristic zero (see e.g. [**5**]).

In this note, we consider certain varieties where the structure of the whole cone of effective one–cycles turns out to be very simple. These are the normal projective varieties where a semisimple group acts with a unique closed orbit. For such a variety X, we show that the cone $NE(X)$ is generated by the closures of positive strata of dimension one, for an appropriate Bialynicki–Birula decomposition of X (recalled in Section 1). These closures are rational curves passing through a common point, the sink of the decomposition; their classes form a basis of the rational vector space $N_1(X)_{\mathbb{Q}}$, and the latter is isomorphic to the rational Chow group of one–cycles. If, in addition, X is nonsingular, then so are our rational curves, and their classes form a basis of the group $N_1(X)$ (Theorem 2 and Corollary 2).

Moreover, the cone of nef divisors is generated by the closures of negative strata of codimension one. These are globally generated Cartier divisors, and their classes in $N^1(X)_{\mathbb{Q}}$ form the dual basis to our basis of $N_1(X)_{\mathbb{Q}}$; finally, $N^1(X)$ is isomorphic to the Picard group. As a consequence, every nef divisor on X is globally generated (Theorem 1).

The simplest examples of projective varieties where a semisimple group acts with a unique closed orbit are of course flag varieties. For these, our results are well–known: the cone of effective one–cycles is freely generated by the classes of Schubert varieties of dimension one, while the classes of opposite Schubert varieties of codimension one form the dual basis of the nef cone. Moreover, numerical and rational equivalence coincide.

These results were generalized in [**3**] and [**4**] to all projective simple spherical varieties (i.e., normal projective varieties where a connected reductive group acts with a unique closed orbit, and where a Borel subgroup acts with a dense orbit), in characteristic zero. There the main tool was the classification of embeddings of spherical homogeneous spaces. In the present note, the use of the Bialynicki–Birula decomposition (for possibly singular varieties) yields more general results.

Another interesting class of examples consists of orbit closures of a finite set of ordered points of the projective line, under the diagonal action of PGL(2). Clearly, these orbit closures are projective varieties where PGL(2) acts with a unique closed orbit; and they turn out to be normal, as shown by Iozzi and Poritz (in characteristic zero; see [**10**]). In the final section of this note, we show that their cone of nef divisors is generated by the pull–backs of points under the various projections to the projective line, while the cone of effective curves is generated by the projective lines mapped to points under all but one projection (Proposition 1). We also obtain another proof of the normality of these orbit closures, and we show that they yield all simple complete embeddings of PGL(2), in arbitrary characteristics.

Note finally that the duality between closures of positive strata of dimension one and of negative strata of codimension one, is very seldom satisfied by (say) nonsingular projective varieties where a torus acts with only finitely many fixed points. One checks, for example, that this duality fails for all toric surfaces having at least 5 fixed points.

Acknowledgments. Many thanks to L. Bonavero and S. Druel for very hepful discussions and comments; their joint work [**2**] (with C. Casagrande and O. Debarre) on a conjecture of Mukai has been the starting point of this note.

1. The cone of numerically effective divisors

We begin by fixing notation and recalling some general results on group actions, to be used throughout this work.

All algebraic varieties and groups are defined over an algebraically closed field K of arbitrary characteristic.

Let G be a connected reductive algebraic group. Let $B, B^- \subseteq G$ be opposite Borel subgroups with common torus T and unipotent radicals $U = R_u(B), U^- = R_u(B^-)$; then $B = TU$ and $B^- = TU^-$. Let $\mathcal{X}^*(T)$ be the character group of T (the weight lattice of G), and

$\mathcal{X}_*(T)$ the group of one–parameter subgroups of T (the dual lattice of $\mathcal{X}^*(T)$).

A *G–module* is a the space of a rational, finite–dimensional G–representation. Important examples of G–modules are the *induced modules* $H^0(\omega)$, where $\omega \in \mathcal{X}^*(T)$: these consist of all regular functions f on G such that $f(gtu) = \omega(t^{-1})f(g)$ for all $g \in G$, $t \in T$ and $u \in U$. The G–action is defined by $(g \cdot f)(h) = f(g^{-1}h)$. Recall that $H^0(\omega)$ is non–zero if and only if ω is dominant; then the set of U–fixed points in $H^0(\omega)$ is a line where T acts with weight ω (see [**11**, II.2] for these results, and for more on induced modules).

A *variety* is a separated integral scheme of finite type over K. For a connected linear algebraic group H, a *H–variety* is a variety X endowed with an algebraic action of H. The fixed point subset of H in X is denoted by X^H and regarded as a reduced closed subscheme of X. An example of a H–variety is the projectivization $\mathbb{P}(M)$ of a H–module M; an H–variety X is *H–quasi–projective* if it admits an equivariant embedding into some $\mathbb{P}(M)$. Equivalently, X admits an ample H–linearizable invertible sheaf. Any normal H–variety X admits a covering by H–quasi–projective open subsets; as a consequence, X is H–quasi–projective if it is quasi–projective, or if it contains a unique closed H–orbit [**12**, Theorem 1].

Let $\mathcal{L}$ be an invertible sheaf on a normal H–variety. If the Picard group of H is trivial, then $\mathcal{L}$ admits a H–linearization; in the general case, there exists a finite covering $\tilde{H} \to H$ of algebraic groups such that $\tilde{H}$ has a trivial Picard group. Thus, $\mathcal{L}$ is $\tilde{H}$–linearizable, and some positive power $\mathcal{L}^n$ is H–linearizable (for all this, see [**19**, I.3] and also [**12**], [**13**]).

Next we obtain a criterion for a G–projective variety to contain a unique closed orbit.

Lemma 1. *The following conditions are equivalent for a G–projective variety X:*

(1) *X contains a unique closed G–orbit.*
(2) *X^U consists of a unique point.*
(3) *There exists a finite equivariant morphism $f : X \to \mathbb{P}(H^0(\omega))$, for some dominant weight ω.*

Then X is fixed pointwise by the connected centre $Z(G)^0$, so that G acts on X via its semisimple quotient $G/Z(G)^0$.

Proof. (1)⇒(2) By assumption, X^B consists of a unique point x. In other words, T acts on X^U with x as its unique fixed point. Since X^U is a T–projective variety, it follows that $X^U = \{x\}$.

(2)⇒(1) follows from Borel's fixed point theorem.

(1)⇒(3) Let $\mathcal{L}$ be a G–linearized, very ample invertible sheaf on X; let $Y \subseteq X$ be the closed G–orbit. Then the G–module $H^0(Y, \mathcal{L})$ contains a B–eigenvector σ. By [**14**, Korollar 2.3], there exist a positive integer n such that $\sigma^n \in H^0(Y, \mathcal{L}^n)$ lifts to a B–eigenvector $\tau \in H^0(X, \mathcal{L}^n)$. Let $N \subseteq H^0(X, \mathcal{L}^n)$ be the G–submodule generated by τ. Then the set of common zeroes in X of all sections in N is closed, G–stable, and does not contain Y. Hence this set is empty, since Y is the unique closed orbit. So the canonical rational map $X - \to \mathbb{P}(N^*)$ is in fact a finite equivariant morphism. Moreover, the G–module N being generated by a B–eigenvector, it is a quotient of a universal highest weight module; thus, its dual N^* is a submodule of some $H^0(\omega)$, by [**11**, II.2.13].

(3)⇒(2) Since U is unipotent connected and $H^0(\omega)^U$ is a line, $\mathbb{P}(H^0(\omega))^U$ consists of a unique point; thus, the set X^U is finite. On the other hand, X^U is connected by [**9**]; hence it consists of a unique point.

Finally, the centre $Z(G)$ acts on $H^0(\omega)$ by scalars [**11**, II.2.8], and hence fixes $\mathbb{P}(H^0(\omega))$ pointwise. Since f is finite, it follows that $Z(G)^0$ fixes X pointwise.

Definition. A G–projective variety satisfying one of the conditions of Lemma 1 is called *simple*.

We will construct generators of the nef cone of a simple G–projective variety, by using the *Bialynicki–Birula decomposition* that we now review, after [**1**] (in the nonsingular case) and [**16**] (in the general case).

Let X be a T–projective variety and let $\lambda \in \mathcal{X}_*(T)$. The multiplicative group $\mathbb{G}_m$ acts on X via $\lambda : \mathbb{G}_m \to T$, and we denote by X^λ the corresponding fixed point set. Given $x \in X$, the map $\mathbb{G}_m \to X$, $t \mapsto \lambda(t) \cdot x$ extends uniquely to a morphism $\mathbb{P}^1 \to X$; this defines $\lim_{t\to 0} \lambda(t) \cdot x$ and $\lim_{t\to\infty} \lambda(t) \cdot x$. Both are λ–fixed points, distinct unless $x \in X^\lambda$.

Recall also that X^λ equals X^T for any $\lambda \in X_*(T)$ outside a finite union of proper subgroups; such a λ is called *regular*.

For any $\lambda \in \mathcal{X}_*(T)$ and any closed subset $Y \subseteq X^\lambda$, let

$$X^+(Y) = \{x \in X \mid \lim_{t\to 0} \lambda(t) \cdot x \in Y\}.$$

Then $X^+(Y)$ is a locally closed T–invariant subset of X, and the map

$$p^+ : X^+(Y) \to Y,\ x \mapsto \lim_{t\to 0} \lambda(t)\cdot x$$

is a surjective affine T–invariant morphism (these facts are easily checked in the ambient projectivization of a T-module). Moreover, X is the disjoint union of the subsets $X^+(Y)$, where Y runs over all connected components of X^T. These subsets are called the *positive strata*.

As a consequence, there exists a unique open positive stratum $X^+(Y)$. Then $Y = Y^+$ is irreducible; it is called the *source* of X for the λ–action.

Likewise, X is the disjoint union of the

$$X^-(Y) = \{x \in X \mid \lim_{t\to\infty} \lambda(t)\cdot x \in Y\},$$

where Y runs over the connected components of X^λ. The unique component $Y = Y^-$ such that $X^-(Y)$ is open in X is called the *sink*.

Note that $X^+(Y)\cap X^-(Y) = Y$ for any $Y \subseteq X^\lambda$, so that $X^-(x) = \{x\}$ for all $x \in Y^+$. Moreover, by [**16**], one has for any $x \in X^\lambda$:

$$(*) \qquad \dim X - \dim_x X^\lambda \leq \dim_x X^+(x) + \dim_x X^-(x).$$

As a consequence, the source of X consists of those $x \in X^\lambda$ such that $X^-(x) = \{x\}$ (indeed, if $X^-(x) = \{x\}$, then $X^-(x') = (p^-)^{-1}(x')$ equals $\{x'\}$ for all x' in a neighborhood of x in X^λ, since $\{x'\}$ is the unique closed $\mathbb{G}_m$–orbit in its fiber under p^-. Now $(*)$ implies that $\dim X = \dim X^+(Y)$ for any irreducible component $Y \subseteq X^\lambda$ containing x.)

If, in addition, X is nonsingular, then so are X^λ and $X^+(Y)$, $X^-(Y)$ for all components Y of X^λ. Moreover, $X^+(Y)$ and $X^-(Y)$ intersect transversally along Y, and the morphisms p^+, p^- are affine bundles, by [**1**].

Next we recall the definition of certain subgroups of G associated with a given $\lambda \in \mathcal{X}_*(T)$; see [**19**, II.2.6] for details. Let

$$G(\lambda) = \{g \in G \mid \lambda(t)g\lambda(t^{-1}) \text{ has a limit in } G \text{ as } t \to 0\},$$

so that

$$G(-\lambda) = \{g \in G \mid \lambda(t)g\lambda(t^{-1}) \text{ has a limit in } G \text{ as } t \to \infty\}.$$

Then $G(\lambda)$ and $G(-\lambda)$ are opposite parabolic subgroups of G, with common Levi subgroup the centralizer of λ. Moreover, the unipotent radical of $G(\lambda)$ is

$$R_uG(\lambda) = \{g \in G \mid \lim_{t\to 0} \lambda(t)g\lambda(t^{-1}) = 1\}.$$

Finally, $G(\lambda)$ equals B if and only if λ lies in the interior of the Weyl chamber associated with B; then $G(-\lambda) = B^-$.

Consider now a G–projective variety X. Then we may choose a regular $\lambda \in \mathcal{X}_*(T)$ such that $G(\lambda) = B$. Hence $X^+(Y)$ is B–invariant for any closed subset $Y \subseteq X^\lambda = X^T$ (since $\lambda(t)g \cdot x = \lambda(t)g\lambda(t^{-1})\lambda(t) \cdot x$). Likewise, $X^-(Y)$ is B^-–invariant. This implies at once that the sink Y^- is fixed pointwise by B. In particular, we obtain

LEMMA 2. *Let X be a simple G–projective variety. Then the sink consists of the unique B–fixed point x^-, and the source consists of the unique B^-–fixed point x^+.*

Put for simplicity $X^- = X^-(x^-)$. Then X^- is an open affine B^-–invariant neighborhood of x^-; one easily checks that such a neighborhood is unique. Moreover, $\mathbb{G}_m$ acts on the algebra of regular functions $R = H^0(X^-, \mathcal{O}_{X^-})$ via $(t \cdot f)(x) = f(\lambda(t^{-1})x)$, and this yields a positive grading of R. By Nakayama's Lemma, any graded invertible R–module is generated by a homogeneous element, unique up to scalar. This implies

LEMMA 3. *With the preceding notation, the group of isomorphism classes of T–linearized invertible sheaves on X^- is isomorphic to $\mathcal{X}^*(T)$, via pull–back to the fixed point x^-.*

We will also need the following variant of a result of Knop [**15**, Lemma 2.2].

LEMMA 4. *Let H be a connected linear algebraic group, X a normal H–variety, and D an effective Weil divisor on X whose support contains no H–orbit. Then D is a globally generated Cartier divisor.*

Proof. Let $i : X^{\mathrm{reg}} \to X$ be the inclusion of the nonsingular locus. Then the sheaf $i^*\mathcal{O}_X(D)$ is invertible. Replacing H by a finite cover, we may assume that this sheaf is H–linearized. Since $\mathcal{O}_X(D) \cong i_*i^*\mathcal{O}_X(D)$(by normality of X), it follows that $\mathcal{O}_X(D)$ is H–linearized as well.

Let $Y \subseteq X$ be the subset of all points where $\mathcal{O}_X(D)$ is not invertible. Then Y is closed, H–invariant, and contained in the support of D; thus, Y contains no orbit. Hence Y is empty, and D is Cartier.

Likewise, the base locus of D is empty, so that D is globally generated.

We now come to our first main result.

THEOREM 1. *Let X be a normal, simple G–projective variety. Let $x^- \in X$ be the B–fixed point, $X^- \subseteq X$ its unique B^-–invariant open affine neighborhood, and $D_1, \ldots, D_r$ the irreducible components of $X \setminus X^-$. Then the following hold:*

1. *$D_1, \ldots, D_r$ are globally generated Cartier divisors. Their linear equivalence classes form a basis of the Picard group of X.*
2. *Every ample divisor on X is linearly equivalent to a unique linear combination of $D_1, \ldots, D_r$ with positive integer coefficients.*
3. *Every nef divisor on X is linearly equivalent to a unique linear combination of $D_1, \ldots, D_r$ with non–negative integer coefficients.*

Proof. (1) The first assertion follows from Lemma 4 (applied to $H = B$). For the second assertion, let D be a Cartier divisor on X. Then the invertible sheaf $\mathcal{O}_X(D)$ admits a T–linearization; by Lemma 3, it follows that the pull–back of $\mathcal{O}_X(D)$ to X^- is trivial (as an invertible sheaf). Thus, D is linearly equivalent to a linear combination of $D_1, \ldots, D_r$ with integer coefficients. These coefficients are unique, since every regular invertible function on X^- is constant.

(2) Let D be an ample divisor on X. There exists a positive integer n such that the invertible sheaf $\mathcal{O}_X(nD)$ is very ample and G–linearized. Then the G–module $H^0(G \cdot x^-, \mathcal{O}_X(nD))$ contains a B^-–eigenvector σ. Note that $\sigma(x^-) \neq 0$, since $\sigma \neq 0$ and $B^- \cdot x^- = B^- B \cdot x^-$ is dense in $G \cdot x^-$. Replacing n by a positive multiple, we may also assume that σ lifts to $\tau \in H^0(X, \mathcal{O}_X(nD))$; then we may further assume that τ is a T–eigenvector. Since $\tau(x^-) \neq 0$, it follows that τ has no zero on X^-, so that nD is linearly equivalent to a linear combination of $D_1, \ldots, D_r$ with non–negative integer coefficients.

Since the divisor $mD - D_1 - \cdots - D_r$ is ample for large m, this shows that D is linearly equivalent to a linear combination of $D_1, \ldots, D_r$ with positive rational coefficients. Together with (1), this implies our statement.

(3) Let E be a nef divisor on X. Since $D_1 + \cdots + D_r$ is ample (by (2)), the divisor $E + D_1 + \cdots + D_r$ is ample as well. Applying again (2) completes the proof.

In particular, the Picard group $\mathrm{Pic}(X)$ is a free abelian group of finite rank. Together with [**6**, Examples 19.1.2, 19.3.3], this implies

COROLLARY 1. *For any normal, simple G–projective variety X, the natural map $\mathrm{Pic}(X) \to N^1(X)$ is an isomorphism. In other words, rational and numerical equivalence coincide for Cartier divisors.*

2. The cone of effective one–cycles

We will construct generators of the cone of effective one–cycles of a simple G–projective variety X, that are dual to the generators of the nef cone obtained in Theorem 1. For this, recall that every effective cycle on X is rationally equivalent to a linear combination of B–invariant subvarieties with positive coefficients, by [**8**]. Thus, we will study the B–invariant curves in X; we begin with the following easy result, generalizing [**3**, Proposition 2.1] to arbitrary characteristics.

LEMMA 5. *Let C be a B–invariant curve in a simple G–projective variety. Then C is fixed pointwise by the radical of a unique minimal parabolic subgroup $P(C) \supset B$, and $C = \overline{B \cdot x}$ for a unique $x \in C^T$. Moreover, the normalization of C is isomorphic to $\mathbb{P}^1$, and the normalization map is bijective.*

Proof. By assumption, C^B consists of a unique point. And since C is T–invariant, $C \setminus C^B$ contains a unique T–fixed point x. The reduced isotropy group B_x equals $U_x T$, where U_x is a closed reduced subgroup of codimension one in U, normalized by T. It follows easily that $U_x = R_u(P)$ for a unique minimal parabolic subgroup $P \supset B$. Then the radical $R(P)$ is a normal subgroup of B contained in B_x. Thus, $R(P)$ fixes pointwise $\overline{B \cdot x} = C$, and $R_u(P)$ is the kernel of the U–action in C.

Since C consists of a dense B–orbit and a unique fixed point, it is rational, and the normalization map is bijective.

REMARK 1. In characteristic zero, C is isomorphic to $\mathbb{P}^1$ by [**3**, 2.2, 2.6]. This does not generalize to arbitrary characteristics: let indeed K be a field of characteristic $p \geq 3$ and let $G = \mathrm{SL}(2)$, with its Borel subgroup B of upper triangular matrices. Let M be the space of homogeneous polynomials of degree $p+2$ in two variables x, y, where G acts by linear substitutions. The line $Ky^{p+2} \subset M$ is B–invariant; let $N = M/Ky^{p+2}$ be the quotient B–module, and let $[x^{p+1}y]$ be the image of $x^{p+1}y$ in $\mathbb{P}(N)$. Then

$$C = \overline{B \cdot [x^{p+1}y]}$$

is a B–invariant curve in $\mathbb{P}(N)$. Since

$$(x+ty)^{p+1}y = x^{p+1}y + tx^p y^2 + t^p x y^{p+1} + t^{p+1}y^{p+2},$$

the curve C is singular at $[xy^{p+1}]$, the unique B–fixed point in $\mathbb{P}(N)$. On the other hand, C is a B–invariant curve in

$$X = G \times^B \mathbb{P}(N),$$

and the latter is a simple, nonsingular G–projective variety.

Another useful observation is the following

LEMMA 6. *Any simple G–projective variety contains only finitely many B–invariant curves.*

Proof. Let X be a simple G–projective variety. By Lemma 5, it suffices to show that $X^{R(P)}$ contains only finitely many B–invariant curves, where $R(P)$ is the radical of a minimal parabolic subgroup $P \supset B$. Then the quotient group $P/R(P)$ is isomorphic to SL(2) or PGL(2); it acts on $X^{R(P)}$ with a unique closed orbit (since $R_u(P) \subset U$, and X^U consists of a unique point). Thus, we may assume that $G = \mathrm{SL}(2)$. Then every $H^0(\omega)$ is multiplicity–free as a T–module, so that the set $\mathbb{P}(H^0(\omega))^T$ is finite. Thus, X^T is finite by Lemma 1. This implies our statement, by Lemma 5 again.

We may now state our second main result.

THEOREM 2. (1) *With the notation of Theorem 1, the sink of each D_i is a unique point x_i^-, isolated in X^T. Moreover, D_i is the closure of $X^-(x_i^-)$, whereas $X^+(x_i^-) = B \cdot x_i^-$; the B–invariant curve $C_i = \overline{B \cdot x_i^-}$ intersects D_i at the unique point x_i^-, and intersects no other D_j. As a consequence, $(D_i \cdot C_i)$ is a positive integer, and $(D_i \cdot C_j) = 0$ for $j \neq i$.*

(2) *The cone $NE(X)$ is generated by the classes of $C_1, \ldots, C_r$, and these form a basis of the rational vector space $N_1(X)_{\mathbb{Q}}$.*

(3) *Any negative stratum having an irreducible component of codimension one is actually irreducible, with closure some D_i. Likewise, any positive stratum having an irreducible component of dimension one is open in some C_i.*

(4) *If, in addition, X is nonsingular, then every D_i intersects transversally C_i at x_i^-. In particular, D_i is nonsingular at x_i^-, and $(D_i \cdot C_i) = 1$. Moreover, all C_i are isomorphic to $\mathbb{P}^1$, and their classes form a basis of the group $N_1(X)$.*

Proof. (1) Let Y_i be the sink of D_i and let $x \in Y_i$. Then $X^+(x)$ is positive–dimensional (since $x \neq x^-$), and $X^+(x) \cap D_i = D_i^+(x) =$

$\{x\}$. Since D_i is a Cartier divisor, it follows that $X^+(x)$ is one–dimensional. As a consequence, $\overline{B \cdot x}$ is a B–invariant curve with T–fixed points x and x^-. Now Lemma 6 implies that $Y_i = \{x\}$. Since $X^+(Y_i)$ is B-invariant, we must have $X^+(Y_i) = B \cdot x$.

Let Y be an irreducible component of X^T through x. Then $X^+(Y) \cap D_i = D_i^+(Y \cap D_i)$ contains x as an isolated point, so that $\dim_x X^+(Y) \leq 1$. On the other hand, $X^+(Y)$ contains the curve $B \cdot y$ for any $y \in Y$; it follows that $Y = \{x\}$. In other words, x is isolated in X^T; hence $X^+(x) = B \cdot x$.

To show that $\overline{X^-(x)} = D_i$, we choose an open affine T–invariant neighborhood $X(x) \subset X$ of x, where D_i is a principal divisor associated to a regular function f, eigenvector of T. Then $X(x)$ contains both $X^-(x)$ and $X^+(x) = B \cdot x$; moreover, the weight of f is negative on λ (since f restricts to a non–zero function on $B \cdot x$). It follows that f vanishes at any point of $X^-(x)$, so that $X^-(x) = D_i^-(x)$.

Since $x^- \notin D_i$, the curve $C_i = \overline{B \cdot x}$ intersects D_i at x only. We show that C_i intersects no D_j for $j \neq i$. Otherwise, $x \in D_j$. Moreover, x is not the sink of D_j by the preceding step, so that $D_j^+(x) \neq \{x\}$. But $D_j^+(x) \subseteq X^+(x) = B \cdot x$, and hence $B \cdot x \subseteq D_j$. Thus, $x^- \in D_j$, a contradiction.

(2) follows from (1) together with Theorem 1.

(3) Let $X^-(Y)$ be a negative stratum having an irreducible component D of codimension one in X. Then $D = D_i$ for some index i. Together with (1), it follows that $x_i^- \in D$, whence $Y = \{x_i^-\}$ and $X^-(Y) = D_i^-$.

Next let $X^+(Y)$ be a positive stratum having an irreducible component C of dimension one. Then C is B–invariant, so that $C = B \cdot x$ for some $x \in Y$. Let Z be an irreducible component of Y containing x, then $C \subseteq B \cdot Z \subseteq X^+(Y)$, and $B \cdot Z$ is irreducible. Hence $C = B \cdot Z$, that is, $Y = \{x\}$. Now every irreducible component of $X^+(x)$ is B–invariant and contains x; thus, $X^+(x) = B \cdot x = C$. By $(*)$, we have $\dim X \leq \dim_x X^+(x) + \dim_x X^-(x)$. Thus, $X^-(x)$ is a divisor at x. Since $X^-(x)$ is disjoint from X^-, it follows that $x = x_i^-$ for some index i, whence $C = C_i$.

(4) Since X is nonsingular and x_i^- is an isolated fixed point, $X^-(x_i^-)$ and $X^+(x_i^-)$ intersect transversally at x_i^-; thus, the same holds for their closures D_i and C_i. Together with (1) and Theorem 1, it follows that D_i restricts to a globally generated Cartier divisor of degree 1 on C_i. But the normalization of C_i is isomorphic to $\mathbb{P}^1$,

so that $C_i \cong \mathbb{P}^1$. Finally, by (2), any $\gamma \in N_1(X)$ decomposes as $\gamma = \sum_{i=1}^r c_i C_i$, and every $c_i = (D_i \cdot \gamma)$ is an integer.

REMARK 2. In particular, any stratum of codimension at most one in a simple G–projective variety contains a unique T–fixed point. But the whole fixed point set may be infinite; in fact, it may have arbitrary irreducible components, as shown by the following construction.

Let $Y \subseteq \mathbb{P}^n$ be a projective variety. Let $G = \mathrm{SL}(3)$ with standard opposite Borel subgroups B, B^-; let ω be the highest root, i.e., the highest weight of the adjoint representation. Then the zero–weight space $H^0(\omega)^T$ has dimension 2, so that $H^0(n\omega)^T$ has dimension $\geq n+1$. Thus, we may regard Y as a subvariety of $\mathbb{P}(H^0(n\omega)^T)$. Now let $X = \overline{G \cdot Y}$ (closure in $\mathbb{P}(H^0(n\omega))$. Clearly, X is a simple G–projective variety, and Y is an irreducible component of X^T. In fact, both $X^+(Y)$ and $X^-(Y)$ have codimension 3 in X, since $U \cdot Y$ (resp. $U^- \cdot Y$) is dense in $X^+(Y)$ (resp. $X^-(Y)$).

Finally, we compare the group $N_1(X)$ with the Chow group $A_1(X)$ of one–cycles modulo rational equivalence.

COROLLARY 2. *For any normal, simple G–projective variety X, the map $A_1(X) \to N_1(X)$ is an isomorphism over the rationals. If, in addition, X is nonsingular, then this map is an isomorphism.*

Proof. We begin with the case where X is nonsingular. Then it suffices to show that the classes of $C_1, \ldots, C_r$ generate the group $A_1(X)$. Decomposing X into the positive strata $X^+(Y)$ and using the long exact sequence for Chow groups [**6**, Proposition 1.8], this reduces to checking the vanishing of $A_1(X^+(Y))$ whenever $\dim X^+(Y) \geq 2$.

Recall that $p^+ : X^+(Y) \to Y$ is an affine bundle. By Lemma 2 and [**6**, Proposition 1.9], it follows that $A_1(X^+(Y)) = 0$ unless $\dim X^+(x) = 1$ for all $x \in Y$. In the latter case, since $B \cdot x \subseteq X^+(x)$ for all $x \in Y$, the fibers of p^+ are B–invariant curves in $X^+(Y)$. By Lemma 6, it follows that Y is a unique point, so that $X^+(Y)$ is one–dimensional.

For arbitrary X, recall that the group $A_1(X)$ is generated by classes of B–invariant curves. Together with Theorem 2 (3), this reduces to checking the following assertion: let $X^+(Y)$ be a positive stratum and let $C \subseteq X^+(Y)$ be a B–invariant curve which is not an irreducible component. Then the class of C is zero in $A_1(X^+(Y))_{\mathbb{Q}}$.

Since C is contained in an irreducible component of $X^+(Y)$ of dimension ≥ 2, we may find a closed irreducible B–stable surface

$S \subseteq X^+(Y)$ containing C. Let $Z = p^+(S)$, then $Z \neq S$ (otherwise $S \subseteq X^T$, whence $S \subseteq X^B$, a contradiction). For any $x \in Z$, the fiber at x of $p^+ : S \to Z$ contains $B \cdot x$. By Lemma 6, it follows that Z consists of a unique point x, the source of S. In particular, S is affine, and its algebra of regular functions is negatively graded. Consider now the normalization $\tilde{S}$ of S. The group B acts on $\tilde{S}$ without fixed points, so that $\tilde{S}$ is nonsingular. On the other hand, the algebra of regular functions of $\tilde{S}$ is negatively graded as well; as a consequence, $\tilde{S}$ is isomorphic to the affine plane. In particular, $A_1(\tilde{S}) = 0$, whence $A_1(S)_{\mathbb{Q}} = 0$. This proves our assertion.

3. The simple complete PGL(2)–embeddings

As an illustration of our results, we describe the cones of effective one–cycles and nef divisors for all *simple complete embeddings of* PGL(2), that is, for the normal complete PGL(2)–varieties containing an open orbit isomorphic to PGL(2) and a unique closed orbit. Along the way, we obtain a realization of these embeddings as PGL(2)–orbit closures in a product of copies of the projective line, suggested by work of Iozzi and Poritz [**10**].

Note that there is a combinatorial classification of all embeddings of PGL(2), presented e.g. in [**17**], as part of the Luna–Vust theory of embeddings of homogeneous spaces; the geometric realization of simple complete PGL(2)–embeddings can be deduced from this classification. The Chow rings of smooth complete SL(2)–embeddings, and their cones of effective one–cycles, are described in [**18**].

We introduce some notation. Let $G = PGL(2)$ and let T be the image in G of the torus of diagonal matrices in SL(2). The map $t \mapsto \begin{pmatrix} t & 0 \\ 0 & t^{-1} \end{pmatrix}$ defines a one–parameter subgroup λ of T, which generates the group $\mathcal{X}_*(T)$. The opposite Borel subgroups $B = G(\lambda)$, $B^- = G(-\lambda)$ are the images in G of the subgroups of upper (resp. lower) triangular matrices in SL(2).

For the standard action of G on $\mathbb{P}^1$, the isotropy group of the sink ∞ (resp. the source 0) is B (resp. B^-). We will consider the diagonal action of G on the product $(\mathbb{P}^1)^r$ of $r \geq 3$ copies of $\mathbb{P}^1$. This is a (very) simple G–projective variety, with closed orbit being the small diagonal, $\operatorname{diag} \mathbb{P}^1$; the sink is ∞^r, with B^-–invariant open affine neighborhood $(\mathbb{P}^1 \setminus \{0\})^r \cong \mathbb{A}^r$ where T acts by scalar multiplication. The irreducible divisors of Theorem 1 are the

$$D_i = (\mathbb{P}^1)^{i-1} \times \{0\} \times (\mathbb{P}^1)^{r-i},$$

and the curves of Theorem 2 are the lines

$$C_i = \infty^{i-1} \times \mathbb{P}^1 \times \infty^{r-i},$$

where $1 \le i \le r$.

PROPOSITION 1. *Let $p_1, \dots, p_r \in \mathbb{P}^1$ be pairwise distinct points and let*

$$X = X(p_1, \dots, p_r) \subseteq (\mathbb{P}^1)^r$$

be the closure of the orbit $G \cdot (p_1, \dots, p_r)$. Then the following statements hold.

(1) *The irreducible components of the boundary $X \backslash G(p_1, \dots, p_r)$ are the divisors*

$$\partial_i X = \{(x^{i-1}, y, x^{r-i}) \mid x, y \in \mathbb{P}^1\} = G \cdot C_i \ (1 \le i \le r).$$

These are isomorphic to $\mathbb{P}^1 \times \mathbb{P}^1$, and they intersect along $\operatorname{diag} \mathbb{P}^1$.

(2) *The open subset X^- is isomorphic to $\mathbb{A}^1 \times \Sigma$, where Σ is the affine cone over the rational normal curve in $\mathbb{P}^{r-2}$. As a consequence, X is a normal, Cohen–Macaulay, simple G–projective variety; it is singular along* $\operatorname{diag} \mathbb{P}^1$, *if $r \ge 4$.*

(3) *The irreducible divisors constructed in Theorem 1 are the pull–backs of 0 under the r projections $X \to \mathbb{P}^1$; they are normal and Cohen–Macaulay. The curves constructed in Theorem 2 are the lines $C_1, \dots, C_r$.*

(4) *The boundary divisors $\partial_1 X, \dots, \partial_r X$ form a basis of the divisor class group of X with rational coefficients; the latter is isomorphic to the Picard group with rational coefficients.*

(5) *A canonical divisor for X is*

$$-K_X = \partial_1 X + \cdots + \partial_r X = \frac{2}{r-2}(D_1 + \cdots + D_r).$$

As a consequence, $-(r-2)K_X$ is very ample, so that X is $\mathbb{Q}$–Fano.

Conversely, any simple complete embedding of G is isomorphic to $X(p_1, \dots, p_r)$, where $r \ge 3$ and $p_1, \dots, p_r \in \mathbb{P}^1$ are uniquely determined up to permutation and diagonal action of G.

Proof. (1) If $r = 3$, then $X = (\mathbb{P}^1)^3$ and our assertions are evident; so we may assume that $r \ge 4$. We may also assume that $p_1 = \infty$ and $p_2 = 0$; then $p_3, \dots, p_r$ are distinct non–zero scalars.

The first projection $z_1 : X \to \mathbb{P}^1 = G \cdot p_1 \cong G/B$ is a locally trivial fibration. It yields an isomorphism $X \cong G \times^B S$, where $S =$

$\overline{B\cdot(p_1,\ldots,p_r)}\subset X$ is an irreducible projective B–invariant surface with a unique fixed point ∞^r. Moreover, the map

$$U^-\times S^-\to X^-,\ (g,s)\mapsto g\cdot s$$

is an isomorphism, where $S^-=S\cap(\mathbb{P}^1\setminus\{0\})^r$ is an irreducible surface in $\mathbb{A}^r$, invariant under scalar multiplication. Since $S^-\subseteq\{\infty\}\times\mathbb{A}^{r-1}$ (as $p_1=\infty$), we will regard S^- in $\mathbb{A}^{r-1}$ via projection.

Since $\begin{pmatrix} t & u\\ 0 & t^{-1}\end{pmatrix}\cdot p_i=t^2p_i+tu$ in $\mathbb{P}^1\setminus\{\infty\}$, we see that S^- is the closure of the image of the rational map

$$\mathbb{A}^2-\to\mathbb{A}^{r-1},\ (t,u)\mapsto(\frac{1}{tu},\frac{1}{t^2p_3+tu},\ldots,\frac{1}{t^2p_r+tu}).$$

For $2\le i\le r$, let $x_i=(t^2p_i+tu)^{-1}$. Equivalently, $tux_2=1$, and $tp_ix_i+u(x_i-x_2)=0$ for $3\le i\le r$. Hence S^- is defined (as a closed subset of $\mathbb{A}^{r-1}$) by the vanishing of all 2×2 minors of the $2\times(r-2)$ matrix

$$A=\begin{pmatrix} p_3x_3 & p_4x_4 & \cdots & p_rx_r\\ x_3-x_2 & x_4-x_2 & \cdots & x_r-x_2\end{pmatrix}.$$

It follows that the boundary $S^-\setminus B\cdot(p_1,\ldots,p_r)$ consists of the diagonal

$$(x_2=x_3=\cdots=x_r)=S^-\cap G\cdot C_1,$$

together with the $r-1$ coordinate lines

$$(x_2=\cdots=x_{i-1}=x_{i+1}=\cdots=x_r=0)=C_i^-.$$

On the other hand, the open subset X^- equals $U^-\cdot S^-$, and intersects all G–orbits in X (since X^- intersects the unique closed orbit). Thus, the boundary $X\setminus G\cdot(p_1,\ldots,p_r)$ is the union of the irreducible divisors $\partial_1X=G\cdot C_1,\ldots,\partial_rX=G\cdot C_r$.

(2) Clearly, X is a simple G–projective variety. We claim that its regular locus intersects all irreducible components of the boundary. To see this, consider the projection $(z_i,z_j,z_k):(\mathbb{P}^1)^r\to(\mathbb{P}^1)^3$ where $1\le i<j<k\le r$. Then the restriction of (z_i,z_j,z_k) to X is birational, and bijective over $G\cdot(C_j\setminus\{\infty^r\})$. Now our claim follows from Zariski's main theorem.

By that claim, X is nonsingular in codimension one. Next we show that X is Cohen–Macaulay, or, equivalently, that S^- is Cohen–Macaulay. For this, consider the closed subscheme $\Sigma\subseteq\mathbb{A}^{r-1}$ associated with the ideal generated by the 2×2 minors of the matrix A. Then S^- is the support of Σ. On the other hand, we may regard Σ as a closed subscheme of the space $M_{2,r-2}$ of $2\times(r-2)$ matrices, an affine space of dimension $2r-4$. Then Σ is the scheme–theoretic intersection of the closed subscheme of matrices of rank at most one (a Cohen–Macaulay variety of dimension $r-1$) with a linear space of dimension $r-1$. Since $\dim\Sigma = 2$, it follows that Σ is Cohen–Macaulay. Moreover, one easily checks that the intersection $\Sigma\cap(x_2=0)\subseteq\mathbb{A}^{r-1}$ is reduced, of dimension one. Thus, x_2 is a non–zero–divisor on Σ, and the latter is reduced. It follows that $\Sigma = S^-$.

Thus, $X^-\cong U^-\times S^-\cong\mathbb{A}^1\times\Sigma$, where Σ is the affine cone over a curve $C\subset\mathbb{P}^{r-2}$. Clearly, C is rational and not contained in any hyperplane; moreover, C is projectively normal, since Σ is Cohen–Macaulay and nonsingular in codimension one. Thus, C is a rational normal curve.

(3) follows from the fact that X contains $C_1,\dots,C_r$, together with Theorem 2.

(4) For the divisor class group $A_2(X)$, we have an exact sequence

$$\bigoplus_{i=1}^{r}\mathbb{Z}\,\partial_i X\to A_2(X)\to A_2(G\cdot(p_1,\dots,p_r))\to 0.$$

Moreover, the group $A_2(G\cdot(p_1,\dots,p_r)) = \mathrm{Pic}(\mathrm{PGL}(2))$ has order 2, and the classes of the $\partial_i X$ are linearly independent in $A_2(X)_{\mathbb{Q}}$ (since every regular invertible function on PGL(2) is constant). Therefore, these classes form a basis of $A_2(X)_{\mathbb{Q}}$. The lattter contains $\mathrm{Pic}(X)_{\mathbb{Q}}$ as a subspace of dimension r; thus, they are equal.

(5) By (4), we have

$$-K_X = a_1\partial_1 X+\cdots+a_r\partial_r X$$

for unique rational coefficients $a_1,\dots,a_r$. If $r=3$, then $X=(\mathbb{P}^1)^3$ and one obtains easily $a_1=a_2=a_3=1$. In the general case, any $\partial_i X$ is mapped isomorphically to its image under some projection to $(\mathbb{P}^1)^3$; it follows that $a_i=1$.

To express $-K_X$ in terms of $D_1,\dots,D_r$, consider the divisors of the rational functions z_i-z_j, where $z_1,\dots,z_r : X\to\mathbb{P}^1$ denote the projections. We obtain

$$\mathrm{div}(z_i-z_j) = -D_i-D_j+\sum_{k,k\notin\{i,j\}}\partial_k X,$$

whence

$$\partial_1 X+\cdots+\partial_r X = \frac{2}{r-2}(D_1+\cdots+D_r)$$

in $A_2(X)_{\mathbb{Q}}$.

For the final assertion, let X be a simple complete embedding of G. Choose $p \in X$ in the open G–orbit. By Lemma 1, there exists a finite equivariant morphism

$$f : X \to \mathbb{P}(H^0(\omega)).$$

Now $H^0(\omega)$ is the space of homogeneous polynomials of degree d in two variables x, y, where G acts by linear substitutions; here d is an even positive integer (see e.g. [**11**, II.2.16]). Let $P(x, y) \in H^0(\omega)$ be a representative of $f(p) \in \mathbb{P}(H^0(\omega))$; let

$$P(x,y) = \prod_{i=1}^{s} (a_i x + b_i y)^{m_i}$$

be a decomposition into a product of pairwise distinct linear forms, with multiplicities. Since f is finite, then so is the isotropy group of the line KP. Thus, $s \geq 3$.

The source of the λ–action on $\mathbb{P}(H^0(\omega))$ consists of the image of y^d, and the corresponding open subset $\mathbb{P}(H^0(\omega))^-$ consists of the images of those homogeneous polynomials where the coefficient of y^d is non–zero. Since f is finite and equivariant, $X^- = f^{-1}(\mathbb{P}(H^0(\omega))^-$. It follows that $(X \setminus X^-) \cap G \cdot p$ consists of s irreducible components. On the other hand, any irreducible component D_i of $X \setminus X^-$ intersects $G \cdot p$ (otherwise, D_i is G–invariant, and hence contains the closed G–orbit, a contradiction). Thus, $s = r$ with the notation of Theorem 1.

Next let

$$f_i : X \to Y_i = \operatorname{Proj} \bigoplus_{n=0}^{\infty} H^0(X, \mathcal{O}_X(nD_i))$$

be the morphism associated with the globally generated divisor D_i. Then Y_i is a normal simple G–projective variety, and f_i is equivariant, separable, with connected fibers. Moreover, the Picard group of Y_i is freely generated by a unique ample divisor, which pulls back to D_i. By considering a finite morphism $Y_i \to \mathbb{P}(H^0(\omega_i))$ and arguing as above, one obtains that $\dim Y_i = 1$. On the other hand, G acts on Y_i with a dense separable orbit and a unique closed orbit, whence Y_i is isomorphic to the projective line with standard G–action.

Consider now the product morphism

$$f = \prod_{i=1}^{r} f_i : X \to \prod_{i=1}^{r} Y_i \cong (\mathbb{P}^1)^r.$$

Then f is finite, since $D_1 + \cdots + D_r$ is ample. Let $f(p) = (p_1, \ldots, p_r)$, then the p_i are pairwise distinct (since so are the D_i), and their number is at least 3. Thus, the restriction $G \cdot p \to G \cdot f(p)$ is an isomorphism. On the other hand, $f(X)$ is normal by (1), so that f is a closed immersion by Zariski's main theorem.

So we have proved that X embeds equivariantly into $(\mathbb{P}^1)^r$, where r is the rank of Pic(X); moreover, the r projections $z_i : X \to \mathbb{P}^1$ are uniquely determined. This implies the remaining uniqueness assertion.

References

[1] A. Bialynicki–Birula: Some theorems on actions of algebraic groups, *Ann. of Math.* **98** (1973), 480–497.

[2] L. Bonavero, C. Casagrande, O. Debarre, S. Druel: Sur une conjecture de Mukai, *Comment. Math. Helv.*, (to appear) (math.AG/0204314).

[3] M. Brion: Variétés sphériques et théorie de Mori, *Duke Math. J.* **72** (1993), 369–404.

[4] M. Brion: Curves and divisors in spherical varieties. In: *Algebraic groups and Lie groups*, 21–34, Austral. Math. Soc. Lect. Ser. **9**, Cambridge Univ. Press, Cambridge, 1997.

[5] H. Clemens, J. Kollar, S. Mori: Higher dimensional complex geometry, *Astérisque* **166** (1988).

[6] W. Fulton: Intersection theory. Second edition, Springer-Verlag, Berlin, 1998.

[7] R. Hartshorne: Ample subvarieties of algebraic varieties, Lecture Notes in Mathematics **156**, Springer-Verlag, Berlin, 1970.

[8] A. Hirschowitz: Le groupe de Chow équivariant, *C. R. Acad. Sci. Paris Sér. I Math.* **298** (1984), 87–89.

[9] G. Horrocks: Fixed point schemes of additive group actions, *Topology* **8** (1969), 233–242.

[10] A. Iozzi, J. Poritz: Boundary compactifications of SL$(2, \mathbb{R})$ and SL$(2, \mathbb{C})$, *Forum Math.* **11** (1999), 385–397.

[11] J. C. Jantzen: Representations of Algebraic Groups, Academic Press, Boston, 1987.

[12] F. Knop, H. Kraft, D. Luna, T. Vust: Local properties of algebraic group actions. In: *Algebraic transformation groups and invariant theory*, 63–75, Birkhäuser Verlag, Basel, 1989.

[13] F. Knop, H. Kraft, T. Vust: The Picard group of a G–variety. In: *Algebraic transformation groups and invariant theory*, 77–87, Birkhäuser Verlag, Basel, 1989.

[14] F. Knop: Über Bewertungen, welche invariant under einer reduktiver Gruppe sind, *Math. Ann.* **295** (1993), 333–363.

[15] F. Knop: The asymptotic behavior of invariant collective motion, *Invent. math.* **116** (1994), 309–328.

[16] J. Konarski: Decompositions of normal algebraic varieties determined by an action of a one–dimensional torus, *Bull. Acad. Polon. Sci. Sér. Sci. Math. Astronom. Phys.* **26** (1978), 295–300.
[17] L. Moser–Jauslin: Smooth embeddings of $SL(2)$ and $PGL(2)$, *J. Algebra* **132** (1990), 384–405.
[18] L. Moser–Jauslin: The Chow rings of smooth complete $SL(2)$–embeddings, *Compositio Math.* **82** (1992), 67–106.
[19] D. Mumford, J. Fogarty, F. Kirwan: Geometric invariant theory. Third edition, Springer-Verlag, Berlin, 1994.

Université de Grenoble I
Département de Mathématiques
Institut Fourier, UMR 5582 du CNRS
38402 Saint-Martin d'Hères
Cedex, France
E-mail address: `Michel.Brion@ujf-grenoble.fr`

Representations of double affine lie algebras

Vyjayanthi Chari and Thang Le

Dedicated to Professor C.S. Seshadri on his 70th *birthday*

Abstract. We study a family of indecomposable integrable modules for the double affine algebra, the so–called Weyl modules. These modules have non–zero level with respect to one of the central elements and zero level with respect to the other. We give a condition for the modules to be irreducible, and use the idea of fusion product of modules to prove that they are generically reducible.

0. Introduction

The representation theory of Kac–Moody algebras, and in particular that of affine Lie algebras, has been extensively studied over the past twenty years. The representations that have had the most applications are the integrable ones, so called because they lift to the corresponding group.

The affine Lie algebra associated to a finite-dimensional complex simple Lie algebra $\mathfrak{g}$ is the universal (one–dimensional) central extension of the Lie algebra of polynomial maps from $\mathbf{C}^* \to \mathfrak{g}$. There are essentially two kinds of interesting integrable representations of this algebra: one is where the center acts as a positive integer, the positive energy representations; and the other is where the center acts trivially, or the level zero representations. Both kinds of representations have interesting applications: representations of the first kind have connections with number theory through the Rogers–Ramanujam identities, the theory of vertex algebras and conformal field theory; representations of the second kind are connected with the six–vertex and XXZ–model (see [**11**] and the references therein), the Kostka polynomials and the fermionic formulas of Kirillov and Reshetikhin,

[**18**]. The study of such level zero representations was begun in [**4**] and continued in [**5**],[**6**]. More recently a geometric approach to such representations was developed in [**21**] for the corresponding quantum algebras.

The affine Lie algebras admit an obvious generalization. Namely, we can consider central extensions of the polynomial maps $(\mathbf{C}^*)^\ell \to \mathfrak{g}$. Not surprisingly, these algebras are a lot more complicated, for instance the central extension is now infinite–dimensional. A systematic study of such algebras can be found in [**2**] and the representation theory has been studied in [**3**], [**9**], [**10**], [**20**]. In general, interesting theories have been found for the quotients of this algebra by a central ideal of finite–codimension.

One such quotient is the double affine algebra, this algebra is obtained from the affine algebra in essentially the same way as the affine algebra is obtained from the finite–dimensional algebra. One can also define a corresponding quantum object, and representations of these have been studied in [**21**], [**22**], [**23**]. However, relatively little is still known about the integrable representations of the double affine algebras.

In this paper we study representations of the double affine algebra $\mathfrak{g}_{\rm tor}$ when one of the centers acts trivially; this is also the situation studied in the quantum case mentioned above. The category of such representations is not semisimple and our interest is in the study of indecomposable integrable representations of $\mathfrak{g}_{\rm tor}$ rather than the irreducible ones. We are motivated by considerations coming from the study of quantum affine algebras [**8**] and modular Lie algebras. Thus, we believe that these indecomposable representations should be the limit as $q \to 1$ of the irreducible representations of the corresponding quantum algebra. In the case of quantum affine algebras this is in fact a conjecture which has been checked in many cases; for double affine algebras, no such conjecture is possible at the moment, since the notion of $q \to 1$ is not well–defined. However, Theorem 4 can be viewed as the classical analog of the result in [**22**].

The paper is organized as follows. In section 1, we set up the notation to be used in the rest of the paper. In section 2, we recall several results on integrable representations of affine Lie algebras. In section 3, we define a family of indecomposable integrable representations for the double affine algebra and identify their irreducible quotients.

Sections 4 and 5 are devoted to the study of the structure of these representations. Thus we give a sufficient condition for the indecomposable modules to be irreducible. We also show that in fact the modules are almost always reducible in Section 5. In the case of affine algebras, this was proved by passing to the quantum situation. In this paper, however, we prove it by using the notion of fusion product representations, an idea introduced recently by Feigin and Loktev in [**12**].

1. Preliminaries

1.1. Let $\mathfrak{g}_{\rm fin}$ be a complex finite-dimensional simple Lie algebra of rank n, and let $\mathfrak{h}_{\rm fin}$ be a Cartan subalgebra of $\mathfrak{g}_{\rm fin}$. Fix a set $\{\alpha_i : 1 \le i \le n\}$ (resp. $\{\omega_i : 1 \le i \le n\}$) of simple roots (resp. fundamental weights) of $\mathfrak{g}_{\rm fin}$ with respect to $\mathfrak{h}_{\rm fin}$. Let $R^+_{\rm fin}$ be the corresponding set of positive roots and let $\theta \in R^+_{\rm fin}$ be the highest root. Given $\alpha \in R^+_{\rm fin}$, let $\mathfrak{g}^{\pm\alpha}_{\rm fin}$ be the corresponding root spaces and fix non–zero elements $x^\pm_\alpha \in \mathfrak{g}^{\pm\alpha}_{\rm fin}$ and $h_\alpha \in \mathfrak{h}_{\rm fin}$ such that the elements $x^\pm_\alpha$, h_α span a subalgebra isomorphic to sl_2. Set $x^\pm_i = x^\pm_{\alpha_i}$, $h_i = h_{\alpha_i}$. Define subalgebras

$$\mathfrak{n}^\pm_{\rm fin} = \oplus_{\alpha\in R^+_{\rm fin}} \mathfrak{g}^{\pm\alpha}_{\rm fin}.$$

We have,

$$\mathfrak{g}_{\rm fin} = \mathfrak{n}^-_{\rm fin} \oplus \mathfrak{h}_{\rm fin} \oplus \mathfrak{n}^+_{\rm fin}.$$

Let $Q^+_{\rm fin}$ (resp. $P^+_{\rm fin}$) be the non-negative root (resp. weight) lattice of $\mathfrak{g}_{\rm fin}$.

1.2. The corresponding untwisted affine Lie algebra $\mathfrak{g}_{\rm aff}$ is the universal central extension of the Lie algebra of Laurent polynomial maps from $\mathbf{C}^* \to \mathfrak{g}_{\rm fin}$. Thus

$$\mathfrak{g}_{\rm aff} = \mathfrak{g}_{\rm fin} \otimes \mathbf{C}[t_1, t_1^{-1}] \oplus \mathbf{C}c_1,$$

with bracket given by

$$[xt_1^r, yt_1^m] = [x,y]t_1^{r+m} + r\delta_{r,-m} < x, y > c_1, \quad [c_1, xt_1^r] = 0, \qquad x, y \in \mathfrak{g}_{\rm fin},\ r, m \in \mathbf{Z}$$

where $<,>$ is the symmetric non–degenerate invariant form on $\mathfrak{g}_{\rm fin}$ which is normalized so that $< h_\theta, h_\theta >= 2$.

Let d_1 be the derivation $t_1 d/dt_1$ of $\mathbf{C}[t_1, t_1^{-1}]$; then d_1 acts on $\mathfrak{g}_{\rm aff}$ in the obvious way. The extended affine algebra $\mathfrak{g}^{\rm e}_{\rm aff}$ is the semi–direct product of $\mathbf{C}d_1$ and $\mathfrak{g}_{\rm aff}$, i.e. $\mathfrak{g}^{\rm e}_{\rm aff} = \mathfrak{g}_{\rm aff} \oplus \mathbf{C}d_1$ with the Lie bracket between d_1 and $\mathfrak{g}_{\rm aff}$ being given by

$$[d_1, c_1] = 0, \quad [d_1, xt_1^m] = mxt_1^m, \quad x \in \mathfrak{g}_{\rm fin}, m \in \mathbf{Z}.$$

The algebra $\mathfrak{g}_{\rm aff}^{\rm e}$ admits a symmetric, invariant bilinear form $< , >_{aff}$ which is defined as follows: for $x, y \in \mathfrak{g}_{\rm fin}$, $r, m \in \mathbf{Z}$

$$< xt_1^r \ , \ yt_1^m >_{aff} = \delta_{r,-m} < x \ , y >$$

and

$$< c_1 \ , \ d_1 >_{aff} = 1, \quad < c_1 \ , \ \mathfrak{g}_{\rm aff} >_{aff} = 0, \quad < d_1 \ , \ xt^r >_{aff} \ = \ 0.$$

Set

$$\mathfrak{h}_{\rm aff}^{\rm e} = \mathfrak{h}_{\rm fin} \oplus \mathbf{C}c_1 \oplus \mathbf{C}d_1, \quad \mathfrak{n}_{\rm aff}^+ = \mathfrak{g} \otimes t_1\mathbf{C}[t_1] \oplus \mathfrak{n}_{\rm fin}^+,$$
$$\mathfrak{n}_{\rm aff}^- = \mathfrak{g} \otimes t_1^{-1}\mathbf{C}[t_1^{-1}] \oplus \mathfrak{n}_{\rm fin}^-.$$

Clearly we can regard $\mathfrak{g}_{\rm fin}$ as the algebra of constant maps in $\mathfrak{g}_{\rm aff}^{\rm e}$. Given any element $\lambda \in (\mathfrak{h}_{\rm fin})^*$ we regard it as an element of $(\mathfrak{h}_{\rm aff}^{\rm e})^*$ by setting $\lambda(c_1) = \lambda(d_1) = 0$. Define $\delta_1 \in (\mathfrak{h}_{\rm aff}^{\rm e})^*$ by

$$\delta_1(d_1) = 1, \quad (\delta_1)|_{\mathfrak{h}_{\rm fin} \oplus \mathbf{C}c_1} = 0.$$

Set $\alpha_0 = \delta_1 - \theta$. The elements $\{\alpha_i : 0 \le i \le n\}$ are called the affine simple roots. The corresponding set of positive affine roots is

$$R_{\rm aff}^+ = \{\pm\alpha + r\delta_1 : \alpha \in R_{fin}^+, r \in \mathbf{Z}, r > 0\} \cup R_{\rm fin}^+ \cup \{r\delta_1 : r \in \mathbf{Z}, r > 0\}.$$

The root space corresponding to $\pm\alpha + r\delta_1$ is $\mathfrak{g}_{\rm fin}^{\pm\alpha} \otimes \mathbf{C}t_1^r$, and that corresponding to $r\delta_1$ is $\mathfrak{h}_{\rm fin} \otimes \mathbf{C}t_1^r$. Fix non–zero elements $x_0^\pm \in \mathfrak{g}_{\rm aff}^{\pm\alpha_0}$ and $h_0 \in \mathfrak{h}_{\rm aff}^{\rm e}$ so that the subalgebra spanned by $x_0^\pm, h_0$ is isomorphic to sl_2. Then,

$$c_1 = h_0 + h_\theta.$$

Let $\omega_0 \in (\mathfrak{h}_{\rm aff}^{\rm e})^*$ be defined by

$$\omega_0(h) = 0, \quad \omega_0(c_1) = 1, \quad \omega_0(d_1) = 0, \quad (h \in \mathfrak{h}_{\rm fin}).$$

The affine root lattice $Q_{\rm aff}^+$ and the affine weight lattice $P_{\rm aff}^+$ in $(\mathfrak{h}_{\rm aff}^{\rm e})^*$ are now defined in the obvious way.

1.3. The double affine algebra $\mathfrak{g}_{\rm tor}$ and the extended algebra $\mathfrak{g}_{\rm tor}^{\rm e}$ are obtained from $\mathfrak{g}_{\rm aff}^{\rm e}$ in essentially the same way that $\mathfrak{g}_{\rm aff}$ and $\mathfrak{g}_{\rm aff}^{\rm e}$ were obtained from $\mathfrak{g}_{fin}$. More precisely, we have

$$\mathfrak{g}_{\rm tor} = \mathfrak{g}_{\rm aff} \otimes \mathbf{C}[t_2, t_2^{-1}] \oplus \mathbf{C}c_2 \oplus \mathbf{C}d_1, \quad \mathfrak{g}_{\rm tor}^{\rm e} = \mathfrak{g}_{\rm aff} \otimes \mathbf{C}[t_2, t_2^{-1}]$$
$$\oplus \mathbf{C}c_2 \oplus \mathbf{C}d_1 \oplus \mathbf{C}d_2,$$

with the Lie bracket given as follows. The element c_2 is central, and

$$[d_2, xt_2^r] = rxt_2^r, \quad [xt_2^r, yt_2^m] = [x, y]t_2^{r+m} + \delta_{r,-m} < x, y >_{\rm aff} c_2,$$
$$(x, y \in \mathfrak{g}_{\rm aff}, r, m \in \mathbf{Z}).$$

Clearly $\mathfrak{g}_{\rm aff}$ is a subalgebra of $\mathfrak{g}_{\rm tor}$. Set

$$\mathfrak{h}^{\rm e}_{\rm tor} = \mathfrak{h}^{\rm e}_{\rm aff} \oplus \mathbf{C}c_2 \oplus \mathbf{C}d_2.$$

Let $\delta_2 \in (\mathfrak{h}^{\rm e}_{\rm tor})^*$ be defined by setting

$$\delta_2(h) = 0, \quad h \in (\mathfrak{h}^{\rm e}_{\rm aff} \oplus \mathbf{C}c_2), \quad \delta_2(d_2) = 0.$$

As before, we regard an element $\lambda \in (\mathfrak{h}^{\rm e}_{\rm aff})^*$ as an element of $(\mathfrak{h}^{\rm e}_{\rm tor})^*$ by setting it to be zero on $\mathbf{C}c_2$ and $\mathbf{C}d_2$. The set of roots of the double affine algebra are then,

$$R_{\rm tor} = \{\pm\alpha + n_2\delta_2 : \alpha \in R^+_{\rm aff}, n_2 \in \mathbf{Z}\} \cup \{n_2\delta_2 : n_2 \in \mathbf{Z}\ \ n_2 \neq 0\}.$$

Unlike the affine case, there is no natural choice of simple roots. However, we will work with the following partition of $R_{\rm tor}$ into mutually disjoint sets,

$$\begin{aligned} R_{\rm tor}(>) &= \{\alpha + n_2\delta_2 : \alpha \in R^+_{\rm aff}, n_2 \in \mathbf{Z}, \} \\ R_{\rm tor}(0) &= \{n_2\delta_2 : n_2 \in \mathbf{Z},\ n_2 \neq 0\}, \\ R_{\rm tor}(<) &= \{-\alpha + n_2\delta_2 : \alpha \in R^+_{\rm aff}, n_2 \in \mathbf{Z}, \}. \end{aligned}$$

The subalgebras $\mathfrak{g}_{\rm tor}(>)$, $\mathfrak{g}_{\rm tor}(0)$ and $\mathfrak{g}_{\rm tor}(<)$ are defined in the obvious way.

Given a Lie algebra $\mathfrak{a}$, let $\mathbf{U}(\mathfrak{a})$ denote its universal enveloping algebra. We let $\mathbf{U}_{fin}$ (resp. $\mathbf{U}^{\rm e}_{\rm aff}$, $\mathbf{U}_{\rm tor}$) denote the enveloping algebra of $\mathfrak{g}_{\rm fin}$ (resp. $\mathfrak{g}^{\rm e}_{\rm aff}$, $\mathfrak{g}_{\rm tor}$). By the Poincare–Birkhoff–Witt theorem, we have

$$\begin{aligned} \mathbf{U}_{\rm fin} &= \mathbf{U}(\mathfrak{n}^-_{\rm fin})\mathbf{U}(\mathfrak{h}_{\rm fin})\mathbf{U}(\mathfrak{n}^+_{\rm fin}), \\ \mathbf{U}^{\rm e}_{\rm aff} &= \mathbf{U}(\mathfrak{n}^-_{\rm aff})\mathbf{U}(\mathfrak{h}^{\rm e}_{\rm aff})\mathbf{U}(\mathfrak{n}^-_{\rm aff}), \\ \mathbf{U}_{\rm tor} &= \mathbf{U}(\mathfrak{g}_{tor}(<))\mathbf{U}(\mathfrak{h}_{\rm tor})\mathbf{U}(\mathfrak{g}_{\rm tor}(0))\mathbf{U}(\mathfrak{g}_{tor}(>)). \end{aligned}$$

For $h \in \mathfrak{h}_{\rm aff}$, let $\Lambda^\pm(h,u)$ be the power series in an indeterminate u with coefficients in $\mathbf{U}_{\rm tor}$, defined by

$$\Lambda^\pm(h,u) = \sum_{r=0}^{\infty} \Lambda(h,r)u^{\pm r} = \exp - \left(\sum_{r=1}^{\infty} \frac{ht_2^{\pm r}u^r}{r}\right). \tag{1.1}$$

It is not hard to see that the elements $\{\Lambda^\pm(h,r) : h \in \mathfrak{h}_{\rm aff}, r \in \mathbf{Z}, r \geq 0\}$ generate the subalgebra $\mathbf{U}(\mathfrak{g}_{\rm tor}(0))$. Further, by taking the formal derivative with respect to u, we see that for all $r \geq 1$, we have

$$\Lambda(h,r) = -\frac{1}{r}\sum_{s=1}^{r}(ht_2^s)\Lambda(h,s-r).$$

For $\alpha \in R^+_{\rm fin}$ and $r \in \mathbf{Z}$, the the elements $\{x^\pm_\alpha t_1^r t_2^s : s \in \mathbf{Z}\}$ generate a subalgebra of $\mathfrak{g}_{\rm tor}$ which is isomorphic to the affine algebra associated to sl_2. The following lemma is a special case of Lemma 7.1 in [**13**].

LEMMA 1.1. *Let $\beta \in R^+_{\text{aff}}$ be of the form $\alpha + r_1\delta_1$ for some $r_1 \in \mathbf{Z}$. For all $s \geq 0$, we have*

$$(x^+_\beta t_2^{\pm 1})^s.(x^-_\beta)^{s+1} = \sum_{m=0}^{s} (x^-_\beta t_2^{\pm m})\Lambda^\pm(h_\beta,\ s-m) +\ X,$$

$$(x^+_\beta t_2^{\pm 1})^{s+1}.(x^-_\beta)^{s+1} = \Lambda^\pm(h_\beta, s+1) + Y,$$

where X and Y are in the left ideal of $\mathbf{U}_{\text{tor}}$ generated by the subalgebra $\mathfrak{g}_{\text{tor}}(>)$.

□

2. Representation theory of $\mathfrak{g}_{\text{fin}}$ and $\mathfrak{g}^{\text{e}}_{\text{aff}}$

In this section, we discuss the representation theory of $\mathfrak{g}_{\text{fin}}$ and $\mathfrak{g}^{\text{e}}_{\text{aff}}$. We shall be interested in the finite–dimensional representations of $\mathfrak{g}_{\text{fin}}$ and their analogues, the integrable representations of $\mathfrak{g}^{\text{e}}_{\text{aff}}$.

2.1. For $\lambda = \sum_{i=1}^n \lambda(h_i)\omega_i \in P^+_{\text{fin}}$, (resp. $\lambda = \sum_{i=0}^n \lambda(h_i)\omega_i \in P^+_{\text{aff}}$), let $V_{\text{fin}}(\lambda)$ (resp. $V_{\text{aff}}(\lambda)$) be the unique irreducible $\mathfrak{g}_{\text{fin}}$–module (resp. $\mathfrak{g}^{\text{e}}_{\text{aff}}$–module) with highest weight λ and highest weight vector v_λ. Thus, $V_{\text{fin}}(\lambda) = \mathbf{U}_{\text{fin}}.v_\lambda$ (resp. $V_{\text{aff}}(\lambda) = \mathbf{U}^{\text{e}}_{\text{aff}}.v_\lambda$) is generated by the element v_λ, subject to the following relations:

$$\mathfrak{n}^+_{\text{fin}}.v_\lambda = 0,\quad h.v_\lambda = \lambda(h).v_\lambda,\quad (x_i^-)^{\lambda(h_i)+1}.v_\lambda = 0,\qquad i = 1,\cdots,n,\ h \in \mathfrak{h}_{\text{fin}},$$

$$(\text{resp. } \mathfrak{n}^+_{\text{aff}}.v_\lambda = 0,\quad h.v_\lambda = \lambda(h).v_\lambda,\quad (x_i^-)^{\lambda(h_i)+1}.v_\lambda = 0,\qquad i = 0,\cdots,n,\ h \in \mathfrak{h}^{\text{e}}_{\text{aff}}.)$$

It is well–known that the set $\{V_{\text{fin}}(\lambda) : \lambda \in P^+_{\text{fin}}\}$ is in one–one correspondence with the set of isomorphism classes of irreducible finite–dimensional representations of $\mathfrak{g}_{\text{fin}}$. Further, any finite–dimensional $\mathfrak{g}_{\text{fin}}$–module is isomorphic to a direct sum of irreducible $\mathfrak{g}_{\text{fin}}$–modules.

2.2. The $\mathfrak{g}^{\text{e}}_{\text{aff}}$–modules $V_{\text{aff}}(\lambda)$ are not finite–dimensional, but are integrable [**17**, Chapter 10] in the following sense.

DEFINITION 2.1. A $\mathfrak{g}^{\text{e}}_{\text{aff}}$–module V is said to be integrable if

$$V = \bigoplus_{\mu \in (\mathfrak{h}^{\text{e}}_{\text{aff}})^*} V_\mu,$$

where $V_\mu = \{v \in V : hv = \mu(h)v\ \ \forall\ h \in \mathfrak{h}^{\text{e}}_{\text{aff}}\}$, and if for all $\alpha \in R^+_{\text{fin}}$, $r \in \mathbf{Z}$ the elements $x^\pm_\alpha t_1^r$ act locally nilpotently on V. □

To describe the isomorphism classes of irreducible integrable representations of $\mathfrak{g}^{\mathrm{e}}_{\mathrm{aff}}$, we need to introduce two more families of modules. The first one is obtained by just taking the restricted dual $V^*_{aff}(\lambda)$ of $V_{\mathrm{aff}}(\lambda)$, i.e., the $\mathfrak{g}^{\mathrm{e}}_{\mathrm{aff}}$–module generated by an element v^*_λ subject to the relations

$$\mathfrak{n}^-_{\mathrm{aff}}.v^*_\lambda = 0, \quad h.v^*_\lambda = -\lambda(h).v^*_\lambda, \quad (x^+_i)^{\lambda(h_i)+1}.v^*_\lambda = 0, \\ i = 0, \cdots, n, \quad h \in \mathfrak{h}^{\mathrm{e}}_{\mathrm{aff}}.$$

Notice that the center c_1 acts as a positive integer on $V_{\mathrm{aff}}(\lambda)$ (if $\lambda \neq 0$) and as a negative integer on $V^*_{aff}(\lambda)$.

The second family that we need are the loop modules which were introduced in [**5**]. For $k \geq 1$, $\lambda_1, \cdots, \lambda_k \in P^+_{\mathrm{fin}}$, $a_1, \cdots, a_k \in \mathbf{C}^*$, $b \in \mathbf{C}$, define a $\mathfrak{g}^{\mathrm{e}}_{\mathrm{aff}}$–module structure on

$$L(V_{\mathrm{fin}}(\lambda_1) \otimes \cdots \otimes V_{\mathrm{fin}}(\lambda_k)) := V_{\mathrm{fin}}(\lambda_1) \otimes \cdots \otimes V_{\mathrm{fin}}(\lambda_k) \otimes \mathbf{C}[t, t^{-1}],$$

as follows:

$$\begin{aligned} c_1(v_1 \otimes \cdots \otimes v_k) \otimes t^s &= 0, \\ d_1(v_1 \otimes \cdots \otimes v_k) \otimes t^s &= (s+b)(v_1 \otimes \cdots \otimes v_k) \otimes t^s, \\ xt_1^r(v_1 \otimes \cdots \otimes v_k) \otimes t^s &= \left(\sum_{i=1}^k a_i^r v_1 \otimes \cdots \otimes v_{i-1} \otimes xv_i \otimes v_{i+1} \right. \\ &\qquad \left. \otimes \cdots \otimes v_k \right) \otimes t^{r+s}, \end{aligned}$$

where $\{v_i \in V_{\mathrm{fin}}(\lambda_i) : 1 \leq i \leq k\}$, $s, r \in \mathbf{Z}$, $x \in \mathfrak{g}_{\mathrm{fin}}$. Denote this module by $V_{\mathrm{aff}}(\boldsymbol{\lambda}, \mathbf{a}, b)$.

The following result was proved in [**5**].

Proposition 2.1. *Let $k \geq 1$, $b \in \mathbf{C}$, $\lambda_1, \cdots, \lambda_k \in P^+_{\mathrm{fin}}$ and assume that $a_1, \cdots, a_k$ are distinct non–zero complex numbers.*

(i) *The $\mathfrak{g}^{\mathrm{e}}_{\mathrm{aff}}$–module $V_{\mathrm{aff}}(\boldsymbol{\lambda}, \mathbf{a}, b)$ is irreducible iff for every $r \in \mathbf{Z}$ there exists an integer $m_r \in \mathbf{Z}$ with $m_r \neq 0 \mod r$ such that*

$$\sum_{i=1}^k a_i^{m_r} \lambda_i \neq 0.$$

(ii) *Suppose that there exists $r > 0$ such that $\sum_{i=1}^k a_i^m \lambda_i = 0$ for all $m \neq 0 \mod r$. Then the $\mathfrak{g}_{\mathrm{aff}}$–module $L(V_{\mathrm{fin}}(\lambda_1) \otimes \cdots \otimes V_{\mathrm{fin}}(\lambda_k))$ is completely reducible. The irreducible submodules are generated by the elements $v_{\lambda_1} \otimes \cdots \otimes v_{\lambda_k} \otimes t^\ell$, $0 \leq \ell < r$.*

□

REMARK. The characters of these modules have been studied in [**14**]. Further, in the case of sl_n a crystal model was obtained for these modules in [**15**].

The next result was proved in [**4**], [**5**].

THEOREM 1. *Let V be an irreducible integrable $\mathfrak{g}^e_{\rm aff}$-module. Assume also that $\dim(V_\mu) < \infty$ for all $\mu \in \mathfrak{h}^e_{\rm aff}$. Let $\ell \in \mathbf{Z}$ be such that $c_1 v = \ell v$ for all $v \in V$. Then:*

(i) *If $\ell > 0$ (resp. $\ell < 0$) there exists $\lambda \in P^+_{aff}$ such that $V \cong V_{\rm aff}(\lambda)$ (resp. $V \cong V^*_{aff}(\lambda)$).*

(ii) *If $\ell = 0$, then there exists $k \geq 1$, $\lambda_1, \cdots, \lambda_k \in P^+_{\rm fin}$, $b \in \mathbf{C}$ and distinct non–zero complex numbers $a_1, \cdots, a_k$ such that V is isomorphic to either $V_{\rm aff}(\boldsymbol{\lambda}, \mathbf{a}, b)$ or to one of its irreducible submodules as described in Proposition 2.1.*

REMARK. In [**16**] a closely related family of bounded admissible modules for $\mathfrak{g}_{\rm aff}$ were studied and a similar classification theorem was obtained.

The category of integrable representations is in general far from semisimple. However, one has the following result, [**17**, Chapter 10, Theorem 10.7].

THEOREM 2. *Let V be an integrable $\mathfrak{g}^e_{\rm aff}$–module such that $\dim V_\mu < \infty$ for all $\mu \in (\mathfrak{h}^e_{\rm aff})^*$. Assume that there exists $\mu_1, \cdots, \mu_s \in P^+_{\rm aff}$ such that*

$$V_\mu \neq 0 \implies \mu \in \mu_i - Q^+_{\rm aff},$$

$$(\text{resp.} V_\mu \neq 0 \implies \mu \in -\mu_i - Q^-_{aff},)$$

for some $1 \leq i \leq s$. Then, V is isomorphic to a direct sum of representations of the form $V_{\rm aff}(\lambda)$ (resp. $V^*_{aff}(\lambda)$) where $\lambda \in P^+_{\rm aff}$. In particular, if $\lambda, \mu \in P^+_{\rm aff}$, the tensor product $V_{\rm aff}(\lambda) \otimes V_{\rm aff}(\mu)$ is a reducible but completely reducible $\mathfrak{g}^e_{\rm aff}$–module.

□

2.3. We give an easy example of an integrable representation $V = \oplus_{\mu \in (\mathfrak{h}^e_{\rm aff})^*} V_\mu$ (with $\dim V_\mu < \infty$ for all $\mu \in (\mathfrak{h}^e_{\rm aff})^*$) of $\mathfrak{g}^e_{\rm aff}$ which is indecomposable and reducible. Such representations are studied in greater detail in [**8**]. Assume that $\mathfrak{g}_{\rm fin} = sl_2$ and that h, x, y is the standard basis of sl_2. Let V be the free module over $\mathbf{C}[t_1, t_1^{-1}]$ of rank 4, with basis v_0, v_1, v_2, w_0. It is not hard to check (using the Chevalley presentation [**17**, Chapter 7]) that the following formulae define an action of $\mathfrak{g}_{\mathrm{aff}}^{\mathrm{e}}$ on V. The center c_1 acts trivially, and for any $v \in V$, $n \in \mathbf{Z}$, $d_1 v t_1^n = n t_1^n v$. In addition,

$$h(v_i t_1^r) = (2-2i)(v_i t_1^r), \quad x(v_i t_1^r) = (3-i)(v_{i-1} t_1^r),$$
$$y(v_i t_1^r) = (i+1)(v_{i+1} t_1^r) \quad (v_3 = 0),$$
$$(x t_1^{-1})(v_0 t_1^r) = 0, \quad (x t_1^{-1})(v_1 t_1^r) = 2(v_0 t_1^{r-1}),$$
$$(x t_1^{-1})(v_2 t_1^r) = (v_1 + w_0) t_1^{r-1},$$
$$\mathfrak{g}_{\mathrm{aff}}.w_0 t_1^r = 0,$$
$$(y t_1)(v_0 t_1^r) = (v_1 + w_0) t_1^{r+1}, \quad (y t_1) v_1 t_1^r = 2 v_2 t_1^{r+1}, \quad (y t_1) v_2 = 0,$$

where we set $v_{-1} = v_3 = 0$. It is clear that this module is generated by the element v_0 and is indecomposable. However for all $r \in \mathbf{Z}$, the elements $w_0 t_1^r$ generate a proper submodule of it.

2.4. We shall need the following result, [**6**] in our study of representations of the double affine algebra.

THEOREM 3. *Let $\lambda \in P_{\mathrm{aff}}^+$ and let $\mu_1, \cdots, \mu_k \in P_{\mathrm{fin}}^+$. Assume also that $a_1, \cdots, a_k$ are distinct non–zero complex numbers and let $b \in \mathbf{C}$. The $\mathfrak{g}_{\mathrm{aff}}^{\mathrm{e}}$–module $V_{\mathrm{aff}}(\lambda) \otimes V_{\mathrm{aff}}(\boldsymbol{\mu}, \mathbf{a}, b)$ is irreducible if*

(i) $\sum_{i=1}^k a_i \mu_i \neq 0$,
(ii) *there exists $\alpha \in R_{\mathrm{fin}}^+$ such that either*

$$(k+1)\lambda(c_1) < (\mu + \lambda)(h_\alpha),$$

or

$$k\lambda(c_1) < (\mu^* - \lambda)(h_\alpha),$$

where $\mu = \sum_{i=1}^k \mu_i$ and μ^* is the highest weight of the $\mathfrak{g}_{\mathrm{fin}}$–module that is dual to $V_{\mathrm{fin}}(\mu)$.

□

Notice that, in particular, the theorem implies the existence of irreducible integrable representations with infinite–dimensional weight spaces. A partial converse to the theorem was proved in [**7**]. More recently, a converse has been proved in [**1**] in the case of sl_2.

3. Representations of $\mathfrak{g}_{\mathrm{tor}}$

The representation theory of $\mathfrak{g}_{\mathrm{tor}}^{\mathrm{e}}$ is substantially more complicated than the affine case, and a classification theorem of the kind in Theorem 1 seems much more difficult; however, see [**9**], [**10**]. The study of $\mathfrak{g}_{\mathrm{tor}}$–modules is closely related to (see Remark 3.3), but somewhat simpler than the study of $\mathfrak{g}^{e}_{tor}$–modules, and so in this paper, we restrict our attention to representations of $\mathfrak{g}_{tor}$. Thus, we are interested in the category of integrable representations of $\mathfrak{g}_{tor}$ on which c_2 acts trivially, but c_1 acts non–trivially. Since this category is not semisimple, we shall discuss both irreducible and indecomposable representations of $\mathfrak{g}_{tor}$.

3.1. We begin with the definition of integrable $\mathfrak{g}_{tor}$–modules.

DEFINITION 3.1. A representation V of $\mathfrak{g}_{tor}$ is called integrable if

$$V = \bigoplus_{\lambda\in(\mathfrak{h}_{tor})^*} V_\lambda$$

where

$$V_\lambda = \{v \in V : h.v = \lambda(h)v \ \forall\ h \in \mathfrak{h}_{tor}\},$$

and if the elements $x_\alpha^{\pm}t_1^{m_1}t_2^{m_2}$ act locally nilpotently on V for all $\alpha \in R^+_{fin}$, $m_1, m_2 \in \mathbf{Z}$. Further, we say that V is admissible if $\dim V_\lambda < \infty$ for all $\lambda \in (\mathfrak{h}_{tor})^*$. □

It is not hard to see that if V is integrable and admissible $\mathfrak{g}^{e}_{tor}$–module, and $0 \neq v \in V_\lambda$, then $\mathbf{U}_{fin}.v$ is a finite–dimensional $\mathbf{U}_{fin}$–module and that $\mathbf{U}^{e}_{aff}.v$ is an integrable $\mathfrak{g}^{e}_{aff}$–module.

3.2. Given $\lambda_1, \cdots, \lambda_k \in (\mathfrak{h}^{e}_{aff})^*$, $a_1, \ldots, a_k \in \mathbf{C}^*$, define an action of $\mathfrak{g}_{tor}$ on $V_{aff}(\lambda_1) \otimes \cdots \otimes V_{aff}(\lambda_k)$ as follows: c_2 acts as zero and for $x \in \mathfrak{g}_{aff}$, $m \in \mathbf{Z}$,

$$\begin{aligned} xt_2^m.(v_1 \otimes \cdots \otimes v_k) = \textstyle\sum_{j=1}^k a_j^m (v_1 \otimes \cdots \otimes v_{j-1} \otimes xv_j \otimes v_{j+1} \\ \otimes \cdots \otimes v_k), \quad v_i \in V_{aff}(\lambda_i), 1 \leq i \leq k. \end{aligned}$$

Denote this module by $V_{tor}(\boldsymbol{\lambda}, \mathbf{a})$.

LEMMA 3.1. *For all $k \geq 1$, $\boldsymbol{\lambda} = (\lambda_1, \cdots, \lambda_k) \in (P^+_{aff})^k$, $\mathbf{a} = (a_1, \cdots, a_k) \in (\mathbf{C}^*)^k$, the module $V_{tor}(\boldsymbol{\lambda}, \mathbf{a})$ is an integrable admissible $\mathfrak{g}_{tor}$–module which is irreducible iff $a_r \neq a_s$ for all $1 \leq r \neq s \leq k$. Let v_{λ_r} be the highest weight vector in $V_{aff}(\lambda_r)$. We have*

$$(3.1)\quad \mathbf{U}_{tor}(>).v_{\lambda_1} \otimes \cdots \otimes v_{\lambda_k} = 0,$$

$$(3.2)\quad \Lambda^{\pm}(h_i, u)(v_{\lambda_1} \otimes \cdots \otimes v_{\lambda_k}) = \prod_{j=1}^{k}(1 - a_j^{\pm 1}u)^{\lambda_j(h_i)}(v_{\lambda_1} \otimes \cdots \otimes v_{\lambda_k}), \forall\ 0 \leq i \leq n.$$

Proof. Since $x_\alpha^\pm t_1^m$ acts locally nilpotently on $V_{\text{aff}}(\lambda_i)$ for all $\alpha \in R_{\text{fin}}^+$, it follows from the definition that $V_{\text{tor}}(\boldsymbol{\lambda}, \mathbf{a})$ is integrable. The module is admissible because the weight spaces of $V_{\text{aff}}(\lambda_i)$ are finite–dimensional. If $a_r \neq a_s$ for all $r \neq s$, notice that for all $x \in \mathfrak{g}_{\text{aff}}$, the element $x \prod_{s \neq r}(t_2 - a_s)$ acts only on the rth component of the tensor product. The irreducibility of $V_{\text{tor}}(\boldsymbol{\lambda}, \mathbf{a})$ now follows from the irreducibility of the $\mathfrak{g}_{\text{aff}}^{\text{e}}$–module $V_{\text{aff}}(\lambda_i)$. Suppose now that $a_r = a_s$ for some pair (r, s); we can assume without loss of generality (by applying a suitable permutation) that $r = 1, s = 2$. It is easy to deduce from the results of [**17**, Chapter 9] that $V_{\text{aff}}(\lambda_1) \otimes V_{\text{aff}}(\lambda_2)$ is a reducible $\mathfrak{g}_{\text{aff}}^{\text{e}}$–module; let W be a proper submodule. It is clear that $W \otimes V_{\text{aff}}(\lambda_3) \otimes \cdots \otimes V_{\text{aff}}(\lambda_k)$ is a proper non–zero submodule of $V_{\text{tor}}(\boldsymbol{\lambda}, \mathbf{a})$.

Finally, note that (3.1) and (3.2) follow from the definition of the $\mathfrak{g}_{\text{tor}}$–action on $V_{\text{tor}}(\boldsymbol{\lambda}, \mathbf{a})$ and (1.1)

3.3. We now construct a family of integrable indecomposable $\mathfrak{g}_{\text{tor}}$–modules. This family will be maximal in a suitable sense. We begin with the following definition.

Given a collection $\mathbf{p} = \{p_r^\pm \in (\mathfrak{h}_{\text{aff}})^* : r \in \mathbf{Z}, r > 0\}$ and $\lambda \in (\mathfrak{h}_{\text{tor}})^*$, let $M_{\text{tor}}(\lambda, \mathbf{p})$ be the left $\mathbf{U}_{\text{tor}}$–module defined as follows. Let $I(\lambda, \mathbf{p})$ be the left ideal in $\mathbf{U}_{\text{tor}}$ generated by

$$\mathbf{U}_{\text{tor}}(>), \quad \Lambda^\pm(h_i, r) - p_r^\pm(h_i), \quad h' - \lambda(h'), \quad (h' \in \mathfrak{h}_{\text{tor}}, 0 \leq i \leq n).$$

Set

$$M_{\text{tor}}(\lambda, \mathbf{p}) = \mathbf{U}_{\text{tor}}/I(\lambda, \mathbf{p}).$$

Clearly $M_{\text{tor}}(\lambda, \mathbf{p})$ is a left $\mathbf{U}_{\text{tor}}$–module. Let $v_{\lambda,\mathbf{p}}$ be the image of 1 in $M_{\text{tor}}(\lambda, \mathbf{p})$. Standard arguments prove that $M_{\text{tor}}(\lambda, \mathbf{p})$ is a free $\mathbf{U}_{\text{tor}}(<)$ module generated by $v_{\lambda,\mathbf{p}}$ and hence that $M_{\text{tor}}(\lambda, \mathbf{p})$ has a unique irreducible quotient, which we denote by $V_{\text{tor}}(\lambda, \mathbf{p})$.

Remark One defines analogues of these modules for $\mathfrak{g}_{\text{tor}}^{\text{e}}$ by replacing $I(\lambda, \mathbf{p})$ by a graded ideal $I(\lambda, \chi_{\mathbf{p}})$ as follows. Thus let $\lambda \in (\mathfrak{h}_{\text{tor}}^{\text{e}})^*$ and let $\chi_{\mathbf{p}} : \mathbf{U}(\mathfrak{g}_{\text{tor}}) \to \mathbf{C}[t, t^{-1}]$ be the graded homomorphism defined by setting,

$$\chi_{\mathbf{p}}(\Lambda^\pm(h_i, r)) = p_r^\pm(h_i)t^r.$$

Then, $M(\lambda, \chi_{\mathbf{p}})$ is the quotient of $\mathbf{U}(\mathfrak{g}_{\text{tor}}^{\text{e}})$ by the left ideal $I(\lambda, \chi_{\mathbf{p}})$ generated by,

$$\mathbf{U}_{\text{tor}}(>), \quad \ker(\chi_{\mathbf{p}}), \quad h' - \lambda(h'), \quad (h' \in \mathfrak{h}_{\text{tor}}^{\text{e}}).$$

One has analogues of the next two results of this section, and these modules will be further studied elsewhere.

3.4. The following result determines a necessary and sufficient condition for $M_{\rm tor}(\lambda, \mathbf{p})$ to have an integrable quotient and identifies the irreducible quotient in this case.

PROPOSITION 3.1. *The module $M_{\rm tor}(\lambda, \mathbf{p})$ has an integrable quotient iff there exists an $(n+1)$–tuple of polynomials (with constant term one) $\boldsymbol{\pi} = (\pi_0, \cdots, \pi_n)$ in an indeterminate u, such that the following conditions hold for all $i = 0, \cdots, n$:*

(i) $\lambda(h_i) = deg\ \pi_i$,
(ii) $\sum_{r\geq 0} p_r^{\pm}(h_i)u^r = \pi_i^{\pm}(u)$, *where* $\pi_i^+(u) = \pi_i$ *and* $\pi_i^-(u) = \frac{u^{deg\pi_i}\pi^+(u^{-1})}{(u^{deg\pi_i}\pi^+(u^{-1}))(0)}$.

Proof. Suppose that $M_{\rm tor}(\lambda, \mathbf{p})$ has an integrable $\mathfrak{g}_{\rm tor}$–quotient W and let $w_{\mathbf{p}}$ be the image of $v_{\lambda,\mathbf{p}}$ in W. By the representation theory of sl_2 it follows that $\lambda(h_i)$ is a non–negative integer for all $0 \leq\leq n$. Further $\lambda(h_i) = r_i$ is the smallest non–negative integer such that

$$(x_i^-)^{r_i+s}.w_{\mathbf{p}} = 0, \quad 0 \leq i \leq n, \quad \forall s > 0.$$

Applying $(x_i^+t_2)^{r_i+s}$ to the preceding equation and using Lemma 1.1, we see that

$$\Lambda^+(h_i, m).w_{\mathbf{p}} = p_m^+(h_i) = 0 \ \text{ if } m \geq r_i + 1.$$

To see that $p_{r_i}^+(h_i) \neq 0$, note that for all $0 \leq i \leq n$ we have,

$$(h_i t_2^{-1})(x_i^-)^{r_i+1}.w_{\mathbf{p}} = (x_i^- t_2^{-1})(x_i^-)^{r_i}.w_{\mathbf{p}} = 0.$$

Since the elements $x_i^{\pm}t_2^{\pm 1}$ generate a subalgebra isomorphic to sl_2 it follows from the representation theory of sl_2 that,

$$(x_i^+t_2)^{r_i}(x_i^-)^{r_i}.w_{\mathbf{p}} \neq 0.$$

Using Lemma 1.1 we see that this means that $p_r^+(h_i) \neq 0$ for all $0 \leq i \leq n$. Hence $\pi_i^+(u) = \sum_s p_s^+(h_i)u^s$ is a polynomial of degree r_i. Similarly one can prove that $\pi_i^-(u) = \sum_s p_s^-(h_i)u^s$ is a polynomial of degree r_i. To see that $\pi_i^{\pm}$ are related as in the proposition, one proceeds as in the proof of Proposition (1.1)(iv),(v) in [**8**], we omit the details.

Conversely, given $\boldsymbol{\pi} = (\pi_0, \cdots, \pi_n)$, consider the set $\{a_1, a_2, \cdots, a_r\}$ of distinct roots of $\pi = \prod_{j=0}^n \pi_j$. Let m_{ij} be the multiplicity with which a_i occurs as a root in π_j. Set

$$\mu_j = \sum_{i=0}^{n} m_{ji}\omega_i, \quad 1 \leq j \leq r.$$

It is not hard to see that there exists a map of $\mathfrak{g}_{\mathrm{tor}}$–modules $M_{\mathrm{tor}}(\lambda, \mathbf{p}) \to V_{\mathrm{tor}}(\boldsymbol{\mu}, \mathbf{a})$. Since $V_{\mathrm{tor}}(\boldsymbol{\mu}, \mathbf{a})$ is integrable, the theorem follows.

3.5. We shall only be interested in the modules $M_{\mathrm{tor}}(\lambda, \mathbf{p})$ which are given by an $(n+1)$–tuple of polynomials as in the preceding proposition. Set $\lambda_{\boldsymbol{\pi}} = \sum_{i=0}^{n}(\deg \pi_i)\omega_i$. Then $\lambda_{\boldsymbol{\pi}} = \lambda$, and we set $M_{\mathrm{tor}}(\boldsymbol{\pi}) = M_{\mathrm{tor}}(\lambda_{\boldsymbol{\pi}}, \mathbf{p})$ and denote by $V_{\mathrm{tor}}(\boldsymbol{\pi})$ the unique irreducible quotient of $M_{\mathrm{tor}}(\boldsymbol{\pi})$. Let $v_{\boldsymbol{\pi}} = v_{\lambda,\mathbf{p}}$. Notice that this immediately proves that $V_{\mathrm{tor}}(\boldsymbol{\pi}) \cong V_{\mathrm{tor}}(\boldsymbol{\mu}, \mathbf{a})$ for a suitable choice of $\boldsymbol{\mu}, \mathbf{a}$. It now follows easily that

$$V_{\mathrm{tor}}(\boldsymbol{\lambda}, \mathbf{a}) \cong V_{\mathrm{tor}}(\boldsymbol{\lambda}', \mathbf{a}')$$

if and only if there exists a permutation σ of $\{1, \cdots, k\}$ such that

$$\lambda_i = \lambda'_{\sigma(i)}, \quad a_i = a'_{\sigma(i)}.$$

We now define the maximal integrable quotient $W_{\mathrm{tor}}(\boldsymbol{\pi})$ of $M_{\mathrm{tor}}(\boldsymbol{\pi})$. Thus, let $W_{\mathrm{tor}}(\boldsymbol{\pi})$ be the quotient of $M_{\mathrm{tor}}(\boldsymbol{\pi})$ by the submodule generated by the elements

$$(x_i^-)^{r_i+1}.v_{\boldsymbol{\pi}}, \quad r_i = \deg \pi_i, \quad 0 \le i \le n.$$

Let $w_{\boldsymbol{\pi}}$ be the image of $v_{\boldsymbol{\pi}}$ in $W_{\mathrm{tor}}(\boldsymbol{\pi})$. The following lemma is immediate.

LEMMA 3.2. *The $\mathbf{U}^{\mathrm{e}}_{\mathrm{aff}}$–submodule of $W_{\mathrm{tor}}(\boldsymbol{\pi})$ generated by $w_{\boldsymbol{\pi}}$ is isomorphic to $V_{\mathrm{aff}}(\lambda_{\boldsymbol{\pi}})$, where $\lambda_{\boldsymbol{\pi}} = \sum_{i=0}^{n}(\deg \pi_i)\omega_i \in P^+_{\mathrm{aff}}$.* □

PROPOSITION 3.2. *The $\mathfrak{g}_{\mathrm{tor}}$–module $W_{\mathrm{tor}}(\boldsymbol{\pi})$ is integrable and admissible. Further, any integrable quotient of $M_{\mathrm{tor}}(\boldsymbol{\pi})$ is a quotient of $W_{\mathrm{tor}}(\boldsymbol{\pi})$.*

Proof. To see that $W_{\mathrm{tor}}(\boldsymbol{\pi})$ is integrable, first observe that $W_{\mathrm{tor}}(\boldsymbol{\pi}) = \mathbf{U}_{\mathrm{tor}}(<).w_{\boldsymbol{\pi}}$. Since the elements $x_\alpha^{\pm} t_1^{m_1} t_2^{m_2}$, $\alpha \in R^+_{\mathrm{fin}}$, $m_1, m_2 \in \mathbf{Z}$, act locally nilpotently (via the adjoint action) on $\mathbf{U}_{\mathrm{tor}}$ it is enough to prove that they act locally nilpotently on $w_{\boldsymbol{\pi}}$. If $m_2 = 0$, the result follows since $V_{\mathrm{aff}}(\lambda)$ is an integrable $\mathfrak{g}^{\mathrm{e}}_{\mathrm{aff}}$–module, and so we have

$$(x_\alpha^- t_1^{m_1})^N.w_{\boldsymbol{\pi}} = 0,$$

for some $N \ge 0$ depending on α and m_1. Applying $h_\alpha t_2^r$ to the preceding equation, we get

$$(x_\alpha^- t_1^{m_1})^{N-1}(x_\alpha^- t_1^{m_1} t_2^r).w_{\boldsymbol{\pi}} = 0.$$

Repeating, we find that for any $r_1, \cdots, r_N \in \mathbf{Z}$

$$(x_\alpha^- t_1^{m_1} t_2^{r_1})(x_\alpha^- t_1^{m_1} t_2^{r_2}) \cdots (x_\alpha^- t_1^{m_1} t_2^{r_N}).w_{\boldsymbol{\pi}} = 0.$$

This proves that the elements $(x_\alpha^- t_1^{m_1} t_2^{m_2})$ act nilpotently on $w_{\boldsymbol{\pi}}$ and hence that $W_{\rm tor}(\boldsymbol{\pi})$ is integrable.

To see that W is admissible, fix a total order on the set $R_{\rm aff}^+$. By the Poincare–Birkhoff–Witt theorem it follows that the weight space $W_{\lambda-\eta}$ for $\eta \in Q_{\rm aff}^+$ is spanned by elements of the form

$$x_{\beta_1}^- t_2^{k_1} \cdots x_{\beta_r}^- t_2^{k_r}.w_{\mathbf{p}},$$

where $r \geq 1$, $\beta_1 \leq \beta_2 \leq \cdots \leq \beta_r$ are elements of $R_{\rm aff}^+$ such that $\sum_s \beta_s = \eta$ and $k_1, \cdots, k_r \in \mathbf{Z}$. Since the number of roots in $R_{\rm aff}^+$ that add up to a fixed $\eta \in Q_{\rm aff}^+$ is finite, to see that $W_{\lambda-\eta}$ is finite–dimensional it suffices to prove that the number of choices for $k_1, \cdots, k_r$ is finite. In fact, it suffices to prove that for every $\beta \in R_{\rm aff}^+$ there exists an integer N_β such that $x_\beta^- t_2^k.w_{\boldsymbol{\pi}}$ is in the span of elements of the form $x_\beta^- t_2^s.w_{\boldsymbol{\pi}}$, with $-N_\beta \leq s \leq N_\beta$. For then an obvious induction on r proves that, for every $\eta \in Q_{\rm aff}^+$, there exists an integer N_η such that $W_{\lambda-\eta}$ is spanned by elements of the form

$$x_{\beta_1}^- t_2^{k_1} \cdots x_{\beta_r}^- t_2^{k_r}.w_{\boldsymbol{\pi}}, \quad -N_\eta \leq k_i \leq N_\eta, \sum_s \beta_s = \eta.$$

If $\beta \in R_{\rm aff}^+ \backslash \{m\delta_1 : m \in \mathbf{Z}, m > 0\}$, it follows from Lemma 1.1 and the fact that $(x_\beta^-)^{\lambda(h_\beta)+1} = 0$ that $x_\beta^- t_2^k.w_{\boldsymbol{\pi}}$ is in the span of $\{x_\beta^- t_2^r.w_{\boldsymbol{\pi}} : -\lambda(h_\beta) \leq r \leq \lambda(h_\beta)\}$. If $\beta = m\delta_1$, $m > 0$, the corresponding negative root vectors in $\mathfrak{g}_{\rm aff}^{\rm e}$ are $\{h_i t_1^{-m} : 1 \leq i \leq n\}$. Take $\beta = \alpha_i + m\delta_1$ in the first equation in Lemma 1.1 and apply x_i^+ to it. This gives,

$$\sum_{r=0}^{s} h_i t_1^{-m} t_2^s \Lambda^\pm(h_i, s-r).w_{\boldsymbol{\pi}} = 0.$$

Again, it follows that the element $h_i t_1^{-m} t_2^{m_2}.w_{\boldsymbol{\pi}}$ is in the span of the elements $\{h_i t_1^{-m} t_2^{s_2}.w_{\boldsymbol{\pi}} : -\lambda(h_i) \leq s_2 \leq \lambda(h_i)\}$. This completes the proof of the proposition.

COROLLARY 3.1. *We have* $W_{\rm tor}(\boldsymbol{\pi}) = \mathbf{U}(\mathfrak{g}_{\rm tor}(<)) \otimes \mathbf{C}[t_2]).w_{\boldsymbol{\pi}}$.

Proof. Since $W_{\rm tor}(\boldsymbol{\pi})$ is an integrable module, it follows from Lemma 1.1 that if $\beta \in R_{\rm aff}^+ \backslash \{r_1\delta : r_1 \in \mathbf{Z}\}$, then

$$\sum_{m=0}^{N_\beta} (x_\beta^- t_2^m) \Lambda^\pm(h_\beta, \quad N_\beta - m).w_{\boldsymbol{\pi}} = 0,$$

where $N_\beta = \lambda_{\boldsymbol{\pi}}(h_\beta)$. Applying $h_\beta t_2^{m_2}$, $m_2 < 0$, to both sides of the equation gives

$$\sum_{m=0}^{N_\beta} (x_\beta^- t_2^{m+m_2}) \Lambda^\pm(h_\beta,\ N_\beta - m).w_{\boldsymbol{\pi}} = 0.$$

This proves that $(x_\beta^- t_2^{m_2}).w_{\boldsymbol{\pi}}$ is in the span of elements $\{(x_\beta^- t_2^m).w_{\boldsymbol{\pi}} : m > m_2\}$ and hence by a simple induction in the span of elements of the form $\{(x_\beta^- t_2^m).w_{\boldsymbol{\pi}} : m \geq 0\}$. A similar statement then follows for roots of the form $\{r_1\delta_1 : r_1 \in \mathbf{Z}\}$, exactly as in the proof of the proposition. The corollary follows.

We conclude this section with some result on the structure of the modules $W_{\rm tor}(\boldsymbol{\pi})$ that we shall need in the last section of the paper.

LEMMA 3.3. *Regarded as a module for $\mathfrak{g}_{\rm aff}^{\rm e}$, we have*

$$W_{\rm tor}(\boldsymbol{\pi}) \cong \bigoplus_{\mu \in P_{\rm aff}^+} m(\mu) V_{\rm aff}(\mu),$$

for some $m(\mu) \in \mathbf{Z}$, $m(\mu) \geq 0$. Further, $m(\lambda_{\boldsymbol{\pi}}) = 1$ and

$$m(\mu) \neq 0 \implies \mu = \lambda_{\boldsymbol{\pi}} - P_{\rm aff}^+.$$

Proof. We note that

$$(W_{\rm tor}(\boldsymbol{\pi}))_\mu \neq 0 \implies \mu \in \lambda_{\boldsymbol{\pi}} - Q_{\rm aff}^+,$$

and $\dim (W_{\rm tor}(\boldsymbol{\pi}))_\mu$ is finite. The Lemma now follows from Theorem 2.

Although, we have so far been interested only in the representations of $\mathfrak{g}_{\rm tor}$, we shall also need the corresponding results for the algebra $\mathfrak{g}_{\rm tor}^+ = \mathfrak{g}_{\rm aff}[t_2] \oplus \mathbf{C}d_1$. One can define, in the obvious way, a family of $\mathfrak{g}_{\rm tor}^+$–modules $M_{\rm tor}^+(\lambda, \mathbf{p}^+)$, where $\lambda \in (\mathfrak{h}_{\rm aff}^{\rm e})^*$, $\mathbf{p}^+ = \{p_r^+ \in (\mathfrak{h}_{\rm aff})^* : r \in \mathbf{Z}, r > 0\}$. It is not hard to see that there exists an injective map of $\mathfrak{g}_{\rm tor}^+$–modules

$$M_{\rm tor}^+(\lambda, \mathbf{p}^+) \to M(\lambda, \mathbf{p}),$$

where $\mathbf{p} = \{p_r : r \in \mathbf{Z}, r \neq 0\}$ is such that $p_r = p_r^+$ for all $r > 0$. Further, one can show as in Proposition 3.1, that $M_{\rm tor}^+(\lambda, \mathbf{p}^+)$ has an integrable quotient if and only if $\mathbf{p}^+$ is given by an $(n+1)$–tuples of polynomials. Let $W_{\rm tor}^+(\boldsymbol{\pi})$ be the corresponding maximal integrable quotient, i.e. the quotient of $M_{\rm tor}^+(\boldsymbol{\pi})$ by the submodule generated by elements of the form $(x_i^-)^{\deg\, \pi_i + 1}.v_{\boldsymbol{\pi}}$.

PROPOSITION 3.3. *Assume that* $\boldsymbol{\pi}$ *is an* $(n+1)$*–tuple of polynomials with constant term one. Then,*

$$W_{\rm tor}^+(\boldsymbol{\pi}) \cong W_{\rm tor}(\boldsymbol{\pi})$$

as $\mathfrak{g}_{\rm tor}^+$*–modules.*

Proof. It is clear that there exists a map of $\mathfrak{g}_{\rm tor}^+$–modules $\iota : M_{\rm tor}^+(\boldsymbol{\pi}) \to W_{\rm tor}(\boldsymbol{\pi})$, which by Corollary 3.1 is surjective. Suppose that $v = gv_{\boldsymbol{\pi}} \in M_{\rm tor}^+(\boldsymbol{\pi})$ maps to zero. This means that $\iota(gv_{\boldsymbol{\pi}}) = g\iota(v_{\boldsymbol{\pi}})$ is in the $\mathfrak{g}_{\rm tor}$–submodule of $M_{\rm tor}(\boldsymbol{\pi})$ generated by the elements $(x_i^-)^{\deg\,\pi_i+1}.\iota(v_{\boldsymbol{\pi}})$. Since

$$\sum_{i=0}^{n}\mathbf{U}(\mathfrak{g}_{\rm tor}(<)).(x_i^-)^{\deg\,\pi_i+1}.\iota(v_{\boldsymbol{\pi}}) = \sum_{i=0}^{n}\mathbf{U}(\mathfrak{n}_{\rm aff}^-\otimes\mathbf{C}[t_2^{-1}]t_2^{-1})\mathbf{U}(\mathfrak{n}_{\rm aff}^- \otimes\mathbf{C}[t_2])(x_i^-)^{\deg\,\pi_i+1}\iota(v_{\boldsymbol{\pi}}),$$

and $M_{\rm tor}(\boldsymbol{\pi})$ is a free $\mathfrak{g}_{\rm tor}(<)$–module, it follows from the PBW–theorem that

$$g \in \sum_{i=0}^{n}\mathbf{U}(\mathfrak{n}_{\rm aff}^-\otimes\mathbf{C}[t_2])(x_i^-)^{\deg\,\pi_i+1}\iota(v_{\boldsymbol{\pi}}).$$

Hence, we find that the induced map $W_{\rm tor}^+(\boldsymbol{\pi}) \to W_{\rm tor}(\boldsymbol{\pi})$ is injective, and the proposition is proved.

The final result of this section is an analog for the modules $W_{\rm tor}(\boldsymbol{\pi})$ of the factorization that holds for the irreducible modules $V_{\rm tor}(\boldsymbol{\mu},\mathbf{a})$. The proof of this proposition is a modification of the proof of Proposition 3.1 in [**8**], the details of the proof can be found in [**19**].

PROPOSITION 3.4. *Let* $\boldsymbol{\pi} = (\pi_1,\cdots,\pi_n)$ *be an* $(n+1)$*–tuple of polynomials with constant term one. Assume that* $a_1,\cdots,a_k$ *are the distinct roots of* $\prod_{i=0}^n \pi_i$*. For* $1\le s\le k$*, let* $\boldsymbol{\pi}_j = (\pi_{j,0},\pi_{j,1},\cdots,\pi_{j,n})$ *be the* $(n+1)$*–tuple of polynomials defined by*

$$\pi_{j,i} = (1-a_j^{-1}u)^{m_{i,j}},$$

where $m_{i,j}$ *is the multiplicity of the root* a_j *in* π_i*. Then,*

$$W(\boldsymbol{\pi}) \cong W(\boldsymbol{\pi}_1)\otimes\cdots\otimes W(\boldsymbol{\pi}_k),$$

as $\mathfrak{g}_{\rm tor}$*–modules.*

4. An irreducibility criterion for $W_{\rm tor}(\boldsymbol{\pi})$

It follows from Proposition 3.2 that the unique irreducible quotient of $W_{\rm tor}(\boldsymbol{\pi})$ is $V_{\rm tor}(\boldsymbol{\pi})$. In this section we give a condition for the quotient map to be an isomorphism.

Recall that θ is the highest root of $R^+_{\rm fin}$ and write $h_\theta = \sum_{i=1}^n m_i h_i$. For $0 \le i \le n$, and $a \in \mathbf{C}^\times$, define an $(n+1)$–tuple of polynomials $\boldsymbol{\pi}_{i,a}$ by

$$\pi_j = 1, \quad j \neq i, \quad \pi_i = (1 - au).$$

THEOREM 4. Assume that either $i = 0$ or that $i \in \{j : m_j = 1, 1 \le j \le n\}$. Then,

$$W_{\rm tor}(\boldsymbol{\pi}_{i,a}) \cong V_{\rm tor}(\boldsymbol{\pi}_{i,a})$$

as $\mathbf{U}^{\rm e}_{\rm aff}$–modules.

Proof. Set $\boldsymbol{\pi} = \boldsymbol{\pi}_{i,a}$. Consider the map

$$\phi_r : (\mathfrak{g}_{\rm aff} t_2^r) \otimes W_{\rm tor}(\boldsymbol{\pi}) \to W_{\rm tor}(\boldsymbol{\pi})$$

given by $\phi_r(xt_2^r, w) = (xt_2^r)w$. This is clearly a map of $\mathfrak{g}_{\rm aff}$–modules, where $\mathfrak{g}_{\rm aff}$ acts on the first factor through the adjoint representation. Let p_μ denote the projection of

$$p_\mu : W_{\rm tor}(\boldsymbol{\pi}) = \bigoplus_{\nu \in P^+_{\rm aff}} V_{\rm aff}(\nu) \to W(\mu)$$

where $W(\mu)$ is the μth isotypical component of $W_{\rm tor}(\boldsymbol{\pi})$. Then p_μ is a map of $\mathfrak{g}_{\rm aff}$–modules, and hence for every $r \in \mathbf{Z}$ we get a map of $\mathfrak{g}_{\rm aff}$–modules

$$\phi_{r,\mu} = p_\mu . \phi_r : (\mathfrak{g}_{\rm aff} t_2^r) \otimes W_{\rm tor}(\boldsymbol{\pi}) \to W(\mu).$$

We show that for all $r \in \mathbf{Z}$ the restriction of $\phi_{r,\mu}$ to $(\mathfrak{g}_{\rm aff} t_2^r) \otimes V_{\rm aff}(\boldsymbol{\pi})$ is zero if $\mu \neq \lambda_{\boldsymbol{\pi}}$. Observe that $\lambda_{\boldsymbol{\pi}} = \omega_i$ where i is such that $m_i = 1$ and hence $\omega_i(c_1) = 1$. This implies that

$$(\lambda_{\boldsymbol{\pi}} + \theta)(h_\theta) > 2, \quad \text{if} \quad i \neq 0$$

and

$$(\theta - \lambda_{\boldsymbol{\pi}})(h_\theta) > 1, \quad i = 0.$$

Further, since $c_1 t_2^r . w_{\boldsymbol{\pi}}$ is a scalar multiple of $w_{\boldsymbol{\pi}}$, it follows that $\phi_{r,\mu}(c_1 t_2^r, v) = 0$ for all $v \in V_{\rm aff}(\lambda_{\boldsymbol{\pi}})$ if $\mu \neq \lambda_{\boldsymbol{\pi}}$. This implies that $\phi_{r,\mu}$ factors through to a map of $\mathfrak{g}_{\rm aff}$–modules

$$\mathfrak{g}_{\rm fin}(\mathbf{C}[t_1, t_1^{-1}]) t_2^r \otimes V_{\rm aff}(\lambda_{\boldsymbol{\pi}}) \to W(\mu).$$

Since

$$\mathfrak{g}_{\mathrm{fin}}(\mathbf{C}[t_1, t_1^{-1}])t_2^r \cong L(V_{\mathrm{fin}}(\theta))$$

as $\mathfrak{g}_{\mathrm{aff}}$–modules, it follows from Theorem 3 that $(\mathfrak{g}_{\mathrm{aff}}t_2^r) \otimes V_{\mathrm{aff}}(\lambda_{\boldsymbol{\pi}})$ is irreducible and hence $\phi_{r,\mu}$ must be either zero or an isomorphism. On the other hand the module $(\mathfrak{g}_{\mathrm{aff}}t_2^r) \otimes V_{\mathrm{aff}}(\lambda_{\boldsymbol{\pi}})$ has infinite–dimensional weight spaces while $W(\mu)$ has finite–dimensional weight spaces and hence in fact we must have, $\phi_{r,\mu} = 0$. This means that as a $\mathfrak{g}_{\mathrm{aff}}^{\mathrm{e}}$–module we have

$$W_{\mathrm{tor}}(\boldsymbol{\pi}_{i,a}) \cong V_{\mathrm{aff}}(\lambda_i).$$

On the other hand the irreducible quotient of $W_{\mathrm{tor}}(\boldsymbol{\pi}_{i,a})$ is the module $V_{\mathrm{aff}}(\lambda_i, a)$ and this by definition is isomorphic to $V_{\mathrm{aff}}(\lambda_i)$ as a $\mathfrak{g}_{\mathrm{aff}}^{\mathrm{e}}$–module. The theorem now follows.

5. Fusion product and reducibility of $W_{\mathrm{tor}}(\boldsymbol{\pi})$

To understand the $\mathfrak{g}_{\mathrm{aff}}^{\mathrm{e}}$–module structure of $W_{\mathrm{tor}}(\boldsymbol{\pi})$ and in particular to prove that it is not isomorphic to $V_{\mathrm{tor}}(\boldsymbol{\pi})$, we need to introduce the notion of the fusion product of representations of $\mathfrak{g}_{\mathrm{aff}} \otimes \mathbf{C}[t_2]$. This was introduced by Feigin and Loktev in [**12**].

Thus, let $\mathfrak{a}$ be any Lie algebra and let $\mathfrak{a}[t]$ be the algebra of polynomial maps $\mathbf{C} \to \mathfrak{a}$. For $k \geq 1$, let $V_1, \cdots, V_k$ be representations of $\mathfrak{a}$ and let $a_1, \cdots, a_k \in \mathbf{C}$ be arbitrary complex numbers. As in the previous sections, one sees that the tensor product $V_1 \otimes \cdots \otimes V_k$ admits a structure of $\mathfrak{a}[t]$–module given by

$$xt^r(v_1 \otimes \cdots \otimes v_k) = \sum_{j=1}^{k} a_j^r v_1 \otimes \cdots \otimes v_{j-1} \otimes xv_j \otimes v_{j+1} \cdots \otimes v_k,$$

for all $x \in \mathfrak{a}$, $r \geq 0$, $v_i \in V_i$, $1 \leq i \leq k$. Denote this module by $V_1(a_1) \otimes \cdots \otimes V_k(a_k)$. The following lemma is easily proved.

LEMMA 5.1. *Assume that there exist elements $v_j \in V_j$ such that $V_j = \mathbf{U}(\mathfrak{a}[t]).v_j$ for all $1 \leq j \leq k$ and that $a_1, \cdots, a_k$ are distinct complex numbers. Then, $V_1(a_1) \otimes \cdots \otimes V_k(a_k)$ is generated as an $\mathfrak{a}[t]$–module by $v_1 \otimes \cdots \otimes v_k$.* □

The Lie algebra $\mathfrak{a}[t]$ and its enveloping algebra $\mathbf{U}(\mathfrak{a}[t])$ are obviously graded by powers of t. Let $\mathbf{U}(\mathfrak{a}[t])_r$ be the rth–graded piece. If $V_1, \cdots, V_k$ are generated by elements $v_1, \cdots, v_k$ and $a_1, \cdots, a_k$ are distinct complex numbers, the $\mathfrak{a}[t]$–module $V = V_1(a_1) \otimes \cdots \otimes V_k(a_k)$ admits an $\mathfrak{a}$ (but not an $\mathfrak{a}[t]$) equivariant filtration. The rth–filtered piece is given by

$$V(r) = \sum_{0 \le s \le r} \mathbf{U}(\mathfrak{a}[t])_s (v_1 \otimes \cdots \otimes v_k).$$

The associated graded vector space $\oplus_{r \ge 0} V(r)/V(r-1)$ is obviously an $\mathfrak{a}$–module. Since $\mathfrak{a}t^s.V(r) \subset V(r+s)$, one can define an $\mathfrak{a}[t]$–module structure on $\oplus_{r \ge 0} V(r)/V(r-1)$. This module, denoted by $V_1(a_1) * V_2(a_2) * \cdots * V_k(a_k)$, is called the fusion product of the modules $V_1, \cdots, V_k$ with respect to $a_1, \cdots, a_k$. Let $v_1 * \cdots * v_k$ be the image of $v_1 \otimes \cdots \otimes v_k$ in $V_1(a_1) * V_2(a_2) * \cdots * V_k(a_k)$.

Assume now that $\mathfrak{a} = \mathfrak{g}_{\mathrm{aff}}$ and that $V_j = V_{\mathrm{aff}}(\lambda_j)$, where $\lambda_j \in P_{\mathrm{aff}}^+$, $1 \le j \le k$. It follows from Theorem 2 that

$$V_1(a_1) * V_2(a_2) * \cdots * V_k(a_k) \cong V_1 \otimes \cdots \otimes V_k, \tag{5.1}$$

as $\mathfrak{g}_{\mathrm{aff}}^{\mathrm{e}}$–modules.

The next lemma is not hard to check.

Lemma 5.2. *Let $k \ge 1$, and let $\lambda_1, \cdots, \lambda_k \in P_{\mathrm{aff}}^+$. Assume that $a_1, \cdots, a_k$ are distinct scalars. The fusion product $V_{\mathrm{aff}}(\lambda_1)(a_1) * \cdots * V_{\mathrm{aff}}(\lambda_k)(a_k)$ is generated as an $\mathfrak{a}[t]$–module by the element $v = v_1 \otimes \cdots \otimes v_k$. Further,*

$$\mathfrak{n}_{\mathrm{aff}}^+[t].v = 0, \quad (\mathfrak{h}_{\mathrm{aff}})t\mathbf{C}[t].v = 0, \quad h.v = \sum_j \lambda_j(h).v, \quad h \in \mathfrak{h}_{\mathrm{aff}}.$$

□

For any $\lambda = \sum_{i=0}^n \lambda_i \omega_i$, where $\lambda(h_i) \ge 0$ for all $0 \le i \le n$, fix distinct complex numbers $\{c_{ij} : 0 \le i \le n, 1 \le j \le \lambda(h_i)\}$. Consider the fusion product

$$W(\lambda) = V_{\mathrm{aff}}(\omega_0)(c_{01}) * \cdots * V_{\mathrm{aff}}(\omega_0)(c_{0\lambda_0}) * \cdots * V_{\mathrm{aff}}(\omega_n)(c_{n1}) \\ * \cdots * \cdots * V_{\mathrm{aff}}(\omega_n)(c_{n\lambda_n}).$$

For any complex number $a \in \mathbf{C}$, let $W(\lambda, a)$ be the ${\mathfrak{g}_{\mathrm{tor}}}^+$–module obtained by pulling back $W(\lambda)$ through the Lie algebra homomorphism $\mathfrak{g}_{tor}^+ \to \mathfrak{g}_{tor}^+$ given by sending $xt_2^r \to x(t_2 - a)^r$, for all $x \in \mathfrak{g}_{\mathrm{aff}}$ and sending $d_1 \to d_1$.

We turn now to the $\mathfrak{g}_{\mathrm{tor}}$-module $W_{\mathrm{tor}}(\boldsymbol{\pi})$. We restrict our attention to the case where there exists $a \in \mathbf{C}^*$ and $n_j \in \mathbf{Z}$, $n_j \ge 0$, such that if, $\boldsymbol{\pi} = (\pi_1, \cdots, \pi_n)$ then,

$$\pi_j(u) = (1 - au)^{n_j}, \quad 1 \le j \le n.$$

THEOREM 5. There exist a surjective map of $\mathfrak{g}_{tor}^{+}$–modules $W_{\text{tor}}(\boldsymbol{\pi}) \to W(\lambda_{\boldsymbol{\pi}}, a)$.

Proof. By Proposition 3.3 it is enough to prove that the element $(v_{\omega_0})^{\otimes(\lambda_0)} \otimes \cdots \otimes (v_{\omega_n})^{\otimes \lambda_n}$ satisfies the defining relations of the element $w_{\boldsymbol{\pi}}$. But this is clear from Lemma 5.2.

It is now not hard to see that $W_{\text{tor}}(\boldsymbol{\pi})$ is generally reducible. This is because we know from the results of Section 3 that the irreducible quotient of $W_{\text{tor}}(\boldsymbol{\pi})$ in this case is $V_{\text{tor}}(\lambda_{\boldsymbol{\pi}}, a)$, and

$$V_{\text{tor}}(\lambda_{\boldsymbol{\pi}}, a) \cong V_{\text{aff}}(\lambda_{\boldsymbol{\pi}})$$

as $\mathfrak{g}_{\text{aff}}^{\text{e}}$–modules. On the other hand the module $W(\lambda_{\boldsymbol{\pi}}, a)$ is isomorphic as a $\mathfrak{g}_{\text{aff}}^{\text{e}}$–module to a tensor product of integrable irreducible $\mathfrak{g}_{\text{aff}}^{\text{e}}$–modules, and hence must be reducible as a $\mathfrak{g}_{\text{aff}}^{\text{e}}$–module.

References

[1] D. Adamovic, *An application of* $\mathbf{U}(\mathfrak{g})$*–bimodules to representation theory of affine Lie algebras*, preprint, 2000.

[2] B. Allison, S. Azam, S. Berman, Y. Gao, A. Pianzola, *Extended affine Lie algebras and their root systems*, Mem. Amer. Math. Soc. **126**, (1997), no. 603.

[3] S. Berman and Y. Billig, *Irreducible representations for toroidal Lie algebras*, Jour. Alg. **221** (1999), 188–231.

[4] V. Chari, *Integrable representations of affine Lie algebras*, Invent. Math. **85** (1986), no. 2, 317–335.

[5] V. Chari and A. Pressley, *New unitary representations of loop groups*, Math. Ann. **275** (1986), no. 1, 87–104.

[6] V. Chari and A. Pressley, *A new family of irreducible integrable modules for affine Lie algebras*, Math. Ann. **277** (1987), no.3, 543–562.

[7] V. Chari and A. Pressley, *An application of Lie superalgebras to affine Lie algebras*, Jour. Alg. **135**, (1990), 203–216.

[8] V. Chari and A. Pressley, *Weyl modules for classical and quantum affine algebras.* Represent. Theory **5** (2001), 191–223

[9] S. E. Rao, *Classification of irreducible integrable modules for multi–loop algebras with finite–dimensional weight spaces*, Jour. Alg. **246**, (2001), 215–225.

[10] S. E. Rao, *Classification of irreducible integrable modules for toroidal Lie algebras with finite–dimensional weight spaces*, preprint, (2001).

[11] K. Fabricius, B. M. McCoy, *Bethe's equation is incomplete for the XXZ model at roots of unity*, Preprint, cond-mat/0009279.

[12] B. Feigin and S. Loktev, *On generalizes Kostka polynomials and the quantum Verlinde rule*, Amer. Math.Soc. Transl. **2** Vol.**194**, (1999), 61–79.

[13] H. Garland,*The arithmetic theory of loop algebras*, J. Algebra, **53**, (1978), 480–551.

[14] J. Greenstein, *Characters of simple bounded modules over an untwisted affine algebra*, Algebr. Represent. Theory, (to appear).

[15] J. Greenstein, *Littelmann's path crystal and bounded modules for* $\widehat{sl_{\ell+1}}$, preprint, 2002.
[16] A. Joseph, *A completion of the quantized enveloping algebra of a Kac–Moody algebra,* J. Alg., **214** (1999), no.1, 235–275.
[17] V. G. Kac, *Infinite dimensional Lie algebras*, Cambridge University Press, Cambridge, United Kingdom, 1992.
[18] A. N. Kirillon and N. Yu. Reshetikhin, *Representations of Yangians and the multiplicities of the inclusion of the irreducible components of the tensor product of representations of simple Lie algebras*, J.Soviet Math. **52** (1990), 3156–3164.
[19] T. Le, *Representations of infinite–dimensional Lie algebra*, UCR thesis, 2002.
[20] R. V .Moody, S. E. Rao and T. Yokonuma, *Toroidal Lie algebras and vertex representations*, Geom. Dedicata, **35**, (1990), 283–307.
[21] H. Nakajima, Quiver varieties and finite-dimensional representations of quantum affine algebras, J. Amer. Math. Soc. **14** (2001), no. 1, 145–238.
[22] Y. Saito, K. Takemura and D. Uglov, *Toroidal Action on level one modules of* $\mathbf{U}_q(\widehat{sl_n})$, Transform. groups, **3**, (1998), 75–102
[23] M. Varagnolo and E. Vasserot, *Double loop algebras and Fock sapce*, Invent. Math. **133** (1998), 133–159.

DEPARTMENT OF MATHEMATICS
UNIVERSITY OF CALIFORNIA
RIVERSIDE, CA 92521.

Poincaré series of line bundles on varieties

STEVEN DALE CUTKOSKY

Dedicated to Professor Seshadri on the occasion of his 70th birthday

Abstract. We associate a Poincaré series to the dimension of global sections of multiples of r line bundles on a proper variety X. We show that this series is rational on a nonsingular characteristic 0 curve, but can be irrational on a singular curve, and can be irrational in higher dimensions. We also show that the series can be irrational in positive characteristic, even on a nonsingular curve.

1. Introduction

Suppose that X is a projective variety over an algebraically closed field of characteristic zero, and that $\mathcal{L}_1, \ldots, \mathcal{L}_r$ are line bundles on X. We consider the function

$$h(\underline{n}) = h^0(X, \mathcal{L}_1^{n_1} \otimes \cdots \otimes \mathcal{L}_r^{n_r})$$

for $\underline{n} = (n_1, \ldots, n_r) \in \mathbf{N}^r$, and the Poincaré series

$$f(\underline{n}) = \sum_{\underline{n} \in \mathbf{N}^r} h(\underline{n}) t^{\underline{n}}$$

where $t^{\underline{n}} = t_1^{n_1} \cdots t_r^{n_r}$ is a monomial in the variables $t_1, \ldots, t_r$.

If X is a nonsingular curve, then $h(\underline{n})$ is a little complicated. An illustrative example is as follows.

Let C be a nonsingular elliptic curve, over an algebraically closed field k.

Fix a point $p_\infty \in C$ as the 0 in the group law on C. For $p \in C$, let $\overline{p}$ be the divisor $\overline{p} - p_\infty$. We then have a group isomorphism

Partially supported by NSF.

$C \to \mathrm{Pic}^0(C)$ given by $p \mapsto \overline{p}$. Let $p \in C$ be a point such that $\overline{p}$ has infinite order in $\mathrm{Pic}^0(C)$.

Let $\mathcal{L}_1 = \mathcal{O}_C(\overline{p})$, $\mathcal{L}_2 = \mathcal{O}_C(-\overline{p})$. Then

$$h(n_1, n_2) = h^0(C, \mathcal{L}_1^{n_1} \otimes \mathcal{L}_2^{n_2}) = \begin{cases} 1 & \text{if } n_1 = n_2 \\ 0 & \text{otherwise} \end{cases}$$

However, this is essentially the most complicated situation that can arise on a nonsingular curve (in characteristic zero), and in fact we have that the Poincaré series

$$f(\underline{n}) = \sum_{\underline{n} \in \mathbf{N}^r} h(\underline{n}) t^{\underline{n}}$$

on a nonsingular curve is a rational series. We prove this in Theorem 4.1 of Section 4.

We also have that the series $f(\underline{n})$ is always rational on a singular curve, as long as $r \leq 2$. When $r = 1$ this is immediate from the Riemann-Roch formula and Theorem 2.2. The case $r = 2$ follows from the method of proof of Theorem 4.1, using Theorem 15 of [**2**] in place of Proposition 2.1. Theorem 15 of [**2**] is a general statement for cyclic subgroups of algebraic groups.

However, if C is a singular curve, $r \geq 3$ and $\mathrm{Pic}^0(C)$ is not semi-abelian, then the Poincaré series may be irrational. We give in Example 4.2 an example of a singular curve C with $\mathrm{Pic}^0(C) \cong \mathbf{C}^2$, and 3 line bundles such that the associated Poincaré series is irrational.

In Section 5 we consider the corresponding problem on a nonsingular surface X. If X is a nonsingular projective surface over a field of characteristic zero, and $r = 1$, we have that the series $f(n)$ is rational, since $h(n)$ is a polynomial with periodic coefficients for large n (Theorem 4 [**2**]). In Example 5.1 of Section 5, we give an example of a nonsingular surface S and two line bundles $\mathcal{L}_1$ and $\mathcal{L}_2$ ($r = 2$) such that the associated Poincaré series is not rational. This example is based on the construction of Example 4 [**2**]. Essential use is made of the fact that $\mathcal{L}_1$ and $\mathcal{L}_2^{-1}$ are ample. This leaves open the question of rationality for the more classical case where the line bundles are associated to effective divisors, that is $h^0(X, \mathcal{L}_i) > 0$ for $1 \leq i \leq r$. In Example 6.1 we give an example of 3 line bundles $\mathcal{L}_1, \mathcal{L}_2, \mathcal{L}_3$ on a nonsingular surface X such that $h^0(X, \mathcal{L}_i) > 0$ for $1 \leq i \leq 3$, and the associated Poincaré series is not rational. The underlying idea of this example is the explict calculation of the Abel-Jacobi map of a singular curve C embedded in X, with $\mathrm{Pic}^0(C) \cong \mathbf{C}^2$ of Section 3. This is also the curve used in Example 4.2. We show that the image of C in $\mathrm{Pic}^0(C)$ is just the curve $y = 2x^3$ (Lemma 3.2), which contains infinitely many rational points.

A related example is constructed in [**3**], where the Poincaré series associated to the exceptional divisors of a resolution of a normal surface singularity is considered.

Over a field of characteristic $p > 0$ the Poincaré series can be irrational if X is a nonsingular curve and $r = 2$ (this follows from an analysis of the line bundles on the supersingular curve constructed in Example 3 of [**2**]). An analysis of the Poincaré series of Example 3 of [**2**] shows that if X is a nonsingular surface of characteristic $p > 0$, then the Poincaré series associated to a singular line bundle can be irrational.

2. Preliminaries

Suppose that X is a nonsingular projective surface over a field k. If $\mathcal{L}$ is a line bundle (or a divisor D) on X and C is an integral curve on X, then $\mathcal{L} \cdot C$ (or $D \cdot C$) will denote the line bundle $\mathcal{L} \otimes \mathcal{O}_C$ (or the linear equivalence class of $\mathcal{O}_C \otimes \mathcal{O}_X(D)$). $(\mathcal{L} \cdot C)$ (or $(C \cdot D)$) will denote the degree of $\mathcal{L} \otimes \mathcal{O}_C$ (or of $\mathcal{O}_X(D) \otimes \mathcal{O}_C$).

If $\mathcal{M}$ is a coherent sheaf on X, then $H^i(X, \mathcal{M})$ has finite length as a k-vector space for $0 \leq i$. We will denote

$$h^i(X, \mathcal{M}) = \dim_k H^i(X, \mathcal{M}).$$

The following result is a direct consequence of Lang's conjecture, proven by McQuillan in [**6**]. Recall that a semi-abelian variety is a commutative algebraic group such that there is an exact sequence

$$0 \to \mathbf{G}_m^a \to G \to A \to 0$$

where A is an abelian variety, and $a \in \mathbf{N}$.

PROPOSITION 2.1. *Suppose that G is a semi-abelian variety over an algebraically closed field k of characteristic 0, H is a finitely generated abelian group, and $\pi : H \to G(k)$ is a group homomorphism. Suppose that $Y \subset G$ is a closed integral subvariety such that $Y(k) \cap \pi(H)$ is Zariski dense in Y. Then there exists a subgroup M of H and $n_0 \in H$ such that*

$$\pi(n) \in Y(k) \text{ iff } n \in n_0 + M.$$

PROOF. Let $\Gamma = \pi(H)$, and

$$\overline{\Gamma} = \{x \in G(k) \mid nx \in \Gamma \text{ for some } n \in \mathbf{N}\}.$$

$Y(k) \cap \overline{\Gamma}$ is Zariski dense in X, so by Lang's conjecture (proven in [**6**]) $Y = b + \Delta$ for some $b \in G(k)$ and semi-abelian subvariety Δ of Γ. By assumption, there exists an $n_0 \in H$ such that $\pi(n_0) \in b + \Delta$, so we can assume that $b = \pi(n_0)$. Let

$$M = \{n \in H \mid \pi(n) \in \Delta(k)\}$$

a subgroup of H. We have

$$n_0 + M = \{n \in H \mid \pi(n) \in Y(k)\}.$$

□

By a *rational convex polyhedral set in* $\mathbf{Q}^r$ we mean a set of the form

$$P = \{\underline{n} \in \mathbf{Q}^r \;/\; L_i(\underline{n}) \geq b_i, \quad 1 \leq i \leq m\} \tag{1}$$

where $m \in \mathbf{N}$ and, for $1 \leq i \leq m$, L_i is an integral linear form on $\mathbf{Q}^r$ and $b_i \in \mathbf{Z}$.

THEOREM 2.2. *Let $P \subseteq \mathbf{Q}^r_{\geq 0}$ be a rational convex polyhedral set in $\mathbf{Q}^r$. Let $\underline{m} \in \mathbf{Z}^r$, $M \subset \mathbf{Z}^r$ be a subgroup and suppose that $q(\underline{n})$ is a polynomial in $\mathbf{Q}[t_1, \ldots, t_r]$. Then, there exist $d_i \in \mathbf{N}$, $\underline{a}_i \in \mathbf{N}^r$ and a polynomial $p(t_1, \ldots, t_r) \in \mathbf{Q}[t_1, \ldots, t_r]$, such that*

$$\sum_{\underline{n} \in P \cap (\underline{m} + M)} q(\underline{n}) \underline{t}^{\underline{n}} = \frac{p(t_1, \ldots, t_r)}{\prod_{i=1}^{s} (1 - \underline{t}^{\underline{a}_i})^{d_i}}.$$

This is proven in Section 7 of [**3**].

3. The Abel-Jacobi map of a special quintic curve

The results of this section are Lemma 9.3 and Lemma 9.4 of [**3**], which we reproduce for the reader's convenience.

LEMMA 3.1. *There exists a rational, complex Gorenstein projective curve C which has an isolated singularity $\tilde{p}$, with local ring*

$$\mathcal{O}_{C,\tilde{p}} \cong \mathbf{C}[t^2, t^5]_{(t^2, t^5)},$$

and the following properties:

1. *C has arithmetic genus $p_a(C) = 2$.*
2. *If $\mathcal{L}$ is a line bundle on C and deg($\mathcal{L}$) ≥ 8 then $\mathcal{L}$ is generated by global sections.*
3. *If $\mathcal{L}$ is a line bundle on C and deg($\mathcal{L}$) ≥ 10 then $\mathcal{L}$ is very ample.*

(4) *If $\mathcal{L}$ is a line bundle on C of negative degree, there exists a nonsingular projective surface S and an embedding $C \subset S$ such that $C \cdot C \sim \mathcal{L}$.*

PROOF. Let $C_0 = V(x_2^3x_1^2 - x_0^5) \subset X_0 = \mathbf{P}^2_{\mathbf{C}}$. C_0 is a rational curve with two singular points, $q_0 = (0:0:1)$ and $q_1 = (0:1:0)$. We will resolve the singularity at q_1. There are regular parameters $y = \frac{x_0}{x_1}$, $z = \frac{x_2}{x_1}$ at q_1.

$$\mathcal{O}_{C_0,q_1} = (\mathbf{C}[y,z]/z^3 - y^5)_{(y,z)}.$$

Let $\pi_1 : X_1 \to X_0$ be the blowup of q_1. Let C_1 be the strict transform of C_0 on X_1. Let $E_1 = \pi_1^{-1}(q_1)$, $q_2 = E_1 \cap C_1$. Let (y_1, z_1) be the regular parameters at q_2 defined by $y = y_1, z = y_1z_1$. $z_1^3 - y_1^2 = 0$ is a local equation of C_1 at q_2. Let $\pi_2 : X_2 \to X_1$ be the blowup of q_2. Let C_2 be the strict transform of C_1 on X_2, $E_2 = \pi_1^{-1}(q_2)$, $q_3 = E_2 \cap C_2$. Let (y_1, z_1) be the regular parameters at q_2 defined by $y_1 = y_2z_2, z_1 = z_2$. $z_2 - y_2^2 = 0$ is a local equation of C_2 at q_3, and C_2 is thus nonsingular at q_2. Set $\pi = \pi_1 \circ \pi_2$. Identify E_1 with its strict transform on X_2. q_0 is the only singular point on C_2.

$$\pi_1^*(C_0) = C_1 + 3E_1.\ \pi^*(C_0) = C_2 + 5E_2 + 3E_1.$$

$$C_2 \cdot C_2 \sim \pi^*(C_0) \cdot C_2 - 5E_2 \cdot C_2 - 3E_1 \cdot C_2.$$

$C_2 \cdot E_2$ and $C_2 \cdot E_1$ are supported at q_2, $y_2 = 0$ is a local equation of E_1 at q_2 and $z_2 = 0$ is a local equation of E_2 at q_2. Thus $C_2 \cdot E_1 = q_2$, $C_2 \cdot E_2 = 2q_2$, and

$$C_2 \cdot C_2 \sim \pi^*(C_0) \cdot C_2 - 13q_2.$$

Let V be a general cubic curve on X_0.

$$V \cdot C_0 = p_1 + \cdots + p_{25}$$

where $p_1, \ldots, p_{25} \in C_0$ are distinct nonsingular points.

$$\pi^*(C_0) \cdot C_2 \sim \pi^*(V) \cdot C_2 = p_1 + \cdots + p_{25}.$$

Thus

$$C_2 \cdot C_2 \sim p_1 + \cdots + p_{25} - 13q_2.$$

Set $D = C_2 \cdot C_2$. $\deg(D) = 12$.

$$\hat{\mathcal{O}}_{C_2,q_0} = \mathbf{C}[[\frac{x_0}{x_2}, \frac{x_1}{x_2}]]/(\frac{x_1}{x_2})^2 - (\frac{x_0}{x_2})^5 \cong \mathbf{C}[[t^2, t^5]] \subset \mathbf{C}[[t]]$$

and $\ell(\mathbf{C}[[t]]/\hat{\mathcal{O}}_{C_2,q_0}) = 2$. Let $C = C_2$.

The arithmetic genus of C is thus

$$p_a(C) = p_a(\mathbf{P}^1) + 2 = 2$$

(c.f. Exercise IV 1.8 [**5**]).

Since C is a local complete intersection, the Riemann-Roch Theorem is applicable on C (c.f Exercise IV 1.9 [**5**]). In particular, there is a canonical bundle ω_C on C such that for any line bundle $\mathcal{L}$ on C, $h^1(C,\mathcal{L}) = h^0(C, \omega_C \otimes \mathcal{L}^{-1})$ (Serre duality) and $\chi(\mathcal{L}) = \deg(\mathcal{L}) + 1 - p_a(C)$. We further have $\deg(\omega_C) = 2p_a(C) - 2 = 2$.

C has an affine cover by opens sets $U_1 = \operatorname{spec}(\mathbf{C}[t^2, t^5])$ and $U_2 = \operatorname{spec}(\mathbf{C}[\frac{1}{t}]$. Let ∞ be the point on C with maximal ideal $(\frac{1}{t})$. ∞ is the point q_2 on C_2. Let $\tilde{p}$ be the point q_1 on C. The maximal ideal of the point $\tilde{p}$ is $m = (t^2, t^5) \subset \mathbf{C}[t^2, t^5])$. Consider the line bundle $\mathcal{L}_1$ defined by

$$H^0(U_1, \mathcal{L}_1) = \mathbf{C}[t^2, t^5]t^2$$
$$H^0(U_2, \mathcal{L}_1) = \mathbf{C}[\tfrac{1}{t}],$$

and the line bundle $\mathcal{L}_2$ defined by

$$H^0(U_1, \mathcal{L}_2) = \mathbf{C}[t^2, t^5]t^5$$
$$H^0(U_2, \mathcal{L}_2) = \mathbf{C}[\tfrac{1}{t}].$$

multiplying $\mathcal{L}_1$ by $\frac{1}{t^2}$ and $\mathcal{L}_2$ by $\frac{1}{t^5}$ we see that $\mathcal{L}_1 \cong \mathcal{O}_C(-2\infty)$ and $\mathcal{L}_2 \cong \mathcal{O}_C(-5\infty)$. We thus have short exact sequences

$$0 \to \mathcal{K}_1 \to \mathcal{O}_C(-2\infty) \oplus \mathcal{O}_C(-5\infty) \to m \to 0 \tag{2}$$

$$0 \to \mathcal{K}_2 \to \mathcal{O}_C(-4\infty) \oplus \mathcal{O}_C(-7\infty) \to m^2 \to 0 \tag{3}$$

of coherent $\mathcal{O}_C$ modules, for some modules $\mathcal{K}_1$ and $\mathcal{K}_2$.

Let $a \in C$ be a closed point, with ideal sheaf m_a. From the exact sequence

$$0 \to \mathcal{L}m_a \to \mathcal{L} \to \mathcal{L}/\mathcal{L}m_a \to 0$$

we see that $\mathcal{L}$ is generated by global sections if $H^1(C, \mathcal{L}m_a) = 0$ for all $a \in C$. By Serre Duality, we have that if $\mathcal{N}$ is a line bundle of degree > 2 then $H^1(C, \mathcal{N}) = 0$. By (3) we see that $\mathcal{L}$ is generated by global sections if $\deg(\mathcal{L}) \geq 8$.

By Proposition II.7.3 [**5**] a line bundle $\mathcal{L}$ on C is very ample if

(1) $\mathcal{L}$ is generated by global sections
(2) $H^0(C, \mathcal{L}m_a) \to H^0(C, \mathcal{L}m_a/\mathcal{L}m_a m_b)$ is surjective for distinct closed point $a, b \in C$
(3) $H^0(C, \mathcal{L}m_a) \to H^0(C, \mathcal{L}m_a/\mathcal{L}m_a^2)$ is surjective for every closed point $a \in C$.

From the exact sequences

$$0 \to \mathcal{L}m_a m_b \to \mathcal{L}m_a \to \mathcal{L}m_a/\mathcal{L}m_a m_b \to 0$$

and

$$0 \to \mathcal{L}m_a^2 \to \mathcal{L}m_a \to \mathcal{L}m_a/\mathcal{L}m_a^2 \to 0,$$

Serre Duality, (2) and (3), we see that $\mathcal{L}$ is very ample if $\deg(\mathcal{L}) \geq 10$.

Suppose that $\mathcal{L} \in \mathrm{Pic}(C_2)$ has degree $-e < 0$. Let $r = e + 12$. $\deg(D - \mathcal{L}) = r > 10$ implies $D - \mathcal{L}$ is very ample, so by Bertini's theorem,

$$D - \mathcal{L} \sim a_1 + \cdots + a_r$$

where $a_1, \ldots, a_r \in C_2$ are distinct nonsingular points in C_2. Let $\lambda : X_3 \to X_2$ be the blowup of $a_1, \ldots, a_r$. Let $F_i = \lambda^{-1}(a_i)$ for $1 \leq i \leq r$. Let $C_3 \cong C$ be the strict transform of C_2. $\lambda^*(C_2) = C_3 + F_1 + \cdots + F_r$.

$$C_3 \cdot C_3 \sim \lambda^*(C_2) \cdot C_3 - a_1 - \cdots - a_r \sim D - a_1 - \cdots - a_r \sim \mathcal{L}.$$

□

Let C be the curve of Lemma 3.1, with singular point $\tilde{p}$. Let $\pi : \mathbf{P}^1 \to C$ be the normalization of C, with function field $\mathbf{C}(\mathbf{P}^1) = \mathbf{C}(t)$ where $t = 0$ is a local equation of $q = \pi^{-1}(\tilde{p})$. Let $\infty \in C$ be the point with local equation $\frac{1}{t} = 0$. Let

$$\mathrm{Div}^0 = \text{ Group of Weil divisors of degree 0 on } C - \tilde{p}.$$

$$\mathrm{Pic}^0(C) \cong \mathrm{Div}^0/\sim$$

where $D_1 \sim D_2$ if $D_1 - D_2 = (f)$ for some $f \in \mathbf{C}(t)$ which is a unit in $\mathcal{O}_{C,\tilde{p}}$ (c.f. II.6 [**5**]).

Suppose that $D \in \mathrm{Div}^0$. There exists $f_D \in \mathbf{C}(t)$ such that $(f_D) = D$ (divisor computed on $\mathbf{P}^1$), and f_D is unique up to multiplication by a nonzero constant in $\mathbf{C}$. Define $\Lambda : \mathrm{Div}^0 \to \mathbf{C}^2$ by

$$\Lambda(D) = \left(\frac{d}{dt}\log(f_D) \mid_{t=0}, \frac{d^3}{dt^3}\log(f_D) \mid_{t=0} \right).$$

$\Lambda(D)$ is a well defined group homomorphism.

For $D \in \mathrm{Div}^0$, we have an expansion

$$f_D = \sum_{i=0}^{\infty} a_i t^i \in \hat{\mathcal{O}}_{\mathbf{P}^1,q} = \mathbf{C}[[t]].$$

$f_D \in \hat{\mathcal{O}}_{C,\tilde{p}}$ if and only if $a_1 = a_3 = 0$. Since $f_D \in \mathbf{C}(t)$ and $\mathcal{O}_{C,\tilde{p}} = \hat{\mathcal{O}}_{C,\tilde{p}} \cap \mathbf{C}(t)$ (c.f. Lemma 2 [**1**]), $f_D \in \mathcal{O}_{C,\tilde{p}}$ if and only if $a_1 = a_3 = 0$. Since $\Lambda(D) = 0$ if and only if $a_1 = a_3 = 0$, $\Lambda(D) = 0$ if and only if $D \sim 0$. By allowing a_1 and a_3 to vary, we see that Λ is onto. Thus Λ is a group isomorphism of $\mathrm{Pic}^0(C)$ with $\mathbf{C}^2$.

We will consider the Abel-Jacobi map

$$AJ : C - \{\tilde{p}\} \to \mathbf{C}^2$$

defined by $AJ(\alpha) = \Lambda(\alpha - \infty)$.

$$\begin{aligned} AJ(\alpha) &= \left(\frac{d\log(t-\alpha)}{dt} \Big|_{t=0}, \frac{d^3\log(t-\alpha)}{dt^3} \Big|_{t=0} \right) \\ &= \left(\frac{1}{-\alpha}, \frac{2}{-\alpha^3} \right). \end{aligned}$$

Define the image of the Abel-Jacobi map to be

$$W = AJ(C - \{\infty\}).$$

W is the subvariety of $\mathbf{C}^2$ defined by $y = 2x^3$.

LEMMA 3.2. *Suppose that $D \in Div^0$. Then*

$$h^1(C, \mathcal{O}_C(D + \infty)) = h^0(C, \mathcal{O}_C(D + \infty)) = \begin{cases} 0 & \text{if } \Lambda(D) \notin W \\ 1 & \text{if } \Lambda(D) \in W \end{cases}$$

PROOF. There exists $f \in \mathbf{C}(t)$ such that $(f) = D$ (divisor computed on $\mathbf{P}^1$).

$$H^0(\mathbf{P}^1, \mathcal{O}_{\mathbf{P}^1}(D + \infty))$$

has $\mathbf{C}$ basis $\frac{1}{f}, \frac{t}{f}$ and

$$H^0(C, \mathcal{O}_C(D + \infty)) = \left\{ \lambda \in H^0(\mathbf{P}^1, \mathcal{O}_{\mathbf{P}^1}(D + \infty)) \mid \lambda \in \mathcal{O}_{C,\tilde{p}} \right\}.$$

There is an expansion

$$\frac{t}{f} = a_1 t + a_2 t^2 + \cdots$$

in $\mathbf{C}[[t]] = \hat{\mathcal{O}}_{\mathbf{P}^1,q}$ where $a_1 \neq 0$, so $\frac{t}{f} \notin \mathcal{O}_{C,\tilde{p}}$. For $0 \neq \alpha \in \mathbf{C}$,

$$\begin{aligned} \tfrac{t-\alpha}{f} \in \mathcal{O}_{C,\tilde{p}} &\Leftrightarrow \Lambda(-D + \alpha - \infty) = 0 \\ &\Leftrightarrow \Lambda(D) = AJ(\alpha) \end{aligned}$$

We further have

$$\frac{1}{f} \in \mathcal{O}_{C,\tilde{p}} \Leftrightarrow \Lambda(-D) = 0 \Leftrightarrow \Lambda(D) = AJ(\infty) = 0.$$

Since $p_a(C) = 2$, the equality of h^0 and h^1 follows from the Riemann-Roch Theorem. □

4. The function $h(\underline{n})$ on a curve

THEOREM 4.1. *Suppose that C is a nonsingular projective integral curve over an algebraically closed field k of characteristic zero, and $\mathcal{L}_1, \ldots, \mathcal{L}_r$ are line bundles on C. Then the Poincaré series*

$$\sum_{n \in \mathbf{N}^r} h(\underline{n}) t^{\underline{n}}$$

is a rational series, where

$$h(\underline{n}) = h^0(C, \mathcal{L}_1^{n_1} \otimes \mathcal{L}_r^{n_r}).$$

PROOF. Let $\chi(\mathcal{N})$ be the Euler characteristic of a line bundle $\mathcal{N}$. Let g be the genus of C. Let $d_i = \deg(\mathcal{L}_i)$ for $1 \le i \le r$.

$$h(\underline{n}) = \chi(\mathcal{L}_1^{n_1} \otimes \cdots \mathcal{L}_r^{n_r}) \tag{4}$$

if $n_1d_1 + \cdots + n_rd_r > 2g-2$. Thus $h(\underline{n})$ is a polynomial for $n_1d_1 + \cdots + n_rd_r > 2g-2$, and

$$\sum_{\underline{n}\in\mathbf{N}^r, n_1d_1+\cdots+n_rd_r>2g-2} h(\underline{n})t^{\underline{n}}$$

is a rational series by Theorem 2.2. For $s \in \mathbf{N}$, let

$$H_s = \{\underline{n} \in \mathbf{Z}^r \mid n_1d_1 + \cdots + n_rd_r = s\}.$$

Since $h(\underline{n}) = 0$ if $n_1d_1 + \cdots + n_rd_r < 0$, we are reduced to proving that the $2g-1$ series

$$\sum_{\underline{n}\in H_s\cap\mathbf{N}^r} h(\underline{n})t^{\underline{n}} \tag{5}$$

are rational for $0 \le s \le 2g-2$.

For the rest of this proof we will fix s with $0 \le s \le 2g-2$, and prove the rationality of (5).

Fix a base point $p_0 \in C$ and set $\overline{\mathcal{L}}_i = \mathcal{L}_i \otimes \mathcal{O}_C(-d_ip_0)$ for $1 \le i \le r$.

We have a group homomorphism

$$\pi : \mathbf{Z}^r \to \mathrm{Pic}^0(C)$$

defined by

$$\pi(\underline{n}) = \overline{\mathcal{L}}_1^{n_1} \otimes \cdots \otimes \overline{\mathcal{L}}_r^{n_r}.$$

Let

$$\Omega_{s,i} = \{\mathcal{M} \in \mathrm{Pic}^0(C) \mid h^1(\mathcal{M} \otimes \mathcal{O}_C(sp_0)) \ge i\}.$$

The $\Omega_{s,i}$ are closed sets in $\mathrm{Pic}^0(C)$ (as shown in the proof of Theorem 8 of [**2**]). There exists ω_s such that $\Omega_{s,l} = \emptyset$ for $l > w_s$.

By Proposition 2.1 there exists an index set $\Delta_{s,i}$ with corresponding $m_\delta \in \mathbf{Z}^r$ and subgroup $M_\delta < \mathbf{Z}^r$ to each $\delta \in \Delta_{s,i}$ such that for $\underline{n} \in \mathbf{Z}^r$,

$$\pi(\underline{n}) \in \Omega_{s,i} \Leftrightarrow \underline{n} \in \cup_{\delta\in\Delta_{s,i}}(m_\delta + M_\delta).$$

Define

$$\Psi_i : \mathbf{Z}^r \to \mathbf{N}$$

by

$$\Psi_i(\underline{n}) = \begin{cases} 1 & \text{if } \underline{n} \in \cup_{\delta\in\Delta_{s,i}}(m_\delta + M_\delta) \\ 0 & \text{otherwise} \end{cases}$$

Now define

$$\Psi_s : \mathbf{Z}^r \to \mathbf{N}$$

by

$$\Psi_s(\underline{n}) = \sum_{i=1}^{\omega_s} \Psi_{s,i}(\underline{n}).$$

we have

$$\Psi_s(\underline{n}) = h^1(C, \overline{\mathcal{L}}_1^{n_1} \otimes \cdots \otimes \overline{\mathcal{L}}_r^{n_r} \otimes \mathcal{O}_C(sp_0))$$

for $\underline{n} \in \mathbf{Z}^r$, and for $\underline{n} \in H_s$,

$$h^1(\mathcal{L}_1^{n_1} \otimes \cdots \otimes \mathcal{L}_r^{n_r}) = \Psi_s(\underline{n}).$$

Thus

$$\sum_{\underline{n}\in H_s\cap\mathbf{N}^r} h^1(\mathcal{L}_1^{n_1} \otimes \cdots \mathcal{L}_r^{n_r}) t^{\underline{n}} = \sum_{i=1}^{\omega_s} \Big(\sum_{\underline{n}\in H_s\cap\mathbf{N}^r} \Psi_{s,i}(\underline{n}) t^{\underline{n}} \Big).$$

Fix i with $1 \leq i \leq \omega_s$. Well order the set $\Delta_{s,i}$. To prove the rationality of (5) we are reduced to proving the rationality of

(6)

$$\sum_{\underline{n}\in H_s\cap\mathbf{N}^r} \Psi_{s,i}(\underline{n}) t^{\underline{n}}$$

$$= \sum_{j\geq 1} (-1)^{j-1} \left(\sum_{\delta_1<\cdots<\delta_j} \Big(\sum_{\underline{n}\in H_s\cap\mathbf{N}^r\cap(\underline{n}_{\delta_1}+M_{\delta_1})\cap\cdots\cap(n_{\underline{\delta}_j}+M_{\delta_j})} t^{\underline{n}} \Big) \right)$$

where $\delta_1, \ldots, \delta_j \in \Delta_{s,i}$.

If $\underline{n}_0 \in (\underline{n}_{\delta_1} + M_{\delta_1}) \cap \cdots \cap (n_{\underline{\delta}_j} + M_{\delta_j})$, then

$$(\underline{n}_{\delta_1} + M_{\delta_1}) \cap \cdots \cap (n_{\underline{\delta}_j} + M_{\delta_j}) = \underline{n}_0 + M_{\delta_1} \cap \cdots \cap M_{\delta_j}.$$

The rationality of (6) now follows from Theorem 2.2. □

If we allow C to be singular, but require C to have semi-abelian Pic^0, then the same proof shows the rationality of the series

$$\sum_{\underline{n}\in\mathbf{N}^r} h(\underline{n}) t^{\underline{n}}.$$

Furthermore, we get rationality in general on a singular curve, as long as $r \leq 2$. If $r = 1$ this is a trivial statment (it is immediate from the Riemann-Roch formula and Theorem 2.2). The case $r = 2$

follows from the method of the proof of Theorem 4.1, using Theorem 15 of [**2**] in place of Proposition 2.1. Theorem 15 of [**2**] is a general statement for cyclic groups on algebraic groups.

However, if $r \geq 3$ and $\mathrm{Pic}^0(C)$ is not semi-abelian, then the Poincaré series may not be rational, as is shown by the following example, where $\mathrm{Pic}^0(C) \cong \mathbf{C}^2$ is not semi-abelian.

EXAMPLE 4.2. *There exists a complex, integral (singular) projective curve C, and line bundles $\mathcal{L}_1$, $\mathcal{L}_2$, $\mathcal{L}_3$ on C such that the Poincaré series*

$$\sum_{\underline{n}\in\mathbf{N}^3} h(\underline{n})t^{\underline{n}}$$

is not rational, where

$$h(\underline{n}) = h^0(C, \mathcal{L}_1^{n_1} \otimes \mathcal{L}_2^{n_2} \otimes \mathcal{L}_3^{n_3}).$$

PROOF. Let C be the curve of Lemma 3.1. Let notation be that preceding Lemma 3.2. Set

$$\mathcal{L}_1 = (1,0), \mathcal{L}_2 = (0,1) \in \mathrm{Pic}^0(C),$$
$$\mathcal{L}_3 = \mathcal{O}_C(\infty).$$

Since $p_a(C) = 2$, we have by Serre duality that

$$h(n_1, n_2, n_3) = \chi(\mathcal{L}_1^{n_1} \otimes \mathcal{L}_2^{n_2} \otimes \mathcal{L}_3^{n_3})$$

if $n_3 > 2$.

$$h(n_1, n_2, 0) = \begin{cases} 0 & \text{if } n_1 \text{ or } n_2 \neq 0 \\ 1 & \text{if } n_1 = n_2 = 0 \end{cases}$$

By Serre duality, $h(n_1, n_2, 2) = 0$ with one possible exception, that is if n_1, n_2 are such that

$$\omega_C \cong \mathcal{L}_1^{n_1} \otimes \mathcal{L}_2^{n_2} \otimes \mathcal{O}_C(2\infty).$$

The interesting case (which follows from Lemma 3.2) is

$$h(n_1, n_2, 1) = \begin{cases} 1 & \text{if } n_2 = 2n_1^3 \\ 0 & \text{otherwise} \end{cases}$$

Thus

$$\sum_{\underline{n}\in\mathbf{N}^r} h(\underline{n})t^{\underline{n}}$$

is irrational if and only if

$$\sum_{n_1,n_2=0}^{\infty} h(n_1, n_2, 1)t_1^{n_1}t_2^{n_2}$$

is irrational.

Set

$$a_{n_1,n_2} = h(n_1, n_2, 1) = \begin{cases} 1 & \text{if } n_2 = 2n_1^3 \\ 0 & \text{otherwise} \end{cases}$$

We will assume that $f = \sum_{m,n=0}^{\infty} a_{m,n} t_1^m t_2^n$ is rational, and derive a contradiction. Assuming that the series is rational, there exist $r > 0$ and a nonzero polynomial

$$Q = \sum_{i,j=0}^{r} b_{ij} t_1^i t_2^j$$

such that fQ is a polynomial. Thus there exists a constant σ such that

$$\sum_{i,j=0}^{\infty} a_{m-i,n-j} b_{ij} = 0$$

whenever $m + n > \sigma$ and $m, n \geq r$.

We make the following simple observation. Given $n > 0$, there exists $m(n) > 0$ such that if $x_0 > m(n)$ is an integer, and we set $y_0 = 2x_0^3$, then

$$[x_0 - n, x_0 + n] \times [y_0 - n, y_0 + n] \cap \mathbf{Z}^2 \cap \{y = 2x^3\} = \{(x_0, y_0)\}.$$

Choose (x_0, y_0) as above (with $n = r$) and x_0 so big that $x_0, y_0 > r$ and $x_0 + y_0 > \sigma$.

For fixed a, b with $0 \leq a \leq r$, $0 \leq b \leq r$, set $m = x_0 + a$, $n = y_0 + b$. Then

$$0 = \sum_{i,j=0}^{r} a_{m-i,n-j} b_{ij} = b_{ab}.$$

Thus $Q = 0$, a contradiction. □

5. Irrationality on Surfaces I

If X is a normal projective surface over a field k of characteristic 0, and $\mathcal{L}$ is a line bundle on X, then it follows from Theorem 4 of [**2**] that $h^0(X, \mathcal{L}^n)$ is a polynomial with periodic coefficients for $n >> 0$. Thus the series

$$\sum_{n=0}^{\infty} h^0(X, \mathcal{L}^n) t^n$$

is a rational series by Theorem 2.2.

We show in the following example that rationality fails for surfaces if $r = 2$.

EXAMPLE 5.1. *There exists a nonsingular (integral) projective surface S and line bundles $\mathcal{L}_1$, $\mathcal{L}_2$ on S such that the series*

$$\sum_{i,j=0}^{\infty} h^0(S, \mathcal{L}_1^i \otimes \mathcal{L}_2^j) t_1^i t_2^j \tag{7}$$

is an irrational series.

We will use a construction of [**2**]. As in Example 4 of [**2**], Let C be an elliptic curve over an algebraically closed field k, and let $S = C \times C$. Let $\Delta \subset S$ be the diagonal, $p \in C$ be a closed point, $A = \pi_1^{-1}(p)$, $B = \pi_2^{-1}(p)$, where $\pi_i : S \to C$, $i = 1, 2$ are the projections.

Let $\mathcal{L}_1 = \mathcal{O}_S(A + 2B + 3\Delta)$, $\mathcal{L}_2 = \mathcal{O}_S(-A - B - \Delta)$,

$$h(i,j) = h^0(S, \mathcal{L}_1^i \otimes \mathcal{L}_2^j).$$

We will show that the associated Poincaré series

$$f = \sum_{i,j=0}^{\infty} h(i,j) t_1^i t_2^j$$

is irrational. Let

$$\tau = 2 - \frac{\sqrt{3}}{3},$$

an irrational number. For $i, j \geq 0$,

$$h(i,j) = \begin{cases} \chi(\mathcal{L}_1^i \otimes \mathcal{L}^j) = \frac{1}{2}\left((\mathcal{L}_1^i \otimes \mathcal{L}_2^j)^2\right) > 0 & \text{if } j < i\tau \\ 0 & \text{if } j > i\tau \\ 1 & \text{if } i = j = 0 \end{cases}$$

as follows from the calculations of Example 4 [**2**].

Suppose that f is rational. Then there exist $r > 0$ and a nonzero polynomial

$$Q = \sum_{i,j=0}^{r} b_{ij} t_1^i t_2^j$$

such that fQ is a polynomial. Thus there is $\sigma > 0$ such that

$$\sum_{i,j=0}^{r} a_{m-i,n-j} b_{i,j} = 0$$

whenever $m + n \geq \sigma$ and $m, n \geq r$.

Let λ be the largest real number such that there exists a nonzero coefficient b_{ij} of Q such that

$$j - i\tau = \lambda.$$

Since τ is irrational, there exists a unique solution in natural numbers (i_0, j_0) to $j - i\tau = \lambda$. We necessarily have $b_{i_0,j_0} \neq 0$. We further have that $b_{ij} = 0$ if $j - i\tau > \lambda$.

LEMMA 5.2. *Suppose that N is a given constant. Then there exists $(\alpha_0, \beta_0) \in \mathbf{N}^2$ such that $\beta_0 < \alpha_0\tau$, $\alpha_0, \beta_0 > N$ and if $(\alpha, \beta) \in \mathbf{N}^2$ is such that*

(1) $\alpha_0 - r \le \alpha \le \alpha_0 + r$
(2) $\beta_0 - \tau\alpha_0 \le \beta - \tau\alpha \le 0$

Then $(\alpha, \beta) = (\alpha_0, \beta_0)$.

PROOF. If x is a real number, $\lceil x \rceil$ will denote the smallest integer $\ge x$. Choose a real number ϵ such that

$$0 < \epsilon < \min \{\lceil i\tau \rceil - i\tau \mid i \in \{-r, -r+1, \dots, -1, 1, 2, \dots, r\}\}.$$

As a consequence of Kroneker's theorem (c.f. Theorem 438 [**4**]) there exist natural numbers (α_0, β_0) such that $\alpha_0 > N$ and $0 < \alpha_0\tau - \beta_0 < \epsilon$. Set $\lambda = \alpha_0\tau - \beta_0$. Suppose there exists $(\alpha, \beta) \in \mathbf{N}^2$ such that $\alpha_0 - r \le \alpha \le \alpha + r$ and $0 < \alpha\tau - \beta < \lambda$. Then there exists a nonzero integer i such that $-r \le i \le r$ and $\alpha = \alpha_0 + i$.

$$(\alpha_0 + i)\tau - \lambda < \beta < (\alpha_0 + i)\tau.$$

We have

$$\begin{aligned} \alpha_0\tau - \lambda + i\tau &< \beta < \alpha_0\tau + i\tau \\ \beta_0 + i\tau &< \beta < \beta_0 + \lambda + i\tau \\ i\tau &< \beta - \beta_0 < i\tau + \lambda. \end{aligned}$$

There can be no integral solutions in β by our choice of ϵ. □

With the notation of the Lemma 5.2, taking $N = \max\{\sigma, r\}$, set

$$m = i_0 + \alpha_0, n = \beta_0 + j_0.$$

Then $(\alpha_0, \beta_0) = (m - i_0, n - j_0)$.

$$\begin{aligned} 0 &= \textstyle\sum_{ij=0}^{r} b_{ij} a_{m-i,n-j} \\ &= \textstyle\sum_{j-i\tau>\lambda} b_{ij} a_{m-i,n-j} + \sum_{j-i\tau<\lambda} b_{ij} a_{m-i,n-j} \\ &\qquad + \textstyle\sum_{j-i\tau=\lambda} b_{ij} a_{m-i,n-j} \\ &= \textstyle\sum_{j-i\tau<\lambda} b_{ij} a_{m-i,n-j} + b_{i_0,j_0} a_{m-i_0,n-j_0} \end{aligned} \quad . \tag{8}$$

Make a change of variables

$$(\alpha, \beta) = (m - i, n - j)$$

and set $(\alpha_0, \beta_0) = (m - i_0, n - j_0)$, $\psi = \beta_0 - \alpha_0\tau$.

$j_0 - i_0\tau = \lambda$ implies $\psi = \beta_0 - \alpha_0\tau = (n - m\tau) - \lambda$.

If $j - i\tau < \lambda$ then

$$\beta - \alpha\tau = (n - m\tau) + (i\tau - j) > (n - m\tau) - \lambda = \psi.$$

$$\sum_{j-i\tau<\lambda} b_{ij}a_{m-i,n-j} = \sum_{\beta-\alpha\tau>\psi} a_{\alpha,\beta}b_{m-\alpha,n-\beta}. \tag{9}$$

We have $0 \le m - \alpha \le r$ implies

$$\alpha_0 - r \le \alpha_0 + i_0 - r = m - r \le \alpha \le m = \alpha_0 + i_0 \le \alpha_0 + r$$

so $a_{\alpha,\beta} = 0$ in this range. Thus (9) is zero, and substituting into (8), we conclude that $b_{i_0,j_0} = 0$, a contradiction to our assumption that Q is nonzero. Thus the series f is irrational.

6. Irrationality on Surfaces II

In Example 5.1 of Section 5, the fact that $\mathcal{L}_1$ was ample and $\mathcal{L}_2^{-1}$ was ample played a critical role in the construction. This leaves open the question if an example can be constructed on a surface of effective divisors with irrational Poincaré series. We construct such an example in this section.

We construct an example of 3 line bundles which have nonzero global sections (they are the line bundles associated to effective divisors) on a nonsingular complex projective surface such that the associated Poincaré series is not rational.

The construction is similar to that of Theorem 9.1 of [**3**], which shows that the Poincaré series of a resolution of singularities of a surface singularity with 3 exceptional components can be irrational.

Example 6.1. *There exists a nonsingular complex projective surface X of and line bundles $\mathcal{N}_1$, $\mathcal{N}_2$, $\mathcal{N}_3$ on X such that*

$$h^0(X, \mathcal{N}_1) > 0,\ h^0(X, \mathcal{N}_2) > 0,\ h^0(X, \mathcal{N}_3) > 0$$

and the series

$$\sum_{i,j,k=0}^{\infty} h^0(X, \mathcal{N}_1^i \otimes \mathcal{N}_2^j \otimes \mathcal{N}_3^k)t_1^i t_2^j t_3^k \tag{10}$$

is an irrational series.

Let C be the curve of Lemma 3.1. Suppose that $\mathcal{M}$ is a line bundle on C of degree $-d$ with $d \ge 3$, and p_1, p_2 are distinct points on $C - \{\tilde{p}\}$. Let $\overline{\mathcal{L}} = \mathcal{M} \otimes \mathcal{O}_C(p_1 + p_2)$. Lemma 3.1 shows that there exists a nonsingular, complex projective surface S and an embedding $C \subset S$ such that $C \cdot C = \overline{\mathcal{L}}$.

Let $X \to S$ be the blowup of S at the points p_1 and p_2, with exceptional divisors F_1 and F_2 (both isomorphic to $\mathbf{P}^1$). Identify C with its strict transform on X. Then $C \cdot C \sim \mathcal{M}$, $C \cdot F_1 = p_1$, $C \cdot F_2 = p_2$, $(F_1^2) = -1$, $(F_2^2) = -1$.

Recall from the proof of Lemma 3.1 that we have a birational morphism $S \to \mathbf{P}^2$ such that C is the strict transform of a curve of degree 5 in $\mathbf{P}^2$. Let H be the pullback on X of a hyperplane section of $\mathbf{P}^2$. We will now fix $d = 3$, and continue to assume that $\mathcal{M}$ is a line bundle of degree $-d = -3$ on C. Set

$$\begin{aligned} \Delta_1 &= H + 3F_1 + 4F_2 + 4C \\ \Delta_2 &= H + 4F_1 + 3F_2 + 4C \\ \Delta_3 &= H + 4F_1 + 4F_2 + 4C \end{aligned}$$

If we set $F_3 = C$, then we have $(\Delta_i \cdot F_j) = \delta_{ij}$. Let $\overline{H} = H \cdot C - 5\infty$, $\overline{p}_1 = p_1 - \infty$, $\overline{p}_2 = p_2 - \infty$, $\mathcal{L} = \mathcal{M} + 3\infty$. Now choose $\mathcal{L}$ to be

$$\mathcal{L} = (1, \frac{9}{4}) - \frac{1}{4}\overline{H},$$

Set $p_1 = \infty$, and choose p_2 so that $\overline{p}_2 = (-1, -2) \in W$. Then

$$\Delta_1 \cdot C = \overline{H} + 4\overline{p}_2 + 4\mathcal{L} = \overline{H} + (-4, -8) + (4, 9) - \overline{H} = (0, 1)$$

$$\Delta_2 \cdot C = \overline{H} + 3\overline{p}_2 + 4\mathcal{L} = \overline{H} + (-3, -6) + (4, 9) - \overline{H} = (1, 3)$$

LEMMA 6.2. *There exists at most 2 divisors $D = a\Delta_1 + b\Delta_2 + c\Delta_3$ on X such that $D \cdot C \sim \omega_C + (F_1 + F_2 + C) \cdot C$ or $D \cdot C \sim \omega_C$, where ω_C is the canonical bundle of C.*

PROOF. Suppose that $D \cdot C \sim \omega_C + (F_1 + F_2 + C) \cdot C$. Then

$$c = (D \cdot C) = \deg \omega_C - 1 = 1.$$

$$D \cdot C \sim a(0, 1) + b(1, 3) + \Delta_3 \cdot C$$

implies

$$(b, a + 3b) \sim \omega_C + (F_1 + F_2 + C) \cdot C - \Delta_3 \cdot C$$

so there is at most one solution in a and b.

If $D \cdot C \sim \omega_C$, we have $c = 2$ and $(b, a + 3b) \sim \omega_C - 2\Delta_3 \cdot C$ so there is at most one solution in a and b. □

LEMMA 6.3. *There exists $\lambda > 0$ such that $a, b, c \geq 0$ and $a+b+c > \lambda$ implies*

$$h^1(X, \mathcal{O}_X(a\Delta_1 + b\Delta_2 + c\Delta_3)) = \begin{cases} 0 & \text{if } c \geq 2 \\ 1 & \text{if } c = 1, a = 2b^3 - 3b - 1 \\ 0 & \text{if } c = 1, a \neq 2b^3 - 3b - 1 \end{cases}$$

PROOF. By Lemma 6.2 we can choose λ so that $a, b, c \geq 0$ and $a+b+c>0$ implies

$$(a\Delta_1 + b\Delta_2 + c\Delta_3) \cdot C \not\sim \omega_C + (F_1 + F_2 + C) \cdot C$$

and

$$(a\Delta_1 + b\Delta_2 + c\Delta_3) \cdot C \not\sim \omega_C.$$

Suppose that $\mathcal{N}$ is a line bundle on X. We have exact sequences

$$\begin{aligned} 0 \to \mathcal{O}_{F_1+F_2}(-n(F_1+F_2+C)-C) \otimes \mathcal{N} &\to \mathcal{O}_{(n+1)(F_1+F_2+C)} \otimes \mathcal{N} \\ &\to \mathcal{O}_{n(F_1+F_2+C)+C} \otimes \mathcal{N} \to 0 \end{aligned} \tag{11}$$

for $n \geq 0$,

$$\begin{aligned} 0 \to \mathcal{O}_C(-n(F_1+F_2+C)) \otimes \mathcal{N} &\to \mathcal{O}_{n(F_1+F_2+C)+C} \otimes \mathcal{N} \\ &\to \mathcal{O}_{n(F_1+F_2+C)} \otimes \mathcal{N} \to 0 \end{aligned} \tag{12}$$

for $n \geq 1$,

$$\begin{aligned} 0 \to \mathcal{O}_{F_1}(-n(F_1+F_2+C)-C) \otimes \mathcal{N} &\to \mathcal{O}_{n(F_1+F_2+C)+F_1+C} \otimes \mathcal{N} \\ &\to \mathcal{O}_{n(F_1+F_2+C)+C} \otimes \mathcal{N} \to 0 \end{aligned} \tag{13}$$

for $n \geq 0$, and

$$\begin{aligned} 0 \to \mathcal{O}_{F_2}(-n(F_1+F_2+C)-C) \otimes \mathcal{N} &\to \mathcal{O}_{n(F_1+F_2+C)+F_2+C} \otimes \mathcal{N} \\ &\to \mathcal{O}_{n(F_1+F_2+C)+C} \otimes \mathcal{N} \to 0 \end{aligned} \tag{14}$$

for $n \geq 0$. Observe that we have

$$\begin{aligned} &H^1(X, \mathcal{O}_{F_1+F_2}(-n(F_1+F_2+C)-C) \otimes \mathcal{N}) = 0 && \text{if } (\mathcal{N} \cdot F_i) \geq 0 \text{ for } i = 1, 2 \\ &H^1(X, \mathcal{O}_{F_1}(-n(F_1+F_2+C)-C) \otimes \mathcal{N}) = 0 && \text{if } (\mathcal{N} \cdot F_1) \geq 0 \\ &H^1(X, \mathcal{O}_{F_2}(-n(F_1+F_2+C)-C) \otimes \mathcal{N}) = 0 && \text{if } (\mathcal{N} \cdot F_2) \geq 0. \end{aligned}$$

We also have

$$\begin{aligned} &H^1(X, \mathcal{O}_C(-n(F_1+F_2+C)) \otimes \mathcal{N}) = 0 \\ &\quad \text{if } n + (\mathcal{N} \cdot C) > 2 \text{ or} \\ &\quad n + (\mathcal{N} \cdot C) = 2 \text{ and } -n(F_1+F_2+C) \cdot C + \mathcal{N} \cdot C \not\sim \omega_C. \end{aligned}$$

Let $\overline{\Delta}_i = \Delta_i - H$ for $1 \leq i \leq 3$. From the exact sequence

$$\begin{aligned} 0 \to \mathcal{O}_X((a+b+c)H) &\to \mathcal{O}_X(a\Delta_1 + b\Delta_2 + c\Delta_3) \\ &\to \mathcal{O}_{a\overline{\Delta}_1 + b\overline{\Delta}_2 + c\overline{\Delta}_3}(a\Delta_1 + b\Delta_2 + c\Delta_3) \to 0 \end{aligned}$$

we see that

$$\begin{aligned} &H^1(X, \mathcal{O}_X(a\Delta_1 + b\Delta_2 + c\Delta_3)) \\ &\qquad \cong H^1(X, \mathcal{O}_{a\overline{\Delta}_1 + b\overline{\Delta}_2 + c\overline{\Delta}_3}(a\Delta_1 + b\Delta_2 + c\Delta_3)). \end{aligned}$$

For $0 \le \alpha \le a$, $0 \le \beta \le b$, $0 \le \gamma \le c$,

$$\begin{aligned} &a\Delta_1 + b\Delta_2 + c\Delta_3 - (\alpha\overline{\Delta}_1 + \beta\overline{\Delta}_2 + \gamma\overline{\Delta}_3) \\ &\qquad = (a-\alpha)\Delta_1 + (b-\beta)\Delta_2 + (c-\gamma)\Delta_3 + (\alpha+\beta+\gamma)H. \end{aligned}$$

Consider the exact sequences, with $0 \le \alpha \le a$, $0 \le \beta \le b$, $0 \le \gamma \le c$

(15)

$$\begin{aligned} 0 \to\ & \mathcal{O}_{\overline{\Delta}_1}((a-\alpha)\Delta_1 + (b-\beta)\Delta_2 + (c-\gamma)\Delta_3 + (\alpha+\beta+\gamma)H) \\ & \to \mathcal{O}_{(\alpha+1)\overline{\Delta}_1+\beta\overline{\Delta}_2+\gamma\overline{\Delta}_3}(a\Delta_1 + b\Delta_2 + c\Delta_3) \\ & \to \mathcal{O}_{\alpha\overline{\Delta}_1+\beta\overline{\Delta}_2+\gamma\overline{\Delta}_3}(a\Delta_1 + b\Delta_2 + c\Delta_3) \to 0 \end{aligned}$$

with $\alpha < a$,

(16)

$$\begin{aligned} 0 \to\ & \mathcal{O}_{\overline{\Delta}_2}((a-\alpha)\Delta_1 + (b-\beta)\Delta_2 + (c-\gamma)\Delta_3 + (\alpha+\beta+\gamma)H) \\ & \to \mathcal{O}_{\alpha\overline{\Delta}_1+(\beta+1)\overline{\Delta}_2+\gamma\overline{\Delta}_3}(a\Delta_1 + b\Delta_2 + c\Delta_3) \\ & \to \mathcal{O}_{\alpha\overline{\Delta}_1+\beta\overline{\Delta}_2+\gamma\overline{\Delta}_3}(a\Delta_1 + b\Delta_2 + c\Delta_3) \to 0 \end{aligned}$$

with $\beta < b$,

(17)

$$\begin{aligned} 0 \to\ & \mathcal{O}_{\overline{\Delta}_3}((a-\alpha)\Delta_1 + (b-\beta)\Delta_2 + (c-\gamma)\Delta_3 + (\alpha+\beta+\gamma)H) \\ & \to \mathcal{O}_{\alpha\overline{\Delta}_1+\beta\overline{\Delta}_2+(\gamma+1)\overline{\Delta}_3}(a\Delta_1 + b\Delta_2 + c\Delta_3) \\ & \to \mathcal{O}_{\alpha\overline{\Delta}_1+\beta\overline{\Delta}_2+\gamma\overline{\Delta}_3}(a\Delta_1 + b\Delta_2 + c\Delta_3) \to 0 \end{aligned}$$

with $\gamma < c$.

Set

$$\mathcal{N} = (a-\alpha)\Delta_1 + (b-\beta)\Delta_2 + (c-\gamma)\Delta_3 + (\alpha+\beta+\gamma)H$$

with $a, b \ge 0$ and $c \ge 1$. $(\mathcal{N} \cdot F_1) = a - \alpha$, $(\mathcal{N} \cdot F_2) = b - \beta$, $(\mathcal{N} \cdot C) = (c-\gamma) + 5(\alpha+\beta+\gamma)$. If $a+b+c > 0$, we get the necessary vanishing in (11)-(14) to conclude that

(18) $H^1(X, \mathcal{O}_{\overline{\Delta}_1}((a-\alpha)\Delta_1 + (b-\beta)\Delta_2 + (c-\gamma)\Delta_3 + (\alpha+\beta+\gamma)H)) = 0$

if $0 \le \alpha < a$, $0 \le \beta \le b$, $0 \le \gamma \le c$, $\alpha + \beta + \gamma > 0$,

(19) $H^1(X, \mathcal{O}_{\overline{\Delta}_2}((a-\alpha)\Delta_1 + (b-\beta)\Delta_2 + (c-\gamma)\Delta_3 + (\alpha+\beta+\gamma)H)) = 0$

if $0 \le \alpha \le a$, $0 \le \beta < b$, $0 \le \gamma \le c$ $\alpha + \beta + \gamma > 0$, and

(20) $H^1(X, \mathcal{O}_{\overline{\Delta}_3}((a-\alpha)\Delta_1 + (b-\beta)\Delta_2 + (c-\gamma)\Delta_3 + (\alpha+\beta+\gamma)H)) = 0$

if $0 \le \alpha \le a$, $0 \le \beta \le b$, $0 \le \gamma < c$, $\alpha + \beta + \gamma > 0$.

From (15) - (20) it follows that

$$H^1(X, \mathcal{O}_X(a\Delta_1 + b\Delta_2 + c\Delta_3)) \cong H^1(X, \mathcal{O}_{\overline{\Delta}_3}(a\Delta_1 + b\Delta_2 + c\Delta_3)).$$

It then follows from (11) - (14) and our choice of λ that

$$H^1(X, \mathcal{O}_X(a\Delta_1 + b\Delta_2 + c\Delta_3)) \cong H^1(X, \mathcal{O}_C(a\Delta_1 + b\Delta_2 + c\Delta_3))$$

if $a, b, c \geq 0$, $c \geq 1$ and $a + b + c > \lambda$. The conclusions of Lemma 6.3 now follow from Lemma 3.2 if $c = 1$, since

$$a\Delta_1 + b\Delta_2 + \Delta_3 \sim (a+1)\Delta_1 + b\Delta_2 + F_1$$

and

$$(a+1)\Delta_1 \cdot C + b\Delta_2 \cdot C = (a+1)(0,1) + b(1,3) = (b, a+3b+1).$$

If $c \geq 2$, we have $H^1(C, \mathcal{O}_C(a\Delta_1 + b\Delta_2 + c\Delta_3)) = 0$ by Serre duality and our choice of λ. □

We will now give the proof of Theorem 6.1. Let notation be as above for the surface X. Set $a_{ijk} = h^1(X, \mathcal{O}_X(i\Delta_1 + j\Delta_2 + k\Delta_3))$. We will show that the series

$$f = \sum_{i,j,k=0}^{\infty} a_{ijk} t_1^i t_2^j t_3^k$$

is not rational.

We first observe that, given $n > 0$, there exists $m(n) > 0$ such that if $x_0 > m(n)$ is an integer, and $y_0 = 2x_0^3 - 3x_0 - 1$, then
(21)

$$\begin{aligned}([x_0 - n, x_0 + n] \times [y_0 - n, y_0 + n]) \cap \mathbf{Z}^2 \cap \{y = 2x^3 - 3x - 1\} \\ = \{(x_0, y_0)\}.\end{aligned}$$

Suppose that f is rational. Then there exist $r > 0$ and a nonzero polynomial

$$Q = \sum_{i,j,k=0}^{r} b_{ijk} t_1^i t_2^j t_3^k$$

such that fQ is a polynomial. Thus there is $\sigma > 0$ such that

$$\sum_{i,j,k=0}^{r} a_{l-i,m-j,n-k} b_{i,j,k} = 0$$

whenever $l + m + n \geq \sigma$ and $l, m, n \geq r$, and

$$0 = \sum_{i,j=0}^{r} \left(\sum_{k=0}^{\alpha} a_{l-i,m-j,\alpha-k} b_{ijk}\right)$$

if $l+m+\alpha \geq \sigma$, $l,m \geq r$, $0 \leq \alpha < r$. Suppose that l,m are such that $l+m \geq \sigma$. Then by lemma 6.3

$$0 = \sum_{ijk=0}^{r} a_{l-i,m-j,r+1-k} b_{ijk} = \sum_{i,j=0}^{r} b_{ijr} a_{l-i,m-j,1}.$$

By (21) and lemma 6.3 we can choose l,m to conclude that $b_{ijr} = 0$ for $0 \leq i,j \leq r$.

Suppose that α is such that $1 \leq \alpha \leq r$ and $b_{ijk} = 0$ for $0 \leq i,j \leq r$ and $\alpha \leq k \leq r$. Suppose that l,m are such that $l+m \geq \sigma$. Then by lemma 6.3

$$\begin{aligned} 0 &= \textstyle\sum_{i,j=0}^{r} \sum_{k=0}^{\alpha} b_{ijk} a_{l-i,m-j,\alpha-k} \\ &= \textstyle\sum_{i,j=0}^{r} a_{l-i,m-j,0} b_{ij\alpha} + \sum_{i,j=0}^{r} a_{l-i,m-j,1} b_{i,j,\alpha-1} \\ &= \textstyle\sum_{i,j=0}^{r} a_{l-i,m-j-1,1} b_{i,j,\alpha-1}. \end{aligned}$$

By (21) and lemma 6.3, we can choose l,m to conclude that $b_{i,j,\alpha-1} = 0$ for $0 \leq i,j \leq r$. Thus we conclude by descending induction on α that $Q = 0$, a contradiction, so that

$$\sum_{i,j,k=0}^{\infty} h^1(X, \mathcal{O}_X(i\Delta_1 + j\Delta_2 + k\Delta_3)) t_1^i t_2^j t_3^k$$

is irrational. Since $\Delta_1, \Delta_2, \Delta_3$ are effective,

$$\sum_{i,j,k=0}^{\infty} h^2(X, \mathcal{O}_X(i\Delta_1 + j\Delta_2 + k\Delta_3)) t_1^i t_2^j t_3^k$$

is a polynomial by Serre Duality. Since

$$\sum_{i,j,k=0}^{\infty} \chi(\mathcal{O}_X(i\Delta_1 + j\Delta_2 + k\Delta_3)) t_1^i t_2^j t_3^k$$

is rational, (where $\chi(\mathcal{O}_X(i\Delta_1 + j\Delta_2 + k\Delta_3)$ is the Euler characteristic and is consequentially a polynomial) we must then have that

$$\sum_{i,j,k=0}^{\infty} h^0(X, \mathcal{O}_X(i\Delta_1 + j\Delta_2 + k\Delta_3)) t_1^i t_2^j t_3^k$$

is irrational.

References

[1] Abhyankar, S., *Local uniformization on algebraic surfaces over ground fields of characteristic $p \neq 0$*, Annals of Math 63, (1956), 491-526.

[2] Cutkosky, S.D. and Srinivas, V., *On a problem of Zariski on dimensions of linear systems*, Annals of Math. 137 (1993), 531-559.

[3] Cutkosky, S.D. and Reguera, A., *Poincaré Series of Resolutions of Singularities*, preprint, available on www.math.missouri.edu/~cutkosky.
[4] Hardy, G.H. and Wright, E.M., *An Introduction to the Theory of Numbers*, Oxford Univ. Press, (1938).
[5] Hartshorne, R., *Algebraic Geometry*, Graduate Texts in Mathematics 52, Springer Verlag, New York, (1977).
[6] McQuillan, M., *Division points on semi-abelian varieties*, Inv. Math. 120 (1995), 143-159.
[7] Zariski, O., *The Theorem of Riemann-Roch for high multiples of an effective divisor on an algebraic surface*, Annals of Math, 96 (1962), 560-616.

Dept. of Math., University of Missouri, Columbia, MO., 65211
E-mail address: cutkoskys@missouri.edu

TRIPLE CANONICAL COVERS OF VARIETIES OF MINIMAL DEGREE

Francisco Javier Gallego
and
Bangere P. Purnaprajna

Dedicated to C.S. Seshadri on his 70th birthday.

Abstract. In this article we study pluriregular varieties X of general type with base-point-free canonical bundle whose canonical morphism has degree 3 and maps X onto a variety of minimal degree Y. We carry out our study from two different perspectives.

First we study in Section 2 and Section 3 the canonical ring of X describing completely the degrees of its minimal generators. We apply this to the study of the projective normality of the images of the pluricanonical morphisms of X. Our study of the canonical ring of X also shows that, if the dimension of X is greater than or equal to 3, there does not exist a converse to a theorem of M. Green that bounds the degree of the generators of the canonical ring of X. This is in sharp contrast with the situation in dimension 2 where such converse exists, as proved by the authors in a previous work.

Second, we study in Section 4, the structure of the canonical morphism of X. We use this to show among other things the non-existence of some a priori plausible examples of triple canonical covers of varieties of minimal degree. We also characterize the targets of flat canonical covers of varieties of minimal degree. Some of the results of Section 4 are more general and apply to varieties X which are not necessarily regular, and to targets Y that are scrolls which are not of minimal degree.

The first author was partially supported by MCT project number BFM2000-0621 and by UCM project number PR52/00-8862. The second author was partially supported by the General Research Fund of the University of Kansas at Lawrence. The first author is grateful for the hospitality of the Department of Mathematics of the University of Kansas at Lawrence.

Typeset by $\mathcal{A}\mathcal{M}\mathcal{S}$-TEX

1. Introduction

The purpose of the article is to study pluriregular varieties (i.e., varieties such that all the intermediate cohomology of its structure sheaf vanishes) of general type X of dimension m with base-point-free canonical divisor K_X and such that the canonical morphism π is a generically $3:1$ morphism onto a variety of minimal degree Y. Flat triple covers of varieties have been studied by Miranda, who analyzes the structure of the algebra associated to these covers. In this paper the morphisms we consider are not necessarily flat, but their restrictions to the intersection of $m-1$ general members of $|K_X|$ are. The varieties of general type whose canonical morphism maps to a variety of minimal degree play an important role in various settings. In dimension 2 they are central in the classification of surfaces of general type with small c_1^2 and in questions about degenerations and the moduli of surfaces of general type. This is illustrated by the works of Horikawa (see [7], [8], [9] and [10]), Catanese (see [2]), Konno (see [12]), Mendes Lopes and Pardini (see [13]), among others. In higher dimensions, pluriregular varieties of general type have been studied by Catanese ((see [1]) and Green (see [6]) who prove results on the canonical ring of these varieties.

We study these varieties from two different perspectives. The first perspective we take is the study of the structure of their canonical ring. This is handled in Section 2 and Section 3. More precisely our main goal is to find out the degrees of the generators. This goal is accomplished in Theorem 2.7 for odd dimensional varieties and Theorem 3.2 for even dimensional varieties, where we obtain a full description of the degrees in terms of the dimension of X and the degree of Y. We put these two results together in the following

Theorem 1. *Let X be a smooth, pluriregular variety of general type of dimension $m \geq 2$. Assume that K_X is base-point-free and that the canonical morphism associated to K_X, $\pi : X \longrightarrow Y$, is onto an m-fold $Y \subset \mathbf{P}^{r+m-1}$ of minimal degree. Assume also that π is a generically finite morphism of degree 3. Then*

1) *if m is odd, the canonical ring of X is generated by its part of degree 1 and $\frac{r}{2}$ linearly indendent generators in degree $\frac{m+1}{2}$;*
2) *if m is even, the canonical ring of X is generated by its elements of degree 1, r linearly independent generators in degree $\frac{m+2}{2}$ and $r-1$ linearly independent generators in degree $m+1$.*

Apart from its intrinsic interest, Theorem 1 has several applications. In the first place the degrees of the generators of the canonical ring are related to the degrees of the generators of the pluricanonical rings and therefore, with the very ampleness and projective normality associated to pluricanonical morphisms on the variety X. In this regard, using Theorem 1 (1) for the odd dimensional case, we give a necessary and sufficient condition for multiples of the canonical bundle K_X on X to give projectively normal embeddings. Precisely (see Theorem 2.10) we obtain

Let X be an odd dimensional variety satisfying the hypothesis of Theorem 1. Assume in addition that K_X is ample. Then the line bundle $K_X^{\otimes n}$ is very ample and the image of the morphism it induces is projectively normal if and only if $n \geq \frac{m+1}{2}$.

In the even dimensional case using Theorem 1 (2) we obtain a sufficient condition for a multiple of K_X to induce a projectively normal embedding (see Theorem 3.4).

Another interesting consequence of Theorem 1 concerns this general question: "what is the minimal degree of the generators of the canonical ring of a variety of general type and dimension m?." Green (see [6]) proved that the canonical ring of a pluriregular variety of general type is generated in degree less than or equal to m if the canonical morphism does not map X onto a variety of minimal degree (for surfaces this result was also obtained independently by Ciliberto, see [3]). In [5] we proved a converse of Green's result in the case of surfaces. For higher dimensions Theorem 1 draws a very different picture. Indeed Theorem 1 implies that a converse to Green's result, that is, a result saying that the canonical ring of a pluriregular variety of general type is generated in degree less than or equal to n *if and only if* the canonical morphism does not map X onto a variety of minimal degree, is false in dimension greater than 2.

The second approach we take is to study the structure of the canonical morphism of varieties X of general type which are triple covers of varieties of minimal degree Y. In dimension 2 these triple covers have been studied by Konno and Horikawa among others and in dimension 3 by Supino (see [15]). We deal with this topic in Section 4. We will mention here some of the interesting consequences of this study. We observe the fact that working with a morphism induced by a canonical subseries imposes strong restrictions on the parity of the dimension of X and Y and of the degree of Y. These results of ours hold for varieties of general type which are more general than those considered in Section 2 and Section 3. For instance, as a consequence of Theorem 4.2, we obtain this result which holds for canonical morphisms of arbitrary degree and for any scroll Y not necessarily fibered over $\mathbf{P}^1$:

(1) An even dimensional variety of general type does not admit a generically finite canonical morphism of odd degree to a scroll.

When the canonical morphism is generically finite of degree 3, we go further in our study and obtain Theorem 4.1 and Theorem 4.3. We mention below some of the consequences of these theorems:

(2) We show that there is a relation between the dimension m of X and the degree r of Y. More precisely, we see that the condition of m being odd forces r to be even (see also [15], Remark 3.1 when Y is a rational normal scroll). In particular there are no odd dimensional varieties of general type which are triple canonical covers of linear $\mathbf{P}^m$.

With the additional hypothesis of flatness on the canonical morphism, one obtains the following stronger results:

(3) A variety of general type which is a canonical triple cover of a variety of minimal degree is pluriregular (see Theorem 4.1).

(4) All triple flat canonical covers $X \xrightarrow{\pi} Y$ of varieties of minimal degree are trisections of a line bundle on Y (see Theorem 4.3).

(5) We determine precisely what are the varieties of minimal degree which occur as targets of canonical morphisms of degree 3. (see Theorem 4.3).

Moreover, the pluriregular varieties of general type mentioned in Theorem 4.3 illustrate the theorems that we prove in this paper.

We will say now a few words on the techniques used to prove the results of Sections 2 and 3. These techniques involve the study of the $\mathcal{O}_Y$-algebra structure on $\pi_*\mathcal{O}_X$ to find the multiplicative structure of the canonical ring of X. Even though we reduce the problem from a complicated variety to a simpler variety, as a variety of minimal degree Y appear to be when compared to the variety X, there are natural difficulties that arise in the process. The proof of Theorem 2.7 and Theorem 3.2 involves the study of multiplication of global sections of line bundles on the m-dimensional variety of general type X. To do so we relate the problem of multiplication of global sections of line bundles on X to that of multiplication of global sections of line bundles on a curve C which is an $m-1$ complete intersection on X. We achieve this through an inductive process that we develop in Lemma 2.3. An impression that things become simple on curves is misleading since by a process such as this one obtains bundles of small degree on C which are difficult to handle in the context of multiplication maps. To circumvent this we transfer the problem of multiplication maps on C to a rational normal curve obtained as $m-1$ complete intersection on Y by means of pushing down from C by π. We have traded one difficulty, that of handling small degree bundles on the curve C of X, to another, namely, the study of certain maps on $\mathbf{P}^1$. These maps involve global sections of vector bundles on $\mathbf{P}^1$ but are not multiplication maps of global sections. The key ingredient to this approach, which allows us to settle the problem, is to use the algebra structure of $\mathcal{O}_C$ over $\mathcal{O}_{\mathbf{P}^1}$ to interpret the maps appearing on $\mathbf{P}^1$ and to see their relation with the multiplication maps on C. Note finally that, unlike double covers, a triple cover is not determined by the branch locus and the algebra structure it induces on $\mathcal{O}_X$ might be quite complicated, as the work of Miranda in [14] shows. This is another difficulty one encounters but we are able to overcome it in the context of this paper.

ACKNOWLEDGEMENTS: We thank N. Mohan Kumar for useful discussions regarding this article which led in particular to an improved exposition of the proof of Theorem 4.3.

2. The canonical ring of triple covers of varieties of minimal degree (odd dimensional case)

Convention. Throughout this article the base field $\mathbf{k}$ has characteristic 0 and $\overline{\mathbf{k}} = \mathbf{k}$. By variety we will mean an irreducible projective variety over $\mathbf{k}$.

The purpose of this section is to study the canonical ring of an odd dimensional, pluriregular variety X of general type with base-point-free canonical bundle K_X and such that its canonical morphism maps

X generically $3:1$ onto a variety of minimal degree. We start with some definitions and notations.

Definition. A variety X of dimension m is said to be *pluriregular* if
$H^1(\mathcal{O}_X) = \cdots = H^{m-1}(\mathcal{O}_X) = 0$.

Notation 2.1. Let X be a smooth variety of general type of dimension m and let K_X be its canonical bundle. Assume that K_X is base-point-free. Let $X_1 \subset \cdots \subset X_{m'} \subset \cdots X_m = X$ be smooth irreducible subsequent m'-dimensional, complete intersections of members of $|K_X|$. We will also denote X_1 by C. We denote $K_X|_{X_{m'}}$ by $L_{m'}$ (in particular $L_m = K_X$) and we will also denote L_1 by θ. Finally let

$$H^0(X_{m'}, L_{m'}^{\otimes s}) \otimes H^0(X_{m'}, L_{m'}^{\otimes t}) \xrightarrow{\alpha(s,t;m')} H^0(X_{m'}, L_{m'}^{\otimes s+t})$$

be the usual multiplication map of global sections.

We will use also some abridged notation:
$\alpha(s,t;m) = \alpha(s,t)$,
$\alpha(s,t;1) = \beta(s,t)$,
$\alpha(s,1;m') = \alpha(s;m')$
$\alpha(s,1;m) = \alpha_s$ and
$\alpha(s,1;1) = \beta_s$.

The idea of the proofs of Theorem 1 is to relate the computation of generators of the canonical ring to the knowledge of the images of certain maps of multiplication of global sections of line bundles on X. We will find out the image of these maps by studying analogous maps of multiplication of global sections of line bundles on the varieties $X_{m'}$ defined above, and eventually, by studying maps on the curve C. In order to realize the link between the maps on $X_{m'}$ and the maps on X_m we need a couple of lemmas.

Lemma 2.2. *Let X be a smooth, pluriregular variety of general type of dimension m, such that its canonical bundle K_X is base-point-free. Then:*
1) $H^b(L_{m'}^{\otimes a}) = 0$ for all $2 \leq m' \leq m$, $1 \leq b \leq m'-1$
2) The natural maps of restriction of global sections

$$H^0(L_{m'}^{\otimes n}) \longrightarrow H^0(L_{m'-i}^{\otimes n})$$

surject for all $2 \leq m' \leq m$, all $1 \leq i \leq m'-1$ and all $n \geq 1$. In particular, the maps

$$H^0(L_{m'}^{\otimes n}) \longrightarrow H^0(L_{m'-1}^{\otimes n}) \text{ and}$$
$$H^0(K_X^{\otimes n}) \longrightarrow H^0(\theta^{\otimes n})$$

surject for all $n \geq 1$.

Proof. If $m' = m$, 1) follows from hypothesis, Kawamata-Viehweg vanishing and Serre duality. Arguing by induction on the codimension and using the long exact sequence of cohomology which arises from

$$0 \longrightarrow L_{m'}^{\otimes n-1} \longrightarrow L_{m'}^{\otimes n} \longrightarrow L_{m'-1}^{\otimes n} \longrightarrow 0$$

we prove 1) for all $2 \leq m' \leq m$. To prove 2) it is enough to show the surjectivity of

$$H^0(L_{m'}^{\otimes n}) \xrightarrow{\mu} H^0(L_{m'-1}^{\otimes n})$$

for all $2 \leq m' \leq m$ and $n \geq 1$, since the maps

$$H^0(L_{m'}^{\otimes n}) \longrightarrow H^0(L_{m'-i}^{\otimes n})$$

are composite of maps like the previous one. But the surjectivity of μ follows at once from the vanishing of $H^1(L_{m'}^{\otimes n-1})$, which follows from 1). □

Lemma 2.3. *Let X be a smooth, pluriregular variety of general type of dimension m, such that its canonical bundle K_X is base-point-free.*
1) Let $s_1, s_2 \geq 1$. If

$$H^0(K_X^{\otimes s_1}) \otimes H^0(K_X^{\otimes s_2}) \xrightarrow{\alpha(s_1,s_2;m)} H^0(K_X^{\otimes s_1+s_2}) \text{ surjects, so does}$$
$$H^0(\theta^{\otimes s_1}) \otimes H^0(\theta^{\otimes s_2}) \xrightarrow{\alpha(s_1,s_2;1)} H^0(\theta^{\otimes s_1+s_2})\ .$$

2) Given $s_1, s_2 \geq 1$, assume that the maps

$$H^0(\theta^{\otimes s'}) \otimes H^0(\theta^{\otimes s_2}) \xrightarrow{\alpha(s',s_2;1)} H^0(\theta^{\otimes s'+s_2})$$

surject for all $1 \leq s' \leq s_1$. Then the maps

$$H^0(L_{m'}^{\otimes s'}) \otimes H^0(L_{m'}^{\otimes s_2}) \xrightarrow{\alpha(s',s_2;m')} H^0(L_{m'}^{\otimes s'+s_2})$$

surject for all $1 \leq s' \leq s_1$ *and all* $1 \leq m' \leq m$.

3) *Given* $s_1, s_2 \geq 1$, *assume that the maps*

$$H^0(\theta^{\otimes s'}) \otimes H^0(\theta^{\otimes s_2}) \xrightarrow{\alpha(s',s_2;1)} H^0(\theta^{\otimes s'+s_2})$$

surject for all $1 \leq s' \leq s_1 - 1$. *Then the image of the map*

$$H^0(L_{m'}^{\otimes s_1}) \otimes H^0(L_{m'}^{\otimes s_2}) \xrightarrow{\alpha(s_1,s_2;m')} H^0(L_{m'}^{\otimes s_1+s_2})$$

is the inverse image of the image of map

$$H^0(\theta^{\otimes s_1}) \otimes H^0(\theta^{\otimes s_2}) \xrightarrow{\alpha(s_1,s_2;1)} H^0(\theta^{\otimes s_1+s_2})$$

by the obvious restriction map

$$H^0(L_{m'}^{\otimes s_1+s_2}) \longrightarrow H^0(\theta^{\otimes s_1+s_2}) .$$

Moreover, the codimension of $\alpha(s_1, s_2; m')$ *in* $H^0(L_{m'}^{\otimes s_1+s_2})$ *is the same for all* $1 \leq m' \leq m$.

Proof. First we prove 1). Consider the following commutative diagram:

$$\begin{array}{ccc} H^0(K_X^{\otimes s_1}) \otimes H^0(K_X^{\otimes s_2}) & \longrightarrow & H^0(\theta^{\otimes s_1}) \otimes H^0(\theta^{\otimes s_2}) \\ \downarrow & & \downarrow \\ H^0(K_X^{\otimes s_1+s_2}) & \twoheadrightarrow & H^0(\theta^{\otimes s_1+s_2}) , \end{array}$$

where the horizontal arrows are induced by restricting global sections from X to C. The surjectivity of the bottom horizontal arrow follows from Lemma 2.2. Then it is clear that the surjectivity of the left hand side vertical arrow implies the surjectivity of right hand side vertical arrow.

Now we prove 2). Assume the maps

$$H^0(\theta^{\otimes s'}) \otimes H^0(\theta^{\otimes s_2}) \xrightarrow{\alpha(s',s_2;1)} H^0(\theta^{\otimes s'+s_2})$$

surject for all $1 \leq s' \leq s_1$. We are going to show the surjectivity of $\alpha(s', s_2; m')$ for all $1 \leq m' \leq m$ and all $0 \leq s' \leq s$. The proof is by induction on both on m'. If $m' = 1$, the result is our hypothesis, the case $s' = 0$ being obvious. Now we assume the result true for $m' - 1$ and we will prove it for m' by induction on s'. If $s' = 0$, the result is again obvious. Now we assume the result to be true for $s' - 1$ and we will prove it for s'. Consider the following commutative diagram:

$$\begin{array}{ccccc}
\begin{array}{c} H^0(L_{m'}^{\otimes s'-1})\otimes \\ H^0(L_{m'}^{\otimes s_2}) \end{array} & \hookrightarrow & \begin{array}{c} H^0(L_{m'}^{\otimes s'})\otimes \\ H^0(L_{m'}^{\otimes s_2}) \end{array} & \twoheadrightarrow & \begin{array}{c} H^0(L_{m'-1}^{\otimes s'})\otimes \\ H^0(L_{m'}^{\otimes s_2}) \end{array} \\
\downarrow & & \downarrow & & \downarrow \\
H^0(L_{m'}^{\otimes s'+s_2-1}) & \hookrightarrow & H^0(L_{m'}^{\otimes s'+s_2}) & \twoheadrightarrow & H^0(L_{m'-1}^{\otimes s'+s_2})\ .
\end{array}$$

The last top and bottom horizonta l arrows are surjective by Lemma 2.2. The left hand side vertical arrow is $\alpha(s'-1, s_2; m')$ and surjects by induction hypothesis on s'. The right hand side vertical arrow is the composition of the map

$$H^0(L_{m'-1}^{\otimes s'}) \otimes H^0(L_{m'}^{\otimes s_2}) \longrightarrow H^0(L_{m'-1}^{\otimes s'}) \otimes H^0(L_{m'-1}^{\otimes s_2})\ ,$$

whose surjectivity follows from Lemma 2.2, and the map of multiplication of global sections on $X_{m'-1}$, $\alpha(s', s_2; m'-1)$, which is surjective by induction hypothesis on m'. Then it follows from chasing the diagram that the middle vertical arrow $\alpha(s', s_2; m')$ surjects.

Finally we prove 3) by induction on m'. If $m' = 1$, there is nothing to prove. Now we assume the result true for $m' - 1$ with $m' \geq 2$ and we will prove it for m'. Consider the following commutative diagram:

$$\begin{array}{ccccc}
\begin{array}{c} H^0(L_{m'}^{\otimes s_1-1})\otimes \\ H^0(L_{m'}^{\otimes s_2}) \end{array} & \hookrightarrow & \begin{array}{c} H^0(L_{m'}^{\otimes s_1})\otimes \\ H^0(L_{m'}^{\otimes s_2}) \end{array} & \twoheadrightarrow & \begin{array}{c} H^0(L_{m'-1}^{\otimes s_1})\otimes \\ H^0(L_{m'}^{\otimes s_2}) \end{array} \\
\downarrow & & \downarrow & & \downarrow \\
H^0(L_{m'}^{\otimes s_1+s_2-1}) & \hookrightarrow & H^0(L_{m'}^{\otimes s_1+s_2}) & \twoheadrightarrow & H^0(L_{m'-1}^{\otimes s_1+s_2}) \\
\downarrow & & \downarrow & & \downarrow \\
0 & \longrightarrow & W_{m'} & = & W_{m'-1}\ .
\end{array}$$

The last top and bottom horizontal arrows are surjective by Lemma 2.2. The left hand side vertical arrow is $\alpha(s_1 - 1, s_2; m')$. This map surjects by part 2) of this lemma. The right hand side first vertical arrow is the composition of the map

$$H^0(L_{m'-1}^{\otimes s_1}) \otimes H^0(L_{m'}^{\otimes s_2}) \longrightarrow H^0(L_{m'-1}^{\otimes s_1}) \otimes H^0(L_{m'-1}^{\otimes s_2})\ ,$$

and the map of multiplication of global sections on $X_{m'-1}$, $\alpha(s_1, s_2; m'-1)$. The first of the two above mentioned maps is surjective by Lemma 2.2. Therefore the image of the right hand side first vertical arrow is equal to the image of $\alpha(s_1, s_2; m'-1)$. By induction hypothesis, the image of $\alpha(s_1, s_2; m'-1)$ is the inverse image of the image of $\alpha(s_1, s_2; 1)$ by the natural map of restriction of global sections

$$H^0(X_{m'-1}, L_{m'-1}^{\otimes s_1+s_2}) \longrightarrow H^0(C, \theta^{\otimes s_1+s_2}) .$$

By chasing the diagram it follows that the image of $\alpha(s_1, s_2, m')$ is the inverse image of the image of $\alpha(s_1, s_2, m'-1)$ by the restriction map

$$H^0(X_{m'}, L_{m'}^{\otimes s_1+s_2}) \longrightarrow H^0(X_{m'-1}, L_{m'-1}^{\otimes s_1+s_2}) .$$

So finally the image of $\alpha(s_1, s_2; m')$ is the inverse image of the image of $\alpha(s_1, s_2; 1)$ by the restriction map

$$H^0(X_{m'}, L_{m'}^{\otimes s_1+s_2}) \longrightarrow H^0(C, \theta^{\otimes s_1+s_2}) .$$

Note finally that the isomorphism between $W_{m'}$ and $W_{m'-1}$ at each step of the inductive process tells us that the codimensions of the images of $\alpha(s_1, s_2; m')$ in $H^0(X_{m'}, L_{m'}^{\otimes s_1+s_2})$ are all equal for all $1 \leq m' \leq m$. $\square$

As we have previously mentioned in order to prove Theorem 1 we need to study certain maps of multiplication of global sections of line bundles on the curve C defined in Notation 2.1. We will prove Theorem 1 first when the dimension of m is odd. On the curve C we need the following

Proposition 2.4. *Let C be a smooth curve, let θ be an ample and base-point-free line bundle on C and let m be an odd natural number such that $\theta^{\otimes m} = K_C$. Let $\varphi : C \longrightarrow Y$ be the morphism induced by a subseries W' of $|\theta|$ without base points, and assume that the degree of φ is 3 and that Y is a (smooth) rational normal curve of degree r. Then*

1) If $m \geq 3$, then φ is in fact induced by the complete linear series of θ.

Let

$$H^0(\theta^{\otimes s}) \otimes H^0(\theta^{\otimes t}) \xrightarrow{\beta(s,t)} H^0(\theta^{\otimes s+t})$$

be the usual map of multiplication of global sections and let $\beta_s = \beta(s, 1)$.

2.1) *The map* β_n *in* $H^0(\theta^{\otimes n+1})$ *surjects if and only if* $n \neq \frac{m-1}{2}, m$;
2.2) *The map* $\beta(\frac{m+1}{2}, l)$ *surjects in* $H^0(\theta^{\otimes \frac{m+1}{2}+l})$ *if* $0 \leq l \leq \frac{m+1}{2}$.
2.3) *The image of* $\beta(s_1, s_2)$ *is the same subspace* U' *of* $H^0(\theta^{\otimes \frac{m+1}{2}})$ *for all* s_1, s_2 *such that* $s_1 + s_2 = \frac{m+1}{2}, s_1, s_2 \geq 1$. *The codimension of* U' *is* $\frac{r}{2}$.
2.4) *The map* $\beta(s_1, s_2)$ *surjects if* $s_1, s_2 \geq 0$ *and* $s_1 + s_2 \leq \frac{m-1}{2}$.

3) *The ring* $R_\theta = \bigoplus_{s=1}^{\infty}(R_\theta)_s$, *where* $(R_\theta)_s = H^0(\theta^{\otimes s})$, *is generated by its part of degree* 1 *and by* $\frac{r}{2}$ *generators in degree* $\frac{m+1}{2}$.

Before proving Proposition 2.4 we need the following result:

Lemma 2.5. *Let* C *be a smooth curve, let* θ *be an ample and base-point-free line bundle on* C *and let* m *be a natural number such that* $\theta^{\otimes m} = K_C$. *Let* $\varphi : C \longrightarrow Y$ *be the morphism induced by a subseries* W' *of* $|\theta|$ *without base points, and assume that the degree of* φ *is* 3 *and that* Y *is a (smooth) rational normal curve of degree* r. *Then*

$$\varphi_*(\mathcal{O}_C) = \mathcal{O}_{\mathbf{P}^1} \oplus \mathcal{O}_{\mathbf{P}^1}(-\frac{mr+2}{2}) \oplus \mathcal{O}_{\mathbf{P}^1}(-mr-2)) .$$

In particular, if m *is odd, then* r *is even.*

Proof. The lemma is an easy consequence of relative duality. For further details, see [4], Theorem 2.1. □

(2.6) *Proof of Proposition 2.4.* We prove 1) first. It follows from Lemma 2.5 and projection formula that

$$H^0(\theta) = H^0(\varphi_*(\theta)) = H^0(\mathcal{O}_{\mathbf{P}^1}(r))$$
$$\oplus \mathcal{O}_{\mathbf{P}^1}(\frac{(2-m)r-2}{2}) \oplus \mathcal{O}_{\mathbf{P}^1}((1-m)r-2))) .$$

Then $H^0(\varphi_*(\theta)) = H^0(\mathcal{O}_{\mathbf{P}^1}(r))$, for $m \geq 3$. Therefore φ is induced by the complete $|\theta|$.

We have seen that $\varphi_*(\mathcal{O}_C)$ and $\mathcal{O}_{\mathbf{P}^1} \oplus \mathcal{O}_{\mathbf{P}^1}(-\frac{mr+2}{2}) \oplus \mathcal{O}_{\mathbf{P}^1}(-mr-2))$ are isomorphic as sheaves of modules over $\mathcal{O}_{\mathbf{P}^1}$. We call $E = \mathcal{O}_{\mathbf{P}^1}(-\frac{mr+2}{2}) \oplus \mathcal{O}_{\mathbf{P}^1}(-mr-2))$. Since $\varphi_*(\mathcal{O}_C)$ is also a sheaf of algebras over $\mathcal{O}_{\mathbf{P}^1}$, then the algebra structure of $\varphi_*(\mathcal{O}_C)$ makes $\mathcal{O}_{\mathbf{P}^1} \oplus \mathcal{O}_{\mathbf{P}^1}(-\frac{mr+2}{2}) \oplus \mathcal{O}_{\mathbf{P}^1}(-mr-2))$ into a sheaf of $\mathcal{O}_{\mathbf{P}^1}$-algebras whose multiplication decomposes in the following way (see [14], [11] or [5]): The map

$$\mathcal{O}_{\mathbf{P}^1} \otimes \mathcal{O}_{\mathbf{P}^1} \longrightarrow \mathcal{O}_{\mathbf{P}^1}$$

is the ring multiplication in $\mathcal{O}_{\mathbf{P}^1}$. The maps

$$\begin{array}{rcl} \mathcal{O}_{\mathbf{P}^1} \otimes \mathcal{O}_{\mathbf{P}^1}(-\frac{mr+2}{2}) & \longrightarrow & \mathcal{O}_{\mathbf{P}^1}(-\frac{mr+2}{2}) \\ \mathcal{O}_{\mathbf{P}^1} \otimes \mathcal{O}_{\mathbf{P}^1}(-mr-2) & \longrightarrow & \mathcal{O}_{\mathbf{P}^1}(-mr+2) \\ \mathcal{O}_{\mathbf{P}^1}(-\frac{mr+2}{2}) \otimes \mathcal{O}_{\mathbf{P}^1} & \longrightarrow & \mathcal{O}_{\mathbf{P}^1}(-\frac{mr+2}{2}) \qquad \text{and} \\ \mathcal{O}_{\mathbf{P}^1}(-mr-2) \otimes \mathcal{O}_{\mathbf{P}^1} & \longrightarrow & \mathcal{O}_{\mathbf{P}^1}(-mr+2) \end{array}$$

are the left and right module multiplication of $\mathcal{O}_{\mathbf{P}^1}(-\frac{mr+2}{2})$ and $\mathcal{O}_{\mathbf{P}^1}(-mr+2)$. Finally there is a map

$$E \otimes E \longrightarrow \varphi_* \mathcal{O}_C$$

whose structure depends on the cover φ itself. We will use all this to prove 2). Since $\theta = \varphi^* \mathcal{O}_{\mathbf{P}^1}(r)$, it follows from Lemma 2.5 and projection formula that

$$H^0(\theta^{\otimes n}) = H^0(\mathcal{O}_{\mathbf{P}^1}(nr)) \oplus H^0(\mathcal{O}_{\mathbf{P}^1}(\frac{(2n-m)r-2)}{2})) \\ \oplus H^0(\mathcal{O}_{\mathbf{P}^1}((n-m)r-2)).$$

Let us denote

$$\begin{array}{rcl} A(n) & = & H^0(\mathcal{O}_{\mathbf{P}^1}(nr)) \\ B(n) & = & H^0(\mathcal{O}_{\mathbf{P}^1}(\frac{(2n-m)r-2)}{2})) \\ C(n) & = & H^0(\mathcal{O}_{\mathbf{P}^1}((n-m)r-2)) \end{array} .$$

The way in which global sections of $H^0(\theta^{\otimes s_1})$ and global sections of $H^0(\theta^{\otimes s_2})$ multiply is governed by the ring multiplication in

$$\varphi_* \mathcal{O}_C = \mathcal{O}_{\mathbf{P}^1} \oplus \mathcal{O}_{\mathbf{P}^1}(-\frac{mr+2}{2}) \oplus \mathcal{O}_{\mathbf{P}^1}(-mr-2)) .$$

Thus the map $\beta(s_1, s_2)$ splits as direct sum of the following maps

$$\begin{gathered} A(s_1) \otimes A(s_2) \longrightarrow A(s_1+s_2) \\ A(s_1) \otimes B(s_2) \longrightarrow B(s_1+s_2) \\ A(s_1) \otimes C(s_2) \longrightarrow C(s_1+s_2) \\ B(s_1) \otimes A(s_2) \longrightarrow B(s_1+s_2) \\ C(s_1) \otimes A(s_2) \longrightarrow C(s_1+s_2) \\ [B(s_1) \oplus C(s_1)] \otimes [B(s_2) \oplus C(s_2)] \longrightarrow \\ A(s_1+s_2) \oplus B(s_1+s_2) \oplus C(s_1+s_2)(2.6.1). \end{gathered}$$

The first map $A(s_1) \otimes A(s_2) \longrightarrow A(s_1 + s_2)$ is induced by the ring multiplication in $\mathcal{O}_{\mathbf{P}^1}$ and it is therefore a map of multiplication of global sections of line bundles on $\mathbf{P}^1$. The maps second to fifth are induced by the left and right module structure of E over $\mathcal{O}_{\mathbf{P}^1}$ and are also multiplcation maps of global sections of line bundles on $\mathbf{P}^1$. The structure of the last map is more complex and depends on the structure of φ. Thus the image of $\beta(s_1, s_2)$ will be the sum of the images of those six maps. We assume $s_1, s_2 \geq 0$. Then the first map is always nonzero and surjective, hence $A(s_1+s_2)$ is always contained in the image of $\beta(s_1, s_2)$. On the other hand the vanishing of the groups $B(s_1), B(s_2), B(s_1 + s_2)$ and $C(s_1), C(s_2), C(s_1 + s_2)$ depends on the values of s_1 and s_2. If $B(s_1)$ does not vanish, that is, if $s_1 \geq \frac{m+1}{2}$, then $B(s_1 + s_2)$ does not vanish either and the map

$$B(s_1) \otimes A(s_2) \longrightarrow B(s_1 + s_2)$$

is surjective. Therefore in this case $B(s_1+s_2)$ is contained also in the image of $\beta(s_1, s_2)$. If $C(s_1)$ does not vanish, that is, if $s_1 \geq m + 1$, then $C(s_1 + s_2)$ does not vanish either and the map

$$C(s_1) \otimes A(s_2) \longrightarrow C(s_1 + s_2)$$

is surjective. Therefore in this case $C(s_1 + s_2)$ is contained in the image of $\beta(s_1, s_2)$. There are analogous statements for $B(s_2)$ and $C(s_2)$.

With this information we prove now 2.1), 2.2), 2.3) and 2.4). We prove first 2.1). The multiplication map $\beta_n = \beta(n, 1)$ splits as shown in (2.6.1). If $0 \leq n < \frac{m-1}{2}$, $B(n), C(n), B(n + 1)$ and $C(n + 1)$ all vanish, therefore $H^0(\theta^{\otimes n+1}) = A(n + 1)$, and the map

$$H^0(\theta^{\otimes n}) \otimes H^0(\theta) \xrightarrow{\beta_n} H^0(\theta^{\otimes n+1})$$

is the map

$$A(n) \otimes A(1) \longrightarrow A(n + 1) \ ,$$

which is surjective. On the other hand, $B(\frac{m-1}{2}) = C(\frac{m-1}{2}) = C(\frac{m+1}{2}) = 0$ but $B(\frac{m+1}{2})$ has dimension $\frac{r}{2} > 0$. Then $H^0(\theta) = A(1)$, $H^0(\theta^{\otimes \frac{m-1}{2}}) = A(\frac{m-1}{2})$ and $H^0(\theta^{\otimes \frac{m+1}{2}}) = A(\frac{m+1}{2}) \oplus B(\frac{m+1}{2})$. Then

$$H^0(\theta^{\otimes \frac{m-1}{2}}) \otimes H^0(\theta) \xrightarrow{\beta_{\frac{m-1}{2}}} H^0(\theta^{\otimes \frac{m+1}{2}})$$

is the map

$$A(\frac{m-1}{2}) \otimes A(1) \longrightarrow A(\frac{m+1}{2}) ,$$

whose image has codimension $\frac{r}{2}$ in $H^0(\theta^{\otimes \frac{m+1}{2}})$. If $\frac{m+1}{2} \leq n \leq m-1$, then $H^0(\theta^{\otimes n}) = A(n) \oplus B(n)$ and $H^0(\theta^{\otimes n+1}) = A(n+1) \oplus B(n+1)$, with $A(n), B(n)$, $A(n+1)$ and $B(n+1)$ all different from 0. Then according to ((2.6.1)) if $\frac{m+1}{2} \leq n \leq m-1$ the map β_n decomposes as direct sum of

$$\begin{aligned} A(n) \otimes A(1) &\longrightarrow A(n+1) \text{ and} \\ B(n) \otimes A(1) &\longrightarrow B(n+1) , \end{aligned}$$

which are surjective maps, therefore the image of β_n is $A(n+1) \oplus B(n+1) = H^0(\theta^{\otimes n+1})$. On the other hand $C(m) = 0$ but $C(m+1)$ has dimension $r-1 > 0$. Thus $H^0(\theta^{\otimes m}) = A(m) \oplus B(m)$ but

$$H^0(\theta^{\otimes m+1}) = A(m+1) \oplus B(m+1) \oplus C(m+1) .$$

The map β_m decomposes as direct sum of

$$\begin{aligned} A(m) \otimes A(1) &\longrightarrow A(m+1) \text{ and} \\ B(m) \otimes A(1) &\longrightarrow B(m+1) , \end{aligned}$$

and then its image is $A(m+1) \oplus B(m+1)$, which has codimension $r-1$ in $H^0(\theta^{\otimes m+1})$. Finally if $n \geq m+1$, $H^0(\theta^n) = A(n) \oplus B(n) \oplus C(n)$, and $A(n), B(n)$ and $C(n)$ are all nonzero. Then β_n decomposes as direct sum of

$$\begin{aligned} A(n) \otimes A(1) &\longrightarrow A(n+1) \\ B(n) \otimes A(1) &\longrightarrow B(n+1) \\ C(n) \otimes A(1) &\longrightarrow C(n+1) , \end{aligned}$$

and its image is $A(n+1) \oplus B(n+1) \oplus C(n+1) = H^0(\theta^{n+1})$. This completes the proof of 2.1).

Now we prove 2.2). Recall that if $\frac{m+1}{2} \leq n \leq m$, then $H^0(\theta^{\otimes n}) = A(n) \oplus B(n)$ and that, if $0 \leq l \leq \frac{m-1}{2}$, then $H^0(\theta^{\otimes l}) = A(l)$. As we argued in the proof of 2.1, the map $\beta(\frac{m+1}{2}, l)$ is governed by the algebra structure of $\varphi_* \mathcal{O}_C$. Then, if $0 \leq l \leq \frac{m-1}{2}$, $\beta(\frac{m+1}{2}, l)$ decomposes as direct sum of

$$\begin{aligned} A(\frac{m+1}{2}) \otimes A(l) &\longrightarrow A(\frac{m+1}{2} + l) \\ B(\frac{m+1}{2}) \otimes A(l) &\longrightarrow B(\frac{m+1}{2} + l) . \end{aligned}$$

Then the image of $\beta(\frac{m+1}{2}, l)$ is $A(\frac{m+1}{2}+l) \oplus B(\frac{m+1}{2}+l)$ $= H^0(\theta^{\otimes \frac{m+1}{2}+l})$.

On the other hand, $H^0(\theta^{\otimes \frac{m+1}{2}}) = A(\frac{m+1}{2}) \oplus B(\frac{m+1}{2})$. Therefore the map $\beta(\frac{m+1}{2}, \frac{m+1}{2})$ decomposes as direct sum of

$$A(\frac{m+1}{2}) \otimes A(\frac{m+1}{2}) \longrightarrow A(m+1)$$
$$B(\frac{m+1}{2}) \otimes A(\frac{m+1}{2}) \longrightarrow B(m+1)$$
$$A(\frac{m+1}{2}) \otimes B(\frac{m+1}{2}) \longrightarrow B(m+1)$$
$$B(\frac{m+1}{2}) \otimes B(\frac{m+1}{2}) \xrightarrow{\eta}$$
$$A(m+1) \oplus B(m+1) \oplus C(m+1) \quad (2.6.2).$$

Since the first three maps are surjective We know that $A(m+1) \oplus B(m+1)$ is contained in the image of $\beta(\frac{m+1}{2}, \frac{m+1}{2})$. Now we need to look more carefully at the last one of the above maps. For this we have to study more closely the ring structure of $\varphi_*(\mathcal{O}_C)$. Let

$$\varphi_*(\mathcal{O}_C) \otimes \varphi_*(\mathcal{O}_C) \xrightarrow{\mu} \varphi_*(\mathcal{O}_C)$$

be the map defined by the ring multiplication in $\varphi_*(\mathcal{O}_C)$. Then the projection to $\mathcal{O}_{\mathbf{P}^1}(-mr-2)$ of the restriction of μ to $\mathcal{O}_{\mathbf{P}^1}(\frac{-mr-2}{2}) \otimes \mathcal{O}_{\mathbf{P}^1}(\frac{-mr-2}{2}))$ is nonzero. Otherwise there would be an integral $\mathcal{O}_{\mathbf{P}^1}$-subalgebra of $\varphi_*(\mathcal{O}_C)$ of rank 2, namely $\mathcal{O}_{\mathbf{P}^1} \oplus \mathcal{O}_{\mathbf{P}^1}(\frac{-mr-2}{2})$. This is impossible because the rank of an integral subalgebra of $\varphi_*\mathcal{O}_C$ must divide the rank of $\varphi_*\mathcal{O}_C$. Now if the projection to $\mathcal{O}_{\mathbf{P}^1}(-mr-2)$ of the restriction of μ to $\mathcal{O}_{\mathbf{P}^1}(\frac{-mr-2}{2}) \otimes \mathcal{O}_{\mathbf{P}^1}(\frac{-mr-2}{2}))$ is nonzero, it is an isomorphism. Then the projection of η to $C(m+1)$ is surjective. Since the image of the three first maps in ((2.6.2)) is $A(m+1) \oplus B(m+1)$ we may conclude that the image of $\beta(\frac{m+1}{2}, \frac{m+1}{2})$ is

$$A(m+1) \oplus B(m+1) \oplus C(m+1) = H^0(\theta^{\otimes m+1}) .$$

We proceed now to prove 2.3). Let s_1, s_2 satisfy $s_1 + s_2 = \frac{m+1}{2}$ and $s_1, s_2 \geq 1$. Recall that

$$H^0(\theta^{\otimes \frac{m+1}{2}}) = A(\frac{m+1}{2}) \oplus B(\frac{m+1}{2}) .$$

However, if $s_1 + s_2 = \frac{m+1}{2}$ and $s_1, s_2 \geq 1$, then $s_1, s_2 \leq \frac{m-1}{2}$, hence $H^0(\theta^{\otimes s_1}) = A(s_1)$ and $H^0(\theta^{\otimes s_2}) = A(s_2)$. Thus the map $\beta(s_1, s_2)$ is

$$A(s_1) \otimes A(s_2) \longrightarrow A(\frac{m+1}{2}) .$$

Thus the image of $\beta(s_1, s_2)$ is, for all $s_1 + s_2 = \frac{m+1}{2}$, $s_1, s_2 \geq 1$, equal to $A(\frac{m+1}{2}) = H^0(\mathcal{O}_{\mathbf{P}^1}(\frac{m+1}{2}r))$ which has codimension $\frac{r}{2}$ in $H^0(\theta^{\otimes \frac{m+1}{2}})$.

Finally we prove 2.4). If $0 \leq s \leq \frac{m-1}{2}$ then $H^0(\theta^{\otimes s}) = A(s)$. Then if $s_1, s_2 \geq 0$ and $s_1 + s_2 \leq \frac{m-1}{2}$, the map $\beta(s_1, s_2)$ is the multiplication map

$$A(s_1) \otimes A(s_2) \longrightarrow A(s_1 + s_2)$$

which is surjective.

Now we prove 3). Note that the vector subspace of $(R_\theta)_n$ generated by the elements of degree smaller than n is the sum of the images of the maps $\beta(s_1, s_2)$ with $s_1, s_2 \geq 1$ and $s_1 + s_2 = n$. We saw in 2.1) that the maps β_l surject if $l < \frac{m-1}{2}$. Then the parts of R_θ of degree less than $\frac{m+1}{2}$ are generated by the elements of degree 1. 2.3) tells us that the elements of degree less than $\frac{m+1}{2}$ generates a subspace of $(R_\theta)_{\frac{m+1}{2}}$ of codimension $\frac{r}{2}$, henceforth we need $\frac{r}{2}$ linearly independent elements of degree $\frac{m+1}{2}$ to generate $(R_\theta)_{\frac{m+1}{2}}$. If $\frac{m+1}{2} \leq l \leq m-1$, we saw also in 2.1) that the maps β_l surject, therefore the parts of degree $\frac{m+1}{2}, \ldots, m$ are generated by the elements of R_θ of degree less than or equal to $\frac{m+1}{2}$. From 2.2) it follows that the map $\beta(\frac{m+1}{2}, \frac{m+1}{2})$ surjects, therefore $(R_\theta)_{m+1}$ is also generated by the elements of $(R_\theta)_{\frac{m+1}{2}}$. Finally, if $l \geq m+1$ we saw in 2.1) that the map β_l surjects, thus the parts of R_θ of degree greater than or equal to $m+2$ are generated by the elements of degree $m+1$. This proves 3). □

Now we can prove Theorem 1 when the dimension of the variety of general type X is odd. Precisely we prove

Theorem 2.7. *Let X be a smooth, pluriregular variety of general type of odd dimension $m \geq 3$ and let $Y \subset \mathbf{P}^{r+m-1}$ be an m-fold of minimal degree r . Assume that K_X is base-point-free and let $\pi : X \longrightarrow Y$ be the canonical morphism of X. Assume that π is* *generically finite morphism of degree* 3. *Then the canonical ring* R *of* X *is generated by its part of degree* 1 *and by* $\frac{r}{2}$ *generators in degree* $\frac{m+1}{2}$.

Proof. Let C be the intersection of $m-1$ general members of $|K_X|$. Then C is a smooth and irreducible curve in X. Let $\theta = K_X|_C$. By adjunction $\theta^{\otimes m} = K_C$. We set $\varphi = \pi|_C$. Since C has been taken general, we may, and actually do, choose C so that φ is finite onto a smooth rational normal curve D of degree r in $\mathbf{P}^r$.

As we have seen in the proof of Proposition 2.4, in order to find out the degrees of the generators of R, we only need to study the images of the maps $\alpha(s,t)$ for all $s,t \geq 1$. In fact, since many of them will happen to be surjective it will suffice, for our purpose, to study only some of them. More precisely we will show the following:

1) $H^0(K_X^{\otimes n}) \otimes H^0(K_X) \xrightarrow{\alpha_n} H^0(K_X^{\otimes n+1})$ surject for all $n \geq 0$, $n \neq \frac{m-1}{2}, m$;
2) $H^0(K_X^{\otimes \frac{m+1}{2}}) \otimes H^0(K_X^{\otimes \frac{m+1}{2}}) \xrightarrow{\alpha(\frac{m+1}{2},\frac{m+1}{2})} H^0(K_X^{\otimes m+1})$ surjects;
3) the images of

$$H^0(K_X^{\otimes s_1}) \otimes H^0(K_X^{\otimes s_2}) \xrightarrow{\alpha(s_1,s_2)} H^0(K_X^{\otimes \frac{m+1}{2}})$$

for all $s_1, s_2 \geq 1, s_1+s_2 = \frac{m+1}{2}$ are equal of codimension $\frac{r}{2}$.

Recall that $\varphi = \pi|_C$ is induced by the complete series of θ and that C, θ and φ satisfy the hypothesis of Proposition 2.4. Now to prove 1), 2) and 3) we will relate the maps appearing in 1), 2) and 3) with the multiplication maps on C studied in Proposition 2.4. To establish the relation between them we will use among other things Lemma 2.3. 1) and 2) follows directly from Lemma 2.3 (2) and Proposition 2.4 (2.1) and 2.4 (2.2). We now prove 3). By Proposition 2.4 (2.4) we know that the maps $\beta(s_1,s_2)$ are surjective if $s_1+s_2 \leq \frac{m-1}{2}$. Let us know fix $s_1, s_2 \geq 1$ such that $s_1+s_2 = \frac{m+1}{2}$. Then by Lemma 2.3 (3), we know that the codimension of the image of $\alpha(s_1,s_2)$ in $H^0(K_X^{\otimes \frac{m+1}{2}})$ is the same as the codimension of the image of $\beta(s_1,s_2)$ in $H^0(\theta^{\otimes \frac{m+1}{2}})$. We also know by Lemma 2.3 (3) that the image of $\alpha(s_1,s_2)$ is the inverse image by the obvious map of restriction of sections

$$H^0(K_X^{\otimes \frac{m+1}{2}}) \longrightarrow H^0(\theta^{\otimes \frac{m+1}{2}})$$

of the image of $\beta(s_1,s_2)$. Finally Proposition 2.4 (2.3) tells the image of $\beta(s_1, s_2)$ in $H^0(\theta^{\otimes \frac{m+1}{2}})$ are the same for all $s_1, s_2 \geq 1$ such that $s_1+s_2 = \frac{m+1}{2}$ and have codimension $\frac{r}{2}$. Then all this implies that the images of $\alpha(s_1, s_2)$ are equal and of codimension $\frac{r}{2}$ in $H^0(K_X^{\otimes \frac{m+1}{2}})$ for all $s_1, s_2 \geq 1$ such that $s_1 + s_2 = \frac{m+1}{2}$.

Now we show how 1), 2) and 3) imply the statement of the theorem. 1) says that the $H^0(K_X^{\otimes n}) \otimes H^0(K_X) \longrightarrow H^0(K_X^{\otimes n+1})$ surjects if $1 \leq n \leq \frac{m-3}{2}$, hence the part of R of degree less than or equal to $\frac{m-1}{2}$ is generated in degree 1. The subspace of $H^0(K_X^{\otimes \frac{m+1}{2}})$ generated in degree less than $\frac{m+1}{2}$ is the sum of the images of $\alpha(s_1, s_2)$, where $s_1, s_2 \geq 1$ and $s_1 + s_2 = \frac{m+1}{2}$. 3) implies that such sum has codimension $\frac{r}{2}$ in $H^0(K_X^{\otimes \frac{m+1}{2}})$, hence in order to generate R we need $\frac{r}{2}$ linearly independent generators in degree $\frac{m+1}{2}$. Now 1) implies that $H^0(K_X^{\otimes n}) \otimes H^0(K_X) \longrightarrow H^0(K_X^{\otimes n+1})$ surjects if $\frac{m+1}{2} \leq n \leq m-1$, hence the parts of R of degree $\frac{m+3}{2}, \ldots, m$ are generated in degree $\frac{m+1}{2}$ or less. 2) proves that $H^0(K_X^{\otimes m+1})$ is generated by elements of degree $\frac{m+1}{2}$ and finally, 1) also proves that the part of R of degree greater than $m+1$ is generated in degree $m+1$ or lower. □

Remark 2.8. *Proposition 2.4 states explicitely that the ring $\bigoplus_{n\geq 0} H^0(C, \theta^{\otimes n})$ has in each degree the same number of generators as the canonical ring of X, $\bigoplus_{n\geq 0} H^0(X, K_X^{\otimes n})$. The way Theorem 2.7 is proved shows that the same is true also for the "intermediate" rings $\bigoplus_{n\geq 0} H^0(X_{m'}, L_{m'}^{\otimes n})$, $2 \leq m' \leq m-1$.*

Knowing the generators of the canonical ring of X tells us about the generators of the pluricanonical rings of X and in particular, about which pluricanonical rings are generated in degree 1. We state the result we obtain in this regard in the next Theorem. Before that we recall the following definition due to M. Green:

Definition 2.9. *Let L be a line bundle on a projective variety X. We say that L satisfies property N_0 if L is very ample and the image of X by the embedding induced by $|L|$ is projectively normal, that is, its homogeneous coordinate ring is integrally closed.*

Theorem 2.10. *Let X be a smooth, pluriregular variety of general type of odd dimension $m \geq 3$ and let $Y \subset \mathbf{P}^{r+m-1}$ be an m-fold of minimal degree r. Assume that K_X is ample and base-point-free and let $\pi : X \longrightarrow Y$ be the canonical morphism of X. Assume that π has* *degree 3. Then the line bundle $K_X^{\otimes n}$ satisfies property N_0 if and only if $n \geq \frac{m+1}{2}$.*

Proof. Let C be as in the proof of Theorem 2.7. We carry out the proof in several steps.

Step 1. $K_X^{\otimes n}$ does not satisfy property N_0 if $n < \frac{m+1}{2}$.

Indeed, if $n < \frac{m+1}{2}$, then

$$H^0(K_X^{\otimes n}) \longrightarrow H^0(K_X^{\otimes n}|_C)$$

surjects by Lemma 2.2 and

$$H^0(K_X^{\otimes n}|_C) = H^0(\theta^{\otimes n}) = H^0(\mathcal{O}_{\mathbf{P}^1}(nr)) \ ,$$

hence the n-pluricanonical morphism of X maps C to a rational normal curve, therefore $K_X^{\otimes n}$ is not very ample and does not satisfy property N_0.

Now to see that $K_X^{\otimes n}$ satisfies property N_0 for given $n \geq \frac{m+1}{2}$, it suffices to prove the surjectivity of the following maps:

$$H^0(K_X^{\otimes ln}) \otimes H^0(K_X^{\otimes n}) \xrightarrow{\alpha(ln,n)} H^0(K_X^{\otimes (l+1)n}), \text{ for all } l \geq 1 \ .$$

Step 2. $K_X^{\otimes n}$ satisfies property N_0 if $n = \frac{m+1}{2}$.

The map $\alpha(\frac{m+1}{2}, \frac{m+1}{2})$ surjects as shown in 2) of the proof of Theorem 2.7. On the other hand, the surjectivity of $\alpha(ln, n)$, $l \geq 2$ and $n = \frac{m+1}{2}$ follows from the surjectivity of $\alpha_{n'}$ for all $n' \geq m+1$, which was also shown in 1) of the proof of Theorem 2.7.

Step 3. $K_X^{\otimes n}$ satisfies property N_0 if $\frac{m+1}{2} < n \leq m$. We first show that $\alpha(n, n)$ surjects. Since $\alpha(\frac{m+1}{2}, \frac{m+1}{2})$ surjects and so do the maps $\alpha_{n'}$ if $n' \geq m+1$, the following multiplication map

$$H^0(K^{\otimes \frac{m+1}{2}}) \otimes H^0(K_X^{\otimes (n-\frac{m+1}{2})}) \otimes H^0(K^{\otimes \frac{m+1}{2}}) \\ \otimes H^0(K_X^{\otimes (n-\frac{m+1}{2})}) \xrightarrow{\gamma} H^0(K_X^{\otimes 2n})$$

also surjects. On the other hand γ factorizes through $\alpha(n, n)$, which is therefore surjective.

Finally the surjectivity of the maps $\alpha(ln, n)$ for all $l \geq 2$ and $\frac{m+1}{2} < n \leq m$ follows from the surjectivity of $\alpha_{n'}$ for all $n' \geq m+1$.

Step 4. $K_X^{\otimes n}$ satisfies property N_0 if $n \geq m+1$. The surjectivity of the maps $\alpha_{n'}$ for all $n' \geq m+1$ implies the surjectivity of the maps $\alpha(ln, n)$ for all $l \geq 1$ and $n \geq m+1$. □

3. The canonical ring of triple covers of varieties of minimal degree (even dimensional case)

In this section we proceed to prove Theorem 1 when the dimension of X is even. As we did in Section 2, we need study first multiplication maps on a curve C, which is a complete intersection on X. Thus we prove a result analogous to Proposition 2.4.

Proposition 3.1. *Let C be a smooth curve, let θ be an ample and base-point-free line bundle on C and let m be an even natural number such that $\theta^{\otimes m} = K_C$. Let $\varphi : C \longrightarrow Y$ be the morphism induced by $|\theta|$ and assume that the degree of φ is 3 and that Y is a (smooth) rational normal curve of degree r. Then*

1) The map φ is induced by the complete linear series of θ.

Let

$$H^0(\theta^{\otimes s}) \otimes H^0(\theta^{\otimes t}) \xrightarrow{\beta(s,t)} H^0(\theta^{\otimes s+t})$$

be the usual map of multiplication of global sections and let $\beta_s = \beta(s,1)$.

2.1) The map β_n surjects if $n \neq \frac{m}{2}, m, m+1$; if $n = m+1$ and $r \geq 2$; and if $n = m$ and $r = 1$.

2.2) The images of the maps $\beta(s_1, s_2)$ are equal and of codimension r in $H^0(\theta^{\otimes \frac{m+2}{2}})$ if $s_1, s_2 \geq 1$ and $s_1 + s_2 = \frac{m+2}{2}$.

2.3) The images of the maps $\beta(s_1, s_2)$ are equal and of codimension $r-1$ in $H^0(\theta^{\otimes m+1})$ if $s_1, s_2 \geq 1$ and $s_1 + s_2 = m+1$.

2.4) The map $\beta(\frac{m+2}{2}, \frac{m+2}{2})$ surjects.

2.5) The map $\beta(s_1, s_2)$ surjects if $s_1, s_2 \geq 0$ and $s_1 + s_2 \leq \frac{m}{2}$.

2.6) The map $\beta(s_1, s_2)$ surjects if $s_1, s_2 \geq 0$, $s_1 \geq \frac{m+2}{2}$, and $\frac{m+2}{2} \leq s_1 + s_2 \leq m$.

3) The ring $R_\theta = \bigoplus_{s=1}^{\infty} (R_\theta)_s$, where $(R_\theta)_s = H^0(\theta^{\otimes s})$, is generated by its part of degree 1, by r generators in degree $\frac{m+2}{2}$ and by $r-1$ generators in degree $m+1$.

Proof. The proof 1) is exactly like the proof of 1) of Proposition 2.4. The proof of 2) is also like the proof of part 2) of Proposition 2.4. Here we will only highlight the differences. As we saw in the proof of Proposition 2.4 (2), the key to know the images of the maps $\beta(s_1, s_2)$ is the decomposition of $H^0(\theta^{\otimes s_1})$ and $H^0(\theta^{\otimes s_2})$ as sum of blocks $A(s_1) \oplus B(s_1) \oplus C(s_1)$ and $A(s_2) \oplus B(s_2) \oplus C(s_2)$, i.e., the crucial information is to know which ones among

$A(s_1), B(s_1), C(s_1), A(s_2), B(s_2), C(s_2)$ are zero. This information is obtained from 1). If $0 \leq n \leq \frac{m}{2}$, then $H^0(\theta^{\otimes n}) = A(n)$. Thus, the map β_n surjects if $0 \leq n \leq \frac{m-2}{2}$. If $\frac{m+2}{2} \leq n \leq m$, $H^0(\theta^{\otimes n}) = A(n) \oplus B(n)$, hence the map β_n also surjects if $\frac{m+2}{2} \leq n \leq m-1$. If $n \geq m+2$, $H^0(\theta^{\otimes n}) = A(n) \oplus B(n) \oplus C(n)$, then the map β_n surjects if $n \geq m+2$. If $r = 1$, $H^0(\theta^{\otimes m+1}) = A(m+1) \oplus B(m+1)$, thus β_m surjects if $r = 1$. If $r > 1$, then $H^0(\theta^{\otimes m+1}) = A(m+1) \oplus B(m+1) \oplus C(m+1)$, therefore β_{m+1} surjects if $r > 1$. This proves 2.1). Now we prove 2.2). Recall that if $0 \leq n \leq \frac{m}{2}$, then $H^0(\theta^{\otimes n}) = A(n)$. On the other hand $H^0(\theta^{\otimes \frac{m+2}{2}}) = A(\frac{m+2}{2}) \oplus B(\frac{m+2}{2})$. Since $B(\frac{m+2}{2}) = H^0(\mathcal{O}_{\mathbf{P}^1}(r-1))$, then, if $s_1, s_2 \geq 0$ and $s_1 + s_2 = \frac{m+2}{2}$ the image of the maps $\beta(s_1, s_2)$ are equal to $A(\frac{m+2}{2})$, which has codimension r in $H^0(\theta^{\otimes \frac{m+2}{2}})$. To prove 2.3) recall that if $\frac{m+2}{2} \leq n \leq m$, $H^0(\theta^{\otimes n}) = A(n) \oplus B(n)$ and that $H^0(\theta^{\otimes m+1}) = A(m+1) \oplus B(m+1) \oplus C(m+1)$. Thus, if $s_1, s_2 \geq 0$ and $s_1 + s_2 = m+1$, then either $s_1 \geq \frac{m+1}{2}$ or $s_2 \geq \frac{m+1}{2}$. Let us say that $s_1 \geq \frac{m+1}{2}$, then $s_2 \leq \frac{m+1}{2}$. Then the image of the maps $\beta(s_1, s_2)$ is $A(m+1) \oplus B(m+1)$. Since $C(m+1) = H^0(\mathcal{O}_{\mathbf{P}^1}(r-2))$, the codimension of the image of $\beta(s_1, s_2)$ in $H^0(\theta^{\otimes m+1})$ is $r-1$. Part 2.4) follows from the fact that $H^0(\theta^{\otimes \frac{m+2}{2}}) = A(\frac{m+2}{2}) \oplus B(\frac{m+2}{2})$ arguing as in the proof of Proposition 2.4 (2.2). Part 2.5) follows from the fact that$H^0(\theta^{\otimes n}) = A(n)$ if $0 \leq n \leq \frac{m}{2}$. Now we prove 2.6). Recall that if $\frac{m+2}{2} \leq n \leq m$, then $H^0(\theta^{\otimes n}) = A(n) \oplus B(n)$ and that if $0 \leq n \leq \frac{m}{2}$, then $H^0(\theta^{\otimes n}) = A(n)$. This implies 2.6).

Finally 3) follows from 2.1), 2.2), 2.3) and 2.4). □

Now from the previous Proposition 3.1 and using Lemma 2.2 and Lemma 2.3 we can prove Theorem 1 for the even dimensional case:

Theorem 3.2. *Let X be a smooth, pluriregular variety of general type of even dimension m with K_X base-point-free. Assume that the canonical morphism π is generically finite of degree 3 onto a variety Y of minimal degree r. Then the canonical ring R of X is generated by its elements of degree 1, r generators in degree $\frac{m+2}{2}$ and $r-1$ generators in degree $m+1$.*

Sketch of proof. The proof uses the same ideas as the proof of Theorem 2.7 so we will just outline the key steps. As in Theorem 2.7 we need to know the images of the maps $\alpha(s_1, s_2)$. Precisely the result follows from the following claims:

1) The map α_n surjects if $1 \leq n \leq \frac{m-2}{2}$.
2) The images of the maps $\alpha(s,t)$, when $s+t = \frac{m+2}{2}$ and $s,t \geq 1$ are equal and of codimension r in $H^0(K_X^{\otimes \frac{m+2}{2}})$.
3) The map α_n surjects if $\frac{m+2}{2} \leq n \leq m-1$.
4) The images of the maps $\alpha(s,t)$, when $s+r = m+1$, $s,t \geq 1$ are equal and of codimension $r-1$ in $H^0(K_X^{\otimes m+1})$.
5) The map $\alpha(\frac{m+2}{2}, \frac{m+2}{2})$ surjects if $r=1$.
6) The map α_{m+1} surjects if $r \geq 2$.
7) The map α_n surjects if $n \geq m+2$.

Before we prove the above claims, we will show how they imply the result. Claim 1) implies that the part of R of degree less than or equal to $\frac{m}{2}$ is generated in degree 1. Claim 2) implies that the subspace of the part of degree $\frac{m+2}{2}$ generated in lower degree has codimension r, hence, in order to generate R, r generators of degree $\frac{m+2}{2}$ are needed. Claim 1) and 3) imply that the part of R of degree less than or equal to m is generated in degree 1 and $\frac{m+1}{2}$. By the same argument as before, Claim 4) implies that, in order to generate R, we need $r-1$ linearly independent elements of degree $m+1$. Claims 5) and 6) imply that R_{m+2} is generated in degree $\frac{m+2}{2}$ or lower. Finally, Claim 7) proves that the part of R of degree greater than $m+2$ is generated in degree $m+2$ or lower. Summarizing, the above claims show that R is generated by its part of degree 1, by r linearly independent elements in degree $\frac{m+2}{2}$ and by $r-1$ linearly independent elements in degree $m+1$.

Now we proceed to prove claims 1) to 7). The proof goes like the proof of Theorem 2.7. Thus Claims 1), 3), 6) and 7) follow from Proposition 3.1 (2.1) and Lemma 2.3 (2). The proof of Claim 2) follows from Proposition 3.1 (2.2) and Proposition 3.1 (2.5), and from Lemma 2.3 (3), arguing like for the proof the 3) in the proof of Theorem 2.7. Likewise Claim 4) follows from Proposition 3.1 (2.3) and 2.10 (2.6) and from Lemma 2.3(3). Finally Claim 5) follows from Proposition 3.1 (2.3), (2.4) and 2.10 (2.6) and from Lemma 2.3 (3). □

Remark 3.3. *Proposition 3.1 states explicitely that the ring* $\bigoplus_{n\geq 0} H^0(C, \theta^{\otimes n})$ *has in each degree the same number of generators as* $\bigoplus_{n\geq 0} H^0(X, K_X^{\otimes n})$. *The way Theorem 3.2 is proven shows that the same is true for the "intermediate" rings* $\bigoplus_{n\geq 0} H^0(X_{m'}, L_{m'}^{\otimes n})$, $2 \leq m' \leq m-1$.

We find now sufficient conditions for $K_X^{\otimes n}$ to satisfy property N_0:

Theorem 3.4. *Let X be a smooth, pluriregular variety of general type of even dimension m with ample and base-point-free canonical bundle. Assume that the canonical morphism has degree 3 and maps X onto a variety Y of minimal degree r.*

1) If $r = 1$, then the line bundle $K_X^{\otimes n}$ satisfies propety N_0 if and only if $n \geq \frac{m+2}{2}$.

2) If $r > 1$, then the line bundle $K_X^{\otimes n}$ fails to satisfy property N_0 if $n \leq \frac{m}{2}$ and satisfies property N_0 if $n \geq m+1$.

Proof. The proof of 1) follows steps similar to those of the proof of Theorem 2.10.

Step 1. $K_X^{\otimes n}$ does not satisfy property N_0 if $n \leq \frac{m}{2}$.
Step 2. $K_X^{\otimes n}$ satisfies property N_0 if $n = \frac{m+2}{2}$.
Step 3. $K_X^{\otimes n}$ satisfies property N_0 if $\frac{m+4}{2} \leq n \leq m+1$.
Step 4. $K_X^{\otimes n}$ satisfies property N_0 if $n \geq m+2$. Step 1 is true by the same reason as Step 1 of the proof of Theorem 2.10. The other steps are proven like in Theorem 2.10 and the key facts to use are the surjectivity of $\alpha(\frac{m+2}{2}, \frac{m+2}{2})$ if $r = 1$ and the surjectivity of $\alpha_{n'}$ for all $n' \geq m+2$. This was shown in 5) and 7) of the proof of Theorem 3.2. We leave the rest of the details to the reader.

The proof of 2) goes along the same lines. The fact that the line bundle $K_X^{\otimes n}$ fails to satisfy property N_0 if $n \leq \frac{m}{2}$ is proven like Step 1 of Theorem 2.10. The fact that $K_X^{\otimes n}$ satisfies property N_0 if $n \geq m+1$ follows from the surjectivity of $\alpha_{n'}$ for all $n' \geq m+1$ and $r > 1$. □

Comparing the previous result with Theorem 2.10 we realize that in the even dimensional case the following question is left open:

Question 3.5. *Let X be a smooth variety of general type and dimension m even and let $X \xrightarrow{\pi} Y$ be a generically finite, canonically induced, degree 3 cover of a variety Y of minimal degree $r > 1$. Does $K_X^{\otimes n}$ satisfy property N_0 if $\frac{m+2}{2} \leq n \leq m$?*

The reason why this question cannot be addressed with the present arguments is that we cannot decide whether the map $\alpha(\frac{m+2}{2}, \frac{m+2}{2})$ surjects. The map $\beta(\frac{m+2}{2}, \frac{m+2}{2})$ does surject, as stated in Proposition 3.1 (2.4), but if we were to use Lemma 2.3 to deduce from the surjectivity of $\beta(\frac{m+2}{2}, \frac{m+2}{2})$ the surjectivity of $\alpha(\frac{m+2}{2}, \frac{m+2}{2})$, we would also need the surjectivity of $\beta(\frac{m+2}{2}, \frac{m}{2})$. The latter map is however non surjective (cf. Proposition 3.1 (2.3)).

4. On the structure of the canonical morphism

In the previous sections we studied the canonical ring of pluriregular varieties of general type X whose canonical morphism π was a triple cover of a variety of minimal degree. In this section we investigate the structure of such morphisms. We start relaxing our hypotheses on X and π and study the structure of the canonical morphism $X \xrightarrow{\pi} Y$. This study also tells us about the structure of X and Y. We do not not assume a priori X to be pluriregular or even regular, neither we do assume π to be induced by the complete canonical series of X. In fact we start only with a base-point-free canonical subseries (we will see later, cf. Theorem 4.3, that assuming π to be flat is sufficient for X to be as the varieties studied in Sections 2 and 3, that is, pluriregular and with π induced by the complete canonical series). Despite these weaker hypotheses, in the next theorem we are able to obtain interesting information relating the degree of Y and the dimension of X:

Theorem 4.1. *Let X be a smooth variety of general type of dimension $m \geq 2$ and let $Y \subset \mathbf{P}^{r+m-1}$ be an m-fold of minimal degree. Assume that K_X is base-point-free and let W be a sublinear series of $|K_X|$ without base points. Let $\pi : X \longrightarrow Y$ be the morphism induced by W. Assume that π is generically finite of degree 3. Then*
1) If m is odd, then the degree r of Y is even (in particular Y is not linear $\mathbf{P}^m$)
2) If $h^1(\mathcal{O}_X) = 0$, then π is induced by the complete canonical series of X.

Proof. Since W has no base points, we may choose $m-1$ general members of W so that its intersection C is a smooth and irreducible curve in X and so that $\varphi = \pi|_C$ is finite onto a smooth rational normal curve D of degree r in $\mathbf{P}^r$. Let $\theta = K_X|_C$. By adjunction $\theta^{\otimes m} = K_C$. Now, if m is odd, then C, θ and φ satisfy the hypothesis of Proposition 2.4. Then according to Proposition 2.4 (Lemma 2.5), r is even. On the other hand, C, θ and φ satisfy the hypothesis of Proposition 2.4 if m is odd, and the hypothesis of Proposition 3.1 if m is even. Hence, since $m \geq 2$, according to Proposition 2.4 (1) and Proposition 3.1 (1), φ is induced by the complete linear series $|\theta|$. Since $H^1(\mathcal{O}_X) = 0$, by Kawamata-Viehweg vanishing we have also

$H^1(\mathcal{O}_{X_{m-1}}) = \cdots = H^1(\mathcal{O}_{X_2}) = 0$. Then the fact that φ is induced by the complete $|\theta|$ implies that π is induced by the complete linear series of the canonical of X. This proves 2). □

Theorem 4.1 tells us among other things that a variety of minimal degree Y with degree an even integer, cannot occur as the image of a canonical morphism $X \xrightarrow{\pi} Y$ of degree 3 if the dimension of X is odd. In the next result we further eliminate possible targets of π. Applied to the setting of Section 2, in which Y was a variety of minimal degree and π has degree 3, Theorem 4.2 tells us that there are no even dimensional canonical, generically triple covers of smooth rational normal scrolls. Theorem 4.2 is more general since it considers generically finite canonical morphisms of arbitrary odd degree and smooth scrolls fibered not necessarily over $\mathbf{P}^1$.

Theorem 4.2. *Let X be a smooth variety of general type of dimension m with base-point-free canonical bundle X. Let W be a linear subseries of $|K_X|$ without base points and let $X \xrightarrow{\pi} Y$ be the morphism induced by W. Assume that π is generically finite of degree n and that Y is a scroll. Then either n is even or m is odd. In particular, if $n = 3$, m is odd.*

Proof. Let G be a fiber of Y. Let H be the class of the hyperplane section of Y. Let g be a straight line in G. Then $g = G \cdot H^{(m-2)}$, i.e, g is the complete intersection on Y of G and $m-2$ suitable hyperplane sections. Likewise, π^*g is the complete intersection on X of the pullback by π of $m-2$ suitable hyperplane sections and π^*G. We choose the $m-2$ hyperplane sections suitably so that, by Bertini, the intersection of the pullback by π of $m-2$ of them is a smooth irreducible surface X'. The image of X' by π is a surface Y', which is also a smooth scroll and has dimension 2. Applying adjunction recursively we see that $K_{X'}$ is the restriction of $K_X + (m-2)\pi^*(H)$ to X'. Then $2t-2 = (K_{X'} + \pi^*g)\cdot\pi^*g$, where t is the sectional genus of π^*g. Now we know that $K_X = \pi^*H$, so we have that

$$\begin{aligned} 2t-2 &= [K_X + (m-2)\pi^*H + \pi^*G]\cdot\pi^*g \\ &= [(m-1)\pi^*H + \pi^*G]\cdot\pi^*g = (m-1)n \ . \end{aligned}$$

Thus, if m is even, $m-1$ is odd, so n has to be even. And if n is odd, $m-1$ has to be even, hence m has to be odd. □

The two previous results eliminate several possibilities for the target of the morphism $X \xrightarrow{\pi} Y$. We end the section characterizing, under the stronger hypothesis of π being flat, the targets that actually do occur. Theorem 4.3 also shows that a variety of general type which admits a flat, triple canonical morphism onto a variety of minimal degree is necessarily pluriregular. Thus the varieties of general type which appear in Theorem 4.3 are pluriregular and satisfy therefore the hypotheses of Theorem 1. The theorem below also describes some features of the morphism π.

Theorem 4.3. *Let X be a smooth variety of general type of dimension $m \geq 2$. Assume that K_X is ample and base-point-free. Let W be a linear subseries of $|K_X|$ without base points. Let π be the morphism induced by W. Assume that π is flat and has degree 3. Let $Y = \pi(X)$ be a variety of minimal degree r. Then Y is one of the following:*

1) *$\mathbf{P}^m$ with $m \geq 2$ even.*
2) *A smooth rational normal scroll of odd dimension $m \geq 3$ and even degree r. In this case X is fibered over $\mathbf{P}^1$, with general fiber a smooth variety of general type of dimension $m-1$. The restriction of π to the general fiber is finite and flat, induced by the complete canonical series, and maps onto $\mathbf{P}^{m-1}$, fitting therefore in case 1) above.*
3) *A smooth quadric hypersurface of odd dimension $m \geq 3$.*

Moreover, all possibilities 1), 2) and 3) do really occur.

In addition, π is in fact induced by the complete canonical linear series of X; X is a trisection in the total space of a line bundle L on Y such that L^{-2} is not effective; $\pi_*\mathcal{O}_X = \mathcal{O}_Y \oplus L^{-1} \oplus L^{-2}$; $K_Y = L^{-2}(1)$ and $H^1(\mathcal{O}_X) = \cdots = H^{m-1}(\mathcal{O}_X) = 0$.

Proof. Since π is flat and X is smooth, Y is also smooth. Thus Y must be one of the following:

a) $\mathbf{P}^m$;
b) a smooth rational normal scroll;
c) a smooth quadric hypersurface.

Since π is finite of degree 3 and flat, $\pi_*\mathcal{O}_X = \mathcal{O}_Y \oplus E$, where E is a locally free sheaf of rank 2 (cf. [11]). Using relative duality (see [4], Theorem 2.1), we obtain

$$\pi_*\mathcal{O}_X = \mathcal{O}_Y \oplus L^{-1} \oplus L^{-2} \tag{4.3.1}$$

for a line bundle L such that L^{-2} is not effective and

$$K_Y = L^{-2}(1) \tag{4.3.2}$$.

From [4], Theorem 2.1, it also follows that X is a trisection in the total space of L.

Now we study further restrictions on 1), 2) and 3). Let Y be first $\mathbf{P}^m$. Since $K_Y = \mathcal{O}_{\mathbf{P}^m}(-m-1)$, by (4.3.2) m has to be even. Let Y be now a rational normal scroll. Let us denote by H the divisor class of its hyperplane section and by F the class of a fiber. Let r be the degree of Y. Then $K_Y = -mH + (r-2)F$. Again by (4.3.2) r has to be even and m has to be odd (see also [15], Remark 3.1). The only case left is when Y is a smooth quadric hypersurface of dimension greater than 2. By Lefschetz Theorem, Pic$Y = \mathbf{Z}$ and is generated by the hyperplane section Y. In addition $K_Y = \mathcal{O}_Y(-m)$ by adjunction. Hence by (4.3.2) m is odd.

Since $L^{-2}(1) = K_Y$ and by projection formula

$$H^0(\pi_*\pi^*(\mathcal{O}_Y(1))) = H^0(\mathcal{O}_Y(1)) \oplus H^0(L^{-1}(1)) \oplus H^0(L^{-2}(1)) ,$$

checking for each case a), b) c) we see that $H^0(\pi_*\pi^*(\mathcal{O}_Y(1)) = H^0(\mathcal{O}_Y(1))$, hence π is in fact induced by the complete canonical series of X. To further study the structure of π in case 2), let us denote by G the inverse image of a general fiber F of Y. We have the following exact sequence:

$$\begin{aligned} 0 \longrightarrow H^0(K_X(-G)) \longrightarrow H^0(K_X) \longrightarrow H^0(K_X \otimes \mathcal{O}_G) \\ \longrightarrow H^1(K_X(-G)) \, . \end{aligned}$$

Since π is finite, and $K_X(-G) = \pi^*(\mathcal{O}_Y(H-F))$,

$$H^1(K_X(-G)) = H^1(\mathcal{O}_Y(H-F) \oplus L^{-1}(H-F) \oplus L^{-2}(H-F)) \, .$$

Since $L^{-1}(H-F) = \mathcal{O}_Y(\frac{-m+1}{2}H + \frac{r-4}{2}F)$ and $L^{-2}(H-F) = \mathcal{O}_Y(-mH + (r-3)F)$, the first cohomology group of the three line bundle vanishes. Hence the complete linear series $|K_X|$ restricts to the complete linear series K_G, which maps onto $F = \mathbf{P}^{m-1}$. G is smooth by Bertini, and since K_X is ample and base-point-free, G must be connected. The finiteness of the morphism from G to F and the fact that its degree is 3 are clear. To see the flatness it suffices to see that $\pi_*(\mathcal{O}_X) \otimes \mathcal{O}_G = (\pi|_G)_*(\mathcal{O}_G)$.

Now to see the vanishings of cohomology we will deal case by case. Recall that $L^{-1} = \mathcal{O}_Y(\frac{1}{2}(K_Y - H))$ and that $\pi_*(\mathcal{O}_X) = \mathcal{O}_Y \oplus L^{-1} \oplus L^{-2}$. If $Y = \mathbf{P}^m$, then $H^i(\mathcal{O}_X) = 0$ for all $1 \leq i \leq m-1$, since intermediate cohomology of line bundles on Y vanishes. If Y is a smooth rational normal scroll, then

$$\begin{aligned} H^i(\mathcal{O}_X) = H^i(\mathcal{O}_Y) \oplus H^i(\mathcal{O}_Y(-\frac{1}{2}(m+1)H + \frac{1}{2}(r-2)F)) \\ \oplus H^i(\mathcal{O}_Y(-(m+1)H + (r-2)F)) = 0 \end{aligned}$$

for all $1 \leq i \leq m-1$. The latter can be seen using Serre duality and pushing down to $\mathbf{P}^1$. Now, for Y smooth quadric hypersurface of dimension m we have the sequence

$$0 \longrightarrow \mathcal{O}_{\mathbf{P}^{m+1}}(-2) \longrightarrow \mathcal{O}_{\mathbf{P}^{m+1}} \longrightarrow \mathcal{O}_Y \longrightarrow 0$$

so the vanishing of $H^i(\mathcal{O}_Y)$ for all $1 \leq i \leq m-1$ follows from the vanishing of the intermediate cohomology of line bundles on projective space.

Finally we show that all the possible varieties of minimal degree in 1), 2) and 3) do actually occur. We show this by way of examples of cyclic triple covers. Those examples are easy to construct but we outline them for the convenience of the reader (for examples of triple covers of threefolds see also [15]).

To construct examples of triple covers on $\mathbf{P}^m$, m even, take the triple cyclic cover $X \xrightarrow{\pi} \mathbf{P}^m$ branched along a smooth divisor of degree $\frac{3(m+2)}{2}$. In such case,

$$K_X = \pi^*(K_{\mathbf{P}^m} \otimes \mathcal{O}_{\mathbf{P}^m}(m+2)) = \pi^*\mathcal{O}_{\mathbf{P}^m}(1) \ .$$

Moreover

$$\begin{aligned} H^0(K_X) &= H^0(\mathcal{O}_{\mathbf{P}^m}(1)) \oplus H^0(\mathcal{O}_{\mathbf{P}^m}(-\frac{m}{2})) \oplus H^0(\mathcal{O}_{\mathbf{P}^m}(-m-1)) \\ &= H^0(\mathcal{O}_{\mathbf{P}^m}(1)) \ , \end{aligned}$$

hence π is induced by the complete canonical series of X. Analogously, to construct examples of triple cyclic covers of a hyperquadric Q of odd dimension $m \geq 3$ we take the cover $X \xrightarrow{\pi} Q$ branched along a smooth divisor in Q, complete intersection of a hypersurface of $\mathbf{P}^{m+1}$ of degree $\frac{3(m+1)}{2}$ and Q. Finally, to construct examples of triple cyclic covers of smooth rational normal scrolls of odd dimension $m \geq 3$ and even degree r, consider a smooth rational scroll S possessing smooth divisors linearly equivalent to $3(\frac{m+1}{2}H - \frac{r-2}{2}F)$, where H is the hyperplane class of S and F is the class of a fiber. Then we take the cyclic triple cover $X \xrightarrow{\pi} S$ branched along a smooth divisor linearly equivalent to $3(\frac{m+1}{2}H - \frac{r-2}{2}F)$. □

Remark 4.4 Theorem 1 and Theorem 4.3 show that if $m \geq 3$, there does not exist a converse to the theorem of M. Green (cf. [6]) which says that if X is a smooth, pluriregular variety of general type with base-point-free canonical bundle, then the canonical ring of X is generated in degree less than or equal to m if the image Y of the canonical morphism π of X is not a variety of minimal degree. Indeed, the canonical ring of the varieties appearing in Theorem 4.3 are generated in degree less than or equal to $\frac{m+1}{2}$ if m is odd and $\frac{m+2}{2}$ if m is even. Therefore, Y not being of minimal degree is a sufficient condition for the canonical ring of X to be generated in degree less than or equal to m but it is not a necessary condition. This is in contrast with the situation in dimension 2, where the converse of the result of Green is true, as it was proved by the authors in [5].

References

[1] F. Catanese, *Equations of pluriregular varieties of general type*, Geometry today (Rome, 1984), 47–67, Progr. Math., **60**, Birkhauser Boston, 1985.

[2] F. Catanese, *Commutative algebra methods and equations of regular surfaces*, Algebraic geometry, (Bucharest, 1982), 68–111, Lecture Notes in Math., **1056**, Springer Berlin, 1984.

[3] C. Ciliberto, *Sul grado dei generatori dell'anello di una superficie di tipo generale*, Rend. Sem. Mat. Univ. Politec. Torino **41** (1983), 83–111.

[4] T. Fujita, *Triple covers by smooth manifolds*, J. Fac. Sci. Univ. Tokyo Sect. IA, Math. **35** (1988), 169–175.

[5] F.J. Gallego and B.P. Purnaprajna, *On the canonical ring of covers of surfaces of minimal degree*, to appear in Transactions of the American Mathematical Society.

[6] M.L. Green, *The canonical ring of a variety of general type*, Duke Math. J. **49** (1982), 1087–1113.

[7] E. Horikawa, *Algebraic surfaces of general type with small c_1^2. I*, Ann. of Math. (2) **104** (1976), 357–387.

[8] E. Horikawa, *Algebraic surfaces of general type with small c_1^2, II*, Invent. Math. **37** (1976), 121–155.

[9] E. Horikawa, *Algebraic surfaces of general type with small c_1^2, III*, Invent. Math. **47** (1978), 209–248.
[10] E. Horikawa, *Algebraic surfaces of general type with small c_1^2, IV*, Invent. Math. **50** (1978/79), 103–128.
[11] D. Hahn and R. Miranda, *Quadruple covers of algebraic varieties*, J. Algebraic Geom. **8** (1999), 1–30.
[12] K. Konno, *Algebraic surfaces of general type with $c_1^2 = 3p_g - 6$*, Math. Ann. **290** (1991), 77–107.
[13] M. Mendes Lopes and R. Pardini, *Triple canonical surfaces of minimal degree*, International J. Math. **11** (2000), 553–578.
[14] R. Miranda, *Triple covers in Algebraic Geometry*, Amer. J. Math. **107** (1985), 1123–1158.
[15] P. Supino, *Triple covers of 3-folds as canonical maps*, Communications in Algebra **26**, (1998), 1475–1487.

Dpto. de Álgebra, Facultad de Matemáticas
Universidad Complutense de Madrid, 28040 Madrid, Spain
E-mail address: FJavier_Gallego@mat.ucm.es

405 Snow Hall, Dept. of Mathematics, University of Kansas
Lawrence, Kansas 66045-2142
E-mail address: purna@math.ukans.edu

Semistability and Semisimplicity in representations of low height in positive characteristic

S. Ilangovan, V.B. Mehta and A.J. Parameswaran

Dedicated to Professor C.S. Seshadri on his 70th *birthday*

Abstract. Let G be a semisimple algebraic group and X be a smooth projective variety both defined over an algebraically closed field k of positive characteristic. Here we define the notion of a *low height* representation and show that bundles induced from a semistable G-bundle by a low height representation continue to be semistable.

1. Introduction

Let G be a connected reductive algebraic group over k, where k is an algebraically closed field of characteristic $p > 0$. Here we show that semistability of vector bundles induced from principal G-bundles by a representation of G is closely related with the heights of the dominant weights occurring in the representation. The main result of the paper is (cf. 2.1 for the definition of low height representations)

Theorem 3.1 *Let* $G \to SL(V)$ *be a low height representation. Then for every normal projective variety* X *with an ample line bundle* M *and for every principal* G*-bundle* E *on* X *which is semistable with respect to* M*, the induced principal bundle* $E(SL(V))$ *is also semistable with respect to* M*.*

In an e-mail message to us J-P. Serre had raised the following question: If V_1 and V_2 are semistable bundles on a smooth curve in characteristic $p > 0$, and rank V_1 + rank $V_2 < p + 2$, then is $V_1 \otimes V_2$ again semistable? The above theorem answers this question in the affirmative (cf. Remark 3.4). It sheds some light on the question raised by M.S. Narasimhan concerning the boundedness of semistable G-bundles on curves in char $p > 0$. Also it generalizes the theorem of Ramanan and Ramanathan (cf. [RR]), proved for rank 2 bundles in char $p > 0$, to arbitrary semisimple groups. We have the following converse to the above theorem with mild restriction on the prime concerned.

Theorem 3.6 *Let G be an almost simple group. Let $p > min\{2ht(\omega_i)\}$, where ω_i's are the fundamental weights of G. Let $G \to SL(V)$ be a representation which is not of low height. Then there exists a curve X and a semistable G-bundle E on X such that the induced principal bundle $E(SL(V))$ is not semistable.*

We would like to thank J-P. Serre for the e-mail message in which he raised the question about the semistability of tensor products of vector bundles in analogy with his theorem on semisimplicity of tensor product of representations (cf. [S1]). We also thank Roger Carter for drawing our attention to the work of D. Testerman. We would also like to thank Jantzen for stimulating conversations and helpful suggestions. We thank the referees for pointing out several obscurities in an earlier version of this paper and for making several suggestions to improve the exposition.

2. Preliminaries

Let k be an algebraically closed field of characteristic $p > 0$. Let G be a connected reductive algebraic group over k. We fix some notations here. Let T be a maximal torus of G, $\mathbf{X} := Hom(T, G_m)$ be the character group of T and $\mathbf{Y} := Hom(G_m, T)$ be the 1-parameter subgroups of T. Let $R \subset \mathbf{X}$ be the root system of G with respect to T. Let $\mathcal{W}$ be the Weyl group of the root system R. Let (,) denote the $\mathcal{W}$-invariant inner product on $\mathbf{X} \otimes \mathbf{R}$ such that $(\alpha, \alpha) = 2$ for all short roots α. For $\alpha \in R$, the corresponding co-root $\alpha^\vee$ is $2\alpha/(\alpha, \alpha)$. Let $R^\vee \subset \mathbf{X} \otimes \mathbf{R}$ be the set of all co-roots. Let $B \subset G$ be a Borel subgroup containing T. This choice defines a base Δ^+ of R called the *simple roots*. Let $\Delta^- = -\Delta^+$. A root in R is said to be *positive* if it is a nonnegative linear combination of simple roots. We take the roots of B to be positive by convention. Let $\Delta^\vee \subset R^\vee$ be the basis for the corresponding dual root system. Then we can define the Bruhat ordering on $\mathcal{W}$. The longest element with respect to this ordering of $\mathcal{W}$ is denoted by w_0. A reductive group is classified by these *root-datum*, namely the character group, 1-parameter subgroups, the root system, co-roots and the $\mathcal{W}$-invariant pairing.

Let V be a G-module, i.e., V is a k-vector space together with a linear representation of G to $\mathrm{Aut}(V)$. Then V can be written as direct sum of eigenspaces for T. On each eigenspace T acts by a character. These are called the *weights* of the representation. A weight λ is called *dominant* if $(\lambda\ ,\ \alpha_i^\vee) \geq 0$ for all simple roots $\alpha_i \in \Delta^+$. A character on the Borel subgroup B is called *dominant* if its restiction to the maximal torus is dominant. A character on a parabolic subgroup P is called dominant if it is dominant for any Borel subgroup contained in P (cf. [RR], Remark 1.11). A weight λ is said to be $\geq$ another weight μ if the difference $\lambda - \mu$ is a nonnegative integral linear combination of simple roots, where the difference is taken with respect to the natural abelian group structure of $\mathbf{X}$. The *fundamental weights*, ω_i are uniquely defined by the criterion $(\omega_i\ ,\ \alpha_j^\vee) = \delta_{ij}$. The element ρ of $\mathbf{X} \otimes \mathbf{R}$ is defined to be half sum of positive roots. It can also seen to be equal to the sum of fundamental weights. The *height* (cf [H], Section 10.1) of a root is defined to be the sum of the coefficients in the expression $\alpha = \Sigma k_i \alpha_i$. We extend this notion of height linearly to the set of weights and denote this function by $ht(\ )$. Note that ht of any weight is not always an integer, but is at worst a half-integer. We extend this notion of height to representations as follows:

DEFINITION 2.1. (i) Given a linear representation V of G, we define the *height* of the representation $h_G(V)$ (also denoted by $h(V)$ if G is understood in the given context) to be the maximum of $2ht(\lambda)$,

where λ runs over dominant weights occurring in V.
(ii) A linear representation V of G is said to be a *low height* representation if $h(V) < p$.

The following lemmas give other interpretations of the height. The proofs are easy and hence are omitted.

LEMMA 2.2. *For any dominant weight* λ

$$2ht(\lambda) \;=\; ht(\lambda - w_0\lambda).$$

Moreover if G *is simply laced, then*

$$2ht(\lambda) \;=\; (\lambda\ ,\ 2\rho).$$

LEMMA 2.3. *If the representation of* G *on* $H^0(\lambda)$ *is of low height, then the nilpotent index of* G *on* $H^0(\lambda)$ *is* $2ht(\lambda)+1$, *i.e., for any nilpotent element* X *in the Lie algebra of* G, *the* n*-th iteration of the action of* X *induces the zero endomorphism of* $H^0(\lambda)$ *for* $n > 2ht(\lambda)$, *and the* n*-th iteration of* X *is non zero for* $n = 2ht(\lambda)$ *for some nilpotent element* X.

The following was pointed out to us by Jantzen.

LEMMA 2.4. *For any weight* λ, *we have*

$$2ht(\lambda) = \sum_{\alpha>0} (\lambda, \alpha^\vee)$$

The right hand side of the above equality is the invariant $n_G(V)$ considered by Serre in [S3].

Using the notion of low height we can reinterpret Propositions 2 and 5 of Serre (cf. [S1]) as follows. In fact Serre has obtained more general results in [S3].

THEOREM 2.5. *Let* V *be a linear representation of* G *of low height. Then* V *is semisimple.*

REMARK 2.6. Given an irreducible G-module $L(\lambda)$, $L(\lambda) = \otimes L(\lambda_i)^{[i]}$ with each λ_i a restricted weight. Here, the superscript $[i]$ denotes the action with Frobenius twist. We define the *reduced height* of λ to be $\sum 2ht(\lambda_i)$ and denote it by $\tilde{h}(\lambda)$. Given a linear representation V of G, we define the *reduced height* of the representation,denoted by $\tilde{h}_G(V)$, to be the maximum of $\tilde{h}(\lambda)$, where λ runs over the highest weights occurring in the composition factors of V. A representation V of G is said to be a *low reduced height* representation if $\tilde{h}(V) < p$. Then one can prove along the lines of the proof of Jantzen (cf. [J2]) that any representation of reduced height less than p is semisimple for groups other than A_1.

Now, We give some definitions and results from Geometric Invariant Theory (GIT, cf. [M]) and about principal G-bundles (cf. [RR]).

Let G be a reductive group and V be a linear representation. A vector $x \in V$ is said to be a *semistable* vector for the action of G on V if there exists a G-invariant function s on V such that $s(x) \neq s(0)$. A vector is *unstable* if it is not semistable. It can be shown that a vector x is unstable if and only if the orbit closure $\overline{G \cdot x}$ contains the zero vector.

Let $x \in V$ be an unstable point. Then Kempf has associated canonically a one parameter subgroup $\lambda(x)$ (abbreviated as 1-PS) and a parabolic subgroup $P(x)$. They are respectively called the *instability 1-PS* and the *instability flag.* These subgroups satisfy $\lambda(g \cdot x) = g\lambda(x)g^{-1}$ and $P(g \cdot x) = gP(x)g^{-1}$ for all $g \in G$. We note that $P(x)$ contains $(G_x)_{\text{red}}$ where G_x is the scheme theoretic stabilizer of x. We call $(G_x)_{\text{red}}$ the reduced stabilizer of x.

Let $\pi : E \to X$ be a principal G bundle. Let F be a quasi-projective G-variety. Then we can form the associated fibre bundle $E(F) \to X$ which is the quotient of $E \times F$ under the action of G. A section $\sigma : X \to E(F)$ is given by a G-equivariant function $\tilde{\sigma} : E \to F$. If $H \to G$ is a homomorphism and E_H is a principal H-bundle, then we can construct the principal G-bundle $E_H(G)$. When H is a subgroup of G and E is a principal G bundle such that $E_H(G)$ is isomorphic (as principal bundles) to E, then E_H is called a reduction of structure group of E to H, and E is said to be the extension of structure group from H to G. Note that there is a bijective correspondence between the sections of $E/H = E(G/H) \to X$ and reductions of structure groups of E to H.

DEFINITION 2.7. A G-bundle $E \to X$ is said to *semistable* (resp. *stable*) if for any reduction of structure group to a parabolic subgroup P, and any nontrivial dominant character χ on P, we have $deg\ L_\chi \leq 0$ (resp. $deg\ L_\chi < 0$) where L_χ is the line bundle associated by the character χ to the reduced P bundle.

The line bundle on G/P associated to a nontrivial dominant character is the inverse of an ample line bundle, when P is a maximal parabolic subgroup of G. Moreover it is enough to check the semistability condition for reductions to maximal parabolic subgroups.

Let G act on a projective variety M, linearly with respect to an ample line bundle L. Then $E(L) \to E(M)$ is a line bundle. For any $x \in X$, $E(G)_x$ acts linearly on $E(M)_x$ with respect to the line bundle $E(L)_x$. Let $K = K(X)$ be the function field of X and $\overline{K}$ be an algebraic closure of K. For a section $\sigma : X \to E(M)$, we define deg σ to be the degree of $\sigma^*(E(L))$. We denote $X_0 = \mathrm{Spec}K$, the generic point of X.

Let E be a principal G-bundle on X. Let $\tilde{G}$ be a reductive group and $G \to \tilde{G}$ be a homomorphism. Let $\sigma(X_0) \in (\tilde{G}/\tilde{P})_{\overline{K}}$ be a K-rational point obtained from the reduction of structure group of $E(\tilde{G})$ to a parabolic subgroup $\tilde{P}$. Note that when X is a curve the K rational points of $(\tilde{G}/\tilde{P})_{\overline{K}}$ are in one-to-one correspondence with reductions of structure group of $E(\tilde{G})$ to $\tilde{P}$. Now $G_{\overline{K}}$ acts on $(\tilde{G}/\tilde{P})_{\overline{K}}$. The following result relates the semistability of the principal $\tilde{G}$-bundle, $E(\tilde{G})$, with the semistability (in the sense of GIT, [M]) of the $\overline{K}$ rational point $\sigma(X_0)$.

PROPOSITION 2.8. *(cf. [RR], Proposition* 3.10 *(i) and Proposition* 3.13*)*
(i) If $\sigma(X_0)$ *is a semistable point in the sense of GIT, then deg* $\sigma(X_0) \geq 0$.

(ii) Suppose E is semistable and the connected component of the center of G acts trivially on $(\tilde{G}/\tilde{P})$ and L. If $\sigma(X_0)$ is unstable and the instability 1-PS $\lambda(\sigma(X_0))$ is defined over K, then deg $\sigma(X_0) \geq 0$.

3. Semistability of low height representations

In this section we restrict ourselves to semisimple groups. This is to avoid repeating the condition that all the actions are trivial on the connected component of identity of the center. Let $G \to SL(V)$ be a representation of G. The main aim of this section is to prove:

THEOREM 3.1. *Let G be a semisimple group and $G \to SL(V)$ be a low height representation. Then for every normal projective variety X with an ample line bundle M and for every principal G-bundle E on X which is semistable with respect to M, the induced principal bundle $E(SL(V))$ is also semistable with respect to M.*

We will postpone the proof to the latter part of this section. Let $Q \subset SL(V)$ be a maximal parabolic subgroup and $\sigma : X \to E(SL(V)/Q)$ be a section, i.e., a reduction of structure group to the maximal parabolic subgroup Q. In order to prove the theorem, it suffices to show that either $\sigma(X_0)$ is GIT-semistable or the instability flag $P(\sigma(X_0))$ is defined over K (cf. proposition 2.8). To do this we need the following lemma about exponentiating (integrating) nilpotent elements of the Lie algebra of G.

LEMMA 3.2. *Let V be a linear representation of G of low height. Let $X \in$ Lie (G) be a nilpotent element. Let $W \subset V$ be a vector subspace such that $X(W) \subset W$, i.e., X belongs to the Lie algebra of the stabilizer of W. Then X belongs to the tangent space of the reduced stabilizer of W.*

Proof: We are given $\sigma : G \to SL(V) = SL(n)$, which is a low-height representation. Denote the differential of σ by $d\sigma$. Then if X is a nilpotent element of $Lie(G)$, $d\sigma(X)$ is also nilpotent in $sl(n) = LieSL(n)$. Denote $d\sigma(X)$ by Y. Then as σ is of low height, by lemma 2.3 the index of nilpotency of Y is $\leq p$ and so we may form the exponential series $t \to exp(tY)$. We thus get a homomorphism $f : G_a \to SL(n)$. On the other hand, by a result of Serre (cf. [S3], Page 25, Corollary 2) we get an embedding $g : G_a \to G$ such that the tangent vector of $g(G_a)$ is X. So, if $Y(W) \subset W$, then $g(G_a)(W) = W$, i.e. $X \in$ Lie $(Stab(W_{\text{red}})$. □

Now we recall some constructions from [RR]. Given an unstable vector x in a representation space V of G, we define

$$x_0 = \lim_{t \to 0} \lambda(x)(t) \cdot x$$

Here $\lambda(x)$ denotes the instability 1-PS and the limit is taken in the projective space and lifted to V. Then it is proved in [RR] (cf. Proposition 1.9) that $\lambda(x) = \lambda(x_0)$ and $P(x) = P(x_0)$, i.e., both x and x_0 have the same instability one parameter subgroup and the same instability flag. Let

$$V = \bigoplus_{i \in \mathbf{Z}} V_i$$

be the weight space decomposition of V with respect to $\lambda(x)$. Let

$$V^j = \bigoplus_{i \geq j} V_i$$

be the induced filtration of V. Then clearly, this filtration is invariant for the parabolic $P(x)$. Note that if $\lambda(x)$ is defined over K, then so is $P(x)$ and hence the filtration $V^\bullet$. Let $x = x_{j_0} + \sum_{i > j_0} x_i$ be the decomposition of x into weight vectors with respect to $\lambda(x)$. Note that $x_{j_0} \neq 0$. Let $M(x)$ denote the scheme theoretic intersection of V^{j_0} with the orbit space $\mathcal{O}(x)$ and let $N(x)$ be the inverse image of $M(x)$ under the orbit map $\phi_x : G \to \mathcal{O}(x)$. Here we consider $\mathcal{O}(x)$ with reduced scheme structure. Then it is shown in [RR] (proof of Lemma 2.4, page 279) that the $\overline{K}$ rational points of $N(x)$ coincide with those of $P(x)$. Hence the reduced variety $N(x)_{red}$ is equal to $P(x)$.

Proposition 3.3. *Let $G \to SL(W)$ be any low height representation. Let $Q \subset SL(W)$ be a parabolic subgroup and $x \in SL(W)/Q$ be a GIT unstable point. Then $N(x)$ is reduced.*

Proof: It can be shown that the tangent space $T_e(N(x))$ of $N(x)$ at the identity $e \in G$ has the following description (cf. [RR], page 279):

$$T_e(N(x)) = \{X \in Lie\ G \mid X(x) \in V^{j_0}\}$$

To show that $N(x)$ is reduced, it suffices to show the following

Claim: $T_e(N(x)) = T_e(P(x))$.

Suppose $T_e(N(x)) \supseteq T_e(P(x))$ is a proper inclusion. Then there exists a nilpotent element $X \in T_e(N(x))$ such that $X = \sum_{i<0} X_i$. Let $i_0 := \min\{i \mid X_i \neq 0\}$.

It can be shown that X_{i_0} belongs to the Lie algebra of the stabilizer of x_0 (cf. [RR], proof of lemma 2.4, page 279, line 21). As x_0 is a point of $SL(W)/Q$, this corresponds to a partial flag of subspaces

$$W_1 \subset W_2 \subset \cdots \subset W_k = W.$$

Now the tangent space at identity to the stabilizer can be described as

$$\{X \in Lie\ (G) \mid X(W_i) \subset W_i\}.$$

Then by lemma 3.2, X_{i_0} belongs to the reduced stabilizer of each W_i and hence of x_0, which is contained in $P(x_0) = P(x)$. But $P(x)$ consists only of nonnegative weight vectors for $\lambda(x)$. This contradicts the assumption that $i_0 < 0$. Hence the claim. □

Proof: (**of Theorem 3.1**) Let Q be a maximal parabolic subgroup of $SL(V)$ and let $m \in SL(V)/Q$ be any K-rational point. If m is GIT-semistable for the action of G, we are through by proposition 2.8(i). If m is unstable, we have to prove that the instability flag $P(m)$ is defined over K. By the theory of Galois descent, it suffices to prove that $P(m)$ is defined over the separable closure K_s of K in $\overline{K}$. Now there exists a $\overline{K}$-rational point $g \in G(\overline{K})$ such that $P(x) = gP(m)g^{-1}$ is defined over K_s where $x = g \cdot m$. Then $M(x)$ is defined over K_s. Consider $N = \phi_m^{-1}(M(x))$. As m is a K_s rational point ϕ_m is defined over K_s, it follows that N is defined over K_s. Now N is a coset of $N(x)$ and hence both N and $N(x)$ are isomorphic to each other. By proposition 3.3, $N(x)$ is absolutely reduced and hence N is absolutely reduced. But any absolutely reduced K_s variety has a K_s rational point (cf. [B], section 13.3). This shows that N has a K_s rational point $h \in N$. Then $\phi_m(h) \in M(x)$ and hence $P(\phi_m(h)) = P(x)$ is defined over K_s. Then $P(m) = hP(\phi_m(h))h^{-1} = hP(x)h^{-1}$ is defined over K_s as well. Now the theorem follows from proposition 2.8(ii). □

REMARK 3.4. We note that the tensor product representation $SL(n_1) \times SL(n_2) \to SL(n_1 n_2)$ is a low height representation if $n_1 + n_2 < p+2$. Hence, if V_1 and V_2 are semistable vector bundles of rank n_1 and n_2 respectively, then $V_1 \otimes V_2$ is semistable if $n_1 + n_2 < p+2$.

We also have a converse to this theorem with mild restrictions on the primes. Before proving the converse theorem we need the following

LEMMA 3.5. *Let G be an almost simple group and λ be a nonzero dominant weight such that $V = H^0(\lambda)$ is a low height representation of G. Then given any maximal parabolic P of G, there is a maximal parabolic $\tilde{P}$ of $SL(H^0(\lambda))$ such that $P = \tilde{P} \cap G$.*

Proof: Let U be the unipotent radical of P. Then by Lie-Kolchin's theorem U fixes some nonzero vector in V. Let V^U denote the set of all vectors fixed by U. It can be easily seen that $V^U \neq V$. Since U is a normal subgroup of P, it follows that V^U is P-invariant. The stabilizer of the subspace V^U in $SL(H^0(\lambda))$ is a parabolic $\tilde{P}$. As $H^0(\lambda)$ is of low height, it is irreducible. $G \cap \tilde{P}$ contains P by construction, hence is reduced by lemma 3.2. Since the representation is irreducible and P is a maximal parabolic, P must be equal to $G \cap \tilde{P}$. □

THEOREM 3.6. *Let G be an almost simple group. Let $p > min\{2ht(\omega_i)\}$, where ω_i's are the fundamental weights of G. Let $G \to SL(V)$ be a representation which is not of low height. Then there exists a curve X and a semistable G-bundle E on X such that the induced principal bundle $E(SL(V))$ is not semistable.*

Proof: Let $H \subset G$ be the image of the principal homomorphism from SL_2 to G which exists by [T], [S2] and [S3]. Then we see that V is not a low height representation for H (cf. [S3], proof of theorem 5, page 25). Gieseker (cf.[G]) has constructed a curve X and a stable principal SL_2 bundle E on X such that $F_X^*(E)$ is not semistable where F_X denotes the Frobenius morphism. Then by an easy argument (cf. [RR] 4.1 and 4.2) it follows that $E(SL(V))$ is not semistable as V is not a low height representation for H. Now to complete the proof it remains to show that $E(G)$ is semistable. Consider the representation $G \to SL(H^0(\lambda))$, where λ is a nontrivial dominant weight with minimum height. Then by the assumption on p, this is a low height representation of G and hence is also a low height representation of H. Then, by Theorem 3.1, $E(SL(H^0(\lambda)))$ is semistable. By lemma 3.5, every reduction of structure group of $E(G)$ to a maximal parabolic arises from a reduction of structure group of $E(SL(H^0(\lambda)))$ to a maximal parabolic. Now the semistability follows from the fact that Pic G/P is $\mathbf{Z}$ for any maximal parabolic $P \subset G$ and the restriction of an ample line bundle on $SL(H^0(\lambda))/\tilde{P}$ is an ample line bundle on G/P. Hence the theorem. □

REMARK 3.7. Let G be a semisimple group with $p > h$, where h is the Coxeter number of G. Let $f : SL(2) \to G$ be the principal homomorphism (cf. [T],[S2],[S3]). Under the above assumptions, the referee suggests that it is likely that every semistable principal $SL(2)$ bundle $E \to X$ on a smooth projective curve induces a semistable principal G bundle $E(G) \to X$. However we are unable to prove this assertion.

REMARK 3.8. Let G be an almost simple group and X be a nonsingular projective curve in char p. Let $\sigma : G \to SL(V)$ be a low height representation. It follows from Theorem 3.1 that for every semistable principal G bundle $E \to X$, the induced vector bundle $E(V)$ is a semistable vector bundle on X. This sheds some light on the boundedness of semistable principal G bundles on X.

References

[B] A. Borel, *Linear Algebraic Groups*, Graduate Texts in Math. 129, (1991) Springer, Berlin.

[G] D. Gieseker, *Stable vector bundles and the Frobenius morphism*, Ann. Sci. Ecol. Nor. Sup., 6 (1973).

[H] J. E. Humphreys, *Introduction to Lie Algebras and Representation Theory*, GTM-9, Springer-Verlag (1972).

[J1] J. C. Jantzen, *Representations of Algebraic Groups*, Pure and Applied Mathematics, 131, Acad. Press (1987).

[J2] J. C. Jantzen, *Low dimensional representations of reductive groups are semisimple*, Algebraic Groups and related subjects. (ed. Lehrer et. al. Cambridge University Press.) 1996.

[M] D. Mumford, *Geometric Invariant Theory*, Springer-Verlag, Berlin - Heidelberg (1965).

[P] K. Pommerening *Ueber die unipotenten Klassen reduktiver Gruppen II*, J. Algebra, 65 (1980), 373-398.

[RR] S. Ramanan and A. Ramanathan, *Some remarks on Instability flag*, Tôhoku Math. Journal, 36, (1984), 269-291.

[S1] J-P. Serre, *Sur la semi-simplicité des produits tensoriels de représentations de groupes*, Invent. Math. 116, (1994), 513-530.

[S2] J-P. Serre, *Exemples de plongements des groupes $PSL_2(\mathbf{F}_p)$ dans des groupes de Lie simples*, Invent. Math. 124, (1996), 525-562.

[S3] J-P. Serre, *Moursund Lectures 1998*. Univ. Oregon, Eugene.

[T] D. M. Testerman, *A_1 type overgroups of elements of order p in semisimple algebraic groups and the associated finite groups*, J. Algebra, 177 (1995), 34-76.

School of Mathematics, Tata Institute of Fundamental Research

Homi Bhabha Road, Bombay 400 005, India.

E-mail address: ig@math.tifr.res.in, vikram@math.tifr.res.in

param@math.tifr.res.in

The Development of Standard Monomial Theory-II

V. Lakshmibai[†]

Dedicated to Professor C.S. Seshadri on his 70th *birthday*

Abstract. This article is the survey of the later stages in the development of Standard Monomial Theory.

In this article, we give a survey of the development of Standard Monomial Theory (abbreviated SMT henceforth) in the 1980's and 1990's, to be very precise, the developement after it was established for fundamental weights of classical type in the series *Geometry of* G/P *I-IV* ([**42, 21, 18, 19**]). Recall that in [**19**], a basis for $H^0(G/P, L_\omega)$ has been constructed, ω being a fundamental weight of classical type, i.e., $\langle\omega, \beta^\vee\rangle \leq 2$, for all positive roots β; further, a "standard monomial basis" for $H^0(G/P, L_{m\omega}), m \in \mathbb{N}$ consisting of certain monomials in the elements of the basis for $H^0(G/P, L_\omega)$, has also been constructed. These constructions are done even for the Schubert varieties in $G/P, P$ being the maximal parabolic subgroup associated to ω. For a survey of these results, refer to [**36**] in this volume. We first fix some notation.

Let G be a semi simple, simply connected algebraic group, defined over an algebraically closed field k. Fix a maximal torus T of G and a Borel subgroup B, $B \supset T$. Let $W = N(T)/T$ ($N(T)$ being the normalizer of T in G) be the Weyl group. The T-fixed points of G/B (for the action of G by left multiplication) are precisely the cosets $wB, w \in W$; in the sequel, we shall denote the point $wB \in G/B$ by e_w. We denote by X_w the Schubert variety associated to $w \in W$, namely the subvariety of G/B, which is the Zariski closure of the

[†] Partially supported by NSF grant DMS 9971295.

B-orbit Be_w, endowed with the canonical reduced structure. If Q is a parabolic subgroup, $Q \supset B$, we can more generally work with Schubert varieties in G/Q. These are indexed by W/W_Q, where W_Q is the Weyl group of Q. We denote by Δ, the root system of G relative to T, and S the system of simple roots relative to B. Let $\{\omega_i, 1 \leq i \leq \ell\}$ be the fundamental weights, ℓ being the rank of G. Then $\{\omega_i, 1 \leq i \leq \ell\}$ is a $\mathbb{Z}$-basis for $X(T) := \mathrm{Hom}_{\mathrm{alg\ gp}}(T, k^*)$ (= the group of characters of T).

Let $\lambda \in X(T)$. Then λ induces a character on B which we shall denote by λ only; λ determines a line bundle L_λ on G/B, namely, the line bundle associated to the principal B-bundle $G \to G/B$ for the action of B on k given by λ. Thus we obtain a map $X(T) \to \operatorname{Pic} G/B$ ($\operatorname{Pic} G/B$ being the group of isomorphism classes of line bundles on G/B), which is in fact an isomorphism (since G is simply connected).

Let $\lambda \in X(T)$, say $\lambda = \sum a_i\omega_i$. The following facts are well known:

(1) $H^0(G/B, L_\lambda) \neq 0 \Leftrightarrow \lambda$ is dominant, i.e., $a_i \geq 0$.
(2) L_λ is ample $\Leftrightarrow L_\lambda$ is very ample $\Leftrightarrow \lambda$ is regular dominant, i.e., $a_i > 0$ for all i.
(3) If $\operatorname{char} k = 0$, then $H^0(G/B, L_\lambda), \lambda$ being dominant, give all the finite dimensional, irreducible representations of G.

For a parabolic subgroup $Q, Q \supset B$, we have the notion of a dominant weight λ being Q-dominant (resp. Q regular dominant) which is equivalent to saying that L_λ "descends" to a line bundle on G/Q without base points (resp. descends to an ample (in fact a very ample line bundle) on G/Q).

1. Conjectures for a general SMT

Recall that a maximal parabolic subgroup $P, P \supset B$ (or the associated fundamental weight ω) is of *classical type* if $\langle \omega, \beta^\vee \rangle \leq 2$, for all positive roots β, and that a parabolic subgroup $Q, Q \supset B$ is of *classical type* if it is the intersection of maximal parabolic subgroups of classical type. Note that G is classical if and only if every parabolic subgroup of G is of classical type. Let us also recall the relationship between parabolic subgroups of classical type and Chevalley multiplicities. Given a dominant weight λ, let Q be the stabilizer of λ. For $\tau, \varphi \in W^Q$ (the set of minimal representatives of W/W_Q) such that $X_Q(\varphi)$ is a Schubert divisor in $X_Q(\tau)(\subset G/Q)$, let $m_\lambda(\tau, \varphi) := \langle \lambda, \beta^\vee \rangle$, the *Chevalley multiplicity* of $X_Q(\varphi)$ in $X_Q(\tau)$, where β is the positive root given by $\tau = \varphi s_\beta$. Then Q is of classical type if and only if $m_\lambda(\tau,\varphi) \leq 2$, for all pairs (τ,φ) as above.

After a SMT was developed in G/P-IV for Schubert varieties in $G/P, P$ being a maximal parabolic subgroup of classical type, the next task was to develope a SMT for Schubert varieties in $G/Q, Q$ being a parabolic subgroup of classical type (refered to as the "mixed case" in G/P-IV). Even though, a sketch of proof for the mixed case is given in G/P-IV, it was soon realized that the proof underlying this sketch was not correct. Besides, around this time, Victor Kac had found a flaw in Demazure's work (cf. [**5**]), namely that Demazure's proof of the normality of Schubert varieties and his character formula (in characteristic zero), were not correct. Essential use of these results was made in G/P-I to IV.

In view of this, together with Seshadri, the author presented in G/P-V (cf.[**22**]) an extensive revision of all the earlier work, as well as a new strategy for proving generation by standard monomials. The proof of linear independence is the same as in G/P-IV.

To illustrate the strategy in G/P-V for proving generation by standard monomials, let us work for simplicity, with G/P, where P is a maximal parabolic subgroup of classical type. Let X_τ be a Schubert variety in $G/P, H_\tau$ the zero set of p_τ in X_τ, p_τ being the extremal weight vector in $H^0(G/P, L_\omega)$ (of weight $\tau w_0(\omega), w_0$ being the element of largest length in W and ω the fundamental weight associated to P). We have, set theoretically $H_\tau = \cup X_{w'}$, the union being taken over all Schubert divisors $X_{w'}$ in X_w; hence, $(H_\tau)_{red} = \cup X_{w'}$, scheme theoretically. In view of linear independence of Standard monomials, we have only to prove that

(*)

$$dim\, H^0(X_\tau, \mathcal{O}_{X_\tau}(n)) = \#\{\text{Standard monomials on } X_\tau \text{ of degree } n\}$$

(note that $\mathcal{O}_{X_\tau}(1)$ is just the restriction of L_ω to X_τ). By induction, we can suppose that (*) holds on $(H_\tau)_{red}$. If we could compute $\dim H^0(H_\tau, \mathcal{O}_{H_\tau}(n))$, then we could use the exact sequence

$$(**) \qquad 0 \longrightarrow \mathcal{O}_{X_\tau}(n-1) \to \mathcal{O}_{X_\tau}(n) \to \mathcal{O}_{H_\tau}(n) \to 0$$

to compute $\dim H^0(X_\tau, \mathcal{O}_{X_\tau}(n))$. Thus, the essential difficulty stems from the fact that H_τ need not be reduced and the problem is to understand the scheme H_τ. For this, we proceed as follows:

Fix a Schubert divisor X_φ in X_τ, $\varphi = s_\alpha\tau, \alpha$ simple. By induction, SMT holds on X_φ; as a consequence, we obtain a nice filtration $\{I_j(\varphi)\}$ of the ideal sheaf $I((H_\varphi)_{red})$ of $(H_\varphi)_{red}$ in $\mathcal{O}_{X_\varphi}$ such that the associated graded is $\simeq \oplus\mathcal{O}_{X(\lambda_j)}(-1)$, λ_j running over all elements in

W^P such that (φ, λ_j) is an admissible pair (for the definition of an admissible pair, refer to [**22**] or see Definition 5.7 below in §5).

Let $Z_\phi \to \mathbb{P}^1$ be the fibre space with fibre X_ϕ, associated to the principal fibration

$$SL(2,\alpha) \to SL(2,\alpha)/B_\alpha = \mathbb{P}^1 \text{ (or } P_\alpha \to P_\alpha/B \simeq \mathbb{P}^1),$$

P_α being the minimal parabolic subgroup associated to α (here, $SL(2,\alpha)$ denotes the copy of SL_2 in G corresponding to α). Let $\psi : Z_\phi \to X_\tau$ be the canonical birational morphism defined by $(p,x) \mapsto p \cdot x$ ($p \in P_\alpha$ and $x \in X_\phi$). We take the sheaves $\widetilde{I}_j(\phi)$ on Z_ϕ "associated to $I_j(\phi)$" in the sense of fibre spaces. Then we get a nice filtration $\widetilde{I}(\phi)$ on Z_ϕ since $Z_\phi \longrightarrow \mathbb{P}^1$ is a fibration. Using $\widetilde{I}(\phi)$, we get the following:

$$(***) \qquad dim\, H^0(Z_\phi, \psi^*(\mathcal{O}_{X_\tau}(n))) = \#\{\text{Standard monomials on } X_\tau \text{ of degree } n\}$$

From this we deduce (*), and obtain that SMT holds on X_τ. As a consequence, we get a filtration $\{I_j(\tau)\}$ of the ideal sheaf $I((H_\tau)_{red})$ of $(H_\tau)_{red}$ in $\mathcal{O}_{X_\tau}$, such that the associated graded is $\simeq \oplus \mathcal{O}_{X(\lambda_j)}(-1)$, λ_j running over all elements in W^P such that (τ, λ_j) is an admissible pair. We also get the vanishing theorems for positive line bundles on Schubert varieties, as well as the validity of Demazure's conjecture. The normality and Demazure's character formulae (for any characteristic) for the class of Schubert varieties in G/Q, Q of classical type, follow as consequences.

Soon after we had the material of G/P-V, Mehta and Ramanathan proved that the Schubert varieties are Frobenius split; the vanishing theorems for positive line bundles on Schubert varieties, normality of Schubert varieties and Demazure's conjecture, all followed as consequences ([**34, 40, 39, 12**]), while these results followed from G/P-V only for the Schubert varieties in the flag varieties of classical type. Though SMT had helped in proving such theorems earlier in many significant cases, the method of Frobenius splitting seemed much better than SMT for proving, in generality, such results. Interestingly these two points of view have come together recently ([**14, 15**]).

Once SMT was established in [**19, 22**] for $G/Q, Q$ being of classical type, it became clear that there should be a general SMT valid for all semi-simple groups, i.e., on G/Q, where Q need not be of classical type. Following the method pursued in G/P-V, one had first to consider the case of G/P, where P is a maximal parabolic subgroup not of classical type. Then G is necessarily an exceptional group.

For P being of classical type, one has that the basis for $H^0(G/P, L_\omega)$ as constructed in [**19, 22**] consists of *certain* pairs of elements of W^P (the so called "admissible pairs"). Hence it is but natural to expect that the indexing set of a basis of $H^0(G/P, L_\omega)$ (where P is a maximal parabolic subgroup not of classical type) would be *certain* n-tuples $(\tau_1, \ldots, \tau_n)$, $\tau_i \in W/W_P$ (since Chevalley multiplicities ≥ 3 can occur in this case). The problem was at first to come up with a precise definition for these "*admissible n-tuples*" as was done for the case of admissible pairs. The simplest case is G_2/P, where P is associated to the right end root. The author first worked out SMT for this case (cf.[**16**]) and later for G_2 and E_6 in collaboration with Rajeswari ([**20**]). Working out SMT for the other exceptional groups in the same manner became quite complicated and the attention turned to formulating good conjectures for a general SMT.

Based on some computations on exceptional groups (and some crucial discussions with Seshadri), the author formulated the following two conjectures for a general SMT (cf. [**23**]). The first conjecture (Conjecture A below) describes a character formula for the T-module Char $H^0(X(\tau), L)$, while the second conjecture (Conjecture B below) describes a basis for $H^0(X(\tau), L)$.

Let λ be a dominant weight, Q the stabilizer of λ, and L_λ the associated line bundle on G/Q. Let $\tau, \sigma \in W^Q$ and a a rational number $0 < a < 1$.

DEFINITION 1.1. *An a-chain connecting τ, σ is a sequence $\underline{\tau} = (\tau = \tau_0 > \tau_1 > \cdots > \tau_s = \sigma), \tau_i \in W^Q$, such that $X_Q(\tau_i)$ is of codimension one in $X_Q(\tau_{i-1}), 1 \leq i \leq s)$ and if $d_i = m_\lambda(\tau_{i-1}, \tau_i)$ (the Chevalley multiplicity of $X_Q(\tau_i)$ in $X_Q(\tau_{i-1})$), then $ad_i \in \mathbb{Z}$, for all i.*

Note that since $a < 1$, we have, $d_i > 1$. Set $e_i = ad_i$.

DEFINITION 1.2. *Let $w \in W^Q$. A standard diagram π of shape λ on $X_Q(w)$ consists of a pair $(\underline{\tau}, \underline{a})$, where*

$$\underline{\tau} = (\tau_0 > \tau_1 > \cdots > \tau_r), \tau_i \in W^Q, w \geq \tau_0,$$

$$\underline{a} = (0 = a_0 < a_1 < \cdots < a_r < a_{r+1} = 1),\ a_i \text{ rational}, 1 \leq i \leq r$$

such that we have an a_i-chain connecting τ_{i-1} and $\tau_i, 1 \leq i \leq r$.

REMARK 1.3. *In the above definition, we allow for $r = 0$, in which case, π will be just an element of W^Q. If Q is a parabolic sub group of classical type, then the standard diagrams of shape λ on G/Q are of the form $\pi = (\underline{\tau}, \underline{a})$, where $\underline{\tau} = (\tau, \varphi), \underline{a} = (0 < \frac{1}{2} < 1)$; thus* *in this case, standard diagrams of shape λ on G/Q are precisely the admissible pairs of* [**19, 22**].

DEFINITION 1.4. *For a standard diagram π of shape λ on G/Q, define*

$$wt\,\pi = \sum_{i=0}^{r} b_i \tau_i(\lambda),\ b_i = a_{i+1} - a_i.$$

Conjecture A There exists a basis $\{p_\pi\}$ of weight vectors of $H^0(X_Q(w), L)$ indexed by standard diagrams $\pi = (\underline{\tau}, \underline{a})$ of shape λ on $X_Q(w)$ such that weight of $p_\pi = -wt\pi$.

REMARK 1.5. *Observe that $\sum b_i = 1$, i.e., wt p_π is in the convex cone spanned by the extremal weights. Also, Conjecture A implies that*

$$\operatorname{Char} H^0(X_Q(w), L_\lambda) = \sum_{\pi \in \Pi_\lambda(w)} e^{-wt\pi},$$

where $\Pi_\lambda(w) = \{\text{standard diagrams of shape } \lambda \text{ on} X_Q(w)\}$. *We see that Conjecture A is equivalent to the above character formula. We refer to this as the SMT character formula.*

A conjectural basis for the Weyl module $V_{\lambda,\mathbb{Z}}$. Let U denote the universal enveloping algebra of Lie G, $U_{\mathbb{Z}}$ the Kostant $\mathbb{Z}$-form of U, and $U_{\mathbb{Z}}^+$ the $\mathbb{Z}$-subalgebra of $U_{\mathbb{Z}}$ generated by $\frac{X_\beta^n}{n!}, n \in \mathbb{N}$, and β any positive root. Let λ be a dominant integral weight. Let $V_{\lambda,\mathbb{Q}}$ denote the irreducible $G_{\mathbb{Q}}$-module with highest weight λ. Fix a highest weight vector $e \in V_{\lambda,\mathbb{Q}}$. For $w \in W$, fix a lift n_w in $N(T)$, and set $e_w = n_w \cdot e$, an *extremal weight vector* of weight $w(\lambda)$ (note that e_w is unique up to scalars). Let $V_{\lambda,\mathbb{Z}} = U_{\mathbb{Z}} \cdot e, V_{\lambda,\mathbb{Z}}(w) = U_{\mathbb{Z}}^+ \cdot e_w$. For any field k, let $V_{\lambda,k} = V_{\lambda,\mathbb{Z}} \otimes k, V_{\lambda,k}(w) = V_{\lambda,\mathbb{Z}}(w) \otimes k$. Then $V_{\lambda,k}$ is called the *Weyl module* and $V_{\lambda,k}(w)$ is called the *Demazure module associated to w*. We have the following well-known facts:

$$H^0(G/Q, L_\lambda) = V_{\lambda,k}^*,$$

$$H^0(X_Q(w), L_\lambda) = (V_{\lambda,k}(w))^*.$$

Given an a-chain connecting τ, σ as in Definition 1.1, define the operator

$$D_{\tau,\sigma} = X_{-\beta_1}^{(e_1)} X_{-\beta_2}^{(e_2)} \cdots X_{-\beta_s}^{(e_s)}$$

where $X_{-\beta_i}^{(e_i)} = \frac{X_{\beta_i}^{e_i}}{e_i!}$, β_i is the positive root given by $\tau_{i-1} = s_{\beta_i}\tau_i$, and $e_i = a\, m_\lambda(\tau_{i-1}, \tau_i)$.

Given $\pi = (\underline{\tau}, \underline{a})$, a standard diagram of shape λ, as in Definition 1.2, we set

$$Q_\pi = D_{\tau_0,\tau_1} \cdot D_{\tau_1,\tau_2} \cdot \cdots \cdot D_{\tau_{r-1},\tau_r} Q_{\tau_r},$$

where Q_{τ_r} is the extremal weight vector in $V_{\lambda,\mathbb{Z}}$ of weight $\tau_r(\lambda)$.
Conjecture B $\{Q_\pi, \pi \in \Pi_\lambda(w)\}$ is a $\mathbb{Z}$-basis (of weight vectors) of $V_{\lambda,\mathbb{Z}}(w)$ (and base change of $\{Q_\pi\}$ gives a basis of $V_{\lambda,k}(w)$).

REMARK 1.6. *Note that corresponding to standard diagrams π with $r = 0$ (notations as in Definition 1.2), we get all the extremal weight vectors in $V_{\lambda,\mathbb{Z}}$. Note also that if $\{P_\pi, \pi \in \Pi_\lambda(w)\}$ denotes the basis of $H^0(X_Q(w), L_\lambda)$ dual to $\{Q_\pi, \pi \in \Pi_\lambda(w)\}$, then the weights of $P_\pi, \pi \in \Pi_\lambda(w)$ are as in Conjecture A above (here, $\Pi_\lambda(w)$ is as in Remark 1.5).*

REMARK 1.7. *When Q is a maximal parabolic subgroup of classical type and $\lambda = \omega$, the associated fundamental weight, if $\pi = (\tau, \mu)$ (an admissible pair), then by Conjecture B above, we have*

$$Q_{\tau,\mu} = X_{-\beta_1} X_{-\beta_2} \cdots X_{-\beta_s} Q_\mu$$

which coincides with the basis constructed in [**22**].

REMARK 1.8. *Even on G/Q, Q being of classical type, the above conjectures are not simple consequences of the SMT of classical type (of* [**19, 22**]*). If Q is a maximal parabolic subgroup of classical type and λ is the associated fundamental weight, then it is not difficult to deduce the SMT character formula of Conjecture A above from the character formula of* [**22**]*, but otherwise, á priori, we get an SMT character formula different from that of* [**22**]*. As for the bases of Conjecture B, even if Q is a maximal parabolic subgroup of classical type and $\lambda = m\omega, m \geq 2$, they are á priori different from those given in* [**22**].

Besides the evidence provided by the already established SMT, another evidence for the above conjectures, was SMT which the author together with Seshadri worked out for the simplest Kac-Moody case, namely $SL(2, \infty)$ (cf.[**24**]). A priori the above conjectures could be formulated for Schubert varieties in the Kac-Moody set up. The importance of the case $SL(2, \infty)$ is that the Chevalley multiplicities even with respect to a fundamental weight tend to ∞. We could work out SMT for this case since the geometry of the Schubert varieties for this case is not hard. Thus the results of [**24**] supported the above conjectures; further, it showed also that SMT would also hold in the Kac-Moody set up.

2. The work of Littelmann

Solving the above conjectures required new ideas. Littelmann proved the conjectures A&B in [**26, 30**] respectively. He also constructed another basis for $H^0(X_Q(w), L_\lambda)$ (cf. [**28**]) suited for SMT, and thus completed the Standard Monomial theory even in the Kac-Moody set-up! We shall briefly describe below Littelmann's results.

The starting point of Littelmann's work was to generalize the Littelwood-Richardson rule for decomposing tensor products of irreducible representations of linear groups into irreducible representations (in characteristic zero). The classical Littelwood-Richardson rule uses the language of the classical Young tableaux. It is but natural to expect that the generalization of this rule would admit of a good answer in the language of the *standard tableaux* of the SMT that was then established for classical groups, G_2 and E_6. He found such an answer in [**25**]. On the other hand the decomposition (into irreducible representations) of a tensor product of irreducible representations has a good answer in terms of *crystal bases and crystal graphs* in the sense of Kashiwara ([**8, 9, 10, 11**]). This led Littelmann to show that the crystal basis of an irreducible representation along with the Kashiwara operators $\widetilde{e}_\alpha$, $\widetilde{f}_\alpha$ have a concrete model in terms of the standard tableaux (of the SMT) endowed with operators e_α, f_α which correspond to $\widetilde{e}_\alpha, \widetilde{f}_\alpha$. This was the starting point for the new methods brought into SMT by Littelmann.

2.1. Littelmann's paths and path operators. Littelmann's idea is to view the standard tableaux as paths and work with these paths.

DEFINITION 2.2. *A path π is a piecewise linear map $\pi : [0,1] \longrightarrow X(T)_\mathbb{R}$ ($X(T)$ being the character group of T, $X(T)_\mathbb{R} = X(T) \otimes_Z \mathbb{R}$).*

DEFINITION 2.3. *Let π_1, π_2 be two paths. Then the concatenation $\pi_1 * \pi_2$ of π_1 and π_2 is the "composite path"*

$$(\pi_1 * \pi_2)(t) = \begin{cases} \pi_1(2t), \ 0 \le t \le \frac{1}{2} \\ \pi_1(1) + \pi_2(2t-1), \frac{1}{2} \le t \le 1. \end{cases}$$

We call $\pi(1)$ the *weight of* π (written $wt\ \pi$).

DEFINITION 2.4. *Let $\lambda \in X(T)$. We denote by π_λ the path joining origin and λ i.e.*

$$\pi_\lambda(t) = t\lambda, \ t \in [0,1]$$

Let $\pi = (\underline{\tau}, \underline{a})$ be a standard diagram of shape λ as in Definition 1.2 (λ being dominant). To this one associates a path denoted by the same letter π as follows:

Let π_i be the path

$$\pi_i(t) = b_i \tau_i(\lambda), b_i = a_{i+1} - a_i, 0 \leq i \leq r.$$

Then π is the concatenation of the π_i's, i.e.,

$$\pi = \pi_1 * \pi_2 * \cdots * \pi_r.$$

Littelmann calls π a Lakshmiabai - Seshadri path (LS path) of shape λ. Note that $wt\ \pi (= \sum_{i=0}^{r} b_i \tau_i(\lambda))$ is just $\pi(1)$.

Root operators e_α, f_α: Littelmann defines the operators e_α, f_α (α simple) on the set of all paths. To be more precise, let $\mathcal{P}$ denote the set of all paths and $\mathbb{Z}[\mathcal{P}]$ the free $\mathbb{Z}$-module whose basis is $\mathcal{P}$. Then e_α, f_α are in $\mathrm{End}_{\mathbb{Z}} \mathbb{Z}[\mathcal{P}]$. For every simple root α, define $s_\alpha(\pi)$ by

$$s_\alpha(\pi)(t) = s_\alpha(\pi(t)),\ t \in [0,1].$$

To define f_α (and e_α), we cut π into a suitable number of smaller paths π_i so that

$$\pi = \pi_1 * \cdots * \pi_m$$

and $f_\alpha(\pi)$ is the concatenation

$$f_\alpha(\pi) = \eta_1 * \cdots * \eta_m$$

where η_i is π_i or $s_\alpha(\pi_i)$. The operator e_α is defined similarly. For the precise definition of the root operators, see [**27**]. One has the property $wt\ f_\alpha(\pi) = wt\ \pi - \alpha$ (resp. $wt\ e_\alpha(\pi) = wt\ \pi + \alpha$). The important point is that f_α, e_α leave stable the set of LS paths. If one views an LS path π as a standard tableau $(\tau_0, \ldots, \tau_r)$ of shape λ, then roughly speaking $f_\alpha(\pi)$ is given by replacing some of these τ_i by $s_\alpha \tau_i$. To give this rule, viewing π as a path, i.e., as a function, is much more convenient.

Denote

$$\mathcal{P}_\lambda = \{\text{the set of all LS paths of shape } \lambda\}.$$

If $\pi = (\underline{\tau}, \underline{a})$ is an LS path, one calls τ_0 *the initial element of* π and it is denoted by $i(\pi)$. Set

$$\mathcal{P}_\lambda(\tau) = \{\text{the set of all LS paths } \pi \text{ of shape } \lambda, \tau \geq i(\pi)\}.$$

Let $\mathcal{A}$ denote the subalgebra of $\mathrm{End}_{\mathbb{Z}}[\mathcal{P}]$ generated by e_α, f_α.

THEOREM 2.5. *(cf. [**27**]) Let λ be dominant, and π_λ be the LS path $\pi_\lambda(t) = t\lambda$.*

(1) *Every* $\pi \in \mathcal{P}_\lambda$ *is of the form*

$$\pi = f_{\alpha_1} \cdots f_{\alpha_s}(\pi_\lambda).$$

(2) $\mathcal{P}_\lambda$ *is a* $\mathcal{A}$*-basis for the* $\mathcal{A}$*-module* $\mathcal{A} \cdot \pi_\lambda$.

(3) $\operatorname{Char} V_{\lambda,k}(\tau) = \sum_{\pi \in \mathcal{P}_\lambda(\tau)} e^{wt\ \pi}$.

(4) *Let* π_1 *and* π_2 *be two paths such that* π_1 *and* π_2 *are in the dominant chamber. Then* $\mathcal{A}_{\pi_1}$ *and* $\mathcal{A}_{\pi_2}$ *are isomorphic as* $\mathcal{A}$*-modules if and only if* $wt\ \pi_1 = wt\ \pi_2$.

REMARK 2.6. *If we take -wt* π *(in the place of wt* π*) in the sum on the RHS of (3), then it is the SMT character formula for* $H^0(X(\tau), L_\lambda)$ *(cf. Remark 1.5).*

REMARK 2.7. *The assertion (2) in Theorem 2.5 gives a nice explanation of LS paths and is very analogous to the properties of the crystal bases. One would expect, of course, that the colored graph* $\mathcal{G}(\mathcal{P}_\lambda)$ *defined by* $\pi_1 \xrightarrow{\alpha} \pi_2$ *if* $f_\alpha(\pi_1) = \pi_2$ *(*π_i *in* $\mathcal{P}_\lambda$*), is isomorphic to Kashiwara's crystal graph; this turns out to be true and is proved in ([**7**]).*

REMARK 2.8. *Using the above results, Littelmann obtains a generalization of the classical Littelwood-Richardson rule and gives a decomposition formula for tensor products of irreducible representations of semi simple algebraic groups into irreducible representations (in characteristic zero); further, he also gives branching rules, i.e., the decomposition of* V_λ *as an L-module, where L is a Levi subgroup of G:*

Decomposition rule: *Let char* $k = 0$. *Let* V_λ, V_μ *be two irreducible G-modules. Consider* $\nu = \lambda + wt\,\pi$, *where* $\pi = (\underline{\tau}, \underline{a})$ *is a LS path of shape* μ, $\underline{\tau} = (\tau_0 > \tau_1 > \cdots > \tau_r), \underline{a} = (0 = a_0 < a_1 < \cdots < a_r < a_{r+1} = 1)$ *such that for each partial sum* $\chi_j := \sum_{i=0}^{j} b_i \tau_i(\lambda)$, $b_i = a_{i+1} - a_i$, $\lambda + \chi_j$ *is dominant, i.e., the shifted path* $\lambda + \pi$ *lies in the dominant Weyl chamber. Let* $Z_{\lambda,\mu}$ *be the set of all such* ν*'s. Then*

$$V_\lambda \otimes V_\mu = \oplus_{\nu \in Z_{\lambda,\mu}} V_\nu.$$

Branching rule: *Let char* $k = 0$. *Let L be a Levi subgroup of G. Call* ν *L-dominant, if* $\langle \nu, \beta^\vee \rangle \geq 0$, *for all simple roots of L. Let* λ *be a dominant weight, and* $\pi = (\underline{\tau}, \underline{a})$, *a LS path of shape* λ, $\underline{\tau} = (\tau_0 > \tau_1 > \cdots > \tau_r), \underline{a} = (0 = a_0 < a_1 < \cdots < a_r < a_{r+1} = 1)$; *further, for each partial sum* $\chi_j := \sum_{i=0}^{j} b_i \tau_i(\lambda)$, $b_i = a_{i+1} - a_i$, *let* χ_j *be L-dominant, i.e., the path* π *lies in the dominant Weyl chamber of L. Let* $Z_\lambda(L)$ *be the set of all such* π*'s. Then as an L-module,* V_λ *has* *the decomposition*

$$V_\lambda = \oplus_{\pi \in Z_\lambda(L)} U_\pi$$

(here, for a L-dominant weight ν, U_ν *denotes the corresponding irreducible L-module). See* [**26**] *for details. If we are in arbitrary characteristic and* V_λ, V_μ *are Weyl modules, then the generalization of decomposition rule for* $V_\lambda \otimes V_\mu$ *is that there is a filtration of* $V_\lambda \otimes V_\mu$ *whose associated graded is a certain direct sum of Weyl modules. This is the property of good filtrations for which Littelmann has given a proof using SMT (*[**29**]*).*

The remaining problem in building a general SMT is the construction of a nice basis $\{p_\pi \mid \pi \in \mathcal{P}_\lambda\}$ for $H^0(G/Q, L_\lambda)$ (as well as for $H^0(X(w), L_\lambda)$). Recall the fact that the extremal weight vectors have no base points (identified as sections of a line bundle on G/Q) so that a non-extremal weight vector is integral over the algebra generated by the extremal weight vectors. (An example of this is the special quadratic relation $p^2_{\tau,\phi} = \pm p_\tau p_\phi$ on $X(\tau)$, a Schubert variety in G/P, P a maximal parabolic subgroup of classical type.) Let $\pi \in \mathcal{P}_\lambda$, say $\pi = (\underline{\tau}, \underline{a})$ (cf. Definition 1.2). Then $wt\ \pi = \sum_{i=0}^r b_i \tau_i(\lambda)$ $(b_i = a_{i+1} - a_i)$. Take l such that $lb_i = n_i$ is an integer. Then $l = \sum n_i$ and we would expect that on the Schubert variety $X(\tau_0)$

$$(*) \qquad p^l_\pi = \prod_{i=0}^{r} p^{n_i}_{\tau_i} \ (\text{ a product of extremal weight vectors}),$$

i.e.,

$$p^l_\pi = \prod_{i=0}^{r} p^{n_i}_{\tau_i} + \text{terms vanishing on } X(\tau_0).$$

Littelmann's construction has this in background. "To extract l-th roots", he uses the Frobenius map of Lusztig for quantum groups at roots of unity (cf. [**32**]). For simplicity, let us suppose that we are in the simply-laced case (his arguments work in the set up of arbitrary symmetrizable Kac-Moody algebras). Using this Frobenius map, one imbeds V_λ into an l-fold tensor product $\otimes^l M_\lambda$ of the Weyl module (over the ring $\mathbb{Z}[v]$) of highest weight λ for the quantum group $U_v(\underline{g})$ at v, an l-th root of unity (here, $\underline{g}$ is the Lie algebra of G). We then get a canonical homomorphism $\otimes M^*_\lambda \longrightarrow V^*_\lambda$. Then one attaches to π a decomposable tensor in $\otimes^l M^*_\lambda$ which is a product of extremal weight vectors in M^*_λ similar to (*) above. We denote by p_π the image of this vector in V^*_λ.

THEOREM 2.9. *(cf.* [**28**]*)*

(1) p_π *is a weight vector of weight -wt* π.
(2) $\{p_\pi, \pi \in \mathcal{P}_\lambda\}$ *is a basis for* $H^0(G/Q, L_\lambda)$.
(3) $p_\pi \mid_{X(\tau)} \not\equiv 0$ *if and only if* $\tau \geq i(\pi)$, *i.e., if and only if* π *belongs to* $\mathcal{P}_\lambda(\tau)$.
(4) $\{p_\pi \mid_{X(\tau)}, \pi \in \mathcal{P}_\lambda(\tau)\}$ *is a basis for* $H^0(X(\tau), L_\lambda)$.

The crucial point of the proof of the above theorem is that $\{p_\pi \mid \pi \in \mathcal{P}_\lambda(\tau)\}$ is linearly independent in $H^0(X(\tau), L_\lambda)$. The proof of linear independence in [**28**] is inspired by a method due to K.N. Raghavan and P. Sankaran ([**38**]).

One can now define standard monomials in $\{p_\pi\}$ of length m (in the spirit of G/P-V) on $X(\tau)$ and show that they form a basis of $H^0(X(\tau), L_\lambda^m)$.

Theorems 2.5, 2.9 above and their consequences constitute a general SMT for all Schubert varieties in the Kac-Moody setup. They subsume all the earlier results, e.g., the results in G/P-V. Also, in the spirit of G/P-V, the above construction of SMT is completely self-contained and doesn't suppose any vanishing theorems for positive line bundles or the truth of Demazure's conjecture. All these follow as consequences. One unexpected consequence, suggested by the proof of Theorem 2.9, is the coming together of the techniques of Frobenius splitting and SMT, to study the geometry of Schubert varieties, in the work of Littelmann and Shrawan Kumar. See [**14, 15**] for details. Littelmann has also constructed bases for Weyl and Demazure modules (as given by conjecture B; see [**30**] for details, appearing in this volume).

3. K-theory and SMT

A very recent interesting development is the connection between SMT and the Grothendieck ring $K(G/B)$. The Chow ring Chow(G/B) has a **Z**-basis consisting of $[X_w]$, the class in Chow(G/B) defined by the cycle represented by the Schubert variety X_w, $w \in W$. Let $\lambda \in \Lambda$ (=the weight lattice), L_λ the associated line bundle. The class $[L_\lambda]$ is defined by a codimension one cycle. We have, the following expression for the product $[X_w] \cdot [L_\lambda]$ in the Chow ring:

$$[X_w] \cdot [L_\lambda] = \sum a_i [X_{w_i}] \tag{1}$$

where the summation runs over all Schubert divisors X_{w_i} in X_w; further, one has an explicit expression for a_i due to Chevalley (cf. [**3**]),namely, $a_i = \langle \lambda, \beta_i^\vee \rangle$, where β_i is the positive root given by $w_{i-1} = w_i s_{\beta_i}$.

Consider $K(G/B)$, the Grothendieck ring of isomorphism classes of coherent sheaves on G/B. Then one knows that the classes $[\mathcal{O}_{X_w}]$ in $K(G/B)$ of the structure sheaves $\mathcal{O}_{X_w}$ of the Schubert varieties X_w form a **Z**-basis of $K(G/B)$. Let

$$(2) \qquad [\mathcal{O}_{X_w}] \cdot [L_\lambda] = \sum_{v \in W} a^\lambda_{w,v} [\mathcal{O}_{X_v}],\ a^\lambda_{w,v} \in \mathbb{Z}.$$

More generally, let $K_T(G/B)$ be the Grothendieck ring of T-equivariant coherent sheaves on G/B. By [**13**] one knows that again the classes $[\mathcal{O}_{X_w}]_T$ of the structure sheaves of the Schubert varieties form a $\mathbb{Z}[\Lambda]$-basis, where $\mathbb{Z}[\Lambda]$ is the group algebra of the weight lattice Λ. Let

$$(3) \qquad [\mathcal{O}_{X_w}]_T \cdot [L_\lambda]_T = \sum_{v \in W} c^\lambda_{w,v} [\mathcal{O}_{X_v}]_T,\ \in \mathbb{Z}[\Lambda].$$

Assume now that $\lambda \in \Lambda^+$. Then we have that in (1) all $a_i \geq 0$ (cf.[**3**]). The integers $a^\lambda_{w,v}$ in (2) were determined by Fulton and Lascoux (cf. [**6**]) for the case $G = SL(n)$; they provide a formula for $a^\lambda_{w,v}$ (using the combinatorics of Grothendieck polynomials). The general case was treated by Mathieu using representaion theory (cf. [**33**]), who shows that $c^\lambda_{w,v}$ are effective i.e., formal sums of characters with positive integral coefficients. Using SMT chraracter formula (cf. Theorem 2.5), Pittie and Ram (cf. [**37**]) have given an explicit determination of $c^\lambda_{w,v}$'s:

$$(4) \qquad [\mathcal{O}_{X_w}]_T \cdot [L_\lambda]_T = \sum [\mathcal{O}_{X_{e(\pi)}}]_T\, e^{\mathrm{wt} p_\pi}$$

where the summation runs over all L-S paths π of shape λ such that $i(\pi) \leq w$, and p_π is the element corresponding to π in the basis as given by Theorem 2.9 (here, for a L-S path $\pi = (\tau_0 > \cdots > \tau_r; a_1 < \cdots < a_r), i(\pi) = \tau_0$ (the initial element), $e(\pi) = \tau_r$ (the end element), and $\mathrm{wt} p_\pi = -\mathrm{wt}\pi$).

Connection between $c^\lambda_{w,v}$ and X^v_w: In [**17**], $c^\lambda_{w,v}$ is shown to be the character of a certain T-module. We briefly describe below this result of [**17**]. For this, we first have to introduce the *Richardson varieties*:

Let $w \in W^Q$. The boundary ∂X_w is given by

$$\partial X_w = \bigcup_{v \in W^Q,\ v<w} X_v,$$

and the irreducible components of ∂X_w are the *Schubert divisors* X_v where $v \in W^Q$, $v < w$ and $\ell(v) = \ell(w) - 1$.

We will also need the *opposite Schubert variety* $X^w := \overline{B^- e_w}$; its boundary ∂X^w is given by

$$\partial X^w = \bigcup_{v \in W^Q, v > w} X^v.$$

DEFINITION 3.1. *Let v, w in W^Q. We call the intersection*

$$X_w^v := X_w \cap X^v$$

a Richardson variety in G/Q. We define its boundaries by

$$(\partial X_w)^v := \partial X_w \cap X^v, (\partial X^v)_w := X_w \cap \partial X^v.$$

REMARK 3.2. *Richardson was the first to look at such varieties (cf. [**41**]) and hence the name! It can be seen easily that X_w^v is non-empty if and only if $v \leq w$. It follows from [**41**] that the Richardson varieties are reduced and irreducible; hence it follows that a union of Richardson varieties (in particular, $(\partial X_w)^v, (\partial X^v)_w$) is reduced. The intersections $Be_w \cap B^- e_w$ were analyzed by Deodhar (cf. [**4**]).*

Set

$$\Pi_w^v = \{ \text{ the set of all LS paths } \pi \text{ of shape } \lambda, w \geq i(\pi) \geq e(\pi) \geq v\}.$$

More generally, for a union of Richardson varieties $Z = \cup_j X_{w_j}^{v_j}$, set

$$\Pi_Z = \{\pi \mid w_j \geq i(\pi) \geq e(\pi) \geq v_j. \text{ for some } j\}.$$

DEFINITION 3.3. *A union of Richardson varieties $Z = \cup_j X_{w_j}^{v_j}$ is called a pointed union if either all the w_j's or all the v_j's are the same*

$(\partial X_w)^v, (\partial X^v)_w$ are examples of pointed unions of Richardson varieties.

The following theorem is proved in [**17**].

THEOREM 3.4. *Let λ be Q-dominant, $w, v \in W^Q, w \geq v$. Let Z be a pointed union of Richardson varieties in G/Q. Then $\{p_\pi, \pi \in \Pi_Z\}$ is a basis for $H^0(Z, L_\lambda)$.*

Determination of $c_{w,v}^\lambda$ as a character: First assume that λ is regular on G/Q. Consider the exact sequence

$$0 \to \mathfrak{I}_{(\partial X^v)_w} \to \mathcal{O}_{X_w^v} \to \mathcal{O}_{(\partial X^v)_w} \to 0$$

where $\mathfrak{I}_{(\partial X^v)_w}$ denotes the ideal sheaf of $(\partial X^v)_w$ in $\mathcal{O}_{X_w^v}$. Tensoring by L_λ, and writing the cohomology exact sequence, we get the exact sequence

$$0 \to H^0(X_w^v, \mathfrak{I}_{(\partial X^v)_w} \otimes L_\lambda) \to H^0(X_w^v, L_\lambda) \to H^0((\partial X^v)_w, L_\lambda) \to 0.$$

(Note that in view of Theorem 3.4, the restriction map $H^0(X_w^v, L_\lambda) \to H^0((\partial X^v)_w, L_\lambda)$ is surjective.) Hence we obtain dim $H^0(X_w^v, \mathfrak{I}_{(\partial X^v)_w} \otimes L_\lambda) = \#\{p_\pi | i(\pi) \leq w, e(\pi) = v\}$. Since by [**17**], these vectors in fact form a basis for the kernel of the restriction map $H^0(X_w^v, L_\lambda) \to H^0((\partial X^v)_w, L_\lambda)$, it follows that $\{p_\pi | i(\pi) \leq w, e(\pi) = v\}$ is actually a basis of $H^0(X_w^v, \mathfrak{I}_{(\partial X^v)_w}) \otimes L_\lambda$. In the non-regular case, we have to work with the λ-boundary $(\partial_\lambda X^v)_w := \cup_{\{v' \in W^Q, w \geq v' > v, v'_\lambda > v_\lambda\}} X_w^{v'}$ in the place of $(\partial X^v)_w$ (here, for $w \in W^Q, w_\lambda$ denotes the element in W^λ representing wW_λ, W_λ being the stabilizer of λ). Thus we obtain

THEOREM 3.5. *With notations as above, let λ be dominant. We have*

$$c_{w,v}^\lambda = \mathrm{char} H^0(X_w^v, L_\lambda \otimes \mathfrak{I}_{(\partial X^v)_w}),$$

$$a_{w,v}^\lambda = \#\{\pi, \text{ an L–S path of shape } \lambda \mid w \geq i(\pi) \geq e(\pi) = v\}.$$

REMARK 3.6. *The above result is also proved in* [**1**].

4. A geometric approach to Standard monomial theory:

In [**31**], Littelmann and Seshadri prove equation (4) in §3 by constructing a filtration $\mathcal{F} := \{\mathcal{F}^i\}$ of $\mathcal{O}_{X_w} \otimes L_\lambda$ by B-equivariant $\mathcal{O}_{G/B}$-modules (using standard monomial basis) such that each subquotient, as a B-equivariant $\mathcal{O}_{G/B}$-sheaf, is isomorphic to $\mathcal{O}_{X_{e(\pi)}} \otimes \chi_{-\mathrm{wt}\pi}$, the structure sheaf $\mathcal{O}_{X_{e(\pi)}}$, twisted by the character $-\mathrm{wt}\pi$ of B (note that $-\mathrm{wt}\pi$ is the weight of p_π), for a suitable π such that $i(\pi) \leq w$. In particular, the above filtration gives rise to a filtration of $H^0(X_w, L_\lambda)$ such that the associated graded $\cong \oplus_{\{\pi, i(\pi) \leq w\}} H^0(X_{e(\pi)}, \mathcal{O}_{X_{e(\pi)}}) \otimes \chi_{-\mathrm{wt}\pi}$.

Two questions naturally arise. First observe that (4) is equivalent to SMT character formula. Also, as mentioned above, the results in [**31**] show that SMT implies a canonical filtration $\{\mathcal{F}^i\}$ for $\mathcal{O}_{X_w} \otimes L_\lambda$ by B-equivariant $\mathcal{O}_{G/B}$-modules.

Question 1 Can one develope a SMT by constructing a nice filtration $\{\mathcal{F}^i\}$ of $\mathcal{O}_{X_w} \otimes L_\lambda$ so that each subquotient is again $\mathcal{O}_{X_v}$ for some v (or a twist of $\mathcal{O}_{X_v}$ by a certain character of B)?

In fact in [**31**], it is shown that for G classical, SMT is equivalent to the existence of a nice filtration as above.

The second question is the following: In the construction of the standard monomial basis in [**28**], as we saw above, the theory of Quantum groups (at a root of unity) is used in a crucial way.

Question 2 Is it possible to develop a SMT purely geometrically (without using any Quantum group theory)?

Together with Brion , the author has answered Questions 1 & 2 in the affirmative in [**2**]. We shall describe below the main steps of SMT as developed in [**2**].

4.1. The flat family: Let $\theta : k^* \to T$ be a regular dominant one parameter sub group. Consider the flat family $\pi : \mathcal{X}_w^v \to \mathbb{A}^1$, where $\mathcal{X}_w^v$ is the closure in $X_w^v \times X_w^v \times \mathbb{A}^1$ of the subset $\{(y, \theta(s)y, s), y \in X_w^v, s \in k^*\}$. The generic fiber of the flat family is $\cong \operatorname{diag}(X_w^v) \subset X_w^v \times X_w^v$, and the special fiber is $\cong \cup_{v \le x \le w} X_x^v \times X_w^x$. We shall denote

$$F = \bigcup_{x \in W^Q} X_x \times X^x, \ F_w^v = \bigcup_{x \in W^Q,\ v \le x \le w} X_x^v \times X_w^x.$$

(here, Q is the stabilizer of λ). Note that $F = F_{w_0}^{id}, w_0$ being the element of largest length in W.

A filtration of $\mathcal{O}_F, \mathcal{O}_{F_w^v}$: We take a total ordering on $W^Q : x_1 \prec x_2 \prec \cdots \prec x_m$ such that $x_i < x_j$ (in the Bruhat order) $\Rightarrow i \le j$. For $x \in W^Q$, let $\mathcal{O}_{F, \prec x}$ (resp. $\mathcal{O}_{F, \succ x}$) be the subsheaf of $\mathcal{O}_F$ consisting of those sections that vanish identically on $X_y \times X^y$ for each $y \succ x$ (resp. $y \prec x$).

Proposition 4.2. *$\mathcal{O}_{F, \prec x}$ (resp. $\mathcal{O}_{F, \succ x}$) are an ascending (resp. descending) filtration of $\mathcal{O}_F$, with associated graded*

$$\bigoplus_{x \in W^Q} \mathcal{O}_{X_x} \otimes \mathcal{O}_{X^x}(-\partial X^x) \,(\text{resp.} \bigoplus_{x \in W^Q} \mathcal{O}_{X_x}(-\partial X_x) \otimes \mathcal{O}_{X^x}).$$

The induced filtrations on the structure sheaf $\mathcal{O}_{F_w^v}$ have associated graded

$$\bigoplus\nolimits_{x \in W^Q,\ v \le x \le w} \mathcal{O}_{X_x^v} \otimes \mathcal{O}_{X_w^x}(-(\partial X^x)_w)$$
$$(\text{resp.} \bigoplus\nolimits_{x \in W^Q,\ v \le x \le w} \mathcal{O}_{X_x^v}(-(\partial X_x)^v) \otimes \mathcal{O}_{X_w^x}).$$

A filtration of $H^0(X_w^v, L_\lambda)$: As a consequence of the above Proposition, we obtain filtrations of $H^0(X_w^v, L_\lambda)$ given as follows: For $v \le x \le w$, let $H^0(X_w^v, L_\lambda)_{\prec x}$ (resp. $H^0(X_w^v, L_\lambda)_{\succ x}$) be the subspace of $H^0(X_w^v, L_\lambda)$ consisting of those sections that vanish identically on X_w^y for all $y \succ x$ (resp. $y \prec x$).

Proposition 4.3. *$\{H^0(X_w^v, L_\lambda)_{\prec x}, v \le x \le w\}$ (resp. $\{H^0(X_w^v, L_\lambda)_{\succ x}, v \le x \le w\}$) defines an ascending (resp. descending) filtration of $H^0(X_w^v, L_\lambda)$ whose associated graded is given by*

$$\bigoplus_{x \in W^Q,\ v \le x \le w} H^0(X_w^x, L_\lambda(-(\partial X^x)_w)).$$

$$(resp. \bigoplus_{x\in W^Q,\, v\le x\le w} H^0(X_x^v, L_\lambda(-(\partial X_x)^v)).)$$

Further, the map

$$H^0(X_w^v, L_\lambda)_{\prec x} \longrightarrow \mathrm{gr}_{\prec x} H^0(X_w^v, L_\lambda) \,(= H^0(X_w^x, L_\lambda(-(\partial X^x)_w)))$$

$$(resp.\ H^0(X_w^v, L_\lambda)_{\succ x} \longrightarrow \mathrm{gr}_{\succ x} H^0(X_w^v, L_\lambda) \,(= H^0(X_x^v, L_\lambda(-(\partial X_x)^v))))$$

is just the restriction to X_w^x.

A filtration of $H^0(X_w^v, L_\lambda(-(\partial X_w)^v))$: A typical sub quotient in the decreasing filtration of $H^0(X_w^v, L_\lambda)$ (as given by Proposition 4.3) being of the form $H^0(X_x^v, L_\lambda(-(\partial X_x)^v))$, we now describe a filtration of $H^0(X_w^v, L_\lambda(-(\partial X_w)^v))$ which will then give similar filtrations for $H^0(X_x^v, L_\lambda(-(\partial X_x)^v)), v \le x \le w$.

For $v \le x \le w$, let $H^0(X_w^v, L_\lambda(-(\partial X_w)^v))_{\prec x}$ be the subspace of $H^0(X_w^v, L_\lambda(-(\partial X_w)^v))$ consisting of those sections that vanish identically on X_w^y for all $y \succ x$.

Proposition 4.4. *$\{H^0(X_w^v, L_\lambda(-(\partial X_w)^v))_{\prec x}, v \le x \le w\}$ defines an ascending filtration of $H^0(X_w^v, L_\lambda(-(\partial X_w)^v))$ whose associated graded is given by*

$$\bigoplus_{x\in W^Q,\, v\le x\le w} H^0(X_w^x, L_\lambda(-(\partial X^x)_w - (\partial X_w)^x))).$$

Further, the map

$$\begin{aligned} H^0(X_w^v, L_\lambda(-(\partial X_w)^v))_{\prec x} &\longrightarrow \mathrm{gr}_x H^0(X_w^v, L_\lambda(-(\partial X_w)^v)) \\ &(= H^0(X_w^x, L_\lambda(-(\partial X^x)_w) - (\partial X_w)^x)) \end{aligned}$$

is just the restriction to X_w^x.

4.5. Construction of a standard basis. In view of Propositions 4.3,4.4, we are reduced to constructing a "nice" basis for $H^0(X_w^v, L_\lambda(-(\partial X_w)^v) - (\partial X^v)_w))$; let us denote this space by $H_w^v(\lambda)$.

Lemma 4.6. *Any element of $H_w^v(\lambda)$ can be lifted to an element of $H^0(G/Q, L_\lambda)$ which vanishes identically on all Schubert varieties $X_y, y \not\ge w$, and all opposite Schubert varieties $X^x, x \not\le v$.*

Let

$$\chi_w^v(\lambda) = \{ \text{ weights of the } T\text{-module } H_w^v(\lambda)\}$$

the weights being counted with the respective multiplicities. Let $\{p_{w,v}^\xi, \xi \in \chi_w^v\}$ be a basis of $H_w^v(\lambda)$ consisting of weight vectors. For any triple $\pi := (w, v, \xi)$ as above, let p_π be a lift of $p_{w,v}^\xi$ in $H^0(G/Q, L_\lambda)$ as given by Lemma 4.6 so that we have p_π vanishes on

$X_y, y \not\geq w$, and on $X^x, x \not\leq v$. For $\pi = (w, v, \xi)$, set $i(\pi) = w, e(\pi) = v$. Then we obtain

PROPOSITION 4.7. *$p_\pi|_{X_y^x} \neq 0 \iff X_{i(\pi)}^{e(\pi)} \subseteq X_y^x \iff x \leq e(\pi) \leq i(\pi) \leq y$.*

Combined with Proposition 4.4 this implies the following

PROPOSITION 4.8. *$\{p_\pi|_{X_w^v} \mid i(\pi) = w, e(\pi) \geq v\}$ is a basis for the T-module $H^0(X_w^v, L_\lambda(-(\partial X_w)^v))$.*

DEFINITION 4.9. *Set*

$$\Pi(\lambda) = \{(v, w, \xi) \mid v, w \in W^\lambda,\ v \leq w,\ \xi \in \chi_w^v(\lambda)\}.$$

For any $v, w \in W^\lambda, v \leq w$, set

$$\Pi_w^v(\lambda) := \{\pi \mid v \leq e(\pi) \leq i(\pi) \leq w\}.$$

In view of Proposition 4.7, we have, $\Pi_w^v(\lambda) = \{\pi \in \Pi(\lambda) \mid p_\pi|_{X_w^v} \neq 0\}$. Propositions 4.3,4.8 imply

THEOREM 4.10. *$\{p_\pi\ |_{X_w^v}, \pi \in \Pi_w^v(\lambda)\}$ is a basis for $H^0(X_w^v, L_\lambda)$, and $\{p_\pi, \pi \in \Pi(\lambda) \setminus \Pi_w^v(\lambda)\}$ is a basis for the kernel of the restriction map $H^0(G/P, L_\lambda) \longrightarrow H^0(X_w^v, L_\lambda)$.*

More generally, for a union Z of Richardson varieties, define

$$\Pi_Z(\lambda) = \{\pi \in \Pi(\lambda) \mid p_\pi|_Z \neq 0\}.$$

THEOREM 4.11. *Let Z be a union of Richardson varieties. Then $\{p_\pi\ |_Z, \pi \in \Pi_Z(\lambda)\}$ is a basis for $H^0(Z, L_\lambda)$, and $\{p_\pi, \pi \in \Pi(\lambda) \setminus \Pi_Z(\lambda)\}$ is a basis for the kernel of the restriction map $H^0(G/P, L_\lambda) \longrightarrow H^0(Z, L_\lambda)$.*

Standard monomials. We also construct a "standard monomial basis" for $H^0(X_w^v, L_{\lambda+\mu}), w, v \in W^P$, and λ, μ, P-dominant and regular (as well as $H^0(Z, L_{\lambda+\mu})$, Z being a union of Richardson varieties), the so-called "mixed case", as described below.

Let λ, μ be dominant weights such that $P_\lambda = P_\mu = P$.

DEFINITION 4.12. *Let $v, w \in W^P, v \leq w$. Let $\varphi \in \Pi_w^v(\lambda)$ and $\psi \in \Pi_w^v(\mu)$. The pair (φ, ψ) is called standard on X_w^v if*

$$v \leq e(\psi) \leq i(\psi) \leq e(\varphi) \leq i(\varphi) \leq w.$$

Then the product $p_\varphi p_\psi \in H^0(G/P, L_{\lambda+\mu})$ is called standard on X_w^v as well.

LEMMA 4.13. *Let $p_\varphi p_\psi$ be a standard product on G/P and let $v \leq w \in W^P$. Then*

$$p_\varphi p_\psi|_{X_w^v} \neq 0 \Longleftrightarrow v \leq e(\psi) \leq i(\varphi) \leq w.$$

THEOREM 4.14. *The standard products on X_w^v form a basis of $H^0(X_w^v, L_{\lambda+\mu})$. The standard products on G/P that are not standard on X_w^v form a basis of the kernel of the restriction map $H^0(G/P, L_{\lambda+\mu}) \longrightarrow H^0(X_w^v, L_{\lambda+\mu})$.*

Theorem 4.14 is proved by first proving similar results for the special fiber F_w^v of the flat family $\mathcal{X}_w^v$ (of which diag (X_w^v) is the generic fiber), and then deduce for X_w^v. See [**2**] for details. In fact, the above Theorem extends to unions of Richardson varieties as follows:

DEFINITION 4.15. *Let $\Pi(\lambda, \mu)$ be the set of all standard pairs (φ, ψ) where $\varphi \in \Pi(\lambda)$ and $\psi \in \Pi(\mu)$. For $v \leq w \in W^P$, let $\Pi_w^v(\lambda, \mu)$ be the subset of standard pairs on X_w^v. In view of Lemma 4.13, we have*

$$\Pi_w^v(\lambda, \mu) = \{(\varphi, \psi) \in \Pi(\lambda, \mu) \mid p_\varphi p_\psi|_{X_w^v} \neq 0\}.$$

For a union Z of Richardson varieties, let

$$\Pi_Z(\lambda, \mu) = \{(\varphi, \psi) \in \Pi(\lambda, \mu) \mid p_\varphi p_\psi|_Z \neq 0\}.$$

We have

THEOREM 4.16. *Let Z be a union of Richardson varieties in G/P. Then the products $p_\varphi p_\psi$, where $(\varphi, \psi) \in \Pi_Z(\lambda, \mu)$, form a basis of $H^0(Z, L_{\lambda+\mu})$. The products $p_\varphi p_\psi$, where $(\varphi, \psi) \in \Pi(\lambda, \mu) \setminus \Pi_Z(\lambda, \mu)$, form a basis of the kernel of the restriction map $H^0(G/P, L_{\lambda+\mu}) \longrightarrow H^0(Z, L_{\lambda+\mu})$.*

Next we consider a family of dominant weights $\lambda_1, \ldots, \lambda_m$ such that $P = P_{\lambda_1} = \cdots = P_{\lambda_m}$. For any union Z of Richardson varieties in G/P, we construct a basis of $H^0(Z, L_{\lambda_1+\cdots+\lambda_m})$, in terms of *standard monomials of degree* m as follows.

DEFINITION 4.17. *Let $\pi_i \in \Pi(\lambda_i)$ for $1 \leq i \leq m$. Then the sequence $\underline{\pi} := (\pi_1, \pi_2, \ldots, \pi_m)$ is standard if*

$$e(\pi_m) \leq i(\pi_m) \leq \cdots \leq e(\pi_1) \leq i(\pi_1).$$

Further, let $v, w \in W^P$ such that $v \leq w$; then $\underline{\pi}$ is standard on X_w^v if

$$v \leq e(\pi_m) \leq i(\pi_m) \leq \cdots \leq e(\pi_1) \leq i(\pi_1) \leq w.$$

$\underline{\pi}$ is standard on $Z = \cup X_{w_i}^{v_i}$ if it is standard on $X_{w_i}^{v_i}$ for some i.

Set

$$\Pi_w^v(\lambda_1,\ldots,\lambda_m) = \{\underline{\pi} = (\pi_1,\pi_2,\ldots,\pi_m) \mid \underline{\pi} \text{ is standard on } X_w^v\},$$
$$\Pi_Z(\lambda_1,\ldots,\lambda_m) = \{\underline{\pi} = (\pi_1,\pi_2,\ldots,\pi_m) \mid \underline{\pi} \text{ is standard on } Z\}.$$

DEFINITION 4.18. *Given* $\underline{\pi}=(\pi_1,\pi_2,\ldots,\pi_m)$, *set* $p_{\underline{\pi}}:=p_{\pi_1}\cdots p_{\pi_m}$. *Note that* $p_{\underline{\pi}} \in H^0(G/P, L_{\lambda_1+\cdots+\lambda_m})$. *If* $\underline{\pi}$ *is standard, then we call* $p_{\underline{\pi}}$ *a standard monomial on* G/P. *If* $\underline{\pi}$ *is standard on* X_w^v *(resp.* Z*), then we call* $p_{\underline{\pi}}$ *a standard monomial on* X_w^v *(resp.* Z*).*

We have

THEOREM 4.19. *Let* Z *be a union of Richardson varieties in* G/P *and let* $\lambda_1,\ldots,\lambda_m$ *be dominant weights such that* $P = P_{\lambda_1} = \cdots = P_{\lambda_m}$. *Then the monomials* $p_{\underline{\pi}}$ *where* $\underline{\pi} \in \Pi_Z(\lambda_1,\ldots,\lambda_m)$ *form a basis of* $H^0(Z, L_{\lambda_1+\cdots+\lambda_m})$. *Further, the monomials* $p_{\underline{\pi}}$ *where* $\underline{\pi} \in \Pi(\lambda_1,\ldots,\lambda_m) \setminus \Pi_Z(\lambda_1,\ldots,\lambda_m)$, *form a basis of the kernel of the restriction map*
$H^0(G/P, L_{\lambda_1+\cdots+\lambda_m}) \longrightarrow H^0(Z, L_{\lambda_1+\cdots+\lambda_m})$.

REMARK 4.20. *Thus, we have a SMT develeped in a geometric way (avoiding Quantum groups); but there is one restriction in the above approach, namely, for the construction of a standard monomial basis for* $H^0(Z, L_{\lambda_1+\cdots+\lambda_m})$, Z *being a union of Richardson varieties in* G/P *(as given by Theorem 4.19), we need all the* λ_i *to be* P-regular. *Thus, if the* λ_i*'s are dominant on* P *(not necessarily regular), then we do not have a basis even for* $H^0(X_w^v, L_{\lambda_1+\cdots+\lambda_m})$ *(in terms of standard monomials). Even if we restrict ourselves to Schubert varieties, the filtration (discussed above) involves Richardson varieties, and unlike the case of Schubert varieties,* $H^0(X_w^v, L_\lambda)$ *and* $H^0(X_{w_\lambda}^{v_\lambda}, L_\lambda)$ *need not be equal! (here,* w_λ, v_λ *are the elements in* W^{P_λ} *representing the cosets* $wW_{P_\lambda}, vW_{P_\lambda}$ *respectively). Thus, even though we have a basis for* $H^0(X_{w_\lambda}^{v_\lambda}, L_\lambda)$ *(as given by Theorem 4.10), it does not determine a basis for* $H^0(X_w^v, L_\lambda)$. *The reason for this is that the image of a Richardson variety need not be a Richardson variety!*

Counter Example. $G = SL(3)$ *with simple reflections* s_1, s_2. *Let* $P = B$, $w = s_2s_1$, $v = s_2$ *and* $\lambda = \omega_1$ *(the fundamental weight fixed by* s_2*). Then* X_w^v *is one–dimensional and mapped isomorphically to its image* $f(X_w^v)$ *in* G/P_λ. *Since the* T*–fixed points in* $f(X_w^v)$ *are* e_{ω_1} *and* $e_{s_2s_{1_{\omega_1}}}$, *it follows that* $f(X_w^v)$ *is not a Richardson variety,* f *being the canonical projection* $G/B \longrightarrow G/P_\lambda$.

Nevertheless, if G *is classical, we do obtain a standard monomial basis for* $H^0(Z, L_{\lambda_1+\cdots+\lambda_m})$, Z *being a union of Richardson varieties* *in G/P, where λ_i's are dominant on P (not necessarily regular) as explained in the next section.*

5. Weights of classical type

In this section, we shall determine the "building blocks"

$$H_w^v(\lambda) = H^0(X_w^v, L_\lambda(-(\partial X_w)^v - (\partial X^v)_w))$$

in the case when the dominant weight λ is of classical type (as introduced in [**22**], cf. the next definition). Along the way, we shall retrieve the results of loc. cit., using our basis $\{p_\pi\}$. In particular, we shall give a geometric characterization of "admissible pairs" of loc. cit. (cf. Definition 5.7 below).

DEFINITION 5.1. *Let λ be a dominant weight. We say, λ is of classical type if $\langle \lambda, \beta^\vee \rangle \leq 2$, for all positive roots β.*

REMARK 5.2.
1. *Any dominant weight of classical type is either fundamental, or a sum of two minuscule fundamental weights (recall that a weight λ is minuscule, if $\langle \lambda, \beta^\vee \rangle \leq 1$, for all positive roots β).*
2. *G is classical if and only if all fundamental weights of G are of classical type.*

For the rest of this section, we fix a dominant weight λ of classical type.

PROPOSITION 5.3. *Let $v, w \in W^\lambda, v \leq w$. Then the T–module $H_w^v(\lambda)$ is at most one–dimensional; further, if non–zero, then it has the weight $-\frac{1}{2}(w(\lambda) + v(\lambda))$.*

As a consequence, the weights of the T–module $H^0(X_w, L_\lambda(-\partial X_w))$ are among the $-\frac{1}{2}(w(\lambda) + x(\lambda))$ where $x \leq w$, and the corresponding weight spaces are one–dimensional.

As a consequence, we have

LEMMA 5.4. *Let $v, w \in W^\lambda, v \leq w$. Further, let $H_w^v(\lambda)$ be non–zero. Then for each divisor X_{ws_β} (resp. X^{vs_γ}) of X_w (resp. X^v) such that $ws_\beta \geq v$ (resp. $vs_\gamma \leq w$), β (resp. γ) being positive roots, we have, $\langle \lambda, \beta^\vee \rangle$ (resp. $\langle \lambda, \gamma^\vee \rangle$) $= 2$.*

We shall denote by $p_{w,v}$ the unique $p_{w,v}^\xi$, if non–zero. By Proposition 5.3, $p_{w,v}$ lifts to a unique T–eigenvector in $H^0(X_w, L_\lambda(-\partial X_w))$; we still denote that lift by $p_{w,v}$. The non–zero $p_{w,v}$'s, where $v \leq w$, form a basis of $H^0(X_w, L_\lambda(-\partial X_w))$.

LEMMA 5.5. *With notations as above, we have* $p_{w,v}^2 = p_w p_v$ *on* X_w*, up to a non–zero scalar.*

A geometric characterization of Admissible pairs: Recall (cf.[**22**]):

DEFINITION 5.6. *Let* $v, w \in W^\lambda$ *such that* X_v *is a divisor in* X_w*. If* $m_\lambda(w,v) = 2$*, then we shall refer to* X_v *as a double divisor in* X_w*.*

DEFINITION 5.7. *A pair* (v,w) *in* W^λ *is called admissible if either* $v = w$ *(in which case, it is called a trivial admissible pair), or there exists a sequence* $w = w_1 > w_2 > \cdots > w_r = v$*, such that* $X_{w_{i+1}}$ *is a double divisor in* X_{w_i}*. We shall refer to such a chain as a double chain.*

THEOREM 5.8. *(cf. [**2**]) Let* $v, w \in W^\lambda, v \leq w$*. Then the pair* (v,w) *is admissible if and only if* $H_w^v(\lambda)$ *is non–zero. In this case, every chain from* v *to* w *is a double chain.*

6. Standard monomials for sums of weights of classical type

In this section, we describe a standard monomial basis for $H^0(X_w^v, L_{\lambda_1+\cdots+\lambda_m})$, where X_w^v is a Richardson variety in G/P, and $\lambda_1, \ldots, \lambda_m$ are dominant weights of classical type of P (in the sense of Definition 5.1). For $w \in W^P$, we shall denote by w_λ the element in W^λ representing the coset wW_λ (here, P_λ is the stabilizer of $\lambda, W_\lambda = W_{P_\lambda}, W^\lambda = W^{P_\lambda}$).

We begin with the case $m = 1$.

Let $v, w \in W^P, v \leq w$. Let λ be a P-dominant weight, not necessarily P–regular. Consider the T–module $H^0(X_w^v, L_\lambda)$, where $v \leq w \in W^P$. Notice that the diagram

$$\begin{array}{ccc} H^0(X_{w_\lambda}, L_\lambda) & \longrightarrow & H^0(X_{w_\lambda}^{v_\lambda}, L_\lambda) \\ \downarrow & & \downarrow \\ H^0(X_w, L_\lambda) & \longrightarrow & H^0(X_w^v, L_\lambda) \end{array}$$

is commutative, where the horizontal (resp. vertical) maps are restrictions (resp. pull–backs). Further, both restrictions are surjective, and the pull–back on the left is an isomorphism, since the natural map $f : X_w \to X_{w_\lambda}$ satisfies $f_*\mathcal{O}_{X_w} = \mathcal{O}_{X_{w_\lambda}}$. Thus, we may regard the T–module $H^0(X_w^v, L_\lambda)$ as a quotient of $H^0(X_{w_\lambda}^{v_\lambda}, L_\lambda)$.

Likewise, by using the commutative diagram

$$\begin{array}{ccc} H^0(X_{w_\lambda}, L_\lambda(-\partial X_{w_\lambda})) & \longrightarrow & H^0(X_{w_\lambda}^{v_\lambda}, L_\lambda(-(\partial X_{w_\lambda})^{v_\lambda})) \\ \downarrow & & \downarrow \\ H^0(X_w, L_\lambda(-\partial_\lambda X_w)) & \longrightarrow & H^0(X_w^v, L_\lambda(-(\partial_\lambda X_w)^v)) \end{array}$$

we may regard the T–module $H^0(X_w^v, L_\lambda(-(\partial_\lambda X_w)^v))$ as a quotient of $H^0(X_{w_\lambda}^{v_\lambda}, L_\lambda(-(\partial X_{w_\lambda})^{v_\lambda}))$ (here, $\partial_\lambda X_w$ (resp. $\partial_\lambda X^v$) denotes $\cup_{\{w'<w \,|\, w'_\lambda<w_\lambda\}} X_{w'}$ (resp. $\cup_{\{v'>v \,|\, v'_\lambda>v_\lambda\}} X^{v'}$). The latter has been described in Proposition 4.8; in particular, it has a basis consisting of the p_{w_λ,x_λ} where $x_\lambda \in W^\lambda$, $v_\lambda \le x_\lambda \le w_\lambda$ and the pair (w_λ, x_λ) is admissible. Now consider the following two facts:

Fact 1: For $\tau \in W^\lambda$ and $v, w \in W^P$, we have, $p_\tau \,|\, X_w^v$ is non zero if and only if there exists a lift τ' for τ in W^P such that $w \ge \tau' \ge v$ (cf. [**2**], Lemma 2).

Fact 2: For an admissible pair (τ, φ) in W^λ,

$$(*) \qquad p_{\tau,\varphi}^2 = p_\tau p_\varphi \text{ on } X_\tau$$

(cf. Lemma 5.5).

Using these two facts, and the above considerations, we prove in [**2**] the following:

THEOREM 6.1. *Let $v \le w \in W^P$ and let λ be a dominant character of classical type of P.*

(1) *The p_{w_λ,x_λ}'s where $x \in W^P$ and $v \le x \le w$, form a basis of $H^0(X_w^v, L_\lambda(-(\partial_\lambda X_w)^v))$.*
(2) *The p_{y_λ,x_λ}'s where $x, y \in W^P$ and $v \le x \le y \le w$, form a basis of $H^0(X_w^v, L_\lambda)$.*

DEFINITION 6.2. *Let $\lambda_1, \ldots, \lambda_m$ be dominant characters of classical type of P. Let $\pi_i = (w_i, v_i)$ where $v_i \le w_i \in W^{\lambda_i}, 1 \le i \le m$ be admissible pairs. Then the sequence $\underline{\pi} = (\pi_1, \ldots, \pi_m)$ is standard if there exist lifts $\tilde{w}_i, \tilde{v}_i$ in W^P for $1 \le i \le m$, such that*

$$\tilde{w}_1 \ge \tilde{v}_1 \ge \cdots \ge \tilde{w}_m \ge \tilde{v}_m.$$

The monomial

$$p_{\underline{\pi}} := p_{w_1,v_1} \cdots p_{w_m,v_m} \in H^0(G/P, L_{\lambda_1+\cdots+\lambda_m})$$

is called standard as well.

Further, let $v, w \in W^P$ such that $v \leq w$; then $\underline{\pi}$ is standard on X_w^v if there exist lifts as above, such that

$$w \geq \tilde{w}_1 \geq \tilde{v}_1 \geq \cdots \geq \tilde{w}_m \geq \tilde{v}_m \geq v.$$

The restriction of $p_{\underline{\pi}}$ to X_w^v is called a standard monomial on X_w^v; it is a T–eigenvector in $H^0(X_w^v, L_{\lambda_1+\cdots+\lambda_m})$.

THEOREM 6.3. *Let $v \leq w \in W^P$ and let $\lambda_1, \ldots, \lambda_m$ be dominant characters of P. If $\lambda_1, \ldots, \lambda_m$ are of classical type, then the standard monomials on X_w^v form a basis for $H^0(X_w^v, L_{\lambda_1+\cdots+\lambda_m})$.*

REMARK 6.4. *In particular, Theorem 6.3 applies to $P = B$ if all fundamental weights are of classical type, that is, if G is classical. Thereby, we retrieve all results of* [**22**].

REMARK 6.5. *The reason why the filtration method of* [**2**] *works for developing a SMT for dominant weights on P that are sums of classical weights is the following: We have for $\tau \in W^\lambda$ and $v, w \in W^P$, $p_\tau \,|\, X_w^v$ is non zero if and only if there exists a lift τ' for τ in W^P such that $w \geq \tau' \geq v$; this together with the fact that for an admissible pair (τ, φ) in W^λ,*

$$(*) \qquad p_{\tau,\varphi}^2 = p_\tau p_\varphi \text{ on } X_\tau$$

implies that if $w_\lambda \geq \tau \geq \phi \geq v_\lambda$, then $p_{\tau,\varphi} \,|\, X_w^v$ is non zero if and only if there exist lifts τ', φ' for τ, φ in W^P such that $w \geq \tau' \geq \varphi' \geq v$. **Thus in order to have a Standard monomial basis for $H^0(X_w^v, L_{\lambda_1+\cdots+\lambda_m})$ (built out of $H^0(X_w^v, L_{\lambda_i})$), we should first have a basis for $H^0(X_w^v, L_{\lambda_i}), 1 \leq i \leq m$); in order to construct a basis for $H^0(X_w^v, L_\lambda), \lambda = \lambda_i$, starting with a basis for $H^0(X_{w_\lambda}^{v_\lambda}, L_\lambda)$, we have to know some precise relations among the basis vectors of $H^0(X_{w_\lambda}^{v_\lambda}, L_\lambda)$, similar to (*).**

REMARK 6.6. *Regarding Question 1 of §4, the results of* [**2**] *show that to develop a SMT, it is enough to have a nice filtration of $H^0(X_w^v, L_\lambda), v, w \in W^\lambda$. Of course, in* [**2**], *a filtration of $H^0(X_w^v, L_\lambda)$ is deduced from a filtration of the structure sheaf of $\cup_{v \leq x \leq w} X_x^v \times X_w^x$ (the special fiber of the flat family of which diag(X_w^v) is the general fiber)*

7. Concluding Remarks:

The main tool introduced in [**2**] towards a geometric construction of a SMT-basis for $H^0(G/P, \mathcal{L}_\lambda)$ is the flat degeneration of the diagonal in $X \times X$ (X being G/P) to the union of all products $X_\tau \times X^\tau$, where the X_τ are the Schubert varieties and the X^τ are the corresponding opposite Schubert varieties. It is interesting to note that while in [**2**], the above degeneration is the starting point towards developing a SMT for Richardson varieties, in [**17**] such a degeneration is obtained as a consequence of SMT ! As seen above, in [**2**], one first constructs a basis for the space of sections of L_λ on X_w^v vanishing on both of the boundaries $(\partial X_w)^v, (\partial X^v)_w$. This basis is then lifted to global sections of L_λ that vanish on all Richardson varieties not containing X_w^v, which in turn give rise to a basis for $H^0(G/P, L_\lambda)$; this basis for $H^0(G/P, L_\lambda)$ is then shown to have compatibility with unions of Richardson varieties. In [**17**], the authors show the compatibility of the basis of [**28**] with unions of Richardson varieties, and the vanishing of a typical basis vector on $(\partial X_w)^v, (\partial X^v)_w$ for suitable w and v. Thus the basis as constructed in [**28**] may be considered as a specially nice basis of the type constructed in [**2**]. In spite of this relationship between the bases of [**2**] and [**28**], there are still some missing links; to be very precise, let $\pi = (\underline{\tau}, \underline{a})$ be an L-S path of shape λ, where $\underline{\tau} = (\tau_0, \ldots, \tau_r), \tau_0 > \tau_1 > \cdots > \tau_r$. The first and last elements describe the smallest Richardson variety $X_{\tau_0}^{\tau_r}$ such that the restriction of the path vector p_π does not vanish identically. The other terms come up naturally in the quantum group construction and have a suggestive algebraic geometric interpretation in [**28**], but they do not have yet an interpretation in the construction in [**2**].

References

[1] M. BRION: Positivity in the Grothendieck group of complex flag varieties, preprint available at math.AG/0105254.

[2] M. BRION and V. LAKSHMIBAI: A geometric approach to Standard Monomial Theory, preprint math.AG/0111054 (2001).

[3] C. CHEVALLEY: Sur les decompositions cellulaires des espaces G/B (with a foreword by A. Borel), *Proc. Sympos., Pure Math.*, Vol.**56**, Part 1, Algebraic Groups and their Generalizations: Classical Methods (University Park, PA; 1991), *Amer. Math. Soc.*, Providence, RI, 1994, 1–23.

[4] V. V. DEODHAR: On some geometric aspects of Bruhat orderings. I. A finer decomposition of Bruhat cells. *Invent. Math.* **79** (1985), 499–511.

[5] M. DEMAZURE, Désingularisation des variétés de Schubert généralisés, *Ann. Sci. Ec. Norm. Sup*, **t.7** (1974), 53-88.

[6] W. FULTON and A. LASCOUX:A Pieri formula in the Grothendieck ring of a flag bundle, *Duke Math. J.*, **76** (1994), 711-729.

[7] A. JOSEPH: Quantum groups and their primitive ideals, Springer, Berlin 1995.

[8] M. KASHIWARA: Bases crystallines, C.R. Acad. Peris, Sŕie I, 311, 277-280 (1990).

[9] M. Kashiwara: Crystalizing the q-analogue of universal enveloping algebras, *Comm. Math. Physics*, **133** (1990), 249-260.
[10] M. Kashiwara: Crystal base and Littelmann's refined Demazure character formula, *Duke Math. J*, 71, 839-858 (1993).
[11] M. Kashiwara: Crystal bases of modified quantized enveloping algebras, *Duke Math. J*, 73, 383-413 (1994).
[12] G. Kempf and A. Ramanathan: Multi-cones over Schubert varieties, *Invent. math.* **87** (1987), 353–363.
[13] B. Kostant and S. Kumar: T-equivariant K-theory of generalized flag varieties, *Proc. Nat. Acad. Sci. USA* **84** (1987), 4351-4354.
[14] S. Kumar and P. Littelmann: Frobenius splitting in characteristic zero and the quantum Frobenius map, *J. Pure and Appl. Algebra* 152 (2000), 201–216.
[15] S. Kumar and P. Littelmann: Algebraization of Frobenius splitting via quantum groups, to appear in *Annals of Mathematics.*
[16] V. Lakshmibai: Standard monomial theory for G_2, *J. Alg.*, **98** (1986), 281-318.
[17] V. Lakshmibai and P. Littelmann: Richardson varieties and equivariant K-theory, math.AG/0201075.
[18] V. Lakshmibai, C. Musili, and C. S. Seshadri: Geometry of $G/P-III$, *Proc. Ind. Acad. Sci.*, **87A** (1978), 93-177.
[19] V. Lakshmibai, C. Musili, and C. S. Seshadri: Geometry of $G/P-IV$, *Proc. Ind. Acad. Sci.*, **88A** Geometry of $G/P-IV$ (1979) 279-362.
[20] V. Lakshmibai and K.N. Rajeswari: Towards a standard monomial theory for exceptional groups, *Contemporary Math., A.M.S.*, **88** (1989),
[21] V. Lakshmibai and C.S. Seshadri: Geometry of G/P - II, *Proc. Ind. Acad. Sci.*, **87A** (1978), 1-54.
[22] V. Lakshmibai and C.S. Seshadri: Geometry of G/P-V, *J. Alg*, **100, no:2** (1986), 462–557.
[23] V. Lakshmibai and C.S. Seshadri: Standard monomial theory, *Proc. Hyderabad Conference on Algebraic Groups*,(S. Ramanan et al.,eds.), Manoj Prakashan, Madras (1991) 279-323.
[24] V. Lakshmibai and C.S. Seshadri: Standard monomial theory for $\widehat{SL}_2$, *Advanced Studies in Mathematical Physics*, Vol. **7** (1989), p.178-234, "Infinite Dimensional Lie Algebras and Groups", Luminy, Proceedings of the conference held at CIRM, Luminy, Marseille, July 4-8, 1988.
[25] P. Littelmann: A generalization of the Littlewood-Richardson rule, *J. Alg.* **130** (1990), 328-368.
[26] P. Littelmann: A Littlewood-Richardson formula for symmetrizable Kac-Moody algebras. *Invent. Math.* **116** (1994), 329-346.
[27] P. Littelmann: Path and root operators in representation theory, *Ann. Math.* **142** (1995), 499-525.
[28] P. Littelmann: Contracting modules and standard monomial theory. *JAMS* **11** (1998), 551–567.
[29] P. Littelmann: The path model, the quantum Frobenius map and standard monomial theory, *Algebraic groups and their representations*, Kluwer Academic publishers (1998), 175-212.

[30] P. Littelmann: Bases for representations, L-S paths and Verma flags (appears in this volume).
[31] P. Littelmann and C.S. Seshadri: A Pieri-Chevalley formula for K(G/B) and Standard Monomial Theory, *The proceedings of the SCHUR MEMORIAM WORKSHOP in Rehovot, December 2000* (to appear), to be published by Birkhäuser.
[32] G. Lusztig: Introduction to quantum groups. *Progress in Mathematics* **110**, Birkhäuser (1993).
[33] O. Mathieu: Positivity of some intersections in $K_0(G/B)$, *J. Pure and Appl. Alg.*, **232** (2000).
[34] V. Mehta and A. Ramanathan: Frobenius splitting and cohomology vanishing for Schubert varieties , *Ann. Math.* **122** (1985), 27–40.
[35] C. Musili: Postulation formula for Schubert varieties , *J. Ind. Math. Soc.*, **36** (1972) 143-171.
[36] C. Musili: The developement of Standard monomial theory-I, preprint (2002).
[37] H. Pittie and A. Ram: A Pieri-Chevalley formula for $K(G/B)$, *Electron. Res. Announc. Amer. Math. Soc.*, Vol. **5** (1999), 102-107.
[38] K.N. Raghavan and P. Sankaran: Monomial bases for representations of classical semisimple Lie algebras, *J. Transformation groups.*
[39] S. Ramanan and A. Ramanathan: Projective normality of flag varieties and Schubert varieties, Inv. Math, 79, (1985), 217-224.
[40] A. Ramanathan: Equations defining Schubert varieties and Frobenius splitting of diagonals, *Pub. Math. IHES* **65** (1987), 61–90.
[41] R. W. Richardson: Intersections of double cosets in algebraic groups, *Indag. Math. (N. S.)* **3** (1992), 69–77.
[42] C.S. Seshadri: Geometry of G/P - I, *C.P. Ramanujam: A Tribute*, **207** (Springer-Verlag), published by Tata Institute (1978), 207–239.

Northeastern University, Mathematics Department
Boston, MA-02115, U.S.A.
E-mail address: lakshmibai@neu.edu

Maximal Subbundles and Gromov-Witten Invariants

H. Lange and P.E. Newstead

To C. S. Seshadri on his 70th birthday

Abstract. Let C be a nonsingular irreducible projective curve of genus $g \geq 2$ defined over the complex numbers. Suppose that $1 \leq n' \leq n-1$ and $n'd - nd' = n'(n-n')(g-1)$. It is known that, for the general vector bundle E of rank n and degree d, the maximal degree of a subbundle of E of rank n' is d' and that there are finitely many such subbundles. We obtain a formula for the number of these maximal subbundles when $(n', d') = 1$. For $g = 2$, $n' = 2$, we evaluate this formula explicitly. The numbers computed here are Gromov-Witten invariants in the sense of a recent paper of Ch. Okonek and A. Teleman (Commun. Math. Phys. 227 (2002), 551-585) and our results answer a question raised in that paper.

1. Introduction

Let C be a non-singular irreducible projective curve of genus $g \geq 2$, defined over the complex numbers, and let E be a vector bundle over C of rank n and degree d. In describing the structure of E, it is important to consider the set of subbundles E' of E of any fixed rank n' and degree d'. In particular, for any n', there is a maximum value of d' for which such subbundles exist; the corresponding subbundles are called *maximal subbundles* of E. It is convenient here to write

$$s(E, E') = n'd - nd'.$$

Supported by DFG Contracts Ba 423/8-1. Both authors are members of the research group VBAC (Vector Bundles on Algebraic Curves), which is partially supported by EAGER (EC FP5 Contract no. HPRN-CT-2000-00099) and by EDGE (EC FP5 Contract no. HPRN-CT-2000-00101).

We can then put

$$s_{n'}(E) = \min s(E, E'),$$

the minimum being taken over all subbundles E' of rank n'. A subbundle E' is then a maximal subbundle if and only if $s(E, E') = s_{n'}(E)$.

Note in particular that the bundle E is stable if and only if $s_{n'}(E) > 0$ for all n', $1 \leq n' \leq n-1$. In general we call the $s_{n'}(E)$ *degrees of stability* and note that, for each n', we can define a stratification of the moduli space $\mathcal{M}(n, d)$ of stable bundles of rank n and degree d by the locally closed subsets

$$U_{n',s}(n, d) = \{E \in \mathcal{M}(n, d) | s_{n'}(E) = s\}.$$

Note that $U_{n',s}(n, d)$ can be non-empty only if $s > 0$ and $s \equiv n'd \bmod n$.

It was proved in [**11**] that $s_{n'}(E) \leq n'(n-n')g$ and in fact it is known [**6**, Théorème 4.4] that the maximum value of s that occurs is given by

$$s = n'(n-n')(g-1) + \varepsilon,$$

where ε is the unique integer $0 \leq \varepsilon \leq n-1$ such that $s \equiv n'd \bmod n$. The corresponding $U_{n',s}(n, d)$ is the unique open stratum of the stratification.

When $s = n'(n-n')(g-1)$, a dimension counting argument shows that one can expect a general $E \in \mathcal{M}(n, d)$ to have a finite number $m_{n'}(E)$ of maximal subbundles of rank n', the number being independent of E provided that it is chosen sufficiently generally. In the simplest case, when $n' = 1$, the number is known; in fact,

$$m_1(E) = n^g$$

(see [**5, 9, 17**] for $n = 2$, [**15**] and [**14**, Proposition 3.9] in general).

Our object in this paper is to obtain a formula for $m_{n'}(E)$ in the case $(n', d') = 1$. Indeed we use Grothendieck-Riemann-Roch and the Porteous formula to express $\mathcal{M}(n', d')$ as the top Chern class of a certain virtual bundle (Theorem 3.1). When $g = 2$ and $n' = 2$, we evaluate this Chern class and give an explicit formula for $m_2(E)$ (Theorem 4.1).

Our result can be interpreted in terms of Gromov-Witten theory. Indeed, in [**14**], the formula for $m_1(E)$ is obtained as an application of a theorem on Gromov-Witten invariants, and the question of computing these invariants when $n' > 1$ is raised. Our results give an answer to this question. Our methods, however, are similar to those of [**15**]. Notice that these invariants were also defined in the last section of [**3**].

It should also be noted that there is a mistake in [**16**], where it is claimed that, when $s \le n'(n-n')(g-1)$, the general bundle in $U_{n',s}(n,d)$ has only one maximal subbundle of rank n'. For $s = n'(n-n')(g-1)$, this is not correct, and indeed the proof given in [**16**] fails also for $s < n'(n-n')(g-1)$. However the result itself is true in this case; we provide a proof of this in Theorem 2.3. It should be emphasised that this does not invalidate any other result in [**16**].

The research for this paper was carried out during a visit by the second author to the Mathematisches Institut der Universität Erlangen-Nürnberg in October/November 2001. He is grateful to the DFG for funding this visit and to the Institut for its hospitality.

2. Uniqueness of Maximal Subbundles

According to [**16**, Theorem 0.1] $U_{n',s}(n,d)$ is locally closed in $\mathcal{M}(n,d)$ and irreducible of dimension

$$\dim U_{n',s}(n,d) = (n^2 - n'(n-n'))(g-1) + s + 1.$$

Hence it makes sense to speak of a general vector bundle in $U_{n',s}(n,d)$.

LEMMA 2.1. *Let E be a general bundle in $U_{n',s}(n,d)$ with $s \le n'(n-n')(g-1)$. Then E possesses only finitely many maximal subbundles E'. For any one of them both E' and $E'' = E/E'$ are general (as points of the moduli spaces $\mathcal{M}(n',d')$ and $\mathcal{M}(n-n',d-d')$ respectively).*

REMARK 2.2. Lemma 2.1 has been proved in [**16**, Claim p. 495 and Lemma 1.4]. Similar statements have been given in [**4, 6, 8**]. We include the proof for the convenience of the reader.

PROOF. We show first that E' and E'' are both general. If not, then either E' or E'' (or both) depends on fewer parameters than the dimension of the corresponding moduli space; this is true irrespective of whether E' and E'' are stable. Note also that $h^0(E''^* \otimes E') = 0$ since E is stable. So the nontrivial extensions

$$0 \longrightarrow E' \longrightarrow E \longrightarrow E'' \longrightarrow 0$$

depend on a number ν of parameters with

$$\begin{aligned}\nu &< \dim \mathcal{M}(n',d') + \dim \mathcal{M}(n-n',d-d') + h^1(E''^* \otimes E') - 1 \\ &= (g-1)(n'^2 + (n-n')^2 + n'(n-n')) + s + 1 \\ &= (g-1)(n^2 - n'(n-n')) + s + 1 \\ &= \dim U_{n',s}(n,d).\end{aligned}$$

This contradicts the generality of E.

If the general E in $U_{n',s}(n,d)$ possesses infinitely many maximal subbundles, then the same inequality holds, again a contradiction. □

If $s < n'(n-n')(g-1)$, we can be more precise:

THEOREM 2.3. *A general vector bundle $E \in U_{n',s}(n,d)$ with $s < n'(n-n')(g-1)$ admits only one maximal subbundle of rank n'.*

Proof. For convenience we write $n'' = n-n'$, $d'' = d-d'$. Passing if necessary to the dual bundle, we may assume that $n' \leq n''$. Moreover assume first, slightly more generally, that $s \leq n'(n-n')(g-1)$, since we want to deduce also something for $s = n'(n-n')(g-1)$.

Fix a maximal subbundle $E' \in \mathcal{M}(n',d')$ of E and write $E'' = E/E' \in \mathcal{M}(n'',d'')$. So

$$s = n'd'' - n''d'.$$

Suppose F is a second maximal subbundle of E. The composed map $\varphi : F \hookrightarrow E \to E''$ is nonzero. Let $G \subset E''$ be its (sheaf theoretic) image. Then we have the following commutative diagram

$$\begin{array}{ccccccccc} 0 & \to & H & \to & F & \to & G & \to & 0 \\ & & \downarrow_j & & \downarrow & & \| & & \\ 0 & \to & E' & \to & \widetilde{E} & \to & G & \to & 0 \\ & & \| & & \downarrow & & \downarrow_i & & \\ 0 & \to & E' & \to & E & \to & E'' & \to & 0 \end{array}$$

where the middle exact sequence is the push out of the upper exact sequence and the pull back of the lower sequence and the composition $F \longrightarrow E''$ is the map φ. Note that i and j are injective (as maps of sheaves). Conversely, given such a diagram with $rk(F) = n'$ and $deg(F) = d'$, then F is a second maximal subbundle of E of rank n'. Let n_G and d_G denote the rank and degree of G, $1 \leq n_G \leq n'$.

Let $A_{n_G,d_G}(E'')$ denote the set of subsheaves of E'' of rank n_G and degree d_G, and define similarly $A_{n_H,d_H}(E')$. Note that $A_{n_G,d_G}(E'')$

and $A_{n_H,d_H}(E')$ can be given a scheme structure by identifying them with Quot schemes. Finally denote by i^* the pull back map

$$H^1(E''^* \otimes E') \to H^1(G^* \otimes E').$$

According to the above remarks, it suffices to show that, if $s < n'(n-n')(g-1)$, then

$$\dim A_{n_G,d_G}(E'') + \dim A_{n_H,d_H}(E') + h^1(G^* \otimes H) + \dim Ker(i^*) < h^1(E''^* \otimes E'). \tag{1}$$

But, by Lemma 2.1 and [**16**, Theorem 0.2],

$$\begin{aligned}\dim A_{n_G,d_G}(E'') &= n_G d'' - n'' d_G - n_G(n'' - n_G)(g-1)\\ \dim A_{n_H,d_H}(E') &= n' d_G - n_G d' - (n' - n_G) n_G (g-1).\end{aligned}$$

Moreover, since i^* is surjective and $h^0(G^* \otimes H) = 0,\ F$ being stable,

$$\begin{aligned}h^1(G^* \otimes H) &= n' d_G - n_G d' + n_G(n' - n_G)(g-1)\\ \dim Ker(i^*) &\leq n'(n'' - n_G)(g-1) + s + n_G d' - n' d_G.\end{aligned}$$

Hence inequality (1) would follow from

$$(n'n'' + n_G(n_G - n' - n''))(g-1) + n_G d'' - n'' d_G + n' d_G - n_G d' + s < n'n''(g-1) + s$$

which is equivalent to

$$n_G(n_G - n' - n'')(g-1) + n_G(d'' - d') + (n' - n'')d_G < 0. \tag{2}$$

But Lemma 2.1 implies that $n_G(n' - n_G)(g-1) \leq s_{n'-n_G}(F) \leq n' d_G - n_G d'$ which gives

$$d_G \geq \frac{n_G}{n'}\left[(n' - n_G)(g-1) + d'\right].$$

Hence the left hand side of (2) is less than or equal to (here we use the fact that $n' \leq n''$)

$$\begin{aligned}\frac{n_G}{n'} & \left[n''(n_G - 2n')(g-1) + d''n' - n''d'\right]\\ &= \frac{n_G}{n'}\left[n''(n_G - n')(g-1) + s - n'n''(g-1)\right]\\ &\leq \frac{n_G}{n'}\left[s - n'n''(g-1)\right] \ \text{(since } n_G \leq n')\\ &< 0 \ \text{ for } \ s < n'n''(g-1).\end{aligned}$$

□

Suppose now $s = n'(n-n')(g-1)$. According to Lemma 2.1 there are only finitely many maximal subbundles for general E. We obtain as a consequence of the proof of Theorem 2.3

PROPOSITION 2.4. *Let E be a general vector bundle on C with $s_{n'}(E) = n'(n-n')(g-1)$ and $n' \leq (n-n')$. If E' and F are 2 maximal subbundles of rank n' of E, the composed map $F \to E \to E/E'$ is injective.*

PROOF. For the proof just note that in the proof of Theorem 2.3 all inequalities have to be equalities. In particular we must have $n_G = n'$, which implies the assertion. □

3. Maximal subbundles of general bundles

Let E be a general stable bundle of rank n and degree d on C and let n' be an integer, $1 \leq n' \leq n-1$, such that

$$s_{n'}(E) = n'(n-n')(g-1) = n'd - nd'.$$

We know by Lemma 2.1 that E has only a finite number of maximal subbundles. Our object in this section is to obtain a formula for this number when $(n', d') = 1$.

We denote by $M_{n'}(E)$ the set of maximal subbundles of E of rank n'. This set can be given a scheme structure by identifying it with Grothendieck's scheme $Quot_{n-n',d''}(E)$ of quotients of E of rank $n-n'$ and degree

$$d'' = d - d' = \frac{1}{n}\left((n-n')d + s_{n'}(E)\right).$$

Now let J denote the Jacobian of C consisting of line bundles of degree 0 on C, and let M_0 denote the moduli space of stable bundles of rank n' and with fixed determinant of degree d' on C. The map

$$\pi: \ J \times M_0 \longrightarrow \mathcal{M}(n', d'): \ (L, F) \longmapsto L \otimes F$$

is an unramified covering of degree n'^{2g}.

Let $\mathcal{L}$ denote a Poincaré bundle on $C \times J$ and $\mathcal{U}$ a universal bundle on $C \times M_0$ (recall that we are assuming $(n', d') = 1$, so $\mathcal{U}$ exists). Let K denote the canonical bundle on C and $q: \ C \times J \times M_0 \longrightarrow J \times M_0$ and $p: \ C \times J \times M_0 \longrightarrow C$ be the canonical projections. We denote the pullbacks of $\mathcal{U}$ and $\mathcal{L}$ to $C \times J \times M_0$ by the same symbols and the restrictions of $\mathcal{U}$ and $\mathcal{L}$ to $\{x\} \times J \times M_0$ by $\mathcal{U}_x$ and $\mathcal{L}_x$.

THEOREM 3.1. *Let E be a general bundle of rank n and degree d, and n' an integer with $1 \leq n' \leq n-1$. Suppose that $n'd - nd' = n'(n-n')(g-1)$ and that $(n', d') = 1$. Then the number of maximal* *subbundles of E of rank n' is given by*

$$m_{n'}(E) = \frac{1}{n'^{2g}} c_{\text{top}}\,(\mathcal{F} - \mathcal{E})\,[J \times M_0]$$

where $\mathcal{E} = q_*(\mathcal{U}^* \otimes \mathcal{L}^* \otimes p^*(E \otimes K))$ and $\mathcal{F} = \bigoplus_{x \in D}(\mathcal{U}_x^* \otimes \mathcal{L}_x^*) \otimes \mathbb{C}^n$, where D is a smooth canonical divisor on C.

Proof. Note first that the conclusion is equivalent to

$$\#\pi^{-1}(M_{n'}(E)) = c_{\text{top}}\,(\mathcal{F} - \mathcal{E})\,[J \times M_0].$$

Consider the exact sequence

$$\begin{aligned} 0 \longrightarrow \mathcal{U}^* \otimes \mathcal{L}^* \otimes p^*E \xrightarrow{\otimes D} \mathcal{U}^* \otimes \mathcal{L}^* \otimes p^*(E \otimes K) \\ \longrightarrow \mathcal{U}^* \otimes \mathcal{L}^*|_{p^*(D)} \otimes \mathbb{C}^n \longrightarrow 0 \end{aligned}$$

where D denotes any smooth canonical divisor. Note that for any maximal subbundle E' of E,

$$\deg(E'^* \otimes E \otimes K) = n'd - nd' + nn'(2g-2) > nn'(2g-2).$$

Since any such E' is stable by Lemma 2.1, $E'^* \otimes E \otimes K$ is semistable, so

$$H^1(E'^* \otimes E \otimes K) = 0.$$

Hence $R^1 q_*(\mathcal{U}^* \otimes \mathcal{L}^* \otimes p^*(E \otimes K)) = 0$ and we have an exact sequence

$$0 \longrightarrow q_*(\mathcal{U}^* \otimes \mathcal{L}^* \otimes p^*E) \longrightarrow \mathcal{E} \xrightarrow{\varphi} \mathcal{F} \longrightarrow R^1 q_*(\mathcal{U}^* \otimes \mathcal{L}^* \otimes p^*E) \longrightarrow 0$$

with $\mathcal{E}$, $\mathcal{F}$ as in the statement of the theorem. Note that, by Riemann-Roch,

$$\begin{aligned} rk\ \mathcal{E} &= n'd - nd' + nn'(g-1) \\ &= 2nn'(g-1) - n'^2(g-1) \\ &= rk\ \mathcal{F} - n'^2(g-1). \end{aligned}$$

Claim: $\pi^{-1}(M_{n'}(E)) = \{(L,F) \in J \times M_0 | rk\ \varphi_{(L,F)} \le rk\ \mathcal{E} - 1\}$.

The claim is a consequence of the following lemma and Lemma 2.1.

LEMMA 3.2. *Let E be a general bundle. Suppose $n'd - nd' = n'(n-n')(g-1)$ and let E' be a stable bundle of rank n' and degree d'. Then any nonzero homomorphism $E' \longrightarrow E$ is an injection onto a subbundle of E. Moreover $h^0(E'^* \otimes E) = 1$.*

PROOF. If $\psi : E' \longrightarrow E$ is injective then $Im\psi$ must be saturated since it has the same degree as a maximal subbundle of E. Suppose that ψ is not injective and let $E_1 = Im\psi$. Since E' is stable, we have

$$\frac{d_1}{n_1} > \frac{d'}{n'} = \frac{d}{n} - \frac{n-n'}{n}(g-1) > \frac{d}{n} - \frac{n-n_1}{n}(g-1),$$

contradicting the generality of E. Finally, if $h^0(E'^* \otimes E) \geq 2$, the morphism

$$\mathbb{P}(H^0(E'^* \otimes E)) \longrightarrow M_{n'}(E)$$

is non-constant, since E' is stable. So $M_{n'}(E)$ is infinite, another contradiction. □

This establishes the claim. It follows by the Porteous formula (see [**1**, p.86]) that the fundamental class of $\pi^{-1}(M_{n'}(E))$ in $H^*(J \times M_0, \mathbb{Z})$ is given by

$$[\pi^{-1}(M_{n'}(E))] = c_{\text{top}}(\mathcal{F} - \mathcal{E})[J \times M_0].$$

To complete the proof of the theorem, we must show that $M_{n'}(E)$ is smooth.

Given the identification of $M_{n'}(E)$ with $Quot_{n-n',d''}(E)$, it is sufficient to prove

LEMMA 3.3. *Let E be a general bundle and E' any subbundle of E. Then*

$$H^1(E'^* \otimes E/E') = 0.$$

PROOF. (due essentially to Laumon [**10**], see also [**15**]). If

$$H^1(E'^* \otimes E/E') \neq 0,$$

then by Serre duality there exists a nonzero homomorphism $E/E' \longrightarrow E' \otimes K$. We thus have a non-zero homomorphism

$$E \longrightarrow E/E' \longrightarrow E' \otimes K \longrightarrow E \otimes K.$$

This homomorphism is clearly nilpotent. But, since E is general, it is very stable (see [**10**, Proposition 3.5]), i.e. it admits no nilpotent homomorphism $E \longrightarrow E \otimes K$ different from zero. This contradiction establishes the lemma and Theorem 3.1. □

REMARK 3.4. It may be of interest to identify the precise generality conditions required on E. In fact we need

(i) $s_{n'}(E) = n'(n-n')(g-1)$;
(ii) $s_{n_1}(E) \geq n_1(n-n_1)(g-1)$ for $1 \leq n_1 \leq n'-1$;
(iii) every maximal subbundle of E of rank n' is stable;
(iv) $M_{n'}(E)$ is finite;

(v) E is very stable.

The application of the Porteous formula, however, requires only the assumption that the relevant degeneracy locus of φ is finite and it then gives the length of this locus in its natural scheme structure. In particular, if we assume only conditions (i), (ii) and (iv), the formula counts (with the appropriate multiplicities) the number of stable maximal subbundles of E of rank n'.

REMARK 3.5. Since $(n', d') = 1$ and $n'd - nd' = n'(n - n')(g - 1)$, it follows that n' divides n and we can then solve for d in terms of n, n', d'. Theorem 3.1 does therefore give an answer to the question of [**14**, section 3.4, Remark 4] whenever $(n', d') = 1$ (or, in the notation of [**14**], $(r, d) = 1$). In [**14**] there is no smoothness assumption on the Quot scheme, so we require only conditions (i)–(iv) of Remark 3.4; the formula of Theorem 3.1 then gives the length of the finite scheme $Quot_{n-n',d''}(E)$.

REMARK 3.6. To apply this theorem, we note that the Chern classes of $\mathcal{U}$ and $\mathcal{L}$ are well known (see [**2**] for $\mathcal{U}$ and [**1**] for $\mathcal{L}$). From this the Chern character of $\mathcal{U}^* \otimes \mathcal{L}^* \otimes p^*(E \otimes K)$ and hence, using Grothendieck-Riemann-Roch, of $\mathcal{E}$ can be computed. The Chern character of $\mathcal{F}$ is also easily computed, hence that of $\mathcal{F} - \mathcal{E}$. Converting this to Chern classes, we obtain $c_{\text{top}}(\mathcal{F} - \mathcal{E})$. We now need to know only the intersection numbers of $J \times M_0$. In principle, these can be calculated. In practice, it may be a difficult computation. In the final section of this paper, we will carry out the computation in the simplest case not previously known.

One could of course have stated the theorem using the cohomology of $M(n', d')$ rather than that of $J \times M_0$. We have chosen the latter, because (at least in the case $n' = 2$) it is easier to work with.

4. The case $g = 2,\ n' = 2$

THEOREM 4.1. *Let E be a general bundle of rank n and degree d on a curve C of genus 2. Suppose that $n \geq 4$ is even and $2d + 4 \equiv 0$ mod n with $\frac{2d+4}{n}$ odd. Then*

$$m_2(E) = \frac{n^3}{48}(n^2 + 2).$$

REMARK 4.2. Note that the right hand side of this expression is always a positive integer for n even. The numerical conditions are precisely what is needed to allow d' to be odd. When they are satisfied we can take d' to be any odd integer and then

$$d = \frac{n}{2}d' + n - 2.$$

Before starting on the proof, we recall some properties of the cohomology ring $H^*(C \times J \times M_0, \mathbb{Z})$. The abelian group $H^1(C, \mathbb{Z})$ has a natural symplectic structure. Let f denote the positive generator of $H^2(C, \mathbb{Z})$. We can then choose a symplectic basis e_1, e_2, e_3, e_4 for $H^1(C, \mathbb{Z})$ such that

$$e_1 e_3 = e_2 e_4 = -f \tag{1}$$

and all the other products $e_i e_j$ for $i < j$ are 0. On $C \times J$ we can normalise $\mathcal{L}$ so that

$$c(\mathcal{L}) = 1 + \xi_1 \tag{2}$$

where $\xi_1 \in H^1(C, \mathbb{Z}) \otimes H^1(J, \mathbb{Z}) \subset H^2(C \times J)$ can be written as

$$\xi_1 = \sum_{i=1}^{4} e_i \otimes \varphi_i \tag{3}$$

with $\varphi_i \in H^1(J, \mathbb{Z})$. We have (see [**1**, p.335])

$$\xi_1^2 = -2\theta f, \quad \theta^2[J] = 2. \tag{4}$$

The variety M_0 is given by the intersection of two quadrics in $\mathbb{P}_5$ (see [**12**]). From this we can see that its cohomology groups are

$$\mathbb{Z},\ 0,\ \mathbb{Z},\ \mathbb{Z}^4,\ \mathbb{Z},\ 0,\ \mathbb{Z}.$$

If α is the positive generator of $H^2(M_0, \mathbb{Z})$, then $\alpha^3[M_0] = 4$, so by (4)

$$\alpha^3\theta^2[J \times M_0] = 8. \tag{5}$$

Taking $d' = 1$ and normalising as in [**13**] we have

$$c_1(\mathcal{U}) = \alpha + f, \quad c_2(\mathcal{U}) = \chi + \xi_2 + \alpha f \tag{6}$$

with $\chi \in H^4(M_0, \mathbb{Z})$ and $\xi_2 \in H^1(C, \mathbb{Z}) \otimes H^3(M_0, \mathbb{Z})$. In the notation of [**13**] and [**7**], $4\chi = \alpha^2 - \beta$ and we write

$$\xi_2^2 = \gamma f, \quad \gamma \in H^6(M_0, \mathbb{Z}). \tag{7}$$

Moreover, when $g = 2$, the relations of the Theorem of [**7**] give

$$\alpha^2 + \beta = 0, \quad \alpha^3 + 5\alpha\beta + 4\gamma = 0.$$

Hence $\beta = -\alpha^2$, $\gamma = \alpha^3$, and we have from (6)

$$c_2(\mathcal{U}) = \frac{\alpha^2}{2} + \xi_2 + \alpha f, \quad \xi_2^2 = \alpha^3 f. \tag{8}$$

We define also $\Lambda \in H^1(J,\mathbb{Z}) \otimes H^3(M_0,\mathbb{Z})$ by

$$\xi_1\xi_2 = \Lambda f. \tag{9}$$

In $H^*(C \times J \times M_0, \mathbb{Z})$, we have for dimensional reasons (noting that $H^5(M_0,\mathbb{Z}) = 0$) that the following classes are all zero:

$$f^2\,,\ \xi_1^3\,,\ \alpha^4\,,\ \xi_1 f\,,\ \xi_2 f\,,\ \alpha\xi_2\,,\ \alpha\Lambda\,,\ \theta^2\Lambda. \tag{10}$$

Finally we need to compute $\theta\Lambda^2$. We can write

$$\xi_2 = \sum_{i=1}^{4} e_i \otimes \psi_i \quad \text{with} \ \ \psi_i \in H^3(M_0,\mathbb{Z}).$$

Using this, (1) and (3), we have

$$\xi_1\xi_2 = (\varphi_1\psi_3 + \varphi_2\psi_4 - \varphi_3\psi_1 - \varphi_4\psi_2)f.$$

So by (9)

$$\Lambda = \varphi_1\psi_3 + \varphi_2\psi_4 - \varphi_3\psi_1 - \varphi_4\psi_2.$$

Also from (3) and (4),

$$\theta = -\varphi_1\varphi_3 - \varphi_2\varphi_4.$$

Hence $\theta^2 = 2\varphi_1\varphi_3\varphi_2\varphi_4$, so by (4)

$$\varphi_1\varphi_3\varphi_2\varphi_4[J] = 1.$$

Now

$$\begin{array}{rcl} \Lambda^2 & = & -2\varphi_1\varphi_2\psi_3\psi_4 + 2\varphi_1\varphi_3\psi_3\psi_1 + 2\varphi_1\varphi_4\psi_3\psi_2 \\ & + & 2\varphi_2\varphi_3\psi_4\psi_1 + 2\varphi_2\varphi_4\psi_4\psi_2 - 2\varphi_3\varphi_4\psi_1\psi_2 \end{array}$$

and

$$\begin{array}{rcll} \theta\Lambda^2 & = & -2\varphi_1\varphi_3\varphi_2\varphi_4\psi_4\psi_2 - 2\varphi_2\varphi_4\varphi_1\varphi_3\psi_3\psi_1 & \\ & = & \theta^2(\psi_1\psi_3 + \psi_2\psi_4) & \\ & = & \frac{1}{2}\theta^2\gamma & \text{(by (7))} \\ & = & \frac{1}{2}\theta^2\alpha^3 & \text{(since } \ \gamma = \alpha^3). \end{array}$$

So by (5)

$$\theta\Lambda^2[J \times M_0] = 4. \tag{11}$$

PROOF OF THEOREM 4.1. Tensoring by a line bundle over X if necessary, we can suppose $d' = 1$. By (2) and (4) we have

$$ch(\mathcal{L}) = 1 + \xi_1 - \theta f$$

while by (6), (8) and (10)

$$ch(\mathcal{U}) = 2 + (\alpha + f) + (-\xi_2) + \left(-\frac{1}{12}\alpha^3 - \frac{1}{4}\alpha^2 f\right).$$

Moreover

$$ch(p^*(E \otimes K)) \cdot p^*td(C) = n + \left(\frac{5}{2}n - 2\right) f.$$

So

$$\begin{aligned} ch(\mathcal{U}^* \otimes \mathcal{L}^* \otimes p^*(E \otimes K)) \cdot p^*td(C) &= 2n + [(4n-4)f - n\alpha - 2n\xi_1] \\ &+ \left[\left(-\left(\frac{5}{2}n - 2\right)\alpha - 2n\theta\right) f + n\alpha\xi_1 - n\xi_2\right] \\ &+ \left[\left(\frac{n}{4}\alpha^2 + n\Lambda + n\alpha\theta\right) f + \frac{n}{12}\alpha^3\right] \\ &+ \left[\left(\frac{5}{24}n - \frac{1}{6}\right)\alpha^3 f - \frac{n}{12}\alpha^3\xi_1\right] + \left[-\frac{n}{12}\alpha^3\theta f\right]. \end{aligned}$$

Hence by Grothendieck-Riemann-Roch

$$\begin{aligned} ch(\mathcal{E}) = 4n - 4 &+ \left[-\left(\frac{5}{2}n - 2\right)\alpha - 2n\theta\right] + \left[\frac{n}{4}\alpha^2 + n\Lambda + n\alpha\theta\right] \\ &+ \left[\left(\frac{5}{24}n - \frac{1}{6}\right)\alpha^3\right] + \left[-\frac{n}{12}\alpha^3\theta\right]. \end{aligned}$$

From (6) and (8) we obtain easily

$$ch(\mathcal{F}) = 4n - 2n\alpha + \frac{n}{6}\alpha^3.$$

So

$$\begin{aligned} ch(\mathcal{F} - \mathcal{E}) = 4 &+ \left[\left(\frac{1}{2}n - 2\right)\alpha + 2n\theta\right] + \left[-\frac{n}{4}\alpha^2 - n\Lambda - n\alpha\theta\right] \\ &+ \left[\left(\frac{1}{6} - \frac{1}{24}n\right)\alpha^3\right] + \left[\frac{n}{12}\alpha^3\theta\right]. \end{aligned}$$

Converting this into Chern classes, we obtain

$$c_{\mathrm{top}}(\mathcal{F} - \mathcal{E}) = c_5(\mathcal{F} - \mathcal{E}) = \left(\frac{1}{24}n^5 - \frac{5}{12}n^3\right)\alpha^3\theta^2 + n^3\theta\Lambda^2.$$

By (5) and (11), we get

$$c_{\mathrm{top}}(\mathcal{F} - \mathcal{E})[J \times M_0] = \frac{1}{3}n^5 + \frac{2}{3}n^3$$

and Theorem 4.1 follows from Theorem 3.1. □

References

[1] Arbarello, E., Cornalba, M., Griffiths, P.A., Harris, J.: Geometry of Algebraic Curves. Grundlehren **267**, Springer-Verlag, 1985.

[2] Atiyah, M.F., Bott, R.: The Yang-Mills equations over Riemann surfaces. Phil. Trans. R. Soc. Lond. A **308** (1982), 523-615.

[3] Behrend, K.: Gromov-Witten invariants in algebraic geometry. Invent. Math. **127** (1997), 601-617.

[4] Brambila-Paz, L., Lange, H.: A stratification of the moduli space of vector bundles on curves. J. Reine Angew. Math. **494** (1998), 173-187.

[5] Ghione, F.: Quelques résultats de Corrado Segre sur les surfaces réglées. Math. Ann. **255** (1981), 77-95.

[6] Hirschowitz, A.: Problèmes de Brill-Noether en rang supérieur. Prépublications Mathématiques n. 91, Nice (1986).

[7] King, A.D., Newstead, P.E.: On the cohomology ring of the moduli space of rank 2 vector bundles on a curve. Topology **37** (1998), 407-418.

[8] Lange, H.: Zur Klassifikation von Regelmannigfaltigkeiten. Math. Ann. **262** (1983), 447-459.

[9] Lange, H.: Höhere Sekantenvarietäten und Vektorbündel auf Kurven. Manuscr. math. **52** (1985), 63-80.

[10] Laumon, G.: Un analogue global du cône nilpotent. Duke Math. J. **57** (1988), 647-671.

[11] Mukai, S., Sakai, F.: Maximal subbundles of vector bundles on a curve. Manuscr. math. **52** (1985), 251-256.

[12] Newstead, P.E.: Stable bundles of rank 2 and odd degree over a curve of genus 2. Topology **7** (1968), 205-215.

[13] Newstead, P.E.: Characteristic classes of stable bundles of rank 2 over an algebraic curve. Trans. Amer. Math. Soc. **169** (1972), 337-345.

[14] Okonek, Ch., Teleman, A.: Gauge theoretical equivariant Gromov-Witten invariants and the full Seiberg-Witten invariants of ruled surfaces. Commun. Math. Phys. **227** (2002), 551-585.

[15] Oxbury, W.M.: Varieties of maximal line subbundles. Math. Proc. Cambridge Phil. Soc. **129** (2000), 9-18.

[16] Russo, B., Teixidor i Bigas, M.: On a conjecture of Lange. J. Alg. Geom. **8** (1999), 483-496.

[17] Segre, C.: Recherches générales sur les courbes et les surfaces réglées algébriques II. Math. Ann. **34** (1889), 1-25.

Mathematisches Institut, Universität Erlangen-Nürnberg
Bismarckstrasse $1\frac{1}{2}$, D-91054 Erlangen, Germany
E-mail address: lange@mi.uni-erlangen.de

Department of Mathematical Sciences, University of Liverpool
Peach Street, Liverpool L69 7ZL, U.K.
E-mail address: newstead@liv.ac.uk

Bases for representations, LS-paths and Verma flags

P. LITTELMANN

Dedicated to Professor Seshadri on his 70th birthday

Abstract. The basic result of this paper is the construction of two kind of bases $\mathbb{B}(\lambda)$ and $\mathbb{M}(\lambda)$ for simple finite dimensional representations $V(\lambda)$ of a semisimple complex Lie algebra $\mathfrak{g}$. The construction combines the combinatorics of LS-paths [**4**], the Weyl group combinatorics related to inclusions of Verma modules [**1**] and the structure of singular vectors in Verma modules [**7**].

Introduction

In this paper we describe the construction of two kinds of bases $\mathbb{B}(\lambda)$ and $\mathbb{M}(\lambda)$ (Theorems 5.2 and 5.3) for simple finite dimensional representations $V(\lambda)$ of a semisimple complex Lie algebra $\mathfrak{g}$. The construction combines the combinatorics of LS-paths [**4**], the Weyl group combinatorics related to inclusions of Verma modules [**1**] and the structure of singular vectors in Verma modules [**7**].

The development of the basis $\mathbb{B}(\lambda)$ was inspired by the similarity of the combinatorial description of inclusions of Verma modules and the combinatorial description of LS-paths. This similarity is used to attach in a canonical way to each LS-path π of shape λ a vector $v_\pi \in V(\lambda)$. The construction of these vectors uses singular vectors in Verma modules which are naturally associated to the LS-path π. By repeating the procedure for all multiples of λ, this basis globalizes to a basis $\mathbb{B}(\lambda, \infty)$ of $U(\mathfrak{n}^-)$ (for λ regular, for a precise statement in

This research has been partially supported by the EC TMR network "Algebraic Lie Representations", contract no. ERB FMRX-CT97-0100.

the non-regular case see Theorem 5.2). This basis is compatible with the surjections $\Psi_m : U(\mathfrak{n}^-) \to V(m\lambda)$, $u \mapsto uv_{m\lambda}$, i.e., the kernel has $\operatorname{Ker}\Psi_m \cap \mathbb{B}(\lambda,\infty)$ as basis, and the basis $\mathbb{B}(\lambda,\infty)$ depends only on the ray $\mathbb{R}_{>0}\lambda$ spanned by λ.

The second basis $\mathbb{M}(\lambda)$ is in some sense a simplified version of the first. It is obtained by attaching to an LS-path π of shape λ a monomial $\mathfrak{m}_\pi$ in $U(\mathfrak{n}^-)$ in the Chevalley generators. The vectors $u_\pi = \mathfrak{m}_\pi v_\lambda$, $v_\lambda \in V(\lambda)$ a highest weight vector, provide a basis $\mathbb{M}(\lambda)$ of the $\mathbb{Z}$-lattice $V_\mathbb{Z}(\lambda)$. $\mathbb{M}(\lambda)$ of $V_\mathbb{Z}(\lambda)$. The monomials used for the construction of $\mathbb{M}(\lambda)$ have first been considered by V. Lakshmibai and C. S. Seshadri in [**3**], where they conjecture that the $u_\pi \in V_\mathbb{Z}(\lambda)$ form a basis, and they give a proof in some special cases. This basis again globalizes nicely to a basis $\mathbb{M}(\lambda,\infty)$ of $U(\mathfrak{n}^-)$ (for λ regular, for a precise statement see Theorem 5.3), and is compatible with the surjections $\Psi_m : U(\mathfrak{n}^-) \to V(m\lambda)$, $u \mapsto uv_{m\lambda}$. The basis $\mathbb{M}(\lambda,\infty)$ has the advantage of being given by monomials, its disadvantage is that it is not anymore canonically defined.

The main tools used in the proof and in the construction are the path model, the quantum Frobenius splitting (to prove linear independence with the same method as in [**5**]), and the theory of Verma modules (mainly its combinatorial part). Since in our construction only real roots are involved in the combinatorics as well as in the computation of singular vectors, all these tools are also available in the case of symmetrizable Kac-Moody algebras. So the construction can be easily adapted to the more general setting of symmetrizable Kac-Moody algebras, and the theorems above hold also in these cases. But to keep the notation simple, we decided to consider in this paper only the case of semisimple Lie algebras.

1. Inclusions of Verma modules

In this section we fix some notation and recall some standard facts on homomorphisms between Verma modules. Let $\mathfrak{g}$ be a semisimple complex Lie algebra and denote $U(\mathfrak{g})$ its universal enveloping algebra. We fix a Cartan subalgebra $\mathfrak{h}$ and a triangular decomposition $\mathfrak{g} = \mathfrak{n}^-\oplus\mathfrak{h}\oplus\mathfrak{n}$ such that $\mathfrak{b} = \mathfrak{h}\oplus\mathfrak{n}$ is a Borel subalgebra of $\mathfrak{g}$ and $\mathfrak{b}^- = \mathfrak{h}\oplus\mathfrak{n}^-$ is the opposite Borel subalgebra. Denote by Φ the root system of $\mathfrak{g}$, let Φ^+ be the corresponding set of positive roots and denote by $\Delta = \{\alpha_1,\dots,\alpha_n\}$ the set of simple roots. For $\beta\in\Phi$ let $H_\beta\in\mathfrak{h}$ be its coroot. If $\lambda\in\mathfrak{h}^*$, then we write $\langle\lambda,\beta\rangle$ for $\lambda(H_\beta)$. We fix a Chevalley basis $\{X_\beta, Y_\beta \mid \beta\in\Phi^+\}\cup\{H_\alpha \mid \alpha\in\Delta\}$ of $\mathfrak{g}$, where $X_\beta\in\mathfrak{g}_\beta$ and

$Y_\beta \in \mathfrak{g}_{-\beta}$. We write often just X_i and Y_i instead of X_{α_i} and Y_{α_i} for $\alpha_i \in \Delta$. Let $\lambda \in \mathfrak{h}^*$, the Verma module $M(\lambda)$ is the $U(\mathfrak{g})$-module

$$M(\lambda) := U(\mathfrak{g}) \otimes_{U(\mathfrak{b})} \mathbb{C}_\lambda,$$

where $\mathbb{C}_\lambda$ denotes the one dimensional $U(\mathfrak{b})$-module obtained by extending λ trivially to $\mathfrak{b}$. By definition, $M(\lambda)$ is a cyclic highest weight module of highest weight λ with highest weight vector $1_\lambda = 1 \otimes 1 \in M(\lambda)$.

For an element w of the Weyl group W and $\mu \in \mathfrak{h}^*$ set $w \circ \mu := w(\mu + \rho) - \rho$, here ρ is the sum of the fundamental weights. We summarize the properties of homomorphisms between Verma modules in the following theorem (due to Verma, Bernstein, Gelfand and Gelfand), see [**1**] or [**2**], Chapter 7.

THEOREM 1.1. *Let $\lambda, \mu \in \mathfrak{h}^*$, $\lambda \neq \mu$.*
i) The vector space $\mathrm{Hom}_{\mathfrak{g}}(M(\mu), M(\lambda))$ *is null or one-dimensional, and every non-zero element of* $\mathrm{Hom}_{\mathfrak{g}}(M(\mu), M(\lambda))$ *is injective.*
ii) If $\dim \mathrm{Hom}_{\mathfrak{g}}(M(\mu), M(\lambda)) = 1$, *then* $\mu \in W \circ \lambda$.
iii) If $\beta \in \Phi^+$, $s_\beta \circ \mu \neq \mu$, then

$$\dim \mathrm{Hom}_{\mathfrak{g}}(M(s_\beta \circ \mu), M(\mu)) = 1 \Leftrightarrow \langle \mu + \rho, \beta \rangle \in \mathbb{N}_+.$$

iv) Every non-trivial homomorphism $\phi : M(\mu) \to M(\lambda)$ factors through maps of the type above: there exists a sequence of positive roots $\beta_1, \ldots, \beta_t$ such that the weights $\lambda_0 = \lambda$, $\lambda_1 = s_{\beta_1} \circ \lambda_0$, ..., $\lambda_t = s_{\beta_t} \circ \lambda_{t-1} = \mu$ are all pairwise different and

$$M(\mu) = M(\lambda_t) \hookrightarrow \ldots \hookrightarrow M(\lambda_2) \hookrightarrow M(\lambda_1) \hookrightarrow M(\lambda_0) = M(\lambda).$$

We fix an ordering of the positive roots, the correspondingly ordered monomials in the Chevalley basis form then a PBW-basis for $U(\mathfrak{n}^-)$. We use the standard filtration on $U(\mathfrak{n}^-)$ respectively $U(\mathfrak{b}^-)$: the elements of $\mathfrak{n}^-$ respectively of $\mathfrak{b}^-$ are all of degree one. The following result is due to Shapovalov. For $\beta = \sum_{\alpha \in \Gamma} \ell_\alpha \alpha \in \Phi^+$ and $m \in \mathbb{N}$ set

PROPOSITION 1.2 ([**8**]). *There exists an element $\theta_{\beta,m} \in U(\mathfrak{n}^- + \mathfrak{h})_{-m\beta}$, unique modulo the left ideal generated by $(H_\beta - m + \langle \rho, \beta \rangle)$, such that*
i) $\theta_{\beta,m} 1_\lambda \in M(\lambda)$ is a highest weight vector of weight $s_\beta \circ \lambda$ for all $\lambda \in \mathcal{H}^\rho_{\beta,m}$,
ii) $\theta_{\beta,m} = \prod_{\alpha \in \Gamma} Y_\alpha^{m\ell_\alpha} +$ elements of the form uh, where $h \in U(\mathfrak{h})$ and $u \in U(\mathfrak{n}^-)_{-m\beta}$ is of degree strictly less than $m \sum \ell_\alpha$.

Note that the condition above on the term of highest degree is independent of the ordering of the product. We call $\theta_{\beta,m}$ a *Shapovalov element* for the pair (β, m). To have uniqueness, it is sometimes more convenient to replace $\theta_{\beta,m}$ by an element $\theta_{\mu,\lambda} \in U(\mathfrak{n}^-)$. By applying $\theta_{\beta,m}$ to 1_λ, one gets by Proposition 1.2:

COROLLARY 1.3. *Let $\lambda, \mu \in \mathfrak{h}^*$ and $\beta = \sum_{\alpha\in\Gamma} \ell_\alpha \alpha \in \Phi^+$ be such that $m := \langle \lambda + \rho, \beta \rangle \in \mathbb{N}$ and $\mu = s_\beta \circ \lambda$. There exists a unique element $\theta_{\mu,\lambda} \in U(\mathfrak{n}^-)_{-m\beta}$ such that*
i) $\theta_{\mu,\lambda} 1_\lambda \in M(\lambda)$ is a highest weight vector of weight μ,
ii) $\theta_{\mu,\lambda} = \prod_{\alpha\in\Gamma} Y_\alpha^{m\ell_\alpha} +$ elements of lesser degree.

2. L-S paths and Verma flags

Let $X \subset \mathfrak{h}^*$ be the lattice of integral weights and denote X^+ the monoid of dominant weights. We discuss now the connection between the combinatorics of L-S paths $\pi = (\underline{\tau}, \underline{a})$ of shape $\lambda \in X^+$ and the combinatorics of inclusions of Verma modules. Let $X_\mathbb{Q} \subset \mathfrak{h}^*$ be the rational span of the weight lattice, we denote $X_\mathbb{Q}^+$ the submonoid $\mathbb{Q}_{\geq 0} X^+$. We define a length function on $X_\mathbb{Q}$ as follows: for $\mu \in X_\mathbb{Q}$ let $\tau \in W$ be the unique element of minimal length such such that $\mu + \rho = \tau(\lambda + \rho)$ for some $\lambda + \rho \in X_Q^+$, we set $\ell(\mu) = \ell(\tau)$. If $\nu \in X_\mathbb{Q}^+$, then we write W_ν for the stabilizer of ν in W. We use the notation "$\geq$" for the Bruhat order on W as well as for the induced order on W/W_ν.

Let $\mu \neq \nu \in X_\mathbb{Q}$ be such that $\dim \operatorname{Hom}_\mathfrak{g}(M(\mu), M(\nu)) = 1$. By Theorem 1.1, there exists a sequence of positive roots $\beta_1, \ldots, \beta_t$ and pairwise different weights $\nu_0 = \nu$, $\nu_1 = s_{\beta_1} \circ \nu_0$, ..., $\nu_t = s_{\beta_t} \circ \nu_{t-1} = \mu$ such that

$$M(\mu) = M(\nu_t) \hookrightarrow \ldots \hookrightarrow M(\nu_1) \hookrightarrow M(\nu_0) = M(\nu), \tag{2.1}$$

and part *iii)* of the theorem implies for the length of the flag: $t \leq \ell(\mu) - \ell(\nu)$.

DEFINITION 2.2. Let $\mu \neq \nu \in X_\mathbb{Q}$. We say that there exists a *maximal Verma flag* for the pair (μ, ν) if $\dim \operatorname{Hom}_\mathfrak{g}(M(\mu), M(\nu)) = 1$ and one can find a flag of length $t = \ell(\mu) - \ell(\nu)$ in (2.1).

We come now to the definition of an L-S path of shape λ. To simplify the notation with the shifts by ρ, we introduce the abbreviation

$$\hat{\mu} = \mu - \rho$$

The definition of an LS-path below is equivalent to the definition given in [**4**], we will comment on the connection in Remark 2.4 and 2.5:

DEFINITION 2.3. An *L-S path* $\pi = (\underline{\tau}, \underline{a})$ *of shape* $\lambda \in X^+$ is a pair of sequences $\underline{\tau} = (\tau_0, \tau_1, \ldots, \tau_r)$ and $\underline{a} = (a_1, \ldots, a_r)$ satisfying the following conditions:
i) $0 < a_1 < \ldots < a_r < 1$ is a strictly increasing sequence of rational numbers,
ii) $\tau_0 > \tau_1 > \ldots > \tau_r$ is a strictly decreasing sequence of cosets in W/W_λ;
iii) for all pairs $(\widehat{\nu_j}, \widehat{\mu_j})$ below there exists a maximal Verma flag:

$$\begin{cases} \nu_0 = a_1\tau_0(\lambda) & \nu_1 = a_2\tau_1(\lambda) & \ldots & \nu_{r-1} = a_r\tau_{r-1}(\lambda) & \nu_r = \tau_r(\lambda) \\ \mu_0 = a_1\tau_1(\lambda) & \mu_2 = a_2\tau_2(\lambda) & \ldots & \mu_{r-1} = a_r\tau_r(\lambda) & \mu_r = \lambda. \end{cases}$$

REMARK 2.4. Note: $s_\beta \circ \hat{\mu} = s_\beta(\hat{\mu} + \rho) - \rho = s_\beta(\mu) - \rho = \widehat{s_\beta(\mu)}$. A simple calculation (using part *iii)* and *iv)* of Theorem 1.1) shows: the existence of a maximal Verma flag for the pairs $(\widehat{\nu_j}, \widehat{\mu_j})$, is equivalent to the existence of an a_j-chain and the integrality conditions on the pairs (τ_j, τ_{j+1}) in the definition of an LS-path of shape λ given in [**4**].

REMARK 2.5. Since $\mu_r = \lambda \in X^+$ and $\nu_r = \tau_r(\lambda) \in X$, there exists always a maximal Verma flag for the pair $(\widehat{\lambda_r}, \widehat{\mu_r})$: Let $\tau_r = s_{i_1} \cdots s_{i_p}$ be a reduced decomposition, then a maximal Verma flag is given by:

$$M(\widehat{\tau_r(\lambda)}) = M((s_{i_1} \cdots s_{i_p}) \circ \widehat{\lambda}) \hookrightarrow \ldots \hookrightarrow M(s_{i_p} \circ \widehat{\lambda}) \hookrightarrow M(\widehat{\mu_r}) = M(\widehat{\lambda})$$

and $\theta_{\widehat{\tau_r(\lambda)}, \widehat{\lambda}}$ is of the form $Y_{i_1}^{\langle s_{i_2} \cdots s_{i_p}(\lambda), \alpha_{i_1} \rangle} \cdots Y_{i_{p-1}}^{\langle s_{i_p}(\lambda), \alpha_{i_{p-1}} \rangle} Y_{i_p}^{\langle \lambda, \alpha_{i_p} \rangle}$.

REMARK 2.6. Set $a_0 = 0$ and $a_{r+1} = 1$. We identify sometimes the path with its "visible" path, i.e., the piecewise linear map $\pi : [0,1] \to X_{\mathbb{R}}$,

$$t \to \sum_{i=1}^{j} (a_i - a_{i-1})\tau_{i-1}(\lambda) + (t - a_j)\tau_j(\lambda) \text{ for } a_j \leq t \leq a_{j+1},\ 0 \leq j \leq r.$$

Having this identification in mind, it makes sense to talk about the endpoint or the *weight* $\pi(1)$ *of the path*. The *LS-path character formula* in [**4**] states: Let $\lambda \in X$ be a dominant weight, let $V(\lambda)$ be the corresponding irreducible finite dimensional representation of $\mathfrak{g}$ and denote $B(\lambda)$ the set of LS-paths of shape λ. Then

$$\text{Char}V(\lambda) = \sum_{\pi\in B(\lambda)} e^{\pi(1)}. \tag{2.7}$$

3. Verma elements

We associate now to an LS-path a unique element in $U(\mathfrak{n}^-)$. Let $\nu, \eta \in X_\mathbb{Q}$ be a pair such that there exists a *maximal Verma flag* of length $t = \ell(\eta) - \ell(\nu)$:

$$M(\eta) = M(\nu_t) \hookrightarrow \ldots \hookrightarrow M(\nu_2) \hookrightarrow M(\nu_1) \hookrightarrow M(\nu_0) = M(\nu) \tag{3.1}$$

and let $\beta_1, \ldots, \beta_t$ be the positive roots such that

$$\eta = \nu_t = s_{\beta_t} \circ \nu_{t-1},\ \ldots,\ \nu_2 = s_{\beta_2} \circ \nu_1,\ \nu_1 = s_{\beta_1} \circ \nu_0,\ \nu_0 = \nu.$$

We associate to the flag (3.1) the monomial $\mathfrak{m}_{\eta,\nu}$ in $U(\mathfrak{n}^-)$ defined by

$$\mathfrak{m}_{\eta,\nu} = \frac{Y_{\beta_t}^{m_t}}{m_t!} \cdots \frac{Y_{\beta_1}^{m_1}}{m_1!}, \quad \text{where } m_j := \langle \nu_{j-1} + \rho, \beta_j \rangle. \tag{3.2}$$

Note that $\mathfrak{m}_{\eta,\nu}$ depends on the choice of the flag in (3.1). To obtain elements which depend only on η and ν, recall that the image of $i : M(\eta) \hookrightarrow M(\nu)$ is independent of the chosen maximal Verma flag because the space of homomorphisms is one-dimensional. By Corollary 1.3, there exists for all $j = 1, \ldots, t$ a unique element $\theta_{\nu_j,\nu_{j-1}}$ in $U(\mathfrak{n}^-)_{-m_j\beta_j}$ such that $\theta_{\nu_j,\nu_{j-1}} 1_{\nu_{j-1}}$ is a highest weight vector in $M(\nu_{j-1})$ of weight ν_j, and

$$\theta_{\nu_j,\nu_{j-1}} = \prod_{\alpha\in\Gamma} Y_\alpha^{m\ell_\alpha} + \text{ elements of lesser degree,}$$

where $\beta_j = \sum_{\alpha\in\Gamma} \ell_\alpha \alpha$. We set $\Theta_{\eta,\nu} = \theta_{\nu_t,\nu_{t-1}} \cdots \theta_{\nu_2,\nu_1} \theta_{\nu_1,\nu_0}$.

LEMMA 3.3. *Set $\nu - \eta = \sum_{\alpha\in\Gamma} q_\alpha \alpha$. The element $\Theta_{\eta,\nu} \in U(\mathfrak{n}^-)$ is uniquely characterized by the following properties: $\Theta_{\eta,\nu} 1_\nu \in M(\nu)$ is a highest weight vector of weight η, and $\Theta_{\eta,\nu} = \prod_{\alpha\in\Gamma} Y_\alpha^{q_\alpha} +$ elements of lesser degree.*

REMARK 3.4. The uniqueness property is independent on the chosen ordering on the product.

PROOF. By construction, we know that $\Theta_{\eta,\nu} 1_\nu$ is a highest weight vector in $M(\nu)$ of weight η, and $\Theta_{\eta,\nu}$ is, up to scalar multiples, uniquely determined by this property in $U(\mathfrak{n}^-)$. Rewriting a product of two monomials in a PBW-basis as a linear combination of the fixed PBW-basis gives a monomial of degree the sum of the two monomials (the product rewritten in the desired order) plus elements of strictly lesser degree. It follows hence that $\Theta_{\eta,\nu}$ has the form described above, and, by fixing the coefficient of the highest degree term, $\Theta_{\eta,\nu}$ is uniquely determined. $\square$

DEFINITION 3.5. Let $\pi = (\underline{\tau}, \underline{a})$ be an LS-path of shape $\lambda \in X^+$, say $\underline{\tau} = (\tau_0, \ldots, \tau_r)$ and $\underline{a} = (a_1, \ldots, a_r)$. By the *Verma element* Θ_π in $U(\mathfrak{n}^-)$ we mean the product

$$\Theta_\pi = \Theta_{\widehat{\nu_0},\widehat{\mu_0}} \Theta_{\widehat{\nu_1},\widehat{\mu_1}} \cdots \Theta_{\widehat{\nu_r},\widehat{\mu_r}}$$

where the weights $\nu_0, \mu_0, \ldots, \nu_r, \mu_r$ are given as in Definition 2.3 by

$$\begin{cases} \nu_0 = a_1\tau_0(\lambda) & \nu_1 = a_2\tau_1(\lambda) & \ldots & \nu_{r-1} = a_r\tau_{r-1}(\lambda) & \nu_r = \tau_r(\lambda); \\ \mu_0 = a_1\tau_1(\lambda) & \mu_2 = a_2\tau_2(\lambda) & \ldots & \mu_{r-1} = a_r\tau_r(\lambda) & \mu_r = \lambda, \end{cases}$$

The Verma elements Θ_π have in general a rather complicated structure, for a general algorithm to compute the $\theta_{\mu,\nu}$ see [**7**]. For this reason it is also useful to define monomials $\mathfrak{m}_\pi$ in a similar way as above:

DEFINITION 3.6. Let $\pi = (\underline{\tau}, \underline{a})$ be an LS-path of shape λ as above in Definition 3.5. For each pair $(\widehat{\nu_j}, \widehat{\mu_j})$ fix a maximal Verma flag and let $\mathfrak{m}_{\widehat{\nu_j},\widehat{\mu_j}}$ be as in (3.2). By the *path monomial* $\mathfrak{m}_\pi$ in $U(\mathfrak{n}^-)$ we mean the monomial

$$\mathfrak{m}_\pi = \mathfrak{m}_{\widehat{\nu_0},\widehat{\mu_0}} \mathfrak{m}_{\widehat{\nu_1},\widehat{\mu_1}} \cdots \mathfrak{m}_{\widehat{\nu_r},\widehat{\mu_r}}$$

REMARK 3.7. $\mathfrak{m}_\pi, \Theta_\pi \in U(\mathfrak{n}^-)$ are $\mathfrak{h}$-eigenvectors of weight $\lambda - \pi(1)$. The Verma element Θ_π depends only on the path π and is hence *canonically* defined, but the path monomial $\mathfrak{m}_\pi$ depends on the choice of the maximal Verma chains necessary for the definition of the $\mathfrak{m}_{\widehat{\nu_j},\widehat{\mu_j}}$.

4. LS-concatenations of shape $m\lambda$

We consider now the LS-paths of shape $m\lambda$ for all $m \in \mathbb{N}$ at once. To "compare" the LS-paths of shape $p\lambda$ with those of shape $q\lambda$, it is convenient to use the following definition which formalizes the usual concatenation of the "visible" paths in Remark 2.6 (see [**4**]):

DEFINITION 4.1. An *L-S concatenation* $\pi = (\underline{\tau}, \underline{a})$ *of shape* $m\lambda \in X^+$ is a pair $\underline{\tau} = (\tau_0, \ldots, \tau_r)$, $\underline{a} = (a_1, \ldots, a_r)$, satisfying the following conditions:

i) $0 < a_1 < \ldots < a_r < m$ is a strictly increasing sequence of rational numbers;

ii) $\tau_0 > \tau_1 > \ldots > \tau_r$ is a strictly decreasing sequence of cosets in W/W_λ;

iii) for all pairs $(\widehat{\nu_j}, \widehat{\mu_j})$ below there exists a maximal Verma flag:

$$\begin{cases} \nu_0 = a_1\tau_0(\lambda) & \nu_1 = a_2\tau_1(\lambda) & \dots & \nu_{r-1} = a_r\tau_{r-1}(\lambda) & \nu_r = m\tau_r(\lambda); \\ \mu_0 = a_1\tau_1(\lambda) & \mu_2 = a_2\tau_2(\lambda) & \dots & \mu_{r-1} = a_r\tau_r(\lambda) & \mu_r = m\lambda, \end{cases}$$

Set $a_0 = 0$ and $a_{r+1} = m$, the weight of π is defined as

$$\pi(1) = \sum_{j=0}^{r}(a_{j+1} - a_j)\tau_j(\lambda)$$

Denote by $B(m\lambda)$ the set of all LS-paths of shape $m\lambda$ and let $C(m\lambda)$ be the set of all LS-concatenations of shape $m\lambda$. We have obviously $C(\lambda) = B(\lambda)$. The next lemma follows immediately from the definitions:

Lemma 4.2. *The following map defines a weight preserving bijection:*

$$\text{(4.3)} \qquad \begin{array}{rcl} \phi : B(m\lambda) & \longrightarrow & C(m\lambda) \\ (\tau_0, \dots, \tau_r; a_1, \dots, a_r) & \mapsto & (\tau_0, \dots, \tau_r; ma_1, \dots, ma_r). \end{array}$$

The advantage of the notation of LS-concatenations is that for $q > p$ we can easily regard $C(p\lambda)$ as a subset of $C(q\lambda)$. The inclusion $i : C(p\lambda) \hookrightarrow C(q\lambda)$ is given by

$$(\tau_0, \dots, \tau_r; a_1, \dots, a_r) \mapsto \begin{cases} (\tau_0, \dots, \tau_r, \mathrm{id}; a_1, \dots, a_r, p) & \text{if } \tau_r \neq \mathrm{id} \\ (\tau_0, \dots, \tau_r; a_1, \dots, a_r) & \text{if } \tau_r = \mathrm{id} \end{cases}$$

and the image is characterized as

$$\text{(4.4)} \qquad i(C(p\lambda)) = \Big\{(\kappa_0, \dots, \kappa_r; b_1, \dots, b_r) \in C(q\lambda) \,\Big|\, \kappa_r = \mathrm{id} \text{ and } b_r \leq p\Big\}$$

The inclusion $i : C(p\lambda) \hookrightarrow C(q\lambda)$ itself is not weight preserving, but note that $p\lambda - \pi(1) = q\lambda - i(\pi)(1)$ for $\pi \in C(p\lambda)$. By identifying the sets with their images, we get a sequence of inclusions

$$\text{(4.5)} \qquad C(\lambda) \subset C(2\lambda) \subset C(3\lambda) \subset C(4\lambda) \subset \dots \subset C(m\lambda) \subset \dots$$

So it makes sense to talk about $C(\lambda, \infty) = \bigcup_{m\in\mathbb{N}} C(\lambda, m)$.

Definition 4.6. We say that $\pi \in C(\lambda, \infty)$ *is of shape* $m\lambda$ if π is an element of $C(m\lambda)$.

Lemma 4.7. *If $\pi \in B(m\lambda)$ and $\pi' \in B(n\lambda)$ are such that $\pi = \pi'$ in $C(\lambda, \infty)$ (via the inclusions in (4.3) and (4.5)), then $\mathfrak{m}_\pi = \mathfrak{m}_{\pi'}$ and $\Theta_\pi = \Theta_{\pi'}$.*

So in the following we will only write π and not make a difference between LS-paths of shape $m\lambda$ and LS-concatenations of shape $m\lambda$. Further, by Lemma 4.7 it makes sense to consider the collection of elements in $U(\mathfrak{m}^-)$:

$$\mathbb{B}(\lambda,\infty)=\{\Theta_\pi \mid \pi\in C(\lambda,\infty)\} \text{ and } \mathbb{M}(\lambda,\infty)=\{\mathfrak{m}_\pi \mid \pi\in C(\lambda,\infty)\}.$$

PROOF (Lemma 4.7). Let $\pi\in C(\lambda,\infty)$, say $\pi=(\underline{\tau},\underline{a})\in C(m\lambda)$. We can associate to π a Verma element and a monomial as in Definition 3.5 and in Definition 3.6, we set

$$\Theta_\pi=\Theta_{\widehat{\nu_0},\widehat{\mu_0}}\Theta_{\widehat{\nu_1},\widehat{\mu_1}}\cdots\Theta_{\widehat{\nu_r},\widehat{\mu_r}} \quad \text{and} \quad \mathfrak{m}_\pi=\mathfrak{m}_{\widehat{\nu_0},\widehat{\mu_0}}\mathfrak{m}_{\widehat{\nu_1},\widehat{\mu_1}}\cdots\mathfrak{m}_{\widehat{\nu_r},\widehat{\mu_r}}$$

where the weights $\nu_0,\mu_0,\ldots,\nu_r,\mu_r$ are given as in Definition 4.1 by $\nu_0=a_1\tau_0(\lambda)$, $\mu_0=a_1\tau_1(\lambda)$, ... It is evident from the definition that Θ_π and $\mathfrak{m}_\pi$ are independent of the fact whether we regard π as an element of $C(m\lambda)$ or $C(n\lambda)$ for $n\geq m$.

Let now $\pi\in B(\lambda)$ be an LS-path of shape $m\lambda$. One verifies easily $\Theta_\pi=\Theta_{\phi(\pi)}$ and $\mathfrak{m}_\pi=\mathfrak{m}_{\phi(\pi)}$ for the map $\phi: B(m\lambda)\to C(m\lambda)$ in (4.3), which finishes the proof. □

5. The basis given by Verma elements

For simplicity, let first λ be a regular integral dominant weight.

THEOREM 5.1.

i) *The Verma elements* $\mathbb{B}(\lambda,\infty)$ *form a basis of* $U(\mathfrak{n}^-)$.
ii) $\mathbb{B}(\lambda,\infty)$ *is compatible with all irreducible representations* $V(m\lambda)$ *of highest weight* $m\lambda$, *i.e.,* $\{\Theta_\pi \mid \pi\in C(\lambda,\infty),\ \pi\notin C(m\lambda)\}$ *is a basis for the kernel of the surjective map*

$$\Psi: U(\mathfrak{n}^-)\to V(m\lambda),\quad u\mapsto uv_{m\lambda},$$

and the set of vectors $\{\Theta_\pi v_{m\lambda} \mid \pi\in C(m\lambda)\}$ *is a basis of* $V(m\lambda)$.

The formulation for a not necessarily regular $\lambda\in X^+$ has to be slightly adapted. Let $\mathfrak{p}$ be the parabolic subalgebra associated to λ, i.e., $\mathfrak{p}=\mathfrak{b}\oplus\mathfrak{n}_\lambda^-$ where $\mathfrak{n}_\lambda^-=\oplus_{\beta\perp\lambda}\mathfrak{g}_{-\beta}$. Here the sum runs over all positive roots orthogonal to λ. Denote by $\mathfrak{n}_\mathfrak{p}^-$ the nilpotent radical of the parabolic subgroup opposite to $\mathfrak{p}$, so $\mathfrak{g}=\mathfrak{p}\oplus\mathfrak{n}_\mathfrak{p}^-$.

Let $\mathcal{I}_\lambda\subset U(\mathfrak{n}^-)$ be the left ideal generated by $\mathfrak{n}_\lambda^-$, then the canonical map $U(\mathfrak{n}_\mathfrak{p}^-)\to U(\mathfrak{n}^-)/\mathcal{I}_\lambda$ is an isomorphism of left $U(\mathfrak{n}_\mathfrak{p}^-)$-modules. Denote by

$$\overline{\mathbb{B}}(\lambda,\infty)\subset U(\mathfrak{n}^-)/\mathcal{I}_\lambda$$

the images $\overline{\Theta}_\pi$ of the Verma elements Θ_π in the quotient module $U(\mathfrak{n}^-)/\mathcal{I}_\lambda$.

THEOREM 5.2. i) *The Verma elements* $\overline{\mathbb{B}}(\lambda,\infty)$ *form a basis of* $U(\mathfrak{n}^-)/\mathcal{I}_\lambda$.

ii) $\overline{\mathbb{B}}(\lambda,\infty)$ *is compatible with all irreducible representations* $V(m\lambda)$ *of highest weight* $m\lambda$, *i.e.,* $\{\overline{\Theta}_\pi \mid \pi \in C(\lambda,\infty),\, \pi \notin C(m\lambda)\}$ *is a basis of the kernel of the surjective map*

$$\Psi : U(\mathfrak{n}^-)/\mathcal{I}_\lambda \to V(m\lambda), \quad u \mapsto uv_{m\lambda},$$

and the set of vectors $\{\overline{\Theta}_\pi v_{m\lambda} \mid \pi \in C(m\lambda)\}$ *is a basis of* $V(m\lambda)$.

Denote by $\overline{\mathbb{M}}(\lambda,\infty) \subset U(\mathfrak{n}^-)/\mathcal{I}_\lambda$ the images $\overline{\mathfrak{m}}_\pi$ of the path monomials $\mathfrak{m}_\pi$ in $U(\mathfrak{n}^-)/\mathcal{I}_\lambda$. The disadvantage of the monomials $\mathfrak{m}_\pi$ is that they are not attached canonically to a path, certain choices have to be made. The advantage is the simpler structure (the elements in $\mathbb{B}(\lambda,\infty)$ are in general not monomials) and they provide a basis already for the Kostant lattice $U_{\mathbb{Z}}(\mathfrak{n}^-)$ and the corresponding lattice $V_{\mathbb{Z}}(m\lambda)$ in the representation. The following theorem has been conjectured by V. Lakshmibai and C. S. Seshadri in [**3**], section 2.

THEOREM 5.3. *i) The path monomials* $\overline{\mathbb{M}}(\lambda,\infty)$ *form a* $\mathbb{Z}$*-basis of the quotient* $U_{\mathbb{Z}}(\mathfrak{n}^-)/\mathcal{I}_\lambda$.
ii) $\overline{\mathbb{M}}(\lambda,\infty)$ *is compatible with all irreducible representations* $V(m\lambda)$ *of highest weight* $m\lambda$, *i.e.,* $\{\overline{\mathfrak{m}}_\pi \mid \pi \in C(\lambda,\infty),\, \pi \notin C(m\lambda)\}$ *is a basis of the kernel of the surjective map*

$$\Psi : U(\mathfrak{n}^-)/\mathcal{I}_\lambda \to V(m\lambda), \quad u \mapsto uv_{m\lambda},$$

and the elements $\{\overline{\mathfrak{m}}_\pi v_{m\lambda} \mid \pi \in C(m\lambda)\}$ *form a* $\mathbb{Z}$*-basis of* $V_{\mathbb{Z}}(m\lambda)$.

We give only the proof of Theorem 5.1 and the proof of Theorem 5.3 in the case of a regular dominant weight, the proof of Theorem 5.2 as well as the proof of Theorem 5.3 in the non-regular case is on the same lines and is left to the reader.

PROOF (Theorem 5.1 and 5.3). The proof is split into several parts, the most difficult part being the proof of the linear independence. In the rest of this section we show how the linear independence implies the rest of the theorems over the complex numbers, the proof of the linear independence (and generating over $\mathbb{Z}$) will be given in section 7.

Let Q be the root lattice and let Q^+ be the submonoide generated by the positive roots. Then $U(\mathfrak{n}^-)$ is a $\mathfrak{h}$-module with weight space decomposition

$$U(\mathfrak{n}^-) = \bigoplus_{\eta \in Q^+} U(\mathfrak{n}^-)_{-\eta}.$$

The map $\Psi : U(\mathfrak{n}^-) \longrightarrow V(m\lambda)$ is surjective, and for a fixed $\eta \in Q^-$ the restriction of Ψ to a weight space is in fact an isomorphism for all $m \gg 0$:

$$\Psi|_{U(\mathfrak{n}^-)_{-\eta}} : U(\mathfrak{n}^-)_{-\eta} \xrightarrow{\sim} V(m\lambda)_{m\lambda-\eta}.$$

By the character formula (Remark 2.6), the dimension of $V(m\lambda)_{m\lambda-\eta}$ is equal to the number of LS-paths of shape $m\lambda$ ending in $m\lambda - \eta$. So for $m \gg 0$:

$$\dim U(\mathfrak{n}^-)_{-\eta} = \dim V(m\lambda)_{m\lambda-\eta} = \sharp\{\pi \in C(m\lambda) \mid \pi(1) = m\lambda - \eta\}.$$

This shows by the (assumed) linear independence that the corresponding Θ_π (respectively $\mathfrak{m}_\pi$) span $U(\mathfrak{n}^-)_{-\eta}$ and hence form a basis for $U(\mathfrak{n}^-)_{-\eta}$.

A similar counting argument proves *ii)* once one has shown that all the Θ_π (respectively $\mathfrak{m}_\pi$) with $\pi \notin C(m\lambda)$ are in the kernel of the map Ψ. So let $\pi \in C(\lambda, \infty)$ and let p be minimal such that $\pi = (\underline{\tau}, \underline{a}) \in C(p\lambda)$. We suppose first $\underline{\tau} = (\tau_0, \ldots, \tau_r)$ is such that $\tau_r \neq \mathrm{id}$. Let $\tau_r = s_{i_1} \cdots s_{i_k}$ be a reduced decomposition. By Remark 2.5 we know that Θ_π (respectively $\mathfrak{m}_\pi$) is of the form $\mathfrak{m} Y_{i_k}^{p\langle \lambda, \alpha_{i_k}\rangle}$ for some element $\mathfrak{m} \in U(\mathfrak{n}^-)$. Now $\pi \notin C(m\lambda)$ can only be if $p > m$, but in this case we have

$$Y_{i_k}^{p\langle \lambda, \alpha_{i_k}\rangle} v_{m\lambda} = 0 \quad \Rightarrow \quad \Theta_\pi v_{m\lambda} = 0 \text{ respectively } \mathfrak{m}_\pi v_{m\lambda} = 0.$$

Next suppose $\tau_r = \mathrm{id}$, then $a_r > m$, otherwise π would be of shape $m\lambda$, see (4.4). Any maximal Verma flag for the pair $(\widehat{a_r\tau_r(\lambda)}, \widehat{a_r\lambda})$ has to be of the form

$$M(\widehat{a_r\tau_r(\lambda)}) \hookrightarrow \ldots \hookrightarrow M(s_\alpha \circ \widehat{a_r\lambda}) \hookrightarrow M(\widehat{a_r\lambda})$$

for some simple root α. So Θ_π is of the form $\mathfrak{m} Y_\alpha^{a_r\langle \lambda, \alpha\rangle}$ and hence $\Theta_\pi v_{m\lambda} = 0$ because $a_r > m$. The arguments for $\mathfrak{m}_\pi$ are again similar: fix a reduced decomposition $\tau_{r-1} = s_{i_1} \ldots s_{i_t}$, then $\mathfrak{m}_\pi$ is of the form $\mathfrak{m} Y_{i_t}^{a_r\langle \lambda, \alpha_{i_t}\rangle}$ and hence $\mathfrak{m}_\pi v_{m\lambda} = 0$ because $a_r > m$. □

6. The elements $\theta_{\mu,\nu}$

We describe in this section in more detail the elements $\theta_{\mu,\nu}$ (Corollary 1.3). Let $\nu \in X_{\mathbb{Q}}$ be a rational weight and let $\beta \in \Phi^+$ be such that $m = \langle \nu + \rho, \beta\rangle$ is a positive integer, set $\mu = s_\beta \circ \nu$. We assume *in addition* that $\ell(\mu) - \ell(\nu) = 1$; in other words: $M(\mu) \subset M(\nu)$ is a maximal Verma flag.

Consider the associated element $\theta_{\mu,\nu}$ of weight $-m\beta$, so $\theta_{\mu,\nu}1_\nu$ is a highest weight vector in $M(\nu)$. We will show that for an appropriate choice of the ordering of the positive roots, the *coefficient of* Y_β^m *is nonzero* in the expression of $\theta_{\mu,\nu}$ as a linear combination of elements of the associated PBW-basis.

Denote $X_{\mathbb{Q}}^+$ the dominant Weyl chamber and set $\phi = X_{\mathbb{Q}}^+ \cap W(\nu+\rho)$. Denote W_ϕ the stabilizer of ϕ and let $W^\phi \subset W$ be the unique representatives of minimal length of the cosets W/W_ϕ. Fix $\kappa \in W^\phi$ such that

(6.1) $\quad \kappa(\phi) = \nu + \rho, \text{ set } s_\beta\kappa = \tau, \text{ so } \tau(\phi) = \nu + \rho - m\beta = \mu + \rho.$

Now $\beta \in N(\tau) = \{\delta \succ 0 \mid \tau^{-1}(\delta) \prec 0\}$ by construction. For a reduced decomposition $\tau = s_{i_1} \cdots s_{i_r}$ one has

(6.2) $\quad N(\tau) = \{\alpha_{i_1}, s_{i_1}(\alpha_{i_2}), s_{\alpha_{i_1}}s_{\alpha_{i_2}}(\alpha_{i_3}), \ldots, s_{\alpha_{i_1}} \cdots s_{\alpha_{i_{r-1}}}(\alpha_{i_r})\},$

so there exists a j such that $\beta = s_{i_1} \cdots s_{i_{j-1}}(\alpha_{i_j})$. Complete the reduced decomposition of τ to a reduced decomposition of the longest word

(6.3) $$w_0 = s_{i_1} \cdot \ldots \cdot s_{i_N}.$$

The reduced decomposition defines a convex enumeration of the positive roots (i.e., if $\beta_i + \beta_\ell = \beta_k$, then either $i > k > \ell$ or $\ell > k > i$), we set:

(6.4) $$\begin{aligned} \beta_1 = \alpha_{i_1},\ \beta_2 &= s_{i_1}(\alpha_{i_2}),\ \beta_3 \\ &= s_{i_1}s_{i_2}(\alpha_{i_3}),\ \ldots,\ \beta_N \\ &= s_{i_1}s_{i_2}\ldots s_{i_{N-1}}(\alpha_{i_N}) \end{aligned}$$

Let $\mathbb{B}$ be the PBW-basis of $U(\mathfrak{n}^-)$ given by the monomials:

(6.5) $\quad \mathbb{B} := \{Y_{\beta_N}^{n_N} \ldots Y_{\beta_{j+1}}^{n_{j+1}} Y_\beta^{n_j} Y_{\beta_{j-1}}^{n_{j-1}} \ldots Y_{\beta_2}^{n_2} Y_{\beta_1}^{n_1} \mid n_1, \ldots, n_N \geq 0\}$

PROPOSITION 6.6. *In the expression of* $\theta_{\mu,\nu} = \sum_{\mathfrak{m}\in\mathbb{B}} a_{\mathfrak{m}}\mathfrak{m}$ *as a linear combination of elements of* $\mathbb{B}$, *the coefficient of* $\mathfrak{m} = Y_\beta^m$ *is a non-zero rational number.*

REMARK 6.7. Since the weights are rational, all coefficients are rational numbers. The important point for later applications is the fact that the coefficient of Y_β^m is non-zero.

The proof of the proposition needs some preparation, it will be given at the end of this section.

Let λ be a dominant integral weight, let $W_\lambda \subset W$ be the stabilizer and let $V(\lambda)$ be the corresponding simple complex $\mathfrak{g}$-module. The weight space $V(\lambda)_{w(\lambda)}$ of weight $w(\lambda)$ is of dimension one for all $w \in W^\lambda$, and a nonzero generator $v_w \in V(\lambda)_{w(\lambda)}$ is called an *extremal weight vector.* The *Demazure submodule* $V(\lambda)_w \subseteq V(\lambda)$ is the $U(\mathfrak{n})$-submodule $U(\mathfrak{n})v_w$, and the *opposite Demazure module* $V(\lambda)_w^- \subseteq V(\lambda)$ is the $U(\mathfrak{n}^-)$-submodule $U(\mathfrak{n}^-)v_w$.

Let $w, u \in W^\lambda$ and $\beta \in \Phi^+$ be such that $s_\beta w = u$ and $\ell(u) - \ell(w) = 1$. Fix $u = s_{i_1} \cdots s_{i_r}$ a reduced decomposition and enumerate correspondingly the positive roots as in (6.4) (so $\beta = \beta_j$ for some $j \leq r$), and let $\mathbb{B}$ as in (6.5).

LEMMA 6.8. *Let $\mathfrak{m} \in \mathbb{B}$ be an element of weight $-k\beta_i$, $i \leq j$. Then either $\mathfrak{m} = Y_\beta^k$ or $\mathfrak{m} = Y_{\beta_N}^{n_N} \ldots Y_{\beta_\ell}^{n_\ell}$ is such that $n_\ell > 0$ for some $\ell < j$. Further, let v_w be an extremal weight vector in $V(\lambda)_{w(\lambda)}$. Then $\mathfrak{m}v_w = 0$ unless $\mathfrak{m} = Y_\beta^k$ for some $0 \leq k \leq \langle w(\lambda), \beta\rangle$.*

As an immediate consequence we get:

COROLLARY 6.9. *Suppose $1 \leq i < j$ and $r \geq 1$ or $i = j$ and $r > \langle w(\lambda), \beta\rangle$. Then $w(\lambda) - r\beta_i$ is not a weight in $V(\lambda)_w^-$, and for $0 \leq r \leq \langle w(\lambda), \beta\rangle$, the weight space of weight $w(\lambda) - r\beta$ is of dimension one in $V(\lambda)_w^-$.*

PROOF (of Lemma 6.8). The proof is by induction on the j such that $\beta_j = \beta$. If $j = 1$, then $\beta = \alpha$ is a simple reflection and the proof is obvious.

Suppose $j > 1$ and that the lemma holds for all $j' < j$. If $\mathfrak{m} = Y_\beta^k$, then $\beta \in N(u)$ implies $\langle u(\lambda), \beta\rangle < 0$ and hence $\langle w(\lambda), \beta\rangle > 0$. It follows by $\mathfrak{sl}_2$-representation theory that $Y_\beta^k v_w = 0$ for $k > \langle w(\lambda), \beta\rangle$ and $Y_\beta^k v_w \neq 0$ for $k \leq \langle w(\lambda), \beta\rangle$. Let $\mathfrak{m} = Y_{\beta_N}^{n_N} \ldots Y_{\beta_\ell}^{n_\ell}$ be such that $n_\ell > 0$ for some $\ell < j$. Recall that w has a reduced decomposition of the form $w = s_{i_1} \ldots s_{i_{j-1}} s_{i_{j+1}} \ldots s_r$. By the choice of the enumeration of the roots, $\ell < j$ implies $\beta_\ell \in N(w)$ and hence $\langle w(\lambda), \beta_\ell\rangle < 0$, so $Y_{\beta_\ell} v_w = 0$. But this implies also $\mathfrak{m}v_w = 0$.

Let $\mathfrak{m} = Y_{\beta_N}^{n_N} \ldots Y_{\beta_\ell}^{n_\ell}$ be of weight $-k\beta_i$ for some $i \leq j$, but $\ell \geq j > 1$ and $\mathfrak{m}$ is not a power of Y_β. Let $\tilde{s}_{i_1} \in \mathrm{Nor}_G(T)$ be a representative of s_{i_1}, set $\delta := s_{i_1}(\beta)$ and $u' := s_{i_1}u$. Note that $\ell(s_\delta u') = \ell(u') - 1$. Consider the reduced expression $w_0 = s_{i_2} \cdot \ldots \cdot s_{i_N} s_{-w_0(\alpha_{i_1})}$ and the associated ordering of the positive roots (compare with (6.4)):

$$\beta_1' := s_{i_1}(\beta_2),\ \beta_2' := s_{i_1}(\beta_3),\ \ldots,\ \beta_{N-1}' := s_{i_1}(\beta_N),\ \beta_N' := -w_0(\alpha_{i_1}),$$

so $\delta = \beta'_{j-1}$. Denote $\mathbb{B}'$ the corresponding PBW–basis given by the monomials of the form $\mathfrak{m}' = Y_{\beta'_N}^{n'_N} \dots Y_{\beta'_1}^{n'_1}$. By induction we know that an element of weight $-k\beta'_q$, $1 \leq q \leq j-1$ of $\mathbb{B}'$ is either equal to Y_δ^k or of the form $Y_{\beta'_N}^{n'_N} \dots Y_{\beta'_t}^{n'_t}$ with $n'_t > 0$ for some $t < j-1$. Consider now the element $\mathfrak{m} = Y_{\beta_N}^{n_N} \dots Y_{\beta_\ell}^{n_\ell} \in \mathbb{B}$ above. Since $\ell \geq j > 1$, the monomial

$$\mathfrak{m}' := \mathbf{Ad}\tilde{s}_{i_1}(\mathfrak{m}) = \mathbf{Ad}\tilde{s}_{i_1}(Y_{\beta_N})^{n_N} \dots \mathbf{Ad}\tilde{s}_{i_1} Y_{\beta_\ell}^{n_\ell}) = cY_{\beta'_{N-1}}^{n_N} \dots Y_{\beta'_{\ell-1}}^{n_\ell}$$

for some $c \in \mathbb{C}^*$. Since $\mathfrak{m}'$ is of weight $-ks_{i_1}(\beta_i) = -k\beta'_{i-1}$ and not a power of Y_δ and an element of $\mathbb{B}'$, we know by induction that $\ell - 1 < j - 1$ and hence $\ell < j$, a contradiction to the assumption. $\square$

Let $\sigma = s_{i_1} \dots s_{i_r}$ be a reduced decomposition of an element in the Weyl group and set $\beta = s_{i_1} \dots s_{i_{r-1}}(\alpha_{i_r}) \in \Phi^+$. Let C be the dominant Weyl chamber and for a natural number m let $\mathcal{H}^\rho_{\beta,m}$ be the affine hyperplane

$$\mathcal{H}^\rho_{\beta,m} := \{\nu \in X_\mathbb{R} \mid \langle \nu + \rho, \beta \rangle = m\}.$$

LEMMA 6.10. *$(s_\beta\sigma \circ C) \cap X \cap \mathcal{H}^\rho_{\beta,m}$ is Zariski dense in $\mathcal{H}^\rho_{\beta,m}$.*

PROOF. We have $s_\beta\sigma = \sigma s_\gamma$, where $\gamma = \alpha_{i_r}$. For a simple root $\alpha \in \Delta$ denote ω_α the corresponding fundamental weight. Now γ is a simple root, so

(6.11)

$$C \cap X \cap \mathcal{H}^\rho_{\gamma,m} = \{\lambda \in X^+ \mid \lambda = (m-1)\omega_\gamma + \sum_{\substack{\alpha \in \Delta - \{\gamma\} \\ n_\alpha \geq 0}} n_\alpha\omega_\alpha\}, n_\alpha\omega_\alpha\},$$

and the latter is obviously Zariski dense in $\mathcal{H}^\rho_{\gamma,m}$. Further, the equation

$$\langle w \circ \nu + \rho, w(\gamma) \rangle = \langle w(\nu + \rho), w(\gamma) \rangle = \langle \nu + \rho, \beta \rangle$$

shows that for all $w \in W$ we have:

$$w \circ \mathcal{H}^\rho_{\gamma,m} = \mathcal{H}^\rho_{w(\gamma),m} \tag{6.12}$$

The lemma follows immediately from (6.11) and (6.12) $\square$

We will use this Zariski dense subset to get an recursive description of the coefficient in question. Let again $\tau, \kappa \in W$ and $\beta \in \Phi^+$ be such that $\tau = s_\beta\kappa$ and $\ell(\tau) = \ell(\kappa) + 1$. Fix a reduced decomposition $\tau = s_{i_1} \cdots s_{i_r}$ and let j be such that $\beta = s_{i_1} \cdots s_{i_{j-1}}(\alpha_{i_j})$. We assume from *now on that* $j > 1$. Set

$$\sigma = s_{i_1} \cdots s_{i_j},\ \alpha = \alpha_{i_1},\ \delta := s_\alpha(\beta) \text{ and } \sigma' := s_\alpha\sigma.$$

Make the decomposition of τ complete to a decomposition of w_0, and let $\mathbb{B}$ be the corresponding PBW–basis of $U(\mathfrak{n}^-)$ (as in (6.3) and (6.5)), and let $\mathbb{B}'$ be the PBW–basis of $U(\mathfrak{n}^-)$ associated to the reduced decomposition of the longest word given by $w_0 = s_{i_2} \dots s_{i_N} s_{i_0}$, where $\alpha_{i_0} = -w_0(\alpha_{i_1})$, so

$$\mathbb{B}' := \{Y_{\beta'_N}^{n_N} \dots Y_{\beta'_2}^{n_2} Y_{\beta'_1}^{n_1} \mid n_1, \dots, n_N \geq 0\}. \tag{6.13}$$

Let m be a positive natural number and consider the Shapovalov elements $\theta_{\beta,m}, \theta_{\delta,m}$ in $U(\mathfrak{n}^- \oplus \mathfrak{h})$. We write the first one as a linear combination of elements in $\mathbb{B}$ (and coefficients in $U(\mathfrak{h})$), the second as a linear combination of elements in $\mathbb{B}'$:

$$\theta_{\beta,m} = \sum_{\mathfrak{m}\in\mathbb{B}, h_{\mathfrak{m}}\in U(\mathfrak{h})} \mathfrak{m} h_{\mathfrak{m}} \quad \text{and} \quad \theta_{\delta,m} = \sum_{\mathfrak{m}'\in\mathbb{B}', h_{\mathfrak{m}'}\in U(\mathfrak{h})} \mathfrak{m}' h_{\mathfrak{m}'}.$$

Let $h_{m\beta} \in U(\mathfrak{h})$ be the coefficient of Y_β^m in $\theta_{\beta,m}$ and let $h_{m\delta} \in U(\mathfrak{h})$ be the coefficient of Y_δ^m in $\theta_{\delta,m}$. Recall that the coefficients are only determined up to the left ideal generated by $(H_\beta - m + \langle \rho, \beta \rangle)$ respectively $(H_\delta - m + \langle \rho, \delta \rangle)$, see Proposition 1.2, and that $s_\alpha \circ \eta \in \mathcal{H}^\rho_{\delta,m}$ for $\eta \in \mathcal{H}^\rho_{\beta,m}$, see (6.12).

LEMMA 6.14. *For all $\eta \in \mathcal{H}^\rho_{\beta,m}$ we have:*

$$\begin{cases} \eta(h_{m\beta}) = \pm(s_\alpha \circ \eta)(h_{m\delta} \prod_{s=1}^{m\langle\beta,\alpha\rangle} (H_\alpha + 1 + s)) & \text{if } \langle \beta, \alpha \rangle \geq 0 \\ (s_\alpha \circ \eta)(h_{m\delta}) = \pm\eta(h_{m\beta} \prod_{s=1}^{-m\langle\beta,\alpha\rangle} (-H_\alpha - s)) & \text{if } \langle \beta, \alpha \rangle < 0 \end{cases}$$

PROOF. We view the elements of $U(\mathfrak{h})$ as polynomial functions on $\mathcal{H}^\rho_{\beta,m}$ respectively $\mathcal{H}^\rho_{\delta,m}$. By Lemma 6.10, $S = (s_\beta \sigma) \circ C \cap X \cap \mathcal{H}^\rho_{\beta,m}$ is Zariski dense in $\mathcal{H}^\rho_{\beta,m}$. Consider $\eta \in S$ and set $\varphi = s_\beta \circ \eta = \eta - m\beta \in \mathcal{H}^\rho_{\delta,m}$. By the choice of η, there exists an integral dominant weight λ such that $\eta = s_\beta \sigma \circ \lambda$. Now $\lambda + \rho$ is a regular integral dominant weight and $\ell(s_\alpha s_\beta \sigma) < \ell(s_\beta \sigma)$ as well as $\ell(s_\alpha \sigma) < \ell(\sigma)$, so it follows that

$$\begin{aligned} \langle \eta + \rho, \alpha \rangle &= \langle s_\beta \sigma(\lambda + \rho), \alpha \rangle = -r < 0, \\ \langle \varphi + \rho, \alpha \rangle &= \langle \sigma(\lambda + \rho), \alpha \rangle = -p < 0. \end{aligned}$$

We have the following square of embeddings of Verma modules:

$$\begin{array}{ccccc} & & M(\eta) & & \\ & \overset{\theta_{\varphi,\eta}}{\underset{\text{weight}-m\beta}{\nearrow}} & & \overset{\theta_{\eta,s_\alpha\circ\eta}}{\underset{\text{weight}-r\alpha}{\searrow}} & \\ M(\varphi) & & & & M(s_\alpha \circ \eta) \\ & \overset{\theta_{\varphi,s_\alpha\circ\varphi}}{\underset{\text{weight}-p\alpha}{\searrow}} & & \overset{\theta_{s_\alpha\circ\varphi,s_\alpha\circ\eta}}{\underset{\text{weight}-m\delta}{\nearrow}} & \\ & & M(s_\alpha \circ \varphi) & & \end{array}$$

Since α is a simple root and because of the normalization we know that

(6.15)

$$\theta_{\varphi,s_\alpha\circ\varphi}=Y_\alpha^p,\quad \theta_{\eta,s_\alpha\circ\eta}=Y_\alpha^r\quad\text{and}\quad \theta_{\varphi,s_\alpha\circ\varphi}\theta_{s_\alpha\circ\varphi,s_\alpha\circ\eta}=\theta_{\varphi,\eta}\theta_{\eta,s_\alpha\circ\eta}.$$

Set $\nu=\lambda+\rho$ and $w=s_\alpha s_\beta\sigma$, then we get for

$$\nu_0=w(\nu),\ \nu_1=s_\alpha w(\nu)=\eta+\rho,\ \nu_2=s_\delta w(\nu),\ \nu_3=s_\beta s_\alpha w(\nu)=\varphi+\rho$$

the following diagram of weights:

$$\begin{array}{ccccc} & & \nu_0=w(\nu) & & \\ & \swarrow r\alpha & & \searrow m\delta & \\ \nu_1=s_\alpha w(\nu) & & & & \nu_2=s_\delta w(\nu) \\ & \searrow m\beta & & \swarrow p\alpha & \\ & & \nu_3=s_\beta s_\alpha w(\nu)=s_\alpha s_\delta w(\nu) & & \end{array}$$

Consider the simple module $V(\nu)$ of highest weight ν. We fix a highest weight vector v_ν. For an extremal weight $\zeta=w(\nu)$, $w\in W$, denote v_ζ the extremal weight vector of weight ζ obtained from v_ν by applying a sequence of maximal divided powers $Y_{\alpha_{j_k}}^{(n_k)}$ according to a reduced decomposition $w=s_{j_1}\cdots s_{j_t}$. Note that this vector is independent of the choice of the reduced decomposition because of the Verma relations. We get by equation (6.15) in the finite dimensional representation $V(\nu)$:

$$Y_\alpha^p\theta_{s_\alpha\circ\varphi,s_\alpha\circ\eta}v_{\nu_0}=\theta_{\varphi,\eta}Y_\alpha^r v_{\nu_0}.$$

Now the triple $(w,s_\delta w,\delta)$ together with the highest weight ν and the PBW-basis $\mathbb{B}'$ as well as the triple $(s_\alpha w,s_\beta s_\alpha w,\beta)$ together with the highest weight ν and the PBW-basis $\mathbb{B}$ satisfy the conditions of Lemma 6.8. So in the expression of $\theta_{\varphi,\eta}$ with respect to $\mathbb{B}$ and the expression $\theta_{s_\alpha\circ\varphi,s_\alpha\circ\eta}$ with respect to $\mathbb{B}'$, only the terms which are a power of Y_β respectively a power of Y_δ give a possibly nonzero contribution when applied to the extremal weight vector v_{ν_1} respectively v_{ν_0}, and hence:

$$(s_\alpha\circ\eta)(h_{m\delta})Y_\alpha^pY_\delta^m v_{\nu_0}\ =\ \eta(h_{m\beta})Y_\beta^mY_\alpha^r v_{\nu_0}$$

Now $Y_\alpha^pY_\delta^m v_{\nu_0}=\pm(m!)(p!)v_{\nu_2}$ and $Y_\beta^mY_\alpha^r v_{\nu_0}=\pm(m!)(r!)v_{\nu_2}$, so

(6.16) $$\eta(h_{m\beta})=\pm(s_\alpha\circ\eta)(h_{m\delta})p!/r!.$$

Note that $p-r=-\langle\varphi+\rho,\alpha\rangle+\langle\eta+\rho,\alpha\rangle=\langle\eta-\varphi,\alpha\rangle=m\langle\beta,\alpha\rangle$. So if $\langle\beta,\alpha\rangle>0$, then we can write (6.16) also as

$$\begin{aligned}\eta(h_{m\beta})&=\pm(s_\alpha\circ\eta)(h_{m\delta})p(p-1)\cdots(r+1)\\&=\pm(s_\alpha\circ\eta)(h_{m\delta}\textstyle\prod_{s=1}^{m\langle\beta,\alpha\rangle}(H_\alpha+1+s)).\end{aligned}$$

If $\langle\beta,\alpha\rangle < 0$, then we can write (6.16) also as

$$\begin{aligned}(s_\alpha \circ \eta)(h_{m\delta}) &= \pm r(r-1)\cdots(p+1)\eta(h_{m\beta}) \\ &= \pm\eta(h_{m\beta}\textstyle\prod_{s=1}^{-m\langle\beta,\alpha\rangle}(-H_\alpha - s)).\end{aligned}$$

If $\langle\beta,\alpha\rangle = 0$, then we have $\eta(h_{m\beta}) = \pm(s_\alpha \circ \eta)(h_{m\delta})$.

Each of the equalities (for either $+$ or $-$) holds for a Zariski dense subset $S' \subset \mathcal{H}_{\beta+\rho,m}$, and hence it holds for all elements η in $\mathcal{H}_{\beta+\rho,m}$. □

PROOF (Proposition 6.6). Let λ, ν, μ and $\tau, \kappa, \sigma, \beta, \alpha = \alpha_{i_1}, \mathbb{B}$ etc. be as in (6.1)–(6.5). The proof of the proposition is by induction on $\ell(\sigma)$. If $\ell(\sigma) = 1$, then β is a simple root and $\theta_{\mu,\nu} = Y_\beta^m$ has obviously the desired properties.

If $\ell(\sigma) \geq 2$, then $\ell(\tau) \geq 2$. Set $\sigma' := s_\alpha\sigma$ and $\delta = s_\alpha(\beta)$, and let $\mathbb{B}'$ be the PBW-basis in (6.13). Now $\ell(\sigma') < \ell(\sigma)$, and $\lambda, s_\alpha\nu, s_\alpha\mu$ and $s_\alpha\tau, s_\alpha\kappa, s_\alpha\sigma, \delta = s_\alpha(\beta)$ etc. are again as in (6.1)–(6.3), so we can apply our induction procedure:

The coefficient $(s_\alpha \circ \nu)(h_m)$ of Y_δ^m in $\theta_{s_\alpha\circ\mu, s_\alpha\circ\nu} = \sum_{\mathfrak{m}'\in\mathbb{B}'} a_{\mathfrak{m}'}\mathfrak{m}'$ is nonzero.

Choose Shapovalov elements

$$\theta_{\delta,m} = \sum_{\mathfrak{m}'\in\mathbb{B}'} \mathfrak{m}' h_{\mathfrak{m}'} \in U(\mathfrak{n}^- \oplus \mathfrak{h}) \quad \text{and} \quad \theta_{\beta,m} = \sum_{\mathfrak{m}\in\mathbb{B}} \mathfrak{m} h_{\mathfrak{m}} \in U(\mathfrak{n}^- \oplus \mathfrak{h})$$

for $\theta_{s_\alpha\circ\mu, s_\alpha\circ\nu}$ respectively $\theta_{\mu,\nu}$ and let $h_{m\delta} \in U(\mathfrak{h})$ be the coefficient of $\mathfrak{m}' = Y_\delta^m$. Since $(s_\alpha \circ \nu)(h_{m\delta})$ is the coefficient of Y_δ^m in $\theta_{s_\alpha\circ\nu, s_\alpha\circ\mu}$, the latter is nonzero.

Let $h_{m\beta} \in U(\mathfrak{h})$ be the coefficient of $\mathfrak{m} = Y_\beta^m$. To prove the proposition, it is sufficient to prove that $\nu(h_{m\beta}) \neq 0$. If $\langle\beta,\alpha\rangle < 0$, then Lemma 6.14 immediately implies $\nu(h_{m\beta}) \neq 0$ because $(s_\alpha \circ \nu)(h_{m\delta}) \neq 0$.

If $\langle\beta,\alpha\rangle \geq 0$, then we know by Lemma 6.14:

$$\nu(h_{m\beta}) = \pm(s_\alpha \circ \nu)(h_{m\delta} \prod_{s=1}^{m\langle\beta,\alpha\rangle} (H_\alpha + 1 + s))$$

Since $(s_\alpha \circ \nu)(h_{m\delta}) \neq 0$, and

$$\begin{aligned}(s_\alpha \circ \nu)(H_\alpha + 1 + s) &= 1 + s + \langle s_\alpha(\nu+\rho) - \rho, \alpha\rangle \\ &= s + \langle s_\alpha\kappa(\lambda), \alpha\rangle > 0\end{aligned}$$

because $\langle s_\alpha\kappa(\lambda), \alpha\rangle \geq 0$. It follows: $\nu(h_{m\beta}) \neq 0$, which finishes the proof. □

7. The linear independence

To prove the linear independence of the Θ_π respectively the $\mathfrak{m}_\pi$, we use the quantum Frobenius splitting. We recall quickly the method developed in [**5**]. Let $A = (a_{i,j})_{1\le i,j\le n}$ be the Cartan matrix of $\mathfrak{g}$ and denote by $A^t = (\overline{a}_{i,j})$, $\overline{a}_{i,j} := a_{j,i}$, the transposed matrix. Let $\underline{d} = (d_1, \ldots, d_n)$, $d_i \in \mathbb{N}$, be minimal such that the matrix $(d_i a_{i,j})$ is symmetric. We denote by d the smallest common multiple of the d_j, and we set $\overline{d} = (\overline{d}_1, \ldots, \overline{d}_n)$, where $\overline{d}_i := d/d_i$. We assume throughout the rest of section 7 that $\ell \in \mathbb{N}$ is divisible by $2d$.

Let Φ^t be the corresponding root system, with basis $\Delta^t = \{\gamma_1, \ldots, \gamma_n\}$, and let $\mathfrak{g}^t$ be the corresponding semisimple Lie algebra. Denote $U_q(\mathfrak{g}^t)$ the quantum group associated to $\mathfrak{g}^t$ over the field $\mathbb{Q}(q)$, with generators $E_{\gamma_i}, F_{\gamma_i}, K_{\gamma_i}$ and $K_{\gamma_i}^{-1}$. We will sometimes just write $E_i, K_i, \ldots$ for the generators $E_{\gamma_i}, K_{\gamma_i}, \ldots$. In addition, we use the following abbreviations:

$$q_i := q^{\overline{d}_i} = q^{\frac{(\gamma_i,\gamma_i)^t}{2}}, \quad [n]_i := \frac{q^{\overline{d}_i n} - q^{-\overline{d}_i n}}{q^{\overline{d}_i} - q^{-\overline{d}_i}}, \quad \begin{bmatrix} n \\ m \end{bmatrix}_i := \frac{[n]_i!}{[m]_i![n-m]_i!},$$

and $E_i^{(k)}$, $F_i^{(k)}$ for the divided powers.

Let $U_{q,A}$ be the Lusztig form of $U_q(\mathfrak{g}^t)$ defined over the ring of Laurent polynomials $\mathcal{A} := \mathbb{Z}[q, q^{-1}]$. We denote by R the ring $\mathcal{A}/I$, where I is the ideal generated by the 2ℓ-th cyclotomic polynomial, and set $U_\xi := U_{q,A} \otimes_A R$.

We can view Φ^t in $\mathfrak{h}^*$ by setting $\gamma_i := \alpha_i/d_i \in \mathfrak{h}^*$. For the weight lattice X^t of $\mathfrak{g}^t$ we get:

$$X^t = \{\lambda \in \mathfrak{h}^* \mid \forall i : \langle \lambda, \gamma_i \rangle \in \mathbb{Z}\} = \{\lambda \in \mathfrak{h}^* \mid \forall i : d_i \langle \lambda, \alpha_i \rangle \in \mathbb{Z}\}.$$

It follows immediately from the definition: $dX^t \subset X \subset X^t$.

Let $\lambda \in X^+$ ($\subset X^t$) be a dominant weight and denote $V_\xi(\lambda)$ the Weyl module for U_ξ of highest weight λ. Fix ℓ such that for all LS-paths $\pi = (\underline{\tau}, \underline{a})$ of shape λ one has $\frac{\ell}{d} a_i \in \mathbb{N}$ for all i. Let m be a positive integer, the module we are interested is the U_ξ-module

$$V = \underbrace{V_\xi(\lambda) \otimes V_\xi(\lambda) \otimes \ldots \otimes V_\xi(\lambda)}_{\frac{m\ell}{d}-\text{times}}$$

and its subspace given by the weight spaces V_μ, $\mu \in X^t$, such that $\frac{d}{\ell}\mu \in X$:

$$V^{\frac{\ell}{d}} = \bigoplus_{\mu \in X^t,\ \frac{d}{\ell}\mu \in X} V_\mu.$$

Set $\ell_i := \frac{\ell d_i}{d}$, the conditions imposed on ℓ imply that we have a Frobenius homomorphism [**6**] between the classical algebra and the quantum enveloping algebra at a 2ℓ-th root of unity:

$$\mathrm{Fr}' : U(\mathfrak{n}^+) \longrightarrow U_\xi(\mathfrak{n}^{t,+}), \quad X_i^{(k)} \mapsto E_i^{(\ell_i k)}.$$

where $X_i^{(k)} = X_i^k/k!$ is the divided power of the generator. We have a similar map $\mathrm{Fr}' : U(\mathfrak{n}^-) \longrightarrow U_\xi(\mathfrak{n}^{t,-})$ for the generators $Y_i^{(k)} \mapsto F_i^{(\ell_i k)}$ of $\mathfrak{n}^-$.

The subspace $V^{\frac{\ell}{d}} \subset V$ becomes via Fr' a $U(\mathfrak{n}^+)$ and a $U(\mathfrak{n}^-)$-submodule. In [**5**] we have shown that these actions can be glued together to make $V^{\frac{\ell}{d}}$ into a $U(\mathfrak{g})$-module. Let $u_\lambda \in V_\xi(\lambda)$ be a highest weight vector (for the action of U_ξ) and let $v_{m\lambda} \in V(m\lambda)$ be a highest weight vector (for the action of $U(\mathfrak{g})$). The map $v_{m\lambda} \mapsto (u_\lambda)^{\otimes \frac{m\ell}{d}}$ extends to a $U(\mathfrak{g})$-equivariant monomorphism

$$V(m\lambda) \hookrightarrow V^{\frac{\ell}{d}}.$$

To prove that the Θ_π respectively the $\mathfrak{m}_\pi$ are linearly independent, we show that for $\pi \in C(m\lambda)$ the vectors $\Theta_\pi \cdot (u_\lambda)^{\otimes \frac{m\ell}{d}} \in V^{\frac{\ell}{d}}$ (respectively the vectors $\mathfrak{m}_\pi \cdot (u_\lambda)^{\otimes \frac{m\ell}{d}}$) are linearly independent.

To do so, we need some more notation. Let $\succ$ be the usual ordering on the weights in X^t, i.e., $\mu \succ \nu$ if the difference is a sum of positive roots. We define an induced partial order on weight vectors in $V_\xi(\lambda)$ by saying $u_\mu \succ u_\nu$ if $\mu \succ \nu$. We extend this to a partial order on weight vectors in $V^{\frac{\ell}{d}}$. Set

$$t = \frac{m\ell}{d}, \quad \underline{u}_\mu = u_{\mu_1} \otimes \ldots \otimes u_{\mu_t}, \text{ and } \underline{u}_\nu = u_{\nu_1} \otimes \ldots \otimes u_{\nu_t},$$

then we say

$$\underline{u}_\mu \succ \underline{u}_\nu \Leftrightarrow \exists\, j : \mu_j \succ \nu_j \text{ and } \mu_{j+1} = \nu_{j+1}, \ldots, \mu_t = \nu_t.$$

We say more precisely that $\underline{u}_\mu$ is greater than $\underline{u}_\nu$ *after* k:

$$\underline{u}_\mu \succ_{\mathbf{k}} \underline{u}_\nu \Leftrightarrow \exists\, j > k : \mu_j \succ \nu_j \text{ and } \mu_{j+1} = \nu_{j+1}, \ldots, \mu_t = \nu_t.$$

For all $\tau \in W/W_\lambda$ fix an extremal weight vector $u_\tau \in V_\xi(\lambda)$ of weight $\tau(\lambda)$. Let $\pi = (\underline{\tau}, \underline{a}) \in C(m\lambda)$ be an LS-concatenation of shape $m\lambda$. We denote $u_\pi \in V^{\frac{\ell}{d}}$ the vector:

$$u_\pi = \underbrace{u_{\tau_0} \otimes \ldots \otimes u_{\tau_0}}_{\frac{a_1\ell}{d} \text{ times}} \otimes \underbrace{u_{\tau_1} \otimes \ldots \otimes u_{\tau_1}}_{\frac{(a_2-a_1)\ell}{d} \text{ times}} \otimes \ldots \otimes \underbrace{u_{\tau_r} \otimes \ldots \otimes u_{\tau_r}}_{\frac{(m-a_r)\ell}{d} \text{ times}}$$

PROPOSITION 7.1. *Let $\pi \in C(m\lambda)$ be a LS-concatenation of shape $m\lambda$. Then there exists a non-zero complex number c such that*

$$\Theta_\pi(u_\lambda)^{\otimes \frac{m\ell}{d}} = cu_\pi + \sum(\text{ terms } \underline{u} \prec\ u_\pi).$$

Since the u_π, $\pi \in C(m\lambda)$ are linearly independent, as an immediate consequence we see that the $\Theta_\pi v_{m\lambda} \in V(m\lambda) \hookrightarrow V^{\frac{\ell}{d}}$ are linearly independent. So we get by the LS-path character formula (Remark 2.6):

COROLLARY 7.2. $\{\Theta_\pi v_{m\lambda} \in V(m\lambda) \mid \pi \in C(m\lambda)\}$ *is a basis of* $V(m\lambda)$.

PROOF (Theorem 5.1, end). Since any weight space in $U(\mathfrak{n}^-)$ maps under $U(\mathfrak{n}^-) \to V(m\lambda)$ isomorphically onto a weight space in $V(m\lambda)$ for $m \gg 0$, to prove the linear independence of the Θ_π it is sufficient to prove that the $\Theta_\pi v_{m\lambda} \in V(m\lambda)$ are linearly independent for $\pi \in C(m\lambda)$ and $m \gg 0$. But this follows from Corollary 7.2. □

PROOF (Proposition 7.1). Let $\pi = (\underline{\tau}, \underline{a})$ be an LS-concatenation of shape $m\lambda$, where $\underline{\tau} = (\tau_0, \ldots, \tau_r)$, $\underline{a} = (a_1, \ldots, a_r)$. The proof is by induction on the length $\ell(\tau_0)$ of the longest element τ_0. To make the induction procedure work, we will prove the following more precise statement:

$$\Theta_\pi(u_\lambda)^{\otimes \frac{m\ell}{d}} = cu_\pi + \sum(\text{ terms } \underline{u} \text{ such that } u_\pi \succ_{\ell\frac{\mathbf{a_1}}{\mathbf{d}}} \underline{u}),$$

If $\ell(\tau_0) = 0$, then $\tau_0 = \text{id}$ and hence $\pi = (\text{id})$, $\Theta_\pi = 1$, and the proposition holds obviously in this case.

Let $\ell(\tau_0) > 0$. If $\ell(\tau_0) - \ell(\tau_1) = 1$, then $\tau_0 = s_\beta\tau_1$ for some positive root, and the pair $\eta = (\tau_1, \ldots, \tau_r; a_2, \ldots, a_r)$ is again a LS-concatenation of shape $m\lambda$ and $\Theta_\pi = \theta_{\widehat{\nu_0}, \widehat{\mu_0}}\Theta_\eta$, where $\mu_0 = a_1\tau_0(\lambda)$ and $\nu_0 = a_1\tau_1(\lambda)$.

Otherwise, there exists for (ν_0, μ_0), a maximal Verma flag:

$$M(\hat{\nu}_0) = M((s_{\beta_t}\cdots s_{\beta_1})\circ\hat{\mu}_0) \hookrightarrow \ldots \hookrightarrow M(s_{\beta_1}\circ\hat{\mu}_0) \hookrightarrow M(\hat{\mu}_0).$$

Let $\kappa = s_{\beta_t}\tau_0$, then $\eta = (\kappa, \tau_1, \ldots, \tau_r; a_1, a_2, \ldots, a_r)$ is an LS-concatenation of shape $m\lambda$ and $\Theta_\pi = \theta_{\widehat{\nu_0}, \widehat{\mu}'_0}\Theta_\eta$, where $\mu'_0 = a_1\kappa(\lambda)$.

Set $\nu = \widehat{\nu_0}$ and $\mu = \widehat{\mu_0}$ in the first case and $\mu = \widehat{\mu_0}'$ in the second case. Then in both cases, the induction hypothesis applies to Θ_η, and $\Theta_\pi = \theta_{\nu,\mu}\Theta_\eta$, and $\nu = s_\beta \circ \mu$, $\langle \mu + \rho, \beta \rangle = p > 0$, and $\ell(\mu) - \ell(\nu) = 1$.

To discuss both cases uniformly, set $\eta = (\kappa_0, \ldots, \kappa_r; b_0, \ldots, b_r)$. We know by induction

$$\Theta_\eta(u_\lambda)^{\otimes \frac{m\ell}{d}} = cu_\eta + \sum(\text{ terms } \underline{u} \text{ such that } u_\eta \succ_{\ell\frac{\mathbf{b_0}}{\mathbf{d}}} \underline{u}).$$

By construction we have $\mu = \widehat{a_0\kappa_0(\lambda)}$, $\nu = \widehat{a_0\tau_0(\lambda)}$, where $a_0 \leq b_0$, $\ell(\tau_0) - \ell(\kappa_0) = 1$ and $s_\beta\tau_0 = \kappa_0$.

Since the tensor components of u_π and u_η coincide for $k > b_0\frac{m\ell}{d}$, i.e., for

$$u_\pi = u_1^\pi \otimes \cdots \otimes u_{\ell m/d}^\pi \text{ and } u_\eta = u_1^\eta \otimes \cdots \otimes u_{\ell m/d}^\eta$$

one has $u_k^\pi = u_k^\eta$ for $k > b_0\frac{m\ell}{d}$, it follows that

$$u_\eta \succ_{\mathbf{b_0}\frac{\ell}{\mathbf{d}}} \underline{u} \Rightarrow u_\pi \succ_{\mathbf{b_0}\frac{\ell}{\mathbf{d}}} \underline{u} \Rightarrow u_\pi \succ_{\mathbf{b_0}\frac{\ell}{\mathbf{d}}} \theta_{\nu,\mu}\underline{u}$$

To prove the proposition, it suffices hence to prove:

$$\theta_{\nu,\mu}u_\eta = cu_\pi + \sum(\text{ terms } \underline{u} \text{ such that } u_\pi \succ_{\ell\frac{\mathbf{a_0}}{\mathbf{d}}} \underline{u})$$

for some complex number $c \neq 0$. By choosing an appropriate ordering of the positive roots, we know by Proposition 6.6 that in the expression of $\theta_{\nu,\mu} = \sum_{\mathfrak{m}\in\mathbb{B}} a_\mathfrak{m}\mathfrak{m}$ as linear combination of elements of the corresponding PBW-basis $\mathbb{B}$, the coefficient of $\mathfrak{m} = Y_\beta^p$ is a non-zero rational number. Further, by Corollary 6.9, if $\mathfrak{m} \neq Y_\beta^p$, then $\mathfrak{m} = Y_{\beta_N}^{n_N} \dots Y_{\beta_x}^{n_x}$ $(n_x > 0)$ is such that in the opposite Demazure module $V(m\lambda)_{\kappa_0}^- = U(\mathfrak{n}^-)v_{\kappa_0}$, $\kappa_0(m\lambda) - s\beta_x$ is never a weight for $s > 0$ and hence $\mathrm{Fr}'(Y_{\beta_x})$ applied to $u_{\kappa_0}^{\otimes\ell/d} \in V(m\lambda)_{\kappa_0(m\lambda)} \subset V^{\frac{\ell}{d}}$ is zero. But this implies of course also $\mathrm{Fr}'(Y_{\beta_x})u_{\kappa_0}^{a_0\otimes\ell/d} = 0$ Since $u_\eta = u_{\kappa_0}^{\otimes a_0\ell/d} \otimes \underline{u}'$, it follows that

$$\mathrm{Fr}'(Y_{\beta_x})u_\eta = \underbrace{\left(\mathrm{Fr}'(Y_{\beta_x})u_{\kappa_0}^{\otimes a_0\ell/d}\right) \otimes \underline{u}'}_{=0} + \sum(\mathfrak{m}_1 u_{\kappa_0}^{\otimes a_0\ell/d}) \otimes (\mathfrak{m}_2\underline{u}')$$

where $\mathfrak{m}_1, \mathfrak{m}_2 \in U_\xi(\mathfrak{n}^{t,-})$ are such that $\mathfrak{m}_2$ has no constant term. Obviously we have $\underline{u}' \succ \mathfrak{m}_2\underline{u}'$, so if $\mathfrak{m}$ is a monomial in the expression of $\theta_{\nu,\mu}$ different from Y_β^p, then

$$\mathfrak{m}u_\eta = \sum(\text{ terms } \underline{u} \text{ such that } u_\pi \succ_{\ell\frac{\mathbf{a_0}}{\mathbf{d}}} \underline{u}).$$

So to finish the proof of the proposition, it remains to show:

$$\mathrm{Fr}'(Y_\beta^p)u_\eta = cu_\pi + \sum(\text{ terms } \underline{u} \text{ such that } u_\pi \succ_{\ell\frac{\mathbf{a_0}}{\mathbf{d}}} \underline{u})$$

for some $c \in \mathbb{C}^*$. As above, since $u_\eta = u_{\kappa_0}^{\otimes a_0\ell/d} \otimes \underline{u}'$, it follows that

$$\mathrm{Fr}'(Y_\beta^p)u_\eta = \left(\mathrm{Fr}'(Y_{\beta_x}^p)u_{\kappa_0}^{\otimes a_0\ell/d}\right) \otimes \underline{u}' + \sum(\mathfrak{m}_1 u_{\kappa_0}^{\otimes a_0\ell/d}) \otimes (\mathfrak{m}_2\underline{u}')$$

where $\mathfrak{m}_1, \mathfrak{m}_2 \in U_\xi(\mathfrak{n}^{t,-})$ are such that $\mathfrak{m}_2$ has no constant term. As above, since $\underline{u}' \succ \mathfrak{m}_2\underline{u}'$, we get:

$$\mathrm{Fr}'(Y_\beta^p)u_\eta = \left(\mathrm{Fr}'(Y_{\beta_x}^p)u_{\kappa_0}^{\otimes a_0\ell/d}\right) \otimes \underline{u}' + \sum(\text{ terms } \underline{u} \text{ such that } u_\pi \succ_{\ell\frac{\mathbf{a_0}}{\mathbf{d}}} \underline{u}).$$

So the proof is finished once we show: $\mathrm{Fr}'(Y_{\beta_x}^p)u_{\kappa_0}^{\otimes a_0\ell/d} = cu_{\tau_0}^{\otimes a_0\ell/d}$ for some nonzero complex number. Since the weight space (for U_ξ) is one-dimensional, it suffices to show that the first is nonzero. But this follows immediately from $\mathfrak{sl}_2$-theory and the following observation. Set

$$W = \underbrace{V_\xi(\lambda) \otimes \ldots \otimes V_\xi(\lambda)}_{\frac{\mathbf{a_0}\ell}{\mathbf{d}}-\text{times}},$$

and let $\bigoplus_{\epsilon\in X^t} W_\epsilon$ be the weight space decomposition. Denote $W(\beta)$ the sum of all U_ξ-weight spaces for weights integral for β:

$$W(\beta) = \bigoplus_{\substack{\epsilon\in X^t \\ \langle\epsilon,\beta\rangle\in\mathbb{Z}}} W_\epsilon.$$

Then $W(\beta)$ is actually a module for the subalgebra of $\mathfrak{g}$ generated by Y_β, X_β via the Frobenius maps Fr'. This can be proved in the same way as $V^{\frac{\ell}{d}}$ becomes a $\mathfrak{g}$-module in [**5**]. Further, $u_{\kappa_0}^{\otimes a_0\ell/d} \in W(\beta)$ is a highest weight vector for this operation, and a lowest weight vector for the simple module generated by this vector is $u_{\tau_0}^{\otimes a_0\ell/d}$. □

PROOF (Theorem 5.3, end). The proof that the

$$\{\mathfrak{m}_\pi v_{m\lambda} \mid \pi \in C(m\lambda)\} \subset V(m\lambda)\}$$

form a $\mathbb{Z}$-basis of $V_\mathbb{Z}(\lambda)$ uses the same arguments as the proof of Proposition 7.1 above, so we give only a sketch of the proof. Again, we embed $V(m\lambda) \hookrightarrow V^{\frac{\ell}{d}}$ in the quantum group representation for $V = V_\xi(\lambda)^{\otimes m\ell/d}$. Let $A = \mathbb{Z}[\mathrm{x}, \mathrm{x}^{-1}]$ be the ring of Laurent polynomials and let R be ring R/I, where I is the ideal generated by the 2ℓ-th cyclotomic polynomial. Let V_R, $U_R(\mathfrak{g})$, $U_{q,R}(\mathfrak{g}^t)$ etc. be the forms obtained from the corresponding $\mathbb{Z}$-form in the classical case and the corresponding A-form in the quantum group case by base change (as in [**5**]). The corresponding Frobenius map and the embedding of representations above is then already defined over R. By applying the same kind of inductive arguments as above, by fixing the u_π

appropriately and by applying $\mathfrak{sl}_2$-theory a little more rigorously, one gets

$$\mathfrak{m}_\pi(u_\lambda)^{\otimes\frac{m\ell}{d}} = \xi^t u_\pi + \sum(\text{ terms } \underline{u} \text{ such that } u_\pi \succ_{\ell\frac{\mathbf{a_1}}{\mathbf{d}}} \underline{u}),$$

for some power of the fixed primitive 2ℓ-th root of unity ξ. By the linear independence of the u_π, this proves the linear independence of the $\mathfrak{m}_\pi(u_\lambda)^{\otimes\frac{m\ell}{d}}$, and by the character formula this proves that they span a R-lattice of maximal rank in $V_R(\lambda)$. Now the u_π can be completed to an R-basis of V_R. Since the u_π are "leading terms" for the $\mathfrak{m}_\pi(u_\lambda)^{\otimes\frac{m\ell}{d}}$, it follows that they form actually a basis for $V_R(\lambda)$. Since they are by definition actually elements in $V_{\mathbb{Z}}(\lambda)$, they form hence a basis of the $\mathbb{Z}$-lattice $V_{\mathbb{Z}}(\lambda)$. □

References

[1] I. N. Bernstein,I. M. Gelfand and S. I. Gelfand: Structure of representations generated by vectors of highest weight, *Funct. Anal. Appl.* **5**, 1-9 (1971)

[2] J. Dixmier: Enveloping Algebra, Graduate Studies in Math, Volume II, AMS (1996)

[3] V. Lakshmibai and C. S. Seshadri: Standard monomial theory, in *Proceedings of the Hyderabad Conference on Algebraic Groups*, Manoj Prakashan, Madras, 279-323, (1991).

[4] P. Littelmann: A Littlewood–Richardson formula for symmetrizable Kac-Moody algebras. *Invent. Math.* **116**, 329-346 (1994).

[5] P. Littelmann: Contracting modules and standard monomial theory. *JAMS* **11**, 551–567 (1998)

[6] G. Lusztig: *Introduction to Quantum Groups,* Progr. Math. **110**, Birkhäuser, Boston, 1993.

[7] F.G. Malikov, B.L. Feigin and D.B. Fuks: Singular vectors in Verma Modules over Kac-Moody algebras. *Funct. Anal. Appl.* **20**, 25-37 (1986)

[8] N.N. Shapovalov: On a bilinear form on the universal enveloping algebra of a complex semisimple Lie algebra, *Funct. Anal. Appl.* **6**, 307-311 (1972)

Fachbereich Mathematik, Universität Wuppertal
Gauss-Strasse 20, 42097 Wuppertal, Germany
E-mail address: littelmann@math.uni-wuppertal.de

INVARIANT TENSOR FIELDS AND ORBIT VARIETIES FOR FINITE ALGEBRAIC TRANSFORMATION GROUPS

Mark Losik, Peter W. Michor and Vladimir L. Popov

To C. S. Seshadri on the occasion of his 70th birthday

Abstract. Let X be a smooth algebraic variety endowed with an action of a finite group G such that there exists a geometric quotient $\pi_X : X \to X/G$. We characterize rational tensor fields τ on X/G such that the *pull back* of τ is regular on X: these are precisely all τ such that $\operatorname{div}_{R_{X/G}}(\tau) \geqslant 0$ where $R_{X/G}$ is the *reflection divisor* of X/G and $\operatorname{div}_{R_{X/G}}(\tau)$ is the $R_{X/G}$*-divisor* of τ. We give some applications, in particular to a generalization of Solomon's theorem. In the last section we show that if V is a finite dimensional vector space and G a finite subgroup of $\mathrm{GL}(V)$, then each automorphism ψ of V/G admits a biregular lift $\varphi : V \to V$ provided that ψ maps the regular stratum to itself and $\psi_*(R_{X/G}) = R_{X/G}$.

1. Introduction

Let X be a smooth algebraic variety endowed with an action of a finite group G. Assume that there exists a geometric quotient $\pi_X : X \to X/G$ (this is always the case if X is quasi-projective, cf. Subsection 2.4).

In this paper we study the interrelations between rational tensor fields on X and X/G. If τ is a rational tensor field on X/G, we

M.L. and P.W.M. were supported by "Fonds zur Förderung der wissenschaftlichen Forschung, Projekt P 14195 MAT".

Typeset by $\mathcal{A}_{\mathcal{M}}\mathcal{S}$-TEX

define a G-invariant rational tensor field $\pi_X^*(\tau)$ on X called the *pull back* of τ. If θ is a G-invariant rational tensor field on X, we define a rational tensor field $\pi_{X*}(\theta)$ on X/G called the *push forward* of θ. We have $\pi_{X*}(\pi_X^*(\tau)) = \tau$ and $\pi_X^*(\pi_{X*}(\theta)) = \theta$.

Given this, we consider the following problem: Let τ be a rational tensor field on X/G. When is the pull back $\pi_X^*(\tau)$ regular on X?

To that end we consider the Luna stratification of X/G. Let $(X/G)_1$ be the union of all codimension 1 strata. We show that $(X/G)_1$ is contained in the smooth locus of X/G. Let $(X/G)_1 = (X/G)_1^1 \cup \ldots \cup (X/G)_1^d$ be the decomposition into irreducible components. We show that for any $l = 1, \ldots, d$ and $z \in (X/G)_1^l$, $x \in \pi_X^{-1}(z)$ the stabilizer of x is a cyclic group whose image under the slice representation is generated by a pseudo-reflection of order r_l depending only on l. We encode this information into the *reflection divisor* $R_{X/G} := r_1 R_1 + \ldots + r_d R_d \in \mathrm{Div}(X/G)$ where R_l is the prime divisor whose support is the closure of $(X/G)_1^l$. Further, for any nonzero rational tensor field τ on X/G and positive divisor $B \in \mathrm{Div}(X/G)$ we define $\mathrm{div}(\tau) \in \mathrm{Div}(X/G)$, the *divisor of* τ, and $\mathrm{div}_B(\tau) \in \mathrm{Div}(X/G)$, the *$B$-divisor of* τ. Locally $\mathrm{div}(\tau)$ is the minimum of divisors of the component functions of τ in the framings of tangent and cotangent bundles. The divisor $\mathrm{div}_B(\tau)$ is obtained from $\mathrm{div}(\tau)$ by means of some "modification along" B (see Subsection 3.6).

Our main result is that $\pi_X^*(\tau)$ is regular on X if and only if $\mathrm{div}_{R_{X/G}}(\tau) \geqslant 0$.

As a corollary we obtain a generalization of Solomon's theorem [So], and prove that the push forward of a G-invariant regular tensor field on X that is skew symmetric with respect to the covariant entries is regular on the smooth locus of X/G.

Another application pertains to the case where X is a vector space V with a linear action of G: we obtain a characterization of G-invariant polynomials on $V^{\oplus s}$ in terms of rational multi-symmetric covariant tensor fields on V/G.

In the last section we prove that any automorphism ψ of the algebraic variety V/G such that $\psi((V/G)_0) \subseteq (V/G)_0$ and $\psi_*(R_{V/G}) = R_{V/G}$ can be lifted to an automorphism of the algebraic variety V. The proof is based on the relevant result in the analytic setting, [KLM], so what we really prove is that analytic lift is actually algebraic.

Throughout in this paper we assume that the base field k is algebraically closed of characteristic 0. In the last section we use the result from [KLM] that is proved for $k = \mathbb{C}$, so passing to the general case is carried out by Lefschetz's principle. However note that our proofs of the results from Sections 2–4 could be extended *mutatis mutandis* to the case when $\operatorname{char} k$ is positive and subject to some non-divisibility and magnitude conditions; for this purpose one should use the relevant replacement of the slice theorem [Lu] proved in [BR].

This paper is the algebraic geometrical companion of paper [KLM] where similar results were obtained for analytic actions of finite groups.

We thank Yu. Neretin for helpful discussions. Thanks are also due to the referee for suggestions and comments which has led to eliminating some inaccuracies and improvement in the exposition.

Notation, terminology and conventions:

- $\mathcal{O}_{x,X}$ and $T_{x,X}$ are respectively the local ring and tangent space at a point x of an algebraic variety X.
- $X_{\rm sm}$ is the smooth locus of X.
- $\operatorname{Div}(X)$ is the Weil divisor group of X.
- $\operatorname{div}(f)$ is the divisor of a rational function f.
- $\operatorname{supp}(D)$ is the support of $D \in \operatorname{Div}(X)$.
- A positive divisor is called *prime* if it not sum of two positive divisors.
- $m_{C,D}$ is the multiplicity of a prime divisor C in a divisor D.
- $d_x\varphi$ is the differential of a morphism φ at a point x.
- $G\cdot z$ and G_z are respectively the orbit and stabilizer of a point z of a set Z endowed with an action of a group G.
- $Z^S := \{z \in Z \mid g \cdot z = z \text{ for all } g \in S\}$.
- $|S|$ is the number of elements of a finite set S.

All group actions considered in this paper are algebraic actions on algebraic varieties.

2. Preliminaries

2.1. Excellent morphisms. Recall that a morphism of algebraic varieties $\varphi : X \to Y$ is called *étale at a point* $x \in X$ if the homomorphism of local rings $\varphi^* : \mathcal{O}_{\varphi(x),Y} \to \mathcal{O}_{x,X}$ induces the isomorphism of their completions. If $x \in X_{\rm sm}$ and $\varphi(x) \in Y_{\rm sm}$, then φ is étale at x iff $d_x\varphi : T_{x,X} \to T_{\varphi(x),Y}$ is an isomorphism. If φ is *étale* at each point of X, it is called *étale*.

Let G be a reductive algebraic group. We refer to [PV] regarding the following basic facts.

Let X be an affine algebraic G-variety. We denote by $X/\!\!/G$ the categorical quotient of X by G, i.e., the affine algebraic variety whose algebra of regular functions is $k[X]^G$ (it is finitely generated by Hilbert's theorem). We denote by $\pi_X = \pi_{X,G} : X \to X/\!\!/G$ the canonical projection (induced by the inclusion $k[X]^G \hookrightarrow k[X]$).

Every fiber of π_X contains a unique closed orbit. For a point $x \in X$, if the orbit $G \cdot x$ is closed in X, then the stabilizer G_x is reductive. If G is finite, then $\pi_X : X \to X/\!\!/G$ is a geometric quotient and we denote $X/\!\!/G$ by X/G.

Let $\varphi : X \to Y$ be a G-morphism of affine G-varieties. Then we have the following commutative diagram

$$\begin{array}{ccc} X & \xrightarrow{\varphi} & Y \\ {\scriptstyle \pi_{X,G}}\downarrow & & \downarrow{\scriptstyle \pi_{Y,G}} \\ X/\!\!/G & \xrightarrow{\varphi/\!\!/G} & Y/\!\!/G, \end{array} \tag{2.1.1}$$

where $\varphi/\!\!/G$ is induced by the restriction of $\varphi^* : k[Y] \to k[X]$ to $k[Y]^G$ (if G is finite, we denote $\varphi/\!\!/G$ by φ/G). If $\varphi/\!\!/G$ is étale and (2.1.1) is the base change diagram (so that $X = X/\!\!/G \times_{Y/\!\!/G} Y$), then φ is called *excellent*. In this case, φ itself is étale. If φ is excellent, then for any $b \in X/\!\!/G$ the restriction of φ yields a G-isomorphism $\pi_X^{-1}(b) \to \pi_Y^{-1}((\varphi/\!\!/G)(b))$. Hence if Y is irreducible and φ is excellent, then G acts faithfully on X iff G acts faithfully on Y.

2.2. Étale slices. The following theorem is proved in [Lu].

Theorem 2.2.1. *Let X be an affine algebraic G-variety for a reductive algebraic group G, and let $x \in X$ be a point such that the orbit $G\cdot x$ is closed in X.*

(i) *There is a G_x-stable locally closed affine subvariety $S \subseteq X$, called étale slice at x, such that $x \in S$ and the G-morphism*

$$\sigma_S : G *_{G_x} S \to X \tag{2.2.1}$$

induced by $G \times S \to X$, $(g, s) \mapsto g \cdot s$, is excellent.

(ii) *If $x \in X_{\text{sm}}$, there exist a π_{S,G_x}-saturated affine open neighborhood U of x in S and a G_x-equivariant excellent morphism*

$$\lambda_x : U \to T_{x,S}. \qquad \square \tag{2.2.2}$$

Here by $G *_{G_x} S$ we denote the homogeneous fiber space over G/G_x with the fiber S, i.e., the categorical quotient of $G \times S$ by the G_x-action $h \cdot (g, s) = (gh^{-1}, h \cdot s)$ (actually in this case it is the geometrical quotient).

Theorem 2.2.1 is crucial for this paper. We will apply it only in the setting of finite group actions, in which case its proof is simpler than in general one. Therefore for making the paper more self-contained, in a short Appendix at the end of the paper we give, following the referee's suggestion, some details of the proof of Theorem 2.2.1 for finite G.

2.3. The Luna stratification. Let G be a reductive algebraic group. Denote by $\mathcal{V}(G)$ the set of isomorphism classes of all algebraic vector G-bundles over homogeneous spaces G/H where H ranges over all reductive subgroups of G.

Let X be an affine algebraic G-variety. For a point $b \in X/\!\!/G$, let $G \cdot x$ be the unique closed orbit in $\pi_X^{-1}(b)$. Since G_x is reductive, there is a G_x-stable direct linear complement N_x to $T_{x,G\cdot x}$ in $T_{x,X}$. The G-bundle $G *_{G_x} N_x \to G/G_x$, called the *normal bundle* of $G \cdot x$, is the representative of some element $\nu(b) \in \mathcal{V}(G)$. In this way we get a mapping $\nu : X/\!\!/G \to \mathcal{V}(G)$, $b \mapsto \nu(b)$.

Let $s \in \mathcal{V}(G)$. Then the sets $(X/\!\!/G)_s := \nu^{-1}(s)$ and $X_s := \pi_X^{-1}((X/\!\!/G)_s)$ are called the *strata* of type s of $X/\!\!/G$ and X respectively.

The following theorem is proved in [Lu].

Theorem 2.3.1. *Retain the above notation and assume that X is smooth. Then*

(i) *The image of ν in $\mathcal{V}(G)$ is finite.*
(ii) *Each stratum $(X/\!\!/G)_s$ is a smooth locally closed subvariety of $X/\!\!/G$.*
(iii) *Each stratum X_s is a locally closed smooth π_X-saturated subvariety of X. There is a G-variety F_s such that each fiber $\pi_X^{-1}(b)$, $b \in (X/\!\!/G)_s$, is G-isomorphic to F_s and the restriction $\pi_X|_{X_s} : X_s \to (X/\!\!/G)_s$ is a fiber bundle with the fiber F_s locally trivial in étale topology (i.e., each point of $(X/\!\!/G)_s$ has an open neighborhood U such that the pull back of this bundle over a suitable étale covering of U becomes trivial).*

If $\varphi : X \to Y$ is an excellent morphism of smooth irreducible affine algebraic G-varieties, then $X_s = \varphi^{-1}(Y_s)$ and $(X/\!\!/G)_s = (\varphi/\!\!/G)^{-1}((Y/\!\!/G)_s)$ for every $s \in \mathcal{V}(G)$.

2.4. Quotients by finite group actions. Now assume that G is a finite group and X an algebraic G-variety (not necessarily affine). Consider the property

(Q) Every G-orbit is contained in an affine open subset of X.

For instance, (Q) holds if X is quasi-projective.

Then one has the following criterion.

Theorem 2.4.1. *The following properties are equivalent:*

(i) (Q) *holds.*

(ii) *There exists a geometric quotient*

$$\pi_X = \pi_{X,G} : X \to X/G. \tag{2.4.1}$$

The morphism π_X *is automatically affine* (*i.e., for any open affine subset* U *of* X/G *the set* $\pi_X^{-1}(U)$ *is affine*).

Proof. See [PV, Theorem 4.14]. □

Assume that there exists a geometric quotient (2.4.1). Take a point $x \in X$. As the group G is finite, the orbit $G \cdot x$ is closed. As the morphism π_X is affine, the orbit $G \cdot x$ is contained in a G-stable affine open subset of X. Hence Theorem 2.2.1 is applicable to the action of G on this subset. This yields

Theorem 2.4.2. *Assume that for an action of a finite group* G *on an algebraic variety* X *there exists a geometric quotient* (2.4.1). *Then statements* (i) *and* (ii) *of Theorem* 2.2 *hold.*

We retain the notation of Subsection 2.3. For any point $b \in X/G$, the fiber $\pi_X^{-1}(b)$ is an orbit $G \cdot x$. As it is finite, $N_x = T_{x,X}$, so $\nu(b)$ is the class of $G *_{G_x} T_{x,X}$. Like in Subsection 2.3, we consider the mapping $\nu : X/G \to \mathcal{V}(G)$, $b \mapsto \nu(b)$, and for any $s \in \mathcal{V}(G)$ define the sets $(X/G)_s := \nu^{-1}(s)$ and $X_s := \pi_X^{-1}((X/G)_s)$. We call them the *strata* of type s of X/G and X respectively. As every orbit is contained in a G-stable affine open subset of X, applying Theorem 2.3.1 we obtain the following

Theorem 2.4.3. *Assume that for an action of a finite group* G *on a smooth algebraic variety* X *there exists a geometric quotient* (2.4.1). *Then statements* (i), (ii) *and* (iii) *of Theorem* 2.3 *hold.*

The definition of strata yields the following description. Let H be the stabilizer of a point in X, and let M be a finite dimensional algebraic H-module over k. Then

$$G\cdot\{\,x\in X^H \mid G_x=H, \text{ and the } H\text{-modules } T_{x,X} \text{ and } M \text{ are isomorphic}\,\}$$

is a (possibly empty) stratum X_s, where s is the class of $G*_H M$, and each stratum of X is obtained in this way. If X_s is nonempty, Z is an irreducible component of X_s, and z is a point of Z, then existence of morphism (2.2.2) implies that $\dim Z=\dim M^H$ and $Z\subseteq X^{G_z}$.

We denote by $(X/G)_i$, resp. X_i, the union of all strata of X/G, resp. X, of codimension i. This is a locally closed smooth subvariety of X/G, resp. X. We have $X_i=\pi_X^{-1}((X/G)_i)$. The subvariety $(X/G)_0$, resp. X_0, is a single stratum of X/G, resp. X. It is called the *principal* stratum and is characterized among all $(X/G)_i$, resp. X_i, by the property that it is dense and open in X/G, resp. X. We have the inclusion $(X/G)_0\subseteq(X/G)_{\rm sm}$. If X is irreducible, X_0 coincides with the open set of all points whose stabilizer is the kernel of action; in particular if the action is faithful, X_0 is the open set of all points with trivial stabilizers. By Theorem 2.4.2, the map $\pi_X|_{X_0}\colon X_0\to(X/G)_0$ is étale.

2.5. Local description of the codimension 1 strata. The following lemma is a basic tool of our analysis of pull backs of tensor fields on quotient varieties. It shows that, in étale topology, a neighborhood of a point of a codimension 1 stratum of any quotient variety coincides with such a neighborhood for a linear action of a finite cyclic group generated by pseudo-reflection.

Lemma 2.5.1. *Assume that for a faithful action of a finite group G on an irreducible smooth algebraic variety X there exists a geometric quotient* (2.4.1). *Let x be a point of X_1. Then:*

(i) *G_x is a finite cyclic group.*
(ii) *There are a character $\chi: G_x\to k^\times$ of order $|G_x|$ and the local parameters $u_1,\dots,u_n$ of X at x such that u_n is a local equation of X_1 at x, each $u_1,\dots,u_{n-1}$ is G_x-invariant, and $g\cdot u_n=\chi(g)u_n$ for all $g\in G_x$.*
(iii) $\pi_{X,G}(x)\in(X/G)_{\rm sm}$.
(iv) $y:=\pi_{X,G_x}(x)\in(X/G_x)_{\rm sm}$.
(v) *There are the local parameters $v_1,\dots,v_n$ of X/G_x at y such that*

(2.5.1) $\pi^*_{X,G_x}(v_i) = u_i$ *if* $1 \leqslant i \leqslant n-1$, *and* $\pi^*_{X,G_x}(v_n) = u_n^{|G_x|}$.

Proof. By Theorem 2.4.2, there is an étale slice S of X at x. As G is finite, S is a G_x-stable neighborhood of x. Hence the set U in (2.2.2) is a G_x-stable neighborhood of x as well. Since X is irreducible, U is dense in X. As the action of G on X is faithful, this implies that the action of G_x on U is faithful as well.

It follows from smoothness of X_1 and X that T_{x,X_1} is a hyperplane in $T_{x,X}$. Since the irreducible component of X_1 containing x lays in X^{G_x}, we have $T_{x,X_1} \subseteq T_{x,X}^{G_x}$. As the action of G_x on U is faithful and morphism (2.2.2) is excellent, the action of G_x on $T_{x,S} = T_{x,X}$ is faithful as well. As there is a one-dimensional G_x-stable direct complement to T_{x,X_1} in $T_{x,X}$, this implies that $T_{x,X_1} = T_{x,X}^{G_x}$, and G_x is a cyclic group of order $|G_x|$ whose generator acts on $T_{x,X}$ as a pseudo-reflection with mirror T_{x,X_1}.

From this we deduce that there are a character $\chi : G_x \to k^\times$ of order $|G_x|$, and a basis $t_1, \ldots, t_n$ of $T^*_{x,X}$ such that $t_1, \ldots, t_{n-1}$ are G_x-invariant and t_n is a G_x-semi-invariant with weight χ. The sequence $t_1, \ldots, t_n$ is a system of local parameters of $T_{x,X}$ at 0, and t_n is a local equation of T_{x,X_1} at 0. Since the morphism (2.2.2) is étale,

$$u_1 := \lambda_x^*(t_1), \ldots, u_n := \lambda_x^*(t_n) \in k[U]$$

is a system of local parameters at x that has the properties stated in (ii).

We have $k[T_{x,X}] = k[t_1, \ldots, t_n]$. Hence $k[T_{x,X}]^{G_x}$ is the polynomial algebra in $t_1, \ldots, t_{n-1}, t_n^{|G_x|}$, and $T_{x,X}/G_x$ is an affine space.

The set $(G *_{G_x} U)/G$ is open in $(G *_{G_x} S)/G$. By Theorem 2.4.2, we have the following étale morphisms

$$\begin{aligned} X/G \xleftarrow{\sigma_S/G} (G \times_{G_x} S)/G &\hookleftarrow (G \times_{G_x} U)/G \\ = U/G_x &\xrightarrow{\lambda_x/G_x} T_{x,X}/G_x. \end{aligned}$$

Thus the following commutative diagram arises:

$$\begin{array}{ccccc} X & \xleftarrow{\text{embedding}} & U & \xrightarrow{\lambda_x} & T_{x,X} \\ {\scriptstyle \pi_{X,G}}\downarrow & & {\scriptstyle \pi_{U,G_x}}\downarrow & & \downarrow{\scriptstyle \pi_{T_{x,X},G_x}} \\ X/G & \xleftarrow{\sigma_S/G} & U/G_x & \xrightarrow{\lambda_x/G_x} & T_{x,X}/G_x \end{array} \tag{2.5.2}$$

As $\pi_{T_{x,X},G_x}(0) \in (T_{x,X}/G_x)_{\mathrm{sm}}$ and $U/G_x = \pi_{X,G_x}(U)$ is a neighborhood of y in X/G_x, this yields $y \in (X/G_x)_{\mathrm{sm}}$ and $\pi_{X,G}(x) = (\sigma_S/G)(y) \in (X/G)_{\mathrm{sm}}$.

By our construction, $k[T_{x,X}/G_x] = k[w_1, \dots, w_n]$ where $w_1 \dots, w_n$ are defined by the equalities

$$\pi^*_{T_{x,X}}(w_i) = t_i \text{ if } 1 \leqslant i \leqslant n-1, \text{ and } \pi^*_{T_{x,X}}(w_n) = t_n^{|G_x|}.$$

Since λ_x/G_x is étale, $v_1 := (\lambda_x/G_x)^*(w_1), \dots, v_n := (\lambda_x/G_x)^*(w_n)$ is a system of parameters of X/G_x at y. It follows from (2.5.2) that

$$\pi^*_{U,G_x}(v_i) = u_i \text{ if } 1 \leqslant i \leqslant n-1, \text{ and } \pi^*_{U,G_x}(v_n) = u_n^{|G_x|}. \quad \square$$

2.6. Reflection divisor of quotient variety. Let X be an irreducible smooth algebraic variety endowed with a faithful action of a finite group G such that there exists a geometric quotient (2.4.1). As X is normal, X/G is normal as well, [PV, §4 and Theorem 3.16].

It follows from the discussion in Subsection 2.4 that if $(X/G)_1 \neq \varnothing$ and

$$(X/G)_1 = (X/G)_1^1 \cup \ldots \cup (X/G)_1^d \tag{2.6.1}$$

is the decomposition into irreducible components, then for each $l = 1, \dots, d$ and $x \in \pi_X^{-1}(b)$, $b \in (X/G)_1^l$ the integer $|G_x|$ does not depend on b and x. Denote it by r_l.

We encode the information about $r_1, \dots, r_d$ into the divisor

$$R_{X/G} := r_1 R_1 + \ldots + r_d R_d \in \mathrm{Div}(X/G), \tag{2.6.2}$$

where $R_l \in \mathrm{Div}(X/G)$ is the prime divisor such that $\mathrm{supp}(R_l)$ is the closure of $(X/G)_1^l$.

If $(X/G)_1 = \varnothing$, we put $R_{X/G} = 0$. Thus

$$R_{X/G} = 0 \iff (X/G)_1 = \varnothing. \tag{2.6.3}$$

We call $R_{X/G}$ the *reflection divisor of* X/G. We remark that

$$(X/G)_1^i \cap (X/G)_1^j = \varnothing \text{ if } i \neq j. \tag{2.6.4}$$

because of smoothness of $(X/G)_1$.

3. Lifting Tensor Fields

3.1. Rational tensor fields on algebraic varieties. Let Y be an irreducible algebraic variety. Denote by $T(Y_{\rm sm})$, resp. $T^*(Y_{\rm sm})$, the tangent, resp. cotangent, bundle of $Y_{\rm sm}$. We define a rational tensor field of type $\binom{p}{q}$ on Y as a rational section of the tensor product of p copies of $T(Y_{\rm sm})$ and q copies of $T^*(Y_{\rm sm})$ whose factors are taken in a fixed order. We call $p+q$ the *valency* of such a section. As changing of the order yields an isomorphic vector bundle, we consider only the following representative of this class of vector bundles:

$$T^p_q(Y_{\rm sm}) := T(Y_{\rm sm})^{\otimes p} \otimes T^*(Y_{\rm sm})^{\otimes q}.$$

The set of points where a rational tensor field τ of type $\binom{p}{q}$ on Y is defined is an open subset of $Y_{\rm sm}$. We denote it by $\operatorname{dom}(\tau)$. We say that τ is *regular* on a subset U of X if $U \subseteq \operatorname{dom}(\tau)$. For a point $y \in \operatorname{dom}(\tau)$, we denote by $\tau(y)$ the value of τ at y.

The set $\Gamma_{\rm rat}(T^p_q(Y_{\rm sm}))$ of rational tensor fields of type $\binom{p}{q}$ on Y has a natural structure of $k(Y)$-module. A rational tensor field ν of type $\binom{1}{0}$ on Y is a rational vector field on $Y_{\rm sm}$ or, equivalently, a derivation of the k-algebra $k(Y)$. A rational tensor field ω on Y of type $\binom{0}{1}$ is a rational differential 1-form on $Y_{\rm sm}$. We denote by $\omega(\nu)$ the rational function on Y whose value at a point $y \in \operatorname{dom}(\omega) \cap \operatorname{dom}(\nu)$ is equal to the value of $\omega(y)$ at $\nu(y)$. Rational sections of $\wedge^q(Y_{\rm sm})$ are rational differential q-forms on $Y_{\rm sm}$. Rational differential q-forms on Y defined in [Sh] are naturally identified with them by means of restriction to $Y_{\rm sm}$.

Let $v_1, \ldots, v_n$ be a system of local parameters of $Y_{\rm sm}$ at a point $y \in Y_{\rm sm}$. Then there is a neighborhood V of y in $Y_{\rm sm}$ such that $v_1, \ldots, v_n \in k[V]$ and $dv_1(a), \ldots, dv_n(a)$ is a basis of $T^*_{a,V}$ for each point $a \in V$, [Sh, Ch. III, §5]. Let $\frac{\partial}{\partial v_1}, \ldots, \frac{\partial}{\partial v_n}$ be the regular vector fields on V defined by conditions

$$dv_i\Big(\frac{\partial}{\partial v_j}\Big) = \delta_{ij}. \tag{3.1.1}$$

Then the (numbered) set of tensors $(\frac{\partial}{\partial v_{i_1}} \otimes \cdots \otimes \frac{\partial}{\partial v_{i_p}} \otimes dv_{j_1} \otimes \cdots \otimes dv_{j_q})(a)$ for each point $a \in V$ is a basis of $T^{\otimes p}_{a,V} \otimes T^{*\otimes q}_{a,V}$.

This implies that for each $\tau \in \Gamma_{\rm rat}(T^p_q(Y_{\rm sm}))$ there is a unique decomposition

$$\tau = \sum\nolimits_{i_1,\ldots,i_p,j_1,\ldots,j_q} \tau^{i_1\ldots i_p}_{j_1\ldots j_q} \frac{\partial}{\partial v_{i_1}} \otimes \cdots \otimes \frac{\partial}{\partial v_{i_p}} \otimes dv_{j_1} \otimes \cdots \otimes dv_{j_q}, \tag{3.1.2}$$

where $\tau^{i_1\dots i_p}_{j_1\dots j_q} \in k(Y)$, and we have

$$V \cap \operatorname{dom}(\tau) := \{a \in V \mid \text{ each } \tau^{i_1\dots i_p}_{j_1\dots j_q} \text{ is regular at } a\}. \tag{3.1.3}$$

From (3.1.2) we also obtain that for any point $a \in V$,

$$\tau(a) = 0 \iff \tau^{i_1\dots i_p}_{j_1\dots j_q}(a) = 0 \ \forall\, i_1, \dots, i_p, j_1, \dots, j_q. \tag{3.1.4}$$

By (3.1.3), (3.1.4), the set of all zeros of τ is a closed subset of $\operatorname{dom}(\tau)$.

Now assume that the variety Y is normal. If $\operatorname{dom}(\tau) \neq Y_{\mathrm{sm}}$ and

$$Y_{\mathrm{sm}} \setminus \operatorname{dom}(\tau) = \textstyle\bigcup_{i=1}^{m} (\tau)^i_\infty \tag{3.1.5}$$

is the decomposition of $Y_{\mathrm{sm}} \setminus \operatorname{dom}(\tau)$ into irreducible components, then by (3.1.3),

$$\operatorname{codim}_Y (\tau)^i_\infty = 1 \text{ for all } i. \tag{3.1.6}$$

Thus if the "variety of poles" of a tensor field (i.e., the subvariety of Y_{sm} where this field is not defined) is nonempty, it is unmixed and of codimension 1. On the contrary, if p or $q \neq 0$, then by (3.1.4), the variety of zeros of a tensor field of type $\binom{p}{q}$ can be mixed and/or of codimension greater than 1.

3.2. Divisor of a rational tensor field. Let $\tau \in \Gamma_{\mathrm{rat}}(T^p_q(Y_{\mathrm{sm}}))$ be a nonzero rational tensor field. We define $\operatorname{div}(\tau) \in \operatorname{Div}(Y)$, the *divisor of* τ, as follows.

Let $C \in \operatorname{Div}(Y)$ be a prime divisor. Since Y is normal, $\operatorname{supp}(C) \cap Y_{\mathrm{sm}} \neq \varnothing$. Consider in Y an open neighborhood U of a point $y \in \operatorname{supp}(C) \cap Y_{\mathrm{sm}}$ where C is given by a local equation t. Then there is a unique integer $\nu(C, \tau)$ such that

(i) $\operatorname{supp}(C) \cap \operatorname{dom}(t^{-\nu(C,\tau)}\tau) \neq \varnothing$;
(ii) $t^{-\nu(C,\tau)}\tau$ does not identically vanish on the set $\operatorname{supp}(C) \cap \operatorname{dom}(t^{-\nu(C,\tau)}\tau)$.

To see this, take V and $v_1, \dots, v_n$ as above assuming moreover that $V \subseteq U$ (this can be attained by replacing V by $V \cap U$). Thus (3.1.2) holds. Then (i) and (ii) hold iff there is a point $z \in \operatorname{supp}(C)$ such that all functions $t^{-\nu(C,\tau)}\tau^{i_1\dots i_p}_{j_1\dots j_q}$ are defined at z but not all of them vanish at z. This unequivocally identifies $\nu(C, \tau)$ as

$$\nu(C, \tau) = \min_{i_1,\dots i_p, j_1, \dots, j_q} m_{C, \operatorname{div}(\tau^{i_1\dots i_p}_{j_1\dots j_q})}. \tag{3.2.1}$$

It is immediately seen that $\nu(C,\tau)$ is well defined (i.e., depends only on C and τ but not on the choice of V and t), and that the set $\{C \mid \nu(C,\tau) \neq 0\}$ is finite (maybe empty). Now the divisor $\operatorname{div}(\tau)$ is defined by

$$m_{C,\operatorname{div}(\tau)} := \nu(C,\tau) \quad \text{for all } C. \tag{3.2.2}$$

For $p = q = 0$ this gives the usual definition of divisor of a rational function.

It follows from the definition and (3.1.5), (3.1.6) that if $m_{C,\operatorname{div}(\tau)} < 0$, then $\operatorname{supp}(C)$ is the closure in Y of one of the $(\tau)^i_\infty$'s from (3.1.5). Whence if τ is a rational tensor field on Y, then

$$\operatorname{div}(\tau) \geqslant 0 \iff \tau \text{ is regular on } Y_{\mathrm{sm}}. \tag{3.2.3}$$

On the other hand, we remark that, unlike in the case of rational functions, in general $\operatorname{div}(\tau) \leqslant 0$ does not imply that rational tensor field τ vanishes nowhere.

3.3. Pull backs and push forwards of rational tensor fields. Consider a dominant rational map $\alpha : Z \to Y$ of irreducible algebraic varieties of the same dimension. As $\operatorname{char} k = 0$, the set Z_α of points of Z_{sm} where α is defined and étale is dense and open in Z, cf. [Sh, Ch. II, §6, no. 2, Lemma 2].

Consider a rational tensor field $\tau \in \Gamma_{\mathrm{rat}}(T^p_q(Y_{\mathrm{sm}}))$. As $d_z\alpha : T_{z,Z} \to T_{\alpha(z),Y}$ for $z \in Z_\alpha$ is an isomorphism, it yields an isomorphism

$$\begin{gathered}\alpha_z^* : T^{\otimes p}_{\alpha(z),Y} \otimes T^{*\otimes q}_{\alpha(z),Y} \longrightarrow T^{\otimes p}_{z,Z} \otimes T^{*\otimes q}_{z,Z}, \\ \text{where } \alpha_z^* := ((d_z\alpha)^{-1})^{\otimes p} \otimes ((d_z\alpha)^*)^{\otimes q}.\end{gathered} \tag{3.3.1}$$

Hence for $W := \alpha^{-1}(\operatorname{dom}(\tau)) \cap Z_\alpha$ we obtain the following section of $T^p_q(W)$:

$$\alpha^*(\tau) : W \to T^p_q(W), \quad \alpha^*(\tau)(z) := \alpha_z^*(\tau(\alpha(z))). \tag{3.3.2}$$

We claim that (3.3.2) is a rational tensor field on Z, regular on W. To show this, take a point $z \in W$, a system of local parameters $v_1, \ldots, v_n$ of Y_{sm} at $y = \alpha(z)$, and a neighborhood V of y in $\alpha(W)$ as in Subsection 3.1. Then conditions (3.1.1) define the rational vector fields $\frac{\partial}{\partial v_j}$ such that decomposition (3.1.2) holds. Put

$$u_i := \alpha^*(v_i). \tag{3.3.3}$$

As α is étale et z, the sequence $u_1, \dots, u_n$ is a system of parameters of Z_{sm} at z. Let U be a neighborhood of z in Z_α such that $\alpha(U) \subseteq V$. Then $u_1, \dots, u_n \in k[U]$ and $du_1(a), \dots, du_n(a)$ is a basis of $T^*_{z,U}$ for each point $a \in U$. Let $\frac{\partial}{\partial u_1}, \dots, \frac{\partial}{\partial u_n}$ be the regular vector fields on U defined by the conditions $du_i(\frac{\partial}{\partial u_j}) = \delta_{ij}$. Then from (3.3.1), (3.3.2) and (3.1.1) we deduce that

$$\alpha^*(dv_i) = du_i \text{ and } \alpha^*(\tfrac{\partial}{\partial v_i}) = \tfrac{\partial}{\partial u_i} \text{ for all } i. \tag{3.3.4}$$

Hence by (3.1.2), there is a unique decomposition

$$\begin{aligned}\alpha^*(\tau) = \textstyle\sum_{i_1,\dots,i_p,j_1,\dots,j_q} \alpha^*(\tau^{i_1\dots i_p}_{j_1\dots j_q}) \tfrac{\partial}{\partial u_{i_1}} \otimes \dots \otimes \tfrac{\partial}{\partial u_{i_p}} \\ \otimes du_{j_1} \otimes \dots \otimes du_{j_q}.\end{aligned} \tag{3.3.5}$$

As $\tau^{i_1\dots i_p}_{j_1\dots j_q} \in k(Y)$, this shows that $\alpha^*(\tau)$ is a rational tensor field on Z, regular on W. We call $\alpha^*(\tau)$ the *pull back* of τ. Thus we have a map

$$\alpha^* : \Gamma_{\mathrm{rat}}(T^p_q(Y_{\mathrm{sm}})) \to \Gamma_{\mathrm{rat}}(T^p_q(Z_{\mathrm{sm}})), \quad \tau \mapsto \alpha^*(\tau).$$

Now we consider the case where Z is an irreducible smooth algebraic G-variety X for a finite group G such that there exists a geometric quotient (2.4.1), and $Y = X/G$, $\alpha = \pi_X$. The fixed point set of the natural action of G on $\Gamma_{\mathrm{rat}}(T^p_q(X_{\mathrm{sm}}))$ is precisely the set of all G-invariant rational tensor fields of type $\binom{p}{q}$ on X.

If $\tau \in \Gamma_{\mathrm{rat}}(T^p_q((X/G)_{\mathrm{sm}}))$, then the definition implies that the field $\pi^*_X(\tau)$ is G-invariant. Thus we obtain a map

$$\pi^*_X : \Gamma_{\mathrm{rat}}(T^p_q((X/G)_{\mathrm{sm}})) \to \Gamma_{\mathrm{rat}}(T^p_q(X_{\mathrm{sm}}))^G, \; \tau \mapsto \pi^*_X(\tau).$$

We claim that π^*_X is bijective. To prove this, take a field $\varphi \in \Gamma_{\mathrm{rat}}(T^p_q(X_{\mathrm{sm}}))^G$. As $\mathrm{dom}(\varphi)$ is a G-stable and open subset of X, and X is an irreducible variety, $W := \pi_X(\mathrm{dom}(\varphi)) \cap (X/G)_0$ is a dense open subset of X/G. Take points $b \in W$ and $x \in \pi_X^{-1}(b)$. As π_X is étale at each point of X_0, we have an isomorphism $(\pi_X)^*_x$ (see (3.3.1)). Since φ is G-invariant, $((\pi_X)^*_x)^{-1}(\varphi(x))$ depends only on b but not on the choice of x in $\pi_X^{-1}(b)$. Hence the following formula defines a section of $T^p_q(W)$:

$$\pi_{X*}(\varphi) : W \to T^p_q(W), \; \pi_{X*}(\varphi)(b) = ((\pi_X)^*_x)^{-1}(\varphi(x)). \tag{3.3.6}$$

We claim that $\pi_{X*}(\varphi)$ is a rational tensor field on X/G, regular on W. We call it the *push forward* of φ. The definitions imply that the maps π_X^* and

$$\pi_{X*}:\Gamma_{\mathrm{rat}}(T^p_q(X_{\mathrm{sm}}))^G\to\Gamma_{\mathrm{rat}}(T^p_q((X/G)_{\mathrm{sm}})),\ \varphi\mapsto\pi_{X*}(\varphi),$$

are inverses of one another.

In order to prove the latter claim, fix a system $v_1,\dots,v_n$ of local parameters of X/G at b, and let V be a neighborhood of b in W such that $v_1,\dots,v_n\in k[V]$ and $dv_1(b),\dots,dv_n(b)$ is a basis of $T^*_{b,V}$ for each point $b\in V$. The set $U:=\pi_X^{-1}(V)$ is an open G-stable neighborhood of x in $\operatorname{dom}(\varphi)\cap X_0$. As $\pi_X|_{X_0}$ is étale, $u_1=\pi_X^*(v_1),\dots,u_n=\pi_X^*(v_n)\in k[U]^G$ is a system of local parameters of X at x such that $du_1(a),\dots,du_n(a)$ is a basis of $T^*_{a,U}$ for each point $a\in U$. Hence there is a unique decomposition

$$\begin{aligned}\varphi=\textstyle\sum_{i_1,\dots,i_p,j_1,\dots,j_q}\varphi^{i_1\dots i_p}_{j_1\dots j_q}\frac{\partial}{\partial u_{i_1}}\otimes\dots\otimes\frac{\partial}{\partial u_{i_p}}\\ \otimes\, du_{j_1}\otimes\dots\otimes du_{j_q},\end{aligned}\tag{3.3.7}$$

where $\varphi^{i_1\dots i_p}_{j_1\dots j_q}\in k(X)$ and $du_i(\frac{\partial}{\partial u_j})=\delta_{ij}$.

As the fields φ, du_i and $\frac{\partial}{\partial u_i}$ are G-invariant, uniqueness of (3.3.7) and the inclusion $\operatorname{dom}(\varphi)\supseteq U$ imply that all coefficients $\varphi^{i_1\dots i_p}_{j_1\dots j_q}$ in (3.3.7) are contained in $k[U]^G$. Hence, as $\pi_X:X\to X/G$ is a geometric quotient, there are functions $\psi^{i_1\dots i_p}_{j_1\dots j_q}\in k[V]$ such that

$$\varphi^{i_1\dots i_p}_{j_1\dots j_q}=\pi_X^*(\psi^{i_1\dots i_p}_{j_1\dots j_q}),\tag{3.3.8}$$

cf. [PV, §4]. On the other hand, the definitions imply that for any point $a\in U$,

$$\begin{aligned}&((\pi_X^*)_a)^{-1}\big((\tfrac{\partial}{\partial u_{i_1}}\otimes\dots\otimes\tfrac{\partial}{\partial u_{i_p}}\otimes du_{j_1}\otimes\dots\otimes du_{j_q})(a)\big)\\ &=\big(\tfrac{\partial}{\partial v_{i_1}}\otimes\dots\otimes\tfrac{\partial}{\partial v_{i_p}}\otimes dv_{j_1}\otimes\dots\otimes dv_{j_q}\big)(\pi_X(a)).\end{aligned}\tag{3.3.9}$$

From (3.3.6)–(3.3.9), we deduce that the following decomposition holds

$$\begin{aligned}\pi_{X*}(\varphi)=\textstyle\sum_{i_1,\dots,i_p,j_1,\dots,j_q}\psi^{i_1\dots i_p}_{j_1\dots j_q}\frac{\partial}{\partial v_{i_1}}\otimes\dots\otimes\frac{\partial}{\partial v_{i_p}}\\ \otimes\, dv_{j_1}\otimes\dots\otimes dv_{j_q}.\end{aligned}\tag{3.3.10}$$

In turn, (3.3.10) yields that $\pi_{X*}(\varphi)$ is a rational tensor field of type $\binom{p}{q}$ on X/G, regular on W.

3.4. Regularity of the pull back of a tensor field on the quotient variety: Example. We retain the notation and conventions of the previous subsection. Our aim in Section 3 is to give a necessary and sufficient condition of regularity of the pull back of a rational tensor field on X/G. According to the discussion in Subsection 3.3, if $\pi_X^*(\tau)$ is regular on X, then τ is automatically regular on $(X/G)_0$. However in general this condition is not sufficient: tensor fields on X/G with "bad singularities" along the irreducible components of $(X/G)_1$ may have pull backs that are not regular on the whole X. The following simplest example illustrates this phenomenon.

Example. Let X be a one dimensional linear space over k. Let G be a finite group acting on X by $g \cdot a := \chi(g)a$ where $\chi : G \to k^\times$ is a character of order r. If u is a nonzero linear function on X, then $k[X] = k[u]$ and $k[X]^G = k[u^r]$. Hence X/G is an affine line with the coordinate function v, and $\pi_X := \pi_{X,G} : X \to X/G$ is given by $\pi_X^*(v) = u^r$.

Each nonzero rational tensor field τ of type $\binom{p}{q}$ on X/G has the form

$$\tau = \alpha v^m \prod\nolimits_{i=1}^l (v-\gamma_i)^{d_i} (\tfrac{\partial}{\partial v})^{\otimes p} \otimes (dv)^{\otimes q},$$

where $m, d_i \in \mathbb{Z}$, $\alpha, \gamma_i \in k \setminus \{0\}$ for all i, and $\gamma_i \neq \gamma_j$ for all $i \neq j$. Since $\pi_X^*(dv) = ru^{r-1}du$ and $\pi_X^*(\frac{\partial}{\partial v}) = \frac{1}{ru^{r-1}}\frac{\partial}{\partial u}$, we have

$$\begin{aligned}\pi_X^*(\tau) &= \pi_X^*(v)^m \prod\nolimits_{i=1}^l (\pi_X^*(v)-\gamma_i)^{d_i} (\pi_X^*(\tfrac{\partial}{\partial v}))^{\otimes p} \otimes (\pi_X^*(dv))^{\otimes q}\\ &= r^{q-p} u^{(r-1)(q-p)+rm} \prod\nolimits_{i=1}^l (u^r-\gamma_i)^{d_i} (\tfrac{\partial}{\partial u})^{\otimes p} \otimes (du)^{\otimes q}.\end{aligned}$$

Therefore the rational tensor field $\pi_X^*(\tau)$ is regular on X if and only if

(i) $d_i \geqslant 0$ for all i,

(ii) $(r-1)(q-p)+rm \geqslant 0$.

For instance, if $r = 3$, $m = -3$, $q-p = 7$, and $d_i = 0$ for all i, then τ is regular on $(X/G)_0$ but not regular on X/G, and $\pi_X^*(\tau)$ is regular on X. If $r = 3$, $m = -5$, $q-p = 7$ and $d_i = 0$ for all i, then τ is regular on $(X/G)_0$, not regular on X/G, and $\pi_X^*(\tau)$ is not regular on X. □

From the general criterion of regularity of the pull back that will be obtained below we will see that this example essentially elucidates the general mechanism governing the phenomenon under investigation.

3.5. Regularity of the pull back of a tensor field on a quotient variety: Case with no strata of codimension 1. The corollary of the following proposition shows that if there are no strata of codimension 1, then the situation is simple and regularity of τ on $(X/G)_0$ is not only necessary but also sufficient for regularity of $\pi_X^*(\tau)$ on X.

Proposition 3.5.1. *Let X be an irreducible smooth algebraic variety endowed with an action of a finite group G. Assume that there exists a geometric quotient* (2.4.1). *Let U be an open subset of X/G such that $U \cap (X/G)_1 = \varnothing$. Consider a tensor field $\tau \in \Gamma_{\mathrm{rat}}(T_q^p((X/G)_{\mathrm{sm}}))$ such that $U \cap (X/G)_0 \subseteq \mathrm{dom}(\tau)$. Then $\pi_X^{-1}(U) \subseteq \mathrm{dom}(\pi_X^*(\tau))$.*

Proof. As $\mathrm{dom}(\tau) \supseteq U \cap (X/G)_0$, we have $\mathrm{dom}(\pi_X^*(\tau)) \supseteq \pi_X^{-1}(U) \cap X_0$. Since $U \cap (X/G)_1 = \varnothing$, the complement of $\pi_X^{-1}(U) \cap X_0$ in $\pi_X^{-1}(U)$ has codimension $\geqslant 2$. Now the claim follows from (3.1.5) and (3.1.6). □

Corollary 3.5.1. *Assume moreover that X contains no strata of codimension* 1. *Then the pull back $\pi_X^*(\tau)$ of a rational tensor field τ on X/G is regular on X if and only if τ is regular on the principal stratum $(X/G)_0$.*

Proof. As $(X/G)_1 = \varnothing$, one can take $U = X/G$. □

3.6. Modifying the divisor of a tensor field by means of a nonnegative divisor. Let Y be an irreducible normal algebraic variety and $B \in \mathrm{Div}(Y)$ a nonnegative divisor. Consider a nonzero rational tensor field $\tau \in \Gamma_{\mathrm{rat}}(T_q^p(Y_{\mathrm{sm}}))$. In this subsection we define a divisor $\mathrm{div}_B(\tau) \in \mathrm{Div}(Y)$ that we call the *B-divisor of τ*. It is obtained from $\mathrm{div}(\tau)$ by means of some "modification along" B. In the next subsection we will use this notion in our criterion of regularity of the pull back of a tensor field τ on the quotient variety X/G: it is formulated in terms of $\mathrm{div}_{R_{X/G}}(\tau)$, where $R_{X/G}$ is the reflection divisor of X/G, see (2.6.2).

If $B = 0$, we define

(3.6.1) $$\mathrm{div}_0(\tau) := \mathrm{div}(\tau).$$

Now assume that $B > 0$,

(3.6.2) $$B = b_1 B_1 + \ldots + b_d B_d,$$

where B_i is a prime divisor and b_i a positive integer for all i. Then $\operatorname{div}_B(\tau)$ is obtained from $\operatorname{div}(\tau)$ by modifying the multiplicities $m_{B_1,\operatorname{div}(\tau)}, \dots, m_{B_d,\operatorname{div}(\tau)}$ as defined below.

To describe this modification, fix some B_l. As Y is normal and $\operatorname{codim}(\operatorname{supp}(B_l)) = 1$,

we have $\operatorname{supp}(B_l) \cap Y_{\mathrm{sm}} \neq \varnothing$. Hence taking local parameters at a point $y \in \operatorname{supp}(B_l) \cap Y_{\mathrm{sm}}$, we can find an open neighborhood V of y in Y_{sm} and a system v of functions $v_1, \dots, v_n \in k[V]$ such that $\operatorname{supp}(B_l) \cap V \neq \varnothing$, the ideal of $\operatorname{supp}(B_l) \cap V$ in $k[V]$ is generated by v_n, and $dv_1(a), \dots, dv_n(a)$ is a basis of $T^*_{a,V}$ for each $a \in V$. Hence there is a unique decomposition

$$\tau = \sum\nolimits_{i_1,\dots,i_p,j_1,\dots,j_q} \delta^{i_1\dots i_p}_{j_1\dots j_q}, \tag{3.6.3}$$

where (see (3.1.2))

$$\begin{gathered}\delta^{i_1\dots i_p}_{j_1\dots j_q} = \tau^{i_1\dots i_p}_{j_1\dots j_q} \frac{\partial}{\partial v_{i_1}} \otimes \dots \otimes \frac{\partial}{\partial v_{i_p}} \otimes dv_{j_1} \otimes \dots \otimes dv_{j_q}, \\ \text{where } \tau^{i_1\dots i_p}_{j_1\dots j_q} \in k(Y).\end{gathered} \tag{3.6.4}$$

With each nonzero tensor field $\delta^{i_1\dots i_p}_{j_1\dots j_q}$ we associate the following three integers m, p' and q' (depending on this field):

- $m := m_{B_l,\operatorname{div}(\tau^{i_1\dots i_p}_{j_1\dots j_q})}$;
- p' (resp. q') is the number of factors $\frac{\partial}{\partial v_n}$ (resp. dv_n) in the right-hand side of equality (3.6.4).

Now define

$$\mu(b_l B_l, \delta^{i_1\dots i_p}_{j_1\dots j_q})_{V,v} := (b_l - 1)(q' - p') + b_l m, \tag{3.6.5}$$

$$\varrho(b_l B_l, \tau)_{V,v} := \min_{i_1,\dots,i_p,j_1\dots j_q} \mu(b_l B_l, \delta^{i_1\dots i_p}_{j_1\dots j_q})_{V,v}. \tag{3.6.6}$$

It easily follows from (3.6.3)–(3.6.6) that for any nonzero $f \in k(Y)$ and $\tau, \varsigma \in \Gamma_{\mathrm{rat}}(T^p_q(Y)_{\mathrm{sm}})$ such that $\tau + \varsigma \neq 0$ we have

$$\begin{gathered}\varrho(b_l B_l, f\tau)_{V,v} = b_l m_{B_l,\operatorname{div}(f)} + \varrho(b_l B_l, \tau)_{V,v}, \\ \varrho(b_l B_l, \tau + \varsigma)_{V,v} \geqslant \min\{\varrho(b_l B_l, \tau)_{V,v}, \varrho(b_l B_l, \varsigma)_{V,v}\}.\end{gathered} \tag{3.6.7}$$

Lemma 3.6.1. *For any nonzero* $\tau \in \Gamma_{\mathrm{rat}}(T_q^p(Y)_{\mathrm{sm}})$, $\varsigma \in \Gamma_{\mathrm{rat}}(T_t^s(Y)_{\mathrm{sm}})$ *we have*

$$\varrho(b_l B_l, \tau \otimes \varsigma)_{V,v} = \varrho(b_l B_l, \tau)_{V,v} + \varrho(b_l B_l, \varsigma)_{V,v}. \tag{3.6.8}$$

Proof. If τ and ς are " monomials in v", i.e., $\tau = f\frac{\partial}{\partial v_{i_1}} \otimes \cdots \otimes \frac{\partial}{\partial v_{i_p}} \otimes dv_{j_1} \otimes \cdots \otimes dv_{j_q}$, $\varsigma = g\frac{\partial}{\partial v_{a_1}} \otimes \cdots \otimes \frac{\partial}{\partial v_{a_s}} \otimes dv_{b_1} \otimes \cdots \otimes dv_{b_t}$ where $f, g \in k(Y)$, then $\tau \otimes \varsigma$ is such a monomial as well, and the number p' (resp. q', m) for $\tau \otimes \varsigma$ is the sum of numbers p' (resp. q', m) for τ and σ. Therefore in this case the claim readily follows from the definitions (3.6.5), (3.6.6).

Consider the general case. Then by (3.1.2) we have $\tau = \sum_i \tau_i$, $\varsigma = \sum_j \varsigma_j$ where each τ_i and ς_j is a monomial in v. From (3.6.6) we obtain $\varrho(b_l B_l, \tau)_{V,v} = \min_i \varrho(b_l B_l, \tau_i)_{V,v}$, $\varrho(b_l B_l, \varsigma)_{V,v} = \min_j \varrho(b_l B_l, \varsigma_j)_{V,v}$. Since $\tau \otimes \varsigma = \sum_{i,j} \tau_i \otimes \varsigma_j$ and each $\tau_i \otimes \varsigma_j$ is a monomial in v, we have

$$\begin{aligned}
\varrho(b_l B_l, \tau \otimes \varrho)_{V,v} &= \min_{i,j} \varrho(b_l B_l, \tau_i \otimes \varrho_j)_{V,v} \\
&= \min_{i,j} \left(\varrho(b_l B_l, \tau_i)_{V,v} + \varrho(b_l B_l, \varsigma_j)_{V,v}\right) \\
&= \min_i \varrho(b_l B_l, \tau_i)_{V,v} + \min_j \varrho(b_l B_l, \varrho_j)_{V,v} \\
&= \varrho(b_l B_l, \tau)_{V,v} + \varrho(b_l B_l, \varrho)_{V,v}. \quad \square
\end{aligned}$$

Lemma 3.6.2. $\varrho(b_l B_l, \tau)_{V,v}$ *depends only on* τ *and* $b_l B_l$ *but not on the choice of* V *and* $v_1, \ldots, v_n$.

Proof. First we remark that if V' is an open subset of V and $\operatorname{supp}(B_l) \cap V' \neq \varnothing$, then clearly $\varrho(b_l B_l, \tau)_{V',v} = \varrho(b_l B_l, \tau)_{V,v}$.

Now consider another open subset $\tilde{V}$ of Y_{sm}, and a system $\tilde{v}$ of functions $\tilde{v}_1, \ldots, \tilde{v}_n \in k[\tilde{V}]$ such that $\operatorname{supp}(B_l) \cap \tilde{V} \neq \varnothing$, the ideal of $\operatorname{supp}(B_l) \cap \tilde{V}$ in $k[\tilde{V}]$ is generated by $\tilde{v}_n$, and $d\tilde{v}_1(a), \ldots, d\tilde{v}_n(a)$ is a basis of $T^*_{a,\tilde{V}}$ for each point $a \in \tilde{V}$. We have to show that $\varrho(b_l B_l, \tau)_{V,v} = \varrho(b_l B_l, \tau)_{\tilde{V},\tilde{v}}$. Since $\operatorname{supp}(B_l)$ is irreducible, $\operatorname{supp}(B_l) \cap V \cap \tilde{V} \neq \varnothing$. Hence using the remark above and replacing V and $\tilde{V}$ by $V \cap \tilde{V}$ we can (and shall) assume that $V = \tilde{V}$. This reduces the problem to proving that

$$\varrho(b_l B_l, \tau)_{V,v} \leqslant \varrho(b_l B_l, \tau)_{V,\tilde{v}}. \tag{3.6.9}$$

To show that (3.6.9) holds, we need the divisibility properties of some functions. Namely, the defining property of $\tilde{v}_1, \dots, \tilde{v}_n$ implies that there are unique decompositions

$$dv_i = \sum_{j=1}^n g_{ij} d\tilde{v}_j, \quad \frac{\partial}{\partial v_i} = \sum_{j=1}^n h_{ij} \frac{\partial}{\partial \tilde{v}_j}, \quad \text{where } g_{ij}, h_{ij} \in k[V]. \tag{3.6.10}$$

Since either of v_n and $\tilde{v}_n$ is a generator of the ideal of $\operatorname{supp}(B_l) \cap V$ in $k[V]$, we have

$$v_n = f\tilde{v}_n, \text{ where } f \text{ and } 1/f \in k[V]. \tag{3.6.11}$$

We claim that

$$\tilde{v}_n \text{ divides } g_{ni} \text{ and } h_{in} \text{ in } k[V] \ \forall\, i = 1, \dots, n-1. \tag{3.6.12}$$

To prove this, take a point $a \in \operatorname{supp}(B_l) \cap V$ and an index i, $1 \leqslant i \leqslant n-1$. As $\tilde{v}_n(a) = 0$, from (3.6.11) we deduce that

$$dv_n(a) = f(a) d\tilde{v}_n(a). \tag{3.6.13}$$

On the other hand, it follows from (3.6.10) that $dv_n(a) = \sum_{j=1}^n g_{nj}(a) d\tilde{v}_j(a)$. Since $d\tilde{v}_1(a), \dots, d\tilde{v}_n(a)$ is a basis of $T^*_{a,V}$, the latter equality and (3.6.13) imply that g_{ni} vanishes on $\operatorname{supp}(B_l) \cap V$, whence it is divisible by $\tilde{v}_n$ in $k[V]$. Further, by (3.1.1), the value of $dv_n(a)$ on $\frac{\partial}{\partial v_i}(a)$ is equal to 0. On the other hand, by (3.6.13) and (3.6.10), it is equal to the value of $f(a) d\tilde{v}_n(a)$ on $\sum_{j=1}^n h_{ij}(a) \frac{\partial}{\partial \tilde{v}_j}(a)$ that in turn is equal to $f(a) h_{in}(a)$. By (3.6.11), this implies that $h_{in}(a) = 0$. Thus h_{in} vanishes on $\operatorname{supp}(B_l) \cap V$, whence it is divisible by $\tilde{v}_n$ in $k[V]$. This completes the proof of (3.6.12).

Now we can proceed to proving (3.6.9). By (3.6.6), for each nonzero $\delta^{i_1 \dots i_p}_{j_1 \dots j_q}$ we have $\varrho(b_l B_l, \delta^{i_1 \dots i_p}_{j_1 \dots j_q})_{V,v} = \mu(b_l B_l, \delta^{i_1 \dots i_p}_{j_1 \dots j_q})_{V,v}$. This and (3.6.7) imply that it suffices to prove (3.6.9) for the case when τ is a monomial in v,

$$\tau = \frac{\partial}{\partial v_{i_1}} \otimes \cdots \otimes \frac{\partial}{\partial v_{i_p}} \otimes dv_{j_1} \otimes \cdots \otimes dv_{j_q}. \tag{3.6.14}$$

So assume that (3.6.14) holds. Then we prove (3.6.9) by induction with respect to the valency $p+q$. First we prove (3.6.9) for $p+q = 1$, i.e., when $\tau = \frac{\partial}{\partial v_i}$ or $\tau = dv_i$ for some i. Use the notation of (3.6.5) for $\delta^{i_1\dots i_p}_{j_1\dots j_q} = \frac{\partial}{\partial v_i}$. We have $m=0$, $q'=0$. If $i<n$, then $p'=0$. If $i=n$, then $p'=1$. So, by (3.6.5), (3.6.6), we have

$$\varrho(b_l B_l, \tfrac{\partial}{\partial v_i})_{V,v} = \mu(b_l B_l, \tfrac{\partial}{\partial v_i})_{V,v} = \begin{cases} 0 & \text{if } i<n, \\ 1-b_l & \text{if } i=n. \end{cases} \tag{3.6.15}$$

Similarly, using the notation of (3.6.5) for $\delta^{i_1\dots i_p}_{j_1\dots j_q} = dv_i$, we have: $m=0$, $p'=0$; if $i<n$, then $q'=0$, if $i=n$, then $q'=1$. By (3.6.5), (3.6.6), this yields

$$\varrho(b_l B_l, dv_i)_{V,v} = \mu(b_l B_l, dv_i)_{V,v} = \begin{cases} 0 & \text{if } i<n, \\ b_l-1 & \text{if } i=n. \end{cases} \tag{3.6.16}$$

According to (3.6.10), (3.6.6) we have

$$\begin{aligned} \varrho(b_l B_l, \tfrac{\partial}{\partial v_i})_{V,\tilde v} &= \min_j \mu(b_l B_l, h_{ij} \tfrac{\partial}{\partial \tilde v_j})_{V,\tilde v}, \\ \varrho(b_l B_l, dv_i)_{V,\tilde v} &= \min_j \mu(b_l B_l, g_{ij} d\tilde v_j)_{V,\tilde v}. \end{aligned} \tag{3.6.17}$$

From (3.6.7) and (3.6.15), (3.6.16) (where v_i, v, i are replaced by $\tilde v_j$, $\tilde v$, j resp.) we obtain

$$\begin{aligned} \mu(b_l B_l, h_{ij} \tfrac{\partial}{\partial \tilde v_j})_{V,\tilde v} &= \begin{cases} b_l m_{B_l,\mathrm{div}(h_{ij})} & \text{if } j<n, \\ b_l m_{B_l,\mathrm{div}(h_{ij})} + 1 - b_l & \text{if } j=n, \end{cases} \\ \mu(b_l B_l, g_{ij} d\tilde v_j)_{V,\tilde v} &= \begin{cases} b_l m_{B_l,\mathrm{div}(g_{ij})} & \text{if } j<n, \\ b_l m_{B_l,\mathrm{div}(g_{ij})} + b_l - 1 & \text{if } j=n. \end{cases} \end{aligned} \tag{3.6.18}$$

By (3.6.10), (3.6.11) we have $m_{B_l,\mathrm{div}(h_{ij})} \geqslant 0$, $m_{B_l,\mathrm{div}(g_{ij})} \geqslant 0$ for all i,j, and $m_{B_l,\mathrm{div}(h_{in})} \geqslant 1$, $m_{B_l,\mathrm{div}(g_{ni})} \geqslant 1$ for all $i<n$. Hence (3.6.18) implies

$$\begin{aligned} \mu(b_l B_l, h_{ij} \tfrac{\partial}{\partial \tilde v_j})_{V,\tilde v} &\geqslant \begin{cases} 0 & \text{if } j<n, \\ 1 & \text{if } j=n, i<n, \\ 1-b_l & \text{if } j=n, i=n, \end{cases} \\ \mu(b_l B_l, g_{ij} d\tilde v_j)_{V,\tilde v} &\geqslant \begin{cases} 0 & \text{if } j<n, i<n, \\ b_l & \text{if } j<n, i=n, \\ b_l-1 & \text{if } j=n. \end{cases} \end{aligned} \tag{3.6.19}$$

Since $b_l \geqslant 1$, it follows from (3.6.17), (3.6.19), (3.6.15), (3.6.16) that

$$\varrho(b_l B_l, \tfrac{\partial}{\partial v_i})_{V,\tilde v} \geqslant \left\{\begin{array}{ll} 0 & \text{if } i < n, \\ 1 - b_l & \text{if } i = n \end{array}\right\} = \varrho(b_l B_l, \tfrac{\partial}{\partial v_i})_{V,v},$$
$$\varrho(b_l B_l, dv_i)_{V,\tilde v} \geqslant \left\{\begin{array}{ll} 0 & \text{if } i < n, \\ b_l - 1 & \text{if } i = n \end{array}\right\} = \varrho(b_l B_l, dv_i)_{V,v},$$

whence (3.6.9) for $p + q = 1$.

Now assume that $p+q > 1$. Then we can write $\tau = \tau' \otimes \tau''$ where τ' and τ'' are the monomials in v that have valencies $< p + q$. By Lemma 3.6.1,

$$\begin{aligned} \varrho(b_l B_l, \tau)_{V,v} &= \varrho(b_l B_l, \tau')_{V,v} + \varrho(b_l B_l, \tau'')_{V,v}, \\ \varrho(b_l B_l, \tau)_{V,\tilde v} &= \varrho(b_l B_l, \tau')_{V,\tilde v} + \varrho(b_l B_l, \tau'')_{V,\tilde v}, \end{aligned} \tag{3.6.20}$$

and by the inductive hypothesis,

$$\begin{aligned} \varrho(b_l B_l, \tau')_{V,v} &\leqslant \varrho(b_l B_l, \tau')_{V,\tilde v}, \\ \varrho(b_l B_l, \tau'')_{V,v} &\leqslant \varrho(b_l B_l, \tau'')_{V,\tilde v}. \end{aligned} \tag{3.6.21}$$

Clearly (3.6.9) follows from (3.6.20), (3.6.21). □

Given Lemma 3.6.2, we denote $\varrho(b_l B_l, \tau)_{V,v}$ by $\varrho(b_l B_l, \tau)$.

Now we define the divisor $\operatorname{div}_B(\tau)$ for a positive divisor B given by (3.6.2) as follows. Let $C \in \operatorname{Div}(Y)$ be a prime divisor, then

$$m_{C,\operatorname{div}_B(\tau)} := \begin{cases} \varrho(b_l B_l, \tau) & \text{if } C = B_l \text{ for some } l, \\ m_{C,\operatorname{div}(\tau)} & \text{if } C \neq B_l \text{ for all } l. \end{cases} \tag{3.6.22}$$

Note that (3.6.22), (3.2.1), (3.2.2) and Lemma 3.6.1 imply

Corollary 3.6.1. *For any positive divisor B and nonzero tensor fields $\tau \in \Gamma_{\mathrm{rat}}(T^p_q(Y)_{\mathrm{sm}})$, $\varsigma \in \Gamma_{\mathrm{rat}}(T^s_t(Y)_{\mathrm{sm}})$ the following equality holds:*

$$\operatorname{div}_B(\tau \otimes \varsigma) = \operatorname{div}_B(\tau) + \operatorname{div}_B(\varsigma).$$

3.7. Behavior at étale morphisms. Let $\alpha : Z \to Y$ be a surjective étale morphism of irreducible smooth algebraic varieties. Let S be a subvariety of codimension 1 in Y, and let f be a local equation of S at a point $y \in S_{\mathrm{sm}}$. Then $df(y) \neq 0$, and as α is étale, $d(\alpha^*(f))(z) \neq 0$ for any point $z \in \alpha^{-1}(y)$. This implies that $\alpha^*(f)$ is a local equation at z of the subvariety $\alpha^{-1}(S)$ of Z. Hence if $C \in \mathrm{Div}(Y)$ and $C' \in \mathrm{Div}(Z)$ are prime divisors, then $m_{C',\alpha^*(C)} = 0$ or 1, depending on whether $\mathrm{supp}(C')$ is an irreducible component of $\alpha^{-1}(\mathrm{supp}(C))$ or not.

Lemma 3.7.1. *Let τ be a nonzero rational tensor field on Y and let $B \in \mathrm{Div}(Y)$ be a positive divisor. If $C \in \mathrm{Div}(Y)$ and $C' \in \mathrm{Div}(Z)$ are prime divisors such that $m_{C',\alpha^*(C)} = 1$, then*

$$m_{C,\mathrm{div}_B(\tau)} = m_{C',\mathrm{div}_{\alpha^*(B)}(\alpha^*(\tau))}. \tag{3.7.1}$$

Proof. We use the notation of Subsections 3.1, 3.2 and 3.3, so that (3.1.2) and (3.3.5) hold, and t is a local equation of C. Then $\alpha^*(t)$ is a local equation of C' in an appropriate neighborhood of a point of $\mathrm{supp}(C')$. Hence

$$m_{C,\mathrm{div}(\tau^{i_1\cdots i_p}_{j_1\cdots j_q})} = m_{C',\mathrm{div}(\alpha^*(\tau^{i_1\cdots i_p}_{j_1\cdots j_q}))} \tag{3.7.2}$$

for all $i_1,\ldots,i_p,j_1,\ldots,j_q$. Then from (3.1.2), (3.3.4), (3.3.5), (3.2.1), (3.2.2) and (3.7.2) we deduce that

$$m_{C,\mathrm{div}(\tau)} = m_{C',\mathrm{div}(\alpha^*(\tau))}. \tag{3.7.3}$$

From (3.7.3) and (3.6.22) it follows that (3.7.1) holds if $C \neq B_i$ for each B_i in decomposition (3.6.2).

Now assume that $C = B_l$ for some l. Then $b_l = m_{C,B} = m_{C',\alpha^*(B)}$. Use the notation of Subsection 3.6. By (3.3.3), the local equations of C and C' in the appropriate neighborhoods are respectively v_n and u_n. Therefore from (3.1.2), (3.3.5), (3.6.5) and (3.6.6) we deduce the equality $\varrho(b_lC,\tau) = \varrho(b_lC',\alpha^*(\tau))$. By (3.6.22), this yields (3.7.1). □

Corollary 3.7.1. *The following properties are equivalent:*

(i) $\mathrm{div}_B(\tau) \geqslant 0$,
(ii) $\mathrm{div}_{\alpha^*(B)}(\alpha^*(\tau)) \geqslant 0$.

3.8. Regularity of the pull back of a tensor field on a quotient variety: General criterion. Let X be an irreducible smooth algebraic variety endowed with a faithful action of a finite group G. Assume that there exists a geometric quotient (2.4.1). Recall that we defined by (2.6.2) the reflection divisor $R_{X/G}$ of X/G,

$$R_{X/G} := r_1R_1 + \ldots + r_dR_d \in \mathrm{Div}(X/G).$$

The following theorem is the main result of Section 3.

Theorem 3.8.1. *Let X be an irreducible smooth algebraic variety endowed with an action of a finite group G. Assume that there exists a geometric quotient* (2.4.1). *Let τ be a nonzero rational tensor field on X/G. Then the following properties are equivalent:*

(i) $\mathrm{div}(\pi_X^*(\tau)) \geqslant 0$, *i.e., the pull back $\pi_X^*(\tau)$ of τ is regular on X;*

(ii) $\mathrm{div}_{R_{X/G}}(\tau) \geqslant 0$.

Proof. We can (and shall) assume that the action is faithful. By the definitions of $\mathrm{div}(\pi_X^*(\tau))$ and $\mathrm{div}_{R_{X/G}}(\tau)$, either of conditions (i) and (ii) implies that τ is regular on $(X/G)_0$ and $\pi_X^*(\tau)$ is regular on X_0. So we can (and shall) assume that τ and $\pi_X^*(\tau)$ share these properties.

If $R_{X/G} = 0$, the claim follows from (2.6.3) and Corollary 3.5.1. So we shall assume that $R_{X/G} > 0$, i.e., $(X/G)_1 \neq \varnothing$. As $\mathrm{dom}(\pi_X^*(\tau)) \supseteq X_0$, regularity of $\pi_X^*(\tau)$ on X is equivalent to regularity of $\pi_X^*(\tau)$ on X_1 (see (3.1.5)–(3.1.6)).

Let x be a point of X_1 and $z := \pi_X(x) \in (X/G)_1$. By (2.6.4), there is a unique irreducible component $(X/G)_1^l$ of $(X/G)_1$ (see (2.6.1)) containing z. As $\pi_X|_W : \pi_X^{-1}(W) \to W$ for any open subset W of X/G is the geometrical quotient for the action of G on $\pi_X^{-1}(W)$, cf. [PV, 4.2], the problem is local, i.e., we have to prove that $\pi_X^*(\tau)$ is regular at x if and only if $\varrho(r_lR_l, \tau) \geqslant 0$, see (3.6.22), (2.6.2). To that end we apply Lemma 2.5.1.

Let U be the G_x-stable neighborhood of x from the proof of this lemma such that the commutative diagram (2.5.2) holds. Then we have

$$\begin{gathered}\pi_X^*(\tau)|_U = ((\sigma_S/G) \circ \pi_{U,G_x})^*(\tau) = \pi_{U,G_x}^*(\theta),\\ \text{where } \theta = (\sigma_S/G)^*(\tau) = (\pi_{U,G_x})_*(\pi_X^*(\tau)).\end{gathered} \tag{3.8.1}$$

Let $D \in \mathrm{Div}(U/G_x)$ be the prime divisor such that $\mathrm{supp}(D)$ is the irreducible component of $(\sigma_S/G)^{-1}((X/G)_1^l)$ containing $y = \pi_{X,G_x}(x)$. Since σ_S/G is étale, we have $m_{D,(\sigma_S/G)^*(R_{X/G})} = r_l$, and by Lemma 3.7.1, our problem is reduced to proving that the following properties are equivalent:

(a) $\pi_X^*(\tau)|_U$ is regular at x,
(b) $\varrho(r_l D, \theta) \geqslant 0$.

Let $u_1, \dots, u_n$ and $v_1, \dots v_n$ be the local parameters from Lemma 2.5.1. Replacing, if necessary, U by a smaller neighborhood U' of x, and then U' by $\cap_{g\in G_x} g\cdot U'$, we can (and shall) assume that

- $U = \pi_{X,G_x}^{-1}(V)$ where V is a neighborhood of y in X/G_x,
- $X_1 \cap U$ is irreducible,
- $u_1, \dots, u_n \in k[U]$,
- $v_1, \dots, v_n \in k[V]$,
- the ideal of $X_1 \cap U$ in $k[U]$ is generated by u_n,
- the ideal of $\pi_{U,G_x}(X_1)$ in $k[V]$ is generated by v_n,
- $du_1(a), \dots, du_n(a)$, is a basis of $T^*_{a,U}$ for each $a \in U$,
- $dv_1(b), \dots, dv_n(b)$ is a basis of $T^*_{b,V}$ for each $b \in V$.

From (2.5.1) and (3.1.1) we deduce that

$$\begin{aligned} \pi_{X,G_x}^*(dv_i) &= du_i \text{ for } 1 \leqslant i \leqslant n-1, \text{ and} \\ \pi_{X,G_x}^*(dv_n) &= r_l u_n^{r_l - 1} du_n, \end{aligned} \tag{3.8.2}$$

$$\begin{aligned} \pi_{X,G_x}^*\left(\frac{\partial}{\partial v_i}\right) &= \frac{\partial}{\partial u_i} \text{ if } 1 \leqslant i \leqslant n-1, \text{ and} \\ \pi_{X,G_x}^*\left(\frac{\partial}{\partial v_n}\right) &= \frac{1}{r_l u_n^{r_l-1}} \frac{\partial}{\partial u_n}. \end{aligned} \tag{3.8.3}$$

By Lemma 2.5.1 and $du_i(\frac{\partial}{\partial u_j}) = \delta_{ij}$, for each $g \in G_x$ we have

$$\begin{aligned} g\cdot du_i &= du_i \text{ if } 1 \leqslant i \leqslant n-1, \text{ and } g\cdot du_n = \chi(g) du_n, \\ g\cdot \frac{\partial}{\partial u_i} &= \frac{\partial}{\partial u_i} \text{ if } 1 \leqslant i \leqslant n-1, \text{ and } g\cdot \frac{\partial}{\partial u_n} = \chi(g)^{-1} \frac{\partial}{\partial u_n}. \end{aligned} \tag{3.8.4}$$

In view of the properties of $u_1, \dots, u_n$, there is a unique decomposition

$$\begin{aligned} \pi_X^*(\tau) = \textstyle\sum_{i_1,\dots,i_p,j_1,\dots,j_q} \psi_{j_1\dots j_q}^{i_1\dots i_p} \frac{\partial}{\partial u_{i_1}} \otimes \dots \otimes \frac{\partial}{\partial u_{i_p}} \\ \otimes du_{j_1} \otimes \dots \otimes du_{j_q}, \end{aligned} \tag{3.8.5}$$

where $\psi^{i_1\dots i_p}_{j_1\dots j_q} \in k(U)$.

Since $\pi_X^*(\tau)$ is G_x-invariant, (3.8.4) and the uniqueness of decomposition (3.8.5) imply that each summand in the right-hand side of (3.8.2) is G_x-invariant.

Consider such a summand

$$(3.8.6) \qquad (\pi_X^*(\tau))^{i_1\dots i_p}_{j_1\dots j_q} := \psi^{i_1\dots i_p}_{j_1\dots j_q}\frac{\partial}{\partial u_{i_1}} \otimes\cdots\otimes \frac{\partial}{\partial u_{i_p}} \otimes du_{j_1} \otimes\cdots\otimes du_{j_q},$$

and assume that in (3.8.6) there are exactly p' factors $\frac{\partial}{\partial u_n}$ and exactly q' factors du_n. Divide $p'-q'$ by $r_l := |G_x|$ (see (2.6.2)) with reminder:

$$(3.8.7) \qquad p'-q' = sr_l + t,\ \ s,t \in \mathbb{Z},\ 0 \leqslant t < r_l.$$

Since the order of χ is $|G_x|$, and the field (3.8.6) is G_x-invariant, (3.8.4) and (3.8.7) yield

$$(3.8.8) \qquad g\cdot\psi^{i_1\dots i_p}_{j_1\dots j_q} = \chi(g)^t\psi^{i_1\dots i_p}_{j_1\dots j_q} \quad \text{for all } g \in G_x.$$

(a)⇒(b): Assume that $\pi_X^*(\tau)|_U$ is regular at x. Shrinking U if necessary we can (and shall) assume that each $\psi^{i_1\dots i_p}_{j_1\dots j_q}$ is contained in $k[U]$. We claim that

$$(3.8.9) \qquad \psi^{i_1\dots i_p}_{j_1\dots j_q} = u_n^t\tilde\psi^{i_1\dots i_p}_{j_1\dots j_q} \quad \text{for some } \ \tilde\psi^{i_1\dots i_p}_{j_1\dots j_q} \in k[U]^{G_x}.$$

Indeed, if $t=0$, then $\psi^{i_1\dots i_p}_{j_1\dots j_q} \in k[U]^{G_x}$ by (3.8.8), so (3.8.9) holds. Let $t>0$. As $t<|G_x|$, there is $g_0 \in G_x$ such that $\chi(g_0)^t \neq 1$. Take a point $a \in X_1 \cap U$. Since $a \in X^{G_x}$, from (3.8.9) we obtain that $g\cdot\psi^{i_1\dots i_p}_{j_1\dots j_q}(a) = \psi^{i_1\dots i_p}_{j_1\dots j_q}(g^{-1}\cdot a) = \psi^{i_1\dots i_p}_{j_1\dots j_q}(a) = \chi(g_0)^t\psi^{i_1\dots i_p}_{j_1\dots j_q}(a)$, whence $\psi^{i_1\dots i_p}_{j_1\dots j_q}(a) = 0$. Thus $\psi^{i_1\dots i_p}_{j_1\dots j_q}$ vanishes on $X_1 \cap U$. Hence $\psi^{i_1\dots i_p}_{j_1\dots j_q} = u_n\tilde\psi^{i_1\dots i_p}_{j_1\dots j_q}$ for some $\tilde\psi^{i_1\dots i_p}_{j_1\dots j_q} \in k[U]$. As $k[U]$ is an integral domain, from this, (3.8.9) and Lemma 2.5.1 we deduce that $g\cdot\tilde\psi^{i_1\dots i_p}_{j_1\dots j_q} = \chi(g)^{t-1}\tilde\psi^{i_1\dots i_p}_{j_1\dots j_q}$ for all $g \in G_x$. Now the same arguments can be applied to $\tilde\psi^{i_1\dots i_p}_{j_1\dots j_q}$, and so on. Whence the claim.

Now, as $\tilde\psi^{i_1\dots i_p}_{j_1\dots j_q} \in k[U]^{G_x}$, we have $\tilde\psi^{i_1\dots i_p}_{j_1\dots j_q} = \pi^*_{X,G_x}(\varepsilon^{i_1\dots i_p}_{j_1\dots j_q})$ for some

$$(3.8.10) \qquad \varepsilon^{i_1\dots i_p}_{j_1\dots j_q} \in k[V].$$

Hence using (3.8.2), (3.8.3) and (3.8.9), we can rewrite the field (3.8.6) as follows

$$(\pi_X^*(\tau))_{j_1\dots j_q}^{i_1\dots i_p} = \pi_{X,G_x}^*(r_l^{p'-q'} v_n^{p'-q'-s} \varepsilon_{j_1\dots j_q}^{i_1\dots i_p} \frac{\partial}{\partial v_{i_1}} \otimes \dots \otimes \frac{\partial}{\partial v_{i_p}} \otimes dv_{j_1} \otimes \dots \otimes dv_{j_q}).$$

This equality and (3.8.5), (3.8.6), (3.8.1) imply that

$$\begin{aligned} \theta &= \textstyle\sum_{i_1,\dots,i_p,j_1,\dots j_q} (\pi_{U,G_x})_*(\pi_X^*(\tau))_{j_1\dots j_q}^{i_1\dots i_p} \\ &= \textstyle\sum_{i_1,\dots,i_p,j_1,\dots j_q} \theta_{j_1\dots j_q}^{i_1\dots i_p} \frac{\partial}{\partial v_{i_1}} \otimes \dots \otimes \frac{\partial}{\partial v_{i_p}} \otimes dv_{j_1} \otimes \dots \otimes dv_{j_q}, \end{aligned} \tag{3.8.11}$$

where

$$\theta_{j_1\dots j_q}^{i_1\dots i_p} = r_l^{p'-q'} v_n^{p'-q'-s} \varepsilon_{j_1\dots j_q}^{i_1\dots i_p}. \tag{3.8.12}$$

Since v_n is the local equation of D at y, it follows from (3.8.12) and (3.8.10) that

$$m_{D,\operatorname{div}(\theta_{j_1\dots j_q}^{i_1\dots i_p})} - p' + q' + s \geqslant 0. \tag{3.8.13}$$

Taking into account the identity

$$\begin{aligned} &(r_l - 1)(q' - p') + r_l m_{D,\operatorname{div}(\theta_{j_1\dots j_q}^{i_1\dots i_p})} \\ &\qquad = r_l\big(m_{D,\operatorname{div}(\theta_{j_1\dots j_q}^{i_1\dots i_p})} - p' + q' + s\big) + t, \end{aligned} \tag{3.8.14}$$

we deduce from (3.8.13), (3.8.7) and $r_l > 0$ that

$$(r_l - 1)(q' - p') + r_l m_{D,\operatorname{div}(\theta_{j_1\dots j_q}^{i_1\dots i_p})} \geqslant 0. \tag{3.8.15}$$

Now (3.8.11), (3.8.14) and definitions (3.6.5), (3.6.6) immediately imply that $\varrho(r_l D, \theta) \geqslant 0$.

(b)⇒(a): Assume that $\varrho(r_l D, \theta) \geqslant 0$. In view of the properties of $v_1, \dots, v_n$, there is a unique decomposition

$$\theta = \textstyle\sum_{i_1,\dots,i_p,j_1,\dots,j_q} \vartheta_{j_1\dots j_q}^{i_1\dots i_p}, \quad \text{where} \tag{3.8.16}$$

$$(3.8.17)\quad \vartheta^{i_1\dots i_p}_{j_1\dots j_q} = \theta^{i_1\dots i_p}_{j_1\dots j_q} \frac{\partial}{\partial v_{i_1}} \otimes \cdots \otimes \frac{\partial}{\partial v_{i_p}} \otimes dv_{j_1} \otimes \cdots \otimes dv_{j_q}$$

for some $\theta^{i_1\dots i_p}_{j_1\dots j_q} \in k(V)$.

Assume that in (3.8.17) there are exactly p' factors $\frac{\partial}{\partial v_n}$ and exactly q' factors dv_n. Let s and t be defined by (3.8.7). As v_n is a local equation of D at y, the definition of $\varrho(r_l D, \theta)$ (see (3.6.5), (3.6.6)) and the inequality $\varrho(r_l D, \theta) \geqslant 0$ imply that (3.8.15) holds. From here, (3.8.14), (3.8.7) and $r_l > 0$ we deduce that (3.8.13) holds. In turn this implies that shrinking V if necessary, we can (and shall) assume that

$$(3.8.18)\qquad \theta^{i_1\dots i_p}_{j_1\dots j_q} = v_n^{p'-q'-s} \tilde{\theta}^{i_1\dots i_p}_{j_1\dots j_q}, \quad \text{where } \tilde{\lambda}^{i_1\dots i_p}_{j_1\dots j_q} \in k[V].$$

Then (3.8.17), (3.8.18), (3.8.2), (3.8.3), (2.5.1) yield

$$\pi^*_{U,G_x}(\vartheta^{i_1\dots i_p}_{j_1\dots j_q}) = r_l^{q'-p'} u_n^t \pi^*_{U,G_x}(\tilde{\theta}^{i_1\dots i_p}_{j_1\dots j_q}) \frac{\partial}{\partial u_{i_1}} \otimes \cdots \otimes \frac{\partial}{\partial u_{i_p}} \otimes du_{j_1} \otimes \cdots \otimes du_{j_q}.$$

As $t \geqslant 0$ and $\tilde{\theta}^{i_1\dots i_p}_{j_1\dots j_q} \in k[V]$, this shows that $\pi^*_{U,G_x}(\vartheta^{i_1\dots i_p}_{j_1\dots j_q})$ is regular on U. Hence by (3.8.16) and (3.8.1), the field $\pi^*_X(\tau)|_U$ is regular on U. □

4. Some Applications

4.1. Generalization of Solomon's theorem. Let V be a finite dimensional vector space over k and G a finite subgroup of $\mathrm{GL}(V)$. It is well known that the following properties are equivalent, e.g., see [PV]:

(i) V/G is smooth.
(ii) V/G is isomorphic to affine space.
(iii) G is generated by pseudo-reflections.

Assume that conditions (i)–(iii) hold. Then Solomon's theorem can be reformulated as the statement that the algebra of all invariant regular differential forms on V is the pull back of the algebra of all regular differential forms on V/G.

In [M] this statement was generalized to proper smooth polar actions of Lie groups on Riemannian manifolds, and in [B] to algebraic actions of reductive groups on smooth affine algebraic varieties with smooth categorical quotients. From Theorem 3.8.1 we deduce the following

Corollary 4.1.1. *Let X be an irreducible smooth algebraic variety endowed with an action of a finite group G. Assume that there exists a geometric quotient* (2.4.1). *Let τ be a rational differential form on X/G. Then $\pi_X^*(\tau)$ is regular on X if and only if τ is regular on $(X/G)_{\mathrm{sm}}$.*

Proof. By Theorem 3.8.1, we have to show that $\operatorname{dom}(\tau) = (X/G)_{\mathrm{sm}}$ is equivalent to $\operatorname{div}_{R_{X/G}}(\tau) \geqslant 0$. As either of these conditions implies $\operatorname{dom}(\tau) \supseteq (X/G)_0$, we can (and shall) assume that this inclusion holds. As $(X/G)_1 \subset (X/G)_{\mathrm{sm}}$ by Lemma 2.5.1, it follows from (3.1.5), (3.1.6) that $\operatorname{dom}(\tau) = (X/G)_{\mathrm{sm}}$ is equivalent to $\operatorname{dom}(\tau) \supseteq (X/G)_1$.

We can (and shall) assume that the action is faithful. We use the notation of Subsection 3.6 for $Y = X$ and $B = R_{X/G}$. The problem is local, i.e., we have to show that $\operatorname{dom}(\tau) \supseteq (X/G)_1^l$ (see (2.6.1)) iff $\varrho(r_l R_l, \tau) \geqslant 0$. For each field (3.6.4) we have $p' = p = 0$ and $q' \leqslant 1$ by the skew symmetry condition. Whence $p' - q' = 0$ or -1. This easily implies that $(r_l - 1)(q' - p') + r_l m \geqslant 0$ is equivalent to $m \geqslant 0$, and we are done. □

4.2. Rational tensor fields of type $\binom{\otimes p}{\wedge q}$. Consider a more general type of rational tensor fields than differential forms. Namely, let Y be an irreducible algebraic variety. Consider the vector bundle $T(Y_{\mathrm{sm}})^{\otimes p} \otimes \wedge^q T^*(Y_{\mathrm{sm}})$ over Y_{sm}. Its rational sections are precisely rational tensor fields of type $\binom{p}{q}$ on Y that are skew symmetric with respect to the covariant entries. We call them *rational tensor fields of type $\binom{\otimes p}{\wedge q}$*. It is clear that pull backs and push forwards of rational tensor fields of type $\binom{\otimes p}{\wedge q}$ are rational tensor field of type $\binom{\otimes p}{\wedge q}$ as well.

Corollary 4.2.1. *Let X be an irreducible smooth algebraic variety endowed with an action of a finite group G. Assume that there exists a geometric quotient* (2.4.1). *Let φ be a G-invariant regular tensor field of type $\binom{\otimes p}{\wedge q}$ on X. Then $\pi_{X*}(\varphi)$ is regular on $(X/G)_{\mathrm{sm}}$.*

Proof. We have $\varphi = \pi_X^*(\tau)$ for $\tau := \pi_{X*}(\varphi)$. Arguing like in the proof of Corollary 4.1.1 we reduce the problem to showing that $\varrho(r_l R_l, \tau) \geqslant 0$ implies $\operatorname{dom}(\tau) \supseteq (X/G)_1^l$. For each field (3.6.4) we have $q' \leqslant 1$ by the skew symmetry condition. Hence $p' - q' \geqslant -1$. Therefore $(r_l - 1)(q' - p') + r_l m \leqslant r_l(m + 1) - 1$. It is easily seen that nonnegativity of the left hand side of the latter inequality implies $m \geqslant 0$, so we are done. □

4.3. Partially symmetric tensor fields. Let Y be an irreducible algebraic variety. Rational sections of $\mathrm{S}^{q_1}(T^*(Y_{\mathrm{sm}})) \otimes \cdots \otimes \mathrm{S}^{q_d}(T^*(Y_{\mathrm{sm}}))$ are called *rational multi-symmetric covariant tensor fields of type* $(q_1, \ldots, q_d)$ on Y. If Y is smooth and σ is such a field, then

$$T(\mathrm{dom}(\sigma))^{\oplus d} \to k, \ v \mapsto (\sigma(a))(v) \text{ for each}$$
$$a \in \mathrm{dom}(\sigma), v \in T_{a,Y}^{\oplus d},$$

is a rational function on the algebraic variety $T(Y)^{\oplus d}$, homogeneous of multi-degree $(q_1, \ldots, q_d)$ with respect to the natural diagonal action of $k^\times$ on $T(Y)^{\oplus d}$. This function is regular on $T(Y)^{\oplus d}$ iff σ is regular on Y. Every such rational function is obtained in this way. Therefore if Y is smooth and endowed with an action of a finite group G, Theorem 3.8.1 provides a characterization of G-invariant regular functions on $T(Y)^{\oplus d}$ in terms of rational multi-symmetric covariant tensor fields on Y/G.

If $Y = V$ is a vector space and G a subgroup of $\mathrm{GL}(V)$, we have $T(V)^{\oplus d} = V^{\oplus(d+1)}$ and the action of G on $T(V)^{\oplus d}$ coincides with the diagonal action on $V^{\oplus(d+1)}$. Thus Theorem 3.8.1 yields a characterization of G-invariant polynomials on $V^{\oplus(d+1)}$ in terms of rational multi-symmetric covariant tensor fields on V/G. Note that algebras $k[V \oplus \ldots \oplus V]^G$ were studied in [Hu] for some groups G generated by pseudo-reflections (i.e., when V/G is smooth).

5. Lifting Automorphisms of Quotients

5.1. Let X be an irreducible algebraic variety endowed with an action of a finite group G such that there exists the geometric quotients (2.4.1). If α is an automorphism of X/G induced by a G-automorphism of X, then $\alpha((X/G)_0) = (X/G)_0$ and $\alpha_*(R_{X/G}) = R_{X/G}$.

The following theorem is a conversion of this statement for linear actions. It can be seen as complementing the results in [B], [Sch], [Lo], and [KLM].

Theorem 5.1.1. *Let V be a finite dimensional complex vector space and let $G \subset \mathrm{GL}(V)$ be a finite group. Let ψ be an automorphism of the algebraic variety V/G such that $\psi(V/G)_0 \subseteq (V/G)_0$ and $\psi_*(R_{V/G}) = R_{V/G}$. Then there is an automorphism φ of the alge-*

braic variety V such that the following diagram is commutative

$$(5.1.1)\qquad \begin{array}{ccc} V & \xrightarrow{\ \varphi\ } & V \\ {\scriptstyle \pi_{V,G}}\big\downarrow & & \big\downarrow{\scriptstyle \pi_{V,G}} \\ V/G & \xrightarrow{\ \psi\ } & V/G \end{array} \quad .$$

Proof. It is proved in [KLM] that there is an analytic automorphism φ of the analytic space V such that the diagram (5.1.1) is commutative. Hence the claim immediately follows from

Lemma 5.1.1. *Let X, Y and Z be complex algebraic varieties. Assume that Y is irreducible and smooth and X is endowed with an action of a finite group H. Let the maps α, β, γ in the commutative diagram*

$$\begin{array}{ccccc} & & X & & \\ & {\scriptstyle \gamma}\nearrow & & \searrow{\scriptstyle \alpha} & \\ Y & & \xrightarrow{\quad \beta \quad} & & Z \end{array}$$

have the properties:

(i) *α and β are morphisms,*
(ii) *every nonempty fiber of α is an H-orbit,*
(iii) *γ is analytic.*

Then γ is a morphism of algebraic varieties.

Proof. Consider in $X \times Y$ the subsets

$$\Delta = \{(x,y) \in X \times Y \mid \alpha(x) = \beta(y)\},$$
$$\Gamma = \{(\gamma(y), y) \in X \times Y \mid y \in Y\}.$$

Clearly Δ is Zariski closed. As Γ is the graph of γ and Y is smooth, (iii) implies that Γ is an analytic subset of $X \times Y$, cf. [He, Ch. IV, no. 5]; clearly it is analytically isomorphic to Y.

Consider the action of H on $X \times Y$ through the first factor. By (ii), we have

$$(5.2.1)\qquad \Delta = \textstyle\bigcup_{h \in H} h(\Gamma).$$

Taking into account that the decompositions of any complex algebraic variety into irreducible components in the categories of algebraic varieties and analytic sets coincide, cf. [GR, Ch. V, B, Proposition 1], we deduce from irreducibility of Y that Γ is an irreducible analytic set. Hence (5.2.1) yields that each $h(\Gamma)$ is an irreducible component of the analytic set Δ. Thereby $h(\Gamma)$ is an irreducible component of the algebraic variety Δ. Thus Γ is a Zariski closed subset of $X \times Y$.

Let $X \xleftarrow{\pi_1} X \times Y \xrightarrow{\pi_2} Y$ be the projections. As $\pi_2|_\Gamma : \Gamma \to Y$ is a bijective morphism of algebraic varieties and Y is normal, Zariski's Main Theorem implies that $\pi_2|_\Gamma$ is an isomorphism of algebraic varieties. Therefore $(\pi_2|_\Gamma)^{-1}$ is a morphism (actually, an isomorphism) of algebraic varieties. Whence $\gamma = \pi_1 \circ (\pi_2|_\Gamma)^{-1}$ is a morphism as well. □

5.2. Remarks. By Lefschetz's principle, see, e.g., [Si, p. 164], Theorem 5.1.1 remains true over any algebraically closed field of characteristic 0. Notice that the proof of the result from [KLM] used in the proof of Theorem 5.1.1 involves lifting of a flat torsion free connection (that could be carried over to the algebraic setting) and the geodesic exponential mapping that in turn involves solving ordinary differential equations.

For general actions and $k = \mathbb{C}$ there are further obstructions for lifting of automorphisms of X/G that are related to the fundamental group of X/G.

6. Appendix

Here we give some details of the proof of Theorem 2.2.1 for finite G.

This proof is based on the following preliminary results all proved in [Lu].

(a) Let X be a normal affine variety with a faithful action of G. Let H be a subgroup of G and let x be a point of X. Then the natural map $\alpha : X/H \to X/G$ is étale at $\pi_{X,H}(x)$ if and only if $H \supseteq G_x$.

(b) Let in addition $\varphi : X \to Y$ be a finite equivariant morphism to another normal affine G–variety. If φ is étale at x, and if $G_{\varphi(x)} = G_x$, then the map $\varphi/G : X/G \to Y/G$ is étale at $\pi_{X,G}(x)$.

(c) If $x \in X_{\rm sm}$, then there exists a G_x-equivariant morphism $\varphi : X \to T_{x,X}$ such that φ is étale at x and $\varphi(x) = 0$.

Statement (a) is [Lu, Lemme 2], proved by references to results of commutative algebra; (b) is [Lu, Lemme 1], proved on p. 92, Case (B); and (c) is [Lu, Lemme, p. 96] (one can assume that X is a G-stable subvariety of a G-module k^n and $x = 0$, cf. [PV, Theorem 1.5];

if $\pi : k^n \to T_{x,X}$ is the projection parallel to a G_x-stable complement to $T_{x,X}$, then clearly one can take $\varphi = \pi|_X$.)

Note that (a) implies that $X \to X/G$ is étale whenever G acts freely.

To prove Theorem 2.2.1 (i), note that $X/G_x \to X/G$ is étale at $\pi_{X,G_x}(x)$, by (a). So there exists a G_x-invariant open subset $S \subseteq X$ such that $S/G_x \to X/G$ is étale. Then its base change $X \times_{X/G} S/G_x \to X$ is étale as well. But the composition of $\psi : G *_{G_x} S \to X \times_{X/G} S/G_x$, $G_x(g, s) \mapsto (g s, G_x s)$, with this base change is just the natural map $G *_{G_x} S \to X$, which is obviously étale (since it factors as the open immersion $G *_{G_x} S \to G *_{G_x} X$, followed by the projection $G *_{G_x} X = G/G_x \times X \to X$). Thus, ψ is étale as well. On the other hand, the (set-theoretic) fiber of ψ at $\tilde{x} = G_x \cdot (e, x) \in G *_{G_x} S$ is the unique point $\tilde{x}$. Thus, shrinking S, we may achieve that ψ is an isomorphism.

To prove Theorem 2.2.1 (ii), consider φ from (c). Then $\varphi/G_x : X/G_x \to T_{x,X}/G_x$ is étale at $\pi_{X,G_x}(x)$, by (b). By the preceding argument, the natural map $X \to X/G_x \times_{T_{x,X}/G_x} T_{x,X}$ is étale at x, and its fiber there is the unique point x. This yields a G_x-invariant open subset $U \subseteq X$ such that $U/G_x \to T_{x,X}/G_x$ is étale, and $U \to U/G_x \times_{T_{x,X}/G_x} T_{x,X}$ is an isomorphism.

References

[BR] P. Bardsley, R. W. Richardson, *Étale slices for algebraic transformation groups in characteristic p*, Proc. Lond. Math. Soc., I, Ser. **51** (1985), 295–317.

[B] M. Brion, *Differential forms on quotients by reductive group actions*, Proc. AMS **126** (1998), 2535–2539.

[GR] R. C. Gunning, H. Rossi, *Analytic Funstions of Several Complex Variables*, Prentice-Hall, Inc., Englewood Cliffs, N. J., 1965.

[He] M. Hervé, *Several Complex Variables*, Oxford University Press, Bombay, 1963.

[Hu] M. Hunziker, *Classical invariant theory for finite reflection groups*, Transformation Groups **2** (1997), no. 2, 147–163.

[KLM] A. Kriegl, M. Losik, P. W. Michor, *Tensor fields and connections on holomorphic orbit spaces of finite groups*, to appear in: J. Lie Theory **13** (2003).

[Lo] M. Losik, *Lifts of diffeomorphisms of orbit spaces for representations of compact Lie groups*, Geom. Dedicata **88** (2001), 21–36.

[Lu] D. Luna, *Slices étales*, Bull. Soc. Math. France, Memoire **33** (1973), 81–105.

[M] P. W. Michor, *Basic differential forms for actions of Lie groups*, Proc. AMS **124** (1996), 1633–1642; II, Proc. AMS **125** (1997), 2175–2177.

[PV] V. L. Popov, E. B. Vinberg, *Invariant Theory*, Algebraic Geometry, IV, Encycl. of Math. Sci., vol. 55, Springer-Verlag, Heidelberg, 1994, pp. 123–284.

[Sch] G. W. Schwarz, *Lifting smooth homotopies of orbit spaces*, Publ. Math. IHES **51** (1980), 37–136.

[Sh] I. R. Shafarevich, *Basic Algebraic Geometry*, Springer-Verlag, Berlin, Heidelberg, 1994.

[Si] J. H. Silverman, *The Arithmetic of Elliptic Curves*, Springer-Verlag, New York, Berlin, 1986.

[So] L. Solomon, *Invariants of finite reflection groups*, Nagoya Math. J. **22** (1963), 57–64.

Saratov State University,
ul. Astrakhanskaya 83, Saratov, 410026, Russia
E-mail address: LosikMV@info.sgu.ru

Institut für Mathematik, Universität Wien
Strudlhofgasse 4, A-1090 Wien, Austria; *and:*
Erwin Schrödinger Institut für Mathematische Physik
Boltzmanngasse 9, A-1090 Wien, Austria
E-mail address: Peter.Michor@esi.ac.at

Steklov Mathematical Institute,
Russian Academy of Sciences,
Gubkina 8, Moscow, 117966, Russia
E-mail address: popov@ppc.msk.ru

Generators of a general ideal

M. Pavaman Murthy

Dedicated to C.S. Seshadri for his seventieth birthday

Abstract. Let R be a commutative ring and $A = R[X_1, ..., X_n]$. Let I_m be the ideal generated by $X_1, ..., X_m$, $0 < m < n+1$, and r a positive integer. We show here that, for all $(m-1) < k < (r+1)$,$GL(k, A)$ acts transitively on the the set of all $(f_1, ..., f_k)$ which generate I_m, if and only if for all $k < (r+1)$, all unimodular elements in A^k are completable to $k \times k$-invertible matrices over A.

1

Let A be a commutative ring. Recall that $\underline{a} = (a_1, \ldots, a_n) \in A^n$ is *unimodular* if $\sum_{i=1}^n Aa_i = A$. We say that $\underline{a} \in A^n$ is completable if there exists an $\alpha \in GL_n(A)$ such that $\underline{a}\alpha = (1, 0, \ldots, 0)$. It is easy to see that all unimodular vectors in A^n are completable if and only if $P \oplus A \approx A^n \Longrightarrow P \approx A^{n-1}$.

Our main result in this note is the following:

Theorem 1.1 Let R be a commutative ring and $A = R[X_1, \ldots, X_m]$. Let r be a positive integer. Then the following two conditions are equivalent.

1) Every unimodular element in A^k is completable for all $1 \leq k \leq r$.
2) For any $n, 1 \leq n \leq m, n \leq k \leq r$, if $(f_1, \ldots, f_k) \in A^k$ is such that $f_1, \ldots, f_k$ generate $I = \sum_{i=1}^n AX_i$, then there exists an $\alpha \in GL_k(A)$ such that $(f_1, \ldots, f_k)\alpha = (X_1, \ldots, X_n, 0, \ldots, 0)$.

In other words, Theorem 1.1. says that the condition 1) is equivalent to the statement: $GL_k(A)$ acts transitively on the set $S = \{(f_1, \ldots, f_k) \in A^k; \sum_{i=1}^n AX_i = \sum_{i=1}^k Af_i\}$ for all $n \leq m; n \leq k \leq r$.

It is known that all unimodular elements in A^r are completable, for all $r \geq 1$, if R is a regular local ring containing a field (in view of results of Lindel [L] and D. Popescu ([P1], [P2]), see also [SW]), or when R is a noetherian ring of dimension ≤ 1 (in view of Bhatwadekar-Roy) [B-R]). Thus, we have

COROLLARY 1.2 Let R be a regular local ring containing a field or a noetherian ring of dimension ≤ 1. Let $A = R[X_1, \ldots, X_n]$. Let $f_1, \ldots, f_r$ generate $I = \sum_{i=1}^n AX_i$. Then there is an $\alpha \in GL_r(A)$ such that $(f_1, \ldots, f_r)\alpha = (X_1, \ldots, X_n, 0, \ldots, 0)$.

At the cost of repetition, we record the following special case.

COROLLARY 1.3 Let k be a field and $A = k[X_1, \ldots, X_n]$. Then $\sum_{i=1}^n Af_i = \sum_{i=1}^n AX_i$ if and only if $(f_1, \ldots, f_n)\alpha = (X_1, \ldots, X_n)$ for some $\alpha \in GL_n(A)$.

Our proof of Theorem 1.1. is obtained by modifying arguments in Sathaye's result [S, Lemma 10(B), p.221].

Proof of Theorem 1.1. 2) $\Longrightarrow$ 1): For $k \leq r$, let $(g_1, \ldots, g_k) \in A^k$ be unimodular, $A = R[X_1, \ldots, X_m]$. Then $g_1X_1, \ldots, g_kX_1$ generate $I = AX_1$. So by the case $n = 1$ of 2), there is an $\alpha \in GL_k(A)$ such that $(g_1X_1, \ldots, g_kX_1)\alpha = (X_1, 0, \ldots, 0)$. We immediately have $(g_1, \ldots, g_k)\alpha = (1, 0, \ldots, 0)$.

1) $\Longrightarrow$ 2): We prove this implication by induction on n. By the nature of conditions 1) and 2), we may assume $k = r$. Suppose $n = 1$. Let $\sum_{i=1}^n Af_i = AX_1$. Then $f_i = g_iX_1, 1 \leq i \leq r$, where $(g_1, \ldots, g_r)$ is unimodular. Then by 1), we have $(g_1, \ldots, g_r)\alpha = (1, 0, \ldots, 0)$ and hence $(f_1, \ldots, f_r)\alpha = (X_1, 0, \ldots, 0)$, for some $\alpha \in GL_r(A)$.

Suppose now that $n > 1$ and assume 1). We simplify notation by replacing R by $R[X_{n+1}, \ldots, X_m]$ and assume $A = R[X_1, \ldots, X_n]$. Let $\sum_{i=1}^r Af_i = \sum_{i=1}^n AX_i = I, f_i = f_i(X_1, \ldots, X_n) \in A$. Setting $X_n = 0$, we get $\sum_{i=1}^r A'f_i' = \sum_{i=1}^{n-1} A'X_i$, where, $A' = R[X_1, \ldots, X_{n-1}]$ and $f_i' = f_i(X_1, \ldots, X_{n-1}, 0)$. Since condition 1) holds for A, it easily follows that it also holds for A'. Hence, by induction hypothesis, there is an $\alpha' \in GL_r(A')$ such that $(f_1', \ldots, f_r')\alpha' = (X_1, \ldots, X_{n-1}, 0, \ldots, 0)$. Now $f_i \equiv f_i'(mod AX_n)$ and hence we have

$$(f_1, \ldots, f_r)\alpha' = (Y_1, \ldots, Y_{n-1}, g_nX_n, \ldots, g_rX_n),$$
$$Y_i \equiv X_i(mod AX_n), 1 \leq i \leq n-1 \qquad (I)$$

We claim that $(Y_1, \ldots, Y_{n-1}, g_n, \ldots, g_r) \in A^r$ is unimodular. For, if not, there is a maximal ideal $M \subset A$ such that

$$I = \sum_{i=1}^{n} AX_i = \sum_{i=1}^{n-1} AY_i + \sum_{i=n}^{r} Ag_iX_n \subset \sum_{i=1}^{n} AY_i + \sum_{i=n}^{r} Ag_i \subset M$$

Then $g_jX_n \in IM, n \leq j \leq r$ and hence $Y_i mod(IM), 1 \leq i \leq n-1$ generate I/IM. Since I/I^2 is a free A/I-module and hence $\dim_K(I/IM) = n, K = A/M$, we get a contradiction.

Thus $(Y_1, \ldots, Y_{n-1}, g_n, \ldots, g_r)$ is unimodular and so we have, for $\lambda_i = \lambda_i(X_1, \ldots, X_n) \in A, 1 \leq i \leq r$,

$$\lambda_1 Y_1 + \ldots + \lambda_{n-1}Y_{n-1} + \lambda_n g_n + \ldots + \lambda_r g_r = 1 \qquad (II)$$

Noting that $\sum_{i=1}^{n} AY_i + AX_n = \sum_{i=1}^{n} AX_i$, we get

$$A = \sum_{i=1}^{n-1} AY_i + \sum_{j=n}^{r} A\lambda_j = \sum_{i=1}^{n-1} AY_i + AX_n + \sum_{j=n}^{r} A\lambda_j = \sum_{i=1}^{n} AX_i + \sum_{j=n}^{r} A\lambda_j$$

Writing $\lambda_j(0)$ for $\lambda_j(0, \ldots, 0)$, we see that $(\lambda_n(0), \ldots, \lambda_r(0)) \in R^{r-n+1}$ is unimodular and $\lambda_j = \lambda_j(0) + \mu_{jn}X_n + \sum_{k=1}^{n-1} \mu_{jk}Y_k$, with $\mu_{jk} \in A, n \leq j \leq r, 1 \leq k \leq n$. Then setting $\lambda'_i = \lambda_i + \sum_{j=n}^{r} \mu_{ji}g_j, 1 \leq i \leq n-1$ and $\lambda'_j = \lambda_j(0) + \mu_{jn}X_n, n \leq j \leq r$, we can re-write equation (II) as

$$\sum_{i=1}^{n-1} \lambda'_i Y_i + \sum_{j=n}^{r} \lambda'_j g_j = 1 \qquad (III)$$

Thus $(\lambda'_n, \ldots, \lambda'_r) \in A^{r-n+1}$ is unimodular modulo $\sum_{i=1}^{n-1} A\lambda'_i$ and is also unimodular modulo AX_n. Hence $(\lambda'_1 X_n, \ldots, \lambda'_{n-1}X_n, \lambda'_n, \ldots, \lambda'_r) \in A^r$ is unimodular. Multiplying (III) by X_n, and setting $h_i = \lambda'_i X_n, 1 \leq i \leq n-1, h_j = \lambda'_j, n \leq j \leq r$, we get

$$X_n = \sum_{i=1}^{n-1} h_i Y_i + \sum_{j=n}^{r} h_j(g_j X_n) \qquad (IV)$$

Since $(h_1, \ldots, h_r) \in A_r$ is unimodular, by condition 1) of Theorem 1.1, there is a $\beta \in GL_r(A)$ such that

$$\beta \begin{bmatrix} 0 \\ \vdots \\ 0 \\ 1 \end{bmatrix} = \begin{bmatrix} h_1 \\ \vdots \\ h_r \end{bmatrix}$$

and hence

$(Y_1, \ldots, Y_{n-1}, g_n X_n, \ldots, g_r X_n)\beta = (p_1, \ldots, p_{r-1}, X_n)$. Hence by (I), we have, $(f_1, \ldots, f_r)\alpha'\beta = (p_1, \ldots, p_{r-1} X_n)$. Let $q_i = p_i(X_1, \ldots, X_{n-1}, 0), 1 \le i \le n-1$. Then by adding suitable multiples of X_n to the each p_i in $(p_1, \ldots, p_{r-1}, X_n)$, we can change p_i to q_i. In other words, there is an elementary matrix $\epsilon \in GL_r(A)$ such that $(p_1, \ldots, p_{r-1}, X_n)\epsilon = (q_1, \ldots, q_{r-1}, X_n)$. Thus $(f_1, \ldots, f_r)\alpha'\beta\epsilon = (q_1, \ldots, q_{r-1}, X_n)$. Since $\sum_{i=1}^{n} AX_i = \sum_{i=1}^{r} Af_i$, it follows that $\sum_{i=1}^{r-1} A'q_i = \sum_{i=1}^{n-1} A'X_i$, where $A' = R[X_1, \ldots, X_{n-1}]$. Now, by induction hypothesis there is a $\gamma' \in GL_{r-1}(A')$ such that $(q_1, \ldots, q_{r-1})\gamma' = (X_1, \ldots, X_{n-1}, 0, \ldots, 0)$. Hence $(q_1, \ldots, q_{r-1}, X_n)\gamma$ $= (X_1, \ldots, X_{n-1}, 0, \ldots, 0, X_n)$ where $\gamma = \begin{pmatrix} \gamma' & 0 \\ 0 & 1 \end{pmatrix}$. Thus for a suitable permutation matrix $\pi \in GL_r(A)$, we have $(q_1, \ldots, q_{r-1}, X_n)\gamma\pi = (X_1, \ldots, X_n, 0, \ldots, 0)$. Thus $(f_1, \ldots, f_r)\alpha = (X_1, \ldots, X_n, 0, \ldots, 0)$, where $\alpha = \alpha'\beta\epsilon\gamma\pi$. This finishes the proof of Theorem 1.

2. Application to cancellation of certain modules

Let A be a commutative ring. We say that an A-module M is *cancellative* if for all finite free A-modules F, $M \oplus F \approx M' \oplus F \Longrightarrow M \approx M'$, for any A-module M'.

Let R be a commutative ring and $A = R[X_1, \ldots, X_n]$. Suppose stably free A-modules of rank $\le n$ are free. As an application of Theorem 1.1, we prove here the following amusing result that the A-module $A^n/A(X_1, \ldots, X_n)$ associated to $T_{\mathbb{P}^{n-1}(R)/R}$ the relative tangent bundle of $\mathbb{P}^{n-1}(R)$ over R is cancellative.

THEOREM 2.1 Let R be a commutative ring and $A = R[X_1, \ldots, X_n]$. Suppose all stably free A-modules of rank $\le n$ are free. Then $M = A^n/A(X_1, \ldots, X_n)$ is cancellative.

Proof. Let $f : M \oplus A^l \xrightarrow{\sim} M' \oplus A^l$ be an isomorphism with M' an A-module and $l \ge 0$. Let $\eta : A^n \twoheadrightarrow M$ be the canonical surjection. Let $\epsilon : A \longrightarrow A^n$ be the map with $\epsilon(1) = (X_1, \ldots, X_n)$. Then we have the exact sequence:

$$0 \longrightarrow A \xrightarrow{(\epsilon,0)} A^n \oplus A^l \xrightarrow{g} M' \oplus A^l \longrightarrow 0$$

where $g = f \circ (\eta \oplus 1_{A^l})$. Setting $P = g^{-1}(M')$, we get exact sequences:

$$0 \longrightarrow A \longrightarrow P \longrightarrow M' \longrightarrow 0$$

and

$$0 \longrightarrow P \longrightarrow A^n \oplus A^l \longrightarrow A^l \longrightarrow 0$$

Thus P is stably free of rank n and by hypothesis $P \approx A^n$. Thus we have an exact sequence:

$$0 \longrightarrow A \xrightarrow{(\epsilon')} A^n \xrightarrow{\eta'} M' \longrightarrow 0$$

Let $\epsilon'(1) = (Y_1, \ldots, Y_n)$. Dualising this sequence, we get

$$0 \longrightarrow (M')^* \xrightarrow{(\eta')^t} A^n \xrightarrow{(\epsilon')^t} A \longrightarrow Ext_A^1(M', A) \longrightarrow 0,$$

where $(M')^* = \mathrm{Hom}_A(M', A)$ and $(\epsilon')^t(e_i) = Y_i$, the e_i being the standard basis for A^n. Thus $Ext_A^1(M', A) \approx A/I', I' = \sum_{i=1}^n AY_i$. The exact sequence $0 \longrightarrow A \xrightarrow{\epsilon} A^n \xrightarrow{\eta} M \longrightarrow 0$ similarly yields $Ext_A^1(M, A) \approx A/I, I = \sum_{i=1}^n AX_i$. Since $M \oplus A^l \approx M' \oplus A^l$, we have $A/I \approx A/I'$ i.e. $I = I'$. Since all stably free A-modules of rank $\leq n$ are free, it follows that all unimodular elements in A^k are completable for $1 \leq k \leq n+1$. Hence by Theorem 1.1., we have $(Y_1, \ldots, Y_n)\gamma = (X_1, \ldots, X_n)$, for some $\gamma \in GL_n(A)$. Clearly γ induces an isomorphism

$$M' \approx A^n/A.(Y_1, \ldots, Y_n) \approx A^n/A.(X_1, \ldots, X_n) = M$$

.

REMARK 2.2 It is an easy exercise that M is reflexive for $n \geq 3$.
If R is a regular local ring containing a field, or R is a noetherian ring of dimension ≤ 1, all stably free $R[X_1, \ldots, X_n]$-modules are free (in view of ([L],[P1],[P2]) and [B-R]. Hence we have the following corollary (cf.Corollary 1.2)

COROLLARY 2.3 Let R be a noetherian ring satisfying one of the following two conditions:

a) R is a regular local ring containing a field
b) dim $R \leq 1$

Let $A = R[X_1, \ldots, X_n]$. Then the A-module $M = A^n/A.(X_1, \ldots, X_n)$ is cancellative.

REMARK 2.4 Ler k be an algebraically closed field of characteristic zero. Let A be a finitely generated integral k-algebra of dimension n. Let M be a reflexive A-module of rank r. Mohan Kumar ([M]) has shown that M is cancellative if i) $r \geq n$, ii) M has finite projective dimension and iii) M is locally free outside a finite set of closed points of Spec A. He has also given examples to show that the conditions on the field k or iii) cannot be relaxed. It is not known if i) can be weakened or if ii) can be dropped. When $A = k[X_1, \ldots, X_n], n \geq 3$, the module $M = A^n/A.(X_1, \ldots, X_n)$ gives an example of a cancellative non-projective reflexive A-module of rank $n-1$.

We end this note with the following question:

Question 2.5 Let k be an algebraically closed field of characteristic zero and $A = k[X_1, \ldots, X_n]$. Let M be a reflexive A-module which is locally free outside finitely many closed points of $SpecA$. Is M cancellative?

Vector bundles over $\mathbb{P}^{n-1}_{\mathbb{C}}$ of rank $\leq n-1$ provide examples of graded reflexive modules over $\mathbb{C}[X_1, \ldots, X_n]$ which are locally free outside $\mathbb{C}^n - \{0\}$. We do not know if they are cancellative.

We thank R. Sridharan profusely for his immense patience and help in the preparation of this note.

References

[B-R] S.M. Bhatwadekar and A. Roy, Some theorems about projective modules over polynomial rings, J. of Algebra, 86 (1983), 150-158.

[L] H. Lindel, On a question of Bass, Quillen and Suslin concerning projective modules over polynomial rings, Invent. Math. 65 (1981), 319-323.

[M] N. Mohan Kumar, A note on cancellation of reflexive modules, J. Ramanujan Math. Soc.17, No.2 (2002) 93-100.

[P1] D. Popescu, General Neron desingularization and approximation, Nagoya Math. J., 104 (1986) 85-115.

[P2] D. Popescu, Letter to the editor, Nagoya Math.J., 118 (1990) 45-53.

[S] A. Sathaye, On the Förster-Eisenbud-Evans Conjecture, Invent. Math., 46 (1978) 211-224.

[SW] R.G. Swan, Neron-popescu desingularization, Proc. Alg. and Geom., (1995) 135-192, International Press (1998).

The University of Chicago
Chicago, IL 60637, U.S.A.
E-mail address: murthy@math.uchicago.edu

The Development of Standard Monomial Theory–I

C. MUSILI

With grateful homage to C.S. Seshadri, on his 70th *birthday.*

Abstract. In this exposition, we will give a comprehensive report on some of the important highlights in the early phase of "standard monomial theory" (SMT) and its first applications as developed by Seshadri (with Musili, Lakshmibai and Littelmann) (built on the ideas of Hodge's enumerative geometry, [**22**]). The later developments are given in [**32**] (appearing in this volume).

Introduction

The concept of "Standard Monomials", introduced in 1940's by Hodge ([**22**], [**23**]) to study the Schubert varieties in the Grassmannian/Flag varieties, was developed in 1970's into a "Theory" by Seshadri, in collaboration with Musili, Lakshmibai, Littelmann, etc., to study the Schubert varieties in the generalised flag varieties G/Q where G is a semi-simple group and Q is a parabolic subgroup of G.

It is well-known that the case of the group $G = \mathrm{SL}(n, k)$ gives rise to the Grassmannians G/P (resp. the Flag varieties G/Q) corresponding to the maximal parabolic subgroups P (resp. arbitrary parabolic subgroups Q). The Grassmannian is the basic object, serving as the motivating and guiding model of this theory. Lots of very nice things happen in the Grassmannian, as can be seen rather explicitly and concretely. It is fairly clear that not many of these properties can be expected to be true in case of the generalised Grassmannians G/P (or the Flag varieties G/Q) for arbitrary semi-simple groups G.

One naturally wonders therefore to what extent one could expect such properties to hold for the Schubert varieties in G/Q, in general. The aim of the Standard Monomial Theory is to answer this as precisely as possible.

The study of the geometry of Schubert varieties in the Grassmannian and Flag varieties gave birth to the Standard Monomial Theory (SMT). The main steps in its subsequent growth are to ascend from the easier to more complex situations like: **"Grassmannian → Minuscule weights → Quasi-minuscule weights → Classical type weights → general weights"**, whatever these words could mean now (but explained in the text below).

In what follows, we will give a comprehensive narration of some of the important initial developments in the birth, growth and some applications of the standard monomial theory. The later developments are given in [**32**]. First we need to recall some terminology and fix notation.

0. Notation

k	=	a field of arbitrary characteristic
$\overline{k}$	=	an algebraic closure of k
G	=	a semi-simple, simply connected, split (Chevalley) group over k, defined over $\mathbb{Z}$
T	=	a maximal torus of G
n	=	the rank of G = the dimension of T
B	=	a Borel subgroup of G containing T
B^-	=	the Borel subgroup of G opposite to B, i.e., $B \cap B^- = T$
Δ	=	the system of roots relative to T
Δ^+	=	the set of positive roots relative to B
Δ^-	=	$-\Delta^+$ = the set of negative roots relative to B
S	=	simple roots α_i, $1 \leq i \leq n$
$\alpha^\vee$	=	the coroot of $\alpha \in \Delta$
W	=	$N_G(T)/T$, $N_G(T)$ being the normaliser of T in G
	=	Weyl group of G generated by the simple reflections $s_i = s_{\alpha_i}$, $1 \leq i \leq n$
w_0	=	the element of maximal length in W
$\mathbf{i}$	=	$-w_0$, the Weyl involution
$(\, , \,)$	=	a W-invariant inner product on $X(T) \otimes_{\mathbb{Z}} \mathbb{R}$
$\langle \alpha, \beta^\vee \rangle$	=	$2\dfrac{(\alpha, \beta)}{(\beta, \beta)}$ for $\alpha, \beta \in \Delta$, i.e., $\beta^\vee = 2\beta/(\beta, \beta)$

ϖ_i = the fundamental weight dual to α_i, i.e., $\langle \varpi_i, \alpha_j^\vee \rangle = \delta_{ij}$, $1 \leq i, j \leq n$

V_{ϖ_i} = the fundamental representation of G of highest weight ϖ_i

U_α = the unipotent group $\approx \mathbb{G}_a$ associated to $\alpha \in \Delta$

$X(T)$ = $\mathrm{Hom}_{\mathrm{alg\ gp}}(T, \mathbb{G}_m)$ = the group of characters of T with a $\mathbb{Z}$-basis $\{\varpi_1, \cdots, \varpi_n\}$

$X(T)^+$ = $\left\{ \sum_{i=1}^n m_i \varpi_i \,\middle|\, m_i \in \mathbb{Z}^+ \right\}$ = the dominant chamber or the positive cone spanned by the ϖ_i's

L_i = the line bundle on G/B, associated to ϖ_i

$L(\lambda)$ = the line bundle on G/B, associated to $\lambda = \sum_i m_i \varpi_i$, $1 \leq i \leq n$
$= L_i^{\otimes m_i} \otimes \cdots \otimes L_n^{\otimes m_n}$; $H^0(G/B, L(\lambda)) \neq (0)$
$\iff \lambda \in X(T)^+$ and $L(\lambda)$ is ample
$\iff L(\lambda)$ is very ample $\iff m_i > 0$, $\forall\, i$

$P = P_i$ = the maximal parabolic subgroup of G omitting α_i i.e., associated to the maximal subset $S_P = S - \{\alpha_i\}$ of S or generated by B and $U_{-\alpha}$ with $\alpha = \sum_{j \neq i} m_j \alpha_j, m_j \in \mathbb{Z}^+$

Q = a parabolic subgroup of G associated to a subset S_Q of S or generated by B and $U_{-\alpha}$ with $\alpha = \sum_{\alpha_j \in S_Q} m_j \alpha_j, m_j \in \mathbb{Z}^+$
$= \cap_{\alpha_i \notin S_Q} P_i$, the intersection of all the maximal parabolic subgroups of G containing Q

W_P = Weyl group of P, i.e., generated by the simple reflections s_j, $j \neq i$

W_P^{max} = the set of *maximal representatives* of W/W_P in W

W_P^{min} = the set of *minimal representatives* of W/W_P in W

Δ_P = $\left\{ \alpha \in \Delta \,\middle|\, \alpha = \sum_{j \neq i} m_j \alpha_j,\ m_j \in \mathbb{Z} \right\}$

Δ_P^+ = $\left\{ \alpha \in \Delta \,\middle|\, \alpha = \sum_{j \neq i} m_j \alpha_j,\ m_j \in \mathbb{Z}^+ \right\}$
$= \Delta^+ \cap \Delta_P$

Δ_P^- = $\left\{ \alpha \in \Delta \,\middle|\, \alpha = \sum_{j \neq i} m_j \alpha_j,\ m_j \in \mathbb{Z}^- \right\}$
$= \Delta^- \cap \Delta_P$

$\mathrm{Pic}(G/P)$	$=$	the Picard group of G/P with ample generator L_i
$C_Q(\theta)$	$=$	$B\theta Q/Q$, the affine Schubert (Bruhat) cell associated to $\theta \in W/W_Q$
$C_Q^-(\theta)$	$=$	$B^-\theta Q/Q$, the affine opposite Schubert cell associated to $\theta \in W/W_Q$
G/Q	$=$	$\bigcup_{\theta\in W/W_Q} C_Q(\theta)$ $=$ the Bruhat cellular decomposition
	$=$	$\bigcup_{\theta\in W/W_Q} C_Q^-(\theta)$ $=$ the opposite cellular decomposition
$X_Q(\theta)$	$=$	$\overline{B\theta Q/Q}$, the Zariski closure of the Schubert cell $C_Q(\theta)$ in G/Q
	$=$	the Schubert variety in G/Q associated to $\theta \in W/W_Q$
$Y_Q(\theta)$	$=$	$\overline{B^-\theta Q/Q}$, the opposite Schubert variety associated to $\theta \in W/W_Q$
$(W, \geq)$	$=$	the Bruhat partial order on W defined by, $\theta \geq \tau$ $\iff X_B(\theta) \supseteq X_B(\tau) \iff Y_B(\theta) \subseteq Y_B(\tau)$
$\mathrm{Ch}(G/Q)$	$=$	the Chow ring of G/Q (generated by the classes $[X_Q(\theta)], \theta \in W/W_Q$
$C_B(w_0)$	$=$	the affine big cell, $C_B(id)$ the point cell
$C_B^-(id)$	$=$	the affine opposite big cell, $C_B(w_0)$ the opposite point cell
$D_Q(\theta)$	$=$	$X_Q(\theta) \cap C_Q^-(id)$ $=$ the "determinantal" variety
$\mathbf{g}$	$=$	the Lie algebra of G
$\{X_\beta \mid \beta \in \Delta\}$	$\cup$	$\{H_\alpha \mid \alpha \in S\}$, a Chevalley basis of $\mathbf{g}$
$\mathbf{g}_{\mathbb{Z}}$	$=$	the $\mathbb{Z}$-submodule of $\mathbf{g}$ spanned by the Chevalley basis of $\mathbf{g}$
$\mathbf{u} = \mathbf{u}_{\mathbb{Z}}$	$=$	a Kostant $\mathbb{Z}$-form of the universal enveloping algebra of G
$\mathbf{u}^\pm = \mathbf{u}^\pm_{\mathbb{Z}}$	$=$	the subalgebra generated by $\left\{ \frac{X_{\pm\beta}}{m!} \,\middle\vert\, \beta \in \Delta^+, m \in \mathbb{Z}^+ \right\}$
$\mathbf{u}^0 = \mathbf{u}^0_{\mathbb{Z}}$	$=$	the subalgebra generated by $\left\{ \binom{H_\alpha}{m} \,\middle\vert\, \alpha \in S, m \in \mathbb{Z}^+ \right\}$

1. Grassmannian - The Role Model

Some references for this section are [**23**], [**45**], [**46**], [**33**], [**54**], etc.

Let us first recall the classical definition of the Grassmannian $G_{d,n}$ and its Schubert varieties. Let k be a field of arbitrary characteristic and V be a k-vector space of dimension $n \in \mathbb{N}$. Let $\{e_1, \cdots, e_n\}$ be a basis of V. Let E_p be the k-span of $\{e_1, \cdots, e_p\}$, $1 \leq p \leq n$. Let d be an integer such that $1 \leq d \leq n-1$. Let

$$I_{d,n} = \{\mathbf{i} = (i_1, \cdots, i_d) \mid 1 \leq i_1 < \cdots < i_d \leq n\}.$$

1.1 The Grassmannian: Let $G_{d,n}$ be the set of all d-dimensional subspaces of V, called the Grassmannian of d-planes in V.

1.2 Schubert Varieties: Let $\mathbf{a} = (a_1, \cdots, a_d) \in I_{d,n}$. Let $X(\mathbf{a}) = \left\{ W \in G_{d,n} \,\middle|\, \dim_k(W \cap E_{a_j}) \geq j, \ \forall\, j \right\}$, called the **Schubert variety** defined by **a**. This depends on the choice of the **flag**: $E_1 \subset E_2 \subset \cdots \subset E_n$ but it is determined uniquely upto isomorphism.

1.3 Partial Order: There is a natural partial order on $I_{d,n}$, namely: $\mathbf{i} \geq \mathbf{j}$ if $i_t \geq j_t$, $1 \leq t \leq d$. Clearly $I_{d,n}$ has the unique **minimal** element $\mathbf{m}in = (1, 2, \cdots, d)$ and the unique **maximal** element $\mathbf{m}ax = (n-d+1, \cdots, n)$. We have

$$\mathbf{i} \geq \mathbf{j} \iff X(\mathbf{i}) \supseteq X(\mathbf{j}) \text{ and } X(\mathbf{m}in) = \text{pt. and } X(\mathbf{m}ax) = G_{d,n}.$$

1.4 The Plücker embedding: For $\mathbf{i} \in I_{d,n}$, let $e_{\mathbf{i}} = e_{i_1} \wedge \cdots \wedge e_{i_d}$ be the **decomposable tensor** in the exterior d^{th} power $\Lambda^d(V)$ of V. Recall that $\{e_{\mathbf{a}} \mid \mathbf{a} \in I_{d,n}\}$ is a natural basis of $\Lambda^d(V)$. Every vector in $\Lambda^d(V)$ can be written as $\sum_{\mathbf{a}} \lambda_{\mathbf{a}} e_{\mathbf{a}}$ for unique $\lambda_{\mathbf{a}} \in k$.

Recall that for $W \in G_{d,n}$, a subset of d vectors $\{w_1, \cdots, w_d\}$ in W is a basis of $W \iff w_1 \wedge \cdots \wedge w_d \neq 0$. Secondly, given two bases $\{w_1, \cdots, w_d\}$ and $\{v_1, \cdots, v_d\}$ of W, we have $w_1 \wedge \cdots \wedge w_d = \alpha v_1 \wedge \cdots \wedge v_d$ for some $\alpha \in k^*$.

The map $\lambda : G_{d,n} \longrightarrow \mathbb{P}(\Lambda^d(V))$, defined by $W \mapsto [w_1 \wedge \cdots \wedge w_d]$, is well-defined and in fact it is an embedding, called the **Plücker embedding** of $G_{d,n}$ into the projective space $\mathbb{P}(\Lambda^d(V))$. Given a non-zero vector $w = \sum_{\mathbf{a}} \lambda_{\mathbf{a}} e_{\mathbf{a}} \in \Lambda^d(V)$, the homogeneous coordinates of $[w] \in \mathbb{P}(\Lambda^d(V))$ are $(\cdots, \lambda_{\mathbf{a}}, \cdots)$.

If $W \in G_{d,n}$, the homogeneous coordinates of $\lambda(W)$ can be seen to be the *maximal minors* of the $n \times d$ matrix $\mathcal{W}$ obtained first by taking a basis $\{w_1, \cdots, w_d\}$ of W, representing the w_js as the column vectors with respect to the chosen basis of V and the w_j constituting the jth column of $\mathcal{W}$. We write $\lambda(W) = (\cdots, p_{\mathbf{a}}, \cdots)$ (these coordinates are called the **Plücker coordinates** of W) where each $p_{\mathbf{a}}$ is the minor of $\mathcal{W}$ corresponding to the row indices $a_1, \cdots, a_d$.

• **The Dual Basis:** The basis of the dual $\Lambda^d(V)^\vee = \mathrm{Hom}_k(\Lambda^d(V), k)$, dual to the natural basis $\{e_{\mathbf{i}} \mid \mathbf{i} \in I_{d,n}\}$ of $\Lambda^d(V)$ is $\{p_{\mathbf{i}} \mid \mathbf{i} \in I_{d,n}\}$ because the Plücker coordinates are k-valued functions on $\Lambda^d(V)$ such that $p_{\mathbf{j}}(e_{\mathbf{i}}) = \delta_{\mathbf{i},\mathbf{j}}$.

1.5 The Quadratic Relations: The Plücker coordinates $p_{\mathbf{a}}$'s of points of $G_{d,n}$ satisfy the following TWO sets of quadratic relations. The first set of relations are, one for every choice of $2d$ integers, $i_1, \cdots, i_{d-1}; j_1 \cdots j_{d+1}$ between 1 and n (not necessarily all distinct):

1.5.1
$$Q^{i_1\cdots i_{d-1}}_{j_1\cdots j_{d+1}} = \sum_{t=1}^{d+1} p_{i_1\cdots i_{d-1}j_t} p_{j_1\cdots j_{t-1}j_{t+1}\cdots j_{d+1}} = 0.$$

The second set of relations are, one for every pair of incomparable elements $\mathbf{i}, \mathbf{j} \in I_{d,n}$, $\mathbf{i} \not\geq \mathbf{j}$, say $i_1 \geq j_1, \cdots, i_{t-1} \geq j_{t-1}$ but $i_t < j_t$ for some t, then

1.5.2
$$Q(\mathbf{i}, \mathbf{j}) = \sum_{\sigma} \mathrm{sign}(\sigma) p_{\mathbf{i}^\sigma} p_{\mathbf{j}^\sigma} = 0$$

where $\mathbf{i}^\sigma = (i_1^\sigma, \cdots, i_t^\sigma, i_{t+1}, \cdots, i_d)$ and $\mathbf{j}^\sigma = (j_1, \cdots, j_{t-1}, j_t^\sigma, \cdots, j_d^\sigma)$ for every permutation

$$\sigma := \begin{pmatrix} i_1 & < \cdots < & i_t & < & j_t & < \cdots < & j_d \\ i_1^\sigma & < \cdots < & i_t^\sigma & ; & j_t^\sigma & < \cdots < & j_d^\sigma \end{pmatrix}$$

of the $d+1$ symbols $\{i_1 \cdots i_t, j_t \cdots j_d\}$. Such permutations are called the **shuffle permutations** of level t (which are $\binom{d+1}{t}$ in number).

• For a shuffle permutation σ as above, we see that

$$\mathbf{i}^\sigma > \mathbf{i} \text{ and } \mathbf{j} > \mathbf{j}^\sigma, \ \forall\, \sigma \neq \mathrm{Id}.$$

i.e., a **one step betterment** takes place, in the sense that given $\mathbf{i}, \mathbf{j} \in I_{d,n}$ with $\mathbf{i} \not\geq \mathbf{j}$, we can write

1.5.3
$$p_{\mathbf{i}} p_{\mathbf{j}} = \sum_{\sigma \neq \mathrm{Id}} \pm p_{\mathbf{i}^\sigma} p_{\mathbf{j}^\sigma} \text{ with } \mathbf{i}^\sigma > \mathbf{i} \text{ and } \mathbf{j} > \mathbf{j}^\sigma.$$

• After finitely many iterations of using the relations of the type (1.5.3) to each pair $\mathbf{i}^\sigma, \mathbf{j}^\sigma$ with $\sigma \neq \mathrm{Id}$ and $\mathbf{i}^\sigma \not\geq \mathbf{j}^\sigma$, we get that

1.5.4
$$p_{\mathbf{i}} p_{\mathbf{j}} = \sum_{m_{(\mathbf{i}'\mathbf{j}')} \in \mathbb{Z}} m_{(\mathbf{i}'\mathbf{j}')} p_{\mathbf{i}'} p_{\mathbf{j}'} \text{ with } \mathbf{i}' \geq \mathbf{j}' \text{ and } \mathbf{i}' \geq \mathbf{i} \text{ and } \mathbf{j} \geq \mathbf{j}'.$$

1.6 The Coordinate Rings: Recall that the homogeneous coordinate ring of the projective space $\mathbb{P}(\Lambda^d(V))$ is the polynomial algebra $k[\cdots, T_{\mathbf{i}}, \cdots]$ where $T_{\mathbf{i}} = T_{i_1,\cdots,i_d}$, $\forall$ $\mathbf{i} \in I_{d,n}$. Let $k\left[T_{ij} \;\middle|\; 1 \leq i \leq n, 1 \leq j \leq d\right]$ be the polynomial algebra over k. For $\mathbf{i} \in I_{d,n}$, let $p_{\mathbf{i}}$ be the maximal minor of the matrix (T_{ij}) with row indices $i_1, \cdots, i_d$. Let $\mathbf{A}$ be the subalgebra of $k[\cdots, T_{ij}, \cdots]$ generated by the $p_{\mathbf{i}}$'s. The natural k-algebra homomorphism:

$$k[\cdots, T_{\mathbf{i}}, \cdots] \rightarrow \mathbf{A} \hookrightarrow k[\cdots, T_{ij}, \cdots],\ T_{\mathbf{i}} \mapsto p_{\mathbf{i}},$$

has its kernel $\mathbf{p}$ (which is a prime ideal) containing the relations of type (1.5.1/2). If each of the $T_{\mathbf{i}}$'s is given the degree d, then this homomorphism becomes one of degree 0. We have

$$k[\cdots, T_{\mathbf{i}}, \cdots]/\mathbf{p} \approx k[\cdots, p_{\mathbf{i}}, \cdots].$$

• The relations of type (1.5.1) generate the ideal $\mathbf{p}$ if char$(k) = 0$ ([**23**]) and they do not generate the ideal if char$(k) \neq 0$ ([**1**]).
• The relations of type (1.5.2) generate the ideal $\mathbf{p}$ regardless of the characteristic of k.
• The graded k-algebra $\mathbf{A}$ is called the homogeneous coordinate ring of the Grassmannian $G_{d,n}$ for the Plücker embedding.
• The **dimension** of $X(\mathbf{a})$ is given by $\sum_{i=1}^{d}(a_i - i) = \sum_{i=1}^{d} a_i - \sum_{i=1}^{d} i$. Hence, dim$X(\mathbf{m}ax) = \dim(G_{d,n}) = d(n-d)$ and dim$X(\mathbf{m}in) = 0$.
• The coordinate ring of a Schubert variety $X(\mathbf{a})$ can be seen to be

$$\mathbf{A}(a) = \mathbf{A}/ < I_{\mathbf{a}} > \text{ where } I_{\mathbf{a}} = \{p_{\mathbf{b}} \mid \mathbf{a} \not\geq \mathbf{b}\},$$

i.e., $X(\mathbf{a})$ is a linear section of $G_{d,n}$. We have

$$\mathbf{A}(max) = \mathbf{A} = k[\cdots, T_{\mathbf{i}}, \cdots]/\mathbf{p} \approx k[\cdots, p_{\mathbf{i}}, \cdots],$$
$$\mathbf{A}(a) = \mathbf{A}/ < I_{\mathbf{a}} > = k[\cdots, T_{\mathbf{i}}, \cdots]/\mathbf{p} + < \mathcal{I}_{\mathbf{a}} >$$

where $\mathcal{I}_{\mathbf{a}} = \{T_{\mathbf{i}} \mid \mathbf{a} \not\geq \mathbf{i}\}$. We have $\mathbf{A}(min) = k[p_{12\cdots d}]$ (which is a polynomial ring in one variable).

• We write the homogeneous decomposition of the graded rings as

$$\mathbf{A} = \bigoplus_{r \geq 0} \mathbf{A}_r \text{ and } \mathbf{A}(a) = \bigoplus_{r \geq 0} \mathbf{A}(a)_r.$$

For $\mathbf{a} \geq \mathbf{b}$, the inclusion $X(\mathbf{a}) \supseteq X(\mathbf{b})$ induces a surjective homomorphism of graded k-algebras $\mathbf{A}(a) \longrightarrow \mathbf{A}(b)$. It is clear that $\mathbf{A}$ is generated as a k-algebra by $\{p_{\mathbf{i}} \mid \mathbf{i} \in I_{d,n}\}$ and hence $\mathbf{A}(a)$ is generated by $\{p_{\mathbf{i}} \mid \mathbf{a} \geq \mathbf{i}\}$.

1.7 Standard Monomials on $G_{d,n}$: A monomial of degree m in $\mathbf{A}$ is an element in $\mathbf{A}_m$ of the form

$$\underbrace{p_{\mathbf{i}} p_{\mathbf{j}} \cdots p_{\mathbf{k}}}_{m \text{ factors}} = \underbrace{p_{i_1 \cdots i_d} p_{j_1 \cdots j_d} \cdots p_{k_1 \cdots k_d}}_{m \text{ factors}}$$

It is said to be a **standard monomial** of degree m if $\mathbf{i} \geq \mathbf{j} \geq \cdots \geq \mathbf{k}$ in $I_{d,n}$, i.e., we have a **Standard Young Tableaux** [**56**], namely,

$$\underbrace{\begin{matrix} i_1 & \geq & j_1 & \geq & \cdots & \geq & k_1 \\ i_2 & \geq & j_2 & \geq & \cdots & \geq & k_2 \\ \vdots & \vdots & \vdots & \vdots & \vdots\vdots\vdots & \vdots & \vdots \\ i_d & \geq & j_d & \geq & \cdots & \geq & k_d \end{matrix}}_{m \text{ columns}}$$

Standard monomials on $G_{d,n}$ are simply standard monomials on $X(\mathbf{m}ax)$ (or relative to $\mathbf{m}ax$) in the following sense.

1.8 Standard Monomials on $X(\mathbf{a})$: A monomial of degree m in $\mathbf{A}(a)$ is a non-zero element in $\mathbf{A}(a)_m$ of the form

$$\underbrace{p_{\mathbf{i}}p_{\mathbf{j}}\cdots p_{\mathbf{k}}}_{m \text{ factors}} = \underbrace{p_{i_1\cdots i_d}p_{j_1\cdots j_d}\cdots p_{k_1\cdots k_d}}_{m \text{ factors}}$$

Recall that a monomial as above is non-zero on $X(\mathbf{a})$ if and only if $\mathbf{a} \geq \mathbf{i}, \mathbf{j}, \cdots, \mathbf{k}$. It is said to be a **standard monomial** on $X(\mathbf{a})$ if $\mathbf{a} \geq \mathbf{i} \geq \mathbf{j} \geq \cdots \geq \mathbf{k}$ in $I_{d,n}$, i.e., it is standard in $\mathbf{A}$ and non-zero in $\mathbf{A}(a)$. In other words, we have a **Standard Young Tableaux** relative to $\mathbf{a}$ ([**54**], [**33**]), namely,

$$\begin{matrix} a_1 & \geq \\ a_2 & \geq \\ \vdots & \vdots \\ a_d & \geq \end{matrix} \underbrace{\begin{matrix} i_1 & \geq & j_1 & \geq & \cdots & \geq & k_1 \\ i_2 & \geq & j_2 & \geq & \cdots & \geq & k_2 \\ \vdots & \vdots & \vdots & \vdots & \vdots\vdots\vdots & \vdots & \vdots \\ i_d & \geq & j_d & \geq & \cdots & \geq & k_d \end{matrix}}_{m \text{ columns}}$$

1.9 Theorem A: With notation as above, we have the following.

(1) *A Plücker coordinate $p_{\mathbf{i}}$ ($\mathbf{i} \in I_{d,n}$) is non-zero on $X(\mathbf{a})$ $\Longleftrightarrow$ $\mathbf{a} \geq \mathbf{i}$.*

(2) *Standard Monomials of degree m on $X(\mathbf{a})$ span the vector space $\mathbf{A}(a)_m$.*

(3) *Standard Monomials on $X(\mathbf{a})$ are linearly independent.*

(4) *The ideal $\mathbf{p}+ < \mathcal{I}_{\mathbf{a}} >$ is prime and it is the largest homogeneous ideal vanishing on $X(\mathbf{a})$.*

Proof: If a non-zero monomial is not standard, a pair of consecutive factors in the monomial is not standard. Replace that quadratic product by a sum of quadratic products each of which is one step better than the original, as seen in (1.5.2) above. Now we proceed by induction on the Young tableaux associated with monomial, with the **lexicographic** order on the set of all Young tableaux. Thus the set of all standard monomials on $X(\mathbf{a})$ generate $\mathbf{A}(a)$ as a k-algebra.

To see the linear independence of standard monomials in $\mathbf{A}(\mathbf{a})_m$, we proceed by induction on the degree m and on the dimension of $X(\mathbf{a})$. "Cancellation" of a suitable factor and/or "restriction" to a Schubert subvariety clinch the issue.

1.10 Theorem B: *Let L denote the tautological line bundle $\mathcal{O}_{\mathbb{P}}(1)$ on $\mathbb{P} = \mathbb{P}(\Lambda^d(V))$ as well as its restriction to any subvariety of $\mathbb{P}$. Let $q = dim(X(\mathbf{a}))$. Then we have:*

- **(1)** *The restriction map $H^0(\mathbb{P}, L^r) \longrightarrow H^0(X(\mathbf{a}), L^r)$ is surjective $\forall\ r \geq 0$, or, $H^0(X(\mathbf{a}), L^r)$ can be identified with $\mathbf{A}(a)_r$, $\forall\ r \geq 0$.*
- **(2)** *$H^j(X(\mathbf{a}), L^r) = (0)$ for $1 \leq j < q$ and all $r \in \mathbb{Z}$.*
- **(3)** *$H^q(X(\mathbf{a}), L^r) = (0)$ for all $r \geq 0$.*
- **(4)** $\mathrm{Pic}(X(\mathbf{a})) \approx \mathbb{Z}$*, generated by L, if $q \geq 1$.*

1.11 Theorem C: *Let $\widehat{X(\mathbf{a})}$ denote the cone over $X(\mathbf{a})$ for the Plücker embedding. Then we have the following.*

- **(1)** *$X(\mathbf{a})$ is projectively normal, i.e., $\widehat{X(\mathbf{a})}$ is normal at its vertex.*
- **(2)** *$X(\mathbf{a})$ is projectively Cohen-Macaulay.*
- **(3)** *$G_{d,n}$ is projectively factorial.*

The third assertion above is a particular case of the following.

1.12 Theorem D: *The following are equivalent for a Schubert subvariety $X(\mathbf{a})$ of the Grassmannian $G_{d,n}$.*

- **(1)** *$X(\mathbf{a})$ is projectively factorial.*
- **(2)** *$X(\mathbf{a})$ has a unique Schubert subvariety of codimension 1.*
- **(3)** *$p_{\mathbf{a}}$ is a prime in $\mathbf{A}(a)$.*
- **(4)** *$a_1, \cdots, a_d$ are all consecutive or consist of at most two blocks of consecutive integers if $a_1 = 1$.*
- **(5)** *$X(\mathbf{a})$ is a Grassmannian is a suitable space.*
- **(6)** *$X(\mathbf{a})$ is non-singular.*
- **(7)** *$X(\mathbf{a})$ is non-singular at the point cell $X(\mathbf{min})$.*
- **(8)** *$X(\mathbf{a})$ is factorial at the point cell $X(\mathbf{min})$.*
- **(9)** *$X(\mathbf{a})$ is locally factorial.*

Proof: Proceeding as indicated above, we see that the results A to D are established easily thanks to the explicit relations between various objects involved. A sequence of simple facts that goes into the proof is that the coordinate ring B of $D(\mathbf{a}) = X(\mathbf{a}) \cap C^-(\mathbf{min})$ (which is a "determinantal" variety), is the homogeneous localisation of $\mathbf{A}(a)$ at $p_{(\mathbf{min})}$. This happens to be a graded ring. The element $p_{(\mathbf{min})}$ is homogeneous and prime in the ring $\mathbf{A}(a)$ for every $\mathbf{a} \in I_{d,n}$. The obvious identity between the usual and homogeneous localisations is $\mathbf{A}(a)_{p_{(\mathbf{min})}} = B[p_{(\mathbf{min})}, p^{-1}_{(\mathbf{min})}]$. Now for example the factoriality of $\mathbf{A}(a)$ and B are equivalent.

2. The Generalised Flag Varieties G/Q

The Grassmannian $G_{d,n}$, the associated set $I_{d,n}$, the partial order $\geq$ on $I_{d,n}$, the Plücker coordinates $p_{\mathbf{i}}$ $(\mathbf{i} \in I_{d,n})$, the standard monomials in the Plücker coordinates, etc., all form a package concerning a very particular case (for a suitable choice of G and Q) of a wide class of the so called "generalised flag varieties G/Q". We shall now begin with explaining some of the notation listed above for G/Q in general and point out what they are/mean in some particular cases.

2.1 The set-up: Let G be a semi-simple, simply connected, split Chevalley algebraic group (defined over $\mathbb{Z}$) over an algebraically closed field k of arbitrary characteristic. Let Q be a parabolic subgroup, containing a Borel subgroup B, which in turn containing a maximal torus T. Let $n := \dim(T)$ be the rank of G. Let $N_Q(T)$ be the normaliser of T in Q and $W_Q = N_Q(T)/T$ be the **Weyl** group of Q. We write W for the Weyl group W_G of G. Let Δ be the system of roots relative to T, $\Delta^{\pm}$ be the set of positive/negative roots relative to B and $S = \{\alpha_i \mid 1 \leq i \leq n\}$ be the set of positive simple roots there of. Let $s_i := s_{\alpha_i}$ be the simple reflection defined by $\alpha_i \in S$. Let ℓ be the length function on W, as a Coxeter group generated by the simple reflections s_i's. Let $w_{0Q} := (w_0)_Q$ be the unique element in W_Q of maximal length. We write w_0 for w_{0G}.

2.2 The Bruhat Decomposition of G/Q: For $\theta \in W_Q$, let $n_\theta \in N_Q(T)$ be a lift for θ. The coset $n_\theta Q$ in G/Q depends only on θ and not on the lift n_θ chosen for θ. It is often denoted by e_θ. For the natural left action of G on G/Q, the B-orbit $C_Q(\theta) := Be_\theta = Bn_\theta Q$ is called the affine Schubert (or Bruhat) cell associated to θ. This depends only on the coset θW_Q in W/W_Q and not on θ. Moreover, $G/Q = \cup_{\theta \in W/W_Q} C_Q(\theta)$ is a disjoint union of the Schubert cells, called the Bruhat (cellular) decomposition of G/Q.

2.3 The Bruhat Order on W/W_Q: The Zariski closure $X_Q(\theta) := \overline{C_Q(\theta)}$ in G/Q, with the canonical reduced structure, is called the

Schubert variety in G/Q. The Bruhat order $\geq$ on W/W_Q is defined as $\theta \geq \theta'$ if $X_Q(\theta) \supseteq X_Q(\theta')$. See [**16**] for other equivalent forms of the Bruhat order.

For the case when $Q = B$, we have $W_B = \{1\}$ and the Bruhat order on W/W_B is called the Bruhat order on W. It is trivial to see that the natural surjective maps $W \to W/W_Q$, $\theta \mapsto \theta W_Q$, are order preserving.

Since the Schubert cell $C_Q(\theta)$ is a B-orbit in the B-stable variety G/Q, it follows that the boundary of its closure $X_Q(\theta)$ is also B-stable and hence is a disjoint union of B-orbits, i.e., we get a **cellular decomposition**, namely, $X_Q(\theta) = \cup_{\theta \geq \tau \in W/W_Q} C_Q(\tau)$ (i.e., a disjoint union of the Schubert cells).

Let $P_1, \cdots, P_r$ be all the maximal parabolic subgroups of G containing Q so that $Q = \cap_{j=1}^r P_j$. Let $\pi_j : G/Q \to G/P_j$ be the natural projection. Given $\theta, \theta' \in W$, we have

$$\theta \geq \theta' (i.e., X_Q(\theta) \supseteq X_Q(\theta')) \iff \pi_j(X_Q(\theta)) \supseteq \pi_j(X_Q(\theta')),\ \forall\, j \geq 1.$$

In fact, it can be seen that $X_Q(\theta) = \cap_{j=1}^r \pi_j^{-1}(\pi_j(X_Q(\theta)))$.

2.4 The sets $W_Q^{\mathbf{m}in}$ and $W_Q^{\mathbf{m}ax}$: Recall ([**7**]) that in each coset θW_Q, there is a *unique* element $\theta_Q^{\mathbf{m}in}$ of **minimal** length (resp. $\theta_Q^{\mathbf{m}ax}$ of **maximal** length). The set $W_Q^{\mathbf{m}in} := \left\{ \theta_Q^{\mathbf{m}in} \in \theta W_Q \,\middle|\, \theta \in W \right\}$ is called the set of *minimal representatives* of W/W_Q. We have

$$\begin{aligned} W_Q^{\mathbf{m}in} &= \left\{ \theta \in W \,\middle|\, \ell(\theta\theta') = \ell(\theta) + \ell(\theta'),\ \forall\, \theta' \in W_Q \right\} \\ &= \left\{ \theta \in W \,\middle|\, \theta(\alpha) > 0,\ \forall\, \alpha \in S_Q \right\}. \end{aligned}$$

Likewise, the collection $W_Q^{\mathbf{m}ax} := \left\{ \theta_Q^{\mathbf{m}ax} \in \theta W_Q \,\middle|\, \theta \in W \right\}$ is called the set of *maximal representatives* of W/W_Q. We have

$$\begin{aligned} W_Q^{\mathbf{m}ax} &= \left\{ \theta \in W \,\middle|\, \ell(\theta\theta') = \ell(\theta) - \ell(\theta'),\ \forall\, \theta' \in W_Q \right\} \\ &= \left\{ \theta \in W \,\middle|\, \theta(\alpha) < 0,\ \forall\, \alpha \in S_Q \right\}. \end{aligned}$$

Furthermore, if $w_{0Q} := (w_0)_Q$ is the unique element of maximal length in W_Q, then we have

$$W_Q^{\mathbf{m}ax} = W_Q^{\mathbf{m}in} w_{0Q} = \left\{ \theta w_{0Q} \middle| \theta \in W_Q^{\mathbf{m}in} \right\}.$$

2.5 Example: For $G = \mathrm{SL}(n,k)$, we can take T to be the subgroup of all diagonal matrices (so that $n-1$ is its rank), B (resp. B^-) to be the subgroup of all upper (resp. lower) triangular matrices and hence $W = S_n$, the full symmetric group on n symbols $\{1,\cdots,n\}$. The maximal parabolic subgroups of G containing B are exactly the P_d's, $1 \le d \le n-1$, where

$$P_d = \left\{ \begin{pmatrix} A & B \\ 0 & C \end{pmatrix} \in \mathrm{SL}(n,k) \;\middle|\; A \in \mathrm{GL}(d,k) \text{ and } C \in \mathrm{GL}(n-d,k) \right\}$$

whose Weyl group is the subgroup $S_d \times S_{n-d}$ with S_d (resp. S_{n-d}) being the permutation group on the first d symbols $\{1,\cdots,d\}$ (resp. on the last $n-d$ symbols $\{d+1,\cdots,n\}$).

• The quotient variety $\mathrm{SL}(n,k)/P_d$ is precisely the Grassmannian $G_{d,n}$ and its Schubert subvarieties are parametrised by $W/W_{P_d} = S_n/(S_d \times S_{n-d})$.

• The set $W_{P_d}^{\mathbf{m}in}$ of *minimal* representatives of $S_n/(S_d \times S_{n-d})$ in S_n is the set of *shuffle* permutations σ of level d, as described in (1.5.2) above, namely,

$$\sigma := \begin{pmatrix} 1 & <\cdots< & d & ; & d+1 & <\cdots< & n \\ 1^\sigma & <\cdots< & d^\sigma & ; & (d+1)^\sigma & <\cdots< & n^\sigma \end{pmatrix}$$

This is canonically identified with $I_{d,n} = \{1 \le 1^\sigma < \cdots < d^\sigma \le n\}$. (Likewise, the set $W_{P_d}^{\mathbf{m}ax}$ of *maximal* representatives of $S_n/(S_d \times S_{n-d})$ in S_n is the set of permutations

$$\sigma := \begin{pmatrix} 1 & <\cdots< & d & ; & d+1 & <\cdots< & n \\ 1^\sigma & >\cdots> & d^\sigma & ; & (d+1)^\sigma & >\cdots> & n^\sigma \end{pmatrix}$$

which is again obviously identifiable with the set $I_{d,n}$.)

• Fix $1 \le d_1 < \cdots < d_r \le n-1$ and take $Q = \cap_{j=1}^r P_{d_j}$. It can be seen that $W_Q = S_{d_1} \times S_{d_2-d_1} \times \cdots \times S_{n-d_r}$ and hence W/W_Q can be identified with $I_{d_1,\cdots,d_r,n}$ where

$$I_{d_1,\cdots,d_r,n} = \left\{ (\mathbf{i}_1,\cdots,\mathbf{i}_r) \in I_{d_1,n} \times \cdots \times I_{d_r,n} \;\middle|\; \mathbf{i}_j \subset \mathbf{i}_{j+1}, \forall 1 \le j \le r-1 \right\}.$$

• The set $W_Q^{\mathbf{m}in}$ can be seen to be all permutations $\sigma :=$

$$\begin{pmatrix} 1 < \cdots < d_1 & ; & d_1+1 < \cdots < d_2 & ; & d_2+1 < & ; & < n \\ 1^\sigma < \cdots < d_1^\sigma & ; & (d_1+1)^\sigma < \cdots < d_2^\sigma & ; & (d_2+1)^\sigma < & ; & < n^\sigma \end{pmatrix}$$

which in turn is obviously identifiable with $I_{d_1,\cdots,d_r,n}$ above.

• The element of maximal length in W_Q is given by $w_{0Q} =$

$$\begin{pmatrix} 1 < \cdots < d_1 & ; & d_1+1 < \cdots < d_2 & ; & d_2+1 < & ; & < n \\ d_1 > \cdots > 1 & ; & d_2 > \cdots > d_1+1 & ; & d_3 > & ; & > d_r+1 \end{pmatrix}$$

3. The Minuscule Weights

3.0 Recall that for the group $G = \mathrm{SL}(n,k)$, the fundamental representations are $V_{\varpi_d} := \Lambda^d(V_1)$ of highest weight ϖ_d, $1 \leq d \leq n-1$, where $V_1 = k^n$ is the natural representation (of highest weight $\varpi_1 = \epsilon_1$). The character group $X(T)$ has a basis of the fundamental weights $\varpi_1, \cdots, \varpi_{n-1}$ where $\varpi_d : T \longrightarrow k^\star$, $\mathrm{diag}(t_1, \cdots, t_n) \mapsto t_1 \cdots t_d$, $1 \leq d \leq n-1$. In fact, we have $\varpi_d = \epsilon_1 + \cdots + \epsilon_d$ where $\epsilon_d : T \longrightarrow k^\star$, $\mathrm{diag}(t_1, \cdots, t_n) \mapsto t_d$, $1 \leq d \leq n$ with $\epsilon_1 + \cdots + \epsilon_n = 0$. The Weyl group of G is the full symmetric group S_n on n symbols $\{1, \cdots, n\}$.

The weight ϖ_d is called **minuscule** in the sense that all the weights of the irreducible representation space V_{ϖ_d} whose highest weight is ϖ_d, namely, $\left\{\epsilon_{i_1} + \cdots + \epsilon_{i_d} \,\middle|\, \mathbf{i} = (i_1, \cdots, i_d) \in I_{d,n}\right\}$ are **extremal**, i.e., the Weyl group translates of the highest weight ϖ_d $= \epsilon_1 + \cdots + \epsilon_d$ (by the coset space $S_n/(S_d \times S_{n-d})$). The maximal parabolic subgroup associated to ϖ_d is the isotropic subgroup at the line through $e_{\mathbf{min}} = e_1 \wedge \cdots \wedge e_d$, namely,

$$P_d = \left\{ \begin{pmatrix} A & B \\ 0 & C \end{pmatrix} \in \mathrm{SL}(n,k) \,\middle|\, A \in \mathrm{GL}(d,k) \text{ and } C \in \mathrm{GL}(n-d,k) \right\}$$

whose Weyl group is the isotropic subgroup at the highest weight ϖ_d, namely, the subgroup $S_d \times S_{n-d}$. A complete set of the extremal weight vectors of V_{ϖ_d} is given by

$$\left\{ e_{\mathbf{i}} = e_{i_1} \wedge \cdots \wedge e_{i_d} \,\middle|\, \mathbf{i} = (i_1, \cdots, i_d) \in I_{d,n} \right\}.$$

The $\mathrm{SL}(n,k)$-orbit through $e_{\mathbf{min}}$ in the projective space $\mathbb{P}(\Lambda^d(V))$ is the Grassmannian $G_{d,n}$, isomorphic to the quotient variety $\mathrm{SL}(n,k)/P_d$. The Schubert subvarieties of $G_{d,n}$ are the Zariski closures of the B-orbits through the extremal weight vectors described above and are parametrised by $W/W_{P_d} = S_n/(S_d \times S_{n-d}) = I_{d,n}$.

Now we shall look at the minuscule weights in other groups G than $\mathrm{SL}(n,k)$ and see to what extent the Theorems A to D above remain true.

3.1 Minuscule Weights: Let V be a minuscule fundamental representation of G of highest weight ϖ, i.e., all the weights of V are extremal (i.e., the Weyl group translates of ϖ) and hence all the weight spaces are 1-dimensional. Consequently, V has a basis, unique upto non-zero scalar multiples. Let P be the associated maximal parabolic subgroup, i.e., P is the isotropy of the line through a *highest weight* vector, say e_ϖ, in V. We also say that P is **minuscule** since ϖ is minuscule. The G-orbit Ge_ϖ through e_ϖ in $\mathbb{P} = \mathbb{P}(V)$ is closed and is isomorphic to G/P.

The restriction L of the tautological line bundle $\mathcal{O}_{\mathbb{P}}(1)$ to G/P is the **ample generator** of $\mathrm{Pic}(G/P)$. It follows that $H^0(G/P, L) = \mathrm{Hom}(V, k)$ is the dual of V. It is an irreducible G-module of highest weight $\mathbf{i}(\varpi)$ where $\mathbf{i}$ is the Weyl involution.

If W is the Weyl group of G, then the isotropy subgroup at the highest weight ϖ is the Weyl group W_P of P. Thus the set of all weight vectors of V of distinct weights is $\{\tau e_\varpi \mid \tau \in W/W_P\}$. It follows that the weight of τe_ϖ is $\tau(\varpi)$ for the natural action of W on $X(T)$ and that the set

$$\left\{ e_\tau := e_{\tau\varpi} := \tau e_\varpi \;\middle|\; \tau \in W/W_P \right\}$$

is a basis of V, unique upto non-zero scalar multiples.

Let $\{p_\tau \mid \tau \in W/W_P\}$ be the basis of the dual $V^\vee = \mathrm{Hom}_k(V, k)$ $= H^0(G/P, L)$, dual to the basis $\{e_\tau \mid \tau \in W/W_P\}$ of V. It follows that $p_\tau = \pm \tau p_{\mathrm{id}}$ for every $\tau \in W/W_P$. The weight of p_τ is $-\tau(\varpi)$ and in particular, (i) p_{id} is the *lowest weight* vector of weight $-\varpi$ and (ii) p_{w_0} is a highest weight vector of weight $\mathbf{i}(\varpi) := -w_0(\varpi)$ in $H^0(G/P, L)$.

3.2 Standard Monomials: Let $X(\theta) = X_P(\theta)$ be the Schubert variety defined by $\theta \in W/W_P$. Let

$$\mathbf{A}(\theta) = \bigoplus_{m=0}^{\infty} \mathbf{A}(\theta)_m \text{ where } \mathbf{A}(\theta)_m = H^0(X(\theta), L^m) \text{ and}$$

$$\mathbf{A} = \mathbf{A}(w_0) := \bigoplus_{m=0}^{\infty} \mathbf{A}_m \text{ where } \mathbf{A}_m = H^0(G/P, L^m).$$

We have the natural restriction map $\mathbf{A} \longrightarrow \mathbf{A}(\theta)$ which is a surjective k-algebra homomorphism for every $\theta \in W/W_P$, whose kernel is the largest homogeneous ideal I_θ in $\mathbf{A}$ vanishing on $X(\theta)$. It is easy to see that $I_\theta = \langle p_\tau \mid \theta \not\geq \tau \rangle$.

The following are natural and straightforward generalisations from the case of the Grassmannian to the minuscule weights.

3.2(a) An element $f \in \mathbf{A}_m$ of the form

$$f = p_{\tau_1} p_{\tau_2} \cdots p_{\tau_m} \text{ with } \tau_1 \geq \tau_2 \geq \cdots \geq \tau_m,\ \tau_j \in W/W_P,$$

is called a **standard monomial** of degree (or length) m in the p_τ's on G/P.

Standard monomials on G/P are nothing but standard monomials on $X(w_0)$ (or relative to w_0) in the following sense.

3.2(b) An element $f \in \mathbf{A}(\theta)_m$ of the form

$$f = p_{\tau_1} p_{\tau_2} \cdots p_{\tau_m} \text{ with } \theta \geq \tau_1 \geq \tau_2 \geq \cdots \geq \tau_m,\ \tau_j \in W/W_P,$$

is called a **standard monomial** of degree (or length) m on $X(\theta)$.

3.3 The Quadratic Relations: Let $p_\tau p_\sigma$ be a non-standard quadratic monomial. Then there exists a unique relation

$$p_\tau p_\sigma = \sum_{\tau_1 \geq \sigma_1} c_{\tau_1\sigma_1} p_{\tau_1} p_{\sigma_1},$$

such that $c_{\tau_1\sigma_1} \in k^\star$ and (τ_1, σ_1) runs over suitable distinct pairs of Weyl group elements satisfying $\tau_1 > \tau, \sigma > \sigma_1$.

These quadratic relations go parallel to the relations (1.5) above, but these are not so constructive as in the case of $G_{d,n}$. The crucial step that goes in the proof of these relations is the verification (using Demazure's character formula [**15**]) that

$$\dim_k(G/P, L^2) = \#\left\{(\tau,\sigma) \in W/W_P \times W/W_P \;\middle|\; \tau \geq \sigma\right\}.$$

3.4 Theorem A: With notation as above, we have the following.

(1) *The coordinate function p_σ on $X(\theta)$ is non-zero $\iff$ $\theta \geq \sigma$.*

(2) *Standard Monomials of degree m on $X(\theta)$ span the vector space $\mathbf{A}(\theta)_m$.*

(3) *Standard Monomials on $X(\theta)$ are linearly independent.*

(4) *The largest homogeneous ideal $\mathcal{I}_\theta$ vanishing on $X(\theta)$ is generated by $\{p_\sigma \mid \theta \not\geq \sigma\}$.*

3.5 Configuration of Schubert Varieties: If P is a minuscule maximal parabolic subgroup of G, the configuration of the Schubert varieties $X_P(\theta)$ ($\theta \in W/W_P$), their intersections, etc., in G/P are exactly as in the case of the Grassmannian $G_{d,n}$ ([**33**]). It is natural to expect the Theorems A to D above for $G_{d,n}$ to remain true for G/P. It is almost true but NOT quite for the critical reason that we do not EXPLICITLY have the ideal of relations on regular functions vanishing on G/P for its embedding into the projective space $\mathbb{P}(V)$. In other words, the counter part to the quadratic relations, (1.5) above, is not in such best form. While the Theorems A, B and C remain true verbatim, the counter part of "Theorem D" above splits into 2 independent parts (Theorems D1 and D2) as follows ([**10**]).

3.6 Theorem D1: *The following are equivalent for a Schubert subvariety $X(\theta)$ of G/P, (P minuscule).*

- **(1)** *$X(\tau)$ is arithmetically factorial.*
- **(2)** *$X(\theta)$ has a unique Schubert subvariety of codimension 1.*
- **(3)** *p_θ is a prime in $\mathbf{A}(\theta)$.*
- **(4)** *$X(\theta)$ is factorial at the point cell $X(id)$.*
- **(5)** *$X(\theta)$ is locally factorial.*

3.7 Theorem D2: *The following are equivalent for a Schubert subvariety $X(\theta)$ of G/P, (P minuscule).*

- **(1)** *$X(\theta)$ is non-singular.*
- **(2)** *$X(\theta)$ is non-singular at the point cell.*
- **(3)** *$X(\theta)$ satisfies Poincaré duality (or symmetry).*

4. The Quasi-Minuscule Weights

4.1 Quasi-Minuscule Weights: A fundamental representation V of G of highest weight ϖ is said to be **quasi-minuscule** if all the non-zero weights of V are extremal (i.e., the Weyl group translates of ϖ) and hence all the non-zero weight spaces are 1-dimensional. Consequently, except for the "zero weight space", V has a basis, unique upto non-zero scalar multiples.

Let P be the associated maximal parabolic subgroup, i.e., P is the isotropy of the line through a *highest weight* vector, say e_ϖ, in V. We also say that P is **quasi-minuscule** since ϖ is quasi-minuscule. The G-orbit Ge_ϖ through e_ϖ in $\mathbb{P} = \mathbb{P}(V)$ is closed and is isomorphic to G/P.

The restriction L of the tautological line bundle $\mathcal{O}_{\mathbb{P}}(1)$ to G/P is the **ample generator** of $\mathrm{Pic}(G/P)$. It follows that $H^0(G/P, L) = \mathrm{Hom}(V, k)$ is the dual of V. It is an irreducible G-module of highest weight $\mathbf{i}(\varpi)$ where $\mathbf{i}$ is the Weyl involution.

If W is the Weyl group of G, then the isotropy subgroup at the highest weight ϖ is the Weyl group W_P of P. Thus the set of all non-zero weight vectors of V of distinct weights is $\{\tau e_\varpi \mid \tau \in W/W_P\}$.

It follows that the weight of τe_{ϖ} is $\tau(\varpi)$ for the natural action of W on $X(T)$.

4.2 Chevalley's Multiplicity Formula: ([**33**], p.120)

Let ϖ be a fundamental weight with the associated maximal parabolic subgroup P. Let $\alpha \in S$ be the simple root dual to ϖ, i.e., $\left\langle \varpi, \beta^\star \right\rangle = \delta_{\alpha,\beta}$ for all $\beta \in S$. Recall that G/P contains a unique Schubert divisor (Schubert subvariety of codimension 1, say H. We know that $H = X(s_\alpha w_0)$ is the moving divisor moved by α. For $\tau \in W/W_P$, let $[X(\tau)]$ denote the class of the Schubert variety $X(\tau)$ in the Chow ring $\mathrm{Ch}(G/P)$. Then we have

$$\boxed{[X(\tau)] \cdot [H] = \sum_{\tau_i} d_{\tau_i}[X(\tau_i)],\ d_i \in \mathbb{Z}^+,}$$

where $\cdot$ on the left denotes the multiplication in the Chow ring of G/P and the summation on the right is taken over all the Schubert divisors $X(\tau_i)$ in $X(\tau)$. For set-theoretic considerations, every Schubert divisor occurs in the sum on the right, i.e., $d_i \geq 1,\ \forall\ i$. The integer d_i is called the **geometric intersection multiplicity** of $X(\tau_i)$ in $[X(\tau)]\cdot[H]$ or simply the multiplicity of $X(\tau_i)$ in $X(\tau)$. We say that a Schubert divisor $X(\tau_i)$ is an **ordinary** (resp.**double**, resp. **multiple** divisor in $X(\tau)$ if $d_i = 1$ (resp. 2, resp. ≥ 2).

The multiplicity d_i of $X(\tau_i)$ in $X(\tau)$ is calculated as follows. Write $\tau_i = s_{\beta_i}\tau$ for some $\beta_i \in \Delta^+$. Now write $\tau_i = s_{\beta_i}\tau = \tau s_{\gamma_i}$ for some $\gamma_i \in \Delta^+$. Then $d_i = \left\langle \varpi, \gamma_i^\star \right\rangle$. Note that we have $\gamma_i = \pm\tau^{-1}(\beta_i)$ and hence

$$\boxed{d_i = \left\langle \varpi, \gamma_i^\star \right\rangle = \left|\left\langle \varpi, \tau^{-1}(\beta_i)^\star \right\rangle\right| = \left\langle \tau(\varpi), \beta_i^\star \right\rangle.}$$

5. The Classical Type Weights

5.1 Theorem: *Let ϖ, $V = V_\varpi$, P and $\alpha \in S$ be as above. Then we have the following.*

(1) (Minuscule weights): *The following are equivalent:*

(a) *The fundamental weight ϖ (or the maximal parabolic subgroup P) is* **minuscule**, *i.e., all the weights of V are extremal.*

(b) $\left\langle \varpi, \beta^\star \right\rangle$ *= 0 or 1, $\forall\ \beta \in \Delta^+$.*

(c) *The multiplicity of a Schubert divisor* $X(\tau_i)$ *in* $X(\tau)$ *is 1 for every Schubert variety* $X(\tau)$ *of* G/P.

(2) (Quasi-minuscule or Distinguished weights): *The following are equivalent:*

(a) *The fundamental weight* ϖ *(or the maximal parabolic subgroup* P*) is* **quasi-minuscule or distinguished, i.e.,** *all the* **non-zero weights** *of* V *are extremal.*

(b) $\left\langle \varpi, \beta^\star \right\rangle = 0$ *or 1 or 2,* $\forall\ \beta \in \Delta^+$ *and it equals 2* $\iff \varpi = \beta$ *(a root).*

(c) *The multiplicity of a Schubert divisor* $X(\tau_i)$ *is 1 or 2 in every Schubert variety* $X(\tau)$ *of* G/P *and 2 occurs when and only when* $dim(X(\tau)) = \dfrac{1 + \dim(G/P)}{2}$ *and then* $X(\tau)$ *has a unique moving divisor* $X(\tau_i)$.

(3) (Classical type weights): *The following are equivalent:*

(a) *The fundamental weight* ϖ *(or the maximal parabolic subgroup* P*) is* **classical type, i.e.,** $\left\langle \varpi, \beta^\star \right\rangle = 0$ *or 1 or 2,* $\forall\ \beta \in \Delta^+$.

(b) *The multiplicity of a Schubert divisor* $X(\tau_i)$ *is 1 or 2 in every Schubert variety* $X(\tau)$ *of* G/P.

Proof: Easy verification or see [**33**], pp. 105-106.

5.2 Admissible Pairs: The concept of an *admissible pair* is fundamental to a generalisation of the Hodge-Young theory of standard monomials (stated above), to pass from the group SL(n, k) to an arbitrary semi-simple group G. This was suggested by the work of DeConcini and Procesi on classical invariant theory ([**37**], [**13**]). In fact, a basis was given by them for the rings of invariants in the *Classical Invariant Theory* by means of certain "double standard tableaux". These double standard tableaux admit a suitable interpretation giving rise to this notion.

It turns out that these rings of invariants are the coordinate rings of suitable affine open subsets of certain Schubert varieties in G/P for the case when G is a classical group and P a maximal parabolic subgroup of G. Consequently, the "double standard tableaux" basis as above becomes a particular case of the standard monomial basis.

Let ϖ be a fundamental weight of classical type and P be the associated maximal parabolic subgroup of G. A pair of elements (σ, σ') in W/W_P is called an **admissible pair** if either $\sigma = \sigma'$ (in which case it is called a **trivial** admissible pair) or $\sigma \neq \sigma'$ and there exists a finite sequence of elements $\sigma = \sigma_1, \cdots, \sigma_s = \sigma'$ in W/W_P such that

(1) $\sigma_1 > \sigma_2 > \cdots > \sigma_s$ in W/W_P,
(2) $X(\sigma_i)$ is a Schubert divisor in $X(\sigma_{i-1})$, $2 \leq i \leq s$ and
(3) $X(\sigma_i)$ is a double divisor in $X(\sigma_{i-1})$, $2 \leq i \leq s$.

We denote the set of all admissible pairs by $\widehat{W/W_P}$. It is clear that the set of all *trivial* admissible pairs can be naturally identified with W/W_P and $\widehat{W/W_P}$ is therefore a disjoint union of W/W_P and the set of all non-trivial admissible pairs. We extend the (Bruhat) partial order $\geq$ on W/W_P to an order $\geq$ on $\widehat{W/W_P}$ by defining $(\sigma, \sigma') \geq (\tau, \tau')$ if $\sigma' \geq \tau$ in W/W_P. We note that this is not a partial order on $\widehat{W/W_P}$ (since reflexivity fails for non-trivial admissible pairs).

5.3 A Nice Basis for $H^0(G/P, L)$: Assume that $\mathrm{char}(k) = 0$. Let $V = V_\varpi$ be the fundamental representation of G of highest weight ϖ. Let e be a highest weight vector so that the isotropy at e is the maximal parabolic subgroup P. We have the natural identification of G/P with the closed orbit $G \cdot e$ in the projective space $\mathbb{P}(V)$. Let L be the restriction of the tautological line bundle $\mathcal{O}_{\mathbb{P}}(1)$ on $\mathbb{P}(V)$ to G/P as well as any closed subvariety of G/P.

The vector space $H^0(G/P, L)$ is dual to $V = V_\varpi$ and hence is a G-module of highest weight $\mathbf{i}(\varpi)$ where $\mathbf{i}$ is the Weyl involution. To construct a basis for V_ϖ or $H^0(G/P, L)$, we proceed as follows (modulo a host of finer technical details).

5.3(a) Use Demazure's character formula for V_ϖ and parametrise the characters of V_ϖ by the set of all admissible pairs $\widehat{W/W_P}$. The weight associated to an admissible pair $(\sigma, \sigma') \in \widehat{W/W_P}$ is $\dfrac{\sigma(\varpi) + \sigma'(\varpi)}{2}$ where $\sigma(\varpi)$ is the extremal weight associated to $\sigma \in W$.

5.3(b) It is well known that $V_\varpi = \mathbf{u} \cdot e = \mathbf{u}^- \cdot e$ where e is a highest weight vector in V_ϖ. For $\sigma \in W$, let $e_\sigma := \sigma \cdot e \in V_\varpi$. We know that $e_\sigma \notin ke'_\sigma \iff \sigma \neq \sigma'$ in W/W_P. For $\theta \in W/W_P$, we define $V_\varpi(\theta) := \mathbf{u}^+ \cdot e_\theta$. We know that

$$V_\varpi = V_\varpi(w_0) \text{ and } V_\varpi(\theta) \subseteq V_\varpi(\theta') \iff \theta \leq \theta'.$$

5.3(c) First we construct a set of weight vectors $\left\{ q_{(\sigma,\sigma')} \;\middle|\; (\sigma, \sigma') \in \widehat{W/W_P} \right\}$, parametrised by the set of admissible pairs such that the weight of $q_{(\sigma,\sigma')}$ is $\dfrac{\sigma(\varpi)+\sigma'(\varpi)}{2}$ for every $(\sigma,\sigma') \in \widehat{W/W_P}$.

5.3(d) Finally show that $\left\{ q_{(\sigma,\sigma')} \;\middle|\; (\sigma,\sigma') \in \widehat{W/W_P} \right\}$ is linearly independent in V_ϖ, hence $\left\{ q_{(\sigma,\sigma')} \;\middle|\; (\sigma,\sigma') \in \widehat{W/W_P} \right\}$ is a basis of weight vectors for V_ϖ. In fact, for every $\theta \in W/W_P$, we get that $\left\{ q_{(\sigma,\sigma')} \;\middle|\; \theta \geq (\sigma,\sigma') \in \widehat{W/W_P} \right\}$ is a basis of weight vectors for $V_\varpi(\theta)$.

5.3(e) Let $\left\{ p_{(\sigma,\sigma')} \;\middle|\; (\sigma,\sigma') \in \widehat{W/W_P} \right\}$ be the dual basis of $H^0(G/P,L)$, dual to the basis $\left\{ q_{(\sigma,\sigma')} \;\middle|\; (\sigma,\sigma') \in \widehat{W/W_P} \right\}$ of V_ϖ. It follows that $p_{(\sigma,\sigma')}$ is a weight vector of weight $\dfrac{\sigma(-\varpi)+\sigma'(-\varpi)}{2}$ for every $(\sigma,\sigma') \in \widehat{W/W_P}$.

5.3(f) For $\theta \in W/W_P$, let $X(\theta) = \overline{B\theta P/P}$ (the Zariski closure with the reduced structure) be the Schubert variety defined by θ. The constructions/proofs above are geared so as to give that

$$\left\{ p_{(\sigma,\sigma')} \;\middle|\; (\sigma,\sigma') \in \widehat{W/W_P},\ \theta \geq \sigma \right\}$$

is a basis of $V_\varpi(\theta)^\vee = H^0(X(\theta),L)$.

5.4 Standard Monomials: Let

$$\mathbf{A}(\theta) = \bigoplus_{m=0}^{\infty} \mathbf{A}(\theta)_m \text{ where } \mathbf{A}(\theta)_m = H^0(X(\theta),L^m) \text{ and}$$

$$\mathbf{A} = \mathbf{A}(w_0) := \bigoplus_{m=0}^{\infty} \mathbf{A}_m \text{ where } \mathbf{A}_m = H^0(G/P,L^m).$$

We have the natural restriction map $\mathbf{A} \longrightarrow \mathbf{A}(\theta)$ which is a surjective k-algebra homomorphism for every $\theta \in W/W_P$, whose kernel is the largest homogeneous ideal I_θ in $\mathbf{A}$ vanishing on $X(\theta)$. It is easy to see that $I_\theta = \left\langle p_{(\tau,\tau')} \;\middle|\; \theta \not\geq \tau \right\rangle$.

5.4(a) An element $f = p_{(\tau_1,\tau_1')}p_{(\tau_2,\tau_2')}\cdots p_{(\tau_m,\tau_m')}$ in $\mathbf{A}_m$ with

$$\boxed{(\tau_1,\tau_1') \geq (\tau_2,\tau_2') \geq \cdots \geq (\tau_m,\tau_m'),\ (\tau_j,\tau_j') \in \widehat{W/W_P},}$$

is called a **standard monomial** of degree (or length) m on G/P.

5.4(b) An element $f = p_{(\tau_1,\tau_1')} p_{(\tau_2,\tau_2')} \cdots p_{(\tau_m,\tau_m')}$ in $\mathbf{A}(\theta)_m$ with

$$\boxed{\theta \geq (\tau_1,\tau_1') \geq (\tau_2,\tau_2') \geq \cdots \geq (\tau_m,\tau_m'),\ (\tau_j,\tau_j') \in \widehat{W/W_P},}$$

is called a **standard monomial** of degree (or length) m on $X(\theta)$ (or relative to $\theta \in W/W_P$).

5.5 The Quadratic Relations: There are two types of quadratic relations satisfied on $X(\theta)$ by the coordinate functions.

5.5(a) The Global Quadratic Relations: Let $p_{(\tau,\tau')}p_{(\sigma,\sigma')}$ be a non-standard quadratic monomial. Then there exists a unique relation

$$\boxed{p_{(\tau,\tau')}p_{(\sigma,\sigma')} = \sum_{\tau_1' \geq \sigma_1} c_{(\tau_1,\tau_1'),(\sigma_1.\sigma_1')} p_{(\tau_1,\tau_1')} p_{(\sigma_1.\sigma_1')},}$$

such that $c_{(\tau_1,\tau_1'),(\sigma_1.\sigma_1')} \in k^\star$ and $(\tau_1,\tau_1'), (\sigma_1,\sigma_1')$ runs over distinct pairs of admissible pairs satisfying $\tau_1' > \tau, \sigma' > \sigma_1$.

5.5(b) The Local Quadratic Relations: For $(\tau,\tau') \in \widehat{W/W_P}$, we have a relation like, $\boxed{p^2_{(\tau,\tau')} = \pm p_\tau p_{\tau'}}$ on the Schubert variety $X(\tau)$.

As before, a crucial step that goes into a proof of these relations is the verification (using Demazure's character formula [**15**])

$$\boxed{\dim_k(G/P, L^2) = \#\left\{\Big((\tau,\tau'),(\sigma,\sigma')\Big) \in \widehat{W/W_P} \times \widehat{W/W_P} \;\Big|\; \tau' \geq \sigma\right\}.}$$

This is conceptually as easy as in the case of quasi-minuscule weights but technically a little more involved. Essentially by the same pattern/scheme of arguments as before, we prove the following.

5.5(c) Theorem A: *With notation as above, we have the following:*

- **(1)** *For $(\sigma,\sigma') \in \widehat{W/W_P}$, the coordinate function $p_{(\sigma,\sigma')}$ on $X(\theta)$ is non-zero if and only if $\theta \geq \sigma$.*
- **(2)** *Standard Monomials of degree m on $X(\theta)$ span the vector space $\mathbf{A}(\theta)_m$.*
- **(3)** *Standard Monomials on $X(\theta)$ are linearly independent.*
- **(4)** *The largest homogeneous ideal $\mathcal{I}_\theta$ vanishing on $X(\theta)$ is generated by $\{p_{(\sigma,\sigma')} \mid \theta \not\geq \sigma\}$.*

5.5(d) Theorem B: *Let* $L = L(\varpi)$ *be the tautological line bundle* $\mathcal{O}_{\mathbb{P}}(1)$ *on* $\mathbb{P} = \mathbb{P}(V_{\varpi})$ *as well as its restriction to any subvariety of* $\mathbb{P}$. *Let* $q = dim(X(\theta))$ *for* $\theta \in W/W_P$. *Then we have the following.*

(1) *The natural restriction map* $H^0(\mathbb{P}, L^r) \longrightarrow H^0(X(\theta), L^r)$ *is surjective, or,* $H^0(X(\theta), L^r)$ *can be identified with* $\mathbf{A}(\theta)_r$, $\forall\, r \geq 0$.

(2) $H^j(X(\theta), L^r) = (0)$ *for* $1 \leq j < q$ *and all* $r \in \mathbb{Z}$.

(3) $H^q(X(\theta), L^r) = (0)$ *for all* $r \geq 0$.

6. The Variety G/Q for Q of Classical Type

6.1 Parabolic subgroups of Classical Type: Let $Q \supseteq B$ be any parabolic subgroup of G. We say that Q is of **classical type** if every maximal parabolic subgroup P containing Q is of classical type (see (4.1) above).

It is a simple matter to verify that G is of classical type (i.e., G is of type A, B, C or D) if and only if B is of classical type. Furthermore, every group G has at least one maximal parabolic subgroup P which is of classical type (and hence other parabolic subgroups Q of classical type).

Let P_i denote the maximal parabolic subgroup of G corresponding to the fundamental weight ϖ_i, $1 \leq i \leq n$, where n is the rank of G. Let $W_i := W_{P_i}$ be the Weyl group of W_{P_i}, $1 \leq i \leq n$. We assume for simplicity of notation that $P_1, \cdots, P_r$, $r \leq n$, are all those which are of classical type. (For example, $r = n$ if G is a classical group, $r = 5$ if G is of type E_6, etc.) Fix an $s \leq r$ and let $Q = \cap_{i=1}^{s} P_i$ (so that Q is of classical type).

6.2 Young Diagrams of type m: Fix a non-negative integral vector $\mathbf{m} = (m_1, \cdots, m_s) \in (\mathbb{Z}^+)^s$. By a **Young diagram of type m** or **multidegree m**, we mean a pair (σ, σ') defined as follows.

$$\sigma = (\sigma_{ij}),\ \sigma' = (\sigma'_{ij}) \text{ with } (\sigma_{ij}, \sigma'_{ij}) \in \widehat{W/W_i},\ 1 \leq j \leq m_i \ \&\ 1 \leq i \leq s$$

where $\widehat{W/W_i}$ is the set of all admissible pairs associated to P_i (as defined in (4.2) above). We denote this symbolically as follows. We have $(\sigma, \sigma') =$

$$\underbrace{(\sigma_{11}, \sigma'_{11}) \cdots (\sigma_{1m_1}, \sigma'_{1m_1})}_{\text{from } \widehat{W/W_1}} \cdots (\underbrace{\cdots}_{\widehat{W/W_i}}) \cdots \underbrace{(\sigma_{m_s1}, \sigma'_{m_s1}) \cdots (\sigma_{sm_s}, \sigma'_{sm_s})}_{\text{from } \widehat{W/W_s}} \cdot$$

6.3 Standard Young Diagrams on $X(\theta)$: We say that a Young diagram (σ, σ') of type **m** is **standard** on $X(\theta)$ (or relative to) $\theta \in W/W_Q$ (and written $\theta \geq (\sigma, \sigma')$) if there exists a pair (τ, τ'), called a **defining pair**, with the following properties:

(1) $\tau = (\tau_{ij})$, $(\tau') = (\tau'_{ij})$, $1 \leq j \leq m_i$, $1 \leq i \leq s$ with $\tau_{ij}, \tau'_{ij} \in W/W_Q$,

(2) each τ_{ij} (resp. τ'_{ij}) is a lift for σ_{ij} (resp. σ'_{ij}) in W/W_Q and

(3) $\theta \geq \tau_{11} \geq \tau'_{11} \geq \cdots \geq \tau_{1m_1} \geq \tau'_{1m_1} \geq \tau_{21} \geq \tau'_{21} \cdots \geq \tau_{ij} \geq \tau'_{ij} \geq \tau_{ij+1} \geq \tau'_{ij+1} \geq \cdots \geq \tau_{im_i} \geq \tau'_{im_i} \geq \tau_{i+11} \geq \tau'_{i+11} \geq \cdots \geq \tau_{sm_s} \geq \tau'_{sm_s}$ in W/W_Q. Denote this symbolically by

$$\theta \geq (\sigma_{11}, \sigma'_{11}) \geq \cdots \geq (\sigma_{1m_1}, \sigma'_{1m_1}) \geq (\cdots) \geq$$

$$\geq (\cdots) \geq (\sigma_{m_s 1}, \sigma'_{m_s 1}) \geq \cdots \geq (\sigma_{sm_s}, \sigma'_{sm_s}).$$

If $m_t = 0$ for any t, $1 \leq t \leq s$, the family of elements $\sigma_{t\star}, \sigma'_{t\star}, \tau_{t\star}, \tau'_{t\star}$ is understood to be empty.

• The most crucial (but somewhat subtle) point that is to be noted here is the meaning of the comparison $\boxed{(\sigma_{im_i}, \sigma'_{im_i}) \geq (\sigma_{i+11}, \sigma'_{i+11})}$ between elements of the unrelated sets $\widehat{W/W_i}$ and $\widehat{W/W_{i+1}}$. The actual comparison that makes this meaningful is that between their respective lifts τ'_{im_i} and τ_{i+11} in W/W_Q for σ'_{im_i} and σ_{i+11}. (See the comment (6.8)(6) below.)

6.4 Standard Monomials of Multidegree m: Let $\varpi = \sum_{i=1}^{s} m_i \varpi_i$ be the **dominant weight** associated to $\mathbf{m} = (m_1, \cdots, m_s) \in \mathbb{Z}^{+s}$ for $1 \leq s \leq r \leq n$. Let $L(\varpi) = L(\varpi_1)^{m_1} \otimes \cdots \otimes L(\varpi_s)^{m_s}$ be the line bundle on G/Q defined by ϖ. To each Young diagram (σ, σ') of multidegree **m**, we define an element $p_{(\sigma,\sigma')} \in H^0(G/Q, L(\varpi))$ as follows (which is a monomial in the $p'_{(\sigma_{ij},\sigma'_{ij})}s$):

$$p_{(\sigma,\sigma')} = \prod_{1 \leq i \leq s} \prod_{1 \leq j \leq m_i} p_{(\sigma_{ij},\sigma'_{ij})}.$$

We note that $p_{(\sigma,\sigma')}$ is a weight vector of weight

$$\chi(\sigma, \sigma') = \sum_{1 \leq i \leq s} \sum_{1 \leq j \leq m_i} \frac{\sigma_{ij}(-\varpi_i) + \sigma'_{ij}(-\varpi_i)}{2}.$$

We say that the monomial $p_{(\sigma,\sigma')}$ is **standard** on $X(\theta)$ (or relative to) $\theta \in W/W_Q$ if (σ, σ') is standard on $X(\theta)$, i.e., $\theta \geq (\sigma, \sigma')$.

The following are the main results for G/Q when Q is of classical type whose proofs follow the same pattern as their predecessors in the case of maximal parabolic subgroups P of classical type. However, the details are technically more involved.

6.5 Theorem A: *Standard monomials $p_{(\sigma,\sigma')}$ on $X(\theta)$ of multidegree $\mathbf{m} = (m_1, \cdots, m_s)$ form a basis of $H^0(X(\theta), L(\varpi))$. Consequently, we have the*

Character formula:

$$\operatorname{char}[H^0(X(\theta), L(\varpi))] = \sum_{\theta \geq (\sigma,\sigma')} \exp(\chi(\sigma, \sigma')).$$

6.6 Theorem B: *With assumptions and notation as above, we have:*

(1) *The natural restriction map*

$$H^0(G/Q, L(\varpi)) \longrightarrow H^0(X(\theta), L(\varpi))$$

is surjective. In other words, if $L(\varpi)$ is ample (i.e., $m_i \geq 1$ for all $1 \leq i \leq s$), then $X(\theta)$ is projectively normal for the embedding $G/Q \hookrightarrow \mathbb{P}(V_\varpi)$, given by $L(\varpi)$.

(2) *Given a dominant weight $\varpi' = \sum_{i=1}^{s} m'_i \varpi_i$, the natural map*

$$H^0(X(\theta), L(\varpi)) \otimes H^0(X(\theta), L(\varpi')) \longrightarrow H^0(X(\theta), L(\varpi + \varpi'))$$

is surjective and

(3) *$H^j(X(\theta), L(\varpi)) = (0)$ for all $j \geq 1$.*

A combination of the standard monomial bases and the methods of **Frobenius Splittings** (see [**42**], [**52**], [**53**], etc.) yield the following.

6.7 Theorem C: *With notation as above, we have the following:*

(1) *$X(\theta)$ is normal, Cohen-Macaulay and has at the most only rational singularities.*

(2) *G/Q is an intersection of quadrics and $X(\theta)$ is linear section of G/Q for every projective embedding of G/Q.*

(3) *$X(\theta)$ is projectively normal and Cohen-Macaulay.*

(4) *For all θ and τ in W/W_Q, the scheme-theoretic intersection $X(\theta) \cap X(\tau)$ is reduced, i.e., the sum of the ideals $\mathcal{I}_\theta + \mathcal{I}_\tau$ is reduced.*

6.8 Some Comments and Remarks: We note the following.

(1) The transition in the study from Grassmannian to G/P for the case when P is a minuscule maximal parabolic subgroup of G is not only natural but also forced upon since a basis of weight vectors for V_ϖ is unique upto scaling if ϖ is minuscule.

(2) For the case when P is quasi-minuscule, the set of weight vectors of non-zero weights in V_{ϖ} is uniquely determined upto scaling. The zero-weight subspace of V_{ϖ} has no unique basis. Demazure's work ([**15**]) comes handy and suggests a procedure to choose a basis of weight vectors for V_{ϖ}. This is simple and best tested for the case when P is quasi-minuscule. The best and simplest examples of quasi-minuscule weights (but not minuscule) are the *adjoint representations* of G on $\mathbf{g} = \mathrm{Lie}(G)$ when G has only one root length, (i.e., G is of type A, D, E_6, E_7 or E_8), in which case the the zero weight space is the Cartan subalgebra $\mathbf{h} = \mathrm{Lie}(T)$.

(3) When P or Q is of classical type, the method of constructing a nice basis for V_{ϖ} goes absolutely the same way as in the case of quasi-minuscule weights (verbatim) but the concerned verifications are technically involved.

(4) The linear independence of standard monomials is proved using the same arguments/techniques in all the cases (Grassmannian to the classical types).

(5) That the standard monomials generate the space of all monomials is proved the same way in all the cases of G/P (P a maximal parabolic subgroup of classical type). The proof presented in [**34**] for G/Q, Q of classical type (but Q not maximal), is *incomplete.* A complete proof is given [**39**].

(6) As a natural extension of what goes on in the Grassmannian, Hodge gave (in [**22**]) a definition of "standardness" of monomials for the case of $\mathrm{SL}(n,k)/Q$ relative to $\theta = w_0$. An obvious and natural imitation of this to other groups does not work for (i) $G \neq \mathrm{SL}(n,k)$ (Q non-maximal parabolic subgroup of G) and (ii) it does not work for $\theta \neq w_0$ even for the case of $G = \mathrm{SL}(n,k)$ (Q non-maximal).

7. Applications of Standard Monomial Theory

As an application of Standard Monomial Theory for Classical Groups, a unified approach is given to an understanding of several families of affine varieties such as *Determinantal / Ladder Determinantal varieties, variety of Complexes, of Idempotents, Quiver varieties*, $\cdots$, by identifying them with certain open subsets of suitable Schubert varieties (or their unions) and hence deduce that these are *normal, Cohen-Macaulay* and have *rational singularities*, etc., since the Schubert varieties have these properties. We first recall some facts on what are called "Opposite cells" in Schubert varieties in G/Q (especially in $SL(n)/Q$).

7.1. The Big and Opposite Big Cells: The B-orbit $C_Q(w_0)$ $:= Be_{w_0}Q/Q$ in G/Q (w_0 being the unique element of maximal length in W) is called the *Big Cell* in G/Q. It is a dense open subset of G/Q and it gets identified with $R_u(Q)$, the *unipotent radical* of Q, namely, the subgroup of B generated by $\{U_\alpha \mid \alpha \in \Delta^+ \setminus \Delta_Q^+\}$ (cf.[**4**]). Let B^- be the Borel subgroup of G opposite to B, i.e., $B \cap B^- = T$ or the subgroup of G generated by T and $\{U_\alpha \mid \alpha \in \Delta^-\}$. The B^--orbit $C^-(\mathrm{id}) := B^-e_{\mathrm{id}}Q/Q$ is called the *Opposite Big Cell* in G/Q. This is again a dense open subset of G/Q and it gets identified with the unipotent subgroup of B^- generated by $\{U_\alpha \mid \alpha \in \Delta^- \setminus \Delta_Q^-\}$. Observe that both the big cell and the opposite big cell can be identified with the affine space $\mathbb{A}^{N_Q}$ of the same dimension $N_Q = \#\{\Delta^+ \setminus \Delta_Q^+\}$.

For $\theta \in W/W_Q$, the affine variety $D_Q(\theta) := (B^-e_{\mathrm{id}}Q/Q) \cap X_Q(\theta)$ is called (by abuse of language) the *Opposite Big Cell* or simply the *Opposite Cell* in the Schubert $X_Q(\theta)$. In general, it is *not* a cell at all (except for $\theta = w_0$). It is a non-empty affine open subvariety of $X_Q(\theta)$ and a closed subvariety of the affine space $\mathbb{A}^{N_Q}$. Clearly, we have $\dim(D_Q(\theta)) = \dim(X_Q(\theta))$.

7.2. The Coordinate Ring of $D_Q(\theta)$: In terms of the embedding $G/Q \hookrightarrow \mathbb{P}(V_\varpi)$ given by a very ample $L(\varpi)$ on G/Q, the big and opposite big cells of G/Q are the *principal affine* open subsets given by

$$C_Q(w_0) = Be_{w_0}Q/Q = p_{w_0} \neq 0 \text{ and } C_Q^-(\mathrm{id}) = B^-e_{\mathrm{id}}Q/Q = p_{\mathrm{id}} \neq 0$$

where p_{w_0} is a highest weight vector and $f := p_{\mathrm{id}}$ is lowest weight vector in $H^0(G/Q, L(\varpi))$. In particular, $D_Q(\theta)$ is the principal affine open subset $f \neq 0$ on $X_Q(\theta)$.

Suppose $\mathbf{A}_Q(\theta)$ is the homogeneous coordinate ring of $X_Q(\theta)$ for the embedding $G/Q \hookrightarrow \mathbb{P}(V(\varpi))$. Then the coordinate ring of $D_Q(\theta)$ is the homogeneous localisation $R := (\mathbf{A}_Q(\theta))_{(f)}$ of $\mathbf{A}_Q(\theta)$ at f. We know that R is a *filtered ring* with the natural filtration given by $R = \cup_{m=0}^{\infty} R^m$ where

$$R^m = \left\{ \frac{x}{f^m} \;\middle|\; x \in (\mathbf{A}_Q(\theta))_m \right\}, \ \forall\, m \in \mathbb{Z}^+.$$

In general, this R is not a graded ring. But in many cases, viz., Grassmannians, minuscule G/P, etc., it is graded and in that case one can see that $D_Q(\theta)$ has some particularly nice properties ([**10**]).

7.3. Ideal Theory of Schubert Varieties: In a nutshell, Standard Monomial Theory (SMT) when established for a G/Q, gives a "nice coordinatisation" for various projective embeddings of G/Q. One of the most important consequences of this nice coordinatisation is to be able to write down explicitly the equations defining the ideal $\mathcal{I}_\theta$ of a Schubert variety $X_Q(\theta)$ in any given projective embeddings of G/Q. In fact, SMT provides a good hold of the ideals of Schubert varieties, their unions, intersections, hyperplane sections, etc. This in turn allows one to write down the equations defining the affine varieties $D_Q(\theta)$ in the affine space $C_Q^-(\text{id})$.

In what follows, we will list a few cases of G, Q and θ for which the ideal and coordinate ring of $D_Q(\theta)$ are explicitly described, leading to an identification of the $D_Q(\theta)$'s with appropriate members of the stated families of the affine varieties.

7.4. Equations defining Schubert Varieties in G/Q. Let $Q = \cap_{j=1}^s P_j$ be the intersection of maximal parabolic subgroups P_j of classical type. Let ϖ_j be the fundamental weight associated to P_j. Let $\sum_{j=1}^s m_j\varpi_j$ be the **dominant weight** associated to $\mathbf{m} = (m_1, \cdots, m_s) \in \mathbb{Z}^{+s}$. Let $L(\sum_{j=1}^s m_j\varpi_j) = L(\varpi_1)^{m_1} \otimes \cdots \otimes L(\varpi_s)^{m_s}$ be the line bundle on G/Q. Consider the projective embedding $G/Q \hookrightarrow \mathbb{P}(H^0(G/Q, L(\varpi_1 + \cdots + \varpi_s)))$ given by the ample line bundle $L(\varpi_1 + \cdots + \varpi_s)$. Let $\theta \in W/W_Q$. Let

$$\mathbf{A} = \bigoplus\nolimits_{\mathbf{m}} H^0\Big(G/Q, L(\textstyle\sum_{j=1}^s m_j\varpi_j)\Big) \text{ and}$$

$$\mathbf{A}(\theta) = \bigoplus\nolimits_{\mathbf{m}} H^0\Big(X_Q(\theta), L(\textstyle\sum_{j=1}^s m_j\varpi_j)\Big).$$

Let $\mathbf{R} = \bigoplus_{\mathbf{m}}\Big(S^{m_1}(H^0(G/Q, L(\varpi_1)))\otimes\cdots\otimes S^{m_s}(H^0(G/Q, L(\varpi_s)))\Big)$. By (6.7) Theorem C above, the natural map $\mathbf{R} \to \mathbf{A}$ is surjective and its kernel is generated by elements of total degree 2. Furthermore, the natural restriction map $\mathbf{A} \to \mathbf{A}(\theta)$ is surjective as well and its kernel is generated by elements of total degree 1. (Here $S^m(V)$ stands for the mth symmetric power of a vector space V.) The significance of the multigraded ring $\mathbf{R}$ becomes clear from the following.

7.5. Multiprojective Embeddings of G/Q. Let the notation be as above. For $1 \leq j \leq s$, consider the projective embeddings $G/P_j \hookrightarrow \mathbb{P}_j := \mathbb{P}\Big(H^0(G/P_j, L(\varpi_j))\Big)$. Since $Q = \cap_{j=1}^s P_j$, we get a multiprojective embedding

$$G/Q \hookrightarrow G/P_1 \times \cdots \times G/P_s \hookrightarrow \mathbb{P}_1 \times \cdots \times \mathbb{P}_s \hookrightarrow \mathbb{P}$$

where the last arrow is the Segre embedding of the product of the projective spaces $\mathbb{P}_j$'s. The multigraded algebra $\mathbf{R}$ is the homogeneous coordinate ring of the product $\mathbb{P}_1 \times \cdots \times \mathbb{P}_s$. For $\theta \in W/W_Q$, the equations defining the Schubert variety $X_Q(\theta) \subseteq G/Q$ for the embedding above can be easily seen to be the kernel of the composite surjective map $\mathbf{R} \to \mathbf{A} \to \mathbf{A}(\theta)$.

To be more explicit, given $\theta \in W$, let $\mathcal{I}_j$ be the ideal of $\pi_j(X_Q(\theta))$ for the embedding $G/P_j \hookrightarrow \mathbb{P}_j$ where $\pi_j : G/Q \to G/P_j$ is the natural projection. It follows that the ideal of $X_Q(\theta)$ for its embedding in the multiprojective spaces above, is generated by $\sum_{j=1}^{s} \mathcal{I}_j$ in $\mathbf{R}$.

7.6. Multicones over Schubert Varieties in $\mathbf{SL}(n,k)/Q$. We keep the notation as in (2.5) above. Let $G = \mathrm{SL}(n,k)$. Fix positive integers $1 \le d_1 < \cdots < d_s \le n-1$ and take $Q = \cap_{j=1}^{s} P_{d_j}$. We know that W/W_Q can be identified with $I_{d_1,\cdots,d_s,n}$ where

$$I_{d_1,\cdots,d_s,n} = \left\{ (\mathbf{i}_1, \cdots, \mathbf{i}_s) \in I_{d_1,n} \times \cdots \times I_{d_s,n} \,\middle|\, \mathbf{i}_j \subset \mathbf{i}_{j+1},\ \forall\, 1 \le j \le s-1 \right\}.$$

Now given a permutation $\theta \in S_n$, say

$$\theta = \begin{pmatrix} 1 & \cdots & d_1 & d_1+1 & \cdots & d_2 & d_2+1 & \cdots & n \\ 1^\theta & \cdots & (d_1)^\theta & (d_1+1)^\theta & \cdots & (d_2)^\theta & (d_2+1)^\theta & \cdots & n^\theta \end{pmatrix},$$

let $\theta_j \in I_{d_j,n}$ be defined by $\theta_j = (1^\theta, \cdots, (d_j)^\theta) \uparrow$ which simply means the set $\{1^\theta, \cdots, (d_j)^\theta\}$ whose elements are arranged in the ascending order. Now we have $\theta_Q := (\theta_1, \cdots, \theta_s) \in I_{d_1,\cdots,d_s,n}$. Consequently, the multigraded ideal of $X_Q(\theta)$ is generated by $\sum_{j=1}^{s} \mathcal{I}_j$ in $\mathbf{R}$ where $\mathcal{I}_j$ is the ideal of the Schubert variety $X_{P_j}(\theta_j)$ in the Grassmannian $G_{d_j,n}$, as given in (1.6) above.

7.7. Determinantal Varieties: Given $r, s \in \mathbb{N}$, let $T = (T_{ij})$ be an $r \times s$ matrix of indeterminates over a field k. Let t be a positive integer not exceeding $min(r,s)$. Let $D_t(T)$ be the determinantal variety defined by the vanishing of all $t \times t$ minors of T, called the **classical determinantal** variety. There are a host of more general "determinantal" varieties. See [**48**], [**46**], etc., for definitions and basic facts. All these are easily found to be in the families $D_P(\theta)$ and $D_B(\theta)$ for the case of $G = \mathrm{SL}(n,k)$. The standard monomial basis for the coordinate rings of $D_P(\theta)$, induced from the standard monomial basis for the homogeneous coordinate rings of $X_P(\theta)$, get interpreted as the **double standard bi-tableaux**, as explained in ([**48**, pp.393-394] and [**37**, pp.36-39]).

7.8. Ladder Determinantal Varieties: Let $r, s, t, p \in \mathbb{N}$. Let $T = (T_{ij})$ be an $r \times s$ matrix of indeterminates over a field k (as above). Let $1 \leq a_1 < \cdots < a_p \leq r$ and $1 \leq b_1 < \cdots < b_p \leq s$. Form the **ladder** or the **stair case** joining the variables in the corners, namely, "start at T_{1b_1}, vertically go down to $T_{a_1 b_1}$, then horizontally across to $\cdots T_{a_p b_p} \cdots$, finally vertically down to T''_{rb_p}. Let L be the **ladder** of all variables lying below these stairs and to the left of their walls (inclusive of those on the floor/walls of the steps of the ladder). Such a ladder is called **lower sided** ladder. Let $L_t(T)$ be the affine variety defined by the vanishing of all $t \times t$ minors lying entirely in L, called the **ladder determinantal variety**.

Given a staircase, one can consider the the **upper sided** ladder consisting of all variables lying above the stairs. Let $L'_t(T)$ be the corresponding ladder determinantal variety. If two stair cases are such that one lies entirely above the other, take the upper sided ladder determinantal variety defined by the lower stair case and the lower sided one defined by the upper stair case and take their intersection. This is called a **two-sided ladder** determinantal variety.

It can be seen that $L_t(T)$ or $L'_t(T)$ or their intersections are of the form $D_Q(\theta)$ for a suitable choice of Q and θ associated to $G = \mathrm{SL}(n, k)$. For details, see [**3**], [**44**], [**48**], etc.

7.9. Invariant Theory: Some of the affine varieties $D_P(\theta)$ for the cases when $G = \mathrm{Sp}_{2n}$ are seen to be **symmetric determinantal varieties**. Likewise, they are seen to be **skew-symmetric determinantal varieties** for the case when $G = \mathrm{SO}_{2n}$. The coordinate rings of these symmetric/skew-symmetric varieties turn out to be precisely the **rings of invariants** described in [**13**]. The standard monomial basis given therein parametrised by the *double standard bi-tableaux* is nothing but the standard monomial basis we have at hand.

7.10. Variety of Idempotents: The set of all linear endomorphisms of a finite n-dimensional vector space (V over a field k) which are idempotents (i.e., projections) and of a fixed rank (r) is of interest in the context of representation theory of groups and invariant theory. We show that it is the complement of a suitable hyperplane in the *Segre embedding* of the product $G_{r,n} \times G_{n-r,n}$ of the Grassmannian $G_{r,n}$ and its dual $G_{n-r,n}$. It is therefore an affine variety. The natural hyperplane in question is the *Laplace expansion* with respect to the first r columns of the determinant of an $n \times n$ matrix. For details, see [**48**, pp.396-399], [**47**], etc. To realise this in the form $D_Q(\theta)$, we need to take the product group $\mathrm{SL}(n,k) \times \mathrm{SL}(n,k)$ and its parabolic subgroup $Q = P_r \times P_{n-r}$, etc.

7.11. Variety of Complexes: Given finitely many finite dimensional vector spaces V_j over a field k with $\dim_k(V_j) = n_j$, $1 \le j \le r+1$, recall that $C :=$

$$\left\{(f_1,\cdots,f_r) \in \prod_{1\le j\le r} \mathrm{Hom}_k(V_j,V_{j+1}) \middle| f_{j+1}\circ f_j = 0, 1 \le j \le r-1\right\}$$

is called a set of **complexes**. It is easy to see that this is a closed subset of the affine space $\mathbb{A} = \prod_{1\le j\le r} \mathrm{Hom}_k(V_j,V_{j+1})$. This is *not* irreducible and its irreducible components C_j are of the form $C_j = D_Q(\theta_j)$ for $G = \mathrm{GL}(n,k)$ with $n = \sum_{j=1}^{r+1} n_j$, $Q = \cap_{j=1}^{r} P_{m_j}$, P_{m_j} being a maximal parabolic subgroup of G where $m_j = n_1 + \cdots + n_j$ and $\theta_j \in S_n/W_Q$. Thus $C = \cup_{j=1}^{r} D_Q(\theta_j)$ is the "opposite open cell" of a union $\cup_{j=1}^{r} X_Q(\theta_j)$ of the Schubert varieties $X_Q(\theta_j)$ in $\mathrm{GL}(n,k)/Q$ for suitable $\theta_j \in S_n/W_Q$. For details, see [**48**, pp.399-401], [**51**], etc.

7.12. Quiver Varieties. This is a class of affine varieties containing the determinantal/ladder determinantal varieties, varieties of complexes, etc., as particular cases. Fix $\mathbf{n} = (n_1,\cdots,n_{r+1}) \in (\mathbb{Z}^+)^{r+1}$ and vector spaces V_j over a field k such that $\dim_k(V_j) = n_j$ for $1 \le j \le r+1$. The variety of **quiver representations**, of dimension $\mathbf{n}$ and of the **equioriented** quivers of type A_{r+1}, is defined to be the affine space $Z :=$

$$\mathrm{Hom}_k(V_1,V_2)\times\mathrm{Hom}_k(V_2,V_3)\times\cdots\times\mathrm{Hom}_k(V_{r-1},V_r)\times\mathrm{Hom}_k(V_r,V_{r+1})$$

of all r-tuples of linear maps $(f_1,\ldots,f_r)$ where

$$V_1 \xrightarrow{f_1} V_2 \xrightarrow{f_2} \cdots \xrightarrow{f_{r-1}} V_r \xrightarrow{f_r} V_{r+1}$$

7.12(a) Quiver Varieties: Let $\mathbf{r} = (r_{ij})_{1\le i\le j\le r+1}$ be an array of non-negative integers r_{ij} with $r_{ii} = n_i$, and define $r_{ij} = 0$ for $i \not\le j$. The set $Z(\mathbf{r}) :=$

$$\left\{(f_1,\cdots,f_r) \in Z \;\middle|\; \mathrm{rank}(f_{j-1}\circ\cdots\circ f_i : V_i \to V_j) \le r_{ij},\ \forall\, i<j\right\}$$

which is a closed subset of the affine space Z, called the **quiver variety** of rank $\mathbf{r}$.

7.12(b) A natural Zariski dense open subset of $Z(\mathbf{r})$ is the set $Z^0(\mathbf{r})$ where $Z^0(\mathbf{r}) :=$

$$\left\{(f_1, \cdots, f_r) \in Z \;\middle|\; \operatorname{rank}(f_{j-1} \circ \cdots \circ f_i : V_i \to V_j) = r_{ij},\ \forall\, i < j\right\}.$$

• The set $Z(\mathbf{r})$ could be empty for a bad choice of $\mathbf{r}$.
• $Z(\mathbf{r})$ is the variety of complexes for the choice when $\mathbf{r} = \mathbf{0}$.
• The group $\boxed{G_{\mathbf{n}} = \mathrm{GL}(V_1) \times \cdots \times GL(V_{r+1})}$ acts on Z by

$$(g_1, \cdots, g_{r+1}) \cdot (f_1, \cdots, f_r) = (g_2 f_1 g_1^{-1}, g_3 f_2 g_2^{-1}, \cdots, g_{r+1} f_r g_r^{-1}).$$

• It is clear that $Z(\mathbf{r})$ and $Z^0(\mathbf{r})$ are $G_{\mathbf{n}}$-stable subsets of Z. In fact, we have the following (see [**2**], [**18**], etc.).

7.12(c) The $G_{\mathbf{n}}$-orbits of $Z(\mathbf{r})$ are precisely $Z^0(\mathbf{r})$ with $\mathbf{r}$ satisfying:

$$r_{ij} - r_{ij+1} - r_{i-1j} + r_{i-1j+1} \geq 0,\ \forall\, 1 \leq i < j \leq r+1.$$

• Proceeding exactly the same way as in the case of the Variety of complexes, the quiver variety $Z(\mathbf{r})$ is realised as the opposite open cell of a suitable union of Schubert varieties $X_Q(\theta_j)$ in $\mathrm{SL}(n,k)/Q$ where $G = \mathrm{GL}(n,k)$, $n = \sum_{j=1}^{r+1} n_j = \dim_k(\oplus_{j=1}^{r+1} V_j)$, $Q = \cap_{j=1}^{r} P_{m_j}$, P_{m_j} being a maximal parabolic subgroup of G with $m_j = n_1 + \cdots + n_j$ and $\theta_j \in S_n/W_Q$.

8. Singular Loci of Schubert Varieties

One of the most far reaching consequences of the Standard Monomial Theory (SMT) for G/Q is to be able to "write down" the equations defining a Schubert variety $X_Q(\theta)$ for a projective embedding $G/Q \hookrightarrow \mathbb{P}$, explicitly describe the "Zariski tangent spaces" at its points, characterise the singular Schubert varieties and determine the singular loci thereof. See [**38**] for the main idea/plan and [**28**],[**40**],[**31**], [**19**], [**36**], [**41**], etc., for its execution.

8.1. Singularities of Schubert Varieties. For $x \in X_Q(\theta)$, let us write $T_Q(\theta, x)$ for the Zariski tangent space to $X_Q(\theta)$ at x. Recall, we have

$$\dim_k(T_Q(\theta, x)) \geq \dim X_Q(\theta) = \ell(\theta^{\min})$$

and equality holds $\iff$ x is a *smooth* point of $X_Q(\theta)$. A point which is not smooth is called a *singular* point. Here, $\ell : W \to \mathbb{Z}^+$ is the *length function* on the Weyl group W with respect to the simple reflections s_i (which generate W as a Coxeter group).

Since a Schubert variety $X_Q(\theta)$ is B-stable, it is clear that $x \in X_Q(\theta)$ is smooth if and only if the B-orbit Bx through x contains only smooth points of $X_Q(\theta)$. But each B-orbit in G/Q is of the form

Be_τ for a unique $\tau \in W/W_Q$. Thus $Bx = Be_\tau$ and x is smooth on $X_Q(\theta)$ if and only if e_τ is so. Finally, the set $\mathrm{Sing}_Q(\theta) := \mathrm{Sing}(X_Q(\theta))$ of all singular points of $X_Q(\theta)$ is a closed B-stable subvariety and hence is a union of suitable Schubert subvarieties. Since the point cell $Be_{\mathrm{id}} = \{e_{\mathrm{id}}\}$ is in every Schubert variety, it is clear that $X_Q(\theta)$ is singular if and only if e_{id} is a singular point of $X_Q(\theta)$. In summary, we find that $\mathrm{Sing}_Q(\theta) = \bigcup_{j=1}^q X_Q(\theta_j)$ for some $\theta_j \in W/W_Q$ with $\theta \geq \theta_j$. In general, $\mathrm{Sing}_Q(\theta)$ need not be irreducible. It is known that Schubert varieties are *normal* and hence the codimension of $\mathrm{Sing}_Q(\theta)$ in $X_Q(\theta)$ is at least 2, i.e., $\ell(\theta^{\min}) - \ell(\theta_j^{\min}) \geq 2$ for $1 \leq j \leq q$.

8.2. A Canonical Affine Neighbourhood of e_τ. The opposite big cell described in §7.1 above is the particular case of this when $\tau = 1$. For $\tau \in W$, let $B_\tau = n_\tau B n_\tau^{-1}$ (the Borel subgroup conjugated by n_τ) and B_τ^- be the Borel subgroup opposite to B_τ, i.e., $B_\tau \cap B_\tau^- = T$. Let U_τ^- be the unipotent part of B_τ^-, i.e., it is the subgroup generated by the set of unipotent (affine) lines $\{U_\alpha \mid \alpha \in \tau(\Delta^-)\}$.

Now it follows that $U_\tau^- e_\tau Q/Q \approx \mathbb{A}^{N_Q} = \mathbb{A}$, say, is an affine space of dimension $N_Q = \dim(G/Q)$ and an affine open subset of G/Q and hence an affine open neighbourhood of e_τ in G/Q. For $\theta \geq \tau \in W/W_Q$, let $Y_Q(\theta, \tau) = (U_\tau^- e_\tau Q/Q) \cap X_Q(\theta)$ which is an affine open subset of $X_Q(\theta)$ and a closed subvariety of the affine space $\mathbb{A}$. The ideal $I_Q(\theta,\tau)$ of $Y_Q(\theta,\tau)$ in $\mathbb{A}$ is easily determined by means of the Standard Monomial Bases available for $H^0(G/Q, L(\varpi_j))$'s where the ϖ_j's are the fundamental (dominant) weights of G that are stabilised by Q (see 5.5(c) Theorem A, above).

8.2(a) Theorem: Let $\mathcal{B}_j$ be a standard monomial basis of $H^0(G/P_j, L(\varpi_j))$ $(= H^0(G/Q, L(\varpi_j)))$, $1 \leq j \leq s$ and $Q = \cap_{j=1}^s P_j$. Then the ideal $I_Q(\theta,\tau)$ (of $Y_Q(\theta,\tau)$) is generated by the restrictions to $Y_Q(\theta,\tau)$ of those elements of $\mathcal{B}_j$ which vanish on $X_Q(\theta)$.

Once the equations defining $Y_Q(\theta,\tau)$ are at hand, the well-known *Jacobian criterion* can be applied to decide the smoothness or otherwise of $Y_Q(\theta,\tau)$ at the point e_τ which is the same as the smoothness or otherwise of $X_Q(\theta)$ at the point e_τ.

8.2(b) The Tangent Space $T_\tau(X_Q(\theta))$: Recall that the tangent space $T_\tau(G/Q)$ to G/Q at e_τ is given by

$$T_\tau(G/Q) = \bigoplus_{\beta \in \tau(\Delta^+ \setminus \Delta_Q^+)} \mathfrak{g}_\beta \text{ where } \mathfrak{g}_\beta = kX_\beta$$

and hence the tangent space $T_\tau(X_Q(\theta))$ to $X_Q(\theta)$ at e_τ with $\theta \geq \tau$ is the (T-stable) subspace of $T_\tau(G/Q)$ spanned by $\{X_\beta \in T_\tau(X_Q(\theta)\}$. Clearly, $\{X_\beta \mid X_\beta \in T_\tau(X_Q(\theta)\}$ is a basis of $T_\tau(X_Q(\theta))$.

Let $N_{\theta,\tau} = \{\beta \in \tau(\Delta^+) \mid X_\beta \in T_Q(\theta,\tau)\}$ so that $|N_{\theta,\tau}| = \dim(T_\tau(X_Q(\theta))$. Let $S(\theta) = \{\tau \in W/W_Q \mid |N_{\theta,\tau}| \neq \dim(X_Q(\theta))\}$. Determination of the sets

$N_{\theta,\tau}$ and hence of the set $S(\theta)$ is tractable by means of the SMT and this gives the irreducible components of the singular loci of Schubert varieties.

8.2(c) Singular Locus of $X_Q(\theta)$: The singular locus of a Schubert variety $X_Q(\theta)$ in G/Q ($\theta \in W/W_Q$) is given by

$$\mathrm{Sing}(X_Q(\theta)) = \bigcup_{\tau \in S(\theta)} X_Q(\tau).$$

9. Hodge Algebras

The coordinate rings of (i) the Grassmannians, (ii) the generalised Grassmannians G/P, P a maximal parabolic subgroup of G, (iii) the generalised Flag varieties G/Q, (iv) the Schubert varieties $X_Q(\theta)$ in G/Q, (v) the Determinantal and Pfaffian varieties, (vi) varieties of Complexes, so on and so forth, have a common underlying structure which can be axiomatized easily for studying all of them in a unified way (see [**17**] and [**12**]). These axiomatized algebras are called the *Hodge algebras* or the *Algebras with Straightening Laws* (ASL's). The following is a non-technical simple version.

Let W be a non-empty finite set with an order $\geq$ satisfying anti-symmetry and transitivity (but not necessarily reflexivity). Let R be a commutative ring and $\mathbf{A} = R[t_w \mid w \in W]$ be a graded R-algebra generated by $\{t_w\}_{w \in W} \subset \mathbf{A}_1$ (= homogeneous elements of degree 1).

• A non-zero monomial $t_{w_1} \cdots t_{w_r} \in \mathbf{A}_r$ (of degree r) is said to be a **standard monomial** (of degree r) if $w_1 \geq \cdots \geq w_r$ in W.

9.1 Hodge Algebras: We say that $\mathbf{A} = R[t_w \mid w \in W]$ is a **Hodge algebra** or an **Algebra** with **Straightening Laws** (ASL) if the following axioms are satisfied:

Hodge-1: $\mathbf{A}$ is a free R-module admitting the set of all standard monomials as an R-basis.

Hodge-2: (**Straightening Laws**:) If $w, w' \in W$ are such that $w \not\geq w'$, then we have $\boxed{t_w t_{w'} = \sum_{w_1 \geq w_1'} c_{w_1 w_1'} t_{w_1} t_{w_1'}}$ with $c_{w_1 w_1'} \in k^\star$ & $(w_1, w_1') \in W \times W$ satisfy $w_1 > w$.

• The straightening laws above are generalisations of the quadratic relations satisfied by the Plücker coordinates and their subsequent counterparts.

• The relation like $t_w t_{w'} = \sum_{w_1 \geq w_1'} c_{w_1 w_1'} t_{w_1} t_{w_1'}$ in Hodge–2 exists by Hodge–1 and the point of Hodge–2 is the condition that $w_1 > w$.

• By Hodge–2, it can be proved that the standard monomials generate $\mathbf{A}$. In other words, it is enough to assume in Hodge–1 that standard monomials are linearly independent over R.

References

[1] S. Abeasis, On the Plücker relations for the Grassmann Varieties, *Adv. in Math.*, **36**(1980), 277–282.

[2] S. Abeasis and A. Del Fra, Degenerations for the Representations of a Quiver of Type A_m, *J. of Alg.*, **93**(1985), 376–412.

[3] S.S. Abhyankar, *Enumerative Combinatorics of Young Tableaux*, Marcel-Dekker, NY (1988).

[4] A. Borel, *Linear Algebraic Groups*, (second enlarged edition), Graduate Texts in Mathematics, Springer (1991).

[5] A. Borel, Linear Representations of Semi-simple Algebraic Groups, *AMS Proc. Symp. Pure Math., Algebraic Geometry (Arcata)*, **29**(1975), 421–440.

[6] N. Bourbaki, *Algèbre Commutative, Eléments de Math.*, **27,28,30,31**, Hermann, Paris (1964–69); English trans: Addison Wesley, London (1972).

[7] N. Bourbaki, *Groupes et Algèbres de Lie, Chapitres 4, 5 et 6*, Hermann, Paris (1968).

[8] J. Carrell, On the Smooth Points of a Schubert Variety, *CMS Conference Proceedings*, **16**(1994), 15–24; *Proceedings of the Conference on " Representations of Groups: Lie, Algebraic, Finite and Quantum"*, Banff, Alberta (1994).

[9] C. Chevalley, Sur les Décompositions Cellulaires des Espaces G/B (with a foreword by A. Borel), *Proc. Symp. Pure Math.*, (*Part 1, Algebraic Groups and their Generalizations: Classical Methods (Univ. Park, PA: 1991)*), **56**(1994), 1–23.

[10] M.S. Datt, The Minuscule Grassmannian, preprint.

[11] C. DeConcini, D. Eisenbud and C. Procesi, Young Diagrams and Determinantal Varieties, *Invent. math.* **56**(1980), 129–165.

[12] C. DeConcini, D. Eisenbud and C. Procesi, *Hodge Algebras, Asterisque* **91**(1982).

[13] C. DeConcini and C. Procesi, A Characteristic free approach to Invariant Theory, *Adv. in Math.*, **21**(1976), 330–354.

[14] C. DeConcini and E. Strickland, On the Variety of Complexes, *Adv. in Math.*, **41**(1981), 57–77.

[15] M. Demazure, Désingularisation des Variétés de Schubert Généralisés, *Ann. Sci. Ec. Norm. Sup.*, **t.7**(1974), 53–88.

[16] V.V. Deodhar, On Some Geometric aspects of Bruhat Orderings, *I. A finer decomposition of Bruhat cells. Invent. math.* **79**(1985), 499–511.

[17] D. Eisenbud, Introduction to Algebras with Straightening Laws (in Ring Theory and Algebra - III), *Proc. of the Third Oklohoma Conf.*, Ed. B.R. McDonald, Marcel Dekker, NY (1980).

[18] P. Gabriel and V. Dlab (Eds.), *Proc. ICRA-II*, Springer LNM **831**, Ottawa (1979).

[19] N. Gonciulea and V. Lakshmibai, Singular Loci of Ladder Determinantal Varieties and Schubert Varieties, *J. Alg.*, **229**(2000), 463–497.

[20] R. Hartshorne, Appendix to Ample Vector Bundles, *Publ. Math. IHES*, **29**(1966), 63–94.

[21] M. Hochster and J.A. Eagon, Cohen-Macaulay Rings, Invariant Theory, and the Generic Perfection of Determinantal Loci, *Amer. J. Math.*, **93**(1971), 1020–1058.

[22] W.V.D. Hodge, Some Enumerative results in the Theory of Forms, *Proc. Camb. Phil. Soc.*, **39**(1943), 22–30.

[23] W.V.D. Hodge and D. Pedoe, *Methods of Algebraic Geometry*, Vol.**II**, Camb. Univ. Press, Cambridge (1952).

[24] J.I. Igusa, On the Arithmetic Normality of the Grassmann Variety, *Proc. Nat. Acad. Sci.*, USA, **40**(1954), 309–313.

[25] G. Kempf and A. Ramanathan, Multi-cones over Schubert Varieties, *Invent. math.*, **87**(1987), 353–363.

[26] S.L. Kleiman, **Problem 15:** Rigourus Foundations of Schubert's Enumerative Calculus, in *Mathematical Developments arising from Hilbert Problems, AMS Proc. Symp. Pure Math.* **28**(1976), 445–482.

[27] B. Kostant, Lie Algebra Cohomology and the Generalised Borel–Weil Theorem, *Ann. Math.*, **74**(1961), 329–387.

[28] V. Lakshmibai, Singular loci of Schubert Varieties for classical groups, *Bull. Amer. Math. Soc.*, **16**(1987), 83–90.

[29] V. Lakshmibai, On Tangent Spaces to Schubert Varieties – I, *J. Alg.*, **224**(2000), 167–197.

[30] V. Lakshmibai, On Tangent Spaces to Schubert Varieties – II, *J. Alg.*, **230**(2000), 222–244.

[31] V. Lakshmibai, Singular Loci of Varieties of Complexes, *J. Pure and Appl. Alg.*, **152**(2000), 217–230.

[32] V. Lakshmibai, The Development of Standard Monomial Theory – II, (this volume).

[33] V. Lakshmibai, C. Musili and C.S. Seshadri, Geometry of $G/P - III$, *Proc. Ind. Acad. Sci. (Math. Sci.)*, **87A**(1978), 93–177.

[34] V. Lakshmibai, C. Musili and C.S. Seshadri, Geometry of $G/P - IV$, *Proc. Ind. Acad. Sci. (Math. Sci.)*, **88A**(1979), 279–362.

[35] V. Lakshmibai, C. Musili and C.S. Seshadri, Geometry of G/P, *Bulletin (New Series) of the AMS*, **1**(1979), 432–435.

[36] V. Lakshmibai and B. Sandhya, Criterion for Smoothness of Schubert Varieties in $SL(n)/B$, *Proc. Ind. Acad. Sci. (Math. Sci.)*, **100**(1990), 45–52.

[37] V. Lakshmibai and C.S. Seshadri, Geometry of $G/P - II$, *Proc. Ind. Acad. Sci. (Math. Sci.)*, **87A**(1978), 1–54.

[38] V. Lakshmibai and C.S. Seshadri, Singular locus of a Schubert variety, *Bull. Amer. Math. Soc.*, **2**(1984), 363–366.

[39] V. Lakshmibai and C.S. Seshadri, Geometry of $G/P - V$, *J. of Alg.*, **100**(1986), 462–557.

[40] V. Lakshmibai and C.S. Seshadri, Standard Monomial Theory, *Proc. Hyderabad Conference on Algebraic Groups*, (Eds., S. Ramanan et al), Manoj Prakashan, Madras (1991), 279–323.

[41] V. Lakshmibai and J. Weyman, Multiplicities of points on a Schubert Variety in a Minuscule G/P, *Adv. Math.*, **84**(1990), 179–208.

[42] V. Mehta and A. Ramanathan, Frobenius Splitting and Cohomology Vanishing for Schubert Varieties, *Ann. Math.*, **122**(1985), 27–40.

[43] V.B. Mehta and A. Ramanathan, Schubert Varieties in $G/B \times G/B$, *Comp. Math.* **67**(1988), 355–358.
[44] S.B. Mulay, Determinantal Loci and the Flag Variety, *Adv. in Math.*, **74**(1989), 1–30.
[45] C. Musili, Postulation Formula for Schubert Varieties, *J. Ind. Math. Soc.*, **36**(1972), 143–171.
[46] C. Musili, Some Properties of Schubert Varieties, *J. Ind. Math. Soc.*, **38**(1974), 131–145.
[47] C. Musili, A note on the Variety of Projectors, *J. Pure and Applied Algebra*, **74**(1991), 73–84.
[48] C. Musili, Applications of Standard Monomial Theory, *Proc. Hyderabad Conference on Algebraic Groups*, (Eds., S. Ramanan et al), Manoj Prakashan, Madras (1991), 381–406.
[49] C. Musili, *Algebraic Geometry for Beginners*, TRIM-**20**, Hindustan Book Agency (India) Ltd., New Delhi (2001).
[50] C. Musili and C.S. Seshadri, Standard Monomial Theory, in *Séminaire d'Algèbre Paul Dubreil et Marie-Paule Malliavin*, Springer LNM, **867**(1981), 441–478.
[51] C. Musili and C.S. Seshadri, Schubert Varieties and the Variety of Complexes, *Arithmetic and Geometry*, Vol.**II**, (Papers dedicated to I.R. Shafarevich on his 60th Birthday), Progress in Mathematics, Birkhäuser, Basel, **36**(1983), 329–359.
[52] S. Ramanan and A. Ramanathan, Projective Normality of Flag Varieties and Schubert Varieties, *Invent. math*, **79**(1985), 217–224.
[53] A. Ramanathan, Equations defining Schubert Varieties and Frobenius Splitting of Diagonals, *Publ. Math. IHES*, **65**(1987), 61–90.
[54] C.S. Seshadri, Geometry of $G/P - I$, *C.P. Ramanujam: A Tribute*, **207** (Springer-Verlag), Published by TIFR (1978), 207–239.
[55] C.S. Seshadri, *Introduction to the Theory of Standard Monomials*, Brandeis Lecture Notes **4**, Waltham (1985).
[56] H. Weyl, *The Classical Groups*, Princeton Univ. Press, Princeton, NJ (1946).

Department of Math. and Stat., University of Hyderabad,
P.O. Central University, Hyderabad 500 046, INDIA.
E-mail address: <cmsm@uohyd.ernet.in> and <musili_c@yahoo.co.in>

Measures on the Circle invariant under $z \to z^2$ and $z \to z^3$

M. G. Nadkarni

Dedicated to Professor C.S. Seshadri on his 70th *birthday*

Abstract. In this paper we discuss the problem, raised by Furstenberg, whether or not there exist non-atomic measures on the circle simultaneously invariant under the maps $z \to z^2$ and $z \to z^3$, other than the Lebesgue measure.

1. Introduction

Let S^1 denote the circle group and let σ and τ be the maps $z \to z^2$, $z \to z^3$. Let C_0 denote the set of points $z \in S^1$ such that the τ orbit of z, $(\tau^n(z))_{n \in \mathbb{N}}$, has uniform distribution. Let N be the set of z whose τ orbit does not admit any distribution what so ever.
The purpose of this article is to give a structure of probability measures μ on the circle group S^1 simultaneously invariant under the maps σ, τ and deduce as a consequence the result that *if there is a continuous σ-invariant probability measure on S^1 assigning zero probability to $C_0 \cup N$ then there is a continuous probability measure μ on S^1 simultaneously invariant under σ and τ and singular to Lebesgue measure.*

The main observation is that σ acts on ergodic decomposition of τ in a measurable and onto manner and that probability measures on S^1 simultaneously invariant under σ and τ correspond one-one to the σ-invariant probability measures on the ergodic decomposition of τ.
The question whether a continuous probability measure simultaneously invariant under σ and τ is necessarily the Lebesgue measure was raised by H. Furstenberg [**1**]. R. Lyons [**5**] proved that if one of σ

or τ is exact then μ is Lebesgue. This was generalized by D. Rudolph [**6**] who showed that if μ is ergodic for the joint action of σ and τ and if one of σ or τ has positive entropy then μ is Lebesgue. In [**2**] B. Host proved that if ν is any σ-invariant ergodic probability measure with positive entropy, then for a.e. z (with respect to ν), $(\tau^n z)_{n\in\mathbb{N}}$ is uniformly distributed. Rudolph's result quoted above follows from this. We also refer to the work of A. Johnson and D. Rudolph [**3**] E. Lindenstrauss [**4**] connected with the theme of this paper. In particular [**4**] gives other papers connected with the topic.

For simplicity we confine to the case of σ and τ defined above rather than consider more general maps of the type $z \to z^p$, $z \to z^q$ for p, q relatively prime. The results here, which are purely descriptive, sheds some light on the zero entropy case. The article is meant as a discussion paper and more details are given than would normally be included.

2. Ergodic Decomposition of τ

(1). A set $C \subset S^1$ is called τ-invariant if $\tau C \subset C$. If C is a τ-invariant subset of S^1, and if $Q = \{z \in C : -z \in C\}$ then Q is also τ-invariant. For if $w \in \tau Q$ then $w = z^3$ for some $z \in C \cap Q$. So $-z \in C \cap Q$ and $(-z)^3 = -z^3 = -w \in C$ since C is τ -invariant. Consequently $z^3 = w$ and $-z^3 = -w$ are both in C, whence $w \in Q$. Thus $\tau Q \subset Q$, i.e., Q is τ-invariant.

(2) If C supports a τ-invariant ergodic probability measure, say λ, and if Q has positive measure, then Q supports λ since λ is ergodic. So given a τ-invariant ergodic probability measure λ the following dichotomy holds: either, for every support C of λ, $C \cap (-C)$ is a support of λ, or, there exists a support C such that $C \cap (-C)$ is empty. In the first case λ is symmetric in the sense that $\lambda(A) = \lambda(-A)$ for every Borel set A. Indeed the measure $\nu(A) = \lambda(-A)$ is also τ invariant and every support of λ is also a support of ν and since λ is ergodic we see that $\mu = \nu$.

(3). Let $A_n, n \in \mathbb{N}$, be a countable algebra of subsets of S^1 which generates the Borel σ-algebra $\mathcal{B}$ of S^1. We will assume that each A_n is a finite union of semi-closed arcs. Let E be the set points z such that for all $i \in \mathbb{N}$,

$$\lim_{n\to\infty} \frac{1}{n} \sum_{k=0}^{n-1} 1_{A_i}(\tau^k(z))$$

exists. Denote this limit by $\lambda_z(A_i)$ and let ϕ be the map from E into $\Omega = [0,1]^{\mathbb{N}}$ defined by $z \to (\lambda_z(A_i))_{i\in\mathbb{N}}$. The set E, which is same as the set $S^1 - N$, is Borel and since the maps $z \to \lambda_z(A_i)$, $i \in \mathbb{N}$, are Borel, the map ϕ is Borel measurable. Let $I = \phi(E)$. For each $\alpha \in I$ let $C_\alpha = \{z \in E : \phi(z) = \alpha\}$. The collection $C_\alpha, \alpha \in I$, is the partition of E given by ϕ . It is countably generated since ϕ is a Borel map. Each C_α is τ- invariant and for all $i \in \mathbb{N}$, $\lambda_z(A_i)$ is constant for $z \in C_\alpha$. We now change notation and write $\lambda_\alpha(A_i)$ for $\lambda_z(A_i)$ whenever $z \in C_\alpha$. Now, λ_α is countably additive on the algebra $\{A_i, i \in \mathbb{N}\}$ and extends to a countably additive measure on $\mathcal{B}$. The probability measures λ_α are τ invariant ergodic and supported on C_α. C_α does not support any τ-invariant probability measure other than λ_α. The collection $(C_\alpha, \lambda_\alpha), \alpha \in I$, is called the ergodic decomposition of τ. If ν is any probability measure on Borel subsets of E, then $m = \nu \circ \phi^{-1}$ is a measure on the Borel subsets of I. Further the measure λ defined by

$$\lambda(A) = \int_E \lambda_z(A) d\nu(z), = \int_I \lambda_\alpha(A) dm(\alpha), \;\; A \in \mathcal{B}$$

is a τ invariant probability measure on S^1 and every τ invariant probability measure is obtained in this way. Indeed if, λ is any τ invariant probability measure on S^1 (which is necessarily supported on E) then

$$\lambda(A) = \int_E \lambda_z(A) d\lambda(z) = \int_I \lambda_\alpha(A) dm(\alpha),$$

where $m = \lambda \circ \phi^{-1}$. Note that C_α's are pairwise disjoint and only one among these, denote it by C_0, has full Lebesgue measure and all C_α's are of Lebesgue measure zero. Indeed the union of C_α over $\alpha \neq 0$ is of Lebesgue measure zero.

(4) For each Borel set A the map $\alpha \to \lambda_\alpha(A), \alpha \in I$, is Borel measurable. Moreover the map g defined on E by $z \to \alpha$ for $z \in C_\alpha$, is also Borel measurable. The set

$$G = \{\alpha \in I : \lambda_\alpha(A_n) = \lambda_\alpha(-A_n), \forall n\}$$

is a Borel set in I and the set $g^{-1}(G) = \cup_{\alpha\in G} C_\lambda$ is Borel in S^1. Clearly the set $H = E - (g^{-1}(G))$ is also measurable and it is the union of all C_α over α such that λ_α is not symmetric. Since Lebesgue measure is symmetric, τ-invariant and ergodic, we see that C_0 is not a subset of H and so H has Lebesgue measure zero.

3. Supplements

(5) We will need the following considerations in the sequel. It gives conditions under which a factor of a uniquely ergodic system is uniquely ergodic. Let T be a Borel map on S^1 (not necessarily invertible) and let $\mathcal{F}$ be a countably generated sub-σ algebra of $\mathcal{B}$ such that $T^{-1}(\mathcal{F}) \subset \mathcal{F}$. Let $A_\alpha, \alpha \in [0,1]$, be the atoms of $\mathcal{F}$ and for each α let m_α be a measure on $\mathcal{B}$ supported on A_α such that the function $\alpha \to m_\alpha(A)$ is measurable for each $A \in \mathcal{B}$. Assume that T is one-one on each A_α and if $TA_\alpha = A_\beta$ then we write $\beta = T\alpha$. Assume that $m_{T_\alpha} = m_\alpha \circ T^{-1}$. Under these conditions:
every T-invariant probability measure P on $\mathcal{F}$ is the restriction of a T invariant probability measure Q on $\mathcal{B}$.

To see this, view P as a measure on [0,1] and write

$$Q(A) = \int_{[0,1]} m_\alpha(A_\alpha \cap A) dP(\alpha).$$

Then Q is T- invariant and agrees with P on $\mathcal{F}$. Indeed,

$$Q(T^{-1}(A)) = \int_{[0,1]} m_\alpha(A_\alpha \cap T^{-1}(A)) dP(\alpha)$$

$$= \int_{[0,1]} m_\alpha(T^{-1}(T(A_\alpha) \cap A) dP(\alpha)$$

$$\int_{[0,1]} m_{T\alpha}(T(A_\alpha) \cap A) dP(\alpha) = \int_{[0,1]} m_\alpha(A_\alpha \cap A) dP(\alpha) = Q(A),$$

where the T-invariance of P is invoked to justify the step before last. If $A \in \mathcal{F}$ then $m_\alpha(A_\alpha \cap A) = 0$ or 1 according as $A_\alpha \cap A = \emptyset$ or $A_\alpha \subset A$ and we see that Q(A) = P(A), so that Q agrees with P on $\mathcal{F}$. Clearly, if T is uniquely ergodic on $\mathcal{B}$ then it is uniquely ergodic on $\mathcal{F}$. (The converse need not hold, the constructed measure Q need not be ergodic on $\mathcal{B}$ even if P is ergodic.)

(6) If ν is a symmetric probability measure on S^1 invariant under σ then ν is the Lebesgue measure. We see this as follows. Since

ν is symmetric, the odd Fourier coefficients of ν vanish, i.e., for all integers n,

$$\int_{S^1} z^{2n-1} d\nu(z) = 0.$$

If n is even and not zero, then $n = 2^k l$ for some odd integer l. By invariance of ν under σ we have

$$\int_{S^1} z^n d\nu(z) = \int_{S^1} z^{2^k l} d\nu(z) = \int_{S^1} z^l d\nu(z) = 0.$$

Thus all non-zero Fourier co-efficients of ν vanish, so ν is the Lebesgue measure.

4. The Action of σ

(7) We now discuss the action of $\sigma : z \to z^2$ on the ergodic decomposition $(C_\alpha, \lambda_\alpha), \alpha \in I$, of τ. Now, $C_\alpha = C$ supports a unique τ-invariant probability measure $\lambda_\alpha = \lambda$. The set σC is τ- invariant and supports the unique τ- invariant probability measure, viz, $\lambda \circ \sigma^{-1}$. That $\lambda \circ \sigma^{-1}$ is τ-ergodic is obvious. We must verify the τ-action on σC is uniquely ergodic. If C is such that $C \cap (-C) = \emptyset$ then σ is one-one on C and there is nothing to prove. If C is symmetric, then we consider the σ-algebra of symmetric Borel sets, denoted by $\mathcal{F}$, whose atoms are the pairs of opposite points $\{z, -z\}, z \in S^1$. Let $m_{\{-z,z\}} = m_z$ denote the uniform distribution on $\{z, -z\}$. Note that $\tau\{z, -z\} = \{z^3, -z^3\}$, whence τ is one-one on atoms. Clearly $m_z \circ \tau^{-1} = m_{\tau z}$. Since τ-action on C is uniquely ergodic we see, in view of para **5**, that the τ-action on $\mathcal{F}$ is uniquely ergodic. Clearly, then, τ-action on σC is uniquely ergodic.

We next show that the σ action on the ergodic components is onto. To this end let (C, λ) be a uniquely ergodic component of τ. Two cases arise. Case 1. Suppose $\sigma^{-1}C$ contains two disjoint τ-invariant *Borel* sets A and B such that $A = -B$ and $\sigma(A) = \sigma(B) = C$ *a.e.* λ. Then $\sigma\,|_A$ and $\sigma\,|_B$ are one-one. Clearly in this case $(A, \lambda \circ \sigma)$ and $(B, \lambda \circ \sigma)$ are measure spaces uniquely ergodic under τ which map onto (C, λ) under σ. Case 2. Assume now that $\sigma^{-1}C$ does not break in this fashion. The function $\lambda \circ \sigma$ when restricted to the symmetric Borel subsets of $\sigma^{-1}C$ defines a τ- invariant ergodic probability measure on the σ-algebra of symmetric Borel subsets of $\sigma^{-1}(C)$. As in para **5** we can construct a τ invariant probability measure, say P, on the Borel subsets of $\sigma^{-1}C$ which agrees with $\lambda \circ \sigma$ on symmetric Borel sets. We show that P is ergodic under τ. If P is not ergodic with respect to τ, then there will be two distinct τ-ergodic components (A, m) and (B, l) of P, with $A, B \subset \sigma^{-1}C$, A, B disjoint. By unique ergodicity of τ on C, we will have $m \circ \sigma^{-1} = l \circ \sigma^{-1} = \lambda$, $\sigma A = C$ *a.e.* λ, $\sigma B = C$ *a.e.* λ. Since A and B are disjoint, after throwing away $\tau -$ *invariant* m and l null sets from A and B respectively if necessary, we see that $A = -B$ which is a contradiction since we have assumed that $\sigma^{-1}C$ does not break in this fashion.

Clearly $\lambda \circ \sigma^{-1} = \lambda_\beta$ for some β and $\sigma C_\beta = C$ *a.e.* λ. Thus σ-action on the ergodic decomposition of τ is onto.

5. Structure of measures invariant under σ and τ.

(8) Let ν be a σ-invariant probability measure on $E = S^1 - N$. Then μ defined by

$$\mu(A) = \int_E \lambda_z(A) d\nu(z) = \int_I \lambda_\alpha(A) dm(\alpha) \qquad (1),$$

where $m = \nu \circ \phi^{-1}$, is surely τ invariant but it is also σ invariant. We see this as follows:

$$\begin{aligned}\mu(\sigma^{-1}(A)) &= \int_E \lambda_z((\sigma^{-1}(A)) d\nu(z) \\ &= \int_E \lambda_{\sigma z}(A) d\nu(z) \\ &= \int_E \lambda_z(A) d\nu(z),\end{aligned}$$

where the last step follows from the σ-invariance of ν. Clearly every probability measure invariant under both σ and τ is obtained in this way.

(9) Let $D = (\cup_{n\in\mathbb{N}} \sigma^{-n}(H)) \cap E$ where H, as in para **4**, is the union of non-symmetric C_α. Write $R = E - D$. The sets D and R are Borel sets invariant under τ. Note that D has Lebesgue measure zero since H has Lebesgue measure zero.

Let I_1 be the set indices α for which $C_\alpha \subset R$, i.e., $I_1 = \phi(R)$ and let I_2 be the set of indices α with $C_\alpha \subset D$, i.e., $I_2 = \phi(D)$. For every $\alpha \in I_1$, λ_α is symmetric. If ν is a σ-invariant probability measure supported on R then the measure μ given by (1) is σ-invariant and symmetric. Hence it is necessarily the Lebesgue measure in view of para **6**. It is clear therefore that ν is necessarily supported on C_0.

We see that $R - C_0$ has null measure with respect to every σ invariant probability measure ν. We do not know if there is a continuous σ-invariant probability measure supported on D. If there is one, say ν, then (1) will define a probability measure invariant under both σ, and τ and singular to the Lebesgue measure. The result mentioned at the beginning follows.
It can be shown that the τ action on members of H admits eigen–values.

The result of Rudolph mentioned earlier may be interpreted to mean that any σ-invariant probability mesure supported on D has entropy zero, while Host's result implies in addition that same hold also for the set N.

The work was begun while I was visiting the Institute of Mathematical Sciences, Chennai (September - December 2001). It is a pleasure to acknowledge discussions, even so brief, with V. S. Sunder, V. Kodiyalam, R. Balsubramaniam. In Mumbai the topic was discussed with S. G. Dani, J. S. Dani, R. V. Gurjar, E. Lindenstrauss, and R. R. Simha.

References

[1] H. Furstenberg, *Disjointness in ergodic Theory, minimal sets, and a problem in Diophantine approximation*, Math. Systems Theory (1967)**1** 1-49.
[2] B. Host, *Nombres normaux, entropie, translations* Israel Journal of Mathematics, (1995) **91**, 419-428.
[3] A. Johnson D. Rudolph, *Convergence under $\times_p$ of $\times_q$ invariant measures on the Circle*, Advances in Mathematics,(1995) **115**117-140.
[4] E. Lindenstrauss, *p-adic foliation and equidistribution* preprint
[5] R. Lyons, *On measures simultaneously invariant under 2- and -3 invariant,* Israel Journal of Mathematics, (1988)**61** 219-224.
[6] D. Rudolph, $\times 2$ *and* $\times 3$ *invariant measures and entropy,* Ergodic Theory and Dynamical Systems,(1990), **10**, 395-406.

Department of Mathematics, Univ. of Mumbai
Kalina, Mumbai, India 400098
E-mail address: nadkarni@math.mu.ac.in

Affine Hecke Algebras, Cyclotomic Hecke algebras and Clifford Theory

Arun Ram and Jacqui Ramagge

Dedicated to Professor C.S. Seshadri on the occasion of his 70th birthday

Abstract. We show that the Young tableaux theory and constructions of the irreducible representations of the Weyl groups of type A, B and D, Iwahori-Hecke algebras of types A, B, and D, the complex reflection groups $G(r,p,n)$ and the corresponding cyclotomic Hecke algebras $H_{r,p,n}$, can be obtained, in all cases, from the affine Hecke algebra of type A. The Young tableaux theory was extended to affine Hecke algebras (of general Lie type) in recent work of A. Ram. We also show how (in general Lie type) the representations of general affine Hecke algebras can be constructed from the representations of simply connected affine Hecke algebras by using an extended form of Clifford theory. This extension of Clifford theory is given in the Appendix.

0. Introduction

Recent work of A. Ram [Ra2,5] gives a straightforward combinatorial construction of the simple calibrated modules of affine Hecke algebras (of general Lie type as well as type A). The first aim of this paper is to show that Young's seminormal construction and all of its previously known generalizations are special cases of the construction in [Ra5]. In particular, the representation theory of

(a) Weyl groups of types A, B, and D,
(b) Iwahori-Hecke algebras of types A, B, and D,
(c) the complex reflection groups $G(r,p,n)$, and
(d) the cyclotomic Hecke algebras $H_{r,p,n}$,

can be derived entirely from the representation theory of *affine Hecke algebras of type A*. Furthermore, the relationship between the affine Hecke algebra and the objects in (a)-(d) always produces a natural set of Jucys-Murphy type elements and can be used to prove the

Research partially supported by the Natioanl Science Foundation (DMS-0097977) and the National Security Agency (MDA904-01-1-0032).

standard Jucys-Murphy type theorems. In particular, we are able to use Bernstein's results about the center of the affine Hecke algebra to show that, in the semisimple case, the center of the cyclotomic Hecke algebra $H_{r,1,n}$ is the set of symmetric polynomials in the Jucys-Murphy elements.

A. Young's seminormal construction of the irreducible representations of the symmetric group dates from 1931 [Yg1]. Young himself generalized his tableaux to treat the representation theory of Weyl groups of types B and D [Yg2]. In 1974 P.N. Hoefsmit [Hf] generalized the seminormal construction to Iwahori-Hecke algebras of types A, B, and D. Hoefsmit's work has never been published and, in 1985, Dipper and James [DJ], Theorem 4.9 and H. Wenzl [Wz] independently treated the seminormal construction for irreducible representations for Iwahori-Hecke algebras of type A. In 1994 Ariki and Koike [AK] introduced (some of) the cyclotomic Hecke algebras and generalized Hoefsmit's construction to these algebras. The construction was generalized to a larger class of cyclotomic Hecke algebras in [Ar2]. For a summary of this work see [Ra1] and [HR].

General affine Hecke algebras are naturally associated to a reductive algebraic group and the size of the commutative part of the affine Hecke algebra depends on the structure of the corresponding algebraic group (simply connected, adjoint, etc.). The second aim of this paper is to show that it is sufficient to understand the representation theory of the affine Hecke algebra in the simply connected case. We describe explicitly how the representation theory of the other cases is derived from the simply connected case.

The machine which allows us to accomplish this reduction is a form of Clifford theory. Precisely, if R is an algebra and G is a finite group acting on R by automorphisms then the representation theory of the ring R^G of fixed points of the G-action can be derived from the representation theory of R and subgroups of G. This is an extension of the approach to Clifford theory given by Macdonald [Mac2].

Let us state precisely what is new in this paper. The main result has three parts:

(1) The Hecke algebras $H_{r,p,n}$ of the complex reflection groups $G(r,p,n)$ can be obtained as fixed point subalgebras of the Hecke algebra $H_{r,1,n}$ associated to the complex reflection group $G_{r,1,n}$ via

$$H_{r,p,n} = (H_{r,1,n})^{\mathbb{Z}/p\mathbb{Z}}.$$

(2) The Hecke algebras $\tilde{H}_L$ of nonadjoint p-adic groups can be obtained as fixed point subalgebras of the Hecke algebra $\tilde{H}_P$ associated to the corresponding adjoint p-adic group, via

$$\tilde{H}_L = (\tilde{H}_P)^{P/L}.$$

(3) There is a form of Clifford theory (to our knowledge new) that allows one to completely determine the representation theory of a fixed point subalgebra R^G in terms of the representation theory of the algebra R and the group G.

We use this method to work out the representation theory of the algebras $H_{r,p,n}$ in detail. Additional results include,

(4) The discovery of the "right" affine braid groups $\mathcal{B}_{\infty,p,n}$ and affine Hecke algebras $H_{\infty,p,n}$ to associate to the cyclotomic Hecke algebras $H_{r,p,n}$. The representation theory of these new groups and algebras is completely determined from the representation theory of the classical affine braid groups $B_{\infty,1,n}$ and the classical affine Hecke algebras $H_{\infty,1,n}$ of type A as a consequence of the results in (3).

Finally,

(5) We show how the classical trick (due to Cherednik) for determining the representation theory of the algebras $H_{r,1,n}$ from that of $H_{\infty,1,n}$ arises from a map from the affine Hecke algebra of type C to the finite Hecke algebra of type C which corresponds to a folding of the Dynkin diagram. This explanation is new. We show that such maps from the affine Hecke algebra to the finite Hecke algebra cannot exist in general type, and we work out the details of the cases where such homomorphisms do arise from foldings.

In a recent paper Reeder [Re] has used our results to prove the Langlands classification of irreducible representations for general affine Hecke algebras. (Previously, this was known only in the simply connected case, see Kazhdan and Lusztig [KL].) This also provides a classification of the irreducible constituents of unramified principal series representations of general split reductive p-adic groups. (The Kazhdan-Lusztig result provides this classification for groups with connected center.)

Acknowledgements. A. Ram thanks P. Deligne for his encouragement, stimulating questions and helpful comments on the results in this paper. We thank D. Passman for providing some useful references about Clifford theory. We are grateful for the generous support of this research by the National Science Foundation, the National Security Agency and the Australian Research Council.

1. Algebras with Young tableaux theories

A. Young invented the theory of standard Young tableaux in order to describe the representation theory of the symmetric group S_n; the group of $n \times n$ matrices such that

(a) the entries are either 0 or 1,
(b) there is exactly one nonzero entry in each row and each column.

Young himself began to generalize the theory and in [Yg1] he provided a theory for the Weyl groups of type B, i.e. the hyperoctahedral groups $WB_n \cong (\mathbb{Z}/2\mathbb{Z}) \wr S_n$ of $n \times n$ matrices such that

(a) the entries are either 0 or ± 1,
(b) there is exactly one nonzero entry in each row and each column.

In the same paper Young also treated the Weyl group WD_n of $n \times n$ matrices such that

(a) the entries are either 0 or ± 1,
(b) there is exactly one nonzero entry in each row and each column,
(c) the product of the nonzero entries is 1.

W. Specht [Sp] generalized the theory to cover the complex reflection groups $G(r,1,n) \cong (\mathbb{Z}/r\mathbb{Z}) \wr S_n$ consisting of $n \times n$ matrices such that

(a) the entries are either 0 or rth roots of unity,
(b) there is exactly one nonzero entry in each row and each column.

In the classification [ST] of finite groups generated by complex reflections there is a single infinite family of groups $G(r,p,n)$ and exactly 34 others, the "exceptional" complex reflection groups. The groups $G(r,p,n)$ are the groups of $n \times n$ matrices such that

(a) the entries are either 0 or rth roots of unity,
(b) there is exactly one nonzero entry in each row and each column,
(c) the (r/p)th power of the product of the nonzero entries is 1.

Though we do not know of an early reference which generalizes the theory of Young tableaux to these groups, it is not difficult to see that the method that Young uses for the Weyl groups WD_n extends easily to handle the groups $G(r,p,n)$.

Special cases of the groups $G(r,p,n)$ are

(a) $G(1,1,n) = S_n$, the symmetric group,

(b) $G(2,1,n) = WB_n$, the hyperoctahedral group (i.e. the Weyl group of type B_n),

(c) $G(2,2,n) = WD_n$, the Weyl group of type D_n,

(d) $G(r,1,n) \cong (\mathbb{Z}/r\mathbb{Z}) \wr S_n = (\mathbb{Z}/r\mathbb{Z})^n \rtimes S_n$.

The order of $G(r,1,n)$ is $r^n n!$. Let E_{ij} be the $n \times n$ matrix with 1 in the (i,j) position and all other entries 0. Then $G(r,1,n)$ can be presented by generators

$$s_1 = \zeta E_{11} + \sum_{j\neq 1} E_{ii}, \text{ and } s_i = E_{i,i+1} + E_{i+1,i} + \sum_{j\neq i,i+1} E_{jj},\ 2 \le i \le n,$$

where ζ is a primitive rth root of unity, and relations

(B1) $s_i s_j = s_j s_i$, if $|i-j| > 1$,

(B2) $s_i s_{i+1} s_i = s_{i+1} s_i s_{i+1}$, for $2 \le i \le n-1$,

(BB) $s_1 s_2 s_1 s_2 = s_2 s_1 s_2 s_1$,

(C) $s_1^r = 1$,

(R) $s_i^2 = 1$, for $2 \le i \le n$.

The group $G(r,p,n)$ is the subgroup of index p in $G(r,1,n)$ generated by

$$a_0 = s_1^p, \qquad a_1 = s_1 s_2 s_1, \qquad a_i = s_i, \quad 2 \le i \le n.$$

(1.1) Cyclotomic Hecke algebras $H_{r,1,n}$. More recently there has been an interest in Iwahori-Hecke algebras associated to reflection groups and there has been significant work generalizing the constructions of A. Young to these algebras. Iwahori-Hecke algebras of types A, B and D were handled by Hoefsmit [Ho] and other aspects of the theory for these algebras were developed by Dipper, James and Murphy [DJ], [DJM], Gyoja [Gy] and Wenzl [Wz]. In 1994, Ariki and Koike [AK] introduced cyclotomic Hecke algebras $H_{r,1,n}$ for the complex reflection groups $G(r,1,n)$ and they generalized the Young tableau theory to these algebras. Theorem 3.18 below shows that the theory of [AK] is a special case of an even more general theory for affine Hecke algebras.

Let $u_1, \ldots, u_r, q \in \mathbb{C}$, $q \neq 0$. The cyclotomic Hecke algebra is the algebra $H_{r,1,n}(u_1, \ldots, u_r; q)$, over $\mathbb{C}$, given by generators $T_1, \ldots, T_n$ and relations

(B1) $T_i T_j = T_j T_i$, if $|i-j| > 1$,

(B2) $T_i T_{i+1} T_i = T_{i+1} T_i T_{i+1}$, for $2 \le i \le n-1$,

(BB) $T_1 T_2 T_1 T_2 = T_2 T_1 T_2 T_1$,

(qC) $(T_1 - u_1)(T_1 - u_2) \cdots (T_1 - u_r) = 0,$

(qR) $(T_i - q)(T_i + q^{-1}) = 0,$ for $2 \leq i \leq n.$

The algebra $H_{r,1,n}(u_1, \ldots, u_r; q)$ is of dimension $r^n n!$ (see [AK]).

(a) $H_{1,1,n}(1; q)$ is the Iwahori-Hecke algebra of type A_{n-1}.
(b) $H_{2,1,n}(q, -q^{-1}; q)$ is the Iwahori-Hecke algebra of type B_n.
(c) If ζ is a primitive rth root of 1 then $H_{r,1,n}(1, \zeta, \ldots, \zeta^{r-1}; 1)$ is the group algebra $\mathbb{C}G(r, 1, n)$.

Fact (c) says that the representation theory of the groups $G(r, 1, n)$ is a special case of the representation theory of the algebras $H_{r,1,n}$.

(1.2) Cyclotomic Hecke algebras $H_{r,p,n}$. The work of Broué, Malle and Michel [BMM] demonstrated that there are cyclotomic Hecke algebras associated to most complex reflection groups (even exceptional complex reflection groups). In particular, there are cyclotomic Hecke algebras $H_{r,p,n}$ corresponding to all the groups $G(r, p, n)$ and Ariki [Ar2] has generalized the Young tableau mechanism to these groups (see also [HR]). Theorems 3.15 and the mechanism of Theorem 2.8 show that the theory of Ariki is a special case of a general construction for affine Hecke algebras.

Let $r, p, n \in \mathbb{Z}_{>0}$ be such that p divides r and let $d = r/p$. Let $x_0, \ldots, x_{d-1} \in \mathbb{C}$ and let ξ be a primitive pth root of 1. For $1 \leq j \leq r$, define

$$u_j = \xi^k x_\ell, \qquad \text{if } j - 1 = \ell p + k, \qquad (0 \leq k \leq p-1,\ 0 \leq \ell \leq d-1)$$

i.e., $u_1, \ldots, u_r \in \mathbb{C}$ are chosen so that

$$(T_1 - u_1)(T_1 - u_2) \cdots (T_1 - u_r) = (T_1^p - x_0^p)(T_1^p - x_1^p) \cdots (T_1^p - x_{d-1}^p).$$

The algebra $H_{r,p,n}(x_0, \ldots, x_{d-1}; q)$ is the subalgebra of $H_{r,1,n}(u_1, \ldots, u_r; q)$ generated by the elements

$$a_0 = T_1^p, \qquad a_1 = T_1^{-1} T_2 T_1, \qquad a_i = T_i, \quad \text{for } 2 \leq i \leq n.$$

Then

(a) $H_{2,2,n}(1; q)$ is the Iwahori-Hecke algebra of type D_n,
(b) If η is a primitive dth root of unity then $H_{r,p,n}(1, \eta, \cdots, \eta^{d-1}; 1)$ is the group algebra $\mathbb{C}G(r, p, n)$.

(1.3) Affine braid groups of type A. There are three common ways of depicting affine braids [Cr], [GL], [Jo]:

(a) As braids in a (slightly thickened) cylinder,
(b) As braids in a (slightly thickened) annulus,
(c) As braids with a flagpole.

See Figure 1. The multiplication is by placing one cylinder on top of another, placing one annulus inside another, or placing one flagpole braid on top of another. These are equivalent formulations: an annulus can be made into a cylinder by turning up the edges, and a cylindrical braid can be made into a flagpole braid by putting a flagpole down the middle of the cylinder and pushing the pole over to the left so that the strings begin and end to its right.

The group formed by the affine braids with n strands is the affine braid group of type A. The affine braid group $\mathcal{B}_{\infty,1,n}$ is presented by generators $T_2, \dots, T_n$ and X^{ε_1} (see Figure 2) with relations

(B1) $T_iT_j = T_jT_i$, if $|i-j|>1$,

(B2) $T_iT_{i+1}T_i = T_{i+1}T_iT_{i+1}$, for $2 \le i \le n-1$,

(BB) $X^{\varepsilon_1}T_2X^{\varepsilon_1}T_2 = T_2X^{\varepsilon_1}T_2X^{\varepsilon_1}$,

(B1′) $X^{\varepsilon_1}T_i = T_iX^{\varepsilon_1}$, for $3 \le i \le n$.

Inductively define $X^{\varepsilon_i} \in \mathcal{B}_{\infty,1,n}$ by

$$X^{\varepsilon_i} = T_iX^{\varepsilon_{i-1}}T_i, \qquad 2 \le i \le n. \tag{1.4}$$

By drawing pictures of the corresponding affine braids one can check that the X^{ε_i} all commute with each other. View the symbols ε_i as a basis of $\mathbb{R}^n$ so that

$$\mathbb{R}^n = \sum_{i=1}^n \mathbb{R}\varepsilon_i, \qquad \text{and let} \qquad L = \sum_{i=1}^n \mathbb{Z}\varepsilon_i. \tag{1.5}$$

The affine braid group $\mathcal{B}_{\infty,1,n}$ contains a large abelian subgroup

$$X = \{X^\lambda \mid \lambda \in L\}, \tag{1.6}$$

where $X^\lambda = (X^{\varepsilon_1})^{\lambda_1}\cdots(X^{\varepsilon_n})^{\lambda_n}$ for $\lambda = \lambda_1\varepsilon_1 + \cdots + \lambda_n\varepsilon_n \in L$. The *pole winding number* of an affine braid $b \in \mathcal{B}_{\infty,1,n}$ is $\kappa(b)$ where $\kappa\colon \mathcal{B}_{\infty,1,n} \to \mathbb{Z}$ is the group homomorphism defined by $\kappa(X^{\varepsilon_1}) = 1$ and $\kappa(T_i) = 0$, $2 \le i \le n$. The affine braid group $\mathcal{B}_{\infty,p,n}$ is the subgroup of $\mathcal{B}_{\infty,1,n}$ of affine braids with pole winding number equal to 0 (mod p),

$$\mathcal{B}_{\infty,p,n} = \{b \in \mathcal{B}_{\infty,1,n} \mid \kappa(b) = 0 \pmod p\}. \tag{1.7}$$

Define

$$Q = \sum_{i=2}^n \mathbb{Z}(\varepsilon_i - \varepsilon_{i-1}) \qquad \text{and} \qquad L_p = Q + \sum_{i=1}^n p\mathbb{Z}\varepsilon_i, \tag{1.8}$$

for each nonnegative integer p. The lattice L_p is a lattice of index p in L. Then

$$\mathcal{B}_{\infty,p,n} = \langle X^\lambda, T_i \mid \lambda \in L_p, 2 \le i \le n\rangle$$

and the group $X^{L_p} = \langle X^\lambda \mid \lambda \in L_p \rangle$ is an abelian subgroup of $\mathcal{B}_{\infty,p,n}$.

(1.9) Affine Hecke algebras of type A. The affine Hecke algebra $H_{\infty,1,n}$ (resp. $H_{\infty,p,n}$) is the quotient of the group algebra $\mathbb{C}\mathcal{B}_{\infty,1,n}$ (resp. $\mathbb{C}\mathcal{B}_{\infty,p,n}$) by the relations

$$T_i^2 = (q - q^{-1})T_i + 1, \qquad 2 \leq i \leq n.$$

Let L and X^λ be as in (1.5) and (1.6). The subalgebra

$$\mathbb{C}[X] = \mathrm{span}\{X^\lambda \mid \lambda \in L\} \quad (\text{resp. } \mathbb{C}[X^{L_p}] = \mathrm{span}\{X^\lambda \mid \lambda \in L_p\}) \tag{1.10}$$

is a commutative subalgebra of $H_{\infty,1,n}$ (resp. $H_{\infty,p,n}$). The symmetric group S_n acts on the lattice L by permuting the ε_i and the lattices L_p are S_n-invariant sublattices of L. Let $s_i = (i, i-1) \in S_n$ and $\alpha_i = \varepsilon_i - \varepsilon_{i-1}$. For $2 \leq i \leq n$ and $\lambda \in L$ (resp. $\lambda \in L_p$),

$$X^\lambda T_i = X^{s_i\lambda} T_i + (q - q^{-1}) \frac{X^\lambda - X^{s_i\lambda}}{1 - X^{-\alpha_i}}, \tag{1.11}$$

as elements of $H_{\infty,1,n}$ (resp. $H_{\infty,p,n}$).

For each element $w \in S_n$ define $T_w = T_{i_1} \cdots T_{i_p}$ if $w = s_{i_1} \cdots s_{i_p}$ is an expression of w as a product of simple reflections s_i such that p is minimal. The element T_w does not depend on the choice of the reduced expression of w [Bou, Ch. IV §2 Ex. 23]. The sets

$$\{X^\lambda T_w \mid \lambda \in L, w \in S_n\} \qquad \text{and} \qquad \{X^\lambda T_w \mid \lambda \in L_p, w \in S_n\}$$

are bases of $H_{\infty,1,n}$ and $H_{\infty,p,n}$, respectively [Lu]. The center of $H_{\infty,1,n}$ is

$$Z(H_{\infty,1,n}) = \mathbb{C}[X^{\varepsilon_1}, \ldots, X^{\varepsilon_n}]^{S_n} = \mathbb{C}[X]^{S_n}, \tag{1.12}$$

and $\mathbb{C}[X^{L_p}]^{S_n}$ is the center of $H_{\infty,p,n}$ (see Theorem 4.12 below).

2. Representation theory transfer

In this section we provide the mechanism for obtaining the representation theory of $H_{\infty,p,n}$ from $H_{\infty,1,n}$ and for obtaining the representation theory of cyclotomic Hecke algebras from affine Hecke algebras. In order to obtain the representation theory of $H_{\infty,p,n}$ from $H_{\infty,1,n}$ we identify $H_{\infty,p,n}$ as the set of fixed points of a certain group G acting on $H_{\infty,1,n}$ by automorphisms. Once this is done, the extended version of Clifford theory given in the Appendix allows one to construct the representations of $H_{\infty,p,n}$ from those of $H_{\infty,1,n}$. The same technique can be applied to obtain the representations of the braid groups $\mathcal{B}_{\infty,p,n}$ from those of $\mathcal{B}_{\infty,1,n}$, of the complex reflection groups $G(r,p,n)$ from those of $G(r,1,n)$, and of the Weyl groups WD_n from those of the Weyl groups WB_n.

(2.1) Obtaining $H_{\infty,p,n}$-modules from $H_{\infty,1,n}$-modules. The following result is what is needed to apply the Clifford theory developed in the Appendix to derive the representation theory of the algebras $H_{\infty,p,n}$ from that of $H_{\infty,1,n}$.

Theorem 2.2. *Let ξ be a primitive pth root of unity. The algebra automorphism $g\colon H_{\infty,1,n} \to H_{\infty,1,n}$ defined by*

$$g(X^{\varepsilon_1}) = \xi X^{\varepsilon_1}, \qquad \text{and} \qquad g(T_i) = T_i, \quad 2 \le i \le n,$$

gives rise to an action of the group $\mathbb{Z}/p\mathbb{Z} = \{1, g, \ldots, g^{p-1}\}$ on $H_{\infty,1,n}$ by algebra automorphisms and

$$H_{\infty,p,n} = (H_{\infty,1,n})^{\mathbb{Z}/p\mathbb{Z}}$$

is the set of fixed points of the $\mathbb{Z}/p\mathbb{Z}$-action.

Proof. Immediate from the definitions of $H_{\infty,1,n}$, $H_{\infty,p,n}$ and (1.6). ∎

The action of $\mathbb{Z}/p\mathbb{Z}$ on $H_{\infty,1,n}$ which is given in Theorem 2.2 induces an action of $\mathbb{Z}/p\mathbb{Z}$ on the simple $H_{\infty,1,n}$-modules, see (A.1) in the Appendix. The stabilizer K of the action of $\mathbb{Z}/p\mathbb{Z}$ on a simple $H_{\infty,1,n}$-module M is the *inertia group* K of M. The action of K commutes with the action of $H_{\infty,p,n}$ on M and we have a decomposition

$$M \cong \bigoplus_{j=0}^{|K|-1} M^{(j)} \otimes K^{(j)}, \tag{2.3}$$

where $K^{(j)}$, $1 \le j \le |K|-1$, are the simple K-modules and $M^{(j)}$ are $H_{\infty,p,n}$-modules. Theorem A.13 of the Appendix shows that the $M^{(j)}$ are simple $H_{\infty,p,n}$-modules and that all simple $H_{\infty,p,n}$-modules are constructed in this way. In Theorem 3.15 we show that this method gives a combinatorial construction of the module $M^{(j)}$ in any case when the Young tableau theory is available. This is a generalization of the method used in [Ar2] and [HR].

(2.4) The surjective algebra homomorphisms Φ and Φ_p. The homomorphisms Φ and Φ_p described below are the primary tools for transferring results from the affine Hecke algebras to cyclotomic Hecke algebras. Many results are easier to prove for affine Hecke algebras because of the large commutative subalgebra $\mathbb{C}[X]$ which is available in the affine Hecke algebra. The homomorphism Φ has also been used by Cherednik [Ch2], Ariki [Ar3] and many others.

Proposition 2.5. *Let $H_{\infty,1,n}$ be the affine Hecke algebra of type A defined in (1.9) and let $H_{r,1,n}(u_1,\ldots,u_r;q)$ denote the cyclotomic Hecke algebra of (1.1).*

(a) Fix $u_1,\ldots,u_r\in\mathbb{C}$, $q\in\mathbb{C}^*$. There is a surjective algebra homomorphism given by

$$\begin{array}{rccccl} \Phi\colon & H_{\infty,1,n} & \longrightarrow & H_{r,1,n}(u_1,\ldots,u_r;q) & & \\ & T_i & \longmapsto & T_i, & & 2\le i\le n,\\ & X^{\varepsilon_1} & \longmapsto & T_1. & & \end{array}$$

(b) Restricting the homomorphism Φ to $H_{\infty,p,n}$ yields a surjective homomorphism defined by

$$\begin{array}{rccccl} \Phi_p\colon & H_{\infty,p,n} & \longrightarrow & H_{r,p,n}(x_0,\ldots,x_{d-1};q) & & \\ & T_i & \longmapsto & a_i, & & 2\le i\le n,\\ & X^{p\varepsilon_1} & \longmapsto & a_0, & & \\ & X^{\varepsilon_2-\varepsilon_1} & \longmapsto & a_1a_2. & & \end{array}$$

Proof. The result follows directly from the definitions of the affine Hecke algebras $H_{\infty,1,n}$ and $H_{\infty,p,n}$ (see (1.9)) and the cyclotomic Hecke algebras $H_{r,1,n}$ and $H_{r,p,n}$ (see (1.1) and (1.2)). ∎

Let L_p be the lattice defined in (1.8) and define $C_r=\{\lambda=\sum_{i=1}^n\lambda_i\varepsilon_i \mid 0\le\lambda_i<r\}$. Ariki and Koike [AK] and Ariki [Ar2] have shown that the sets

$$\begin{array}{lc} \{\Phi(X^\lambda T_w)\mid \lambda\in C_r, w\in S_n\} & \text{and}\\ \{\Phi_p(X^\lambda T_w)\mid \lambda\in C_r\cap L_p, w\in S_n\} & \end{array}\tag{2.6}$$

are bases of $H_{r,1,n}$ and $H_{r,p,n}$, respectively.

(2.7) Relating $H_{\infty,1,n}$-modules and $H_{r,1,n}$-modules. The representation theory of the affine Hecke algebra $H_{\infty,1,n}$ is equivalent to the representation theory of the cyclotomic Hecke algebras $H_{r,1,n}$ (considering all possible $u_1,\ldots,u_r\in\mathbb{C}$). The elementary constructions in the following theorem allow us to make $H_{\infty,1,n}$-modules into $H_{r,1,n}$-modules *and vice versa*.

Theorem 2.8. *Let $H_{\infty,1,n}$ be the affine Hecke algebra of type A defined in (1.9) and let $H_{r,1,n}(u_1,\ldots,u_r;q)$ be the cyclotomic Hecke algebra of (1.1).*

(a) *Fix $u_1,\ldots,u_r\in\mathbb{C}$, $q\in\mathbb{C}^*$, and let Φ be the surjective homomorphism of Proposition 2.5. If M is a simple $H_{r,1,n}(u_1,\ldots,u_r;q)$-module then defining*

$$hm=\Phi(h)m,\qquad \text{for all } h\in H_{\infty,1,n} \text{ and all } m\in M,$$

makes M into a simple $H_{\infty,1,n}$-module.

(b) *Let M be a simple $H_{\infty,1,n}$-module and let $\rho\colon H_{\infty,1,n} \to \mathrm{End}(M)$ be the corresponding representation. Let $u_1, \ldots, u_r \in \mathbb{C}$ be such that the minimal polynomial $p(t)$ of the matrix $\rho(X^{\varepsilon_1})$ divides the polynomial $(t-u_1)\cdots(t-u_r)$. Define an action of $H_{r,1,n}(u_1, \ldots, u_r; q)$ on M by*

$$T_1 m = X^{\varepsilon_1} m \qquad \text{and} \qquad T_i m = T_i m, \quad 2 \le i \le n,$$

for all $m \in M$. Then M is a simple $H_{r,1,n}(u_1, \ldots, u_r; q)$-module.

Proof. The Theorem follows directly from the definitions of $H_{r,1,n}(u_1, \ldots, u_r; q)$ and $H_{\infty,1,n}$, and the construction of the surjective homomorphism Φ. ∎

Remark 2.9. The same translations work for arbitrary finite dimensional modules; in particular, they work for indecomposable modules and *preserve composition series.*

3. Standard Young tableaux, representations and Jucys-Murphy elements

In this section we review the generalization of standard Young tableaux in [Ra5] which is used to construct representations of the affine Hecke algebras $H_{\infty,1,n}$. Then we show how this theory can be transported to provide combinatorial constructions of simple modules for the cyclotomic Hecke algebras $H_{r,1,n}(u_1, \ldots, u_r; q)$ and $H_{r,p,n}(x_0, \ldots, x_{d-1}; q)$. This approach shows that Jucys-Murphy type elements in the cyclotomic Hecke algebras arise naturally as images of the elements X^{ε_i} in the affine Hecke algebra. The standard Jucys-Murphy type theorems then follow almost immediately from standard affine Hecke algebra facts.

(3.1) Skew shapes and standard tableaux. A partition λ is a collection of n boxes in a corner. We shall conform to the conventions in [Mac] and assume that gravity goes up and to the left.

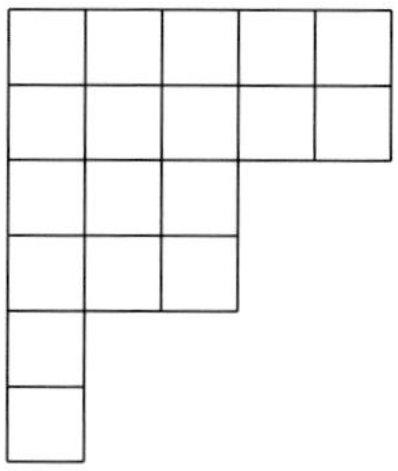

Any partition λ can be identified with the sequence $\lambda = (\lambda_1 \geq \lambda_2 \geq \ldots)$ where λ_i is the number of boxes in row i of λ. The rows and columns are numbered in the same way as for matrices. In the example above we have $\lambda = (553311)$. If λ and μ are partitions such that $\mu_i \leq \lambda_i$ for all i we write $\mu \subseteq \lambda$. The *skew shape* λ/μ consists of all boxes of λ which are not in μ. Any skew shape is a union of connected components. Number the boxes of each skew shape λ/μ along major diagonals from southwest to northeast and

write box_i to indicate the box numbered i.

Let λ/μ be a skew shape with n boxes. A *standard tableau of shape* λ/μ is a filling of the boxes in the skew shape λ/μ with the numbers $1, \ldots, n$ such that the numbers increase from left to right in each row and from top to bottom down each column.

(3.2) Placed skew shapes. Let $\mathbb{R} + i[0, 2\pi/\ln(q^2)) = \{a + bi \mid a \in \mathbb{R}, 0 \leq b \leq 2\pi/\ln(q^2)\} \subseteq \mathbb{C}$. If q is a positive real number then the function

$$\begin{array}{ccc} \mathbb{R} + i[0, 2\pi/\ln(q^2)) & \longrightarrow & \mathbb{C}^* \\ x & \longmapsto & q^{2x} = e^{\ln(q^2)x} \end{array}$$

is a bijection. The elements of $[0,1) + i[0, 2\pi/\ln(q^2))$ index the $\mathbb{Z}$-cosets in $\mathbb{R} + i[0, 2\pi/\ln(q^2))$.

A *placed skew shape* is a pair $(c, \lambda/\mu)$ consisting of a skew shape λ/μ and a *content function*

$$c\colon \{\text{boxes of } \lambda/\mu\} \longrightarrow \mathbb{R} + i[0, 2\pi/\ln(q^2)) \qquad \text{such that} \tag{3.3}$$

$$\begin{array}{ll} c(\text{box}_j) - c(\text{box}_i) \geq 0, & \text{if } i < j \text{ and } c(\text{box}_j) - c(\text{box}_i) \in \mathbb{Z}; \\ c(\text{box}_j) = c(\text{box}_i) + 1, & \text{if box}_i \text{ and box}_j \text{ are on adjacent diagonals;} \\ c(\text{box}_i) = c(\text{box}_j), & \text{if box}_i \text{ and box}_j \text{ are on the same diagonal.} \end{array}$$

This is a generalization of the usual notion of the content of a box in a partition (see [Mac] I §1 Ex. 3).

Suppose that $(c, \lambda/\mu)$ is a placed skew shape such that c takes values in $\mathbb{Z}$. One can visualize $(c, \lambda/\mu)$ by placing λ/μ on a piece of infinite graph paper where the diagonals of the graph paper are indexed consecutively (with elements of $\mathbb{Z}$) from southeast to northwest. The *content* of a box b is the index $c(b)$ of the diagonal that b is on. In the general case, when c takes values in $\mathbb{R} + i[0, 2\pi/\ln(q^2))$, one imagines a book where each page is a sheet of infinite graph paper with the diagonals indexed consecutively (with elements of $\mathbb{Z}$) from southeast to northwest. The pages are numbered by values $\beta \in [0,1) + i[0, 2\pi/\ln(q^2))$ and there is a skew shape $\lambda^{(\beta)}/\mu^{(\beta)}$ placed on page β. The skew shape λ/μ is a union of the disjoint skew shapes on each page,

$$\lambda/\mu = \bigsqcup_{\beta} \left(\lambda^{(\beta)}/\mu^{(\beta)}\right), \qquad \beta \in [0,1) + i[0, 2\pi/\ln(q^2)), \tag{3.4}$$

and the content function is given by

$$\begin{aligned} c(b) = &\ (\text{page number of the page containing } b) \\ &+ (\text{index of the diagonal containing } b). \end{aligned} \tag{3.5}$$

for a box $b \in \lambda/\mu$.

(3.6) Example. The following diagrams illustrate standard tableaux and the numbering of boxes in a skew shape λ/μ.

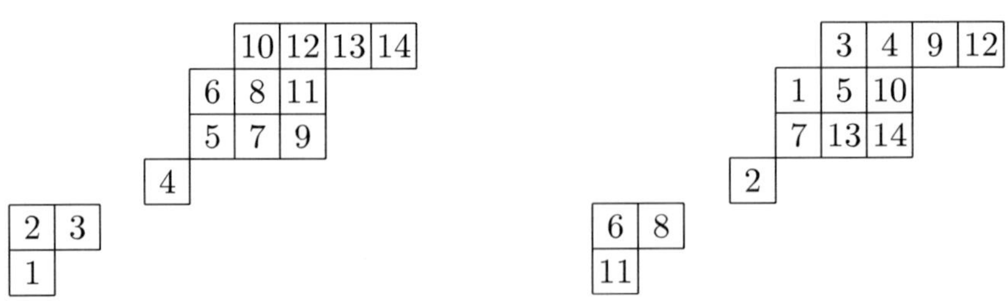

λ/μ with boxes numbered   A standard tableau L of shape λ/μ

The following picture shows the contents of the boxes in the placed skew shape $(c, \lambda/\mu)$ such that the sequence $(c(\text{box}_1), \ldots, c(\text{box}_n))$ is $(-7, -6, -5, -2, 0, 1, 1, 2, 2, 3, 3, 4, 5, 6)$.

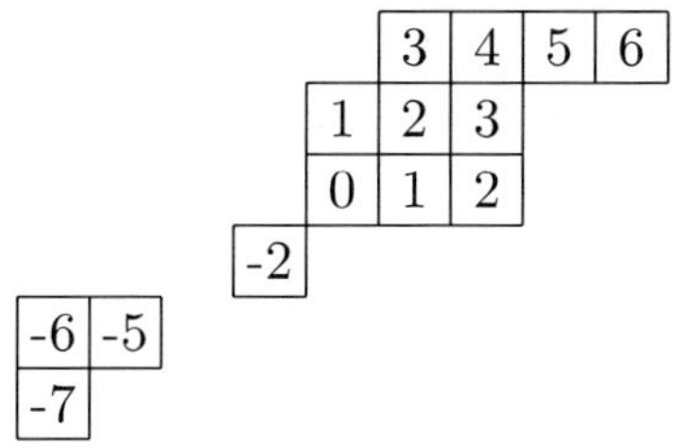

Contents of the boxes of $(c, \lambda/\mu)$

The following picture shows the contents of the boxes in the placed skew shape $(c, \lambda/\mu)$ such that
$(c(\text{box}_1), \ldots, c(\text{box}_n)) = (-7, -6, -5, \frac{-3}{2}, \frac{1}{2}, \frac{3}{2}, \frac{3}{2}, \frac{5}{2}, \frac{5}{2}, \frac{7}{2}, \frac{7}{2}, \frac{9}{2}, \frac{11}{2}, \frac{13}{2})$.

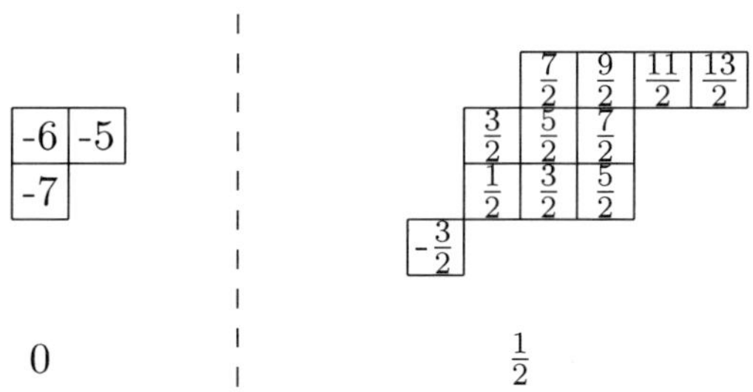

This "book" has two pages, with page numbers 0 and $\frac{1}{2}$. ∎

(3.7) Calibrated $H_{\infty,1,n}$-modules. A finite dimensional $\tilde{H}_{\infty,1,n}$-module M is *calibrated* if it has a basis $\{v_t\}$ such that for each $\lambda \in L$ and each v_t in the basis

$$X^\lambda v_t = t(X^\lambda) v_t, \qquad \text{for some } t(X^\lambda) \in \mathbb{C}.$$

This is the class of representations of the affine Hecke algebra for which there is a good theory of Young tableaux [Ra2].

The following theorem classifies and constructs all irreducible calibrated representations of the affine Hecke algebra $H_{\infty,1,n}$. The construction is a direct generalization of A. Young's classical "seminormal construction" of the irreducible representations of the symmetric group [Yg2]. Young's construction was generalized to Iwahori-Hecke algebras of type A by Hoefsmit [Hf] and Wenzl [Wz] independently, to Iwahori-Hecke algebras of types B and D by Hoefsmit [Hf] and to cyclotomic Hecke algebras by Ariki and Koike [AK]. In (3.7) and (3.11) below we show how all of these earlier generalizations of Young's construction can be obtained from Theorem 3.8. Some parts of Theorem 3.8 are originally due to I. Cherednik, and are stated in [Ch1, §3].

Theorem 3.8. ([Ra, Theorem 4.1]) *Let* $(c, \lambda/\mu)$ *be a placed skew shape with* n *boxes. Define an action of* $H_{\infty,1,n}$ *on the vector space*

$$\tilde{H}^{(c,\lambda/\mu)} = \mathbb{C}\text{-span}\{v_L \mid L \text{ is a standard tableau of shape } \lambda/\mu\}$$

by the formulas

$$\begin{aligned} X^{\varepsilon_i} v_L &= q^{2c(L(i))} v_L, \\ T_i v_L &= (T_i)_{LL} v_L + (q^{-1} + (T_i)_{LL}) v_{s_i L}, \end{aligned}$$

where $s_i L$ *is the same as* L *except that the entries* $i-1$ *and* i *are interchanged,*

$$\begin{aligned} (T_i)_{LL} &= \frac{q - q^{-1}}{1 - q^{2(c(L(i-1)) - c(L(i)))}}, \\ v_{s_i L} &= 0, \qquad \text{if } s_i L \text{ is not a standard tableau,} \end{aligned}$$

and $L(i)$ *denotes the box of* L *containing the entry* i.

(a) $\tilde{H}^{(c,\lambda/\mu)}$ *is a calibrated irreducible* $H_{\infty,1,n}$*-module.*
(b) *The modules* $\tilde{H}^{(c,\lambda/\mu)}$ *are non-isomorphic.*
(c) *Every irreducible calibrated* $H_{\infty,1,n}$*-module is isomorphic to* $\tilde{H}^{(c,\lambda/\mu)}$ *for some placed skew shape* $(c, \lambda/\mu)$.

Remark 3.9. All of the irreducible modules for the affine Hecke algebra have been classified and constructed by Kazhdan and Lusztig [KL]. The construction in [KL] is geometric and noncombinatorial. It is nontrivial (but not very difficult) to relate the construction of Theorem 3.8 and the classification in [KL].

(3.10) Calibrated $H_{\infty,p,n}$-modules. A finite dimensional $H_{\infty,p,n}$-module M is *calibrated* if it has a basis $\{v_t\}$ such that for each $\lambda \in L_p$ and each v_t in the basis

$$X^\lambda v_t = t(X^\lambda) v_t, \qquad \text{for some } t(X^\lambda) \in \mathbb{C}.$$

Let us show how Theorem 3.8, Theorem 2.2 and Theorem A.13 provide explicit constructions of simple calibrated $H_{\infty,p,n}$-modules. The resulting construction is a generalization of the construction of $H_{r,p,n}(x_0, \ldots, x_{d-1}; q)$-modules given by Ariki [Ar2] (as amplified and applied in [HR]). Comparing the following machinations with those in [HR, §3] (where more pictures are given) will be helpful.

The $(\mathbb{Z}/p\mathbb{Z})$-action on $H_{\infty,1,n}$ induces an action of $\mathbb{Z}/p\mathbb{Z}$ on the simple $H_{\infty,1,n}$-modules, as in (A.1) of the Appendix, and this action takes calibrated modules to calibrated modules since the $(\mathbb{Z}/p\mathbb{Z})$-action on $H_{\infty,1,n}$ preserves the subalgebra $\mathbb{C}[X]$. The $\mathbb{Z}/p\mathbb{Z}$ action on simple calibrated modules can be described combinatorially as follows.

If $(c, \lambda/\mu)$ is a placed skew shape with n boxes and $g \in \mathbb{Z}/p\mathbb{Z}$ define

$$g(c, \lambda/\mu) = (c - i\alpha/p, \lambda/\mu), \qquad \text{where } \alpha = 2\pi/\ln(q^2), \tag{3.11}$$

and $c - i\alpha/p$ denotes the content function defined by $(c - i\alpha/p)(b) = c(b) - i\alpha/p$, for all boxes $b \in \lambda/\mu$. To make this definition we are identifying the set $[0, \alpha)$ with $\mathbb{R}/\alpha\mathbb{Z}$. One can imagine the placed skew shape as a book with pages numbered by values $\beta \in [0, 1) + i(\mathbb{R}/\alpha\mathbb{Z})$ and a skew shape $\lambda^{(\beta)}/\mu^{(\beta)}$ on each page. The action of

$$g \text{ cyclically permutes the pages numbered } \beta + i(k/p)\alpha, \quad 0 \leq k < p.$$

If L is a standard tableau of shape $(c, \lambda/\mu)$ let gL denote the same filling of λ/μ as L but viewed as a standard tableaux of shape $g(c, \lambda/\mu)$.

Let ${}^g\tilde{H}^{(c,\lambda/\mu)}$ be the $H_{\infty,1,n}$-module $\tilde{H}^{(c,\lambda/\mu)}$ except twisted by the automorphism g, see (A.1) in the Appendix. It follows from the formulas in Theorem 3.8 that the map

$$\begin{array}{cccc} \phi\colon & {}^g\tilde{H}^{(c,\lambda/\mu)} & \longrightarrow & \tilde{H}^{g(c,\lambda/\mu)} \\ & v_L & \longmapsto & v_{gL} \end{array} \tag{3.12}$$

is an $H_{\infty,1,n}$-module isomorphism. Indeed, since $g^{-1}(T_j) = T_j$ and $(T_j)_{gL,gL} = (T_j)_{LL}$,

$$\begin{aligned} \phi(T_j \circ v_L) &= T_j\phi(v_L), \quad \text{and} \\ \phi(X^{\varepsilon_j} \circ v_L) &= \phi(g^{-1}(X^{\varepsilon_j})v_L) = \phi(\xi^{-1}X^{\varepsilon_j}v_L) \\ &\qquad = q^{2(c(L(j))-i\alpha/p)}v_{gL} = X^{\varepsilon_j}v_{gL} = X^{\varepsilon_j}\phi(v_L), \end{aligned}$$

where $\circ$ denotes the $H_{\infty,1,n}$-action on ${}^g\tilde{H}^{(c,\lambda/\mu)}$ as in (A.1). Identify ${}^g\tilde{H}^{(c,\lambda/\mu)}$ with $\tilde{H}^{g(c,\lambda/\mu)}$ via the isomorphism in (3.12).

Let $(c, \lambda/\mu)$ be a placed skew shape with n boxes and let $K_{(c,\lambda/\mu)}$ be the stabilizer of $(c, \lambda/\mu)$ under the action of $\mathbb{Z}/p\mathbb{Z}$. The cyclic group $K_{(c,\lambda/\mu)}$ is a realization of the inertia group of $\tilde{H}^{(c,\lambda/\mu)}$ and

$$K_{(c,\lambda/\mu)} = \{(g^\kappa)^\ell\colon \tilde{H}^{(c,\lambda/\mu)} \to \tilde{H}^{(c,\lambda/\mu)} \mid 0 \leq \ell \leq |K_{c,\lambda/\mu)}| - 1\},$$

where κ is the smallest integer between 1 and p such that $g^\kappa(c, \lambda/\mu) = (c, \lambda/\mu)$ and $|K_{(c,\lambda/\mu)}|$ is the order of $K_{(c,\lambda/\mu)}$. The elements of $K_{(c,\lambda/\mu)}$ are all $H_{\infty,p,n}$-module isomorphisms. Since $K_{(c,\lambda/\mu)}$ is a cyclic group the irreducible $K_{(c,\lambda/\mu)}$-modules are all one-dimensional and the characters of these modules are given explicitly by

$$\begin{array}{rccc} \eta_j\colon & K_{(c,\lambda/\mu)} & \longrightarrow & \mathbb{C} \\ & g^\kappa & \longmapsto & \xi^{j\kappa}, \end{array} \qquad 0 \le j \le |K_{(c,\lambda/\mu)}| - 1,$$

since ξ^κ is a primitive $|K_{(c,\lambda/\mu)}|$-th root of unity. The element

$$p_j = \sum_{\ell=0}^{|K_{(c,\lambda/\mu)}|-1} \xi^{-j\ell\kappa} g^{\ell\kappa}. \tag{3.13}$$

is the minimal idempotent of the group algebra $\mathbb{C}K_{(c,\lambda/\mu)}$ corresponding to the irreducible character η_j. It follows (from a standard double centralizer result, [Bou2]) that, as an $(H_{\infty,p,n}, K_{(c,\lambda/\mu)})$-bimodule,

$$\tilde{H}^{(c,\lambda/\mu)} \cong \bigoplus_{j=0}^{|K_{(c,\lambda/\mu)}|-1} \tilde{H}^{(c,\lambda/\mu,j)} \otimes K^{(j)}, \text{ where } \tilde{H}^{(c,\lambda/\mu,j)} = p_j \tilde{H}^{(c,\lambda/\mu)}, \tag{3.14}$$

and $K^{(j)}$ is the irreducible $K_{(c,\lambda/\mu)}$-module with character η_j. The following theorem now follows from Theorem A.13 of the Appendix.

Theorem 3.15. *Let $(c, \lambda/\mu)$ be a placed skew shape with n boxes and let $\tilde{H}^{(c,\lambda/\mu)}$ be the simple calibrated $H_{\infty,1,n}$-module constructed in Theorem 3.8. Let $K_{(c,\lambda/\mu)}$ be the inertia group of $\tilde{H}^{(c,\lambda/\mu)}$ corresponding to the action of $\mathbb{Z}/p\mathbb{Z}$ on $H_{\infty,1,n}$ defined by (A.1). If p_j is the minimal idempotent of $K_{(c,\lambda/\mu)}$ given by (3.13) then*

$$\tilde{H}^{(c,\lambda/\mu,j)} = p_j \tilde{H}^{(c,\lambda/\mu)}$$

is a simple calibrated $H_{\infty,p,n}$-module.

Theorem 3.15 provides a generalization of the construction of the $H_{r,p,n}$-modules which was given by Ariki [Ar2] and extended and applied in [HR].

(3.16) Simple $H_{r,1,n}(u_1,\ldots,u_r;q)$-modules. Many (usually all) of the simple $H_{r,1,n}(u_1,\ldots,u_r;q)$-modules can be constructed with Theorems 2.8 and 3.8.

If λ/μ is a skew shape define

$$NW(\lambda/\mu) = \{\text{northwest corner boxes of } \lambda/\mu\},$$

so that $NW(\lambda/\mu)$ is the set of boxes $b \in \lambda/\mu$ such that there is no box of λ/μ immediately above or immediately to the left of b.

Theorem 3.17. *Fix $u_1, \ldots, u_r \in \mathbb{C}^*$ and let $(c, \lambda/\mu)$ be a placed skew shape with n boxes. If*

$$\{q^{2c(b)} \mid b \in NW(\lambda/\mu)\} \subseteq \{u_1, \ldots, u_r\}$$

then the $H_{\infty,1,n}$-module $\tilde{H}^{(c,\lambda/\mu)}$ is a simple $H_{r,1,n}(u_1, \ldots, u_r; q)$-module (via Theorem 2.8).

Proof. Let $(c, \lambda/\mu)$ be a placed skew shape with n boxes and let $\tilde{H}^{(c,\lambda/\mu)}$ be the simple $H_{\infty,1,n}$-module of Theorem 3.8. Let $\rho^{(c,\lambda/\mu)} \colon H_{\infty,1,n} \to \mathrm{End}(\tilde{H}^{(c,\lambda/\mu)})$ be the representation corresponding to $\tilde{H}^{(c,\lambda/\mu)}$. By the formulas in Theorem 3.8 the matrix $\rho(X^{\varepsilon_1})$ is diagonal with eigenvalues $q^{2c(L(1))}$, for $L \in \mathcal{F}^{(c,\lambda/\mu)}$, where $\mathcal{F}^{(c,\lambda/\mu)}$ is the set of standard tableaux of shape λ/μ.

The boxes $L(1)$, $L \in \mathcal{F}^{(c,\lambda/\mu)}$ are exactly the northwest corner boxes of λ/μ and so the minimal polynomial of $\rho(X^{\varepsilon_1})$ is

$$p(t) = \prod_{b \in NW(\lambda/\mu)} (t - q^{2c(b)}).$$

Thus the condition $\{q^{2c(b)} \mid b \in NW(\lambda/\mu)\} \subseteq \{u_1, \ldots, u_r\}$ is exactly what is needed for $\tilde{H}^{(c,\lambda/\mu)}$ to be an $H_{r,1,n}(u_1, \ldots, u_r; q)$-module via Theorem 2.8. ∎

Theorem 3.18. *If the cyclotomic Hecke algebra $H_{r,1,n}(u_1, u_2, \ldots, u_r; q)$ is semisimple, then its simple modules are the modules $\tilde{H}^{(c,\lambda)}$ constructed in Theorem 3.17, where $\lambda = (\lambda^{(1)}, \ldots, \lambda^{(r)})$ is an r-tuple of partitions with a total of n boxes and c is the content function determined by*

$$q^{2c(b)} = u_i, \quad \text{if } b \text{ is the northwest corner box of } \lambda^{(i)}.$$

Proof. The equations $q^{2c(b)} = u_i$, for $b \in NW(\lambda^{(i)})$, determine the values $c(b)$ for all boxes $b \in \lambda$ and thus the pair (c, λ) defines a placed skew shape. By a Theorem of Ariki [Ar1], $H_{r,1,n}(u_1, \ldots, u_r; q)$ is semisimple if and only if

$$[n]_q! \neq 0 \qquad \text{and} \qquad u_i u_j^{-1} \notin \{1, q^2, \ldots, q^{2n}\},$$

where $[n]_q! = [n]_q[n-1]_q \cdots [2]_q[1]_q$ and $[k]_q = (q^k - q^{-k})/(q - q^{-1})$. These conditions guarantee that (c, λ) is a placed skew shape and that the $\tilde{H}_{\infty,1,n}$-module $\tilde{H}^{(c,\lambda)}$ defined in Theorem 3.8 is well defined and irreducible. The reduction in Theorem 2.8 makes $\tilde{H}^{(c,\lambda)}$ into a simple $H_{r,1,n}(u_1, \ldots, u_r; q)$-module and a count of standard tableaux (using binomial coefficients and the classical identity for the symmetric group case) shows that

$$\sum_{\lambda=(\lambda^{(1)},\ldots,\lambda^{(r)})} \dim(\tilde{H}^{(c,\lambda)})^2 = r^n n! = \dim(H_{r,1,n}),$$

where the sum is over all r-tuples $\lambda = (\lambda^{(1)}, \ldots, \lambda^{(r)})$ with a total of n boxes. Thus the $\tilde{H}^{(c,\lambda)}$ are a complete set of simple $H_{r,1,n}(u_1, \ldots, u_r; q)$-modules. ∎

Theorem 3.18 demonstrates that the construction of simple modules for $H_{r,1,n}$ by Ariki and Koike [AK, Theorem 3.7], for $H_{2,1,n}(p, -p^{-1}; q)$ by Hoefsmit [Hf], for $H_{1,1,n}(1; q)$ by Hoefsmit [Hf] and Wenzl [Wz] (independently), and for $\mathbb{C}S_n = H_{1,1,n}(1; 1)$ and $\mathbb{C}WB_n = H_{2,1,n}(1, 1; 1)$ by Young [Yg1-2], are all special cases of Theorem 3.8.

(3.19) Jucys-Murphy elements in cyclotomic Hecke algebras. The following result is well known, but we give a new proof which shows that the cyclotomic Hecke algebra analogues of the Jucys-Murphy elements which have appeared in the literature (see [BMM], [Ra1], [DJM] and the references there) come naturally from the affine Hecke algebra $H_{\infty,1,n}$.

Corollary 3.20. *Let $H_{r,1,n}(u_1, \ldots, u_r; q)$ and $H_{r,p,n}(x_0, \ldots, x_{d-1}; q)$ be the cyclotomic Hecke algebras defined in (1.1) and (1.2).*

(a) The elements

$$M_i = T_i \cdots T_2 T_1 T_2 \cdots T_i\,, \qquad 1 \le i \le n,$$

generate a commutative subalgebra of $H_{r,1,n}(u_1, \ldots, u_r; q)$.

(b) If $H_{r,1,n}(u_1, \ldots, u_r; q)$ is semisimple then every simple $H_{r,1,n}(u_1, \ldots, u_r; q)$-module has a basis of simultaneous eigenvectors of the elements M_i.

(c) The elements

$$M_1^p = a_0, \quad \text{and} \quad M_i M_1^{-1} = a_i \cdots a_3 a_2 a_1 a_3 \cdots a_i\,, \quad 2 \le i \le n,$$

generate a commutative subalgebra of $H_{r,p,n}(x_0, \ldots, x_{d-1}; q)$.

(d) If $H_{r,p,n}(x_0, \ldots, x_{d-1}; q)$ is semisimple then every simple $H_{r,p,n}(x_0, \ldots, x_{d-1}; q)$-module has a basis of simultaneous eigenvectors of the elements M_1^p and $M_i M_1^{-1}$, $2 \le i \le n$.

Proof. (a) The elements X^{ε_i}, $1 \le i \le n$, generate the subalgebra $\mathbb{C}[X] \subseteq H_{\infty,1,n}$. Inductive use of the relation (1.4) shows that $M_i = \Phi(X^{\varepsilon_i})$, where $\Phi\colon H_{\infty,1,n} \to H_{r,1,n}(u_1,\dots,u_r;q)$ is the homomorphism in Proposition 2.5. Thus the subalgebra of $H_{r,1,n}(u_1,\dots,u_r;q)$ generated by the element M_i is the image of $\mathbb{C}[X]$ under the homomorphism Φ.

(b) is an immediate consequence of Theorem 3.18, the construction described in Theorem 3.8, and the fact that $M_i = \Phi(X^{\varepsilon_i})$.

(c) The elements $X^{p\varepsilon_1}$ and $X^{\varepsilon_i-\varepsilon_1}$, $2 \le i \le n$, generate the subalgebra $\mathbb{C}[X^{L_p}] \subseteq H_{\infty,p,n}$ and the images of these elements under the homomorphism $\Phi_p\colon H_{\infty,p,n} \to H_{r,p,n}(x_0,\dots,x_{d-1};q)$ are the elements M_1^p and $M_i M_1^{-1}$.

The proof of part (d) uses the construction described in Theorem 3.15 and is analogous to the proof of part (b). ∎

(3.21) The center of $H_{r,1,n}(u_1,\dots,u_r;q)$. It is an immediate consequence of (1.12) and the proof of Corollary 3.20 that

$$\mathbb{C}[M_1,\dots,M_n]^{S_n} \subseteq Z(H_{r,1,n}(u_1,\dots,u_r;q)),$$

where $M_i = T_iT_{i-1}\cdots T_2T_1T_2\cdots T_{i-1}T_i$. The following proposition shows that this inclusion is an equality.

Proposition 3.22. *If the cyclotomic Hecke algebra $H_{r,1,n}(u_1,\dots,u_r;q)$ is semisimple then its center*

$$Z(H_{r,1,n}(u_1,\dots,u_r;q)) = \mathbb{C}[M_1,\dots,M_n]^{S_n},$$

where $M_i = T_iT_{i-1}\cdots T_2T_1T_2\cdots T_{i-1}T_i$ and $\mathbb{C}[M_1,\dots,M_n]^{S_n}$ is the ring of symmetric polynomials in $M_1,\dots,M_n$.

Proof. By (1.12) $Z(H_{\infty,1,n}) = \mathbb{C}[X]^{S_n}$. Thus, since $M_i = \Phi(X^{\varepsilon_i})$, where $\Phi\colon H_{\infty,1,n} \to H_{r,1,n}(u_1,\dots,u_r;q)$ is the surjective homomorphism of Proposition 2.5, it follows that

$$\begin{aligned}\mathbb{C}[M_1,\dots,M_n]^{S_n} &= \Phi(\mathbb{C}[X]^{S_n}) = \Phi(Z(H_{\infty,1,n}))\\ &\subseteq Z(H_{r,1,n}(u_1,\dots,u_r;q)).\end{aligned}$$

For the reverse inclusion we need to show that the action of the elements $\mathbb{C}[M_1,\dots,M_n]^{S_n}$ distinguishes the simple $H_{r,1,n}(u_1,\dots,u_r;q)$-modules.

Let $\lambda = (\lambda^{(1)},\dots,\lambda^{(r)})$ be an r-tuple of partitions with a total of n boxes and let $\tilde{H}^{(c,\lambda)}$ be the corresponding simple $H_{r,1,n}(u_1,\dots,u_r;q)$-module as constructed by Theorem 3.18 (and Theorem 3.8). If L is a standard tableau of shape λ then $M_1,\dots,M_n$ act on v_L by the multiset of values $(q^{2c(L(1))}, \ldots, q^{2c(L(n))})$. The elementary symmetric functions $e_i(M_1, \ldots, M_n)$ act on $\tilde{H}^{(c,\lambda)}$ by the values $a_i = e_i(q^{2c(L(1))}, \ldots, q^{2c(L(n))})$. Note that a_i does not depend on the choice of the standard tableau L (since $e_i(M_1, \ldots, M_n) \in Z(H_{r,1,n})$). We show that the simple module $\tilde{H}^{(c,\lambda)}$ is determined by the values $a_1, \ldots, a_n$. This shows that the simple modules are distinguished by the elements of $\mathbb{C}[M_1, \ldots, M_n]^{S_n}$.

Let us explain how the values $a_1, \ldots, a_n$ determine the placed skew shape (c, λ). There is a unique (unordered) multiset of values $b_1, \ldots, b_n$ such that $a_i = e_i(b_1, \ldots, b_n)$ for all $1 \le i \le n$. The b_i are determined by the equation

$$(t - b_1) \cdots (t - b_n) = t^n - a_1 t^{n-1} + a_2 t^{n-2} - a_3 t^{n-3} + \cdots \pm a_n.$$

From the previous paragraph, the multiset $\{b_1, \ldots, b_n\}$ must be the same as the multiset $\{q^{2c(L(1))}, \ldots, q^{2c(L(n))}\}$. In this way the values $a_1, \ldots, a_n$ determine the multiset $S = \{c(L(1)), \ldots, c(L(n))\}$.

The multiset S is a disjoint union $S = S_1 \sqcup \cdots \sqcup S_r$ of multisets such that each S_i is in a single $\mathbb{Z}$-coset of $\mathbb{R} + i[0, 2\pi/\ln(q^2))i$ (see (3.4)). These $\mathbb{Z}$ cosets are determined by the values $u_1, \ldots, u_r$ and thus there is a one-to-one correspondence between the u_i and the multisets S_i. Then $|\lambda^{(i)}| = \mathrm{Card}(S_i)$, the nonempty diagonals of $\lambda^{(i)}$ are determined by the values in S_i, and the lengths of the diagonals of $\lambda^{(i)}$ are the multiplicities of the values of the elements of S_i. This information completely determines $\lambda^{(i)}$ for each i. Thus (c, λ) is determined by the values $a_1, \ldots, a_n$. ∎

Remark 3.23. In the language of affine Hecke algebra representations (see [Ra2]) the proof of Proposition 3.22 shows that (when $H_{r,1,n}(u_1, \ldots, u_r; q)$ is semisimple) the simple $H_{r,1,n}(u_1, \ldots, u_r; q)$-modules $\tilde{H}^{(c,\lambda)}$ all have different central characters (as $H_{\infty,1,n}$ modules).

Remark 3.24. The elements M_1^p and $M_i M_1^{-1}$ from Corollary 3.20 cannot be used to obtain a direct analogue of Proposition 3.22 for $H_{r,p,n}(x_0, \ldots, x_{d-1}; q)$. This is because all of the $H_{\infty,p,n}$-modules V_j appearing in the decomposition (3.14) will have the same central character. However, *when n is odd*, the natural analogue of Proposition 3.22 does hold for Iwahori-Hecke algebras of type D_n, $HD_n(q) = H_{2,2,n}(1; q)$. In that case, every simple $H_{2,1,n}$-module has trivial inertia group and the decomposition in (3.14) has only one summand.

4. Affine Hecke algebras of general type

Let R be a reduced root system and let $R^\vee$ be the root system formed by the coroots $\alpha^\vee = 2\alpha/\langle\alpha,\alpha\rangle$, for $\alpha \in R$. Let W be the Weyl group of R and fix a system of positive roots R^+ in R. Let $\{\alpha_1, \ldots, \alpha_n\}$ be the corresponding simple roots and let $s_1, \ldots, s_n$ be the corresponding simple reflections in W. The fundamental weights are defined by the equations $\langle \omega_i, \alpha_j^\vee\rangle = \delta_{ij}$ and the lattices

$$P = \sum_{i=1}^{n} \mathbb{Z}\omega_i \quad \text{and} \quad Q = \sum_{i=1}^{n} \mathbb{Z}\alpha_i,$$

are the weight lattice and the root lattice, respectively. The Dynkin diagrams and the corresponding extended Dynkin diagrams are given in Figure 3. If Γ is a Dynkin diagram or extended Dynkin diagram define

$$m_{ij} = \begin{cases} 2, & \text{if } \overset{i}{\circ} \qquad \overset{j}{\circ}\,, \\ 3, & \text{if } \overset{i}{\circ}\!\!-\!\!\!-\!\!\!-\!\!\overset{j}{\circ}\,, \end{cases} \qquad \text{and} \qquad m_{ij} = \begin{cases} 4, & \text{if } \overset{i}{\circ}\!\!=\!\!=\!\!=\!\!\overset{j}{\circ}\,, \\ 6, & \text{if } \overset{i}{\circ}\!\!\equiv\!\!\equiv\!\!\equiv\!\!\overset{j}{\circ}\,. \end{cases}$$

(4.1) Affine Weyl groups. The extended affine Weyl group is the group

$$\tilde{W} = W \ltimes P = \{wt_\lambda \mid w \in W,\ \lambda \in P\}, \quad \text{where} \quad wt_\lambda = t_{w\lambda}w,$$

for $w \in W$ and $\lambda \in P$ where t_λ corresponds to translation by $\lambda \in P$. Define $s_0 \in \tilde{W}$ by the equation

$$s_0 s_{\phi^\vee} = t_\phi, \qquad \text{where } \phi^\vee \text{ is the highest root of } R^\vee, \tag{4.2}$$

see [Bou Ch. IV §1 no. 2.1]. The subgroup $W_{\text{aff}} = W \ltimes Q$ of $\tilde{W}$ is presented by generators $s_0, s_1, \ldots, s_n$ and relations

$$s_i^2 = 1, \quad 0 \le i \le n, \qquad \text{and} \qquad \underbrace{s_i s_j \cdots}_{m_{ij}} = \underbrace{s_j s_i \cdots}_{m_{ij}}, \quad i \ne j,$$

where the m_{ij} are determined from the extended Dynkin diagram of the root system $R^\vee$. Define

$$\Omega = \{g_i \mid \omega_i \text{ is minuscule}\}, \qquad \text{where} \quad g_i w_0 w_{0,i} = t_{\omega_i}, \tag{4.3}$$

w_0 is the longest element of W and $w_{0,i}$ is the longest element of the group $\langle s_j \mid 1 \le j \le n, j \ne i\rangle$, see [Bou Ch. IV §2 Prop. 6]. Then $\Omega \cong P/Q$ and each element $g \in \Omega$ corresponds to an automorphism of the extended Dynkin diagram of $R^\vee$, in the sense that

$$\text{if } g \in \Omega \quad \text{then} \quad g s_i g^{-1} = s_{\sigma(i)}, \tag{4.4}$$

where σ is the permutation of the nodes determined by the automorphism. Equation (4.4) means that $\tilde{W} = W_{\mathrm{aff}} \rtimes \Omega$. The usual length function on the Coxeter group W_{aff} is extended to the group $\tilde{W}$ by

$$\ell(wg) = \ell(w), \qquad \text{for } w \in W_{\mathrm{aff}} \text{ and } g \in \Omega.$$

Let L be a lattice such that $Q \subseteq L \subseteq P$. View L/Q as a subgroup of $\Omega \cong P/Q$ and let

$$\tilde{W}_L = W \ltimes L = W_{\mathrm{aff}} \rtimes (L/Q).$$

Then $W_{\mathrm{aff}} = \tilde{W}_Q$, $\tilde{W} = \tilde{W}_P$, and $\tilde{W}_L$ is a subgroup of $\tilde{W}$.

(4.5) Affine braid groups. Let L be a lattice such that $Q \subseteq L \subseteq P$. The affine braid group $\tilde{\mathcal{B}}_L$ is the group given by generators T_w, $w \in \tilde{W}_L$, and relations

$$T_w T_{w'} = T_{ww'}, \qquad \text{if } \ell(ww') = \ell(w) + \ell(w').$$

Let $\tilde{\mathcal{B}}_{\mathrm{aff}} = \tilde{\mathcal{B}}_Q$. View L/Q as a subgroup of $\Omega \cong P/Q$. Then $\mathcal{B}_L = \mathcal{B}_{\mathrm{aff}} \rtimes L/Q$ is presented by generators $T_i = T_{s_i}$, $0 \leq i \leq n$, and relations

$$\underbrace{T_i T_j \cdots}_{m_{ij}} = \underbrace{T_j T_i \cdots}_{m_{ij}}, \qquad \text{and} \qquad g T_i g^{-1} = T_{\sigma(i)}, \quad \text{for } g \in \Omega, \tag{4.6}$$

where σ is as in (4.4), and the m_{ij} are specified by the extended Dynkin diagram of $R^\vee$.

Let $P^+ = \sum_{i=1}^n \mathbb{Z}_{\geq 0}\omega_i$ be the dominant weights in P. Define elements X^λ, $\lambda \in P$ by

$$\begin{aligned} X^\lambda &= T_{t_\lambda}, \text{ if } \lambda \in P^+, \text{ and} \\ X^\lambda &= X^\mu (X^\nu)^{-1}, \quad \text{if } \lambda = \mu - \nu \text{ with } \mu, \nu \in P^+. \end{aligned} \tag{4.7}$$

By [Mac3, 3.4] and [Lu], the X^λ are well defined and do not depend on the choice $\lambda = \mu - \nu$, and

$$X^\lambda X^\mu = X^\mu X^\lambda = X^{\lambda+\mu}, \quad \text{for } \lambda, \mu \in P. \tag{4.8}$$

Then $X^\lambda \in \tilde{\mathcal{B}}_L$ if and only if $\lambda \in L$.

(4.9) Affine Hecke algebras. Fix $q \in \mathbb{C}^*$. The affine Hecke algebra $\tilde{H}_L$ is the quotient of the group algebra $\mathbb{C}\tilde{\mathcal{B}}_L$ by the relations

$$(T_i - q)(T_i + q^{-1}) = 0, \qquad \text{for } 0 \leq i \leq n. \tag{4.10}$$

In $\tilde{H}_L$ (see [Mac, 4.2]),

$$X^\lambda T_i = T_i X^{s_i\lambda} + (q-q^{-1})\frac{X^\lambda - X^{s_i\lambda}}{1-X^{-\alpha_i}}, \qquad \text{for } \lambda \in L,\ 1 \le i \le n. \tag{4.11}$$

The Iwahori-Hecke algebra H is the subalgebra of $\tilde{H}$ generated by $T_1, \dots, T_n$.

To our knowledge, the following theorem is due to Bernstein and Zelevinsky in type A, and to Bernstein in general type (unpublished). Lusztig has given an exposition in [Lu]. We give a new proof which we believe is more elementary and more direct.

Theorem 4.12. (Bernstein, Zelevinsky, Lusztig [Lu]) *Let L be a lattice such that $Q \subseteq L \subseteq P$, where Q is the root lattice and P is the weight lattice of the root system R. Let $\tilde{H} = \tilde{H}_L$ be the affine Hecke algebra corresponding to L and let $\mathbb{C}[X] = \operatorname{span}\{X^\lambda \mid \lambda \in L\}$. Let W be the Weyl group of R. Then the center of $\tilde{H}$ is*

$$Z(\tilde{H}) = \mathbb{C}[X]^W = \{f \in \mathbb{C}[X] \mid wf = f \text{ for every } w \in W\}.$$

Proof. Assume

$$z = \sum_{\lambda \in L, w \in W} c_{\lambda,w} X^\lambda T_w \in Z(\tilde{H}).$$

Let $m \in W$ be maximal in Bruhat order subject to $c_{\gamma,m} \neq 0$ for some $\gamma \in L$. If $m \neq 1$ there exists a dominant $\mu \in L$ such that $c_{\gamma+\mu-m\mu,m} = 0$ (otherwise $c_{\gamma+\mu-m\mu,m} \neq 0$ for every dominant $\mu \in L$, which is impossible since z is a finite linear combination of $X^\lambda T_w$). Since $z \in Z(\tilde{H})$ we have

$$z = X^{-\mu} z X^\mu = \sum_{\lambda \in L, w \in W} c_{\lambda,w} X^{\lambda-\mu} T_w X^\mu.$$

Repeated use of the relation (4.11) yields

$$T_w X^\mu = \sum_{\nu \in L, v \in W} d_{\nu,v} X^\nu T_v$$

where $d_{\nu,v}$ are constants such that $d_{w\mu,w} = 1$, $d_{\nu,w} = 0$ for $\nu \neq w\mu$, and $d_{\nu,v} = 0$ unless $v \le w$. So

$$z = \sum_{\lambda \in L, w \in W} c_{\lambda,w} X^\lambda T_w = \sum_{\lambda \in L, w \in W} \sum_{\nu \in L, v \in W} c_{\lambda,w} d_{\nu,v} X^{\lambda-\mu+\nu} T_v$$

and comparing the coefficients of $X^\gamma T_m$ gives $c_{\gamma,m} = c_{\gamma+\mu-m\mu,m}d_{m\mu,m}$. Since $c_{\gamma+\mu-m\mu,m} = 0$ it follows that $c_{\gamma,m} = 0$, which is a contradiction. Hence $z = \sum_{\lambda\in L} c_\lambda X^\lambda \in \mathbb{C}[X]$.

The relation (4.11) gives

$$zT_i = T_i z = (s_i z)T_i + (q - q^{-1})z'$$

where $z' \in \mathbb{C}[X]$. Comparing coefficients of X^λ on both sides yields $z' = 0$. Hence $zT_i = (s_i z)T_i$, and therefore $z = s_i z$ for $1 \le i \le n$. So $z \in \mathbb{C}[X]^W$. ∎

(4.13) Deducing the $\tilde{H}_L$ representation theory from $\tilde{H}_P$. Although we have not taken this point of view in our presentation, the affine Hecke algebras defined above are naturally associated to a reductive algebraic group G over $\mathbb{C}$ [KL] or a p-adic Chevalley group [IM]. In this formulation, the lattice L is determined by the group of characters of the maximal torus of G. It is often convenient to work only with the adjoint version or only with the simply connected version of the group G and therefore it seems desirable to be able to derive the representation theory of the affine Hecke algebras $\tilde{H}_L$ from the representation theory of the affine Hecke algebra $\tilde{H}_P$. The following theorem shows that this can be done in a simple way by using the extension of Clifford theory in the Appendix. In particular, Theorem A.13, can be used to construct all of the simple $\tilde{H}_L$-modules from the simple $\tilde{H}_P$-modules.

Theorem 4.14. *Let L be a lattice such that $Q \subseteq L \subseteq P$, where Q is the root lattice and P is the weight lattice of the root system R. Let $\tilde{H} = \tilde{H}_L$ be the affine Hecke algebra corresponding to L. Then there is an action of a finite group K on $\tilde{H}_P$, acting by automorphisms, such that*

$$\tilde{H}_L = (\tilde{H}_P)^K,$$

is the subalgebra of fixed points under the action of the group K.

Proof. There are two cases to consider, depending on whether the group $\Omega \cong P/Q$ is cyclic or not.

Type	A_{n-1}	B_n	C_n	D_{2n-1}	D_{2n}
Ω	$\mathbb{Z}/n\mathbb{Z}$	$\mathbb{Z}/2\mathbb{Z}$	$\mathbb{Z}/2\mathbb{Z}$	$\mathbb{Z}/4\mathbb{Z}$	$\mathbb{Z}/2\mathbb{Z}\times\mathbb{Z}/2\mathbb{Z}$

Type	E_6	E_7	E_8	F_4	G_2
Ω	$\mathbb{Z}/3\mathbb{Z}$	$\mathbb{Z}/2\mathbb{Z}$	1	1	1

In each case we construct the group K and its action on $\tilde{H}_P$ explicitly. This is necessary for the effective application of Theorem A.13 on examples.

Case 1. If Ω is a cyclic group Ω then the subgroup L/Q is a cyclic subgroup. Suppose

$$\Omega = \{1, g, \ldots, g^{r-1}\} \qquad \text{and} \qquad L/Q = \{1, g^d, \ldots, g^{d(p-1)}\},$$

where $pd = r$. Let ζ be a primitive pth root of unity and define an automorphism

$$\begin{array}{rccc} \sigma: & \tilde{H}_P & \longrightarrow & \tilde{H}_P \\ & g & \longmapsto & \zeta g\,, \\ & T_i & \longmapsto & T_i\,, \qquad 0 \le i \le n. \end{array}$$

The map σ is an algebra isomorphism since it preserves the relations in (4.6) and (4.10). Furthermore, σ gives rise to a $\mathbb{Z}/p\mathbb{Z}$ action on $\tilde{H}_P$ and

$$\tilde{H}_L = (\tilde{H}_P)^{\mathbb{Z}/p\mathbb{Z}}. \tag{4.15}$$

Case 2. If the root system $R^\vee$ is of type D_n, n even, then $\Omega \cong \mathbb{Z}/2\mathbb{Z} \times \mathbb{Z}/2\mathbb{Z}$ and the subgroups of Ω correspond to the intermediate lattices $Q \subseteq L \subseteq P$. Suppose

$$\Omega = \{1, g_1, g_2, g_1g_2 \mid g_1^2 = g_2^2 = 1, g_1g_2 = g_2g_1\}.$$

and define automorphisms of $\tilde{H}_P$ by

$$\begin{array}{rccc} \sigma_1: & \tilde{H}_P & \longrightarrow & \tilde{H}_P \\ & g_1 & \longmapsto & -g_1\,, \\ & g_2 & \longmapsto & g_2\,, \\ & T_i & \longmapsto & T_i\,, \end{array} \qquad \text{and} \qquad \begin{array}{rccc} \sigma_2: & \tilde{H}_P & \longrightarrow & \tilde{H}_P \\ & g_1 & \longmapsto & g_1\,, \\ & g_2 & \longmapsto & -g_2\,, \\ & T_i & \longmapsto & T_i\,. \end{array}$$

Then

$$\tilde{H}_{L_1} = (\tilde{H}_P)^{\sigma_1}, \qquad \tilde{H}_{L_2} = (\tilde{H}_P)^{\sigma_2}, \qquad \text{and} \qquad \tilde{H}_Q = (\tilde{H}_P)^{\langle \sigma_1, \sigma_2 \rangle}, \tag{4.16}$$

where L_1 and L_2 are the two intermediate lattices strictly between Q and P. ∎

5. Where does the homomorphism Φ come from?

The homomorphism $\Phi\colon H_{\infty,1,n} \to H_{r,1,n}$ of Proposition 2.5 is a powerful tool for transporting results about the affine Hecke algebra of type A to the cyclotomic Hecke algebras. In this section we show how this homomorphism arises naturally, from a folding of the

Dynkin diagram of $\tilde{B}_n$, and we give some generalizations of the homomorphism Φ to other types.

Example 1. Type C_n. The root system R of type C_n can be realized by

$$R = \{\pm 2\varepsilon_i, \pm(\varepsilon_j - \varepsilon_i) \mid 1 \le i, j \le n\},$$

where ε_i are an orthonormal basis of $\mathbb{R}^n$. The simple roots and the fundamental weights are given by

$$\begin{aligned} \alpha_1 &= 2\varepsilon_1, \qquad \alpha_i = \varepsilon_i - \varepsilon_{i-1}, \qquad 2 \le i \le n, \\ \omega_i &= \varepsilon_n + \varepsilon_{n-1} + \cdots + \varepsilon_i, \qquad 1 \le i \le n. \end{aligned}$$

If $\phi^\vee$ is the highest root of $R^\vee$ then $\phi = \varepsilon_n + \varepsilon_{n-1}$, and

$$s_{\phi^\vee} = (s_{n-1} \cdots s_2 s_1 s_2 \cdots s_{n-1}) s_n (s_{n-1} \cdots s_2 s_1 s_2 \cdots s_{n-1}). \tag{5.1}$$

Then $\omega_n = \varepsilon_n$ is the only miniscule weight,

$$\begin{aligned} w_0 &= (1,-1)(2,-2)\cdots(n,-n), \\ w_{0,n} &= (1,-1)(2,-2)\cdots(n-1,-(n-1)), \qquad \text{and} \\ w_0 w_{0,n} &= (n,-n) = s_n \cdots s_2 s_1 s_2 \cdots s_n. \end{aligned} \tag{5.2}$$

Thus, from (4.2), (4.3) and (4.6), $\Omega = \{1, g_n\} \cong \mathbb{Z}/2\mathbb{Z}$,

$$\begin{aligned} g_n &= X^{\varepsilon_n} T_n^{-1} \cdots T_2^{-1} T_1^{-1} T_2^{-1} \cdots T_n^{-1}, \\ T_0 &= X^{\varepsilon_n + \varepsilon_{n-1}} (T_n^{-1} \cdots T_2^{-1} T_1^{-1} T_2^{-1} \cdots T_n^{-1}) T_n^{-1} \\ &\qquad (T_n^{-1} \cdots T_2^{-1} T_1^{-1} T_2^{-1} \cdots T_n^{-1}), \end{aligned} \tag{5.3}$$

and

$$g_n T_0 g_n^{-1} = T_n, \qquad \text{and} \qquad g_n T_n g_n^{-1} = T_0. \tag{5.4}$$

The braid group $\tilde{\mathcal{B}}_P(C_n)$ is generated by $T_0, T_1, \ldots, T_n$ and g_n which satisfy relations in (4.6), where the m_{ij} are given by the extended Dynkin diagram $\tilde{B}_n$, see Figure 3. The braid group $\mathcal{B}(C_n)$ is the subgroup generated by $T_1, \ldots, T_n$. These elements satisfy the relations in (4.6), where the m_{ij} are given by the Dynkin diagram C_n. A straightforward check verifies that the map defined by

$$
\begin{array}{llcll}
\Phi_{\tilde{C}C}\colon & \tilde{\mathcal{B}}_P(B_n) & \longrightarrow & \mathcal{B}(C_n) & \\
 & g_n & \longmapsto & 1, & \\
 & T_0 & \longmapsto & T_n, & \\
 & T_i & \longmapsto & T_i, & 1 \le i \le n.
\end{array}
\tag{5.5}
$$

extends to a well defined surjective group homomorphism. From the identity (4.3),

$$
\begin{aligned}
\Phi_{\tilde{C}C}(X^{\varepsilon_n}) &= \Phi_{\tilde{C}C}(X^{\omega_n}) \\
&= \Phi_{\tilde{C}C}(g_n T_n T_{n-1} \cdots T_2 T_1 T_2 \cdots T_{n-1} T_n) \\
&= T_n T_{n-1} \cdots T_2 T_1 T_2 \cdots T_{n-1} T_n.
\end{aligned}
$$

By inductively applying the relation $X^{\varepsilon_i} = T_i X^{\varepsilon_{i-1}} T_i$ we get

$$
\Phi_{\tilde{C}C}(X^{\varepsilon_i}) = T_i T_{i-1} \cdots T_2 T_1 T_2 \cdots T_{i-1} T_i\,, \qquad \text{for all } 1 \le i \le n. \qquad \blacksquare
$$

Example 2. Type A_{n-1}. Since the weight lattice P for the root system of type C_n is the same as the lattice L defined in (1.5) we have an injective homomorphism

$$
\begin{array}{llcll}
\Phi_{\tilde{A}\tilde{C}}\colon & \mathcal{B}_{\infty,1,n} & \longrightarrow & \tilde{\mathcal{B}}_P(C_n) & \\
 & T_i & \longmapsto & T_i, & 2 \le i \le n, \\
 & X^{\varepsilon_i} & \longmapsto & X^{\varepsilon_i}. &
\end{array}
$$

The composition of $\Phi_{\tilde{A}\tilde{C}}$ and the map $\Phi_{\tilde{C}C}$ from (5.5) is the surjective homomorphism defined by

$$
\begin{array}{llcll}
\Phi\colon & \mathcal{B}_{\infty,1,n} & \longrightarrow & \mathcal{B}(C_n) & \\
 & T_i & \longmapsto & T_i, & 2 \le i \le n, \\
 & X^{\varepsilon_i} & \longmapsto & T_i T_{i-1} \cdots T_2 T_1 T_2 \cdots T_{i-1} T_i, & 1 \le i \le n.
\end{array}
$$

In fact, it follows from the defining relations of $\mathcal{B}_{\infty,1,n}$ and $\mathcal{B}(C_n)$ that the map Φ is an isomorphism!

The cyclotomic Hecke algebras $H_{r,1,n}(u_1, \ldots, u_r; q)$ are quotients of $\mathbb{C}\mathcal{B}(C_n)$ and in this way the group homomorphism Φ is the source of the algebra homomorphism

$$
\Phi\colon H_{\infty,1,n} \longrightarrow H_{r,1,n}(u_1, \ldots, u_r; q)
$$

which was used extensively in Section 3 to relate the representation theory of the cyclotomic Hecke algebras $H_{r,1,n}(u_1, \ldots, u_r; q)$ to the affine Hecke algebra of type A. $\blacksquare$

Example 3. Type D_n. Let R be the root system of type D_n. Then $R^\vee$ is also of type D_n and inspection of the Dynkin diagrams of types

$\tilde{D}_n$ and D_n yields a surjective algebra homomorphism defined by

$$\begin{array}{rcccl} \Phi_{\tilde{D}D}\colon & \tilde{\mathcal{B}}_Q(D_n) & \longrightarrow & \mathcal{B}(D_n) & \\ & g_n & \longmapsto & 1, & \\ & T_0 & \longmapsto & T_n, & \\ & T_i & \longmapsto & T_i, & 1 \leq i \leq n. \end{array}$$ ∎

Examples 1, 2, and 3 show that, for types A, B and D, there exist surjective homomorphisms from the affine Hecke algebra to the corresponding Iwahori-Hecke subalgebra. The following example shows that this is not a general phenomenon: there does *not* exist a surjective algebra homomorphism from the affine Hecke algebra of type G_2 to the corresponding Iwahori-Hecke subalgebra of type G_2.

Example 4. Type G_2. If R is the root system of type G_2 then $P = Q$ and $\Omega = \{1\}$.

Proposition 5.6. *Let $\tilde{H}(G_2)$ be the affine Hecke algebra of type G_2 as given by (4.6) and (4.10) and let $H(G_2)$ be the Iwahori-Hecke subalgebra of type G_2 generated by T_1 and T_2. There does not exist an algebra homomorphism $\Phi\colon \tilde{H}(G_2) \to H(G_2)$ such that $\Phi(T_i) = T_i$, for $1 \leq i \leq 2$.*

Proof. There is an irreducible representation of $\tilde{H}(G_2)$ given by

$$\rho(T_1) = \begin{pmatrix} q & 0 \\ 0 & -q^{-1} \end{pmatrix}, \text{ and}$$

$$\rho(T_2) = \frac{1}{q+q^{-1}} \begin{pmatrix} 2-q^{-2} & q^2-1+q^{-2} \\ 3 & q^2-2 \end{pmatrix},$$

(see [Ra1, Theorem 6.11]). We show that there does not exist a 2×2 matrix N which satisfies

$$N^2 = (q-q^{-1})N+1,\ N\rho(T_2)N = \rho(T_2)N\rho(T_2) \text{ and } N\rho(T_1) = \rho(T_1)N.$$

If N exists then N must be diagonal since N commutes with $\rho(T_1)$ and $\rho(T_1)$ is a diagonal matrix with distinct eigenvalues. The first equation shows that N is invertible and the second equation shows that N is conjugate to $\rho(T_2)$. It follows that N must have one eigenvalue q and one eigenvalue $-q^{-1}$. Thus, either

$$N = \begin{pmatrix} q & 0 \\ 0 & -q^{-1} \end{pmatrix} \qquad \text{or} \qquad N = \begin{pmatrix} -q^{-1} & 0 \\ 0 & q \end{pmatrix}.$$

However, neither of these matrices satisfies the relation $N\rho(T_2)N = \rho(T_2)N\rho(T_2)$. This contradiction shows that the representation ρ cannot be extended to be a representation of $\tilde{H}(G_2)$. ∎

In spite of the fact, demonstrated by the previous example, that there does not always exist a surjective algebra homomorphism from the affine Hecke algebra onto its Iwahori-Hecke subalgebra, there *are* interesting surjective homomorphisms from affine Hecke algebras of exceptional type.

Example 5. Type E_6. For the root system of type E_6, $P/Q \cong \mathbb{Z}/3\mathbb{Z}$. Let $\Omega = \{1, g, g^2\}$ where g is as given by (4.3) for the minuscule weight ω_1 (see [Bou, p. 261]). There are surjective algebra homomorphisms

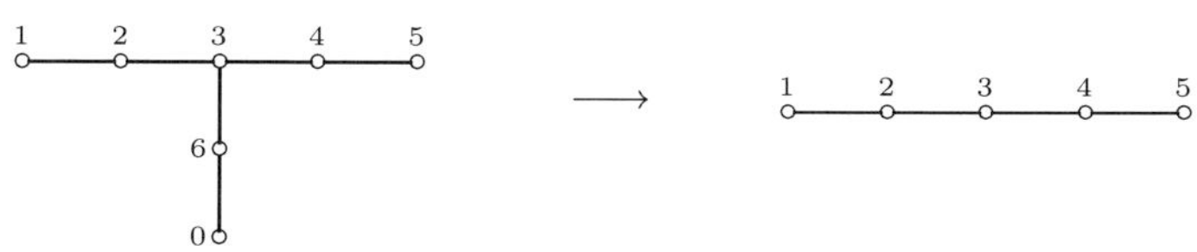

$$\begin{array}{rccc} \Phi\colon & \tilde{\mathcal{B}}_Q(E_6) & \longrightarrow & \mathcal{B}(A_5) \\ & T_0 & \longmapsto & T_5, \\ & T_6 & \longmapsto & T_4, \\ & T_i & \longmapsto & T_i, \qquad 1 \le i \le 5, \end{array}$$

and

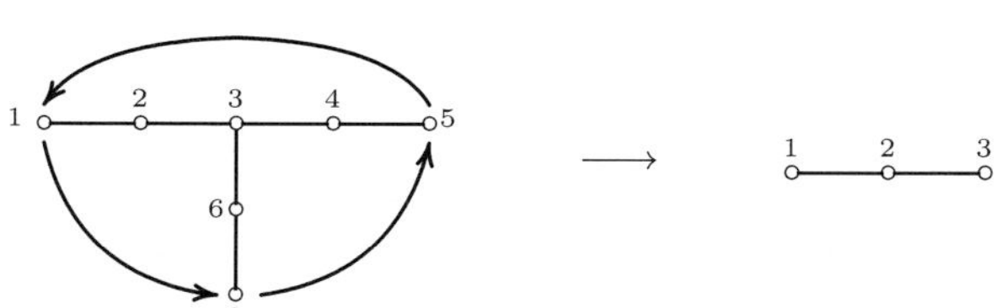

$$\begin{array}{rccccl}
\Phi'\colon & \tilde{\mathcal{B}}_P(E_6) & \longrightarrow & \mathcal{B}(A_3) & & \\
& g & \longmapsto & 1, & & \\
& T_0 & \longmapsto & T_1, & & \\
& T_6 & \longmapsto & T_2, & & \\
& T_5 & \longmapsto & T_1, & & \\
& T_4 & \longmapsto & T_2, & & \\
& T_i & \longmapsto & T_i, & & 1 \le i \le 3.
\end{array}$$

Example 6. Type F_4. For the root system R of type F_4 we have $P = Q$ and $\Omega = \{1\}$. There is a surjective homomorphism

$$\begin{array}{rccccl}
\Phi\colon & \tilde{\mathcal{B}}(F_4) & \longrightarrow & \mathcal{B}(A_2 \times A_3) & & \\
& T_i & \longmapsto & T_i\,, & & 0 \le i \le 4.
\end{array}$$

∎

Appendix: Clifford theory

Let R be an algebra over $\mathbb{C}$ and let G be a finite group acting by automorphisms on R. The *skew group ring* is

$$R \rtimes G = \left\{ \sum_{g \in G} r_g g \;\middle|\; r_g \in R \right\}$$

with multiplication given by the distributive law and the relation

$$gr = g(r)g, \qquad \text{for } g \in G \text{ and } r \in R.$$

Let N be a (finite dimensional) left R-module. For each $g \in G$ define an R-module ${}^gN$, which has the same underlying vector space N but such that

$$ {}^gN \text{ has } R\text{-action given by} \qquad r \circ n = g^{-1}(r)n, \tag{A.1}$$

for $r \in R$, $n \in N$. If W is an R-submodule of N then ${}^gW$ is an R-submodule of ${}^gN$ and so ${}^gN$ is simple if and only if N is simple. Thus there is an action of G on the set of simple R-modules.

Let R^λ be a simple R-module. The inertia group of R^λ is

$$H = \{h \in G \mid R^\lambda \cong {}^hR^\lambda\}. \tag{A.2}$$

If $h \in H$ then Schur's lemma implies that the isomorphism $R^\lambda \cong {}^hR^\lambda$ is unique up to constant multiples (since both R^λ and ${}^hR^\lambda$ are simple). For each $h \in H$ fix an isomorphism $\phi_h\colon R^\lambda \to {}^{h^{-1}}R^\lambda$. Then, as operators on R^λ,

$$\phi_h r = h(r)\phi_h, \quad \text{and} \quad \phi_g\phi_h = \alpha(g,h)\phi_{gh}, \tag{A.3}$$

where $\alpha(g,h) \in \mathbb{C}^*$ are determined by the choice of the isomorphisms ϕ_h. The resulting function $\alpha\colon H \times H \to \mathbb{C}^*$ is called a *factor set* [CR, 8.32].

Let $(\mathbb{C}H)_{\alpha^{-1}}$ be the algebra with basis $\{c_h \mid h \in H\}$ and multiplication given by

$$c_g c_h = \alpha(g,h)^{-1} c_{gh}, \qquad \text{for } g,h \in H. \tag{A.4}$$

Let H^μ be a simple $(\mathbb{C}H)_{\alpha^{-1}}$-module. Then putting

$$rh(m \otimes n) = r\phi_h m \otimes c_h n, \qquad \text{for } r \in R,\ h \in H,\ m \in R^\lambda,\ n \in H^\mu, \tag{A.5}$$

defines an action of $R \rtimes H$ on $R^\lambda \otimes H^\mu$.

Theorem A.6. (Clifford theory) *Let R^λ be a simple R-module and let H be the inertia group of R^λ. Let H^μ be a simple $(\mathbb{C}H)_{\alpha^{-1}}$-module where $\alpha\colon H \times H \to \mathbb{C}^*$ is the factor set determined by a choice of isomorphisms $\phi_h\colon R^\lambda \to {}^hR^\lambda$. Define an action of $R \rtimes H$ on $R^\lambda \otimes H^\mu$ as in (A.5) and define*

$$RG^{\lambda,\mu} = \mathrm{Ind}_{R\rtimes H}^{R\rtimes G}(R^\lambda \otimes H^\mu) = (R \rtimes G) \otimes_{R\rtimes H} (R^\lambda \otimes H^\mu).$$

Then

(a) *$RG^{\lambda,\mu}$ is a simple $R \rtimes G$-module.*
(b) *Every simple $R \rtimes G$-module is obtained by this construction.*
(c) *If $RG^{\lambda,\mu} \cong RG^{\nu,\gamma}$ then R^λ and R^ν are in the same G-orbit of simple R-modules and $H^\mu \cong H^\gamma$ as $(\mathbb{C}H)_{\alpha^{-1}}$-modules.*

Proof. The proof of this theorem is as in [Mac2] except that the consideration of the factor set $\alpha\colon H \times H \to \mathbb{C}^*$ is necessary to correct an error there. We thank P. Deligne for pointing this out to us. A sketch of the proof is as follows.

Let M be a simple $R \rtimes G$-module and let R^λ be a simple R-submodule of M. Then $gR^\lambda \cong {}^gR^\lambda$ as R-modules and $M = \sum_{g\in G} gR^\lambda$ since the right hand side is an $R \rtimes G$-submodule of M. Then

$$M = \sum_{g_i \in G/H} g_i N = \mathrm{Ind}_{R\rtimes H}^{R\rtimes G}(N), \qquad \text{where} \quad N = \sum_{h\in H} hR^\lambda,$$

and the first sum is over a set $\{g_i\}$ of coset representatives of the cosets G/H.

The R-module N is semisimple and by [Bou2]

$$N \cong R^\lambda \otimes H^\mu, \tag{A.7}$$

where $H^\mu = \mathrm{Hom}_R(R^\lambda, N)$. It can be checked that the vector space H^μ has a $(\mathbb{C}H)_{\alpha^{-1}}$-action given by

$$(c_h\psi)(m) = \alpha(h, h^{-1})h\psi(\phi_{h^{-1}}(m)), \quad \text{for } h \in H,\ \psi \in \mathrm{Hom}_R(R^\lambda, N),$$

where c_h is as in (A.4). Then, with $R \rtimes H$-action on $R^\lambda \otimes H^\mu$ given by (A.5), the isomorphism in (A.7) is an isomorphism of $R \rtimes H$-modules (see [CR, Thm. (11.17) (ii)]).

If P is an $(\mathbb{C}H)_{\alpha^{-1}}$-submodule of H^μ then $R^\lambda \otimes P$ is an $R \rtimes H$-submodule of $R^\lambda \otimes H^\mu$ and $\mathrm{Ind}_{R\rtimes H}^{R\rtimes G}(R^\lambda \otimes P)$ is an $R \rtimes G$-submodule of M. Thus H^μ must be a simple $(\mathbb{C}H)_{\alpha^{-1}}$-module.

This argument shows that every simple $R \rtimes G$-module is of the form $RG^{\lambda,\mu}$. The uniqueness follows as in [Mac2, App.]. ∎

Remark A.8. A different choice $\psi_h \colon R^\lambda \to {}^hR^\lambda$ of the isomorphisms in (A.3) may yield a factor set $\beta \colon H \times H \to \mathbb{C}^*$ which is different from the factor set α. However, the algebras $(\mathbb{C}H)_{\beta^{-1}}$ and $(\mathbb{C}H)_{\alpha^{-1}}$ are always isomorphic (a diagonal change of basis suffices).

Lemma A.9. *Define $R^G = \{r \in R \mid g(r) = r$ for all $g \in G\}$ and let*

$$e = (1/|G|)\sum_{g\in G} g \in R \rtimes G.$$

(a) *The map*

$$\begin{array}{cccc} \theta\colon & R^G & \longrightarrow & e(R \rtimes G)e \\ & s & \longmapsto & se \end{array}$$

is a ring isomorphism.

(b) *Left multiplication by elements of R and the action of G by automorphisms make R into a left $R \rtimes G$-module. Right multiplication makes R a right R^G-module. The rings $R \rtimes G$ and $e(R \rtimes G)e$ act on $(R \rtimes G)e$ by left and right multiplication, respectively. The map*

$$\begin{array}{cccc} \psi\colon & R & \cong & (R \rtimes G)e \\ & r & \longmapsto & re \end{array}$$

is an isomorphism of $(R \rtimes G, R^G)$-bimodules.

Proof. (a) If $r \in R^G$ then

$$ere = \frac{1}{|G|}\sum_{g\in G} g(r)ge = \frac{1}{|G|}\sum_{g\in G} re = re.$$

Thus the map θ is well defined and if $r, s \in R^G$ then $rese = rse$, so θ is a homomorphism. If $re = se$ then $r = s$ since $R \rtimes G$ is a free R-module with basis G. Thus θ is injective. If $\sum_{g\in G} r_g g$ is a general element of $R \rtimes G$ then

$$e\left(\sum_{g\in G} r_g g\right)e = \sum_{g,h\in G} h(r_g)hge = \left(\sum_{g,h\in G} h(r_g)\right)e,$$

and, for each $g \in G$, $\sum_{h\in G} h(r_g) \in R^G$. So θ is surjective.

The proof of (b) is straightforward. ∎

Let $(\mathbb{C}H)_\alpha$ be the algebra with basis $\{b_h \mid h \in H\}$ and multiplication given by

$$b_g b_h = \alpha(g,h)b_{gh}, \qquad \text{for } g,h \in H,$$

and let $(\mathbb{C}H)_{\alpha^{-1}}$ be as in (A.4). Let M be a $(\mathbb{C}H)_\alpha$-module. The dual of M is the $(\mathbb{C}H)_{\alpha^{-1}}$-module given by the vector space $M^* = \mathrm{Hom}(M,\mathbb{C})$ with action

$$(c_h\psi)(m) = \alpha(h,h^{-1})^{-1}\psi(b_{h^{-1}}m), \qquad \text{for } h \in H,\ \psi \in M^*.$$

This is a $(\mathbb{C}H)_{\alpha^{-1}}$ action since, for all $g, h \in H$, $\psi \in M^*$,

$$\begin{aligned}
&(c_gc_h\psi)(m)\\
&= \alpha(h,h^{-1})^{-1}\alpha(g,g^{-1})^{-1}\psi(b_{h^{-1}}b_{g^{-1}}m)\\
&= \alpha(h,h^{-1})^{-1}\alpha(g,g^{-1})^{-1}\alpha(h^{-1},g^{-1})\psi(b_{(gh)^{-1}}m)\\
&= \alpha(h,h^{-1})^{-1}\alpha(g,g^{-1})^{-1}\alpha(h^{-1},g^{-1})\alpha(gh,h^{-1}g^{-1})(c_{gh}\psi)(m)\\
&= \alpha(g,h)^{-1}(c_{gh}\psi)(m),
\end{aligned}$$

where the last equality follows from the associativity of the product $b_gb_hb_{h^{-1}}b_{g^{-1}}$ in $(\mathbb{C}H)_\alpha$. If $\rho\colon (\mathbb{C}H)_\alpha \to \mathrm{End}(M)$ is the representation corresponding to M then the representation $\rho^*\colon (\mathbb{C}H)_{\alpha^{-1}} \to \mathrm{End}(M^*)$ corresponding to M^* is

$$\rho^*(c_h) = \alpha(h,h^{-1})^{-1}\rho(b_{h^{-1}})^t = (\rho(b_h)^{-1})^t. \tag{A.11}$$

If M is a $(\mathbb{C}H)_\alpha$-module and N is a $(\mathbb{C}H)_{\alpha^{-1}}$-module then $M\otimes N$ is an $\mathbb{C}H$-module with action defined by

$$h(m\otimes n) = b_hm \otimes c_hn, \qquad \text{for } h \in H,\ m \in M \text{ and } n \in N. \tag{A.12}$$

The following lemma is a version of Schur's lemma which will be used in the proof of Theorem A.13.

Lemma A.12. *Suppose that M and N are simple $(\mathbb{C}H)_\alpha$-modules and let N^* be the $(\mathbb{C}H)_{\alpha^{-1}}$-module which is the dual of N. Let $e_H = (1/|H|)\sum_{h\in H} h$. Then*

$$\dim(e_H(M\otimes N^*)) = \begin{cases} 1, & \text{if } M\cong N,\\ 0, & \text{otherwise.}\end{cases}$$

Proof. Identify $M\otimes N^*$ with $\mathrm{Hom}(N,M)$. Then, by (A.10) and (A.11), the action of $\mathbb{C}H$ on $\mathrm{Hom}(N,M)$ is given by,

$$hA = \rho(b_h)A\rho(b_h)^{-1}, \qquad \text{for } h\in H \text{ and } A\in \mathrm{Hom}(N,M),$$

where $\rho\colon (\mathbb{C}H)_\alpha \to \mathrm{End}(M)$ is the representation corresponding to M. If $A\in\mathrm{Hom}(N,M)$ and $g\in H$ then

$$e_HA = ge_HA = g(e_HA) = \rho(b_g)(e_HA)\rho(b_g)^{-1},$$

and so $\rho(b_g)(e_HA) = (e_HA)\rho(b_g)$ for all $g\in H$. Then, by Schur's lemma, $e_HA = 0$ if $M\not\cong N$ and e_HA is a constant if $M=N$. ∎

Theorem A.13. *Let R^λ be a simple R-module and let H be the inertia group of R^λ. The ring R^G acts on R^λ (by restriction) and $(\mathbb{C}H)_\alpha$ acts on R^λ (by the R-module isomorphisms $\phi_h\colon R^\lambda \cong {}^hR^\lambda$ of (A.3)) and these two actions commute. Thus there is a decomposition*

$$R^\lambda \cong \bigoplus_{\nu\in\hat{H}_\alpha} R^{\lambda,\nu}\otimes (H^\nu)^*,$$

where $\hat{H}_\alpha$ is an index set for the simple $(\mathbb{C}H)_\alpha$-modules, $(H^\nu)^*$ is the dual of the simple $(\mathbb{C}H)_{\alpha^{-1}}$-module H^ν, and $R^{\lambda,\nu}$ is an R^G-module.

(a) *If $R^{\lambda,\mu}\neq 0$ then it is a simple R^G-module.*
(b) *Every simple R^G-module is isomorphic to some $R^{\lambda,\mu}$.*
(c) *The nonzero $R^{\lambda,\mu}$ are pairwise nonisomorphic.*

Proof. The setup of Lemma A.9(b) puts us in the situation of [Gr, §6.2]. If e is the idempotent used in Lemma A.9 then the functor

$$\begin{array}{ccc} R\rtimes G\text{-modules} & \longrightarrow & R^G\text{-modules}\\ M & \longmapsto & eM\end{array}$$

is an exact functor such that if M is a simple $R\rtimes G$-module then eM is either 0 or a simple R^G-module. Furthermore, every simple R^G-module arises as eM for some simple $R\rtimes G$-module M.

Let $RG^{\lambda,\mu}$ be a simple $R \rtimes G$-module as given by Theorem A.6. From the definition of $RG^{\lambda,\mu}$ we obtain

$$\begin{aligned} eRG^{\lambda,\mu} &= e(R \rtimes G) \otimes_{R\rtimes H} (R^\lambda \otimes H^\mu) \\ &= e \otimes (R^\lambda \otimes H^\mu) = ee_H \otimes (R^\lambda \otimes H^\mu) \\ &= e \otimes e_H(R^\lambda \otimes H^\mu), \end{aligned}$$

where $e_H = (1/|H|)\sum_{h\in H} h$. Using the decomposition in the statement of the Theorem, we conclude that, as R^G-modules,

$$\begin{aligned} eRG^{\lambda,\mu} &= e \otimes e_H \left(\bigoplus_{\nu\in\hat{H}} R^{\lambda,\nu} \otimes (H^\nu)^* \otimes H^\mu \right) \\ &= e \otimes e_H \left(\bigoplus_{\nu\in\hat{H}} R^{\lambda,\nu} \otimes e_H((H^\nu)^* \otimes H^\mu) \right) \\ &\cong R^{\lambda,\mu}. \end{aligned}$$

The last isomorphism is a consequence of Lemma A.12. The statement of the Theorem now follows from the results of J.A. Green quoted above.

Remark A.14. It follows from Theorem A.13 that R^λ is semisimple as an R^G-module and the action of $(\mathbb{C}H)_\alpha$ on R^λ generates $\mathrm{End}_{R^G}(R^\lambda)$.

References

[Ar1] S. Ariki, *On the semisimplicity of the Hecke algebra of* $(\mathbb{Z}/r\mathbb{Z}) \wr S_n$, J. Algebra **169** (1994), 216–225.

[Ar2] S. Ariki, *Representation theory of a Hecke algebra of* $G(r,p,n)$, J. Algebra **177** (1995), 164–185.

[Ar3] S. Ariki, *On the decomposition numbers of the Hecke algebra of* $G(m,1,n)$, J. Math. Kyoto Univ. **36** (1996), 789–808.

[AK] S. Ariki and K. Koike, *A Hecke algebra of* $(\mathbb{Z}/r\mathbb{Z}) \wr S_n$ *and construction of its irreducible representations*, Adv. in Math. **106** (1994), 216–243.

[AM] S. Ariki and A. Mathas, *On the number of simple modules of the Hecke algebras of type* $G(r,p,n)$, Math. Zeitschrift **233** (2000), no. 3, 601–623.

[Bou1] N. Bourbaki, *Groupes et algèbres de Lie, Chapitres 4,5 et 6*, Elements de Mathématique, Hermann, Paris 1968.

[Bou2] N. Bourbaki, *Algèbre, Chapitre 8*, Elements de Mathématique, Hermann, Paris 1958.

[BMM] M. Broué, G. Malle, and J. Michel, *Représentations unipotents génériques et blocs des groupes réductifs finis*, Astérique **212**, 1993.

[Ch1] I. Cherednik, *A new interpretation of Gel'fand-Tzetlin bases*, Duke Math. J. **54** (1987), 563–577.

[Ch2] I. Cherednik, *Monodromy representations for generalized Knizhnik-Zamolodchikov equations and Hecke algebras*, Publ. RIMS, Kyoto Univ. **27** (1991), 711–726.

[Cr] J. Crisp, Ph.D. Thesis, University of Sydney, 1997.

[CR] C. Curtis and I. Reiner, *Methods of representation theory–with applications to finite groups and orders* Vol. I, J. Wiley and Sons, 1981.

[DJ] R. Dipper and G.D. James, *Blocks and idempotents of Hecke algebras of general linear groups*, Proc. London Math. Soc. (3) **54** (1987) 57–82.

[DJM] R. Dipper, G.D. James and G.E. Murphy, *Hecke algebras of type B_n at roots of unity*, Proc. London Math. Soc. (3) **70** (1995) 505–528.

[GL] M. Geck and S. Lambropoulou, *Markov traces and knot invariants related to Iwahori-Hecke algebras of type B*, J. Reine Angew. Math. **482** (1997), 191–213.

[Gr] J.A. Green, *Polynomial representations of GL_n*, Lecture Notes in Mathematics **830**, Springer-Verlag, New York, 1980.

[Gy] A. Gyoja, *A q-analogue of Young symmetrizer*, Osaka J. Math. **23** (1986), 841–852.

[HR] T. Halverson and A. Ram, *Murnaghan-Nakayama rules for characters of Iwahori-Hecke algebras of the complex reflection groups $G(r,p,n)$*, Canadian J. Math. **50** (1998), 167–192.

[Hf] P.N. Hoefsmit, *Representations of Hecke algebras of finite groups with BN-pairs of classical type*, Ph.D. Thesis, University of British Columbia, 1974.

[IM] N. Iwahori and H. Matsumoto, *On some Bruhat decomposition and the structure of the Hecke rings of p-adic Chevalley groups*, Publ. Math. I.H.E.S. **25** (1965), 5–48.

[Jo] V.F.R. Jones, *A quotient of the affine Hecke algebra in the Brauer algebra*, Enseign. Math. (2) **40** (1994), 313–344.

[KL] D. Kazhdan and G. Lusztig, *Proof of the Deligne-Langlands conjecture for Hecke algebras*, Invent. Math. **87** (1987), 153–215.

[Lu] G. Lusztig, *Affine Hecke algebras and their graded version*, J. Amer. Math. Soc. **2** (1989), 599–635.

[Mac] I.G. Macdonald, *Symmetric functions and Hall polynomials*, Second edition, Oxford University Press, New York, 1995.

[Mac2] I.G. Macdonald, *Polynomial functors and wreath products*, J. Pure and Appl. Algebra **18** (1980) 173–204.

[Mac3] I.G. Macdonald, *Affine Hecke algebras and orthogonal polynomials*, Séminaire Bourbaki, 47ème année, n° 797, 1994–95, Astérisque **237** (1996), 189–207.

[Ra1] A. Ram, *Seminormal representations of Weyl groups and Iwahori-Hecke algebras*, Proc. London Math. Soc. (3) **75** (1997), 99-133.

[Ra2] A. Ram, *Calibrated representations of affine Hecke algebras*, preprint 1998, results to appear in J. Algebra under the title *Affine Hecke algebras and generalized standard Young tableaux*.

[Ra3] A. Ram, *Standard Young tableaux for finite root systems*, preprint 1998, results to appear in J. Algebra under the title *Affine Hecke algebras and generalized standard Young tableaux*.

[Ra4] A. Ram, *Irreducible representations of rank two affine Hecke algebras*, in *Advances in Algebra and Geometry* (University of Hyderabad Conference 2001), Hindustan Book Agency, New Delhi, 2002 pp. 57–91.

[Ra5] A. Ram, *Skew shape representations are irreducible*, to appear in Contemp. Math., S.-J. Kang and K.-H. Lee eds., Amer. Math. Soc. 2003.

[Re] M. Reeder, *Isogenies of Hecke algebras and a Langlands correspondence for ramified principal series representations*, Represent. Theory **6** (2002), 101–126.

[Sp] W. Specht, *Eine Verallgemeinerung der symmetrischen Gruppe*, Schriften Math. Seminar Berlin **1** (1932), 1–32.

[ST] G. C. Shephard and J.A. Todd, *Finite unitary reflection groups*, Canadian Journal of Math. **6** (1954), 274–304.

[Wz] H. Wenzl, *Hecke algebras of type A_n and subfactors*, Invent. Math. **92** (1988), 349–383.

[Yg1] A. Young, *On quantitative substitutional analysis* (fifth paper), Proc. London Math. Soc. (2) **31** (1929), 273–288.

[Yg2] A. Young, *On quantitative substitutional analysis* (sixth paper), Proc. London Math. Soc. (2) **34** (1931), 196–230.

Department of Mathematics, University of Wisconsin–Madison
Madison, WI 53706, USA.
E-mail address: `ram@math.wisc.edu`

Department of Mathematics, and University of Newcastle
NSW 2308, Australia
E-mail address: `jacqui@maths.newcastle.edu.au`

The Helgason Fourier Transform for Symmetric spaces

Rudra P. Sarkar and Alladi Sitaram

Dedicated to Professor C.S. Seshadri on his 70th *birthday*

Abstract. We discuss the Helgason Fourier transform for L^1 and L^p functions on Riemannian symmetric spaces of the noncompact type.

1. Introduction

Let f be a sufficiently "nice" function on $\mathbb{R}^n$. Then one has the pair of formulae:

$$\widehat{f}(\lambda,\omega) = \frac{1}{(2\pi)^{\frac{n}{2}}} \int_{\mathbb{R}^n} f(x) e^{-i\lambda\langle x,w\rangle} dx \text{ and}$$

$$f(x) = \frac{1}{(2\pi)^{\frac{n}{2}}} \int_{\mathbb{R}^+} \int_{S^{n-1}} \widehat{f}(\lambda,\omega) e^{i\lambda\langle x,w\rangle} d\omega \lambda^{n-1} d\lambda,$$

where $x \in \mathbb{R}^n$, $\lambda \in \mathbb{R}^+ = [0,\infty)$, $\omega \in S^{n-1}$, the unit sphere in $\mathbb{R}^n$ and $d\omega$ is the surface measure on S^{n-1}.

Notice that for fixed λ and ω, $x \mapsto e^{i\lambda\langle x,\omega\rangle}$ are *oscillatory* eigenfunctions of the Laplacian which are constant on the family of parallel hyperplanes $\{\langle \omega, x\rangle = \mathit{constant}\}$. Thus the second formula above can be interpreted as saying that f is a superposition of eigenfunctions of the Laplacian which are also *plane waves*. Thanks to the work of S. Helgason, these ideas have been generalized to symmetric spaces of the noncompact type. To keep the exposition simple we consider only symmetric spaces X of rank 1, i.e. $X = G/K$ where G is a semisimple Lie group of real rank 1 and K is a fixed maximal compact subgroup of G. If Δ is the Laplace-Beltrami operator of X, then the analogues of plane waves are certain eigenfunctions of Δ which are constant on *horocycles.* (These horocycles play the role of hyperplanes in $\mathbb{R}^n$ - see [**4**] for details.) For fixed k and λ, these are the functions: $x \mapsto e^{i\lambda(H(x^{-1}k))}$. Then one defines for sufficiently nice functions, say for $f \in C_c^\infty(X)$, the Helgason Fourier transform on $\mathfrak{a}^* \times K$ by

$$\tilde{f}(\lambda, k) = \int_G f(x) e^{(i\lambda - \rho)(H(x^{-1}k))} dx \tag{1. 1}$$

and one has the inversion formula due to Helgason:

$$f(x) = |\mathcal{W}|^{-1} \int_{\mathfrak{a}^*} \int_K \tilde{f}(\lambda, k) e^{(i\lambda - \rho)(H(x^{-1}k))} dk d\mu(\lambda). \tag{1. 2}$$

Here $d\mu(\lambda)$ is essentially Harish-Chandra's Plancherel measure restricted to the spherical principal series and $|\mathcal{W}|$ is the order of the Weyl group of G/K. (See section 2 for notation.) While the properties of the Helgason Fourier transform have been extensively studied for functions in the Schwartz space and for C_c^∞-functions, very little has been said about general L^1 and L^p functions on X, with the exception of the technical report by Helgason et. al. [**6**]. See the extensive bibliography in [**4**]. Of course, there do exist some isolated results for L^1 and L^p functions, some of them with the added restriction that the functions are also left K-finite (i.e. functions on G that are right K-invariant and left K-finite). In this note we extend some of the results in [**6**] to general L^p-functions on X. In a subsequent paper, we use the Helgason Fourier transform as a basic tool to study questions of the Wiener-Tauberian type.

2. Notation and Preliminaries

Let G be a connected noncompact semisimple Lie group with finite centre and let K be a fixed maximal compact subgroup of G. Let $X = G/K$ be the corresponding Riemannian symmetric space of the noncompact type. Let $G = KAN$ be an Iwasawa decomposition of G and let $\mathfrak{a}$ be the Lie algebra of A. Let $\mathfrak{a}^*$ be the real dual of $\mathfrak{a}$ and $\mathfrak{a}_{\mathbb{C}}^*$ be its complexification. Then for any $g \in G$, $g = K(g) \exp H(g) N(g)$ where $K(g) \in K$, $N(g) \in N$ and $H(g) \in \mathfrak{a}$. Let M be the centralizer of A in K. Let db be a K-invariant normalized measure on K/M. Let $C_c^\infty(X)$ be the set of compactly supported smooth functions on X. As usual we can consider functions on X as right K-invariant functions on G. For $f \in C_c^\infty(X)$, $b \in K$ and $\lambda \in \mathfrak{a}^*$ we define the

Helgason Fourier transform by

$$(2.\ 1)\quad \tilde{f}(\lambda, b) = \int_G f(x) e^{(-i\lambda+\rho)(A(bx))} dx, \text{ where } A(g) = -H(g^{-1}).$$

In the above equality, ρ is the half sum of the positive restricted roots of G/K, dx the Haar measure on G. (Actually $\tilde{f}(\lambda, .)$ is a function on K which is right invariant under M, i.e. a function on K/M; however we choose to ignore this point in this exposition.)

We will assume throughout that X is of rank 1, and hence $\dim \mathfrak{a}^* = 1$. Therefore we can identify $\mathfrak{a}^*_{\mathbb{C}}$ with $\mathbb{C}$ by identifying $\lambda.\rho$ with $\lambda \in \mathbb{C}$. Under this identification $\mathfrak{a}^* = \mathbb{R}$. For each $\lambda \in \mathbb{C}$, let ϕ_λ be the elementary spherical function of Harish-Chandra given by:

$$\phi_\lambda(g) = \int_K e^{(-i\lambda+1)(A(kg))} dk.$$

Here, by $A(kg)$, we actually mean $\rho(A(kg))$. For $\lambda \in \mathbb{C}$ one has the spherical principal series representation π_λ of G on $L^2(K/M)$ defined by:

$$(\pi_\lambda(x)V)(b) = e^{(i\lambda-1)(H(x^{-1}b))} V(K(x^{-1}b)),\ b \in K.$$

(see [**3**] for details). Again, by $H(x^{-1}b)$, we mean $\rho(H(x^{-1}b))$. One also knows that π_λ is unitary and irreducible if λ is real. It can be shown that for $f \in L^1(X)$ and $\lambda \in \mathbb{R}$:

$$\lambda \mapsto \pi_\lambda(f) = \int_G f(x)\pi_\lambda(g) dg$$

maps $\mathbb{R}$ into the space of bounded linear operators on $L^2(K/M)$, where

$$(\pi_\lambda(f)V)(b) = \left(\int_K V(u) du\right) \int_G e^{(i\lambda-1)H(x^{-1}b)} f(x) dx.$$

Thus

$$(\pi_\lambda(f))V(b) = \left(\int_K V(u) du\right) \tilde{f}(\lambda, b).$$

The integral above is over K/M, but as mentioned earlier, we slur over the difference.

3. The Helgason Fourier transform of L^1 and L^p functions

Suppose $\mathcal{S}$ is the Helgason-Johnson strip ([**5**]), i.e. $\mathcal{S} = \{\lambda \in \mathbb{C} : |\text{Im } \lambda| \leq 1\}$. Let $\mathcal{S}_p$ be the corresponding p-strip, i.e. $\mathcal{S}_p = \{\lambda \in \mathbb{C} : |\text{Im } \lambda| \leq (\frac{2}{p} - 1)\}$, where $p \in [1, 2)$. Let $\mathcal{S}^o$ and $\mathcal{S}^o_p$ be the interiors of $\mathcal{S}$ and $\mathcal{S}_p$ respectively. Note that $\mathcal{S}_1 = \mathcal{S}$. It is well known that the elementary spherical function ϕ_λ is bounded if and only if $\lambda \in \mathcal{S}$

([**5**]). It is also known that if $\lambda \in S_p^0$, then $\phi_\lambda \in L^q$, where $\frac{1}{p} + \frac{1}{q} = 1$. This easily follows from the basic estimates for ϕ_0 and ϕ_λ given in [**3**] (Proposition 4.6.1 and Theorem 4.6.4).

There seems to be some misunderstanding regarding the domain of definition of the Helgason Fourier transform for L^1-functions on X ([**1**], p. 319). However it is proved in [**6**] that the *natural domain* of definition of the Helgason Fourier transform for an L^1 functions is the strip $\mathcal{S}_1$. In fact, more precisely we have from [**6**]:

THEOREM 3.1. *Let $f \in L^1(X)$. Then there exits $B = B(f) \subset K$, B a set of full Haar measure in K, such that $\tilde{f}(\lambda, k)$ exists for all $k \in B$, $\lambda \in \mathcal{S}$. For each fixed $k \in B$, $\tilde{f}(\lambda, k)$ is holomorphic for $\lambda \in \mathcal{S}^o$ and continuous on $\mathcal{S}$. Further $\tilde{f}(\lambda, .) \in L^1(K)$ and $\|\tilde{f}(\lambda, .)\|_{L^1(K)} \leq \|f\|_1$, for all $\lambda \in \mathcal{S}$. Finally for each $k \in B$, $\tilde{f}(\lambda, k) \longrightarrow 0$ as $|\lambda| \longrightarrow \infty$ in $\mathcal{S}$.*

Actually a little more is true:

PROPOSITION 3.2. *Let f be as in the above theorem. Then $\|\tilde{f}(\lambda, .)\|_{L^1(K)} \longrightarrow 0$ as $|\lambda| \longrightarrow \infty$, uniformly in $\mathcal{S}$.*

Proof. As $C_c^\infty(X)$ is dense in $L^1(X)$, we find a sequence of functions f_n in $C_c^\infty(X)$ such that $f_n \longrightarrow f$ in $L^1(X)$. Then by the estimates of the Paley-Wiener Theorem ([**4**], III Theorem 5.1), it is easy to see that for each fixed n, $\|\tilde{f}_n(\lambda, .)\|_{L^1(K)} \longrightarrow 0$ uniformly on $\mathcal{S}$ as $|\lambda| \longrightarrow \infty$. Since $\int_K |\tilde{f}_n(\lambda, .) - \tilde{f}(\lambda, .)| dk \leq \|f_n - f\|_1$ for all $\lambda \in \mathcal{S}$, the statement follows.

REMARK 3.3. It must be emphasized that in [**6**], no restriction on the rank of X is made. Although to keep the exposition simple we have assumed rank $X = 1$, it is possible to use the techniques from [**6**] to extend the new result of this section to symmetric spaces of arbitrary rank.

REMARK 3.4. Both the last part of Theorem 3.1 and Proposition 3.2 can be viewed as versions of the classical Riemann-Lebesgue lemma.

REMARK 3.5. Since $\pi_\lambda(f)e_0(k) = \tilde{f}(\lambda, k)$ as L^2 functions on K, it is clear that for $f \in L^1(X)$ and $\lambda \in \mathfrak{a}^*$,

$$\tilde{f}(\lambda, .) \in L^2(K) \quad \text{and} \quad \|\tilde{f}(\lambda, .)\|_{L^2(K)} \leq \|f\|_{L^1(X)}.$$

We now generalize the results in [**6**] to $L^p, 1 \leq p < 2$, although for simplicity we only consider the case rank $X = 1$.

THEOREM 3.6. *Let $f \in L^p(X)$, $1 < p < 2$. Then there exits $B = B(f) \subset K$, B a set of full Haar measure in K, such that $\tilde{f}(\lambda, k)$ exists for all $k \in B$, $\lambda \in \mathcal{S}_p^o$ and $\tilde{f}(\lambda, .)$ is in $L^1(K)$ for every fixed $\lambda \in \mathcal{S}_p^o$. For each fixed $k \in B$, $\tilde{f}(\lambda, k)$ is holomorphic for $\lambda \in \mathcal{S}_p^o$. Finally if*

$$\mathcal{F} = \{\lambda \in \mathbb{C} : |Im\lambda| \leq \delta\},$$

where $\delta < \frac{2}{p} - 1$, then $\|\tilde{f}(\lambda, .)\|_{L^1(K)} \longrightarrow 0$ as $|\lambda| \longrightarrow \infty$ uniformly in $\mathcal{F}$.

Proof. Let $\{i\lambda_n\}$ with λ_n increasing and positive, be a sequence of imaginary points in $\mathcal{S}_p$ which converges to the boundary point $i(\frac{2}{p} - 1)$. Let $\lambda = \mu + i\nu$ be an interior point of $\mathcal{S}_p$. There exists an $n \in \mathbb{Z}^+$ such that $|\nu| \leq \lambda_n$. It follows that

$$\begin{aligned} |e^{(-i\lambda+1)A(kx)}| &= |e^{-(-i\lambda+1)H(x^{-1}k)}| = e^{(\nu-1)H(x^{-1}k)} \\ &\leq e^{(\lambda_n-1)H(x^{-1}k)} + e^{(-\lambda_n+1)H(x^{-1}k)}. \end{aligned} \tag{3. 1}$$

Now,

$$\begin{aligned} \textstyle\int_K |\tilde{f}(\lambda, k)|dk &\leq \textstyle\int_K \int_G |e^{(-i\lambda+1)(A(x,k))}|\, |f(x)|dx\, dk \\ &\leq \textstyle\int_K \int_G (e^{(\lambda_n-1)H(x^{-1}k)} + e^{(-\lambda_n+1)H(x^{-1}k)}) \\ &\qquad\qquad |f(x)|dx\, dk. \end{aligned}$$

Applying Fubini's theorem and using the definition of ϕ_λ, we get

$$\begin{aligned} &\textstyle\int_G \int_K e^{(\lambda_n+1)(A(kx))}|f(x)|dkdx + \int_G \int_K e^{(-\lambda_n+1)(A(kx))}|f(x)|dkdx \\ \leq\ &\textstyle\int_G |f(x)||\phi_{i\lambda_n}(x)|dx + \int_G |f(x)||\phi_{-i\lambda_n}(x)|dx. \end{aligned}$$

Since $\phi_{i\lambda_n} = \phi_{-i\lambda_n}$ is an L^q-function, the above is

$$\leq \|f(x)\|_p \|\phi_{i\lambda_n}\|_q + \|f(x)\|_p \|\phi_{-i\lambda_n}\|_q < \infty. \tag{3. 2}$$

Hence this shows that there exist two sets $C_n, C_{-n} \subset K$, each of full Haar measure in K such that

$$\int_G e^{(\lambda_n+1)(A(kx))}|f(x)|dx \text{ and } \int_G e^{(-\lambda_n+1)(A(kx))}|f(x)|dx$$

exist for every k in C_n and in C_{-n} respectively. Therefore, in view of (3. 1), $\tilde{f}(\lambda, k)$ exists for every $k \in B_n = C_n \cap C_{-n}$.

We now take $B = \cap_n B_n$. Then $\tilde{f}(\lambda, k)$ exists for all $\lambda \in \mathcal{S}_p^o$ and $k \in B$. The above also shows that $\tilde{f}(\lambda, .)$ is in $L^1(K/M)$ for every fixed $\lambda \in \mathcal{S}_p^o$. Now an easy argument along the same lines as Theorem 1 in [**6**] shows that for each fixed $k \in B$, $\lambda \longrightarrow \tilde{f}(\lambda, k)$ is holomorphic in $\mathcal{S}_p^o$. Finally (3. 2) combined with an argument exactly as in Proposition 3.2 yields the last part of this theorem.

4. Concluding remarks

The Helgason Fourier transform is particularly well suited to formulate analogues of the Wiener-Tauberian type results for symmetric spaces. While it is well known that the exact analogue of Wiener's theorem is false for symmetric spaces (see [**2**] and [**7**]), modified versions of the theorem do hold (see [**10**], [**8**] etc.). In a forthcoming paper [**9**] we shall prove:

THEOREM 4.1. *Let G be a noncompact, connected, semisimple Lie group of real rank 1, K a fixed maximal compact subgroup and $X = G/K$ the corresponding symmetric space of rank 1. Let f be a nontrivial integrable function and further assume that $\int_G |f(x)|e^{\varepsilon d(xK,K)}dx < \infty$, for some $\varepsilon > 0$ where d is the Riemannian metric on X. (It follows from the condition on f that there exists a set B of full measure in K such that $\tilde{f}(\lambda, k)$ is defined for $k \in B$, $\lambda \in T = \{\lambda \in \mathbb{C} : |Im\lambda| \leq 1 + \delta\}$, where δ is some fixed but small positive number.) Let $Z = \{\lambda \in T : \tilde{f}(\lambda, .) \equiv 0\}$. Also assume that f vanishes on a set of positive Haar measure.*

If Z is empty, then the left G-translates of f span a dense subspace of $L^1(X)$.

Even in [**8**], [**10**] etc the "Fourier transform" is assumed to exist in an *enlarged* strip like T above. This almost amounts to demanding f to be a little more than in L^1, i.e. it belongs to some sort of weighted L^1-space (as in the condition of the theorem).

References

[1] Bray, William O.; Pinsky, Mark A. *Pointwise Fourier inversion on rank one symmetric spaces and related topics.* J. Funct. Anal. 151 (1997), no. 2, 306-333. MR 99d:43016

[2] Ehrenpreis, L.; Mautner, F. I. *Some properties of the Fourier transform on semi-simple Lie groups. I.* Ann. of Math. (2) 61, (1955). 406-439. MR 16,1017b

[3] Gangolli, Ramesh; Varadarajan, V. S. *Harmonic analysis of spherical functions on real reductive groups.* Ergebnisse der Mathematik und ihrer Grenzgebiete, 101, Springer-Verlag, Berlin, (1988). MR 89m:22015

[4] Helgason, Sigurdur *Geometric analysis on symmetric spaces.* Math. Surveys and Monographs, 39. AMS, Providence, RI, (1994). MR 96h:43009

[5] Helgason, Sigurdur; Johnson, Kenneth *The bounded spherical functions on symmetric spaces.* Advances in Math. 3 (1969) 586-593 MR 40 #2787

[6] Helgason, S.; Rawat, R.; Sengupta, J.; Sitaram A. *Some Remarks on the Fourier Transform on a symmetric space* Indian Statistical Institute-Bangalore Centre Technical Report (1998).

[7] Leptin, Horst *Ideal theory in group algebras of locally compact groups.* Invent. Math. 31 (1975/76), no. 3, 259-278. MR 53 #3189

[8] Sarkar, Rudra P. *Wiener Tauberian theorem for rank one symmetric spaces.* Pacific J. Math. 186 (1998), no. 2, 349-358. MR 99k:43004

[9] Sarkar, Rudra P.; Sitaram, Alladi *The Helgason Fourier Transform for Symmetric Spaces (II),* under preparation.

[10] Sitaram, Alladi *On an analogue of the Wiener Tauberian theorem for symmetric spaces of the noncompact type.* Pacific J. Math. 133 (1988), no. 1, 197-208. MR 89e:43024

STAT-MATH UNIT, INDIAN STATISTICAL INSTITUTE
203 B. T. RD., CALCUTTA 700108, INDIA
E-mail address: rpsarkar@yahoo.com

STAT-MATH UNIT, INDIAN STATISTICAL INSTITUTE
8 TH MILE, MYSORE RD., R. V. COLLEGE POST,
BANGALORE 560059, INDIA
E-mail address: sitaram@isibang.ac.in

On representations of special orthogonal groups over fields of positive characteristics.

S. Senthamarai Kannan and K. V. Subrahmanyam

Dedicated to Prof. C. S. Seshadri on his 70th birthday

Abstract. Let G be the special orthogonal group. Let T be a maximal torus in G and B be a Borel subgroup of G containing T. Let λ be a dominant weight with respect to T. Let L_λ denote the line bundle on G/B corresponding to the dominant weight weight λ. We determine bounds and constraints on p for which $H^0(G/B, L_\lambda)$ is irreducible as a G module. We solve the problem in the case when $\lambda = m\omega_1, m > 0$, and the cases when $\lambda = 2\omega_2, 3\omega_2$. For our proof we make use of ideas from standard monomial theory developed by Seshadri and his school.

1. Introduction

A fundamental problem in the representation theory of reductive algebraic groups is that of determining the formal characters of irreducible modules [**2**]. In [**11**] Lusztig conjectured the characters of certain key modules from which the character of an arbitrary irreducible module can be determined, provided p is greater than or equal to $2h - 3$, where h is the Coxeter number of the group in question. From the works of Kashiwara-Tanisaki, [**7**], Kazhdan-Lustig, [**8**] and Anderson-Jantzen-Soergel, [**1**], the conjecture is now known to be true for $p >> 0$: for every semi-simple, simply connected group with root system Φ defined over an algebraically closed field of characteristic $p > n(\Phi)$, the conjecture holds. This bound on p is a function of the root system Φ. However no explicit bound for $n(\Phi)$ is known.

In this paper we address a closely related problem. Let G be a semi-simple, simply connected Chevalley group over an algebraically closed field of characteristic p. Let T be a maximal torus of G and let $B \supseteq T$ be a Borel subgroup B of G. Let λ be a dominant weight with respect to T, B. Let L_λ denote the line bundle on G/B corresponding to the dominant weight weight λ (see section 2 for definitions). The problem we address is that of determining bounds and constraints on p for which the dual of the Weyl module, $H^0(G/B, L_\lambda)$ is irreducible as a G module.

In this paper we give constraints on p when G is the special orthogonal group and λ is $m\omega_1$, $m > 0$ and when λ is $2\omega_2$, $3\omega_2$. For a precise statement of the result refer to Theorem 6.

The organisation of the paper. In section 2 we recall some definitions and set up some notation. In section 3 we describe our choice of a Borel subgroup for the special orthogonal group $SO(2n+1, K)$ and show some consequences of this choice. In section 4 we compute the Chevalley basis for the special orthogonal group. In section 5 we give constraints of p implying the irreducibility of the $SL(2n+1, K)$-modules $H^0(SL(2n+1, K)/B, L_{m\omega_1})$, $H^0(SL(2n+1, K)/B, L_{2\omega_2})$ and $H^0(SL(2n+1, K)/B, L_{3\omega_2})$. We also state a lemma which relates the irreducibility of the $SO(2n+1, K)$ modules $H^0(SO(2n+1, K)/B, L_{m\omega_j})$, for $j = 1, 2$ to the irreducibility of the corresponding $SL(2n+1, K)$ modules $H^0(SL(2n+1, K)/B, L_{m\omega_j})$. In section 6 we state conditions on p for which $H^0(SO(2n+1, K)/B, L_{m\omega_1})$ is irreducible. We also state conditions on p for which $H^0(SO(2n+1, K)/B, L_{2\omega_2})$ and $H^0(SO(2n+1, K)/B, L_{3\omega_2})$ are irreducible. A proof of the main theorem for these cases is then described, modulo certain computations which are carried out in Appendix B. In Appendix A we give proofs of certain identities from standard monomial theory. These identities express non-standard monomials in terms of standard monomials. In Appendix B we carry out all the computations needed in the $2\omega_2$ case and the $3\omega_2$ case. Finally in Appendix C we give a different proof of our results. This proof was given to us by the referee.

Due to constraints on space we have not included Appendix A and B in this paper. Interested readers may download this from `http://www.cmi.ac.in/~kv/papers/so2n/Appendices-A-B.ps`.

2. Preliminaries

Let K be an algebraically closed field of characteristic $p > 0$. Let G be a semi-simple, simply connected algebraic group over K. We fix a maximal torus T of G. Let B be a Borel subgroup of G containing T. Let $X(T)$ denote the set of characters of T and let $W = N(T)/T$ denote the Weyl group of G with respect to T. Let w_0 denote the longest element of the Weyl group. Let Φ denote the set of roots of G with respect to T and let Φ^+ the set of all positive roots with respect to B. Let $\Delta = \{\alpha_1, \dots, \alpha_n\}$ denote the set of simple roots in Φ^+ and let ω_i denote the fundamental weight of G corresponding to the simple root α_i. For a dominant weight $\lambda \in X(T)$ let $V(\lambda)_K$ (resp. $L(\lambda)_K$) denote the universal Weyl module (resp. irreducible G module) with highest weight λ. We briefly recall this construction [**14**].

Recall that any such G arises as a universal Chevalley group. So there is split group scheme $G_{\mathbb{Z}}$ over $\mathbb{Z}$ such that base change of $G_{\mathbb{Z}}$ by K is G. Let $G_{\mathbb{Q}}$ be the split group scheme over $\mathbb{Q}$ obtained from $G_{\mathbb{Z}}$ by base change to $\mathbb{Q}$. Let $\mathcal{G}$ be the Lie algebra of $G_{\mathbb{Q}}$ and let $\mathcal{U}_{\mathbb{Q}}(\mathcal{G})$ denote the universal enveloping algebra of $G_{\mathbb{Q}}$. X_α, Y_α, $\alpha \in \Phi^+$, and H_{α_i}, $\alpha_i \in \Delta$, denote a Chevalley basis of $\mathcal{U}_{\mathbb{Q}}(\mathcal{G})$. Let $\mathcal{U}_{\mathbb{Z}}(\mathcal{G})$ denote the $\mathbb{Z}$ sub-algebra of $\mathcal{U}_{\mathbb{Q}}(\mathcal{G})$ generated by $X_\alpha^n/n!, Y_\alpha^n/n!$. Let $\mathcal{V}(\lambda)$ denote the unique, irreducible, finite dimensional $\mathcal{U}_{\mathbb{Q}}(\mathcal{G})$ module with highest weight λ. Let v^+ denote a highest weight vector of $\mathcal{V}(\lambda)$. The module $\mathcal{U}_{\mathbb{Z}}(\mathcal{G})v^+$ is a $G_{\mathbb{Z}}$ module and $V(\lambda)_K = \mathcal{U}_{\mathbb{Z}}(\mathcal{G})v^+ \otimes K$ is a G module, the Weyl module with highest weight λ.

It is well known that the dual of the Weyl module with highest weight λ can be realized as the space of sections of a line bundle over G/B [**6**]. We follow the convention in [**13**] and briefly recall how this is done. For a dominant weight λ consider the action of B on $G \times \mathbb{A}^1_K$ given by $b(g,z) = (gb, \lambda(b^{-1})z)$. The quotient space modulo this action is the total space F for the line bundle L_λ on G/B. Let $\sigma : F \to G/B$ the canonical map $\sigma(g,c) = gB$ and let $H^0(G/B, L_\lambda) = \{s : G/B \to F | \sigma \cdot s = id_{G/B}\}$, be the space of sections of the line bundle. Given a section in $H^0(G/B, L_\lambda)$ we get a regular function f on G by composing it with the natural map $G \mapsto G/B$ followed by second projection on F. It can be verified that f satisfies $f(gb) = \lambda(b)f(g)$. Conversely, any regular function satisfying this identity gives us a section in $H^0(G/B, L_\lambda)$. So as a left G module we identify $H^0(G/B, L_\lambda)$ with regular functions satisfying $f(gb) = \lambda(b)f(g)$. Let f be the B-fixed line in $H^0(G/B, L_\lambda)$ with highest weight μ. So f (as a regular function) cannot be zero on the open cell w_0 in G. For t belonging to the maximal torus we have $(t \cdot f)(w_0) = f(t^{-1}w_0) = \mu(t)f(w_0)$. On the other hand $f(t^{-1}w_0) = f(w_0 w_0^{-1} t^{-1} w_0) = \lambda(w_0^{-1}t^{-1}w_0)f(w_0) = w_0(\lambda)(t^{-1})f((w_0)$. This implies that $\mu = -w_0(\lambda)$. So $H^0(G/B, L_\lambda)$ is a G module with highest weight $-w_0(\lambda)$ and $L(-w_o(\lambda))_K$ is a G submodule of $H^0(G/B, L_\lambda)$.

So there is a natural map, $\phi_{G,\lambda}$, from $V(-w_0(\lambda))_K$ to $H^0(G/B, L_\lambda)$ whose image is $L(-w_0(\lambda))_K$.

The main concern of this paper is to determine conditions on p for which this map is an isomorphism.

3. Notation and some observations

For the rest of the paper K will denote an algebraically closed field of characteristic $p > 2$. G will denote $\mathrm{SL}(2n+1, K)$. We fix a basis $\{e_1, \ldots, e_{2n+1}\}$ of a vector space V of dimension $2n+1$ over K and we let $x_1, \ldots, x_{2n+1}$ denote the dual basis. Let $H \subseteq G$ denote $\mathrm{SO}(2n+1, K)$ the special orthogonal group defined as $\{h \in G | hh^t = h^t h = id\}$. Let $T^{'}$ denote the maximal torus in G consisting of diagonal matrices, with respect to the basis $\{e_1, \ldots, e_{2n+1}\}$ and let $B^{'} \supseteq T^{'}$ denote the Borel subgroup of G consisting of upper triangular matrices. Let $T^{''}$ denote the maximal torus of H consisting of matrices of the form

$$O = \begin{pmatrix} a_1 & 0 & 0 & \cdots & 0 & 0 & a_{2n+1} \\ 0 & a_2 & 0 & \cdots & 0 & a_{2n} & 0 \\ \vdots & & & & & & \\ 0 & 0 & \cdots & 1 & \cdots & 0 & 0 \\ \vdots & & & & & & \\ 0 & -a_{2n} & 0 & \cdots & 0 & a_2 & 0 \\ -a_{2n+1} & 0 & 0 & \cdots & 0 & 0 & a_1 \end{pmatrix}$$

such that $a_i^2 + a_{2n+2-i}^2 = 1, \forall i$. It can be verified that the vectors $e_i + (\sqrt{-1})e_{2n+2-1}$, $e_i - (\sqrt{-1})e_{2n+2-1} 1 \leq i \leq n$ and e_{n+1} are weight vectors of V with respect to $T^{''}$. Hence we can select a maximal torus T of G containing $T^{''}$ (namely, the diagonal matrices in G with respect to the basis $e_i + (\sqrt{-1})e_{2n+2-1}, e_i - (\sqrt{-1})e_{2n+2-1}, 1 \leq i \leq n$ and e_{n+1}), and a Borel subgroup B of G containing T [**9**]. Then it can be verified that $B^{''} = B \cap H$ is a Borel subgroup of H containing $T^{''}$, and the following diagram commutes .

$$\begin{array}{ccc} T^{''} & \longrightarrow & B^{''} \\ \downarrow & & \downarrow \\ T^{'} & \longrightarrow & B^{'} \end{array}$$

(Alternatively one can start with an inner product (,) such that $(u_i, u_j) = 0$ unless $i + j = 2n + 2$, with $(u_{n+1}, u_{n+1}) = 2$ and $(u_i, u_{2n+2-i}) = 1$ and use this to define our H; We could then define

T'' and B''. An orthonormal basis constructed from the u_i's can then be used to define T and B.)

Let Φ be the root system of G with respect to T and let Φ^+ denote the set of all positive roots with respect to B. The simple roots will be denoted by $\Delta = \{\alpha_1, \alpha_2, \ldots \alpha_{2n}\}$ and we denote by $\{\omega_1, \omega_2, \ldots, \omega_{2n}\}$, the fundamental weights determined by them. We use the same notation for the line bundle L_{ω_i} on G/B, determined by the character ω_i of T, and its pull back to $H/B \cap H$. Likewise we use the same notation for sections of these two bundles. It will be clear from the context whether we are referring to them as sections over G/B or $H/B \cap H$.

Denoting by $L_{m\omega_i}$ the tensor product of L_{ω_i} taken m times, we have

LEMMA 1. *There is a natural H equivariant surjection ϕ from $H^0(G/B, L_{m\omega_i})$ to $H^0(H/B \cap H, L_{m\omega_i})$. When $m = 1$, $1 \leq i \leq n-1$ this is an isomorphism if the characteristic is not equal to 2.*

PROOF. Consider the following commutative diagram :

$$\begin{array}{ccccc} H^0(G/B, L_{\omega_i})^{\otimes m} & \longrightarrow & H^0(G/B, L_{m\omega_i}) & & \\ \downarrow & & \downarrow & & \\ H^0(H/B \cap H, L_{\omega_i})^{\otimes m} & \longrightarrow & H^0(H/B \cap H, L_{m\omega_i}) & \longrightarrow & 0 \end{array}$$

For $i = 1, \ldots n-1$, both $H^0(G/B, L_{\omega_i})$ and $H^0(H/B \cap H, L_{\omega_i})$ are dual to the i-th exterior power of the standard representation of G and H on V. For a proof see [**4**], Thm. 5.1.6, 5.1.7. So $H^0(G/B, L_{\omega_i}) \simeq H^0(H/B \cap H, L_{\omega_i}), \forall 1 \leq i \leq n-1$ i.e. the left vertical arrow is an isomorphism. The bottom horizontal arrow is a surjection from [**12**] Thm. 1(ii). Hence the right vertical arrow is also a surjection. □

Let s_i (resp. $\overline{s_i}$) denote the section $x_i + (\sqrt{-1})x_{2n+2-i}$ (resp. $x_i - (\sqrt{-1})x_{2n+2-i}$) of $H^0(H/B \cap H, L_{\omega_1})$, $i = 1, n$ and let x_{n+1} denote the (unique, upto scalar) zero weight vector of $H^0(H/B \cap H, L_{\omega_1})$ with respect to T''. It is clear that the section $x_1 + (\sqrt{-1})x_{2n+1}$ is the highest weight vector with respect to T (resp. T'') for the G (resp. H) action on $H^0(G/B, L_{\omega_1})$ (resp. $H^0(H/B \cap H, L_{\omega_1})$) [**9**]. Since this section is non zero on $H/B \cap H$, and this is the only vector of weight ω_1 it follows that the section $(x_1 + (\sqrt{-1})x_{2n+1})^m$ is the highest weight vector of $H^0(H/B \cap H, L_{m\omega_1})$.

Crucial to our proof is the fact that H being the fixed point set of an involution of $\mathrm{SL}(2n+1,K)$ sending g to $(g^t)^{-1}$, G/H is a spherical variety. For our choice of H, B and B' we note the following

OBSERVATION 2.
(a) HB' is dense in G.
(b) There exists a section $s \in H^0(G/B, L_{m\omega_i})$, $i = 1,2$, such that s is a weight vector with respect to both T and T'. Moreover s is an extremal weight vector w.r.t T' and it is a zero weight vector with respect to T''.
(c) The Weyl group W of G with respect to T' is a subgroup (upto signature) of H.
(d) For $1 \le i \le n$, the transformation sending the section s_i to $\overline{s_i}$ and fixing all other sections $s_j, \overline{s_j}, j \ne i$ is in the orthogonal group $O(2n+1,K)$.

PROOF.
(a) This follows from the observation that $H \cap B'$ is finite and $\dim(H) + \dim(B') = \dim\,(G)$.
(b) For $m\omega_1$, the section $s = x_{n+1}^m$ satisfies the requirements of the lemma. For $m\omega_2$, the section $s = (s_n \wedge \overline{s_n})^m$ satisfies the requirements of the lemma.
(c) The Weyl group of G with respect to T' consists of permutation matrices. Permutation matrices satisfy the equation $gg^T = g^Tg = id$. If the determinant of a permutation matrix g is 1 then it belongs to H. Otherwise the matrix $-g$ has determinant 1 and belongs to H.
(d) This transformation is given by the diagonal matrix (w.r.t. the standard basis) with -1 at position $2n+2-i$ and 1's elsewhere and this is clearly in the orthogonal group $O(2n+1,K)$. So the negative of this transformation has determinant 1 and belongs to $\mathrm{SO}(2n+1,K)$. In fact all of $(\mathbb{Z}/2)^{2n+1}$ sits inside $O(2n+1,K)$. □

We let $\sigma_{i,j}, i \ne j$, denote the permutation sending i to j and $\bar{\imath}$ to $\bar{\jmath}$. ¿From observation 2(c) above either $\sigma_{i,j}$ or $-\sigma_{i,j}$ is in H. Let μ_i denote the transformation sending s_i to $\overline{s_i}$. Again from observation 2(d) above -$\mu_i \in \mathrm{SO}(2n+1,K)$.
Remark We are interested in proving that certain H representations V are irreducible. If W is a subspace of V then an H-stable subspace containing W will include all vectors of the form $T(w)$ where $T \in H$ and $w \in W$. From observations (c) and (d) above we are guaranteed that for all $w \in W$, and all $i,j, i \ne j$, $\mu_i(w)$ and $\sigma_{i,j}(w)$ will belong to

the H span of W. So in computing the H span of a given subspace W we may freely use the operators μ_i and $\sigma_{i,j}$.

4. A Chevalley basis for H

For $1 \le i \le n$, denote by ϵ_i the character sending elements in T'' to $a_i + \sqrt{-1}a_{2n+2-i}$, and let ϵ_{n+1} denote the character sending elements in T'' to 1. It can be easily verified that the simple roots are $\epsilon_i - \epsilon_{i+1}$, $1 \le i \le n-1$ and ϵ_{n+1}. We denote these simple roots by $\alpha_i, 1 \le i \le n$.

For $1 \le j \le n-1$ let H_j denote the anti-diagonal matrix with entries $-\sqrt{-1}$, $\sqrt{-1}$, $-\sqrt{-1}$ and $\sqrt{-1}$ at positions $(j, 2n+2-j)$, $(j+1, 2n+1-j)$, $(2n+1-j, j+1)$, $(2n+2-j, j)$ respectively. Let $X_{\alpha_j} = \frac{1}{2}[E_{j,j+1} - \sqrt{-1}E_{j,2n+1-j} - E_{j+1,j} - \sqrt{-1}E_{j+1,2n+2-j} + \sqrt{-1}E_{2n+1-j,j} + \sqrt{-1}E_{2n+2-j,j+1} - E_{2n+1-j,2n+2-j} + E_{2n+2-j,2n+1-j}]$, $1 \le j \le n-1$, where $E_{k,j}$ denotes the matrix with 1 at position (k,j) and zeroes elsewhere. For $1 \le j \le n-1$ let Y_{α_j} denote the matrix $\frac{1}{2}[E_{j,j+1} + \sqrt{-1}E_{j,2n+1-j} - E_{j+1,j} + \sqrt{-1}E_{j+1,2n+2-j} - \sqrt{-1}E_{2n+1-j,j} - \sqrt{-1}E_{2n+2-j,j+1} - E_{2n+1-j,2n+2-j} + E_{2n+2-j,2n+1-j}]$. Let X_{α_n} denote the matrix $E_{n,n+1} + \sqrt{-1}E_{n+2,n+1} - E_{n+1,n} - \sqrt{-1}E_{n+1,n+2}$ and $Y_{\alpha n}$ denote the matrix $E_{n,n+1} - \sqrt{-1}E_{n+2,n+1} - E_{n+1,j} + \sqrt{-1}E_{n+1,2n+2-j}$. It can be easily verified that this can be completed to a Chevalley basis of H. For the simple roots α_j, it can be checked that action of the Chevalley basis on the sections s_i is given by $Y_{\alpha_i}s_j = 0$ if $|j-i| \ge 2$, $Y_{\alpha_i}s_i = -s_{i+1}, 1 \le i \le n-1$, $Y_{\alpha_n}s_n = -x_{n+1}$, $Y_{\alpha_n}x_{n+1} = 2\overline{s_n}$, $Y_{\alpha_i}\overline{s_{i+1}} = \overline{s_i}$, $1 \le i \le n-1$. Likewise the action of X_{α_i} on the sections is given by $X_{\alpha_i}s_j = 0$ if $|j-i| \ge 2$, $X_{\alpha_i}s_{i+1} = s_i, 1 \le i \le n-1$, $X_{\alpha_n}x_{n+1} = 2s_n$, $X_{\alpha_n}\overline{s_n} = -x_{n+1}$, $X_{\alpha_i}\overline{s_i} = -\overline{s_{i+1}}, 1 \le i \le n-1$.

5. Irreducibility of SL$(2n+1, K)$ modules - implications for SO$(2n+1, K)$

We begin the proof of the main theorem by an observation. For G we have

OBSERVATION 3. *If $p > m$ then $H^0(G/B, L_{m\omega_i})$ is irreducible as a G module. If $p > 3$ then $H^0(G/B, L_{2\omega_2})$ and $H^0(G/B, L_{3\omega_2})$ are irreducible as G modules.*

PROOF. In the $m\omega_1$ case the proof is well known, see for example Thm. 2.3 in [**3**]. We sketch a short proof to illustrate our methods. It is clear that $H^0(G/B, L_{m\omega_i})$ consists of homogeneous polynomials of degree m in the variables $x_1, \ldots, x_{2n+1}$. It suffices to show that when

$p > m$ all monomials of degree m in these variables are in the $\mathcal{U}_{\mathbb{Z}}(\mathcal{G})$ span of the highest weight section x_{2n+1}^m. Starting with x_{2n+1}^m it is clear that by applying appropriate elements of the Chevalley basis we get all monomials of degree m in the variables $x_1, \ldots, x_{2n+1}$ so long as $p \not| m$.

The other two cases can be deduced from the general irreducibility criterion for type A in [**6**] II.8.21.I. Again we sketch an elementary proof to illustrate our methods. We first identify the sections of $H^0(G/B, L_{\omega_2})$ with the two by two sub-determinants $p_{i,j} = x_{i,1}x_{j,2} - x_{i,2}x_{j,1}, 1 \leq i, j \leq 2n+1$, of the first two columns of a generic matrix in G. Then $H^0(G/B, L_{2\omega_2})$ is the set of all homogeneous polynomials of degree two in the $p_{i,j}$ and $H^0(G/B, L_{3\omega_2})$ is the set of all homogeneous polynomials of degree 3 in the $p_{i,j}$. Again, it is not difficult to show that provided $p > 3$ all monomials of degree two (resp. three) in the $p_{i,j}$ are in the $\mathcal{U}_{\mathbb{Z}}(\mathcal{G})$ of the highest weight vectors in $H^0(G/B, L_{2\omega_2})$ (resp. $H^0(G/B, L_{3\omega_2})$. We omit the proof since it is similar to the proofs which follow. □

Combining observations 2 and 3 we have

THEOREM 4. *Let $j = 1, 2$. Suppose $H^0(G/B, L_{m\omega_j})$ is irreducible as a G module. Let s be the section as defined in observation 2(b). Suppose $s \in L(-w_0(m\omega_j))_K$. Then $H^0(H/B \cap H, L_{m\omega_j})$ is irreducible as an H module.*

PROOF. Let τ be an element of the the Weyl group of G (w.r.t. T') such that $\tau(s)$ is a highest weight vector, w^+, of $H^0(G/B, L_{m\omega_j})$ with respect to B'. Such a τ exists since s is extremal with respect to T'. Letting $\langle\ \rangle$ denote linear span we have

$$\begin{aligned}\langle Hw^+ \rangle &= \langle H\tau(s) \rangle \\ &\supseteq \langle HB'\tau(s) \rangle \\ &= \langle G\tau(s) \rangle \\ &= H^0(G/B, L_{m\omega_i}).\end{aligned}$$

The second line above follows from the fact that $\tau(s)$ is a highest weight vector with respect to B' and hence is fixed by B'. The third line follows from observation 2 that HB' is Zariski dense in G and from the fact that the linear span of a set of vectors is Zariski closed. The last line follows from the assumption that $H^0(G/B, L_{m\omega_j})$ is irreducible.

Now since the map in lemma 1 is surjective and H equivariant it follows from the above containment that $\langle Hs\rangle = H^0(H/B\cap H, L_{m\omega_j})$. So $H^0(H/B\cap H, L_{m\omega_j}) = L(-w_0(m\omega_j))_K$, is irreducible as an H module. □

We make use of the above theorem to prove our irreducibility theorems. In the next section we prove the following: (All modules in the statement of the lemma are H modules.)

LEMMA 5.
(1) If $p > m$, $p \not| 2(m+n-1)-i$, $1 \le i \le m-1$, then $x_{n+1}^m \in L(-w_0(m\omega_1))_K$.
(2) If $p > 3$ and $p \not| 2n-1$, $p \not| n$, then $(s_n \wedge \overline{s_n})^2 \in L(-w_0(2\omega_2))_K$.
(3) If $p > 3$, $p \not| 2n+1$, $p \not| n+1$, then $(s_n \wedge \overline{s_n})^3 \in L(-w_0(3\omega_2))_K$.

From the above we conclude

THEOREM 6.
(1) If $p > m$, $p \ne 2$, $m \ne 1$ and $p \not| 2(m+n-1)-i$, $1 \le i \le m-1$ then $H^0(H/B\cap H, L_{m\omega_1})$ is irreducible as an H module.
(2) If $p > 3$ and $p \not| 2n-1$, $p \not| n$, then $H^0(H/B\cap H, L_{2\omega_2})$ is irreducible as an H module.
(3) If $p > 3$, $p \not| 2n+1$, $p \not| n+1$, then $H^0(H/B\cap H, L_{3\omega_2})$ is irreducible as an H module.

PROOF. Follows from lemma 5, theorem 4 and observation 3. □

6. Proof of Lemma 5.

6.1. The $m\omega_1$ case.

PROOF. In this case it is clear that $v^+ = s_1^m$. We show by induction that for the chosen range of p, $L(m\omega_1)_K$ includes ${s_n}^{m-j}x_{n+1}^j$, for all $0 \le j \le m$. For $j = m$ this implies that the $L(m\omega_1)_K$ contains x_{n+1}^m completing the proof.
Applying σ_{1n} to s_1^m we have ${s_n}^m \in L(m\omega_1)_K$. Applying Y_{α_n}, we observe,

$$Y_{\alpha_n}{s_n}^m = -m{s_n}^{m-1}x_{n+1}.$$

Since m is invertible in K it follows that ${s_n}^{m-1}x_{n+1} \in L(m\omega_1)_K$. Assume by induction that the section ${s_n}^{m-j}x_{n+1}^j$ is in the $L(m\omega_1)_K$. We have

$$\begin{aligned} Y_{\alpha_n}{s_n}^{m-j}x_{n+1}^j &= -(m-j){s_n}^{m-j-1}x_{n+1}^{j+1} + 2j{s_n}^{m-j}x_{n+1}^{j-1}\overline{s_n} \\ &= {s_n}^{m-j-1}x_{n+1}^{j-1}(-(m-j)x_{n+1}^2 + 2js_n\overline{s_n}). \end{aligned} \tag{6.1}$$

Now applying the operator $Y_{\alpha_{n-1}}$ to the last obtained section above we get

$$Y_{\alpha_{n-1}} {s_n}^{m-j-1} x_{n+1}^{j-1}(-(m-j)x_{n+1}^2 + 2j s_n \overline{s_n}) \\ = {s_n}^{m-j-1} x_{n+1}^{j-1}[2j s_n \overline{s_{n-1}}].$$

Since $p \geq m$ we can invert j in this field and obtain the section ${s_n}^{m-j} x_{n+1}^{j-1}[\overline{s_{n-1}}]$.

Now applying the operator $X_{\alpha_{n-1}}$ to this section we get

$$X_{\alpha_{n-1}} {s_n}^{m-j} x_{n+1}^{j-1}[\overline{s_{n-1}}] \\ = x_{n+1}^{j-1}[-{s_n}^{m-j}\overline{s_n} + (m-j){s_n}^{m-j-1} s_{n-1}\overline{s_{n-1}}] \\ = -{s_n}^{m-j-1} x_{n+1}^{j-1}[s_n\overline{s_n} - (m-j)s_{n-1}\overline{s_{n-1}}].$$

So we also have ${s_n}^{m-j-1} x_{n+1}^{j-1}[s_n\overline{s_n} - (m-j)s_{n-1}\overline{s_{n-1}}] \in L(m\omega_1)_K$. This section along with the sections obtained by applying permutations $\sigma_{n-2,k}, 1 \leq k \leq (n-2)$ shows that ${s_n}^{m-j-1} x_{n+1}^{j-1}[s_n\overline{s_n} - (m-j)s_k\overline{s_k}] \in L(-w_0(m\omega_1))_K, \forall k, 1 \leq k \leq (n-1)$. Adding these $n-1$ sections we see that

(6.2)

$${s_n}^{m-j-1} x_{n+1}^{j-1}((n-1)s_n\overline{s_n} - (m-j)\sum_{k=1}^{n-1} s_k\overline{s_k}) \in L(-w_0(m\omega_1))_K.$$

We show in appendix A.1 that $4\sum_{i=1}^{i=n} s_i * \overline{s_i} + x_{n+1}^2 = 0$. So the last sections obtained above can be rewritten to give

$${s_n}^{m-j-1} x_{n+1}^{j-1}[4(n-1+m-j)s_n\overline{s_n} + (m-j)x_{n+1}^2].$$

Adding the section in equation 6.1 we see that $L(-w_0(m\omega_1))_K$ contains the section $(4(m+n-1)-2j){s_n}^{m-j-1} x_{n+1}^{j-1} s_n\overline{s_n}$. Since the coefficient of this section is invertible in the given field we see that ${s_n}^{m-j-1} x_{n+1}^{j-1} s_n\overline{s_n} \in L(-w_0(m\omega_1))_K$. From equation 6.1 it follows that the section $(m-j){s_n}^{m-j-1} x_{n+1}^{j+1} \in L(-w_0(m\omega_1))_K$. Since $m-j$ is invertible in the given field we are done. □

6.2. The $2\omega_2$ and $3\omega_2$ case. The technique of proof here is similar to the $m\omega_1$ case. We show in the $2\omega_2$ case, by applying a carefully chosen set of operators, that under the assumption $p \not| 2n-1$, $(s_{n-1} \wedge \overline{s_n})((s_n \wedge \overline{s_n}) - (s_{n-1} \wedge \overline{s_{n-1}})) \in L(-w_0(2\omega_2))_K$. Then we show that, under the additional assumption $p > 3, p \not| n$, $(s_n \wedge \overline{s_n})^2 \in L(-w_0(2\omega_2))_K$.

A key problem that we encounter in this case is that there are algebraic dependencies among the sections. So it becomes essential to choose a nice basis in which to represent sections. We choose for our basis standard monomials. Standard monomial theory was developed in a series of papers by Lakshmibai and Seshadri, see for example [**10**]. We recall the necessary facts needed very briefly. Let L be the ample generator of $PicH/P$. Let W_P denote the Weyl group of P and W that of H. Standard monomial theory provides a nice basis for $H^0(G/P, L^{\otimes m})$. The basis is indexed by "admissible pairs" in W/W_P (see the above mentioned papers for details). We confine ourselves to the case when P is the parabolic subgroup corresponding to ω_2.

In appendix A.2 we show that the following relations hold among sections in $H^0(H/B \cap H, L_{2\omega_2})$. To keep the notation short and readable we denote section $(s_i \wedge s_j)$ by the pair (i, j) and the section $\overline{s_j}$ by $\bar{\jmath}$. $n+1$ denotes the section x_{n+1}.
The monomials appearing on the right hand side of the equations below are standard in the sense of [**10**]. The terms on the left are non-standard and the equations given below express these non-standard monomials in terms of the standard basis.

LEMMA 7.

a) $(n, n+1)(n-1, n+1) =$
$-2(n-1,n)((n,\overline{n})-(n-1,\overline{n-1}))+2[(n-2,n)(n-1,\overline{n-2})$
$+ (n-2, n-1)(n, \overline{n-2})] + 2[(n-3,n)(n-1,\overline{n-3})$
$+ (n-3, n-1)(n, \overline{n-3})]$
$\vdots$
$+ 2[(1,n)(n-1,\overline{1}) + (1, n-1)(n, \overline{1})]$.

b) $(1,4)(2,3) = -(1,2)(3,4) + (1,3)(2,4)$. *In general if* $a < b < c < d$ *then,* $(a,d)(b,c) = -(a,b)(c,d) + (a,c)(b,d)$.

c) $2(n,\overline{n})^2 = -2\sum_{i=1}^{n-1}[(i,n)(\overline{n},\bar{\imath}) + (i,\overline{n})(n,\bar{\imath})] + \sum_{i=1}^{n-1}(i, n+1)(n+1,\bar{\imath})$.

d) $2(n-1,\overline{n-1})^2 = -2\sum_{i=1}^{n-2}[(i,n-1)(\overline{n-1},\bar{\imath})+(i,\overline{n})(n,\bar{\imath})] - 2[(n-1,n)(\overline{n},\overline{n-1}) + (n-1,\overline{n})(n,\overline{n-1})] + \sum_{i=1}^{n-2}(i,n+1)(n+1,\bar{\imath}) + (n,n+1)(n+1,\overline{n})$.

6.3. Proof of the $2\omega_2$ case. In this case $v^+ = (1,2)^2$. Applying the permutations $\sigma_{1,n-1}$ and $\sigma_{2,n}$ we see that $(n-1,n)^2 \in$

$L(-w_0(2\omega_2))_K$. We have

$$\begin{aligned}
Y_{\alpha_{n-1}}&\frac{Y^2_{\alpha_n}}{2}((n-1,n)^2)\\
&= -Y_{\alpha_{n-1}}Y_{\alpha_n}((n-1,n)(n-1,n+1))\\
&= -Y_{\alpha_{n-1}}(-(n-1,n+1)^2+2(n-1,n)(n-1,\overline{n}))\\
&= -[2(n-1,n+1)(n,n+1)\\
&\quad -2(n-1,n)((n,\overline{n})-(n-1,\overline{n-1}))].
\end{aligned}$$

Since p is not equal to 2 we can divide the last obtained section by 2 and we get the section

$$(n-1,n+1)(n,n+1)-(n-1,n)((n,\overline{n})-(n-1,\overline{n-1})).$$

Applying lemma 7(a) to this section we get

$$\begin{aligned}
(6.3)\qquad &-3(n-1,n)((n,\overline{n})-(n-1,\overline{n-1}))\\
&+2\sum_{i=1}^{n-2}((i,n)(n-1,\overline{\imath})+(i,n-1)(n,\overline{\imath}))\in L(-w_0(2\omega_2))_K.
\end{aligned}$$

On the other hand starting with $(1,2)^2$ and applying the operators $Y_{\alpha_2}, Y_{\alpha_3}, \ldots, Y_{\alpha_{n-1}}, Y_{\alpha_1}$ it can be verified that

$$Y_{\alpha_1}Y_{\alpha_{n-1}}\ldots Y_{\alpha_2}(1,2)^2 = 2(1,2)(2,n)\in L(-w_0(2\omega_2))_K.$$

Dividing by 2, and applying the operators $\frac{Y^2_{\alpha_2}}{2}\frac{Y^2_{\alpha_3}}{2}\ldots\frac{Y^2_{\alpha_{n-2}}}{2}\frac{Y^2_{\alpha_n}}{2}\frac{Y^2_{\alpha_{n-1}}}{2}$ to the above section it can be seen that

$$\begin{aligned}
\frac{Y^2_{\alpha_{n-1}}}{2}&\frac{Y^2_{\alpha_n}}{2}\frac{Y^2_{\alpha_{n-2}}}{2}\cdots\frac{Y^2_{\alpha_3}}{2}\frac{Y^2_{\alpha_2}}{2}((1,2)(2,n))\\
&=\frac{Y^2_{\alpha_{n-1}}}{2}\frac{Y^2_{\alpha_n}}{2}((1,n-1)(n-1,n))\\
&=-\frac{Y^2_{\alpha_{n-1}}}{2}((1,n-1)(n-1,\overline{n}))\\
&=2(1,n)((n,\overline{n})-(n-1,\overline{n-1}))-2(1,n-1)(n,\overline{n-1}).
\end{aligned}$$

Applying the permutation $\sigma_{1,n-2}$ to the negative of the last section above we see, $(n-2,n)((n,\overline{n})-(n-1,\overline{n-1}))-(n-2,n-1)(n,\overline{n-1})\in L(-w_0(2\omega_2))_K$. Applying $Y_{\alpha_{n-2}}$ to this section we get the section

$$\begin{aligned}
&-(n-1,n)((n,\overline{n})-(n-1,\overline{n-1}))-(n-2,n)(n-1,\overline{n-2})\\
&-(n-2,n-1)(n,\overline{n-2}).
\end{aligned}$$

Now applying the permutation $\sigma_{n-2,i}$, $1 \leq i \leq n-2$ we have

$$(n-1,n)((n,\overline{n})-(n-1,\overline{n-1}))+(i,n)(n-1,\overline{i})+(i,n-1)(n,\overline{i}) \in L(-w_0(2\omega_2))_K, \forall i \; 1 \leq i \leq n-2.$$

Adding these $n-2$ sections we see that

$$(6.4) \qquad (n-2)(n-1,n)((n,\overline{n})-(n-1,\overline{n-1})) + \sum_{i=1}^{n-2}((i,n)(n-1,\overline{i})+(i,n-1)(n,\overline{i})) \in L(-w_0(2\omega_2))_K.$$

Since $p \not| 2n-1$ it follows from equations 6.3 and 6.4 that

$$(6.5) \qquad (n-1,n)((n,\overline{n})-(n-1,\overline{n-1})) \in L(-w_0(2\omega_2))_K.$$

Applying μ_n to the above section it follows that

$$(n-1,\overline{n})((n,\overline{n})+(n-1,\overline{n-1})) \in L(-w_0(2\omega_2))_K.$$

Clearly $(n-1,\overline{n})^2 \in L(-w_0(2\omega_2))_K$ since it is an extremal weight vector. Since

$$(6.6) \qquad Y_{\alpha_{n-1}}(n-1,\overline{n})^2 = -2(n-1,\overline{n})((n,\overline{n})-(n-1,\overline{n-1}))$$

it follows from equation 6.5 and equation 6.6 that $(n-1,\overline{n})(n,\overline{n})$, $(n-1,\overline{n})(n-1,\overline{n-1}) \in L(-w_0(2\omega_2))_K$. It follows by applying μ_n that

$$(6.7) \qquad (n-1,n)(n,\overline{n}), (n-1,n)(n-1,\overline{n-1}) \in L(-w_0(2\omega_2))_K.$$

Starting with the last of these sections we get

$$\begin{aligned} X_{\alpha_{n-2}}Y_{\alpha_{n-2}}&(n-1,n)(n-1,\overline{n-1}) \\ &= X_{\alpha_{n-2}}(n-1,n)(n-1,\overline{n-2}) \\ &= (n-2,n)(n-1,\overline{n-2}) \\ &\quad + (n-1,n)((n-2,\overline{n-2})-(n-1,\overline{n-1})). \end{aligned}$$

¿From equation 6.7, $(n-1,n)(n-1,\overline{n-1}) \in L(-w_0(2\omega_2))_K$. And so

$$(6.8) \qquad (n-2,n)(n-1,\overline{n-2})+(n-1,n)(n-2,\overline{n-2}) \in L(-w_0(2\omega_2))_K.$$

Applying lemma 7(b) to the second term of the above section and simplifying we see that

$$(6.9) \qquad 2(n-2,n)(n-1,\overline{n-2})-(n-2,n-1)(n,\overline{n-2}) \in L(-w_0(2\omega_2))_K.$$

Applying $\sigma_{n,n-1}$ to the section in equation 6.8 we have

$$(n-2,n-1)(n,\overline{n-2})-(n-1,n)(n-2,\overline{n-2})\in L(-w_0(2\omega_2))_K.$$

Applying lemma 7(b) to the second term of the above section and simplifying we see that

$$\begin{aligned}(6.10)\qquad & -(n-2,n)(n-1,\overline{n-2})+2(n-2,n-1)(n,\overline{n-2})\\ & \in L(-w_0(2\omega_2))_K.\end{aligned}$$

Since $p>3$ it follows from equation 6.9 and equation 6.10 that that $(n-2,n)(n-1,\overline{n-2})$ and $(n-2,n-1)(n,\overline{n-2})$ are both in $L(-w_0(2\omega_2))_K$. ¿From equation 6.8 it follows that $(n-1,n)(n-2,\overline{n-2})$ is also in $L(-w_0(2\omega_2))_K$. Now apply the permutation $\sigma_{n-2,i}$, $1\leq i\leq n-2$ to $(n-1,n)(n-2,\overline{n-2})$. Using the sections from equation 6.7 it follows that

$$(6.11)\qquad \forall i,1\leq i\leq n,(n-1,n)(i,\bar{\imath})\in L(-w_0(2\omega_2))_K.$$

Next we show that $(n-1,\overline{n-1})(n-2,\overline{n-2})$ and $(n,\bar{n})(n-2,\overline{n-2})$ are both in $L(-w_0(2\omega_2))_K$. Starting with the section $(n-1,n)(n-2,\overline{n-2})\in L(-w_0(2\omega_2))_K$ we get

$$\begin{aligned}& Y_{\alpha_n}Y_{\alpha_{n-1}}Y_{\alpha_n}(n-1,n)(n-2,\overline{n-2})\\ &\quad=-Y_{\alpha_n}Y_{\alpha_{n-1}}(n-1,n+1)(n-2,\overline{n-2})\\ &\quad=Y_{\alpha_n}(n,n+1)(n-2,\overline{n-2})\\ &\quad=2(n,\bar{n})(n-2,\overline{n-2})\\ &\qquad\quad\in L(-w_0(2\omega_2))_K.\end{aligned}$$

Then applying the permutation $\sigma_{i,n-2}$, $1\leq i\leq n-2$ to the last section above, we have

$$(6.12)\qquad \forall i,1\leq i\leq n-2,(n,\bar{n})(i,\bar{\imath})\in L(-w_0(2\omega_2))_K.$$

Using lemma 7(b) we rewrite this as

$$(6.13)\ \forall i,1\leq i\leq n-2,-(i,n)(\bar{n},\bar{\imath})+(i,\bar{n})(n,\bar{\imath})\in L(-w_0(2\omega_2))_K.$$

Applying the transformation μ_n to the section $(n-1,n)(n-1,\overline{n-2})$ in equation 6.11 we get the section

$$Y_{\alpha_{n-1}}(n-1,\bar{n})(n-2,\overline{n-2})=[(n,\bar{n})-(n-1,\overline{n-1})](n-2,\overline{n-2}).$$

Now from equation 6.12 it follows that $(n-1,\overline{n-1})(n-2,\overline{n-2})\in L(-w_0(2\omega_2))_K$. From this using equation 6.7 and using permutations $\sigma_{n-2,i}$, $1\leq i\leq n-2$ we have

$$(6.14)\qquad \forall i,2\leq i\leq n,(i,\bar{\imath})(i-1,\overline{\imath-1})\in L(-w_0(2\omega_2))_K.$$

Applying μ_n to equation 6.11, we see that $(n-1,\overline{n})(n,\overline{n}) \in L(-w_0(2\omega_2))_K$. And so we get the section

$$\begin{aligned}&Y_{\alpha_{n-1}}(n-1,\overline{n})(n,\overline{n})\\&\quad= -(n,\overline{n})^2 + (n-1,\overline{n-1})(n,\overline{n}) + (n-1,\overline{n})(n,\overline{n-1}).\end{aligned}$$

¿From equation 6.14 it follows that

$$2(n,\overline{n})^2 - 2(n-1,\overline{n})(n,\overline{n-1}) \in L(-w_0(2\omega_2))_K. \tag{6.15}$$

Using lemma 7(c) we rewrite this as

$$\begin{aligned}&-2\sum_{i=1}^{n-2}((i,n)(\overline{n},\overline{\imath}) + (i,\overline{n})(n,\overline{\imath})) + \sum_{i=1}^{n-1}(i,n+1)(n+1,\overline{\imath})\\&-4(n-1,\overline{n})(n,\overline{n-1}) - 2(n-1,n)(\overline{n},\overline{n-1}).\end{aligned} \tag{6.16}$$

¿From equation 6.14 $(n-1,\overline{n-1})(n,\overline{n}) \in \mathcal{U}_{\mathbb{Z}}(\mathcal{H})$. Using lemma 7(b) we rewrite this as $-(n-1,n)(\overline{n},\overline{n-1}) + (n-1,\overline{n})(n,\overline{n-1})$. Using this in conjunction with 6.16, we have

$$\begin{aligned}&2\sum_{i=1}^{n-2}[(i,n)(\overline{n},\overline{\imath}) + (i,\overline{n})(n,\overline{\imath})] - \sum_{i=1}^{n-1}(i,n+1)(n+1,\overline{\imath})\\&+6(n-1,\overline{n})(n,\overline{n-1}) \in L(-w_0(2\omega_2))_K.\end{aligned} \tag{6.17}$$

Adding the $n-2$ sections in 6.13 (for $1 \leq i \leq n-2$) to the section in equation 6.17 we see that

$$\begin{aligned}&4\sum_{i=1}^{n-2}(i,\overline{n})(n,\overline{\imath}) + 6(n-1,\overline{n})(n,\overline{n-1})\\&-\sum_{i=1}^{n-1}(i,n+1)(n+1,\overline{\imath}) \in L(-w_0(2\omega_2))_K.\end{aligned} \tag{6.18}$$

Using the permutation $\sigma_{1,n-1}$ on the above section we have

(6.19)

$$\begin{aligned}&6(1,\overline{n})(n,\overline{1}) + 2\sum_{i=2}^{n-2}(i,\overline{n})(n,\overline{\imath})\\&+2(n-1,\overline{n})(n,\overline{n-1}) - \sum_{i=1}^{n-1}(i,n+1)(n+1,\overline{\imath}) \in L(-w_0(2\omega_2))_K.\end{aligned}$$

Subtracting the above section from the section in equation 6.18 we see that since $p > 2$, $(n-1,\overline{n})(n,\overline{n-1}) - (1,\overline{n})(n,\overline{1}) \in L(-w_0(2\omega_2))_K$.

Using the permutation $\sigma_{1,j}, 2 \leq j \leq n-2$ we see that the following sections are all in $L(-w_0(2\omega_2))_K$.

$$-(j,\overline{n})(n,\overline{j}) + (n-1,\overline{n})(n,\overline{n-1}), \forall j, 1 \leq j \leq n-2$$

Adding these $n-2$ sections we have

$$-\sum_{i=1}^{n-2}(i,\overline{n})(n,\overline{i})+(n-2)(n-1,\overline{n})(n,\overline{n-1}) \in L(-w_0(2\omega_2))_K. \tag{6.20}$$

Starting with $(n-1,n)(n-1,\overline{n-1}) \in L(-w_0(2\omega_2))_K$ (see equation 6.7), we have

$$\begin{aligned}
&Y_{\alpha_n}Y_{\alpha_{n-1}}Y_{\alpha_n}(n-1,n)(n-1,\overline{n-1}) \\
&\quad = -Y_{\alpha_n}Y_{\alpha_{n-1}}(n-1,n+1)(n-1,\overline{n-1}) \\
&\quad = Y_{\alpha_n}[(n,n+1)(n-1,\overline{n-1}) + (n-1,n+1)(n,\overline{n-1})] \\
&\quad = 2(n,\overline{n})(n-1,\overline{n-1}) + 2(n-1,\overline{n})(n,\overline{n-1}) \\
&\qquad - (n-1,n+1)(n+1,\overline{n-1}) \\
&\quad \in L(-w_0(2\omega_2))_K.
\end{aligned}$$

Since the first term of the above section belongs to $L(-w_0(2\omega_2))_K$ we get

$$2(n-1,\overline{n})(n,\overline{n-1}) - (n-1,n+1)(n+1,\overline{n-1}) \in L(-w_0(2\omega_2))_K.$$

Using the permutations $\sigma_{j,n-1}, 1 \leq j \leq n-2$ we see that

$$\forall j, 1 \leq j \leq n-1, 2(j,\overline{n})(n,\overline{j}) - (j,n+1)(n+1,\overline{j}) \in L(-w_0(2\omega_2))_K.$$

Adding these $n-1$ sections we see

$$\begin{aligned}
&2\sum_{i=1}^{n-2}(i,\overline{n})(n,\overline{i}) + 2(n-1,\overline{n})(n,\overline{n-1}) - \sum_{i=1}^{n-1}(i,n+1)(n+1,\overline{i}) \\
&\qquad \in L(-w_0(2\omega_2))_K.
\end{aligned} \tag{6.21}$$

The equations 6.18, 6.20, 6.21 involve the same set of sections. The 3 by 3 coefficient matrix of these 3 equations has determinant $-2n$. Since $p \not| 2n$ it follows that each of $\sum_{i=1}^{n-2}(i,\overline{n})(n,\overline{i})$, $(n-1,\overline{n})(n,\overline{n-1})$ and $\sum_{i=1}^{n-1}(i,n+1)(n+1,\overline{i})$ appears in $L(-w_0(2\omega_2))_K$. ¿From equation 6.15 it follows that $(n,\overline{n})^2 \in L(-w_0(2\omega_2))_K$.

6.4. Proof of the $3\omega_2$ case. This proof proceeds along the same lines as the above proof. We prove the following in the Appendix.

LEMMA 8. *If $p > 3$, $p \not| 2n+1$, the following sections are in $L(-w_0(3\omega_2))_K$.*

(i) $(n-1,n)[-2\sum_{i=1}^{n-2}((i,n)(\overline{n},\overline{\imath}) + (i,\overline{n})(n,\overline{\imath}))$
$\quad + \sum_{i=1}^{n-1}(i,n+1)(n+1,\overline{\imath}) - 5(n-1,n)(\overline{n},\overline{n-1})]$.

(ii) $(n-1,n)[-(n-1,n+1)(n+1,\overline{n-1})$
$\quad + (n-1,n)(\overline{n},\overline{n-1})]$.

(iii) $(n-1,n)[3(i,n+1)(n+1,\overline{\imath})$
$\quad - 2(n-1,n+1)(n+1,\overline{n-1})], \forall i,\ 1 \le i \le n-2$.

(iv) $(n-1,n)[-(i,n)(\overline{n},\overline{\imath}) + (i,\overline{n})(n,\overline{\imath})], \forall i, 1 \le i \le n-2$.

(v) $(n-1,n)[6(n-1,\overline{n})(n,\overline{n-1}) - 7(n-1,n)(\overline{n},\overline{n-1})]$.

(vi) $(n-1,n)[3\sum_{i=1}^{n-2}(i,n+1)(n+1,\overline{\imath}) - 4\sum_{i=1}^{n-2}(i,n)(\overline{n},\overline{\imath})]$.

Assuming the lemma we claim

CLAIM 9. *If $p > 3$, $p \not| 2n+1$, $p \not| n+1$, the following sections are in $L(-w_0(3\omega_2))_K$.*

(a) $(n-1,n)(n-1,\overline{n-1})(n,\overline{n})$, $(n-1,n)(n-1,\overline{n})(n,\overline{n-1})$,
$\quad$ *and* $(n-1,\overline{n})^2(n,\overline{n-1})$.

(b) $(n-1,n)(i,n+1)(n+1,\overline{\imath}), \forall i, 1 \le i \le n-2$.

(c) $(n-1,n)[\sum_{i=1}^{n-2}((i,n)(\overline{n},\overline{\imath}) + (i,\overline{n})(n,\overline{\imath}))]$.

(d) $(n-1,n)(n,\overline{n})^2$.

PROOF. ¿From lemma 8(i), (iv) it follows that the following section is in $L(-w_0(3\omega_2))_K$.

$$(n-1,n)[-4\sum_{i=1}^{n-2}(i,n)(\overline{n},\overline{\imath}) + \sum_{i=1}^{n-1}(i,n+1)(n+1,\overline{\imath})$$
$$- 5(n-1,n)(\overline{n},\overline{n-1})].$$

¿From this and the section in lemma 8(vi) we get

(6.22)
$$(n-1,n)[2\sum_{i=1}^{n-2}(i,n+1)(n+1,\overline{\imath}) - (n-1,n+1)(n+1,\overline{n-1})$$
$$+ 5(n-1,n)(\overline{n},\overline{n-1})] \in L(-w_0(3\omega_2))_K.$$

Adding the $n-2$ sections in lemma 8(iii) it follows that

$$(n-1,n)[3\sum_{i=1}^{n-2}(i,n+1)(n+1,\bar{i})$$
$$-2(n-2)(n-1,n+1)(n+1,\overline{n-1})] \in L(-w_0(3\omega_2))_K.$$

¿From these last two sections we get

$$(n-1,n)[-15(n-1,n)(\bar{n},\overline{n-1})$$
$$+(-4n+11)(n-1,n+1)(n+1,\overline{n-1})] \in L(-w_0(3\omega_2))_K.$$

¿From the above sections and and the section in lemma 8(ii) it follows that the sections $(4n+4)(n-1,n)(n-1,n)(\bar{n},\overline{n-1})$ and $(4n+4)(n-1,n)(n-1,n+1)(n+1,\overline{n-1})$ are both in $L(-w_0(3\omega_2))_K$. Since $p \not| n+1$ and $p \neq 2$ we see that the sections $(n-1,n)(n-1,n+1)(n+1,\overline{n-1})$ and $(n-1,n)^2(\bar{n},\overline{n-1})$ are both in $L(-w_0(3\omega_2))_K$. By applying μ_n to the section $(n-1,n)(n-1,n)(\bar{n},\overline{n-1})$, we get $(n-1,\bar{n})^2(n,\overline{n-1})$. Now using lemma 8(v) it follows that since $p \not| 2,3$ we get $(n-1,n)(n-1,\bar{n})(n,\overline{n-1})$ proving part (a). Applying lemma 7(b) to this we get $(n-1,n)(n-1,\overline{n-1})(n,\bar{n})$. Parts (b) and (c) follows easily using lemma 8(iii), (iv) and (vi). Now multiply the left and right hand sides of lemma 7(c) with $(n-1,n)$. Since each term on the right hand side of the product is in $L(-w_0(3\omega_2))_K$, (d) follows. □

From the above claim (d), using the transformation μ_n, it follows that

$$(n-1,\bar{n})(n,\bar{n})^2 \in L(-w_0(3\omega_2))_K.$$

Applying the operator $Y_{\alpha_{n-1}}$ to this section we get the section

(6.23)
$$-((n,\bar{n})-(n-1,\overline{n-1}))(n,\bar{n})^2+2(n-1,\bar{n})(n,\bar{n})(n,\overline{n-1}).$$

Starting with section $(n-1,n)(n-1,\overline{n-1})(n,\bar{n})$ in claim 9(a), and applying the transformation μ_n to this section we see that $(n-1,\bar{n})(n-1,\overline{n-1})(n,\bar{n}) \in L(-w_0(3\omega_2))_K$. Applying the operator $Y_{\alpha_{n-1}}$ we get the section

(6.24)
$$-((n,\bar{n})-(n-1,\overline{n-1}))(n-1,\overline{n-1})(n,\bar{n})$$
$$-(n-1,\bar{n})(n,\overline{n-1})(n,\bar{n})+(n-1,\bar{n})(n-1,\overline{n-1})(n,\overline{n-1}).$$

Apply the transformation μ_n to the section in claim 9(a) we see $(n-1,n)(n-1,\overline{n})(\overline{n},\overline{n-1}) \in L(-w_0(3\omega_2))_K$. Now applying the operator $Y_{\alpha_{n-1}}$ to this we get
(6.25)

$$-(n-1,n)[(n,\overline{n})-(n-1,\overline{n-1})](\overline{n},\overline{n-1}) \in L(-w_0(3\omega_2))_K.$$

Applying the transformation μ_n we have

$$(n-1,\overline{n})[(n,\overline{n})+(n-1,\overline{n-1})](n,\overline{n-1}) \in L(-w_0(3\omega_2))_K. \tag{6.26}$$

Starting with the section $(n-1,\overline{n})^2(n,\overline{n-1})$ in claim 9(a) and applying the operator $Y_{\alpha_{n-1}}$, we see that

$$(n-1,\overline{n})[(n,\overline{n})-(n-1,\overline{n-1})](n,\overline{n-1}) \in L(-w_0(3\omega_2))_K. \tag{6.27}$$

¿From equations 6.26, 6.27 we see that $(n-1,\overline{n})(n,\overline{n})(n,\overline{n-1}) \in L(-w_0(3\omega_2))_K$. So, from equation 6.23 we get

$$((n,\overline{n})-(n-1,\overline{n-1}))(n,\overline{n})^2 \in L(-w_0(3\omega_2))_K. \tag{6.28}$$

Finally subtracting from the above section the section obtained by applying the transformation μ_n to it we get

$$(n,\overline{n})^3 \in L(-w_0(3\omega_2))_K.$$

Acknowledgements. We thank Prof. C. S. Seshadri for suggesting this problem and for many discussions we had with him. We would also like to thank Prof. S. Ramanan and Prof. R. Sridharan for fruitful discussions.

We would like to thank Prof J. C. Jantzen for going through a preliminary version of this paper and pointing out some major errors. He also referred us to the summation formula and suggested that we could corroborate our results by some programs written by Prof Niels Lauritzen (http://ww.imf.au.dk/niels).

We would like to thank Dr. S. Raghavan for the keen interest he took in our results. He sat through a series of lectures on these results and painstakingly went through all the calculations, which helped us in identifying and correcting some errors.

We thank the referee for generously providing us with an alternative proof of our results. This appears in Appendix C.

References

[1] H. H. Andersen, J. C. Jantzen, W. Soergel, *Representations of quantum groups at a pth root of unity and of semisimple groups in characteristic p*, *Asterisque*, 220, 1994.

[2] S. DONKIN, *An introduction to the Lusztig conjecture*, in R. Carter, M. Geck (eds), *Representations of Reductive Groups*, Publications of the Newton Institute, Cambridge Univ. Press, 1998.
[3] S. DOTY, *The submodule structure of certain Weyl modules for groups of type* A_n, *J. Algebra* **95** (1985), pp 373-283.
[4] R. GOODMAN, N. WALLACH,*Representations and invariants of the classical groups*, Encyclopedia of Mathematics and its applications, *Vol 68*, Cambridge Univ. Press, 1998.
[5] S. HELGASON,*Differential geometry, Lie groups and symmetric spaces, Prentice-Hall*, 1993).
[6] J. C. JANTZEN, *Representations of Algebraic groups*, *Pure and Applied Mathematics*, 9, Springer, 1972.
[7] M. KASHIWARA, T. TANISAKI, *Characters of the negative level highest weight modules for affine Lie algebras*,*Inter.Math.Res.Notices*,3, 1994, pp 151-161.
[8] D. KAZHDAN, G.LUSZTIG,*Tensor structures arising from affine Lie algebras I*, *Jl. Amer. Math. Soc.*, 6, 1993, 905-947.
[9] A. KNAPP,*Lie groups beyond an introduction*, *PM, 140, Birkhauser.*
[10] V. LAKSHMIBAI, C. S. SESHADRI, *Geometry of G/P V*, *Jl. of Algebra*, 100, 1986, pp 462-557.
[11] G. LUSZTIG,*Some problems in the representation theory of Finite Chevalley groups*, in B. Cooperstein, G. Mason (eds), *The Santa Cruz Conference on Finite Groups (1979)*, *Proc. Symp. Pure Math.*, 37, Providence R.I. 1980.
[12] S. RAMANAN, R. RAMANATHAN, *Projective normality of flag varieties and Schubert varieties*, *Invent. Math. 79*, 1985. pp 217-224.
[13] C. S. SESHADRI*Geometry of G/P I*, in K. G. Ramanathan (ed), *C. P. Ramanajum - A tribute*, Springer 1978, pp 207-239.
[14] R. STEINBERG,*Lectures on Chevalley Groups*, mimeographed lecture notes, New Haven: Yale Univ., 1967.

Appendix A. Identities for standard monomials

Appendix B. Calculations for the $3\omega_2$ case

Details can be found in
`http://www.cmi.ac.in/~kv/papers/so2n/Appendices-A-b.ps`

Appendix C. An alternative proof based on the sum formula

In this section we reproduce verbatim an alternative proof of our results given to us by the referee.

First some general remarks. Let G be a connected reductive group over an algebraically closed field of characteristic $p > 0$. For each dominant weight μ let $V(\mu)$ denote the Weyl module and $L(\mu)$ denote the simple G–module with highest weight μ. We denote by $\chi(\mu)$ the formal character of $V(\mu)$ and extend the definition of $\chi(\mu)$ also not not necessarily dominant μ as in [**6**] II.5.7. Let ρ be half the sum of the positive roots. Then one has $\chi(w(\mu+\rho)-\rho) = \det(w)\chi(\mu)$ for all $w \in W$ (the Weyl group); if there exists a root α with $\langle \mu+\rho, \alpha^\vee\rangle = 0$, then $\chi(\mu) = 0$.

Denote the p–adic valuation of the integers by ν_p (such that $\nu_p(p^n) = n$) and set for each dominant weight μ

$$sf(\mu) = -\sum_{\alpha} \sum_{i=1}^{\langle \mu+\rho,\alpha^\vee\rangle -1} \nu_p(i)\,\chi(\mu - i\alpha) \tag{1}$$

where α runs over the positive roots. Then $sf(\mu)$ is the right hand side of the sum formula as in [**6**], II.8.19, cf. [**6**], II.8.17. Let S denote the set of simple roots. For each subset $I \subset S$ set $sf(\mu, I)$ equal to the part of the sum in (1) where we take only positive roots α such that *the support of* α *is equal to* I. So $sf(\mu)$ is the sum of all $sf(\mu, I)$ with $I \subset S$.

A main property of the sum formula is that $sf(\mu) = 0$ if and only if $V(\mu) = L(\mu)$ if and only if $V(\mu)$ is irreducible. (If $sf(\mu) \neq 0$, then it gives some information on possible composition factors of $V(\mu)$, but that is not the point here.) Now in the cases at hand it will be enough to consider $sf(\mu, S) = 0$ (which is easier to compute, obviously). Let me explain, why.

For each subset I of the set of simple roots, let G_I denote the corresponding Levi subgroup of G (generated by our maximal torus and the root subgroups corresponding to roots in the span of I over the integers). For each dominant weight μ let $V_I(\mu)$ denote the Weyl module and $L_I(\mu)$ denote the simple G_I–module with highest weight μ. Now one has:

(2) *If $V(\mu)$ is simple, then $V_I(\mu)$ is simple for all I.*

This follows from comparing [**6**], II.2.11(b) and [**6**], II.5.21(1). (There we actually consider not the Weyl modules, but their duals. The transition, however, is obvious.)

On the other hand, we get:

(3) *Suppose that $V_I(\mu)$ is simple for all* proper *subsets $I \subset S$. Then $sf(\mu) = sf(\mu, S)$.*

This says in particular: Under this assumption $V(\mu)$ is simple if and only if $sf(\mu, S) = 0$.

In order to prove (3) one has to observe that $V_I(\mu)$ is simple if and only if the sum formula for the G_I–module $V_I(\mu)$ is 0 (obvious) if and only if the sum of all $sf(\mu, J)$ with $J \subset I$ is 0 (requires a small argument) if and only if all $sf(\mu, J)$ with $J \subset I$ are 0 (applying (2) to all $G_J \subset G_I$ and using induction). So the assumption in (3) implies that $sf(\mu, I) = 0$ for all proper subsets I of S. Therefore we get $sf(\mu) = sf(\mu, S)$.

We can refine the statement above and say:

(4) *Suppose that $V_I(\mu)$ is simple for all* connected proper *subsets $I \subset S$. Then $sf(\mu) = sf(\mu, S)$.*

Here connected refers to the topology of the Coxeter graph. The point is that the support of each root is connected. So we have $sf(\mu, I) = 0$ for all non-connected I.

Note also that we can always replace G_I by its derived group and μ by its restriction to the maximal torus of that derived group when checking for simplicity.

C.1. The $m\omega_1$ case. After these general considerations we now turn to the case at hand and assume that G has type B_n. I am going to use the notations as in Bourbaki's tables. So we have

(5) $$\rho = \sum_{i=1}^{n} (n + \frac{1}{2} - i)\varepsilon_i \qquad \text{and} \qquad \varpi_k = \sum_{i=1}^{k} \varepsilon_i \text{ if } k < n.$$

The set of positive roots with support equal to S is equal to

(6) $$\{\varepsilon_1\} \cup \{\varepsilon_1 + \varepsilon_k \mid 2 \le k \le n\}.$$

Now consider $\mu = m\varpi_1$ with $m > 0$. If $p > m$, then each $V_I(\mu)$ with $I \subset S$ connected and proper is simple: If $\alpha_1 \notin I$, then μ is 0 for the derived group of G_I, hence $V_I(\mu)$ one dimensional and simple. If $\alpha_1 \in I$, then G_I is of type A and $V_I(\mu)$ is the m-th symmetric power of the natural module of the derived group of G_I, hence simple for $p > m$.

So the assumption in (4) is satisfied for $p > m$ and we get $sf(\mu) = sf(\mu, S)$. We have by (5)

$$\mu + \rho = (m + n - \frac{1}{2})\varepsilon_1 + \sum_{i=2}^{n}(n + \frac{1}{2} - i)\varepsilon_i. \tag{7}$$

This implies $\langle \mu + \rho, \varepsilon_1^\vee \rangle = 2m + 2n - 1$ and $\langle \mu + \rho, (\varepsilon_1 + \varepsilon_k)^\vee \rangle = m + 2n - k$ for $k > 1$, hence

$$sf(\mu, S) = -\sum_{j=1}^{2m+2n-2} \nu_p(j)\,\chi(\mu - j\varepsilon_1) \tag{8}$$

$$-\sum_{k=2}^{n}\sum_{j=1}^{m+2n-k-1} \nu_p(j)\,\chi(\mu - j(\varepsilon_1 + \varepsilon_k))$$

Now a lot of these summands are 0. We have $\mu + \rho - j\varepsilon_1 = (m + n - (1/2) - j)\varepsilon_1 + \sum_{i=2}^{n}(n + (1/2) - i)\varepsilon_i$. If $m + 1 \leq j \leq m + n - 1$, then $i = j + 1 - m$ satisfies $2 \leq i \leq n$ and

$$\langle \mu + \rho - j\varepsilon_1, (\varepsilon_1 - \varepsilon_i)^\vee \rangle = m + n - \frac{1}{2} - j - (n + \frac{1}{2} - (j + 1 - m)) = 0,$$

hence $\chi(\mu - j\varepsilon_1) = 0$. Similarly, if $m + n \leq j \leq m + 2n - 2$, then $i = m + 2n - j$ satisfies $2 \leq i \leq n$ and

$$\langle \mu + \rho - j\varepsilon_1, (\varepsilon_1 + \varepsilon_i)^\vee \rangle = m + n - \frac{1}{2} - j + (n + \frac{1}{2} - (m + 2n - j)) = 0,$$

hence $\chi(\mu - j\varepsilon_1) = 0$.

Since we assume $p > m$, all $\nu_p(j)$ with $j \leq m$ are 0. So the only $\chi(\mu - j\varepsilon_1)$ of interest to us are those with $m + 2n - 1 \leq j \leq 2m + 2n - 2$. For these j we get (writing s_α for the reflection corresponding to a root α)

$$\begin{aligned} s_{\varepsilon_1}(\mu + \rho - j\varepsilon_1) &= (j - n - m + \frac{1}{2})\varepsilon_1 + \sum_{i=2}^{n}(n + \frac{1}{2} - i)\varepsilon_i \\ &= \rho + (j - 2n - m + 1)\varpi_1 \end{aligned}$$

with $0 \le j - 2n - m + 1 \le m - 1$. This implies (for $p > m$)

(9)
$$-\sum_{j=1}^{2m+2n-2} \nu_p(j)\,\chi(\mu - j\varepsilon_1) = \sum_{j=0}^{m-1} \nu_p(2n + 2m - 2 - j)\,\chi((m - j)\varpi_1).$$

We have for $2 \le k \le n$ and $1 \le j \le m + 2n - k - 1$

$$\mu+\rho-j(\varepsilon_1+\varepsilon_k) = (m+n-\frac{1}{2}-j)\varepsilon_1+(n+\frac{1}{2}-k-j)\varepsilon_k+\sum_{i\neq 1,k}(n+\frac{1}{2}-i)\varepsilon_i.$$

If $n + m \le j \le m + 2n - k - 1$, then $\langle \mu + \rho - j(\varepsilon_1 + \varepsilon_k), (\varepsilon_1 + \varepsilon_i)^\vee \rangle$ for $i = 2n + m - j$, if $m + 1 \le j \le m + n - 1$ and $j \neq m - 1 + k$, then $\langle \mu + \rho - j\varepsilon_1, (\varepsilon_1 - \varepsilon_i)^\vee \rangle = 0$ for $i = j + 1 - m$. So in all these cases we get $\chi(\mu - j(\varepsilon_1 + \varepsilon_k)) = 0$. So for $p > m$ the only possibly non-zero contribution from $\varepsilon_1 + \varepsilon_k$ to (8) occurs for $j = m - 1 + k$.

Looking at $\langle \mu + \rho - j(\varepsilon_1 + \varepsilon_k), (\varepsilon_k - \varepsilon_i)^\vee \rangle$ and $\langle \mu + \rho - j(\varepsilon_1 + \varepsilon_k), (\varepsilon_k + \varepsilon_i)^\vee \rangle$ one gets similarly that $\chi(\mu - j(\varepsilon_1 + \varepsilon_k)) = 0$ if $1 \le j \le 2n - k - 1$, $j \neq 2n - 2k + 1$. If $j = m - 1 + k = 2n - 2k + 1$, one gets $\chi(\mu - j(\varepsilon_1 + \varepsilon_k)) = 0$ from $\langle \mu + \rho - j(\varepsilon_1 + \varepsilon_k), (\varepsilon_1 + \varepsilon_k)^\vee \rangle = 0$.

We are left with $j = m - 1 + k \ge 2n - k - 2$. In this case

$$\mu+\rho-j(\varepsilon_1+\varepsilon_k) = (n+\frac{1}{2}-k)\varepsilon_1+(n+\frac{3}{2}-2k-m)\varepsilon_k+\sum_{i\neq 1,k}(n+\frac{1}{2}-i)\varepsilon_i.$$

Applying $s_{\varepsilon_1-\varepsilon_k}s_{\varepsilon_k}$ to this element we get

$$(m + 2k - n - \frac{3}{2})\varepsilon_1 + \sum_{i=2}^{n}(n + \frac{1}{2} - i)\varepsilon_i = \rho + (m - 1 - 2n + 2k)\varpi_1.$$

This shows that (for $p > m$)

$$\begin{aligned}
-\sum_{k=2}^{n}\sum_{j=1}^{m+2n-k-1} & \nu_p(j)\,\chi(\mu - j(\varepsilon_1 + \varepsilon_k)) \\
&= -\sum_{0\le l\le (m-1)/2} \nu_p(n + m - 1 - l)\,\chi((m - 1 - 2l)\varpi_1).
\end{aligned}$$

Assuming in addition $p > 2$, we have $\nu_p(n + m - 1 - l) = \nu_p(2n + 2m-2-2l)$. So adding the last line to (9) we get (for $p > \max(m, 2)$)

(10) $$sf(m\varpi_1) = \sum_{0\le j\le (m-2)/2} \nu_p(2n+2m-3-2j)\,\chi((m-2-2j)\varpi_1).$$

So $V(m\varpi_1)$ is simple if and only if p does not divide any $2(n + m - 1) - (2j + 1)$ with $1 \le 2j + 1 \le m - 1$. These are the same bounds as in Theorem 6(1)

C.2. $2\omega_2$ **and** $3\omega_2$ **cases.** Assume from now on that $n > 2$. We now consider $\mu = 2\varpi_2$ and $\mu = 3\varpi_2$ and assume $p > 3$. Look at all $I \subset S$ connected and proper. If $\alpha_2 \notin I$, then μ is 0 for the derived group of G_I, hence $V_I(\mu)$ one dimensional and simple. If $\alpha_2 \in I$ and if G_I is of type A, then μ restricts to the maximal torus of the derived group of G_I as 2 or 3 times one of the first two fundamental weights; now $p > 3$ implies that $V_I(\mu)$ is simple. The only case left is where G_I is of type B_{n-1}. Then μ restricts to the maximal torus of the derived group of G_I as 2 or 3 times the first fundamental weight. So (10) applied to $n-1$ and $m \in \{2,3\}$ shows that $V_I(2\varpi_2)$ is simple if and only if p does not divide $2n-1$, and that $V_I(3\varpi_2)$ is simple if and only if p does not divide $2n+1$. In particular:

(11) *Suppose that $p > 3$. If p does not divide $2n-1$, then $sf(2\varpi_2) = sf(2\varpi_2, S)$. If p does not divide $2n+1$, then $sf(3\varpi_2) = sf(3\varpi_2, S)$.*

We have

$$2\varpi_2 + \rho = (n + \frac{3}{2})\varepsilon_1 + (n + \frac{1}{2})\varepsilon_2 + \sum_{i=3}^{n}(n + \frac{1}{2} - i)\varepsilon_i.$$

This shows that $\langle 2\varpi_2 + \rho, \varepsilon_1^\vee \rangle = 2n+3$ and $\langle 2\varpi_2 + \rho, (\varepsilon_1 + \varepsilon_2)^\vee \rangle = 2n+2$ and $\langle 2\varpi_2 + \rho, (\varepsilon_1 + \varepsilon_k)^\vee \rangle = 2n+2-k$ for $3 \le k \le n$. One now writes down an expression analogous to (8). As above many summands turn out to be zero; that is checked by the same methods as before. Here is a list of the non-zero summands (for $p > 3$, so I ignore all $-\nu_p(j)\,\chi(2\varpi_2 - j\alpha)$ with $j \le 4$):

$$\begin{aligned}
-\nu_p(2n)\,\chi(2\varpi_2 - 2n\varepsilon_1) &= -\nu_p(n)\,\chi(\varpi_1)\\
-\nu_p(2n+1)\,\chi(2\varpi_2 - (2n+1)\varepsilon_1) &= -\nu_p(2n+1)\,\chi(\varpi_2)\\
-\nu_p(2n)\,\chi(2\varpi_2 - 2n(\varepsilon_1+\varepsilon_2)) &= \nu_p(n)\,\chi(0)\\
-\nu_p(2n+1)\,\chi(2\varpi_2 - (2n+1)(\varepsilon_1+\varepsilon_2)) &= \nu_p(2n+1)\,\chi(\varpi_2)\\
-\nu_p(n)\,\chi(2\varpi_2 - n(\varepsilon_1+\varepsilon_{n-1})) &= \nu_p(n)\,\chi(\varpi_1) \quad \text{if } n > 3.
\end{aligned}$$

So we get for $p > 3$

$$sf(2\varpi_2, S) = \nu_p(n)\,\chi(0). \tag{12}$$

(Note that $\nu_p(n) = 0$ if $n = 3$, so the exception in the last line above does not matter.)

We have

$$3\varpi_2 + \rho = (n + \frac{5}{2})\varepsilon_1 + (n + \frac{3}{2})\varepsilon_2 + \sum_{i=3}^{n}(n + \frac{1}{2} - i)\varepsilon_i.$$

This shows that $\langle 3\varpi_2 + \rho, \varepsilon_1^\vee \rangle = 2n+5$ and $\langle 3\varpi_2 + \rho, (\varepsilon_1+\varepsilon_2)^\vee \rangle = 2n+4$ and $\langle 3\varpi_2+\rho, (\varepsilon_1+\varepsilon_k)^\vee \rangle = 2n+3-k$ for $3 \le k \le n$. Proceeding as above one gets the following non-zero contributions to $sf(3\varpi_2, S)$ for $p > 3$:

$$\begin{aligned}
-\nu_p(2n+1)\,\chi(3\varpi_2 - (2n+1)\varepsilon_1) &= -\nu_p(2n+1)\,\chi(2\varpi_1)\\
-\nu_p(2n+2)\,\chi(3\varpi_2 - (2n+2)\varepsilon_1) &= -\nu_p(n+1)\,\chi(\varpi_1+\varpi_2)\\
-\nu_p(2n+3)\,\chi(3\varpi_2 - (2n+3)\varepsilon_1) &= -\nu_p(2n+3)\,\chi(2\varpi_2)\\
-\nu_p(2n+1)\,\chi(3\varpi_2 - (2n+1)(\varepsilon_1+\varepsilon_2)) &= \nu_p(2n+1)\,\chi(0)\\
-\nu_p(2n+2)\,\chi(3\varpi_2 - (2n+2)(\varepsilon_1+\varepsilon_2)) &= \nu_p(n+1)\,\chi(\varpi_2)\\
-\nu_p(2n+3)\,\chi(3\varpi_2 - (2n+3)(\varepsilon_1+\varepsilon_2)) &= \nu_p(2n+3)\,\chi(2\varpi_2)\\
-\nu_p(n+1)\,\chi(3\varpi_2 - (n+1)(\varepsilon_1+\varepsilon_{n-1})) &= \nu_p(n+1)\,\chi(\varpi_1+\varpi_2)
\end{aligned}$$

The last equation holds for $n > 3$. So we get for $p > 3$

(13) $sf(3\varpi_2, S) = \nu_p(n+1)\,\chi(\varpi_2) + \nu_p(2n+1)\,(-\chi(2\varpi_1) + \chi(0)).$

(Note that $\nu_p(n+1) = 0$ if $n = 3$.)

Combining (11)–(13) we get:

(14) *Suppose that $p > 3$. Then $V(2\varpi_2)$ is simple if and only if p does not divide $(2n-1)n$, and $V(3\varpi_2)$ is simple if and only if p does not divide $(2n+1)(n+1)$.* These bounds are the same as those in Theorem 6(2),(3). One can also use (10) to compute $sf(2\varpi_2, I)$ and $sf(3\varpi_2, I)$ for I of type B_{n-1}. One gets as final result for $p > 3$

(15) $$sf(2\varpi_2) = \nu_p(n)\,\chi(0) + \nu_p(2n-1)\,\chi(2\varpi_1)$$

and

(16) $$sf(3\varpi_2) = \nu_p(n+1)\,\chi(\varpi_2) + \nu_p(2n+1)\,(\chi(2\varpi_1+\varpi_2) - \chi(2\varpi_1) + \chi(0)).$$

Chennai Mathematical Institute, Chennai 600 017, India
E-mail address: kannan@cmi.ac.in, kv@cmi.ac.in

Equivariant Analogue of Grothendieck's Theorem for Vector Bundles on $\mathbb{P}^1$

Shrawan Kumar

To Professor C.S. Seshadri on his seventieth birthday

Abstract. Let T be a complex torus acting algebraically on $\mathbb{P}^1(\mathbb{C})$. In this note we prove a T-equivariant analogue of a theorem of Grothendieck. More specifically, we show that any T-equivariant vector bundle on $\mathbb{P}^1(\mathbb{C})$ is a direct sum of T-equivariant line subbundles.

The aim of this note is to prove the following equivariant analogue of Grothendieck's theorem, existence of which was asked by W. Fulton.

Let T be a complex (connected) torus acting algebraically on $X = \mathbb{P}^1(\mathbb{C})$.

Theorem. *Let E be a T-equivariant algebraic vector bundle on X. Then there exist T-equivariant line subbundles $L_1, \cdots, L_m \subset E$ such that we have a T-equivariant vector bundle isomorphism:*

$$\begin{array}{ccc} L_1 \oplus \cdots \oplus L_m & \xrightarrow{\sim} & E \\ & \searrow \quad \swarrow & \\ & \mathbb{P}^1, & \end{array}$$

under $v_1 \oplus \cdots \oplus v_m \mapsto v_1 + \cdots + v_m$.

Proof. As is well known (cf. [2; Exercise 6.6, Chapter I]), the group of algebraic automorphisms of X can be identified with the projective linear group PGL(2) via $\begin{pmatrix} a & b \\ c & d \end{pmatrix} \cdot [z_1, z_2] = [az_1 + bz_2, cz_1 +$

The author was supported by NSF.

$dz_2]$. Let $D \subset \mathrm{PGL}(2)$ be the standard diagonal subgroup. Since D fixes the point at infinity $[1, 0]$, the action of D on X lifts to give a D-equivariant line bundle structure on $\mathcal{O}_X(1)$ (and hence on any $\mathcal{O}_X(n), n \in \mathbb{Z}$). Thus $\mathcal{O}_X(n)$ acquires a T-equivariant line bundle structure.

By virtue of Grothendieck's theorem (cf. [1; Théorème 2.1]), one can decompose $E \cong \mathcal{O}_X(n_1) \oplus \cdots \oplus \mathcal{O}_X(n_m)$ with $n_1 \geq n_2 \geq \cdots \geq n_m$. Tensoring E with $\mathcal{O}_X(-n_1)$ and putting a T-equivariant line bundle structure on $\mathcal{O}_X(-n_1)$, we can assume that $E \simeq \mathcal{O}_X \oplus \mathcal{O}_X(n_2) \oplus \cdots \oplus \mathcal{O}_X(n_m)$ with each $n_2, \cdots, n_m \leq 0$. Thus any $0 \neq \sigma \in H^0(X, E)$ is nowhere vanishing. Further, $H^0(X, E) \neq 0$. Of course, $H^0(X, E)$ is a T-module. Take a T-eigenvector $0 \neq \sigma \in H^0(X, E)$. Then $\{\mathbb{C}\sigma(x)\}_{x \in \mathbb{P}^1}$ is a T-equivariant line subbundle, denoted L_1, of E.

We get an exact sequence of T-equivariant bundles:

$$(*) \qquad 0 \to L_1 \to E \to E/L_1 \to 0.$$

We claim that this sequence is T-equivariantly split:

Consider the T-module $\mathrm{Hom}_{\mathcal{O}}(E/L_1, E)$ of bundle morphisms. Then we get a natural T-module map

$$\pi : \mathrm{Hom}_{\mathcal{O}}(E/L_1, E) \twoheadrightarrow \mathrm{Hom}_{\mathcal{O}}(E/L_1, E/L_1).$$

Observe that all the line bundles, occurring in E/L_1 as direct summands, have degrees ≤ 0. This can be seen, e.g., by tensoring the sequence $(*)$ with $\mathcal{O}_X(-1)$ and then considering the associated long exact cohomology sequence. Thus $(*)$ splits as vector bundles by [2; Propositions 6.3 and 6.7, Chap. III] (without regarding the T-equivariance). In particular, π is surjective. Thus π induces a surjective map

$$\pi^T : \mathrm{Hom}_{\mathcal{O}}(E/L_1, E)^T \twoheadrightarrow \mathrm{Hom}_{\mathcal{O}}(E/L_1, E/L_1)^T,$$

where the superscript T means T-invariants. Take a preimage f of the identity homomorphism $\mathbf{I}$ under π^T. This f provides a T-equivariant splitting of $(*)$. Thus $E \simeq L_1 \oplus E/L_1$ as T-equivariant bundles. So, by induction on rank E, the theorem follows. □

References

[1] Grothendieck, A.: Sur la classification des fibres holomorphes sur la sphere de Riemann, *Amer. J. Math.* *79* (1957), 121–138.

[2] Hartshorne, R.: *Algebraic Geometry*, GTM 52, Springer-Verlag, 1977.

Department of Mathematics, UNC at Chapel Hill
Chapel Hill, NC 27599–3250, USA.

Vector Bundles and Connections in Physics and Mathematics: Some Historical Remarks

V. S. Varadarajan

To Seshadri, for his seventieth birthday

Abstract. The concept of smooth vector bundles on manifolds and the theory of connections on them has played a fundamental role in physics going back to the discovery of Hermann Weyl around 1918 that electromagnetism can be viewed as a connection on a real line bundle on spacetime. Under the influence of quantum mechanics this idea was modified and the real line bundle was replaced by a hermitian complex line bundle, allowing the discovery, by Dirac, of magnetic monopoles. In mathematics the idea of holomorphic vector bundles goes back to the work of Riemann on multivalued solutions of regular singular differential equations like the hypergeometric equation. This paper examines briefly these historical origins, discusses the evolution of these two streams of thoughts, leading to the surprising climax in which they merge in the theory of moduli spaces.

1. Introduction. I have known Seshadri for many years. I still remember the first time when I came across his name in a Séminaire Chevalley volume on Variétés de Picard. I was at the Indian Statistical Institute at that time and was browsing in the library one day when I noticed that a certain Conjeevaram S. Seshadri had given a series of talks in that seminar. I myself had been born in Conjeevaram (or Kancheepuram, to give the Sanskrit name for this ancient and historic South Indian city), and so the fact that someone, who had grown up in the same milieu as I, had reached a high level in the world of mathematics was not only a thrill but a big source of inspiration for me. Of course it was only many years later that I came to understand a little more of what he had done and the reason why he is regarded with a lot of respect and admiration.

Although a large part of Seshadri's work belongs to the theory of vector bundles on curves, his total work is much broader and has had a substantial impact in many areas. Concerning the vector bundle theory, his famous papers, with M. S. Narasimhan[1] and by himself [2], inaugurated the modern theory of holomorphic vector bundles on compact Riemann surfaces. He then extended the whole theory to the algebraic context in any characteristic. He was the first one to give substantial evidence to the Mumford conjecture on geometric reductivity of reductive algebraic groups by proving it for $GL(2)$ (it was fully proved by Haboush a little later). His work with Lakshmibai and others on flag manifolds of the classical groups extended the classical results on the grassmannians and pioneered the so-called standard monomial theory. His lectures and articles and influential scholarship have had a deep impact on his colleagues and students. After he left the Tata Institute of Fundamental Research where he had been a major force for many decades, he came back to Madras to organize a strong group of young and active mathematicians, working under the auspices of the SPIC, a private consortium of industrial companies. Unlike many of us who went abroad, he remained in India and tried to exert himself in creating and maintaining a visible and active Indian school of mathematics. All of these achievements have been recognized by many awards and honors, both from India and from other countries. It is a privilege to be asked to contribute to this volume in celebration of his seventieth birthday and I am very happy that I have an opportunity to say something on this occasion.

I had much hesitation when Sridharan asked me to contribute an article to this volume because my own work does not have much overlap with Seshadri's. In the end I decided that it may be of some interest to explain, in a manner partly historical and partly pedagogical, how the concept of vector bundles and connections on them evolved from the point of view of both physics and mathematics. Although the matters I propose to discuss are perhaps only tangential to the main themes in Seshadri's work, they do touch upon them in a surprising manner as we shall see at the end.

Before I proceed it is important to have at least an informal understanding of the nature of the objects we will be dealing with.

Roughly speaking, a *vector bundle* is a vector space depending on a parameter which varies on a manifold M; it is then said to be defined *over* M. It is assumed that we can, over small pieces of M, take bases that vary nicely with the parameter; here nicely refers to some smoothness, for instance C^∞ or holomorphic, depending on the context, and the vector bundle will be correspondingly called C^∞ or holomorphic.

C^∞ vector bundles arise naturally in modern physics, especially when the manifold M is spacetime or some extension of it. When M is spacetime, its points may be thought of as representing the locations of particles. These particles obey the laws of quantum physics and therefore have a much richer internal structure than their classical counterparts. For instance they have properties such as spin, isospin, color, charm, etc. These internal properties are encoded by a vector space equipped with an action by a unitary group. When the particle moves, it takes this internal vector space with itself. It is then clear that a mathematical treatment involving these particles has to be in the context of vector bundles over M. However the appearance of vector bundles dates back to a few years before the discovery of quantum theory, when such internal structures of particles were not known. So tracing the development of this concept has some interest.

Even in the informal approach that we have taken above, it must be admitted that the concept of a vector bundle is not very intuitive, although the physical example given above certainly does make it more acceptable. In differential geometry a manifold has vector bundles naturally defined over it - the bundle of tangent vectors, tensors, especially exterior differential forms, and so on. But there are also other situations where vector bundles arise that have nothing to do with these geometrically derived bundles. One of the most interesting examples of a "non-geometric "vector bundle is obtained when we consider a system of linear first order partial differential equations which satisfy the Frobenius condition of integrability. The equations are of the form

$$\frac{\partial u}{\partial x^\mu} + A_\mu u = 0 \qquad (1 \le \mu \le n)$$

where the A_μ are smooth $r \times r$ matrix valued functions defined on an open set $M \subset \mathbf{R}^n$ satisfying the Frobenius condition

$$F_{\mu\nu} := A_{\nu,\mu} - A_{\mu,\nu} + [A_\mu, A_\nu] = 0 \qquad (1 \le \mu, \nu \le n)$$

on M, and u is the unknown function whose values are column vectors having r components. Here, as usual,

$$A_{\mu,\nu} = \frac{\partial A_\mu}{\partial x^\nu}, \quad [A_\mu, A_\nu] = A_\mu A_\nu - A_\nu A_\mu.$$

The holomorphic version with M an open set in $\mathbf{C}^n$ and A_μ holomorphic is even more important for us later. The Frobenius condition is necessary and sufficient that for any point $m \in M$ and any r-vector $u_0 \in \mathbf{C}^n$ there is a solution u of this system in a neighborhood of m with $u(m) = u_0$. At each point of M we can then attach the r-dimensional vector space of initial values of the local solutions at that point and obtain a natural vector bundle on M. Clearly this vector bundle is quite different in nature from the geometric bundles of tangent vectors and tensors. Notice that when M has dimension 1, we have ordinary linear differential equations and there is no need of any integrability condition.

To use this notion of a vector bundle in a free and flexible manner it is necessary to have an apparatus of differential calculus on them. For instance one should be able to differentiate a vector that varies with the vector space in a smooth manner. It is not enough to differentiate the components of this vector in a basis that one is allowed to choose because these bases do not have any intrinsic significance. In fact, any intrinsic definition of derivative of a varying vector implies that one is able to compare the vectors attached to distinct but neighboring points of the manifold M. Thus what is needed to construct a coordinate invariant differential calculus is the existence of isomorphisms between the vector spaces at neighboring points, indeed, points that are infinitely near each other. Such a structure is called a *connection* on the vector bundle.

This article develops two themes. The first, taking up §2-§5, explains how the attempts to understand fundamental issues in physics played a truly important role in the evolution of the idea of a C^∞ vector bundle on a manifold and the concept of connections on such bundles. These ideas eventually led to the modern concept of a *gauge field* which dominates much of the high energy physics of today. Indeed, it is now universally accepted that a consistent field theory of elementary particles and their interactions must be a gauge theory. But in the early days this was not so obvious, and some very remarkable ideas of Weyl, Dirac, Aharanov-Bohm, and Yang-Mills were responsible for the decisive emergence of this view point[3], as I shall explain in §2-§5. The second theme is taken up in §6 where I discuss the evolution of the concept of *holomorphic* vector bundles whose genesis is in function theory and differential equations on Riemann surfaces. On the surface these two themes, one coming from differential geometry and physics and the other from function theory, appear to have no common aspect. But remarkably they do have a deep interaction and I shall explain at the end of the article how the physics and mathematics come together in surprising ways.

I shall try to give a short account of these matters; however the reader should be warned that this account is far from being complete. Indeed, the topic can be treated adequately only in a course of several lectures. I should also put in a disclaimer to the effect that I am not a professional historian of science and so the reader should look upon this effort only as a pedagogical one aiming to trace briefly the development of the basic concepts of vector bundles and connections. To keep my discussion as self-contained and elementary as possible and to make the ideas accessible to a wider audience, especially to students, I have gone over many things that are familiar to experts. I hope I will be forgiven for this.

2. The work of Weyl and his discovery that the electromagnetic vector potential is a connection on a suitable bundle on spacetime and the electromagnetic field is the curvature of this connection. The creation of the theory of special and general relativity by Einstein in the years 1905-1916 had a profound impact on mathematics and mathematicians. The astonishing fact that gravitation is just a manifestation of the curvature of spacetime made a deep impression on mathematicians like Elie Cartan and Hermann Weyl. This was especially the case with Weyl. Weyl's initial work, as is appropriate for a student of Hilbert at that time, was concerned with analysis and spectral theory. By 1916 he had made profound contributions to the spectral theory of differential operators, the asymptotic distribution of their eigenvalues, and to questions of uniform distribution mod 1. But he had also pursued simultaneously a geometrically motivated line of thought, as can be inferred from the fact that his epoch-making book *Die Idee der Riemanschen Fläschen* was published in 1913. It is clear that his mind, which was universal in its scope and tremendously attracted to fundamental questions of mathematics and physics that had a deep philosophical component to them, was fascinated by the role of differential geometry in the understanding of nature. Geometers such as Levi-Civita, Ricci, Schouten, and others had been gaining great insights on the foundations of the differential geometry of Riemannian manifolds. The pseudo Riemannian case, whose theory does not differ in many essential aspects from the Riemannian case, became important after the work of Minkowski and Einstein because of the fact that spacetime is not Riemannian but pseudo Riemannian of signature (+,-,-,-), i.e., a manifold which has a smooth bilinear form on the tangent spaces having this signature at each point. In what follows the term Riemannian is used to include the pseudo Riemannian case also.

Clearly there were two streams of thought - on the one hand, the ideas of Einstein on general relativity and their implications for the foundations of the theory of pseudo Riemannian manifolds as developed by the geometers, and on the other hand, the analysis of the nature of space and spacetime which became possible because of the advances in differential geometry. In my opinion Weyl was the only one who attempted this unification.

From the very beginning it was clear that the basic difficulty in doing physics or mathematics on manifolds that do not possess an intrinsically defined coordinate system was that of making sure that the results that one obtains by working in a single coordinate system remained valid in all of them. The traditional way to get around this issue is by insisting that the fundamental laws should be formulated in the language of tensors and their derivatives. However we have seen above, somewhat informally to be sure, that doing coordinate-invariant differential calculus of tensors requires the existence of a connection, or, isomorphisms between the tangent spaces at neighboring points. Such isomorphisms can be constructed on a Riemannian manifold by what is called *parallel transport,* as was first discovered by Levi-Civita. Thus, a connection may also be thought of as a way of moving tangent vectors along curves on the manifold in a canonical manner. If $t \mapsto \gamma(t)$ $(0 \leq t \leq 1)$ is a curve in M and Y is a tangent vector to M at the point $p = \gamma(0)$, we can find vectors $Y(t)$ tangent to M at $\gamma(t)$, uniquely determined by Y and depending linearly and isomorphically on Y; $Y(t)$ is the parallel transport of Y along the curve γ. These isomorphisms $Y \mapsto Y(t)$ between the tangent spaces at $p = \gamma(0)$ and $\gamma(t)$, allow one to define differentiation of vector and tensor fields along the curve γ. This is the idea of *covariant differentiation of vectors and tensor fields.* In fact, the three concepts *connection, parallel transport, covariant differentiation*

are essentially equivalent.

For defining this connection, known as the *Levi-Civita connection,* Levi-Civita originally assumed that the Riemannian manifold M was imbedded in a higher dimensional euclidean space. Let M be a Riemannian manifold whose metric comes from the flat metric of an ambient space E which is Euclidean. Let us be given a vector field Y defined on M near p, where p be a point of M and let X a tangent vector to M at p. Then the covariant derivative at p of Y in the direction X can be defined as follows. First extend Y to a vector field Y' defined in a neighborhood of p in E, compute the directional derivative $(\partial(X)Y')_p$ (in E) at p in the direction of X, the directional derivative being applied to each component of Y'; the covariant derivative of Y at p in the direction X is then the orthogonal projection of $(\partial(X)Y')_p$ on the tangent space to M at p. Since only the direction X is involved, it is enough if Y is specified just on a curve γ with $\gamma(0) = p$ and $\dot{\gamma}(0) = X$. If $Y(t)$ are vectors tangent to M at $\gamma(t)$ (notation as above), the condition that they are parallel transports of $Y(0)$ is that the covariant derivative of $Y(t)$ in the direction of $\dot{\gamma}(t)$ is 0 for all t. It is not difficult to show that these conditions translate to linear first order ordinary differential equations for $Y(t) = (Y^\mu(t))$ of the form

$$\frac{dY^\mu}{dt} + \sum_{\lambda\nu} \Gamma^\mu_{\lambda\nu}(\gamma(t))\dot{\gamma}^\lambda(t)Y^\nu(t) = 0, \qquad Y^\mu(0) = Y^\mu. \qquad (PT)$$

The functions $\Gamma^\mu_{\lambda\nu}$ which appear as coefficients of these differential equations, are the Christoffel symbols defined in terms of the first order derivatives of the coefficients of the metric tensor. In the geometrical picture described above, differentiation in E which is related to parallel movement of vectors is adjusted to produce a movement of tangent vectors on M, hence the name parallel transport.

It was Weyl who apparently made the observation[4] that the formulae for the parallel transport make sense on *abstract* (as opposed to *imbedded*) Riemannian manifolds. Perhaps one should not put too much emphasis on this point because most classical geometers never worried about global aspects of differential geometry; however, for Weyl, who was one of the pioneers in the global view of manifolds as illustrated by his work in the theory of Riemann surfaces, such a line of thinking was natural.

Another discovery of his, namely that it was not necessary to require that a manifold have a metric in order to define parallel transport on it, was much more fundamental. Weyl's observation was that the notion of parallel transport on a manifold could be taken as an *axiomatic starting point*. He thus introduced the concept of an *affinely connected manifold* on which an (affine) connection was defined axiomatically. He formulated this as the statement that parallel transport was to be defined on the manifold by the equations (PT). The functions $\Gamma^{\mu}_{\lambda\nu}$ were prescribed *a priori* together with their law of transformation under change of coordinates. The transformation law is in fact determined uniquely by the requirement that the parallel transport defined by (PT) is independent of the choice of coordinates. If we write the equations (PT) heuristically as

$$dY^{\mu} = -\sum_{\lambda\nu} \Gamma^{\mu}_{\lambda\nu} Y^{\nu} dx^{\lambda}$$

it suggests itself that we introduce the *matrix* of 1-forms given by

$$A = \left(\sum_{1\leq\lambda\leq n} \Gamma^{\mu}_{\lambda\nu} dx^{\lambda} \right)_{1\leq\mu,\nu\leq n} \qquad (n = \dim(M)).$$

We then say that the matrix A together with its transformation law *defines the connection* on M. Although the elements of A have been indexed as if they are tensors, a connection is *not* a tensor, that is, the $\Gamma^{\mu}_{\lambda\nu}$ are not the components of a tensor; but the *difference* of two connections, in particular, any *infinitesimal variation* of a connection, is a tensor. So the connections form an *affine space* rather than a linear space. The condition

$$\Gamma^{\mu}_{\lambda\nu} = \Gamma^{\mu}_{\nu\lambda}$$

is independent of the coordinate system used since it is manifestly equivalent to the vanishing of a difference of two connections and so is a tensor equation. In this case the connection is said to be *torsionless*. The Levi-Civita connection is rediscovered in two ways: by fiat, giving the formulae for $\Gamma^{\mu}_{\lambda\nu}$ in terms of the metric tensor, and intrinsically, through the important result that *it is the only connection which is torsionless and has the property that parallel transport preserves the length of the vector.* For the general affine connections the curvature tensor can be defined imitating what is done in the Riemannian case. In fact, in terms of the matrices A_{μ} the curvature can be viewed as a matrix F of 2-forms given by

$$F = (1/2)\sum_{\rho\tau} F_{\rho\tau} dx^\rho dx^\tau = \sum_{\rho<\tau} F_{\rho\tau} dx^\rho dx^\tau$$

where

$$F_{\rho\tau} = A_{\tau,\rho} - A_{\rho,\tau} + [A_\rho, A_\tau].$$

For the Levi-Civita connection of a Riemannian manifold the curvature is the usual Riemann-Christoffel curvature tensor and its vanishing is the necessary and sufficient condition that the metric is flat, namely that in a suitable coordinate system it has the form

$$(dx^1)^2 + \cdots + (dx^p)^2 - (dx^{p+1})^2 - \cdots - (dx^{p+q})^2.$$

For a general manifold with an affine connection as described above, the vanishing of F is the necessary and sufficient condition that the manifold equipped with the connection is locally isomorphic to the affine space $\mathbf{R}^n$ where parallel transport is just the usual one; this is equivalent to saying that there is a coordinate system locally in which covariant differentiation is just ordinary differentiation.

This liberation of the concept of a connection from its metric origins, although extremely simple and natural from hindsight (which we know has always 20/20 vision!), turned out to be one of the most important things that Weyl did in differential geometry[5]. Indeed, once the idea of axiomatically introducing connections is accepted, it is easy to realize that one should not be limited to transporting just tangent vectors and tensors along curves, and that one should also think of *axiomatically* attaching vectors to points of the manifold and transporting them using connections *axiomatically defined* in the appropriate way. But I am running ahead of the story.

One cannot be certain at what point in Weyl's thinking his ideas about the foundations of differential geometry merged with his reflections on their role in the foundations of general relativity. It is quite possible that these two strands of thought coexisted in his mind from the beginning. Indeed, after his 1918 paper[5] Weyl generally made no serious conceptual distinction between the differential geometric constructs and their physical interpretation. The next element in the story was Weyl's observation that requiring parallel transport to preserve the lengths of vectors was *not natural* either from the mathematical or the physical point of view. Mathematically it was not natural because one is already assuming that parallel transport can change the direction of the vector and so one can ask why it should keep its length unchanged[6]. On the other hand, it was not natural from the physical point of view either. In fact, the numerical values of the lengths of vectors can be determined only after one chooses a unit of length. Therefore, if we suppose that the base manifold is space or spacetime and that observers are located at various points of it, the assumption of length invariance of vectors under parallel transport meant that *all observers agree on what a unit of length is, even though they may be widely separated in space and time.* Weyl insisted that this was too strong an assumption, and that the proper assumption should be that observers can choose the unit of length only at their location, and that, while they can communicate their choice along paths to other observers, this transference of the unit of length may not be path-independent. Indeed, this should not be surprising, especially since transference of the direction of vectors was already assumed to be path-dependent. It was by reasoning in this manner that Weyl arrived at his remarkable observation[7]: Riemannian geometry cannot be considered as a completely pure infinitesimal geometry because the (metric) enables us to compare, with respect to their length, not only two vectors at the same point, but also the vectors at any two points. *A truly infinitesimal geometry must recognize only the principle of transference of a length from one point to another point infinitely near to the first* (italics as in the original).

So, from Weyl's point of view, the metric of spacetime was determined only up to a scale at each point and that the possible scales at a point of M is represented by a copy of $\mathbf{R}_{>0}$, the multiplicative group of positive real numbers. Weyl made the assumption that after choosing a unit of length at a point, an observer can transfer this choice along curves to other observers. Clearly this transport of *scales* was of a very different nature from parallel transport in Riemannian geometry where only tangent vectors and tensors were transported. It is here that his work on connections that are defined without any reference to a metric became a guiding principle for him. Using his experience with affinely connected manifolds Weyl realized that such a transport of scales could be introduced in the same way; all he had to do was to introduce the matrix of 1-forms analogous to the $(\sum_{\lambda\nu} \Gamma^{\mu}_{\lambda\nu} dx^{\lambda})$ of his affine geometry, and use them to define the differential equations that describe the transport process. Since the set of scales has dimension 1, one has to define a single 1-form A locally on M, given in local coordinates by

$$A = \sum_{\mu} A_\mu dx^\mu$$

for suitable functions $A_\mu(\mu = 0, 1, 2, 3)$. The transport of scales $s(t)$ along a curve $\gamma(t \mapsto \gamma(t))(0 \leq t \leq 1, \gamma(0) = p, \gamma(1) = q)$ connecting two points p and q of the manifold (assumed for simplicity to be in the same coordinate neighborhood) is then determined by the differential equation

$$\frac{ds}{dt} + \sum_{\mu} A_\mu(\gamma(t))\dot{\gamma}^\mu(t)s(t) = 0$$

which is solved by

$$s(t) = \exp\Big\{ - \oint_{\gamma_t} \sum_{\mu} A_\mu dx^\mu \Big\} = \exp\Big\{ - \oint_{\gamma_t} A \Big\}$$

where γ_t is the path γ from 0 to t and $\oint_{\gamma_t}$ is the line integral of A along γ_t. It is important to realize that A is a 1-form as far as coordinate transformations are concerned, but it would change when one considers *transformations that come from a change of scales.* Let the scales be changed locally by a function

$$g : x \longmapsto g(x) > 0.$$

Then in the new units the function s changes over to s' where

$$s'(t) = s(t)g(\gamma(t)).$$

If we require that the differential equation for s' has the same form as the equation for s with A replaced by A', then we must require that A should change to the 1-form A' where

$$A' = A - d(\log g).$$

Weyl described this process as a *change of gauge,* and was led to his *principle of gauge invariance* which asserted that *the fundamental physical laws should be invariant, not only under coordinate transformations, but also, in addition, under changes of gauge,* otherwise called *gauge transformations.*

Notice that the exterior derivative dA of A is independent of the gauge and is a *globally defined* 2-form. Indeed, as the change of gauge changes A to $A + d(\log g)$, the 2-form $F = dA$ is independent of change of gauge and so is globally defined on M. The components of F are

$$F_{\mu\nu} = \frac{\partial A_\nu}{\partial x^\mu} - \frac{\partial A_\mu}{\partial x^\nu} = A_{\nu,\mu} - A_{\mu,\nu}.$$

Clearly F is the *curvature* of this connection.

As already observed, these ideas of Weyl represented a big departure from classical geometry where only vectors like tangent vectors, cotangent vectors, tensors and so on, which are geometrically associated to the underlying manifold, were transported. Here Weyl was breaking new ground by considering the possible scales at the various points as a *bundle* on spacetime, introducing connections on this bundle, and their curvatures as 2-forms defined on spacetime. *If I am not mistaken, this is the first occurrence in physics and differential geometry of a non-geometric vector bundle and connections on it;* the fact that scales form a multiplicative group rather than a vector space is not important since the map $x \mapsto \log x$ converts this multiplicative group into the additive group of real numbers. In modern terms the scale bundle may be thought of either as a principal bundle for the group of positive real numbers or a vector bundle of rank 1. Thus one can describe the situation as follows. There is a natural bundle, the *scale bundle,* on M with group $\mathbf{R}_{>0}$, and the communication of the choice of scales along a curve γ on M is simply a curve γ' in this bundle that lies above γ. Moreover this lifting of γ is nothing but parallel transport in this bundle, described by a connection on it. The curvature of this connection is then a well-defined 2-form on M.

Before proceeding further let me mention as an aside that the generalization of these ideas to any vector bundle in place of the scale bundle is immediate although Weyl did not take this step. The connection is defined by a *matrix-valued* 1-form

$$A = \sum_\mu A_\mu dx^\mu \qquad A_\mu = \left(\Gamma^a_{\mu b}\right)_{1 \leq a,b \leq r}.$$

The vectors attached to a point of the manifold form a vector space of dimension r. Unlike the case of the tangent bundle these vectors may have nothing to do with the tangent space and so we index them by the symbols a, b. If the vector spaces attached to the points of M are equipped with an action of a compact Lie group G, one makes the further assumption that the A_μ take values in the Lie algebra of G, i.e., the matrices $(\Gamma^a_{\mu b})$ lie in the Lie algebra of G. Parallel transport is defined by

$$\frac{ds^a}{dt} + \sum_b \Gamma^a_{\mu b}(\gamma(t)) s^b(t) = 0 \qquad (PT)$$

and covariant differentiation is given by the formula

$$(\nabla_{\partial_\mu} s)^a = \frac{\partial s^a}{\partial x^\mu} + \sum_b \Gamma^a_{\mu b} s^b$$

where (s^a) is a local section of the bundle. This can also be written concisely in the form

$$\nabla_{\partial_\mu} = \partial_\mu + A_\mu \qquad A_\mu = (\Gamma^a_{\mu b})_{1 \le a,b \le r}$$

where the operator on the right acts on sections as

$$s \longmapsto \partial_\mu s + A_\mu s.$$

However, when $\dim(M) \ge 2$, the operators ∇_{∂_μ} and ∇_{∂_ν} do not in general commute for distinct μ, ν. More precisely, let

$$F_{\mu\nu b}{}^a = (\nabla_{\partial_\mu}\nabla_{\partial_\nu} - \nabla_{\partial_\nu}\nabla_{\partial_\mu})^a_b \, .$$

The matrix $F_{\mu\nu}$, with entries $F_{\mu\nu b}{}^a$ can be calculated to be

$$F_{\mu\nu} = A_{\nu,\mu} - A_{\mu,\nu} + [A_\mu, A_\nu].$$

One should think of the $F_{\mu\nu}$ as the coefficients of a matrix-valued 2-form F, the *curvature (form) of the connection,* defined by

$$F = (1/2)\sum_{\mu\nu} F_{\mu\nu} dx^\mu dx^\nu = \sum_{\mu<\nu} F_{\mu\nu} dx^\mu dx^\nu$$

where $A_{\mu,\nu}$ is $\partial A_\mu / \partial x^\nu$ and $[C, D] = CD - DC$. Its vanishing is equivalent to the statement that parallel transport is *locally* path independent, or, equivalently, there is a coordinate system locally in which covariant differentiation is the same as ordinary differentiation. These local data go hand in hand with the analog of Weyl's gauge transformations, namely the maps

$$x \longmapsto g(x)$$

of the part of the manifold being described locally into the group G. Obviously these should be called *gauge transformations* also. Finally, under the gauge transformation g the matrix A changes to the matrix A' where

$$A' = gAg^{-1} - g^{-1}dg.$$

The presence of the term $g^{-1}dg$ shows that this is *not* a linear transformation and so A is not a tensorial object, but the difference of two connections is tensorial. The formula connecting A and A' is of course obtained from the requirement that the equations (PT) describing parallel transport have intrinsic significance. Moreover, if $F' = dA' + A' \wedge A'$, we have,

$$F' = gFg^{-1}.$$

Thus F is tensorial. In the case of the scale bundle we have

$$A' = A - d(\log g), \qquad F' = F$$

as we have seen earlier. In this case F is a globally defined 2-form. In the general case, to get globally defined forms we must take *traces* to get rid of the factors involving g. More precisely, let us take the *characteristic polynomial* of F,

$$\det\left(TI - \frac{1}{2\pi i}F\right) = T^r + c_1(F)T^{r-1} + \cdots + c_r(F).$$

We must remember that the exterior forms of *even* degree form a *commutative algebra* under exterior multiplication, and so F may be viewed as a matrix with entries from this commutative algebra; thus there is no difficulty in computing its characteristic polynomial. The coefficient $c_j(F)$ is an exterior form of degree $2j$. It is *globally defined* on M. It is a remarkable fact (the theorem of Chern-Weil) that the c_j are *closed, and that their cohomology classes do not depend on the connection.* Thus they are invariants of the vector bundle. These constructions make sense for real as well as complex vector bundles but have to be modified a little when we work with a bundle with group G. For a complex vector bundle the cohomology classes of the c_j are called *Chern classes.* For a bundle with structure group as $U(r)$ the cohomology classes of the c_j are actually *integral classes*. This is a far-reaching generalization of the Dirac monopole quantization where one is dealing with $c_1(F) = F$, as we shall see in §4 below. Because of the gauge theoretic origin of these notions the Chern classes are important in physics where they often arise as *topological charges.*

Let us now return to the original discussion. Weyl had therefore been led to a natural generalization of Riemannian geometry from physical considerations. Sometimes this geometry is called a *Weyl geometry,* and the manifolds with a Weyl geometry, *Weyl manifolds.* Weyl geometry is based on two things. First, a metric, which is meaningful only up to scale changes that could be spacetime dependent, i.e., a *conformal metric* in modern terminology[8]; second, a connection on the scale bundle of the manifold. In his calculations which were couched in local terms Weyl simply used the product bundle

$M \times \mathbf{R}_{>0}$. But this is not a restriction, as all principal $\mathbf{R}_{>0}$-bundles on M are trivial, i.e., isomorphic to the product bundle. Indeed, the $\mathbf{R}_{>0}$-bundles are essentially the same as $\mathbf{R}$-bundles, and the latter are classified by $H^1(M, \mathcal{S})$ where $\mathcal{S}$ is the sheaf of real smooth functions on M; and this sheaf is a so-called fine sheaf and so all of its cohomologies in degrees ≥ 1 are 0.

Just as in the case of Riemannian geometry, in Weyl geometry there is a canonical connection on the base manifold, the so-called *Weyl connection,* which is the analogue of the Levi-Civita connection of a Riemannian manifold. The Weyl connection is determined by requiring that it is torsionless and that the lengths of vectors is preserved under parallel transport, *after adjusting for scale changes.* This means that

$$\frac{d}{dt}\left(s(t)^{-1}\sum_{\mu\nu} g_{\mu\nu}(\gamma(t))Y^{\mu}(t)Z^{\nu}(t)\right) = 0$$

where $(Y^{\mu}(t))$ and $(Z^{\nu}(t))$ are tangent vectors at $\gamma(t)$ that remain parallel as t varies, and $s(t)$ is the scale factor at $\gamma(t)$ defined by parallel transport of scales governed by the differential equations described earlier. From this it follows easily that the coefficients of the Weyl connection are given by the formulae

$$\Gamma^{\mu}_{\lambda\nu} = {}^{0}\Gamma^{\mu}_{\lambda\nu} + \frac{1}{2}\left(\delta^{\mu}_{\nu}A_{\lambda} + \delta^{\mu}_{\lambda}A_{\nu} - g_{\lambda\nu}A^{\mu}\right)$$

where ${}^{0}\Gamma^{\mu}_{\lambda\nu}$ are the components of the Levi-Civita connection of the metric g and $A^{\mu} = g^{\mu\tau}A_{\tau}$ as usual (see[4], 206 or[7], 125, equations (49) and (50)).

All this was quite interesting but what excited Weyl tremendously was that he saw in his new geometry *a way to unify electromagnetism with gravitation.* Let me now explain this in a little more detail. At the time we are talking about, there were just *two* field theories in nature; the *electromagnetic field,* which obeyed the Maxwell equations in flat Minkowski spacetime, and the *gravitational field,* which was described by the Einstein equations involving the curvature tensor on an underlying Riemannian manifold.

In the electromagnetic theory, spacetime is $\mathbf{R}^{1,3}$ with the Minkowski flat metric. The electromagnetic field, which is usually described by a skew-symmetric matrix containing the electric and magnetic fields, is an exterior 2-form F

$$F = (1/2)\sum_{\mu\nu} F_{\mu\nu}dx^\mu dx^\nu = \sum_{\mu<\nu} F_{\mu\nu}dx^\mu dx^\nu.$$

The Maxwell equations are then written concisely as

$$dF = 0, \qquad d * F = 0$$

where d is the exterior derivative and $*$ is the Hodge $*$-operator. In $\mathbf{R}^{1,3}$ we can write $F = dA$ for a 1-form

$$A = \sum_\mu A_\mu dx^\mu$$

which is uniquely determined up to an additive term of the form df for a scalar function f; A is called the *vector potential*. This change from F to A gets rid of the equation $dF = 0$. It is then well known that the second equation is the Euler equation for the action

$$\mathcal{A} = \int F \wedge *F dm = (1/4)\int \sum_{\mu\nu} F_{\mu\nu}F^{\mu\nu}$$

where the electromagnetic field F is the 2-form

$$F = dA = (1/2)\sum_{\mu\nu} F_{\mu\nu}dx^\mu dx^\nu, \qquad F_{\mu\nu} = A_{\nu,\mu} - A_{\mu,\nu}.$$

Since the field equations involve only F the vector potential has no physical significance.

In the Einstein theory spacetime is a pseudo Riemannian manifold M of dimension 4 and its metric itself is a dynamical object satisfying the equations in empty space which are concisely written as

$$R_{\mu\nu} = 0$$

where $R_{\mu\nu}$ is the Ricci tensor. They are the Euler equations for the Einstein-Hilbert action

$$\mathcal{A} = \int_M \mathrm{Rsc}\; dm = \int_M \mathrm{Rsc}\, \sqrt{-\det g}\; d^4x$$

where Rsc is the Ricci scalar. From a purely aesthetic point of view it was natural to ask whether one can unify these two theories; this was the starting point of the long but still unfinished quest for a *unified field theory* begun by Einstein and carried on by Weyl and many others.

Let me return now to the scale bundle. The fact that the parallel transport of scales can be defined by a connection expressed locally by a 1-form suggested to Weyl the possibility of identifying this connection with the electromagnetic potential; more precisely, if

$$A = \sum_{\mu} A_\mu dx^\mu$$

is the 1-form in local coordinates, then the A_μ could be identified with the components of the vector potential. This suggestion is reinforced by the observation, mentioned earlier, that the 2-form dA is globally defined and is the local description of the curvature of the connection. Clearly the identification of the connection with the vector potential goes hand in hand with the identification of its curvature with the electromagnetic field. This was what Weyl did.

The dynamical state of the Weyl universe is described by the conformal metric together with the connection on the scale bundle, the *connection* being identified with the electromagnetic vector potential and the *curvature* of this connection with the electromagnetic field. The equations of motion of the Weyl universe were required to be invariant not only under coordinate transformations as in Einstein's theory, but under gauge transformations of scales also. How can one write such equations of motion? The work of Einstein and Hilbert on gravitation gave the clue to Weyl and he wrote down a Lagrangian that was based on the curvature of the Weyl connection. The Euler equations derived from the Weyl Lagrangian gave coupled gravitational and electromagnetic equations, and the gauge invariance of the equations followed from the invariance of the Lagrangian under the gauge transformations determined by smooth positive functions g that represent the scale changes at the various points of M. The presence of the gauge transformations meant that along with tensors one should also consider *tensor densities,* and organize them according to their *weights,* where a tensor density is said to be of *weight* e if it is a tensor that gets multiplied by g^e under the gauge transformation g (e is an integer but can be negative). For instance, the metric tensor is of weight 1, and the curvature tensor $R_{\lambda\mu\nu\rho}$ of the Weyl connection is also of weight 1, while the mixed curvature tensor $R^{\lambda}{}_{\mu\nu\rho}$ is of weight 0, i.e., gauge invariant. Weyl took the Lagrangian W given by

$$W = R^{\mu}{}_{\nu\rho\sigma} R_{\mu}{}^{\nu\rho\sigma}$$

and defined the corresponding action by

$$\mathcal{A} = \int_M W dm = \int W \sqrt{(-1)^q g}\, dx \qquad (g = \det(g_{\mu\nu}))$$

where q is the number of negative terms in the diagonal form of the metric. Since $\sqrt{(-1)^q g}$ has weight $n/2$ where n is the dimension of M, and W has weight -2, it follows that this Lagrangian is gauge invariant only if $-2+n/2=0$, i.e., when $n=4$! Using this Lagrangian Weyl wrote down the corresponding (Euler) equations of motion. It turns out that electromagnetism is still described in Weyl's theory by the Maxwell equations for the curvature F of the connection on the scale bundle, namely,

$$dF = 0, \qquad d * F = 0.$$

However, the fact that the curvature of the Weyl connection combined both the metric and the electromagnetic connection, led to gravitational equations that were more complicated than Einstein's. Weyl took the point of view that only in the absence of the electromagnetic field it is true that gravitation is described by the Einstein equations, and that when both electromagnetism and gravitation are present, the Einstein equations are only an approximation to the more complicated but exact gravitational equations because the gravitational constant is enormously smaller than the electron radius.

The theory of Weyl described above was the first important example of a unified field theory. Weyl developed his ideas in three papers published during the years 1918-1920[9]. From his comments in his various expositions of these ideas it is obvious that Weyl was extremely happy with his ideas that suggested a *differential geometric* origin of electromagnetism, and the possibility of using this to solve the problem of a unified field theory. However, there were serious objections to the Weyl theory, especially from Einstein. Einstein pointed out that if the scale transference is assumed to be path-dependent (which is necessary in order to have electromagnetic phenomena), then an object changes its size when transported around a closed curve. In particular, if the object is a clock, it will keep time differently after a transport along a closed curve. In other words, Einstein argued that the period of a clock will depend on its past history, making objective measurements impossible even for a single observer. Thus Einstein's objections cast doubts on scale transference as the source of electromagnetism[10]. Weyl answered these objections to some extent but he eventually gave up the idea that his geometry was the answer to the problem of unifying electromagnetism and gravitation. Nevertheless, because of the beauty and simplicity of his theory, Weyl remained committed to the concept of the gauge principle on bundles over spacetime and the related concept that electromagnetism should be viewed as a connection on such bundles[11]. I wish to note here that it may be of interest to explore Einstein's objections further by installing the clocks on the *loop space* of M.

3. The rise of quantum theory. Electromagnetism as a connection on the phase bundle. The birth of quantum mechanics in 1925 introduced new themes into physics at the foundational level and Weyl had a major role in the identification and elucidation of some of these. Of particular interest for the present discussion is the modification of the nature of electromagnetism that he proposed as a consequence of quantum mechanics. He realized that quantum theory, with its introduction of the Planck's constant $\hbar$, *made it possible to choose a universal unit of length.* So far as he was concerned, this was the decisive argument in abandoning scale transfer as the source of electromagnetism. But in quantum theory the electron wave functions are undetermined only up to a phase. This phase however was the same at all spacetime points. Weyl had the idea to *make the phase depend on the spacetime points.* Thus he was led to the idea that the scale bundle can be replaced by the *phase bundle* and that one should do *phase transfer* along closed curves. The phases at a point form a group isomorphic to $U(1)$, the multiplicative group of complex numbers of absolute value 1. His proposal was to view the electromagnetic vector potential as a *connection on the phase bundle.* The phase bundle is just $\mathbf{R}^{1,3} \times U(1)$ viewed as a principal bundle for the group $U(1)$. The connection is defined locally by the 1-form

$$-i \sum_{\mu} A_\mu dx^\mu$$

(here $i = \sqrt{-1}$, and we use natural units with $e = \hbar = c = 1$). The appearance of i in the above formula indicates clearly that this is a *quantum* theory. The parallel transport of the phases corresponding to this connection is now defined by the differential equation

$$\frac{ds}{dt} - i \sum_{\mu} A_\mu(\gamma(t))\dot{\gamma}^\mu(t)s(t) = 0$$

which is solved by

$$s(t) = \exp\Big\{ i \oint_{\gamma_t} \sum_{\mu} A_\mu dx^\mu \Big\}.$$

This is the same as the earlier formula on the scale bundle except for the factor i in front of the line integral. The gauge transformations are still defined by functions on M but now with values in $U(1)$:

$$g : x \longmapsto g(x) \in U(1).$$

Locally on M (and globally too, if M is simply connected) we can write

$$g = e^{i\beta}$$

for smooth real functions β which is how Weyl denoted them. The connection form A changes to A' under the gauge transformation g by

$$A' = A - d(\log g)$$

(since $\log g = i\beta$ is defined locally up to an additive constant, $d(\log g)$ is defined without any ambiguity). He discussed all of these matters thoroughly in a remarkable paper[12] in 1929. Apart from what we have talked about, his treatment of electromagnetism allowed Weyl to give a gauge-theoretic explanation of the conservation of electric charge. However phases are quantum mechanical quantities and so Weyl's description of electromagnetism is essentially quantum mechanical. In particular, classical gravitation does not come into this picture at all. So the problem of unification became open again. I do not have the time to go into the various attempts to construct unified field theories, by Weyl himself and many others, since then (see however the end of §5).

Interestingly enough, a $U(1)$-bundle is topologically more subtle than the scale bundle which, as we have noted already, is always trivial. Indeed, the same sheaf-theoretic argument given earlier can be adapted to show that the principal $U(1)$-bundles on a manifold M are classified by $H^2(M, \mathbf{Z})$. In modern terminology the element $d(P)$ of $H^2(M, \mathbf{Z})$ corresponding to a $U(1)$-bundle P is its *degree* or its *Chern class;* its image in $H^2(M, \mathbf{R})$ is determined as the cohomology class of $(2\pi i)^{-1}\Omega$ where Ω is the curvature of an arbitrary connection on P, by the Chern-Weil theorem. If M is a vector space or more generally contractible, this is again 0. But there are M such that $H^2(M, \mathbf{Z})$ is not 0 (see below); new electromagnetic phenomena then become possible because of the *topology* of M. Weyl himself did not consider these *topological effects*. They were first considered by Dirac in his theory of magnetic monopoles in 1931 and by Aharanov-Bohm in 1959 when they proposed their famous experiment to test whether the electromagnetic potentials have physical significance (see the following sections). In both cases the topology of M created the circumstances for new electromagnetic phenomena. The interaction between quantum physics and topology started by Dirac and Aharanov-Bohm has become much more thorough nowadays. For instance the Chern classes give rise to *topological charges* in nonabelian gauge theory which are direct generalizations of the Dirac charge of the magnetic monopole.

Weyl's vision, that *the electromagnetic potential is a connection on a $U(1)$-bundle on spacetime whose curvature is the electromagnetic field, and that electromagnetism is a gauge field theory with structure group $U(1)$*, has proved to be prophetic. Weyl's ideas have survived to this day, through the many generalizations and modifications of the foundations of physics demanded by the quantum revolution.

4. Dirac's theory of monopoles. Weyl's idea that the wave function of the electron allowed changes in its phase dependent on the spacetime points attracted Dirac and he explored this theme. He replaced the principal $U(1)$-bundle by its associated *complex* line bundle and noticed that it acquired a hermitian structure in a natural manner. Let us call this line bundle L. For Dirac the wave functions were now *sections* of L; of course he did not express his ideas in this language but what he was doing amounted exactly to this. Since the fibers have a hermitian structure, the absolute values of the sections at the points of spacetime are well defined independently of the gauge transformations. This led Dirac to introduce what we would now call the Hilbert space $\mathcal{H}(L)$ of *square integrable sections* of L and identify its unit vectors with the states of the system as usual.

Dirac used this more subtle framework to discuss *magnetic monopoles* in one of the most famous papers he ever wrote[13] in 1931. His starting point was the observation that the conventional description of electromagnetism was not completely symmetric between electricity and magnetism. Thus, while an electric charge localized at a point is a perfectly viable concept, both theoretically and experimentally, the same is not true for a magnetic pole localized at a point. All magnets in nature are dipoles; moreover, any treatment of a magnetic monopole using Maxwell equations in the conventional way immediately runs into a contradiction. Indeed, if $\mathbf{B}$ is the magnetic field, and we assume that we have a vector potential that gives rise to the field, the Maxwell equations imply that $\mathbf{B} = d'A$ where d' is exterior differentiation in space so that the *magnetic flux* through any closed 2-dimensional surface is 0. *Thus there are no magnetic charges.* However there is a 2-form on $\mathbf{R}^3 \setminus \{0\}$ which is a candidate for the field of a stationary magnetic monopole located at the origin, namely,

$$F_g = \mathbf{B}_g = \frac{g}{4\pi r^3}(xdydz + ydzdx + zdxdy) \tag{$*$}$$

where g is a nonzero constant. It is easy to check that F is the unique *closed* 2-form up to the scalar factor g, which is invariant under rotations, and that it satisfies the equation $d * F_g = 0$, so that it is a solution of the Maxwell equations

$$dF = 0, \qquad d * F_g = 0.$$

One can now verify that the flux of the field through any closed surface is 0 if the surface does not enclose 0 but is equal to g if the surface encloses 0, so that F can be taken as the magnetic monopole field of charge equal to g located at the origin. The earlier discussion shows that this field cannot come from a vector potential defined in the conventional way. But Dirac discovered that things are different if we consider the Maxwell equations on a *nontrivial* $U(1)$-bundle on $M = \mathbf{R}^3 \setminus \{0\}$, the natural space for treating a monopole located at the origin (we consider the stationary situation so that time can be ignored), and ask whether there is a connection on such a line bundle whose curvature is the field $(*)$. First of all nontrivial principal $U(1)$-bundles exist on M; indeed M is contractible to S^2 and so $H^2(\mathbf{R}^3 \setminus \{0\}, \mathbf{Z}) \simeq H^2(S^2, \mathbf{Z})$ and the latter is isomorphic to $\mathbf{Z}$. Thus the principal $U(1)$-bundles (or line bundles with group $U(1)$) on M are parametrized by *integers*. Let L_n be the bundle corresponding to the nonzero integer n; we can think of it as a principal bundle or a line bundle. Dirac showed that there is a connection on L_n whose curvature Ω_n satisfies

$$\frac{1}{2\pi i}\Omega_n = F_n = \mathbf{B}_n \tag{$MMPF$}$$

and so can appropriately be called the magnetic monopole of charge n. Moreover, as we have remarked earlier, for *any* connection on L_n its curvature Ω has the property that its cohomology class is the same as that of $\mathbf{B}_n$. In other words, only the magnetic monopole fields of *integral* charge n can be obtained from L_n. This is the modern version of the arguments that allowed Dirac to conclude that *the magnetic monopole charge is quantized.* The Dirac quantization of magnetic charge is a consequence of the classification of principal $U(1)$-bundles on M by $\mathbf{Z}$ and the Chern-Weil theorem which permits the detrmination of the Chern class as the class of $(2\pi i)^{-1}\Omega$ where Ω is the curvature of any connection on the bundle. Dirac's work marks the first emergence of Chern classes as topological charges. More remarkably, Dirac showed further that this magnetic charge quantization implied that *the electric charge is also quantized.* No other explanation of the quantization of the electric charge has been given yet.

Dirac proceeded to study the quantum mechanical motion of an electron in the field of a magnetic monopole defined above. The Hilbert space is the space $\mathcal{H}(L)$ defined above where L is the nontrivial hermitian line bundle on M with the monopole connection defined on it. To get a gauge invariant Schrödinger equation Dirac replaced the usual partial derivatives ∂_μ by the *covariant partial derivatives* ∇_{∂_μ}:

$$\partial_\mu \longrightarrow \nabla_{\partial_\mu} = \partial_\mu - iA_\mu .$$

The problem that Dirac considered was to determine the energy states of the energy operator in which the covariant Laplacian replaced the ordinary Laplacian. There is no space here to go in any more detail into this question[14].

Although magnetic monopoles have yet to be discovered, Dirac's monopole idea has proved to be surprisingly persistent. It is believed (as already pointed out by Dirac in his 1931 paper) that the difficulty of finding monopoles is due to the enormous energies needed to separate the dipoles, and that at an early time in the evolution of the universe they must have been around. Indeed, monopoles have to be admitted in many GUT's (Grand Unification Theories), and have recently played a fundamental role in the celebrated Seiberg-Witten theory.

5. Yang-Mills and the gauge principle. The Aharanov-Bohm idea. The next major step in the development of the gauge concept was the 1954 paper of Yang-Mills[15]. Yang and Mills were trying to understand the conservation of isotopic spin in interactions of nuclei and wanted to see if this can be done in a manner analogous to charge conservation in electrodynamics. Their starting point was the observation that the proton and neutron are just two states of a single entity (called the nucleon, an idea already proposed by Heisenberg in 1932). There are many reasons why this is a good idea - for instance, their masses are nearly identical, they have the same spin, and so on. To formulate this idea mathematically one introduces an internal space for the nucleon, called the isotopic spin space, which is a 2-dimensional complex Hilbert space with the standard action of $SU(2)$. When electromagnetic effects are neglected the observer cannot distinguish between the proton and neutron and so is free to choose any state in the isotopic spin space as defining the charged state. This freedom of choice for the observer led Yang and Mills to the principle that physical laws should be invariant under all the rotations of the isotopic spin space. But what made their work a major breakthrough was their insistence, that *to be compatible with the field concept in physics, these rotations of the isotopic spin space should be allowed to be different for observers situated at different points of spacetime.* They were thus led to define an *isotopic gauge transformation* as an arbitrary smooth $SU(2)$-valued function g defined on Minkowski spacetime ($= \mathbf{R}^{1,3}$), and to formulate the principle that all physical laws (when the electromagnetic field is absent) should be invariant under all isotopic gauge transformations g and the corresponding transformations of the nucleon wave functions

$$\psi \longmapsto \psi' = g\psi.$$

From this assumption they proceeded to obtain a theory which, for all intents and purposes, was completely analogous to the Weyl theory except that $SU(2)$ had replaced $U(1)$, i.e., one had now a *nonabelian gauge theory.*

Expressed in modern language, Yang and Mills worked with the $SU(2)$-bundles $\mathbf{R}^{1,3} \times SU(2)$ on Minkowski space $\mathbf{R}^{1,3}$ and the associated Hermitian vector bundles corresponding to unitary representations of $SU(2)$. (Since the base manifold is a vector space and hence contractible, all the principal bundles are product bundles anyway.) The gauge concept then led to the introduction of connections on these bundles described by 1-forms on $\mathbf{R}^{1,3}$ with values in the Lie algebra of $SU(2)$,

$$A = -i \sum_{\mu} A_\mu dx^\mu \qquad (A_\mu = A_\mu^\dagger)$$

where $\dagger$ denotes the adjoint and the A_μ are 2×2 complex hermitian matrix functions. The corresponding field F was defined by them as the curvature of this connection,

$$iF = iF_A = dA + A \wedge A.$$

The additional term $A \wedge A$ by which F differs from the formula for the field in electromagnetism is dictated for gauge invariance. In local coordinates

$$F_{\mu\nu} = A_{\nu,\mu} - A_{\mu,\nu} + i[A_\mu, A_\nu].$$

The dynamical equations are now written in the gauge context starting from a gauge invariant Lagrangian. The bundle under consideration has an adjoint operator analogous to the Hodge $*$-operator, also denoted by $*$, and one takes the action

$$\mathcal{A} = \int_M F \wedge *F dm = \int_M \sum_{\mu\nu} F_{\mu\nu} F^{\mu\nu} dm.$$

It is easy to see that this is gauge invariant and so the corresponding Euler equations are also gauge invariant. These are the celebrated Yang-Mills equations (in the absence of matter). Globally they take the following form:

$$D_A * F_A = 0 \qquad (iF = dA + A \wedge A) \qquad (YM)$$

where D_A is *the covariant exterior derivative defined by the connection A itself.* One has automatiucally

$$D_A F_A = 0$$

which is the *Bianchi identity;* it is the analog of the equation $dF = 0 (F = dA)$ in the Maxwell theory. Thus the Yang-Mills equations generalize the Maxwell equations in empty space which are

$$dF = 0, \qquad d * F = 0, \qquad (F = dA) \qquad (M)$$

where d is the usual exterior derivative.

It is indeed remarkable how similar this whole theory is to that of Weyl and Dirac. However there are major differences because the Yang-Mills theory is a nonabelian gauge theory. For electromagnetism on a principal $U(1)$-bundle, we have $D_A = d$ because the 1-forms take values in the Lie algebra of $U(1)$ which is $\simeq \mathbf{R}$. It is only when we go over to nonabelian gauge groups that the covariant exterior derivative depends on the connection so that the connection itself explicitly enters the field equations. The Yang-Mills field equations for the potentials A_μ are:

$$\sum_\nu (F^{\mu\nu}_{,\nu} + [A_\nu, F^{\mu\nu}]) = 0. \qquad (YM)$$

In electromagnetism the second term is 0 because the commutators are 0, as we observed just a little earlier. But they do not vanish in nonabelian gauge theory. In other words, *the field equations* (YM) *contain the potentials in an essential manner; it is no longer true that the fields contain all the physical information.* This is in striking contrast with *classical electromagnetism* where the Maxwell equations (M) involve only the electromagnetic field. It is also noteworthy that the equations are *nonlinear* in the potentials.

This is the place to make some remarks on what is nowadays called the *gauge principle.* This principle says that once the group of gauge transformations is given, everything in the theory is automatically determined. If we view the gauge group as the symmetry group of the system, this is very similar to the principle in classical geometry, first enunciated by Felix Klein in his Erlangen Program, that any geometry is completely determined by the group of congruent transformations. To make this more precise, let us consider the two basic features of the gauge theory - the wave functions ψ, and the gauge transformations $\psi \mapsto \psi' = g\psi$. Since

$$\frac{\partial(g\psi)}{\partial x^\mu} = g\left(\frac{\partial\psi}{\partial x^\mu} + g^{-1}g,_{\mu}\right) \qquad (g,_{\mu} = \partial g/\partial x^\mu)$$

we can say that under the gauge transformation g the derivative ∂_μ changes to

$$\partial_\mu + g^{-1}g,_{\mu}.$$

So, in order to obtain gauge covariant quantities, it is natural to *counteract* the effects of the gauge transformation by introducing fields A_μ and work with $\partial_\mu + A_\mu$ instead of ∂_μ. Such a choice will be covariant if

$$g(\partial_\mu + A_\mu) = (\partial_\mu + A'_\mu)g$$

where A'_μ is the counteracting term in the new gauge. If $A = \sum_\mu A_\mu dx^\mu$ this is just the condition

$$A' = gAg^{-1} - g^{-1}dg$$

which says that A transforms as a connection. *Thus the specification of the gauge group automatically forces one to introduce gauge potentials(connections) in the associated bundle and use covariant derivatives with respect to this connection instead of the usual derivatives.* Now the wave equations satisfied by the particle (Klein-Gordon, Dirac, Weyl, etc) usually come from a Lagrangian. Moreover, the same reasoning shows that one must make the change

$$\partial_\mu \longrightarrow \nabla_{\partial_\mu} = \partial_\mu - iA_\mu$$

in the Lagrangian of the electron wave function ψ mentioned earlier to have a gauge invariant Lagrangian. The new expression for the Lagrangian with this change is then gauge invariant; but what is more remarkable, it contains the terms that tell us how the field A interacts with the wave function ψ. In other words, the specification of the gauge group and the requirement of gauge invariance already force one to introduce the gauge potentials or connections as well as the form of the interaction of the gauge field with the matter field. As an example, let us consider the free Dirac field which is a spinor field ψ(a 4-component column vector) whose equation of motion are

$$i\left(\sum_\mu \gamma^\mu \partial_\mu + m\right)\psi = 0 \quad \gamma^\mu\gamma^\nu + \gamma^\nu\gamma^\mu = 2\delta_{\mu\nu}$$

arising out of the Lagrangian

$$\mathcal{L}_{\text{Dirac}}(\psi) = -i\psi^\dagger\left(\sum_\mu \gamma^\mu \partial_\mu + m\right)\psi$$

where the γ^μ are Dirac's γ-matrices and $\dagger$ is the complex conjugate. Now,

$$\mathcal{L}_{\text{Dirac}}(e^{i\beta}\psi) = \mathcal{L}_{\text{Dirac}}(\psi) + \left(\sum_\mu \psi^\dagger\gamma^\mu\beta_{,\mu}\psi\right).$$

In other words, $\mathcal{L}_{\text{Dirac}}$ by itself is not gauge invariant; but if we introduce compensating fields A_μ which change to $A_\mu - \beta_{,\mu}$ when ψ changes to $e^{i\beta}\psi$, and simultaneously replace ∂_μ by $\partial_\mu - iA_\mu$, then the new Lagrangian is gauge invariant. A simple calculation then shows that the new Lagrangian is

$$-\psi^\dagger\left(\sum_\mu i\gamma^\mu(\partial_\mu - iA_\mu) + m\right)\psi = \mathcal{L}_{\text{Dirac}}(\psi) - \psi^\dagger\sum_\mu \gamma^\mu A_\mu\psi.$$

The second term in this expression gives the term that corresponds to the interaction of the gauge field A with the matter field ψ. The final expression for the Lagrangian with matter fields present is

$$\mathcal{L} = -(1/2)F\wedge *F - i\psi^\dagger\left(\sum_\mu \gamma^\mu\partial_\mu + m\right)\psi + \psi^\dagger\sum\gamma^\mu A_\mu\psi.$$

In physics this entire set of ideas is called the *gauge principle.* It was quite explicit in the works of Weyl and Dirac on electromagnetism but was elevated to a universal principle after the work of Yang and Mills was fully understood.

In 1959, Aharanov and Bohm wrote a remarkable paper[16] in which they discussed the question *whether the potentials have physical significance.* They suggested that this is true even in electromagnetism. More precisely they predicted that *when M is not simply connected, electromagnetic effects could be present even if the electromagnetic field is* 0. To test this hypothesis they proposed an experiment, the famous *Aharanov-Bohm experiment.* In this experiment, $M = \mathbf{R}^3 \setminus L$ where L is a straight line in space which represents a solenoid which is ultra thin so that its magnetic field does not leak outside of L. Electrons are sent around the solenoid and then allowed to interact, producing interference patterns in a screen. Although in the region in which the electrons travel, namely M, the electromagnetic field is 0, nevertheless, they predicted that *the interference pattern on the screens would change when the flux inside the solenoid is varied.* This experiment was performed and this prediction verified. I cannot go into this in more detail here; see my book with Sundararaman[3] for a more detailed discussion[17]. The fact that Yang-Mills equations involve the potentials in an essential manner should be understood from this perspective.

There is one interesting point worth mentioning. In both the theories of Weyl and Yang-Mills the physical description is through the connections on a principal bundle on spacetime M with group G which is a compact Lie group ($G = U(1), SU(2)$). However, because of the possibility of making gauge transformations, two connections which are gauge equivalent represent the same physical situation. Indeed, as is explained in [17], this is exactly the case in the Aharanov-Bohm experiment. In addition the connections must satisfy the appropriate equations, the Maxwell equations in the Weyl case and the Yang-Mills equation in the theory treated by Yang-Mills. So if $\mathcal{E}$ is the set of connections satisfying the appropriate equations and $\mathcal{G}$ is the group of gauge transformations, the states of the system at the classical level are represented by the points of the *quotient space*

$$\mathcal{M} = \mathcal{E}/\mathcal{G}.$$

Of course one has to quantize this situation before an adequate gauge theory can be developed; but even this classical description is very interesting because it associates with any G-bundle a geometric object $\mathcal{M}$ which is important physically. $\mathcal{M}$ is a *moduli space* of great mathematical interest also. Indeed, it is precisely by studying $\mathcal{M}$ that new discoveries in the topology and geometry of 4-manifolds were made, a striking confirmation of the deep nature of the interaction between mathematics and physics in recent decades[18].

The Yang-Mills theory is a beautiful and natural generalization of the theories of Maxwell, Weyl and Dirac. It must be pointed out however that at the time of their discovery its relationship to the earlier themes was not very clear. It was only subsequently, in the 1970's, that the relationship of gauge theories to the differential geometry of principal and vector bundles, and their natural evolution starting from the themes explored by Weyl and Dirac, were finally understood[19].

The revolution started by Weyl, Dirac, and Yang-Mills, eventually led, in conjunction with a huge development of quantum field theory, to a unification of three of the basic forces in nature, electromagnetic, weak, and strong. However gravitation had remained outside this process. Nevertheless, for the past several years an intensive effort is being made to obtain a quantum theory of gravitation that will then include all forces and will be the unified field theory that has been the holy grail of theoretical physics since the discovery of general relativity by Einstein.

One starting point of a possible unified view of things is the same as Weyl's, namely a principal bundle on spacetime. The group of the bundle however is much bigger than $U(1)$ because *we have to admit many additional types of particles.* Moreover one has to consider not only gauge fields with this structure group but also matter fields. Now in Minkowski space, the typical matter field satisfies the Dirac equation in the massive case and the Weyl equation in the massless case, and so they are *spinor fields.* So to define them in the case when M is curved due to gravitation, we have to introduce spinor fields on curved manifolds. This imposes a topological restriction on M (it should be what is called a spin manifold); the matter fields then satisfy the natural generalization of the Dirac or the Weyl equation to curved spin manifolds. There are additional restrictions on M also because in the classical limit of the vacuum state the theory should describe spacetime with no matter and so M should be Ricci flat (recall that the Einstein equations for matterless spacetime are that the Ricci tensor vanishes). The phenomenology of particle spectra also imposes additional restrictions and difficulties which have been overcome by assuming that M is coupled to a *very small* compact manifold (*compactification* in the terminology of the physicists) of 6 dimensions which is the underlying real manifold of a complex Calabi-Yau manifold of complex dimension 3. This is not the place even to begin talking about these matters but the reader should refer to the address of Witten to the International Congress of Mathematicians at Berkeley in 1986 for an inspiring discussion[20].

6. Holomorphic vector bundles and the meeting of the physics and the mathematics. I mentioned at the beginning that I was going to talk about matters that are tangential to the interests of Seshadri. Tangential means at least some contact and so far there is no hint that the ideas that I have discussed so far have anything to do with *holomorphic* vector bundles on Riemann surfaces! I would like to address this point, at least briefly, now. In keeping with the historical and pedagogical stance of this article let me start by giving a very brief discussion of the *holomorphic* vector bundles.

Although the concept of a holomorphic vector bundle is quite a modern one, the word modern should be taken to refer only to the view point from which the subject is looked at and the language in which its main results are formulated. This use of the modern view point and language is of course natural and appropriate because it is impossible, given what we know about topology and geometry in modern times, to continue to work in the confining paradigm of the ancients. But it is good to know, at least at the starting point, what the subject is from a classical perspective.

Given a compact Riemann surface, one wants to study not only the (meromorphic) functions and differential forms on the surface but also those which are multiple-valued on the given surface but become single-valued when we go over to a covering surface. Thus, together with a given compact Riemann surface, one wants to have the entire tower of its covering surfaces in view. Weyl, one of the founders of the modern theory of Riemann surfaces, emphasized the deep analogy of this situation with the tower of extensions of an algebraic number field. The extensions of a number field are governed by the Galois groups and their representations, and classical number theorists had already obtained a profound understanding of the *abelian* part of the tower, namely the *subtower* of extensions with an *abelian* Galois group, otherwise called the *classfields.* In function theory the analog of the Galois group (at least when the extension is unramified) is the fundamental group. But the classical theory of Riemann surfaces, with its emphasis on the *line integrals* on such surfaces, had access only to the covering surface whose covering group is the *homology* of the surface, which is the largest abelian quotient of the fundamental group. Weyl called the corresponding covering surface the *class surface,* reinforcing the arithmetic analogy with classfields. The multiple-valued functions that become single-valued on the class surface and transform according to *characters* of the homology group were an object of study in the classical theory. These characters constitute a complex torus, the so-called classical Jacobian variety, and the function theory on this torus is the theory of *abelian functions.*

Already Riemann had begun thinking of some *nonabelian* generalizations of this theme, at least on $\mathbf{P}^1$, in his work on the *monodromy* of regular singular differential equations. He considered linear ordinary differential equations of arbitrary order n on the extended complex plane $X = \mathbf{P}^1$ with rational coefficients. Let F be the set of singular points of the coefficients of the equation; it is convenient to assume that $\infty \in F$. At a point $x_0 \notin F$, the local (germs of) solutions to the equation around x_0 form a vector space V of dimension n. By the linearity of the equation one can continue the elements of V along any path not meeting F and the continuation depends only on the homotopy class of the path. Taking closed paths at x_0 it is now clear that we have an action of the fundamental group $\pi_1(X \setminus F, x_0)$ on V; the image of the fundamental group inside $GL(V)$ is the *monodromy group* of the equation in classical terminology. Riemann studied the hypergeometric equation from this point of view. Here the differential equation is of the second order, $F = \{0, 1, \infty\}$; the fundamental group of $X \setminus F = \mathbf{C} \setminus \{0, 1\}$ (with respect to some base point) is then the free group on two generators. In unpublished work that did not become available for some years after he died, Riemann considered also the generalizations to equations of higher order. However, in the general case, one has to impose some restrictions on the orders of the coefficients at their poles so that the solutions of the equations are linear combination of functions of the form $z^\lambda(\log z)^r$. This is equivalent to the requirement that the solutions, although possibly multiple-valued around the singularities, are nevertheless of *moderate growth* (in a suitable sense) at these points. One is thus led to the class of equations with rational coefficients having only *regular singularities;* these are also called *Fuchsian equations* in honor of Fuchs who treated them formally for the first time, Riemann's work coming to light only posthumously.

But the hypergeometric equation, and more generally, the Fuchsian equations, touch the theory of compact Riemann surfaces at a deeper level also. For instance, the periods of elliptic functions satisfy the hypergeometric equation, a fact that is seen from the Legendre formula for the periods of elliptic integrals in terms of the hypergeometric series:

$$K := \int_0^{\pi/2} \frac{d\varphi}{\sqrt{1 - k^2 \sin^2 \varphi}} = \frac{\pi}{2} F(1/2, 1/2, 1; k^2).$$

Fuchs extended this result to hyperelliptic integrals and showed that if one considers a family of hyperelliptic integrals, their periods satisfy a Fuchsian equation. This was in fact a major discovery of Fuchs. The generalizations of these equations are called *Picard-Fuchs equations* and they have become important in physics recently.

Two questions about Fuchsian equations come up naturally. The first is whether the Fuchsian assumption imposes any restriction on the monodromy representation. Given any representation ρ of $\pi_1(X \setminus x_0)$ on $\mathbf{C}^n$, is there a Fuchsian equation with ρ as its monodromy representation? The second question is more delicate. If $y \in F$ is a singular point, the *local monodromy group* is the fundamental group of a punctured disk at y and so is isomorphic to $\mathbf{Z}$. Thus we have an action of $\mathbf{Z}$ on V which is determined by an automorphism L of V. Generically this action is determined by the eigenvalues

$$\alpha_1, \alpha_2, \ldots, \alpha_n$$

of L which are traditionally called the *local exponents* of the equation. The second question is then the following. Suppose one is given all the local exponents at the singularities (these must satisfy certain natural relations), what is the structure of the space of all Fuchsian equations (up to isomorphism) with these singularities and local exponents? For equations of second order with $F = \{0, 1, \infty\}$ Riemann knew that the hypergeometric equation is the only one for fixed local exponents, and he developed the theory of hypergeometric functions from this point of view (the Riemann P-functions). The parameters that describe the equations with given local data are called *ancillary parameters* in classical language.

Hilbert formulated the first question as his twentyfirst problem in his famous 1900 address; this is nowadays called the Riemann-Hilbert(R-H) problem, not only in Hilbert's original formulation but also in its myriad modifications and variations. There is a little vagueness in the Hilbert formulation whether the differential equation is to be sought as a Fuchsian ordinary differential equation or as a first order system

$$\frac{du}{dz} = Au, \qquad A = \sum_k \frac{A_k}{z - a_k}$$

where the A_k are *constant matrices.* As for the second question, Riemann himself had calculated the number of ancillary parameters in his unpublished work, namely, the moduli in his language, for the equations with fixed singularities and local exponents.

The functions envisioned by Riemann and Hilbert are not algebraic as in the classical theory but transcendental; they were called the *Riemann transcendentals.* The local solutions of the system generate a *differential field* and one can keep the arithmetic analogy going by thinking about these fields as forming a tower of *differential field extensions* with the Galois groups replaced by the differential Galois groups; however this point of view did not come in till quite a bit later. It is interesting to note that neither Riemann nor Hilbert made any effort to formulate these questions when the base manifold is taken to be any compact Riemann surface; in terms of the function fields, this is the relative point of view in which the field of rational functions is replaced by the field of meromorphic functions on a compact Riemann surface. That Hilbert, who was one of the first to recognize the importance of relative extensions in number theory (relative means over an arbitrary number field, as opposed to just the rational numbers), and who was also a champion of the analogy between number fields and function fields, did not formulate the R-H problem on an arbitrary compact Riemann surface, is striking.

In its classical formulation over $\mathbf{P}^1$, the R-H problem was studied and solved by several people under varied conditions - Schlessinger, Lappo-Danilevsky, Plemelj, Birkhoff, and so on. But further progress of the R-H problem on other Riemann surfaces had to await the modern development of differential geometry and topology. The formulation and the solution of the R-H problem for arbitrary compact Riemann surfaces was given by Röhrl[21] in 1957. Then Deligne took up this theme in 1970, and in a very influential monograph[22], gave the formulation of the R-H problem in the language of vector bundles and connections. But he went considerably beyond the original context by formulating the R-H problem in *all* dimensions and solving it completely, as well as treating the other classical themes such as the differential equations satisfied by the period matrices, in the more general context.

I shall briefly explain Deligne's formulation of the R-H problem in dimension 1 on an arbitrary compact Riemann surface X. One introduces the sheaf of local solutions of the differential equations which gives rise to a holomorphic vector bundle V on $X \setminus F$ where F is the set of singularities of the equations. There is a natural *holomorphic* connection ∇ on V whose horizontal sections are the local solutions. One can extend the vector bundle to the whole of X. One should add a condition about the moderate growth of the sections of the bundle at the singular points to make sure that the system is regular singular. Deligne's solution is that the assignment that takes a pair (V, ∇) with regular singularities at F to the monodromy action of $\pi_1(X \setminus F, x_0)$ is a functor that gives an equivalence of categories. However some care has to be exercised before deducing the solution to the R-H problem in the classical setting of Hilbert from this result[23].

It must be noted that in dimension 1 all connections are *flat* while this is not so in higher dimensions and one has to restrict oneself to flat connections to formulate the question; in classical language, the partial differential equations must satisfy the Frobenius integrability conditions. These equations in higher dimensions are therefore very special. Such differential equations go back to the papers of Appel, Picard, and others where hypergeometric functions in several variables were first introduced[23].

However one cannot restrict oneself to the regular singular context all the time. Already in dimension 1, many physical problems led to meromorphic differential equations with *irregular singularities* - Bessel, Airy, confluent hypergeometric, and so on. Their systematic treatment was begun in the late nineteenth century by Fabry and Poincaré. Now a characteristic feature of Fuchsian equations is the fact that the local solutions can be computed formally - this is the essence of the well-known Frobenius method of generalized power series and indicial exponents. This approach fails completely in the presence of irregular singularities. So the local theory in the irregular case is already considerably richer than in the regular case. Ponicaré began the process of bridging the gap between the formal and analytic theories with his theory of *asymptotic expansions* of solutions at an irregular singularity. For instance, in the theory of the Bessel equation, the singularity at infinity is irregular while the one at the origin is regular, and the central fact of the theory is the asymptotic structure at infinity of the solutions with given local exponents at the origin. For more general equations the asymptotic aspects are completely captured in the theory of the *Stokes phenomenon.* The modern point of view of these equations began with the work of Malgrange, followed by that of Deligne (expressed in a letter of Deligne to Malgrange), and then continued by Sibuya's work on the Stokes phenomenon. The work of Deligne, Malgrange, and Sibuya revealed that the Stokes phenomenon is really cohomological and is governed by the H^1 of a certain sheaf of noncommutative groups of flat solutions of the equations, called the *Stokes sheaf.* The analysis of the Stokes phenomenon in the classical setting by Balser, Jurkat, and Lutz[24] led to a construction of the *local* moduli space, namely, the moduli space of meromorphic connections with an irregular singularity at $z = 0$. The construction of the moduli *scheme* starting from the Stokes sheaf was accomplished by Babbitt and Varadarajan[25]. Many questions in the global theory of moduli are still open, although a substantial amount of work has been done[25].

Let me now return to the original theme of *algebraic functions* and their nonabelian aspects. If I am not mistaken, it was Weil's paper[26] in 1938 which brought into the foreground the algebraic theory together with the arithmetic analogy and began the search for a better understanding of the *nonabelian* phenomena that govern the systems of functions that transform according to representations of the fundamental group. This is very similar to the R-H context except that there are no singularities. Here not only the classification of the systems corresponding to the various representations of the fundamental group was an objective, but also function theory on the space that parametrizes such systems, the "moduli space ".

Even though Weil had obtained deep insights in his studies, the time was not ripe for making decisive progress. It was only after the vector bundle aspect was placed at the center of the investigations that progress could be made. The major breakthrough came in the papers[1,2] of Narasimhan and Seshadri where the concept of *stability* of an algebraic vector bundle, introduced by Mumford in his geometric invariant theory, was clarified in terms of the *unitarity* of the corresponding representation of the fundamental group of the Riemann surface. The importance of stable vector bundles arises from the fact that they are generic, and to get a "good" moduli space one has to restrict oneself to the set of stable bundles (up to isomorphism of course) and compactify this set when necessary. In the new point of view the moduli space of vector bundles on a compact Riemann surface, properly defined, became the central object of study, emerging as a far-reaching generalization of the classical Jacobians of curves.

The moduli space of the bundles (in the refined sense I have tried to explain above) is a difficult object to apprehend. In particular understanding its topology already requires substantial effort. The Indian school of geometers who followed Seshadri and Narasimhan, as well as several others, made major contributions to this study. However, in a paper published in 1982, Atiyah and Bott[27] *used the theory of Yang-Mills equations and Morse theory to study the topology of these moduli spaces.* This coming together of physics and mathematics, here as well as in the gauge theoretic approach to questions of the geometry of 4-manifolds that I have mentioned earlier where the moduli space of Yang-Mills connections plays a fundamental role, is quite remarkable. The Atiyah-Bott approach to the cohomology of the moduli space of vector bundles, with its origins in differential geometry and the physics of gauge theories, has also proved very fruitful in higher dimensional generalizations of the vector bundle theory.

Acknowledgement. I wish to express my deep gratitude to my friends Robert J. Finkelstein and Donald G. Babbitt for conversations spread over many years on all the topics discussed in this article. These conversations were tremendously inspiring and contributed greatly to my understanding. I am also grateful to Professor Erhard Scholz whose comments have helped me remove many of the ambiguities, errors and misconceptions that were present in the first version of this paper.

NOTES AND REFERENCES

[1] Ann. Math., 82(1965), 540.

2 Ann. Math., 85(1967), 303.

3 See the beautiful account of Yang in *Hermann Weyl, 1885-1985,* K. Chandrasekharan (Ed), Springer Verlag, 1986. I have profited a great deal by reading Yang's paper and have made use of it freely in writing this essay. A more detailed discussion is given in the forthcoming book *An Introduction to Gauge Theories,* by D. Sundararaman and V. S. Varadarajan. See also the articles in the Hermann Weyl International Congress volume *Exact Sciences and their Philosophical Foundations,* Wolfgang Deppert/Kurt Hübner, (Eds), Verlag Peter Lang, 1985, as well as the various chapters in the book *Hermann Weyl's Raum-Zeit-Materie and a General Introduction to his Scientific Work*, Erhard Scholz (Ed), Basel, Birkhäuser 2001.

4 See Weyl's book *Space, Time, and Matter,* Dover, Note 10 to Chapter II.

5 See Sections 14 and 15 as well as the Notes 9 and 10 to Chapter II of Weyl's book *Space, Time, and Matter,* Dover. These ideas were first developed in his paper *Reine Infinitesimalgeometrie*, Math. Zeit., 2(1918), 384. A convenient place to look up Weyl's papers is the *Hermann Weyl Gesammelte Abhandlungen,* K. Chandrasekharan (Ed), Springer Verlag 1968, hereafter abbreviated as GA, I-IV. The paper referred to here is GA, II, 1.

6 In an article reviewing 50 years of relativity, GA, IV, 421, Weyl says the following: *Beim Herumfahren eines Vektors längs einer geschlossenen Kurve durch fortgesetzte infinitesimale Parallelverscheibungkehrt dieser im allgemeinen in einer andern Lage zurück; seineRichtung hat sich geändert. Warum nicht auch seine Länge?*

7 See page 203 of the paper *Gravitation and Electricity* in *The Principles of Relativity,* Dover, 201. The latter is a collection of reprints of the fundamental papers in the theory of relativity, containing the paper of Weyl, the papers of Einstein on special and general relativity, and the famous address of Minkowski in which spacetime occurs for the first time as a mathematical and physical construct.

8 It is not surprising therefore that Weyl had a deep interest in *conformal geometry.* He constructed the analogue of Riemann's curvature tensor in conformal geometry which is nowadays called the *Weyl tensor.* Its vanishing is the necessary and sufficient condition that the manifold is *conformally flat, i.e., its metric can be transformed into a flat metric by a coordinate change followed by a rescaling* (theorem of Weyl-Schouten).

9 The basic papers are GA II, 1; 29; 55. The second of these papers is the most important one from the point of view of physics. Its translation into English is the paper which is a part of the reprint volume referred to in [7]. In addition

Weyl came back to this theme in several subsequent papers of which the most important are GA, III, 217; 229; 245.

10 See Yang's discussion in reference [3].

11 See the discussion in Dirac's paper, Proc. Roy. Soc. Lond., A333(1973), 403.

12 This is the paper GA, III, 245; the papers GA, III, 217, and 229 are variations of this.

13 Proc. Roy. Soc. Lond., A133(1931), 60.

14 The problem here is to obtain the spectral decomposition of the covariant Laplacian acting on the Hilbert space of square integrable sections of the monopole line bundle. See[15], 493. For a treatment of this as well as the spectral analysis of the covariant Dirac operator on this line bundle using the point of view of induced representations see R. P. Langlands in Pac. Jour. Math., 126(1987), 145.

15 See Phys. Rev., 96(1954), 191; also in *Selected Papers, 1945-1980, with Commentary, Chen Ning Yang,* W. H. Freeman and Company, 19; 172.

16 See Aharanov-Bohm, Phys. Rev., 115(1959), 485. The experiment was performed by Chambers in 1960, see Phys. Rev. Lett., 5(1960), 3.

17 Here the manifold is contractible to the circle and so its second integer cohomology is trivial. So only product bundles and globally defined vector potentials arise. As there is no electromagnetic field the connection has vanishing curvature, i.e., it is *flat.* So there is no phase change coming from parallel transport of phases along *small* closed paths but there may be phase changes if the loop is *big.* This is the familiar concept of *holonomy* of a flat connection in differential geometry; the holonomy around a closed path depends only on the homotopy class of the path and leads to a homomorphism of the fundamental group (integers) of the space into the group (circle group) of the bundle. So these homomorphisms are parametrized by a real number mod 1 and correspond to the various interference patterns obtained by the experimenter. Furthermore the connection is *determined up to gauge equivalence* by this homomorphism. So the *physical states of the system are really not the connections per se, but gauge equivalence classes of the connections* which are the points of the moduli space of connections satisfying Maxwell equations[15], 460.

18 The gauge theoretic approach to the geometry of 4-manifolds is the theme of the book of Donaldson and Kronheimer, Oxford, 1990.

19 See C. N. Yang[15], 457; 460; 493; 509; 519 and Yang's article in the *Chern Symposium,* Springer Verlag, 1979, 247.

20 Proceedings of the Int. Cong. Math., Berkeley, 1986, I, 267. I remember that Witten began his lecture at the Congress with the remark that Weyl's attempted unification of electricity and gravitation failed because the forms of matter included in it were not diverse enough. It is of interest to note that Weyl himself had come to the same conclusion. In his preface to the first American printing of his book *Space, Time, and Matter,* he says: *Since then, a unitary field theory, so it seems to me, should encompass at least three fields: electromagnetic, gravitational and electronic. Ultimately the wave fields of other elementary particles will have to be included too - unless quantum physics succeeds in interpreting them all as different quantum states of one particle.*

21 Math. Ann., 133(1957), 1.

22 See Deligne's Springer Lecture Notes 163(1970).

23 For expositions of some of these questions see my articles in Exp. Math., 9(1991), 97 and Bull. Amer. Math. Soc., 33(1996), 1, both of which are reviews. For recent work on the R-H problem see A. A. Bolibruch, Proc. Int. Cong. Math., 2(1994), 1159, Birkhäuser, Basel, 1995.

24 See W. Balser, Jour. für reine und ang. Math., 318(1980), 51; W. Balser, W. Jurkat, and D. Lutz, Funkc. Ekvac., 22(1979), 257; W. Jurkat, Springer Lecture Notes in Math., 637(1978).

25 See Malgrange, Springer Lecture Notes in Mathematics 712(1979), and Sibuya, Bull. Amer. Math. Soc., 83(1977). For the moduli problem see D. G. Babbitt and V. S. Varadarajan, Pac. Jour. Math., 108(1983), 1; Mem. Amer. Math. Soc., 55(1985), no. 325; Asterisqué, 169-170 (1989) as well as my review papers[23]. Roughly speaking, the moduli space is the first cohomology of the Stokes sheaf which is a sheaf of *complex unipotent nonabelian groups* and it is not immediately clear why it should have a nice structure as a variety nor why such a structure, if it exists, should be independent of the (Cěch) covering used to compute it. The basic idea in the Babbitt-Varadarajan approach to the moduli at an irregular singularity (which was based on a suggestion of Deligne) is to view the Stokes sheaf as a sheaf of complex unipotent group *schemes,* defined on the unit circle in the complex plane, so that its first cohomology can be viewed as a functor from commutative complex algebras with units into the pointed sets of the first cohomology of sheaves of groups on the unit circle; the existence of a scheme structure on the moduli space then becomes the question whether this functor is *representable* by a scheme. It turns out, and this is compatible with the Balser-Jurkat-Lutz theory, that this functor is actually represented by *affine space.* For global moduli the literature is too vast and I give only some scattered references. For the Fuchsian case, see K. Iwasaki, Pac. Jour. Math., 155(1992), 319, and the references therein.

The subject of special functions from a modern viewpoint is the theme of the book *From Gauss to Painlevé* by K. Iwasaki, S. Shimomura, and M. Yoshida. For some other aspects of Fuchsian connections and the Stokes sheaf, see the book *Déformations isomonodromiques et variétés de Frobenius,* Savoirs Actuels, EDP Sciences/CNRS Éditions, 2002, by C. Sabbah.

[26] See *Andre Weil Oeuvres Scientifiques,* Springer Verlag, 1979, I, 185.

[27] See Atiyah's *Collected Works,* Oxford, 1988, 5, 265.

DEPARTMENT OF MATHEMATICS
UNIVERSITY OF CALIFORNIA
LOS ANGELES, CA 90095-1555
E-mail address: vsv@math.ucla.edu